MW00716971

New Directions in Civil Engineering

Series Editor
W. F. CHEN
Purdue University

Zdeněk P. Bažant and Jaime Planas
Fracture and Size Effect in Concrete and Other Quasibrittle Materials

W.F. Chen and Seung-Eock Kim
LRFD Steel Design Using Advanced Analysis

W.F. Chen and E.M. Lui
Stability Design of Steel Frames

W.F. Chen and K.H. Mossallam
Concrete Buildings: Analysis for Safe Construction

W.F. Chen and S. Toma
Advanced Analysis of Steel Frames: Theory, Software, and Applications

W.F. Chen and Shouji Toma
Analysis and Software of Cylindrical Members

Y.K. Cheung and L.G. Tham
Finite Strip Method

Hsai-Yang Fang
Introduction to Environmental Geotechnology

Yuhshi Fukumoto and George C. Lee
Stability and Ductility of Steel Structures under Cyclic Loading

Ajaya Kumar Gupta
Response Spectrum Method in Seismic Analysis and Design of Structures

C.S. Krishnamoorthy and S. Rajeev
Artificial Intelligence and Expert Systems for Engineers

Boris A. Krylov
Cold Weather Concreting

Pavel Marek, Milan Guštar and Thalia Anagnos
Simulation-Based Reliability Assessment for Structural Engineers

N.S. Trahair
Flexural-Torsional Buckling of Structures

Senol Utku
Theory of Adaptive Structures: Incorporating Intelligence into Engineered Products

Jan G.M. van Mier
Fracture Processes of Concrete

S. Vigneswaran and C. Visvanathan
Water Treatment Processes: Simple Options

Limit Analysis and Concrete Plasticity

Second Edition

M. P. Nielsen

Department of
Structural Engineering and Materials
Technical University of Denmark

CRC Press

Boca Raton London New York Washington, D.C.

Library of Congress Cataloging-in-Publication Data

Nielsen, Mogens Peter, 1935–
Limit analysis and concrete plasticity / M.P. Nielsen. -- 2nd ed.
p. cm. -- (New directions in civil engineering)
Includes bibliographical references and index.
ISBN 0-8493-9126-1 (alk. paper)
1. Reinforced concrete construction. 2. Plastic analysis (Engineering) 3. Concrete--Plastic properties. I. Title. II. Series.
TA683.N54 1998
624.1′8341—dc21

98-23303
CIP

International Standard Book Number 0-8493-9126-1
Library of Congress Card Number 98-23303
Printed in the United States of America 1 2 3 4 5 6 7 8 9 0
Printed on acid-free paper

Preface
to
2nd edition

This second edition of a book first published in 1984 sets out, like its predecessor, to explain the basic principles of plasticity theory and its application to the design of reinforced and prestressed concrete structures. It is intended for use by advanced students and design engineers who wish to understand the subject in depth rather than simply apply current design codes. A scientific understanding of the subject has the benefit of allowing one to attack real problems much more effectively and safely. Once in a while it also permits a simple solution that does not call for solving a large number of equations on the computer: simplicity is one of the key features of plasticity design methods.

Since 1984 our understanding of plasticity theory as applied to reinforced and prestressed concrete structures has greatly increased. Plasticity theory covers the whole field of reinforced concrete -- that is, it can be applied to any structural element and to any structure. It is now possible to provide rational methods superior to the hitherto dominant empirical methods, and to unify calculations for reinforced and prestressed concrete. In addition, quick control of more sophisticated computerized solutions is now possible.

In this edition I have therefore placed much more emphasis on practical design. Almost all elementary concrete mechanics problems have now been treated in such a way that the solutions may be directly applied by the designer.

Second, the fundamental problems associated with so-called effectiveness factors are explained in considerable detail. These factors reflect softening of the concrete and the fact that in an actual structure the concrete is cracked. Cracks may reduce the strength by up to 50 percent, sometimes more. We still cannot predict these factors except in very special cases, but we understand much better why they exist.

A number of new solutions of specific problems that are important in design are also included, covering concentrated forces, shear walls and deep beams, beams with normal forces and torsional moments, and new solutions dealing with membrane effects in slabs.

Probably the most significant progress has occurred in the treatment of shear in beams and slabs without shear reinforcement or with only a small amount of shear reinforcement. For decades this problem has bewildered researchers. It is shown that a theory based on cracking followed by yielding in the most dangerous crack gives both a clear mental picture about what physically takes place in a shear failure and also provides rather accurate solutions. Their simplicity makes them directly applicable in practical design.

The chapters on joints and bond strength have been extended so it is now fair to say that plastic theory provides reasonable solutions for most structural problems in reinforced concrete.

For almost a decade I took part in the development of Eurocode 2, the future common concrete standard in the European Union. I had the pleasant and difficult task of convincing my colleagues from the other member states, supported to a large extent by our Swiss colleagues, that plasticity theory had to be the basis of ultimate limit state analysis. Eurocode 2 contains some of the more spectacular solutions of plasticity theory which means that the whole Europe is now going to adopt it. In practice this will, of course, not happen overnight. The local traditions are too different, but no doubt it will be the future trend as it has been in Denmark since the beginning of this century and in Switzerland since Professor Thürlimann took over the concrete chair at ETH in Zürich.

Eurocode 2 has already been used in practice in the design of the Öresund Link (submerged tunnel and cable stayed bridge), the cross way from Denmark to the old Danish part of Sweden named Skåne.

Denmark is too small to be a significant research country and so research grants are correspondingly small. This book therefore could not have been accomplished without the work of a large number of students. It is because of my students, rather than the research councils, that most of the progress has taken place. It is impossible to mention them all here. However, I would like to call attention to my former Ph.D. student Bent Feddersen. He is one of the few engineers I know outside the University environment who deep in his heart believes in science. He has undertaken the task of making the technical specifications used in the firm, RAMBØLL, a major consultant, solely based on plasticity theory. This is not an easy job. The many suggestions he has made through the years have been a constant encouragement and inspiration for making progress in the application of plasticity theory.

I would like to mention also my former Ph.D. student Jin-Ping Zhang. Besides providing significant research contributions she has undertaken the difficult job of proofreading the whole text. Due to her efforts the book is much more readable than it would otherwise be. Of course, I am the one to be blamed for any errors that may still be present.

The word processing has mainly been done by my loyal secretary through the years, Bente Jensen. She has corrected my bad English and done an excellent job of finding a proper style of presentation. Chapter 6 was done by Lars Væver Petersen, Gunnar Guttesen and Thomas Jantzen. Most of the drawings are from the 1st edition, new ones have been prepared by Esther Martens.

The book is thus dedicated to my students examplified in the present research group in concrete mechanics. Their names are Junying Liu, Linh Cao Hoang, Morten Bo Christiansen, Lars G. Hagsten and Thomas Jantzen.

Finally, I would like to thank the CRC staff. Particular thanks are due to Felicia Shapiro, Navin Sullivan, Suzanne Lassandro and Mimi Williams, who did a great job in both improving the style and the accuracy of the text.

The Technical University of Denmark
Lyngby, 1998

M. P. Nielsen

Preface
to
1st edition

The purpose of this book is to serve as an introductory text on applications of plastic theories to the design of concrete structures. This subject has a long history in Denmark, where early theories of beams, frames, and slabs have been an everyday tool for many years. Newer theories, as those of disks, beam shear, and joints, are in recent years on their way to being adopted in practical design, while the latest developments, as the theories of punching shear and bond strength, are still awaiting several years of research activity.

Besides providing a useful tool for design work, the plastic theories, compared to the hitherto dominating empirical methods, have the advantage of leading to a thorough understanding of the failure mechanisms in concrete. Therefore, qualitative reasoning in the early stages of the design work of a structure or a structural element may help in selecting possible designs, and in excluding designs with unwanted properties. Furthermore, rational theories have the noteworthy advantage of being far easier to extend into applications within related areas not directly covered by cases studied, while such an extension is more or less impossible or at least extremely difficult and uncertain when dealing with empirical methods.

Research workers in Denmark and a few other, mainly small, countries have for a long time been wondering why the plastic theories of reinforced concrete have had such little impact on the research work in the larger countries. The reason is probably that the concrete world in the early days of concrete history was forced to develop its own methodology with sparse connections to classical mechanics with its roots in linear elastic theories. Since only little progress can be made by identifying concrete with a linear elastic body, the world of mechanics in its classical sense and the world of concrete have been and still are very separate. It is sincerely hoped that this book might serve as a modest contribution to a unification of the two worlds.

The more classical parts of the book, of course, rest heavily on the work of Danish pioneers like Ingerslev, Suenson, and Johansen. Other main parts of the text are based on my own work during the 1960s. I am grateful to the Danish Academy of Technical Sciences, the publisher of my Ph.D. dissertation and my doctoral thesis, for permission to use the material from these works.

The newer areas are mainly results from the work of research groups I have headed at The Engineering Academy in Aalborg and at the Structural Research Laboratory at The Technical University of Denmark in Lyngby. I am indebted to many colleagues and students for their collaboration. Only a few can be mentioned: A. Rathkjen and L. Pilegaard Hansen in Aalborg; and M. Bræstrup, Finn Bach, B. C. Jensen, J. F. Jensen, H. Exner, Uwe Hess, and Jens Kærn in Lyngby. The first draft of Chapters 7, 8, 9 and Section 3.6.5 were written by M. Bræstrup, B. C. Jensen, Uwe Hess, and H. Exner, respectively. For the translation, Mrs. Pauline Katborg and Mrs. Kirsten Aakjær are responsible. The manuscript was typed by Mrs. Kaja Svendsen, and the drawings were made by Mrs. Esther Martens.

I would like to thank all for their contribution to the book.

M. P. Nielsen
Lyngby, Denmark

INTRODUCTION

The theory of plasticity is a branch of the study of the strength of materials which can be traced back at least to Galileo [1638.1], who determined the failure moment of a beam composed of a material with infinite compression strength. In its simplest form this theory deals with materials that can deform plastically under constant load when the load has reached a sufficiently high value. Such materials are called *perfectly plastic materials*, and the theory dealing with the determination of the load-carrying capacity of structures made of such materials is called *limit analysis*.

A general formulation of a complete theory for perfectly plastic materials was given in 1936 by Gvozdev (see [38.1]), but his work was not known in the Western world until the 1950s, where previously, mainly in works of the Prager school at Brown University (see Drucker et al. [52.1] and Prager [52.2]), a very similar theory had been developed.

One of the most important improvements in the development of the plastic theory was undoubtedly the establishment of the *upper* and *lower bound theorems*. The contents of these theorems had indeed been known by intuition long before Gvozdev's work and those of the Prager school appeared, but a complete and precise formulation as given by Gvozdev and by Drucker, Greenberg, and Prager, proved very valuable. These important principles were also stated by Hill [51.1; 52.3].

A modern and exhaustive treatment of the theory of plasticity has been given by Martin [75.1] (see also Prager [59.3] and Hodge [59.4]). Textbooks in other principal languages include the monographs by Kachanov [69.1] (Russian), Masonnet and Save [63.1] (French), and Reckling [67.1] (German).

The use of the plastic properties of reinforced concrete structures goes back to the turn of the century. In the 1908 Danish code of reinforced concrete we find the first traces of a theory of plasticity in the principles given for calculation of continuous beams.

The early applications of plasticity to structural concrete consist of cases where the strength is governed mainly by the reinforcement, e.g., flexure of beams and slabs, and for such problems, the use of a plastic approach has become standard. Prominent examples are the yield hinge method for beams and frames (see Baker [56.1]) and the yield line theory for slabs.

The development of the theory for reinforced concrete slabs was initiated by Ingerslev [21.1; 23.1]. He suggested the calculation of homogeneously reinforced slabs on the assumption of a constant bending moment along certain lines, called yield lines, and he gave several examples of the application of the method.

Later, Ingerslev's work was continued by K. W. Johansen. In his works [31.1; 32.1; 43.1; 62.1] the yield lines have statical as well as geometrical significance as lines along which plastic rotation is taking place at the collapse load. It was thus made possible to estimate yield line patterns by purely geometrical considerations and to calculate upper bounds for the load-carrying capacity by the work equation, an essential extension of Ingerslev's method. Concurrently with K. W. Johansen's work in Denmark, corresponding work was carried out in the U.S.S.R. by Gvozdev (see [49.1]), among others.

One of the most important theoretical problems left unsolved by K. W. Johansen was the establishment of a yield condition for orthotropic slabs, whereby reinforced concrete slabs of greater practical importance than isotropic slabs can be dealt with theoretically. K. W. Johansen's work did include a proposal for the calculation of orthotropic slabs, but for a long time it was necessary to regard the proposal as a purely intuitive suggestion for a practical solution. However, it was proved that the yield conditions derived by the author [63.2; 63.3] for orthotropic slabs lead to Johansen's method as a special case.

Another unsolved problem was a deeper understanding of the nodal force concept. Johansen's presentation left much to be desired as far as clarity was concerned, and it turned out that it was not entirely correct, since limitations to the applicability of the theory had to be introduced (Wood [61.5] and Nylander [60.2; 63.4]).

During the 1960s there was growing interest outside Denmark in the plastic theory for reinforced concrete slabs. Wood [61.5] increased our understanding of the membrane effect, which was studied previously by Ockleston [55.1]; an effect that results in higher load-carrying capacities than those calculated according to the simple theory. Further, Sawczuk and Jaeger [63.5] presented the plastic theory for slabs in general, and similarly, the yield line theory.

The development and application of plastic theory have also lead to attempts to create a lower bound method for slabs. Hillerborg's works [56.2; 59.1; 74.1] represent such an attempt. However, this method has not been given the same general character as the upper bound method. Moreover, even though upper bound solutions are always unsafe from a theoretical point of view, one can generally keep the membrane effect in reserve, and this actually often renders upper bound solutions safe. The strip method has been further developed by Hillerborg [74.1] and by others [68.1; 68.2; 75.7; 78.22].

By the middle 1960s the slab theory had reached almost final form and at that time it appeared as a special and useful case of the general theory of perfectly plastic materials. The developments in slab theory since then have been concerned with three main subjects.

First, the theory as it was developed up to the middle of the 1960s had, as mentioned, taken into account only bending and twisting moments (i.e., the in-plane forces were neglected). This is a more severe restriction in the theory of reinforced concrete slabs than in the classical theory of plates, since forces develop in the middle plane in a reinforced concrete slab not only because of second-order strain effects and restrained edges but also because of the fact that as soon as the concrete cracks, the neutral axis seldom lies in the middle plane. Therefore, the cracking leads to in-plane forces even if the slab edges are not restrained. Several papers have been published on the subject (see Chapter 6), but a general, practical design method has only recently been developed.

Second, the general development of the optimization theory has also touched the theory of reinforced concrete slabs. The first results were reported by Wood [61.5] and Morley [66.1], who gave an exact solution for the simply supported square slab. Since then considerable progress has taken place and a great number of exact solutions exist (see Chapter 6).

Third, the rapid development of automatic data processing has led to a formulation of automatic design methods in reinforced forced concrete slab theory. One of the first contributions in this field was that of Wolfensberger [64.1]. The subject is undergoing rapid development (see Chapter 6) and programs for reinforced concrete slabs based on the theory of plasticity are now available.

The theory of reinforced concrete disks has not been subject to the same interest. Certain attempts to develop convenient formulas for the reinforcement necessary to carry given stress resultants were made as early as the 1920s.

The first attempts dealing with orthogonal reinforcement were made by Leitz [23.2; 25.1; 25.2; 26.1; 30.1] and Marcus [26.2]. Rosenbleuth [55.2], Falconer [56.3], and Kuyt [63.6; 64.2] treated skew reinforcement in accordance with Leitz's principles, the first by a graphic method without trying to solve the problem of optimizing the reinforcement. An attempt to solve this problem was made by K. W. Johansen [57.1], but he only obtained Leitz's formulas. The problem was also treated by Hillerborg [53.1]. The complete set of formulas for orthogonal reinforcement was set up by the author in 1963 [63.3] and the complete set of formulas for skew reinforcement in 1969 [69.2]. The model accepted for reinforced concrete in these works appears to have been used for the first time by N. J. Nielsen [20.1] in his investigations of the stiffness of slabs with different arrangements of reinforcement.

Whereas the plasticity theory of Gvozdev was formulated with explicit reference to structural concrete, the works of the Prager school and those of Hill were concerned primarily with metallic bodies, and plain concrete was long regarded as a brittle material, generally unfit for plastic analysis. The implications of applying rigorous limit analysis to reinforced concrete structures were discussed by Drucker [61.1].

In reinforced beams, slabs, and disks plastic behavior may be attributed essentially to reinforcement. A plastic theory for plain concrete and for reinforced concrete, where the properties of the concrete play an important role, is much more difficult because concrete is not a perfectly plastic material but exhibits a significant strain softening.

Shells generally belong to this group of problems. A practical, plastic method, therefore, has been formulated only for cylindrical shells acting as beams (Lundgren [49.2]).

Within the last decades, the plastic theory has been applied to a number of nonstandard cases, principally shear in plain and reinforced concrete, by research groups at The Technical University of Denmark (Nielsen et al. [78.1]; Bræstrup et al. [77.1]; Nielsen [84.11]). Similar research has been carried out at various other institutions, notably the Swiss Federal Institute of Technology in Zürich, (Müller [78.9]; Marti [80.6]). In May 1979 a Colloquium on Plasticity in Reinforced Concrete was organized in Copenhagen, sponsored by the International Association for Bridge and Structural Engineering. Most of the results obtained so far were collected in the conference reports [78.2; 79.1].

In U.S. important work has been carried out by W. F. Chen [82.8] and T. T. C. Hsu [93.9]. In Canada the experimental and theoretical work by Vecchio and Collins [82.4] has increased our understanding of the behavior of concrete in shear to a large extent. Japanese and Australian research has contributed as well. Recently I became aware of an important plasticity school in Ukraine headed by V. P. Mitrofanov. The developments in plastic theory have also influenced code work and teaching in many places. Eurocode 2, the future common concrete standard in Europe, has adopted many plastic solutions. Influence is also found in Japanese codes.

In this period of time much research is devoted to unravel the constitutive properties of concrete and reinforced concrete on a basic level. Accurate constitutive equations in finite element codes or similar are necessary. In this respect one might ask whether plastic theory has a place in future developments. The author believes that the answer is affirmative. Even at a time in future when advanced computer programs are accurate and reliable, which they are not yet, simple methods will be required for preliminary designs, for checking so-called advanced numerical calculations and for the purpose of formulating code rules. In all these respects the plastic theory, which is the only general theory we have, will be superior to the completely empirical approach.

CONTENTS

PREFACE to 2nd edition

PREFACE to 1st edition

INTRODUCTION

1	**THE THEORY OF PLASTICITY**	1
1.1	**CONSTITUTIVE EQUATIONS**	1
1.1.1	Von Mises's Flow Rule	1
1.2	**EXTREMUM PRINCIPLES FOR RIGID-PLASTIC MATERIALS**	9
1.2.1	The Lower Bound Theorem	10
1.2.2	The Upper Bound Theorem	11
1.2.3	The Uniqueness Theorem	12
1.3	**THE SOLUTION OF PLASTICITY PROBLEMS**	14
1.4	**REINFORCED CONCRETE STRUCTURES**	16
2	**YIELD CONDITIONS**	21
2.1	**CONCRETE**	21
2.1.1	Failure Criteria	21
2.1.2	Failure Criteria for Coulomb Materials and Modified Coulomb Materials	23
2.1.3	Failure Criteria for Concrete	32
2.1.4	Structural Concrete Strength	52
2.2	**YIELD CONDITIONS FOR REINFORCED DISKS**	70
2.2.1	Assumptions	70
2.2.2	Orthogonal Reinforcement	75
	The reinforcement degree	75
	Tension and compression	76
	Pure shear	77
	The yield condition in the isotropic case	80
	The yield condition in the orthotropic case	86
2.2.3	Skew Reinforcement	90
2.2.4	Uniaxial Stress and Strain	93
2.2.5	Experimental Verification	98
2.3	**YIELD CONDITIONS FOR SLABS**	98
2.3.1	Assumptions	98

2.3.2 Orthogonal Reinforcement 98
Pure bending 98
Pure torsion 100
Combined bending and torsion 105
Analytical expressions for the yield conditions 115
Effectiveness factors 117
2.3.3 An Alternative Derivation of the Yield Conditions for Slabs 118
2.3.4 Arbitrarily Reinforced Slabs 119
2.3.5 Experimental Verification 120
2.3.6 Yield Conditions for Shells 120
2.4 REINFORCEMENT DESIGN 120
2.4.1 Disks with Orthogonal Reinforcement 120
2.4.2 Examples 127
Pure tension 128
Shear 131
2.4.3 Disks with Skew Reinforcement 132
2.4.4 Slabs 135
2.4.5 Shells 137
2.4.6 Three-dimensional Stress Fields 137
2.4.7 Reinforcement Design According to the Elastic Theory 138
2.4.8 Stiffness in the Cracked State 140
2.4.9 Concluding Remarks 141

3 THE THEORY OF PLAIN CONCRETE 143

3.1 STATICAL CONDITIONS 143
3.2 GEOMETRICAL CONDITIONS 144
3.3 VIRTUAL WORK 145
3.4 CONSTITUTIVE EQUATIONS 145
3.4.1 Plastic Strains in Coulomb Materials 145
3.4.2 Dissipation Formulas for Coulomb Materials 149
3.4.3 Plastic Strains in Modified Coulomb Materials 153
3.4.4 Dissipation Formulas for Modified Coulomb Materials 157
3.4.5 Planes and Lines of Discontinuity 160
Strains in a plane of discontinuity 160
Plane strain 162
Plane stress 165
3.5 THE THEORY OF PLANE STRAIN FOR COULOMB MATERIALS 168
3.5.1 Introduction 168
3.5.2 The Stress Field 168
3.5.3 Simple, Statically Admissible Failure Zones 175
3.5.4 The Strain Field 178
3.5.5 Simple Geometrically Admissible Strain Fields 181
3.6 APPLICATIONS 187

3.6.1 Pure Compression of a Prismatic Body 187
3.6.2 Pure Compression of a Rectangular Disk 190
3.6.3 A Semi-infinite Body 191
3.6.4 A Slope with Uniform Load 195
3.6.5 Strip Load on a Concrete Block 198
Loading far from the edge 198
Loading near the edge 202
3.6.6 Point Load on a Cylinder or Prism 207
3.6.7 Design Formulas for Concentrated Loading 210
Approximate formulas 210
Semi-empirical formulas 216
Comparison with tests 223
Conclusion 230
Effect of reinforcement 231
Edge and corner loads 232
Group action 235
Size effects 237

4 DISKS 239

4.1 STATICAL CONDITIONS 239
4.2 GEOMETRICAL CONDITIONS 241
4.3 VIRTUAL WORK 241
4.4 CONSTITUTIVE EQUATIONS 242
4.4.1 Plastic Strains in Disks 242
4.4.2 Dissipation Formulas 244
4.5 EXACT SOLUTIONS FOR ISOTROPIC DISKS 246
4.5.1 Various Types of Yield Zones 246
4.5.2 A Survey of Known Solutions 248
4.5.3 Illustrative Examples 253
4.5.4 Comparison with the Elastic Theory 262
4.6 THE EFFECTIVE COMPRESSIVE STRENGTH OF REINFORCED DISKS 262
4.6.1 Strength Reduction due to Internal Cracking 263
4.6.2 Strength Reduction due to Sliding in Initial Cracks 274
4.6.3 Implications of Initial Crack Sliding on Design 278
4.6.4 Plastic Solutions taking into Account Initial Crack Sliding 281
4.6.5 Concluding Remarks 285
4.7 GENERAL THEORY OF LOWER BOUND SOLUTIONS 285
4.7.1 Statically Admissible Stress Fields 285
4.7.2 A Theorem of Affinity 289
4.7.3 The Stringer Method 290
4.7.4 Shear Zone Solutions for Rectangular Disks 300
Distributed load on the top face 300

Distributed load along a horizontal line 304
Arbitrary loads 305
Effectiveness factors 306
4.8 STRUT AND TIE MODELS 306
4.8.1 Introduction 306
4.8.2 The Single Strut 306
4.8.3 Strut and Tie Systems 311
4.8.4 Effectiveness Factors 320
4.8.5 More Refined Models 325
4.9 SHEAR WALLS 328
4.9.1 Introduction 328
4.9.2 Strut Solutions Combined with Web Reinforcement 329
4.9.3 Diagonal Compression Field Solution 336
4.9.4 Effectiveness Factors 342
4.9.5 Test Results 343
4.10 HOMOGENEOUS REINFORCEMENT SOLUTIONS 351
4.10.1 Loads at the Top Face 351
4.10.2 Loads at the Bottom Face 355
4.10.3 A Combination of Homogeneous and Concentrated Reinforcement 355
4.10.4 Very Deep Disks 357
4.11 DESIGN ACCORDING TO THE ELASTIC THEORY 359

5 BEAMS 365

5.1 BEAMS IN BENDING 365
Load-carrying capacity 365
Effectiveness factors 368
5.2 BEAMS IN SHEAR 373
5.2.1 Maximum Shear Capacity, Transverse Shear Reinforcement 373
Lower bound solutions 373
Upper bound solutions 381
5.2.2 Maximum Shear Capacity, Inclined Shear Reinforcement 386
Lower bound solutions 386
Upper bound solutions 388
5.2.3 Maximum Shear Capacity, Beams without Shear Reinforcement 390
Lower bound solutions 390
Upper bound solutions 391
5.2.4 The Influence on Shear Capacity of Longitudinal Reinforcement 393
Beams with shear reinforcement 393

Beams without shear reinforcement 394
5.2.5 Effective Concrete Compressive Strength for Beams in Shear 397
Beams with shear reinforcement 397
Beams without shear reinforcement 400
5.2.6 Theory of Beams without Shear Reinforcement 401
5.2.7 Design of Shear Reinforcement in Beams 422
Beams with constant depth and arbitrary transverse loading 422
Beams with normal forces 428
Beams with variable depth 435
Beams with bent-up bars or inclined prestressing reinforcement 435
Variable θ solutions 436
Lightly reinforced beams 441
Beams with strong flanges 450
Beams with arbitrary cross section 450
5.3 BEAMS IN TORSION 452
Reinforcement design 452
Corner problems 457
Torsion capacity of rectangular sections 459
Effectiveness factors 465
5.4 COMBINED BENDING, SHEAR, AND TORSION 466

6 SLABS 473

6.1 STATICAL CONDITIONS 473
6.1.1 Internal Forces in Slabs 473
6.1.2 Equilibrium Conditions 473
6.1.3 Lines of Discontinuity 475
6.2 GEOMETRICAL CONDITIONS 476
6.2.1 Strain Tensor in a Slab 476
6.2.2 Conditions of Compatibility 478
6.2.3 Lines of Discontinuity, Yield Lines 479
6.3 VIRTUAL WORK, BOUNDARY CONDITIONS 480
6.3.1 Virtual Work 480
6.3.2 Boundary Conditions 482
6.4 CONSTITUTIVE EQUATIONS 487
6.4.1 Plastic Strains in Slabs 487
6.4.2 Dissipation Formulas 490
6.5 EXACT SOLUTIONS FOR ISOTROPIC SLABS 493
6.5.1 Various Types of Yield Zones 493
Yield zone of type 1 493
Yield zone of type 2 494
Yield zone of type 3 496
Yield lines 497

	The circular fan	498
6.5.2	Boundary Conditions	499
	Boundary conditions for yield lines	499
	Boundary conditions for yield zones	501
6.5.3	A survey of Exact Solutions	503
6.5.4	Illustrative Examples	506
	Simple statically admissible moment fields	506
	Simply supported circular slab subjected to uniform load	511
	Simply supported circular slab with circular line load	512
	Semi circular slab subjected to a line load	514
	Rectangular slab subjected to two line loads	516
	Hexagonal slab subjected to uniform load	518
	Concentrated force at a corner	519
	Ring-shaped slab under torsion	522
	Rectangular slab subjected to uniform load	523
6.6	**UPPER BOUND SOLUTIONS FOR ISOTROPIC SLABS**	525
6.6.1	The Work Equation Method and the Equilibrium Method	525
6.6.2	The Relationship between the Work Equation Method and the Equilibrium Method	526
	Bending and torsional moments in the neighborhood of yield lines	526
6.6.3	Nodal Forces	530
	Nodal forces of type 1	531
	Nodal forces of type 2	533
6.6.4	Calculations by the Equilibrium Method	537
6.6.5	Geometrical Conditions	540
6.6.6	The Work Equation	540
6.6.7	Examples	541
	Square slab supported on two adjacent edges	541
	Rectangular slab supported along all edges	546
	Triangular slab with uniform load	551
	Line load on a free edge	552
	Concentrated load	553
	Simply supported square slab with concentrated load	556
6.6.8	Practical Use of Upper Bound Solutions	556
6.7	**LOWER BOUND SOLUTIONS**	571
6.7.1	Introduction	571
6.7.2	Rectangular Slabs with Various Support Conditions	572
	A slab supported on four edges	573
	A slab supported on three edges	577
	A slab supported on two adjacent edges	580
	A slab supported along one edge and on two columns	584
	A slab supported on two edges and on a column	586

Other solutions 587
6.7.3 The Strip Method 587
Square slab with uniform load 589
One-way slab with a hole 590
Triangular slab with a free edge 592
Angular slab 593
Line load on a free edge 595
Slabs supported on a column 598
Concentrated force on simply supported slab 602
Flat slab 603
6.7.4 Some Remarks Concerning the Reinforcement Design 607
6.8 ORTHOTROPIC SLABS 608
6.8.1 The Affinity Theorem 608
6.8.2 Upper Bound Solutions 614
6.9 ANALYTICAL OPTIMUM REINFORCEMENT SOLUTIONS 622
6.10 NUMERICAL METHODS 623
6.11 MEMBRANE ACTION 625
6.11.1 Membrane Effects in Slabs 625
6.11.2 Unreinforced One-way Slabs 629
6.11.3 Work Equation 632
6.11.4 Unreinforced Square Slabs 635
6.11.5 Unreinforced Rectangular Slabs 640
6.11.6 The Effect of Reinforcement 642
6.11.7 Comparison with Tests 643
6.11.8 Conclusion 644

7 PUNCHING SHEAR OF SLABS 649

7.1 INTRODUCTION 649
7.2 INTERNAL LOADS 650
7.2.1 Concentric Loading, Upper Bound Solution 650
The failure mechanism 650
Upper bound solution 651
Analytical results 656
7.2.2 Experimental Verification, Effectiveness Factors 660
Failure surface 660
Ultimate load 661
7.2.3 Practical Applications 664
7.2.4 Eccentric Loading 668
7.2.5 The Effect of Counterpressure and Shear Reinforcement 676
7.3 EDGE AND CORNER LOADS 683
7.3.1 Introduction 683
7.3.2 Corner Load 686
7.3.3 Edge Load 695
7.3.4 General Case of Edge and Corner Loads 696

7.3.5 Eccentric Loading 702
7.4 CONCLUDING REMARKS 708

8 SHEAR IN JOINTS 711

8.1 INTRODUCTION 711
8.2 ANALYSIS OF JOINTS BY PLASTIC THEORY 711
8.2.1 General 711
8.2.2 Monolithic Concrete 712
8.2.3 Joints 716
8.2.4 Statical Interpretation 717
8.2.5 Axial Forces 718
8.2.6 Effectiveness Factors 719
8.2.7 Skew Reinforcement 723
8.2.8 Compression Strength of Specimens with Joints 726
8.3 STRENGTH OF DIFFERENT TYPES OF JOINTS 730
8.3.1 General 730
8.3.2 The Crack as a Joint 730
8.3.3 Construction Joints 740
8.3.4 Butt Joints 750
8.3.5 Keyed Joints 753

9 THE BOND STRENGTH OF REINFORCING BARS 765

9.1 INTRODUCTION 765
9.2 THE LOCAL FAILURE MECHANISM 766
9.3 FAILURE MECHANISMS 772
9.3.1 Review of Mechanisms 772
9.3.2 Splice Strength versus Anchor Strength 773
9.3.3 The Most Important Mechanisms 775
9.3.4 Lap Length Effect 776
9.3.5 Development Length 777
9.4 ANALYSIS OF FAILURE MECHANISMS 781
9.4.1 General 781
9.4.2 Corner Failure 781
9.4.3 *V*-notch Failure 788
9.4.4 Face Splitting Failure 796
9.4.5 Concluding Remarks 797
9.5 ASSESSMENT OF ANCHOR AND SPLICE STRENGTH 802
9.6 EFFECT OF TRANSVERSE PRESSURE AND SUPPORT REACTION 812
9.7 EFFECT OF TRANSVERSE REINFORCEMENT 828
9.7.1 General 828
9.7.2 Transverse Reinforcement does not Yield 830

9.7.3 Transverse Reinforcement Yields 839
9.8 CONCLUDING REMARKS 846

REFERENCES 849

INDEX 891

Chapter 1

THE THEORY OF PLASTICITY

1.1 CONSTITUTIVE EQUATIONS

1.1.1 Von Mises's Flow Rule

A *rigid-plastic material* is defined as a material in which no deformations occur (at all) for stresses up to a certain limit, the yield point. For stresses at the yield point, arbitrarily large deformations are possible without any change in the stresses. In the uniaxial case, a tensile or compressive rod, this corresponds to a stress-strain curve as shown in Fig. 1.1.1. The stress, the *yield stress*, for which arbitrarily large strains are possible, is denoted f_Y. In the figure the yield stresses for tensile and compressive actions are assumed equal.

A rigid-plastic material does not exist in reality. However, it is possible to use the model when the plastic strains are much larger than the elastic strains.

To render the following independent of the actual structure considered, we define a set of *generalized stresses* $Q_1, Q_2, \ldots, Q_n$ which exhibit the property that a product of the form $Q_1 q_1 + \ldots + Q_n q_n$, where $q_1, q_2, \ldots, q_n$ are the corresponding *generalized strains*, defines the virtual work per unit volume, area, or length of the structure. In a three-dimensional continuum, the Q_i are the six components of the stress tensor and the q_i are the corresponding six components of the strain tensor. For a plane beam the Q_i may be selected as the bending moment M, the normal force N, the shear force V, and the q_i are selected as the corresponding strain components.

For arbitrary stress fields the yield point is assumed to be determined by a *yield condition*, for example,

$$f(Q_1, Q_2, \ldots, Q_n) = 0 \tag{1.1.1}$$

Values of $Q_1, Q_2, \ldots$ satisfying (1.1.1) give combinations of the generalized stresses, rendering possible arbitrarily large strains without any change in the stresses. The strains occuring in rigid-plastic bodies are assumed to be plastic deformations (i.e., permanent deformations).

We assume that stresses rendering $f < 0$ correspond to stresses that can be sustained by the material, and thus these stress combinations give no strains. It is assumed that stresses giving $f > 0$ cannot occur.

We now consider a weightless body with a homogeneous strain field characterized by the strains $q_1, q_2, \ldots, q_n$. We pose the question: Which stresses correspond to this strain field in a rigid-plastic body, and which work D must be performed to deform a rigid-plastic body to the given strains?

When the stresses are known, the reply to the latter question is

$$D = \int_V (Q_1 q_1 + \ldots) dV = \int_V W dV \qquad (1.1.2)$$

where W is the work per unit volume, area, or length. In the following, W denotes the dissipation per unit volume, area, or length and D denotes the dissipation.

The reply to the question posed above can be verified or invalidated only by experiment; and for materials with a tensile-compressive stress-strain curve such as that shown in Fig. 1.1.1 (e.g., valid for mild steel) the reply is assumed to be given by *von Mises's hypothesis on maximum work* [28.1]. In accordance with this hypothesis, *the stresses corresponding to a given strain field assume such values that W becomes as large as possible.* The principle implies that of all stress combinations satisfying the yield condition (1.1.1), we should find the stress field rendering the greatest possible work W (i.e., the greatest possible resistance against the deformation in question).

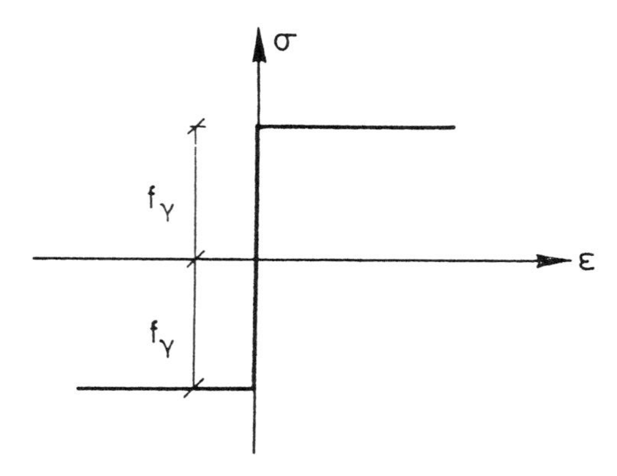

Figure 1.1.1 Uniaxial stress-strain relation for a rigid-plastic material.

Let us imagine that the yield condition (1.1.1) is drawn in a Q_1, $Q_2,\ldots, Q_n$-coordinate system. The surface $f = 0$ denotes the yield surface. Let us imagine the strains represented in the same coordinate system by a vector

$$\bar{\epsilon} = (q_1,q_2,\ldots,q_n) \tag{1.1.3}$$

Setting

$$\bar{\sigma} = (Q_1,Q_2,\ldots,Q_n) \tag{1.1.4}$$

W is equal to the scalar product

$$W = \bar{\sigma}\cdot\bar{\epsilon} \tag{1.1.5}$$

If $\bar{\epsilon}$ is assumed given, $\bar{\sigma}$ is to be determined so that W becomes as large as possible, subject to the condition

$$f(\bar{\sigma}) = 0 \tag{1.1.6}$$

Let us make the provisional assumption that the yield surface is differentiable without plane surfaces or apexes. Furthermore, we assume that the yield surface is *convex*. Finally, the yield surface is assumed to be a closed surface containing the point $(Q_1,\ldots) = (0,\ldots)$. If the variation of W is required to be zero when the stress field is varied from that which is sought, we have

$$\delta W = \delta Q_1 q_1 + \ldots = 0 \tag{1.1.7}$$

Since the stress field $Q_1 + \delta Q_1,\ldots$ also satisfies the condition $f = 0$ (the stress field is varied on the yield surface), we have

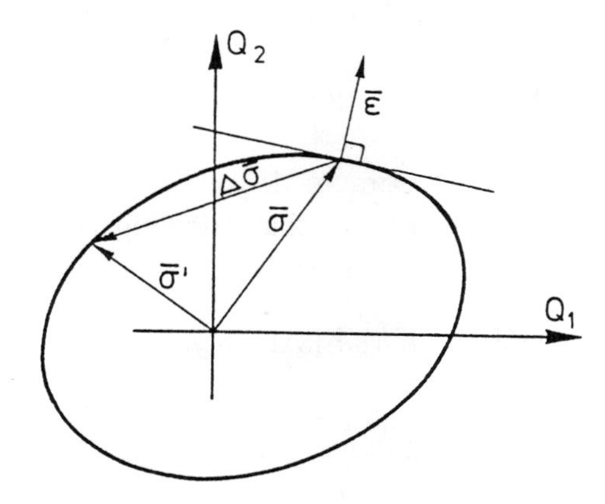

Figure 1.1.2 Maximum work hypothesis.

$$\frac{\partial f}{\partial Q_1}\delta Q_1 + \ldots = 0 \tag{1.1.8}$$

As (1.1.7) and (1.1.8) apply to any variation $\delta Q_1,\ldots$, it is seen then that W is stationary ($\delta W = 0$) when and only when

$$q_i = \lambda \frac{\partial f}{\partial Q_i} \quad , \qquad i = 1,2,\ldots,n \tag{1.1.9}$$

where λ is an indeterminate factor. As known, $\lambda(\partial f/\partial Q_1,\ldots)$ is a normal to the yield surface.

Thus we have shown that when W is stationary, $\bar{\epsilon}$ must be a normal to the yield surface. Equation (1.1.9) is therefore called the *normality condition*. When $f < 0$ for stresses within the yield surface, $(\partial f/\partial Q_1,\ldots)$ is an outward-directed normal. As $W = \bar{\sigma}\cdot\bar{\epsilon}$ is assumed to be nonnegative, it is seen that $\lambda \geq 0$. *Thus* $\bar{\epsilon}$ *becomes an outward-directed normal to the yield surface.*

Under the given assumptions $\bar{\epsilon}$ uniquely determines a point $\bar{\sigma} = (Q_1,\ldots)$ on the yield surface, that is, the point where $\bar{\epsilon}$ is a normal to the yield surface. To show that apart from a stationary value of W, the normality condition also leads to a maximum value, another arbitrary stress field $\bar{\sigma}' = (Q_1', Q_2',\ldots) = \bar{\sigma} + \Delta\bar{\sigma} = (Q_1 + \Delta Q_1,\ldots)$ on the yield surface is considered. If $\bar{\sigma}'$ were the stress field corresponding to the given strain vector $\bar{\epsilon}$, the work would be

$$W' = Q_1' q_1 + \ldots = (Q_1 + \Delta Q_1) q_1 + \ldots = \bar{\sigma}\cdot\bar{\epsilon} + \Delta\bar{\sigma}\cdot\bar{\epsilon} \tag{1.1.10}$$

Since the yield surface is convex, the scalar product $\Delta\bar{\sigma}\cdot\bar{\epsilon}$ will be negative (see Fig. 1.1.2, which shows a two-dimensional case). Thus we have

$$W \geq W' \tag{1.1.11}$$

that is, the work is at a maximum. If the stress field $\bar{\sigma}' = \bar{\sigma} + \Delta\bar{\sigma}$ is entirely within the yield surface, the following applies:

$$W > W' \tag{1.1.12}$$

Equation (1.1.9) is denoted *von Mises's flow rule.*

Let us consider a beam with rectangular cross section ($b \cdot h$) of rigid-plastic material. The cross section is assumed loaded by a bending moment M and a normal force N, which are referred to the center of gravity. The load-carrying capacity is determined from the stress distribution shown in Fig. 1.1.3. It is seen that

$$\left(h - 2y_0\right) b f_Y = N \tag{1.1.13}$$

that is,

$$\frac{y_0}{h} = \frac{1}{2}\left(1 - \frac{N}{N_p}\right) \tag{1.1.14}$$

where the load-carrying capacity in pure tension (tension yield load),

$$N_p = bhf_Y \tag{1.1.15}$$

has been introduced. We then have

$$M = y_0 b f_Y \left(h - y_0\right) = \frac{1}{4} bh^2 f_Y \left[1 - \left(\frac{N}{N_p}\right)^2\right] = M_p \left[1 - \left(\frac{N}{N_p}\right)^2\right] \tag{1.1.16}$$

where the load-carrying capacity in pure bending (yield moment in pure bending),

$$M_p = \frac{1}{4} bh^2 f_Y = \frac{1}{4} h N_p \tag{1.1.17}$$

has been introduced. Setting

$$m = \frac{M}{M_p}, \qquad n = \frac{N}{N_p} \tag{1.1.18}$$

the yield condition is found to be

$$f(m,n) = m + n^2 - 1 = 0 \tag{1.1.19}$$

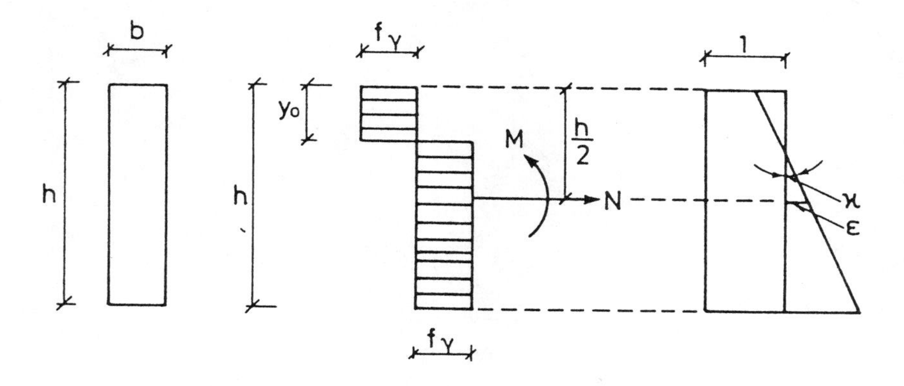

Figure 1.1.3 Stress and strain distribution in a rectangular beam of rigid-plastic material subjected to bending moment and normal force.

The formula applies to positive moments, as well as to tensile and compressive normal forces. It is seen that $f < 0$ for stress resultants which can be carried by the beam.

With the determined position of the neutral axis, and if the curvature of the beam is κ, the strain ε in the center of gravity is equal to

$$\epsilon = \kappa\left(\frac{1}{2}h - y_0\right) = \frac{1}{2}\kappa h \frac{N}{N_p} = \frac{1}{2}\kappa h n \tag{1.1.20}$$

(see Fig. 1.1.3).

Introducing the quantities

$$E = N_p \epsilon , \qquad K = M_p \kappa \tag{1.1.21}$$

we see that

$$E = \frac{1}{2}\kappa h N = 2\kappa M_p n = 2Kn \tag{1.1.22}$$

Thus if we have

$$K = \lambda , \qquad E = 2\lambda n \tag{1.1.23}$$

the dissipation per unit length of the beam is

$$W = M\kappa + N\epsilon = mK + nE \tag{1.1.24}$$

If we apply the flow rule (1.1.9), we get

$$E = \lambda \frac{\partial f}{\partial n} = 2\lambda n , \quad K = \lambda \frac{\partial f}{\partial m} = \lambda \tag{1.1.25}$$

which is seen to comply with the expressions found for E and K.

The quantities (m, n) and (K, E) are the generalized stresses and strains in the case considered.

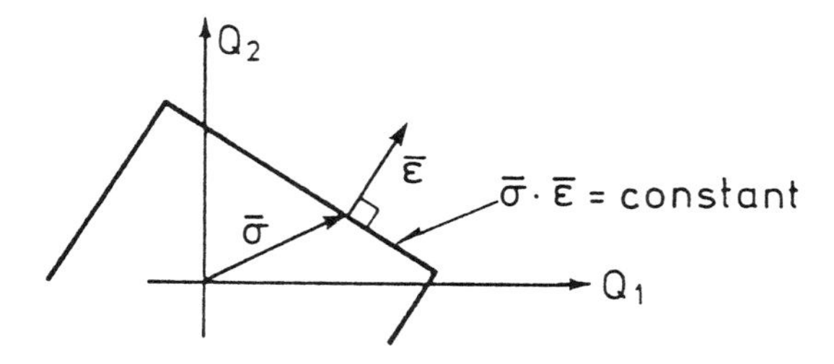

Figure 1.1.4 Flow rule at a straight part of a yield condition.

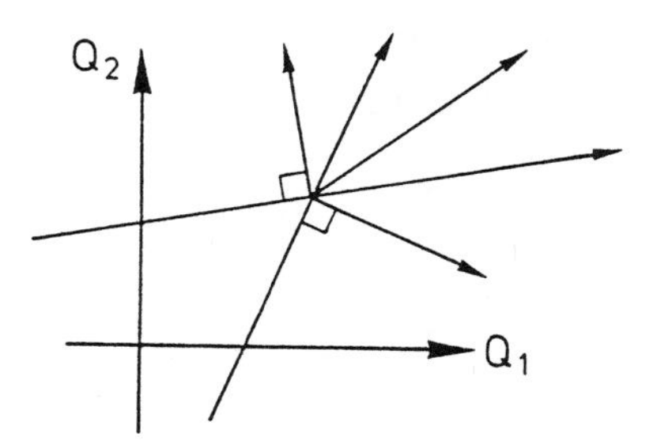

Figure 1.1.5 Flow rule at an apex.

It is important to note that the flow rule (1.1.9) only determines the strains, with the exception of a nonnegative factor λ in correspondence with the properties of the material in the uniaxial case. In other words, only the ratios between the strains can be determined.

If the yield surface contains plane parts, there is no unique relation between $\bar{\epsilon}$ and $\bar{\sigma}$. However, it is seen that W is uniquely determined from $\bar{\epsilon}$, since the scalar product $\bar{\sigma} \cdot \bar{\epsilon}$ has the same value all along the plane part when $\bar{\epsilon}$ is a normal to the plane part (see Fig. 1.1.4, showing a two-dimensional case).

If the yield surface contains an apex or an edge, $\bar{\epsilon}$ is assumed in a point of the apex or edge to lie arbitrarily in the angle determined by the limit positions of $\bar{\epsilon}$ when the point is approached in all possible ways. Thus (1.1.11) will still be valid. In Fig. 1.1.5 the strains possible in an apex are illustrated in a two-dimensional case.

The flow rule described here applies to an isotropic as well as to an anisotropic material.

Let us again consider the beam with a rectangular cross section $(b \cdot h)$ of rigid-plastic material. When this beam is subjected to the load-carrying capacity in pure tension $N_p = bhf_Y$, the strain distribution over the cross section may be any of the strain distributions corresponding to positive strains in the entire cross section. It means (see Fig. 1.1.6) that

$$-\frac{2\epsilon}{h} \leq \kappa \leq \frac{2\epsilon}{h} \tag{1.1.26}$$

or with the notation introduced earlier,

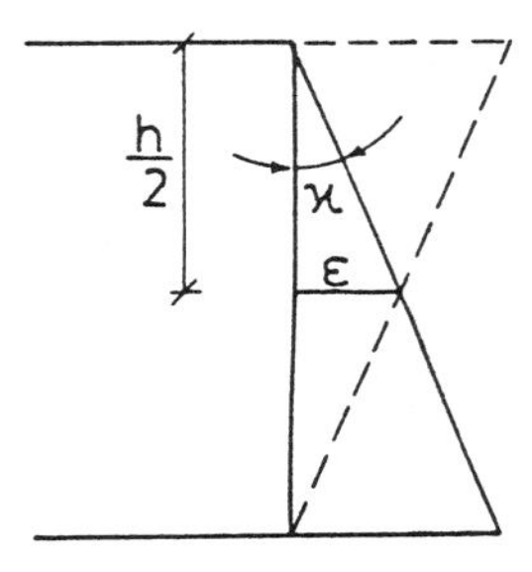

Figure 1.1.6 Limiting strain distributions in a beam subjected to the tension yield load.

$$-\frac{1}{2}E \leq K \leq \frac{1}{2}E \tag{1.1.27}$$

For $N = bhf_Y$ or, equivalently, $n = 1$, we have from (1.1.25)

$$E = 2\lambda, \qquad K = \lambda \tag{1.1.28}$$

(i.e., $K/E = 1/2$). This is the limit position for the normal when $n \rightarrow 1$ through positive moments.

Correspondingly, it is found that the limit position when n → 1 through negative moments is equal to $K/E = -1/2$. Thus it is seen that the foregoing rule on the possible strains in an apex corresponds to the conditions in the beam since (1.1.27) expresses the same conditions as those of Fig. 1.1.5. The conditions for $n = 1$ are illustrated in Fig. 1.1.7.

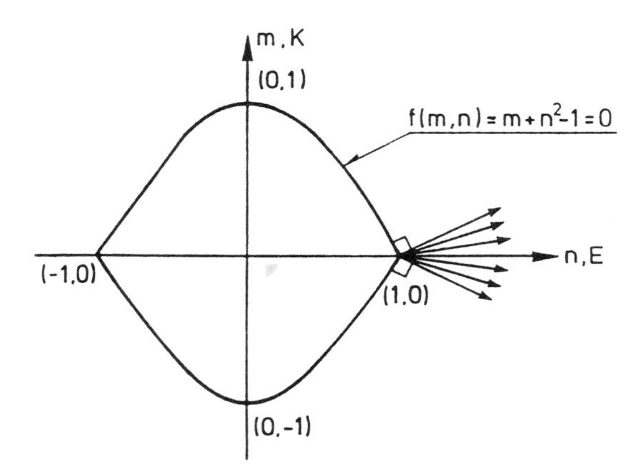

Figure 1.1.7 Flow rule at the apex of the yield condition for a rectangular beam subjected to bending moments and normal forces.

It should be strongly emphasized that the strains defined above for a rigid-plastic material will be for a real material the *strain increments* which might, but would not necessarily, occur when the stress vector reaches the yield surface. Therefore, if we want to verify a flow rule on the basis of experiments, we will have to measure the plastic strain increments at yielding, not the total strains. In many expositions of the plastic theory, either the strain differentials dq_1, dq_2,... or the strain velocities $\partial q_1/\partial t$, $\partial q_2/\partial t$,... are therefore used instead of the strains ε_1, ε_2,... when dealing with the flow rule.

The flow rule (1.1.9) has proved to accord well with reality when applied to metals with perfectly plastic properties, and it is now generally recognized that fundamental equations of this type (the increment type) should be preferred to other types of flow rules used previously.

It is, however, questionable if the application of this flow rule to reinforced concrete structures will give a reliable prediction of the plastic deformations for two reasons: first, because the tensile strength of the concrete will cause a discontinuous crack distribution instead of a continuous distribution as assumed in the theory, and second, because concrete is not a perfectly plastic material. However, as demonstrated in the following chapters, the plastic theory is in many cases able to predict accurately the load-carrying capacity for concrete structures. We shall have to put up with the fact that the prediction of plastic deformations is not necessarily accurate. Therefore, the application of the flow rule is in many cases quite formal and justified only because it enables us to use the upper bound technique (see Section 1.2) in a way that is consistent with the yield conditions.

1.2 EXTREMUM PRINCIPLES FOR RIGID-PLASTIC MATERIALS

As long as the stresses in a body of rigid-plastic material are below the yield point, no deformations occur. This idealization implies that we cannot determine the stress field in such a body when the stresses are below the yield point. When the loading increases to a point where it can be carried only by stresses at the yield point, unlimited deformations are possible without changing the load, if the strains (determined by the normality condition) correspond to a geometrically possible displacement field. The body is then said to be subject to *collapse by yielding*. The corresponding load is called the *collapse load* or the *load-carrying capacity* of the body. The terms *yield load* and *failure load* will also be used. The theory of collapse by yielding is termed *limit analysis*.

For determination of the load-carrying capacity of rigid-plastic bodies the following extremum principles are very useful. The theorems were first formulated by Gvozdev in 1936 [38.1], and independently by Drucker et al. in 1952 [52.1].

1.2.1 The Lower Bound Theorem

First, we show *that if the load has such a magnitude that it is possible to find a stress distribution corresponding to stresses within the yield surface and satisfying the equilibrium conditions and the statical boundary conditions for the actual load, then this load will not be able to cause collapse of the body.* A stress distribution such as this is denoted a *safe and statically admissible stress distribution.*

The proof is made indirectly. Let us assume that for the external load we can find a statically admissible stress distribution $\bar{\sigma} = (Q_1, Q_2, \ldots)$ which in the body or part of the body corresponds to stresses on the yield surface, and that the stresses correspond to strains $\bar{\epsilon} = (q_1, q_2, \ldots)$ in accordance with a displacement field that is geometrically possible for the body. Thus the principle of virtual work reads

$$\Sigma P_i u_i = \int_V \bar{\sigma} \cdot \bar{\epsilon} dV \tag{1.2.1}$$

where P_i and u_i are the external forces and the corresponding displacements, and dV is a volume element, an area element, or a length element. Now, according to the assumption, we can find a safe, statically admissible stress distribution. If this is denoted $\bar{\sigma}' = (Q_1', \ldots)$, we have

$$\Sigma P_i u_i = \int_V \bar{\sigma}' \cdot \bar{\epsilon} dV \tag{1.2.2}$$

using P_i and Q_i' as static quantities in the principle of virtual work, and u_i and $\bar{\epsilon}$ as deformation quantities (i.e., the same as used above).

Now, according to (1.1.12),

$$\bar{\sigma}' \cdot \bar{\epsilon} < \bar{\sigma} \cdot \bar{\epsilon} \tag{1.2.3}$$

leads to a contradiction.

It is possible, of course, that statically admissible stress distributions corresponding to stresses above the yield point also exist, but we have shown that if we can only find one corresponding to stresses below the yield point, the load cannot cause collapse.

If the external load is determined by one parameter $\mu > 0$ in such a way that the individual loading components are proportional to μ, we have a *proportional loading*. The theorem can then be used to find values of the load that are lower than the collapse load corresponding to $\mu = \mu_p$, hence the name the *lower bound theorem*. For all loads where a safe and statically admissible stress distribution can be found, we have

$$\mu < \mu_p \tag{1.2.4}$$

Equation (1.2.4) also applies if only part of the load is proportional to μ and the rest of the load (e.g., the weight) is constant.

In the following sections it is sometimes more convenient to define a safe stress field as a stress field *on* or within the yield surface. Then (1.2.4) reads $\mu \leq \mu_p$.

1.2.2 The Upper Bound Theorem

We now consider a geometrically possible displacement field u_i, which is assumed to correspond to the strains $\bar{\epsilon}$ that are possible in accordance with the normality condition.[1] The work that has to be performed to deform the body corresponding to this strain field is

$$D = \int_V W(q_1,\ldots)dV = \int_V \bar{\sigma} \cdot \bar{\epsilon}\, dV \tag{1.2.5}$$

where $\bar{\sigma}$ is the stresses corresponding to the strains $\bar{\epsilon}$.

A load P_i for which

$$\Sigma P_i u_i > \int_V W dV = \int_V \bar{\sigma} \cdot \bar{\epsilon}\, dV \tag{1.2.6}$$

that is, a load performing work greater than D, cannot be carried by the body.

This theorem is also proved indirectly. Let us assume that the load can be carried by the body. If so, a statically admissible stress distribution Q_i' corresponding to stresses on or within the yield surface can be found for the load P_i.

According to the principle of virtual work, we have

$$\Sigma P_i u_i = \int_V \bar{\sigma}' \cdot \bar{\epsilon}\, dV \tag{1.2.7}$$

[1] In the following the designation "geometrically possible failure mechanism" is also used.

It is, however, not certain that Q_i' according to the flow rule corresponds to the strains $\bar{\epsilon}$. Therefore, according to (1.1.11),

$$\int_V \bar{\sigma} \cdot \bar{\epsilon} dV \geq \int_V \bar{\sigma}' \cdot \bar{\epsilon} dV \tag{1.2.8}$$

Due to (1.2.7) and (1.2.8), we then have

$$\Sigma P_i u_i \leq \int_V \bar{\sigma} \cdot \bar{\epsilon} dV \tag{1.2.9}$$

in conflict with (1.2.6), whereby the theorem has been proved.

By proportional loading this theorem can be used to find values of the load that are greater than or equal to the collapse load. That is, if values of μ are determined that make (1.2.6) an equality (i.e., the work of the load is equal to the resistance against the displacement field), then even the smallest increase of this load cannot, according to the theorem, be carried by the body. From this it is seen that the load found is greater than or equal to the collapse load.

The equation for the determination of μ is

$$\mu \Sigma P_i u_i = \int_V W(q_1,...) dV \tag{1.2.10}$$

corresponding to the loading μP_i, where the P_i are now fixed quantities.

The equation yields

$$\mu = \frac{\int_V W(q_1,...) dV}{\Sigma P_i u_i} \geq \mu_p \tag{1.2.11}$$

Equation (1.2.10) is called the *work equation*. It is noted that the stresses corresponding to the geometrically possible strain field need not satisfy the equilibrium conditions.

We can conclude that if various geometrically possible strain fields are considered, the work equation can be used to find values of the load-carrying capacity that are greater than or equal to the true one: thus the name *upper bound theorem*.

1.2.3 The Uniqueness Theorem

According to the foregoing two theorems, for proportional loading only one load can be found to which it applies:

1. There is a statically admissible stress distribution corresponding to stresses on or within the yield surface.

2. The strains corresponding to the stresses according to the flow rule can be derived from a geometrically possible displacement field.

This is true because loads satisfying condition 1 are smaller than or equal to the collapse load, and loads satisfying condition 2 are greater than or equal to the collapse load. When condition 1 as well as 2 are satisfied, the load found is therefore equal to the collapse load, which is thus uniquely determined.

All of the body does not always "participate" in the collapse. It frequently occurs that only part of the body is deformed at collapse. In the remaining part of the body, the *rigid part*, the stresses cannot be uniquely determined in this case. We know only that they correspond to points within or on the yield surface.

It also occurs that several geometrically possible strain fields lead to the same load-carrying capacity. It can be shown that when we are dealing with two geometrically possible strain fields corresponding to the same external load, the stresses are identical in the parts of the body where in both cases strains different from zero occur. That is, neither the failure mechanism nor the stress field is uniquely determined for a rigid-plastic body; only the load-carrying capacity is.

The proofs of the limit analysis theorems rest on the convexity of the yield surface and the normality of the strain rate vector. For a material, Drucker [50.1] derived these conditions as consequences of the postulate that the work done on the increments of strain by the corresponding increments of stress is nonnegative. This means that it is impossible to extract mechanical energy by a loading cycle, which may be regarded as the definition of a stable material.

It is tempting to consider Drucker's stability postulate as a consequence of the second law of thermodynamics, and according to Ziegler [63.7] it is a special case of the principle of maximum entropy production. On the other hand, Green and Naghdi [65.1] have shown that the postulate implies restrictions on the flow rule which do not follow from the thermodynamic laws. Hence it must be regarded as a constitutive assumption, describing a certain class of materials (cf. Drucker [64.12]). If the stability postulate is valid for elements of materials described by stresses and strains, Ziegler [61.2] has shown that it also holds for a structure of the same materials, described by generalized stresses and strain rates.

The number of generalized variables for a body may be reduced by kinematical constraints or statical conditions. The corresponding yield surface is obtained from the original one by projecting on (respectively, intersecting with) a suitable subspace. The validity of the conditions of convexity and normality for such derived yield surfaces has been demonstrated by Sawczuk and Rychlewski [60.1] and Save [61.3].

The rigid-plastic model is, of course, a drastic idealization of reality. In fact, strains will occur in a body for stresses below the yield point. When the load is increased, the stresses will at some time reach the yield point at one or more points. Here plastic deformations might occur, but generally a further increase of the load will be possible since the stresses will be able to grow in the remaining parts of the body, and in this case there will still be a unique relationship between loading and strains. Only when the yield point has been reached in such a part of the body that plastic strains may occur corresponding to a geometrically possible displacement field will the load-carrying capacity be exhausted, as the strains then will be able to grow without any increase of the load. That is, only plastic strains occur at collapse, and it is these strains that are dealt with in the rigid-plastic model.

1.3 THE SOLUTION OF PLASTICITY PROBLEMS

The basic equations of the plastic theory are for the statical and geometrical parts the same as those used in the elastic theory. Only the constitutive equations are different.

In plastic solutions we often find that the displacement and/or the stress field are discontinuous. The geometrical discontinuities are treated in Chapter 3. The statical discontinuities can be illustrated as follows.

We consider a plane stress field in a disk. To satisfy the law of action and reaction, along a curve ℓ only the following conditions have to be fulfilled:

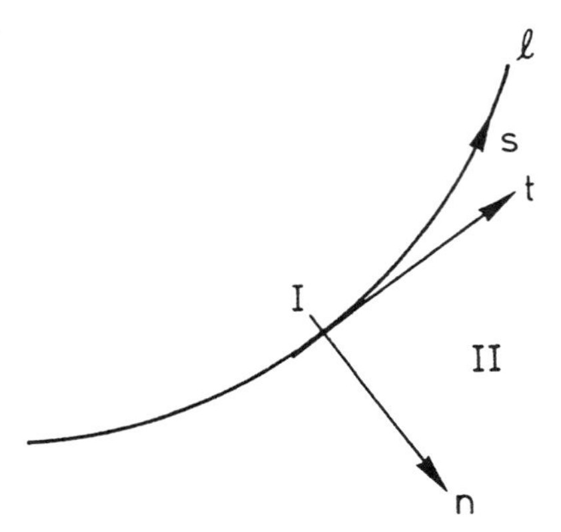

Figure 1.3.1 Coordinate system along a stress discontinuity line in a disk.

$$\sigma_n^I = \sigma_n^{II}, \qquad \tau_{nt}^I = \tau_{nt}^{II} \tag{1.3.1}$$

The notation refers to a local rectangular n, t-coordinate system on ℓ, where t is the tangent in the actual point, and superscripts I and II refer to the two parts I and II into which ℓ divides the body (see Fig. 1.3.1).

If only the stress field in the two parts I and II satisfies the equilibrium conditions, no claims for the sake of equilibrium are made on the stresses σ_t. Therefore, there might be a discontinuity in σ_t along ℓ, which in this case is called a *line of stress discontinuity*. This is illustrated in Fig. 1.3.2.

It is significant to note that a line of stress discontinuity does not give any contribution to the virtual work, which is seen by setting up equations separately for the two parts I and II, and then adding the equations. When (1.3.1) is satisfied, the usual result is attained.

On the other hand, a line of displacement discontinuity gives a contribution. If along ℓ there is a discontinuity in the displacement u_n and u_t of

$$\Delta u_n = u_n^{II} - u_n^I, \qquad \Delta u_t = u_t^{II} - u_t^I \tag{1.3.2}$$

we have the contribution

$$\Delta W_I = \int_\ell W_\ell ds = \int_\ell (\sigma_n \Delta u_n + \tau_{nt} \Delta u_t) b ds \tag{1.3.3}$$

to the internal work W_I.

Here b is the thickness of the disk and s is the arc length along ℓ. This result is obtained directly by setting up the virtual work equation for each separate part I and II and then adding the equations found. Discontinuities in other cases are treated correspondingly.

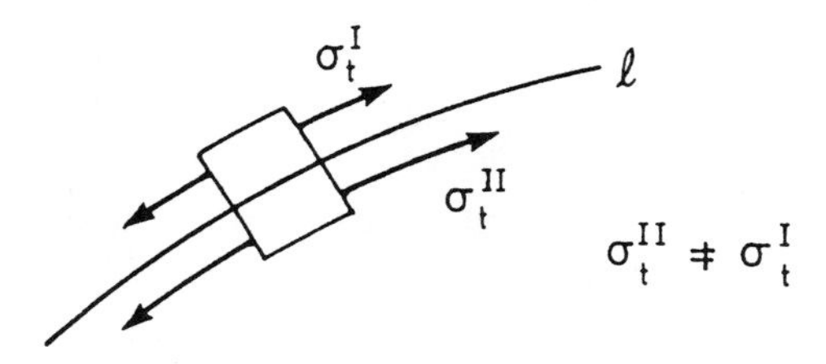

Figure 1.3.2 Example of stress discontinuities in a disk.

There is no standard method to solve load-carrying capacity problems in the plastic theory. Among other things this is due to the fact that the problem cannot be reduced to one single set of differential equations, which is again caused by the fact that the part of the body participating in the collapse is not known beforehand.

Upper and lower bound solutions for proportional loading can be found by the upper and lower bound theorems developed. An *upper bound solution* is found by considering a geometrically possible failure mechanism and by solving the work equation. A *lower bound solution* is found by considering a statically admissible stress field corresponding to stresses within or on the yield surface.

An *exact solution* demands construction of a statically admissible stress field corresponding to stresses within or on the yield surface in the whole body, as well as verification that a geometrically possible strain field, satisfying the constitutive equations, corresponds to this stress field.

1.4 REINFORCED CONCRETE STRUCTURES

Methods based on the lower bound theorem as well as the upper bound theorem have been developed for concrete structures (see the following chapters).

The most obvious application consists of using the lower bound theorem in the design of the reinforcement. If a statically admissible stress field is selected, the necessary reinforcement may be calculated on the basis of the selected stress field. Provided the concrete stresses corresponding to this stress field can also be carried, we have, according to the lower bound theorem, a safe structure. This applies to any concrete structure.

For a beam structure it is particularly simple to determine the complete group of statically admissible stress fields. If the beam structure is n times statically indeterminate, n moments or forces may be chosen arbitrarily and then the entire stress field may be calculated using statics only.

Now, of course, the question arises: How to select the redundant moments and forces? Traditionally, this is done by minimizing the total reinforcement consumption. The procedure rests on the theorem that for a beam structure with no or small normal forces, the moment field minimizing the total amount of bending reinforcement is the same as the moment field in the linear elastic range for a fully cracked structure.

This property of the solution minimizing the reinforcement consumption makes it two-fold attractive. Reinforcement is saved and one gets the best possible stress distribution in the serviceability limit state, i.e., a stress distribution with constant reinforcement stress. If the serviceability limit state load is, say, scaled down from the failure load to half this load, the reinforcement stress will be half the reinforcement stress (normally the yield stress) used to determine the reinforcement.

The proof of the theorem runs in the following way.

The moment field in the linear elastic range is determined as the one minimizing the elastic energy, i.e.,

$$\int \frac{M^2}{EI} dx = min \tag{1.4.1}$$

M is the bending moment, EI the bending stiffness and the integration is performed along the beam axis x of the whole structure.

The optimized plastic solution is the one minimizing the integral

$$\int \frac{|M|}{h_i} dx = min \tag{1.4.2}$$

h_i being the internal lever arm.

For a fully cracked member the reinforcement stress

$$\sigma_s \cong \frac{|M|}{h_i A_s} \tag{1.4.3}$$

A_s being the area of the tensile reinforcement.

Then the curvature will be

$$|\kappa| \cong \frac{\epsilon_s}{h_i} = \frac{|M|}{E_s h_i^2 A_s} \tag{1.4.4}$$

ε_s being the strain in the tensile reinforcement and E_s Young's modulus for the reinforcement. This means that the bending stiffness is

$$EI \cong E_s h_i^2 A_s = E_s |M| h_i / f_Y \tag{1.4.5}$$

since, if f_Y is the yield stress of the reinforcement, $A_s = |M|/h_i f_Y$.

Assuming f_Y = const. and E_s = const. we see that

$$\int \frac{M^2}{EI} dx = \frac{f_Y}{E_s} \int \frac{M^2}{|M| h_i} dx = \frac{f_Y}{E_s} \int \frac{|M|}{h_i} dx = \min \tag{1.4.6}$$

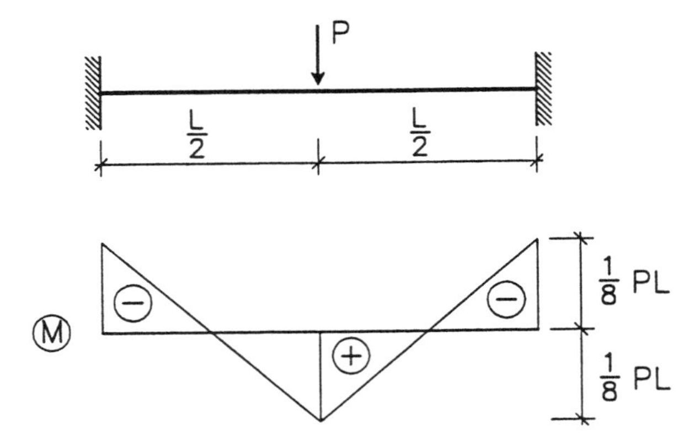

Figure 1.4.1 Beam fixed in both ends and loaded in the middle by a concentrated force.

i.e., the optimized plastic solution and the linear elastic solution for a fully cracked structure are identical. The proof of this theorem was given by the author in 1970 [70.4], but other proofs are found in the literature.

The optimized solution normally leads to curtailed reinforcement, but constant reinforcement will often be chosen in a practical situation.

For other kinds of structures similar theorems have not been established, but it is believed that by approaching the optimal plastic solution, one also approaches the best solution with regard to the serviceability limit state for any reinforced concrete structure where the main part of the elastic energy is absorbed in the reinforcement.

Fortunately, it may be shown that even rather strong deviations from the optimized plastic solution only change the reinforcement stresses a little.

Consider as an example a beam with constant depth fixed in both ends and loaded by a concentrated force P in the middle as shown in Fig. 1.4.1.

The optimized solution is shown in the figure. The absolute values of the bending moments in the ends equal the bending moment in the middle point. The linear elastic solution corresponding to constant bending stiffness will be the same in this case.

Now it may be shown that if the reinforcements in the top side and the bottom side are constant but differ in value, then if the top-side reinforcement is put to say half the value of the bottom reinforcement, the reinforcement stress in the top reinforcement in the ends is only increased by 24% compared to the reinforcement stress in the optimal solution, and the reinforcement stress in the bottom reinforcement is only decreased by 12% [96.3]. In [96.6] other solutions may be found, and methods are developed to determine in an approximate way the reinforcement stresses in beam structures with different top-side and bottom-side reinforcement.

Many examples of this kind and practical experience through decades have led to very liberal rules in the Danish reinforced concrete code [73.2] for the allowable deviations from the optimized plastic solution as well as solutions based on linear elastic, uncracked concrete. Just to mention one rule, it has for many years been allowed to reduce a bending moment calculated on the basis of linear elastic, uncracked concrete to a value ⅓ of the elastic value. Of course, when changing some moments the whole moment field must be changed to satisfy equilibrium. This may be compared with the small deviations of 10-15% often allowed in many countries.

Chapter 2

YIELD CONDITIONS

2.1 CONCRETE

2.1.1 Failure Criteria

Since our knowledge of the structure and composition of materials does not yet enable us to develop the failure criteria based on known natural laws, most failure criteria appear as hypotheses whose application to various materials will have to be evaluated from tests.

In 1776, Coulomb [1776.1] advanced the *frictional hypothesis*. It is based on the observation that failure often occurs along certain *sliding planes* or *yield planes*, the resistance of which is determined by a parameter termed the *cohesion* and an internal *friction*, the magnitude of which depends on the normal stress in the sliding plane. Coulomb's work has been described by Heyman [72.1].

The frictional hypothesis was formulated with the stresses as parameters. Of course, it is an obvious possibility to use strains. This had been suggested by Mariotte in 1682, and the theory was later (about 1840) elaborated by Saint-Venant and Poncelet. The hypothesis postulates that failure will occur when the greatest or smallest *principal strain*, respectively, assumes certain values characteristic for the material (the *principal strain hypothesis)*.

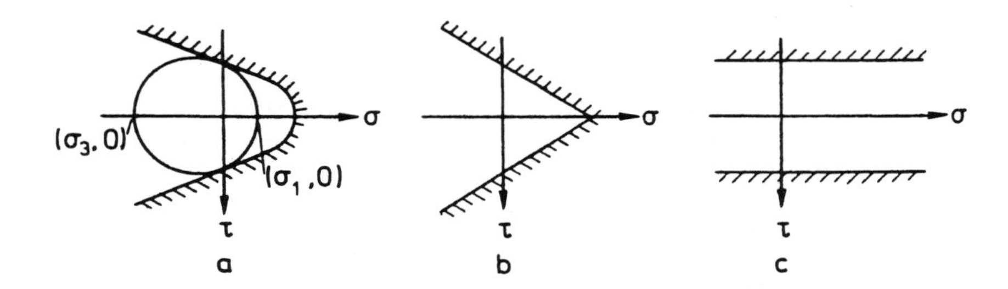

Figure 2.1.1 Mohr's failure envelope (a), Coulomb's frictional hypothesis (b), and Tresca's shear stress criterion (c).

About the same time, Rankine and Lamé advanced the hypothesis that failure occurs when the greatest or smallest *principal stress*, respectively, assumes certain characteristic values (the *principal stress hypothesis)*.

In 1868, Tresca suggested that for mild steel a failure condition could be used that requires only knowledge of the maximum value of the *shear stress*. This was verified experimentally by Guest in about 1900.

In 1882, a more general theory was advanced by Mohr, who assumed that failure occurs when the stresses in a section satisfy the condition

$$f(\sigma, \tau) = 0 \tag{2.1.1}$$

where $f(\sigma, \tau)$ is a function characteristic of the material, and where σ and τ are the normal stress and the shear stress, respectively, in the section. If this condition is illustrated in a σ, τ-coordinate system, a curve, *Mohr's failure envelope*, is obtained (see Fig. 2.1.1a). That failure occurs, meaning that (2.1.1) is satisfied, can be illustrated so that Mohr's circle intersecting the points $(\sigma_1, 0)$ and $(\sigma_3, 0)$, corresponding to the greatest and smallest principal stress, just touches the limiting curve (Fig. 2.1.1). The figure also shows Coulomb's frictional hypothesis (Fig. 2.1.1b) and Tresca's shear stress criterion (Fig. 2.1.1c), both of which can be considered as special cases of Mohr's theory.

Many suggestions have been made for the shape of the Mohr failure envelope. One of the earliest is a parabola (cf. Leon [35.1]), reflecting the experimental fact that the angle of friction (i.e., the slope of the curve) decreases with increasing compression stress. The parabola and the Coulomb hypothesis are defined by two parameters only, which generally are not sufficient to fit the experimental data. One more parameter is gained if the Coulomb hypothesis is combined with an extra limitation on the greatest principal stress σ_1, a *tension cutoff*. This hypothesis leads to the *modified Coulomb criterion*, which has the attractive feature that the tensile strength may be varied independently of the parameters which determine the sliding resistance.

For cast iron, the idea of combining the criteria of maximum shear stress and maximum normal stress appears to be due to Dorn [48.1]. For concrete, the combination of Coulomb sliding failure and Rankine separation was suggested by Cowan [53.2], Johansen [58.1], and Paul [61.4]. It is characteristic of any Mohr criterion that it does not involve the intermediate principal stress, which does have some influence on some materials, according to modern investigations.

Contrary to the hypotheses described above, where the absolute values of the stresses or strains are decisive for the occurrence of failure, in 1903 Beltrami suggested basing a failure criterion on energy considerations. Beltrami formulated his theory from the total internal energy, while Huber (1904) and von Mises (1913) formulated failure criteria that include only the distortion energy.

If the strength is calculated from the forces acting between the individual atoms in the material, values are found which may very well be up to 100 times larger than those that can be measured by testing the usual test specimens. The reduction is caused by a variety of imperfections in the material. Examples include impurities, special conditions in the contact areas between individual crystals (Prandtl, 1928), irregularities in the composition of the crystal lattices, dislocations (Taylor, 1934), and the occurrence of microcracks (Griffith, 1921). Taking such imperfections into consideration when using statistical methods, one can develop general failure conditions, but so far the results have been rather scanty. That the interatomic forces can be used at all as a basis for calculations has been verified by tests on single crystals.

2.1.2 Failure Criteria for Coulomb Materials and Modified Coulomb Materials

For a large group of materials it appears that reasonable failure conditions are attained by combining Coulomb's frictional hypothesis with a bound for the maximum tensile stress. The resulting failure criterion makes it natural to distinguish between two failure modes, *sliding failure* and *separation failure*. In both cases the name refers to what we imagine the relative motion between particles on each side of the failure surface to be. At sliding failure there is motion parallel to the failure surface, while motion at the separation failure is perpendicular to the failure surface. By sliding failure, motion along the failure surface is normally combined with motion off the failure surface.

Sliding failure is assumed to occur in a section when the Coulomb frictional hypothesis is fulfilled; that is, the shear stress $|\tau|$ in the section exceeds the *sliding resistance*, which, as mentioned, can be determined by two contributions. One contribution is *cohesion*, denoted c. The other contribution stems from a kind of internal friction and equals a certain fraction μ of the normal stress σ in the section. The parameter μ is called the *coefficient of friction*. If σ is a compressive stress, it gives a positive contribution to the sliding resistance; if σ is a tensile stress, it gives a negative contribution.

The condition for sliding failure is therefore

$$|\tau| = c - \mu\sigma \tag{2.1.2}$$

where c and μ are positive constants and σ is counted positive as a tensile stress. A material complying with the failure condition (2.1.2) is called a *Coulomb material*.

Separation failure occurs when the tensile stress σ in a section exceeds the *separation resistance* f_A, that is, when

$$\sigma = f_A \tag{2.1.3}$$

A material complying with (2.1.2) and (2.1.3) is called a *modified Coulomb material*. As is clear, three material constants, c, μ, and f_A, must be known for a modified Coulomb material.

If conditions (2.1.2) and (2.1.3) are illustrated in a σ,τ-coordinate system, we have the straight lines shown in Fig. 2.1.2 dividing the plane into two regions. When the stresses in a section correspond to a Mohr's circle lying within the boundary lines, no failure will occur, while stresses corresponding to circles touching the lines represent stress combinations which in fact involve failure. The failure mode depends on whether the contact point lies on the lines $|\tau| = c - \mu\sigma$, which involve sliding failure, or on the line $\sigma = f_A$, which involves separation failure. An angle φ given by $\tan\varphi = \mu$ is called the *angle of friction*.

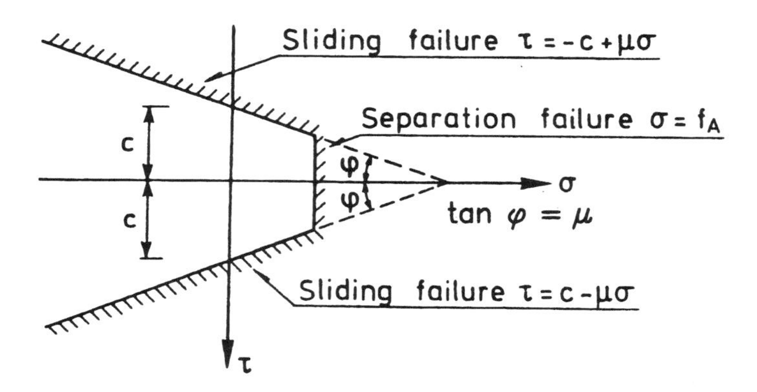

Figure 2.1.2 Failure criterion for a modified Coulomb material.

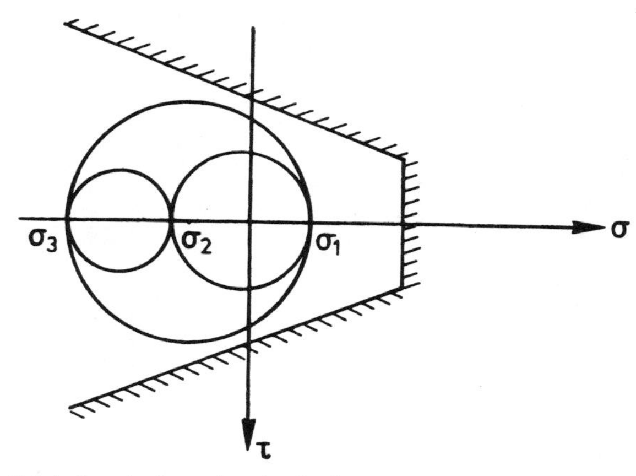

Figure 2.1.3 Mohr's circles of principal stresses.

By Fig. 2.1.3 it is easily discovered whether the stress field in a point, given by the principal stresses σ_1, σ_2, and σ_3, where $\sigma_1 > \sigma_2 > \sigma_3$, will cause failure. Drawing Mohr's circles corresponding to the stress field (see Fig. 2.1.3), we see that the points closest to the boundary lines lie on the circle with $\sigma_1 - \sigma_3$ as the diameter, which means that we have to focus only on the points on this circle. These points represent the stresses in sections parallel to the direction of the intermediate principal stress, and therefore any failure surfaces through the point will be parallel to this direction. As can be seen, the magnitude of the intermediate principal stress has no influence on the failure.

If the circle with diameter $\sigma_1 - \sigma_3$ lies within the boundary lines, failure will not occur. If the circle touches the boundary lines corresponding to sliding failure, this will for reasons of symmetry always occur at two points (see Fig. 2.1.4a). Sliding failure may therefore occur in the two sections which form the angle $90° - \varphi$ with each other. If the circle touches the line corresponding to separation failure, a separation failure will take place (see Fig. 2.1.4b).

Equations (2.1.2) and (2.1.3) can be transformed into relations between the principal stresses σ_1 and σ_3. From Fig. 2.1.4a we find, by perpendicular projection on one of the lines corresponding to sliding failure,

$$\frac{1}{2}(\sigma_1 - \sigma_3) = c\cos\varphi - \frac{1}{2}(\sigma_1 + \sigma_3)\sin\varphi \tag{2.1.4}$$

Introducing $\mu = \tan\varphi$, we have

$$\left(\mu + \sqrt{1+\mu^2}\right)^2 \sigma_1 - \sigma_3 = 2c\left(\mu + \sqrt{1+\mu^2}\right) \tag{2.1.5}$$

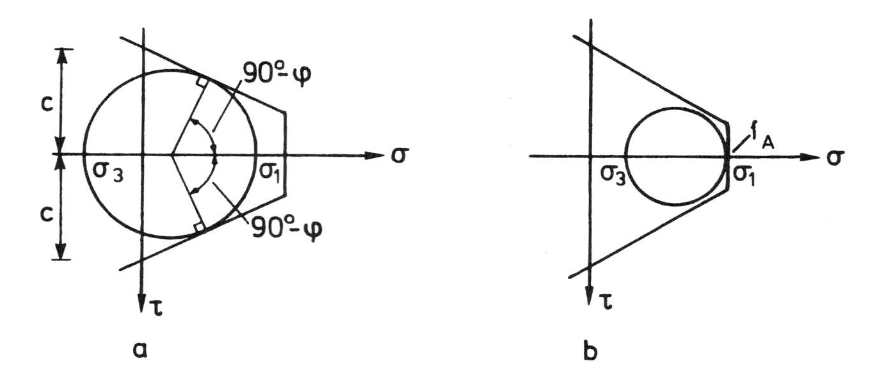

Figure 2.1.4 Mohr's circles at sliding failure (a) and separation failure (b).

If a parameter k is defined by

$$k = \left(\mu + \sqrt{1 + \mu^2}\right)^2 \tag{2.1.6}$$

the conditions for sliding failure can be written

$$k\sigma_1 - \sigma_3 = 2c\sqrt{k} \tag{2.1.7}$$

The condition for separation failure is (cf. Fig. 2.1.4b)

$$\sigma_1 = f_A \tag{2.1.8}$$

The compressive strength f_c of a material is determined by a test, where the stress field at failure is defined by $\sigma_1 = \sigma_2 = 0$ and $\sigma_3 = -f_c$. Since the compression test will always involve sliding failure (see Fig. 2.1.5), we have by application of (2.1.7),

$$-\sigma_3 = f_c = 2c\sqrt{k} \tag{2.1.9}$$

whereupon (2.1.7) can be written

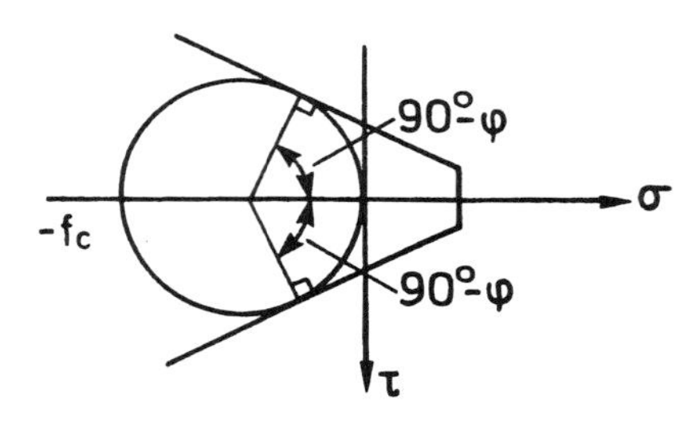

Figure 2.1.5 Mohr's circle at pure compression.

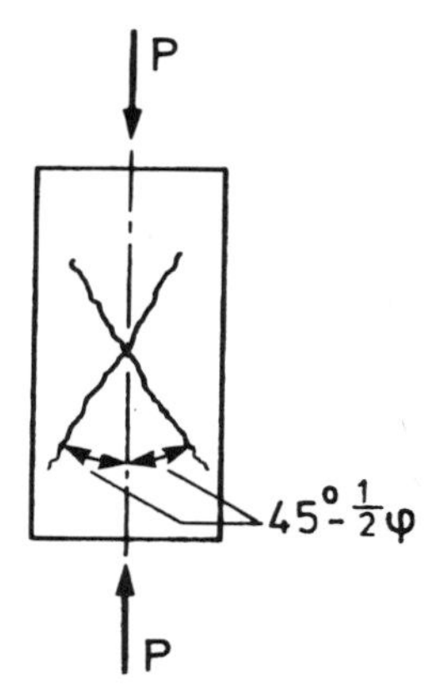

Figure 2.1.6 Failure sections at pure compression.

$$k\sigma_1 - \sigma_3 = f_c \tag{2.1.10}$$

By compression failure we may get failure in two sets of sections forming the angle 90° − φ with each other and, as seen by Mohr's circle, forming the angle 45° − $\varphi/2$ with the direction of force (see Fig. 2.1.6).

The failure condition will be satisfied in all sections that are tangent planes to the set of conical surfaces having the top angle 90° − φ and the axis parallel with the direction of force. A corresponding "conical failure" is often experienced in tests using cylindrical test specimens.

The tensile strength of the material f_t is determined by a test, where the stress field at failure is defined by $\sigma_1 = f_t$, $\sigma_2 = \sigma_3 = 0$. As seen from Fig. 2.1.7, the tensile test holds the possibility of sliding failure as well as separation failure.

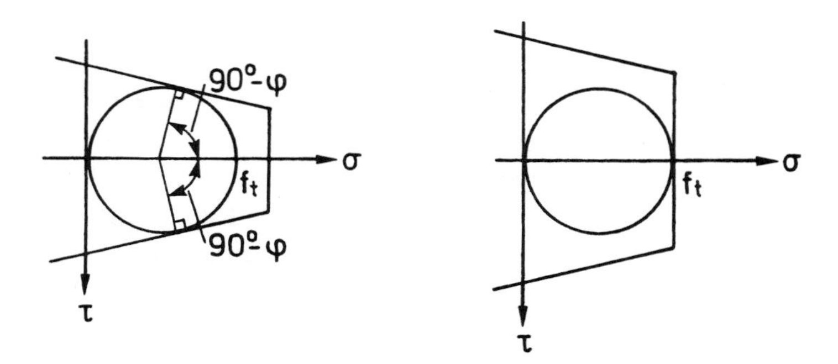

Figure 2.1.7 Mohr's circle at pure tension.

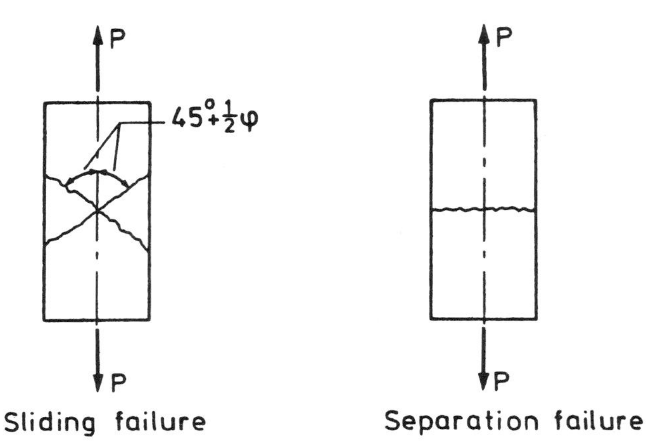

Figure 2.1.8 Failure sections at pure tension.

In the case of sliding failure we have, applying (2.1.10),

$$k\sigma_1 = kf_t = f_c \tag{2.1.11}$$

or

$$f_t = \frac{1}{k} f_c \tag{2.1.12}$$

In the case of separation failure we have, applying (2.1.8),

$$f_t = f_A \tag{2.1.13}$$

This means that the tensile failure is a sliding failure when

$$\frac{1}{k} f_c < f_A \tag{2.1.14}$$

and a separation failure when

$$f_A < \frac{1}{k} f_c \tag{2.1.15}$$

By the sliding failure we may get failure in two sets of sections forming the angle $90^{\circ} - \varphi$ with each other and the angle $45^{\circ} + \varphi/2$ with the direction of force. By separation failure the failure section is perpendicular to the direction of force (see Fig. 2.1.8).

The shear strength of the material f_v is determined by a test where the stress field at failure is defined by $\sigma_1 = -\sigma_3 = f_v$, $\sigma_2 = 0$. As seen from Fig. 2.1.9, this case holds the possibility of sliding as well as separation failure.

In the case of sliding failure we find, applying (2.1.10),

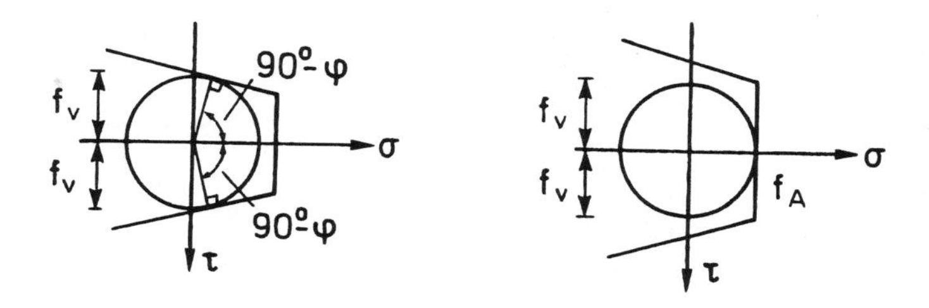

Figure 2.1.9 Mohr's circle at pure shear.

$$k\sigma_1 - \sigma_3 = (k+1)f_v = f_c \tag{2.1.16}$$

or

$$f_v = \frac{1}{1+k} f_c \tag{2.1.17}$$

If the failure is a separation failure, we have

$$f_v = f_A \tag{2.1.18}$$

Thus the shear failure is a sliding failure when

$$\frac{1}{1+k} f_c < f_A \tag{2.1.19}$$

and a separation failure when

$$f_A < \frac{1}{1+k} f_c \tag{2.1.20}$$

If in the shear test we denote the normals for the sections where the maximum shear stresses occur the x and y directions (see Fig. 2.1.10), the first principal direction will form an angle of 45° with these directions. Therefore, at sliding failure the failure sections will form an angle $\varphi/2$ with the x and y directions, respectively, and at separation failure the failure will take place in the section that forms an angle of 45° with the x and y directions.

A plane stress field is defined as a stress field in which the stresses in sections parallel to a plane are zero; this plane is a principal section, and the normal to the plane is a principal direction with the corresponding principal stress equal to zero. Denoting the principal stresses in sections perpendicular to the plane as σ_I and σ_{II}, the failure conditions can be written down.

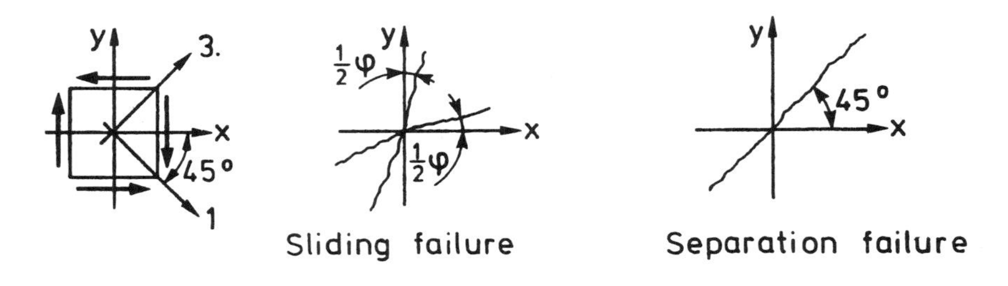

Figure 2.1.10 Failure sections at pure shear.

For $\sigma_I > \sigma_{II} > 0$ we have $\sigma_1 = \sigma_I$ and $\sigma_3 = 0$, which inserted into (2.1.10) gives

$$k\,\sigma_I = f_c \tag{2.1.21}$$

as the condition for sliding failure, while

$$\sigma_I = f_A \tag{2.1.22}$$

is the condition for separation failure.

For $\sigma_I > 0 > \sigma_{II}$ we have $\sigma_1 = \sigma_I$ and $\sigma_3 = \sigma_{II}$, and the condition for sliding failure is

$$k\,\sigma_I - \sigma_{II} = f_c \tag{2.1.23}$$

while the condition for separation failure is the same as in (2.1.22).

For $0 > \sigma_I > \sigma_{II}$ we have $\sigma_1 = 0$ and $\sigma_3 = \sigma_{II}$. Only sliding failure is possible, and the condition is

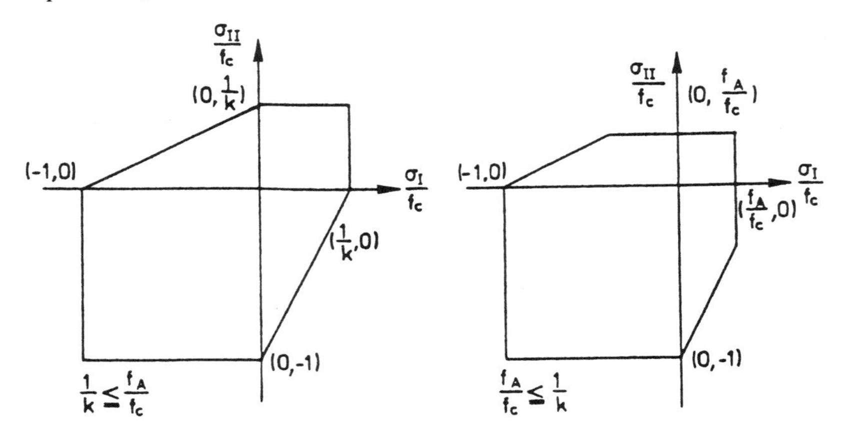

Figure 2.1.11 Failure criteria for a modified Coulomb material at plane stress.

$$k = \frac{1+\sin\varphi}{1-\sin\varphi} = \tan^2\left(45 + \frac{\varphi}{2}\right) = \left(\frac{1+\sin\varphi}{\cos\varphi}\right)^2 = \left(\frac{\cos\varphi}{1-\sin\varphi}\right)^2$$

$$\sin\varphi = \frac{k-1}{k+1} \qquad \cos\varphi = \frac{2\sqrt{k}}{k+1} \qquad \tan\varphi = \frac{k-1}{2\sqrt{k}} = \mu$$

$$f_c = 2c\sqrt{k} = (k+1)\cos\varphi \cdot c = \frac{2\cos\varphi}{1-\sin\varphi} \cdot c$$

$$k-1 = \frac{2\sin\varphi}{1-\sin\varphi} \qquad k = \left(\mu + \sqrt{1+\mu^2}\right)^2$$

$$k+1 = \frac{2}{1-\sin\varphi}$$

$$1-\sin\varphi = \frac{2}{k+1} \qquad 1+\sin\varphi = \frac{2k}{k+1}$$

$$\tan\left(45 + \frac{\varphi}{2}\right) = \frac{1+\sin\varphi}{\cos\varphi} = \sqrt{k}$$

$$\tan\left(45 - \frac{\varphi}{2}\right) = \frac{1-\sin\varphi}{\cos\varphi} = \frac{1}{\sqrt{k}}$$

$$\cos\left(45 + \frac{\varphi}{2}\right) = \frac{1}{\sqrt{k+1}} \qquad \cos\left(45 - \frac{\varphi}{2}\right) = \frac{\sqrt{k}}{\sqrt{k+1}}$$

$$\sin\left(45 + \frac{\varphi}{2}\right) = \frac{\sqrt{k}}{\sqrt{k+1}} \qquad \sin\left(45 - \frac{\varphi}{2}\right) = \frac{1}{\sqrt{k+1}}$$

Box 2.1.1 Useful formulas for manipulating the Coulomb failure criterion (kindly submitted to the author by P. Sandbye).

$$-\sigma_{II} = f_c \tag{2.1.24}$$

In Fig. 2.1.11 the conditions have been drawn in a σ_I, σ_{II}-coordinate system corresponding to the two cases $f_A/f_c > 1/k$ and $f_A/f_c < 1/k$. In the former case the failure will always be a sliding failure.

If we maintain that σ_I is greater than σ_{II}, only the part of the figure that lies below the angular bisector through the first and third quadrant belongs to the failure condition. As a rule this distinction is not made.

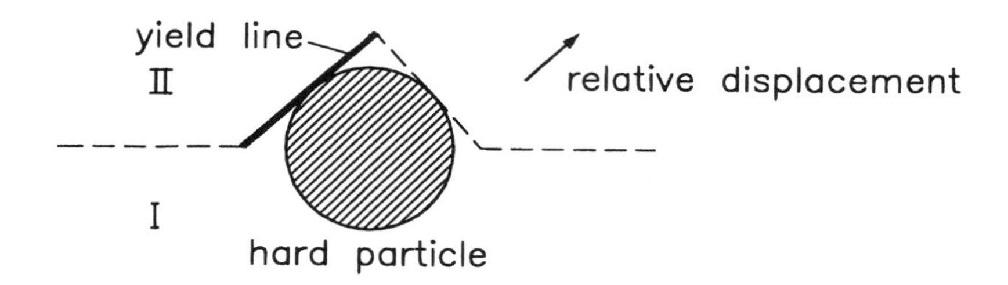

Figure 2.1.12 Yield lines in cement paste.

For later use we shall give some alternative expressions relating the parameters introduced. Equation (2.1.6) might be given the alternative forms

$$k = \left(\frac{1 + \sin \varphi}{\cos \varphi} \right)^2 = \left(\frac{\cos \varphi}{1 - \sin \varphi} \right)^2 = \frac{1 + \sin \varphi}{1 - \sin \varphi} = \tan^2 \left(\frac{\pi}{4} + \frac{\varphi}{2} \right) \tag{2.1.25}$$

Therefore, according to (2.1.9),

$$f_c = \frac{2c \cos \varphi}{1 - \sin \varphi} = 2c \frac{1 + \sin \varphi}{\cos \varphi} = 2c \tan \left(\frac{\pi}{4} + \frac{\varphi}{2} \right) \tag{2.1.26}$$

or

$$c = \frac{1}{2} f_c \frac{1 - \sin \varphi}{\cos \varphi} = \frac{1}{2} f_c \frac{\cos \varphi}{1 + \sin \varphi} \tag{2.1.27}$$

Some useful formulas related to the Coulomb failure condition have been summarized in Box 2.1.1.

2.1.3 Failure Criteria for Concrete

To make accurate predictions about stiffness and strength of concrete, the material must at least be treated as a two-phase material, cement paste and aggregate particles.

Normally the stiffness of the aggregate particles is much higher than the stiffness of the cement paste. The stress field in the two-phase material therefore is rather complicated with stress concentrations around the aggregate particles. In a uniaxial compression stress field, the stress concentrations may lead to cracking at a stress level far below the compressive strength. The cracks are often in the microcrack range, i.e., with crack widths less than $10\mu m$. They are mainly found in the interface between the cement paste and the aggregate particles, and their average direction is in the loading direction. Strictly speaking this means that concrete cannot be treated as an isotropic material, but must be treated as a material with load induced anisotropy.

Microcracks are present even before loading due to shrinkage of the cement paste.

Application of a simple isotropic failure condition like the modified Coulomb criterion is approximate.

Micromechanical modelling of the failure of concrete is still rare. A model based on the theory of plasticity has been formulated by Jin-Ping Zhang [97.2]. We shall shortly describe the main features of this model.

The basic assumption is that cement paste may be treated as an isotropic plastic material with the friction angle $\varphi = 0$. That $\varphi = 0$ for cement paste has been demonstrated by tests carried out at the Technical University of Denmark by K. K. B. Dahl [92.16] and T. C. Hansen [94.12]. These tests showed extremely high ductility of cement paste under triaxial conditions. However, it is well known that cement paste behaves in a rather brittle way in uniaxial tests and other tests with small confinement. In the model this is explained by the presence of hard unhydrated cement particles. Sliding failure may be prevented by the hard particles leading to the formation of yield lines in front of these particles. This is illustrated in Fig. 2.1.12, where part I is assumed at rest while part II is moving in the direction of the yield line formed in front of a hard particle.

There will also be some resistance from other parts of the yield line, shown with dotted lines in the figure, dependent on the amount of microcracking and the amount of load induced cracking. According to the model, the failure in cement paste is always ductile for compression stress fields and the apparent brittleness in cases of small confinement is due to the relatively short lengths of the yield lines formed in front of the hard particles. The failure condition in τ,σ-space for cement paste will be of the type shown in Fig. 2.1.13. Along *AB* the failure is governed by yield lines meeting the hard particles and will be apparently brittle. Along *BC* the failure will be ductile and the yield lines will be formed outside the hard particles.

In triaxial tests with circular-cylindrical test specimens we have $\sigma_1 = \sigma_2 = -p$, p being the pressure applied on the sides as a hydraulic pressure, and $\sigma_3 = -\sigma$, σ being the pressure applied on the ends of the cylinder. According to (2.1.10) we find $\sigma = f_c + p$ for $\varphi = 0$. However, the compressive strength f_c will normally be determined by the part *AB* of the failure condition shown in Fig. 2.1.13. Therefore, in triaxial tests σ as a function of p will normally be of the form shown in Fig. 2.1.14. There will be a transition curve *DE* leading from the uniaxial strength $\sigma = f_c$ for $p = 0$ (point *D*) to the σ-value at *E* where the transition curve intersects the straight line $\sigma = f_{cc} + p = 2c + p$, $f_{cc} = 2c$ being the compressive strength for a Coulomb material with $\varphi = 0$ and the cohesion c, cf. formula (2.1.9).

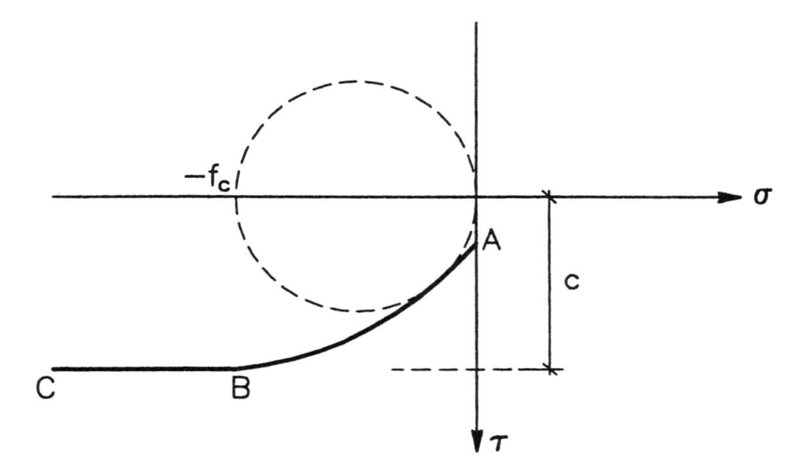

Figure 2.1.13 Failure condition for cement paste in τ,σ-space.

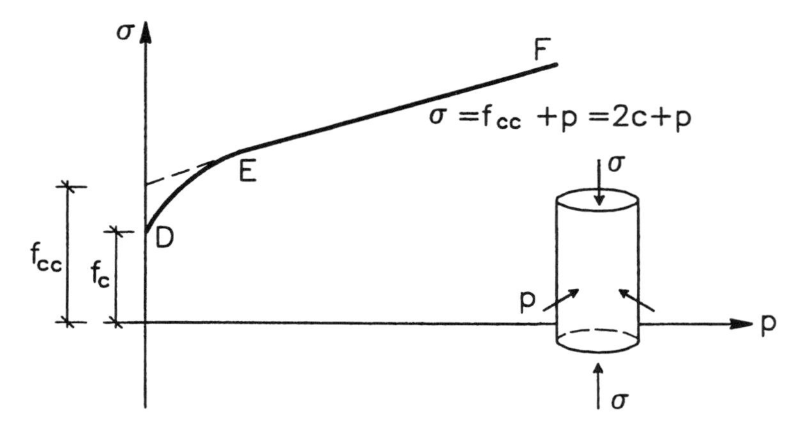

Figure 2.1.14 Failure condition for cement paste in σ,p-space.

In Fig. 2.1.15 some of the Danish test results for cement paste just referred to are shown. Results are shown for compressive strengths f_c = 10/14, 40 and 70 MPa. The test results clearly show that the uniaxial strength f_c of cement paste is lower than the value f_{cc} which would be predicted by prolonging the straight Coulomb line to intersect the σ-axis. The tests also clearly show that the Coulomb line corresponds to the friction angle $\varphi = 0$. The predictions by the model are not shown. The reader is referred to the original paper [97.2]. The predictions give a somewhat smaller strength than the tests which may be attributed to difficulties in calculating the number of hard particles which are strongly depending on the hydration degree.

Now if cement paste is combined with aggregate particles, things get more complicated.

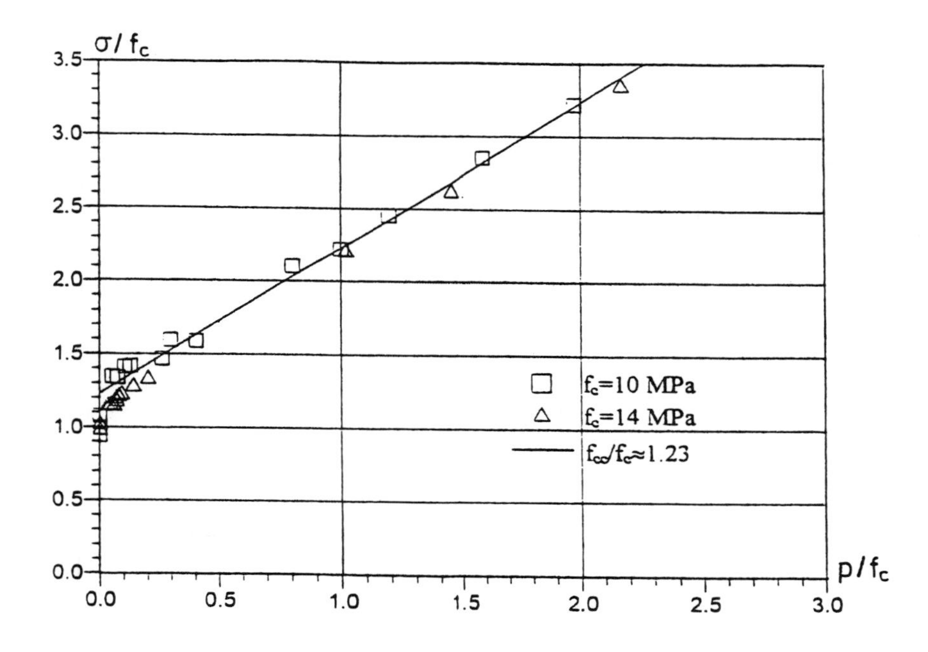

Figure 2.1.15 Triaxial test results for cement paste.

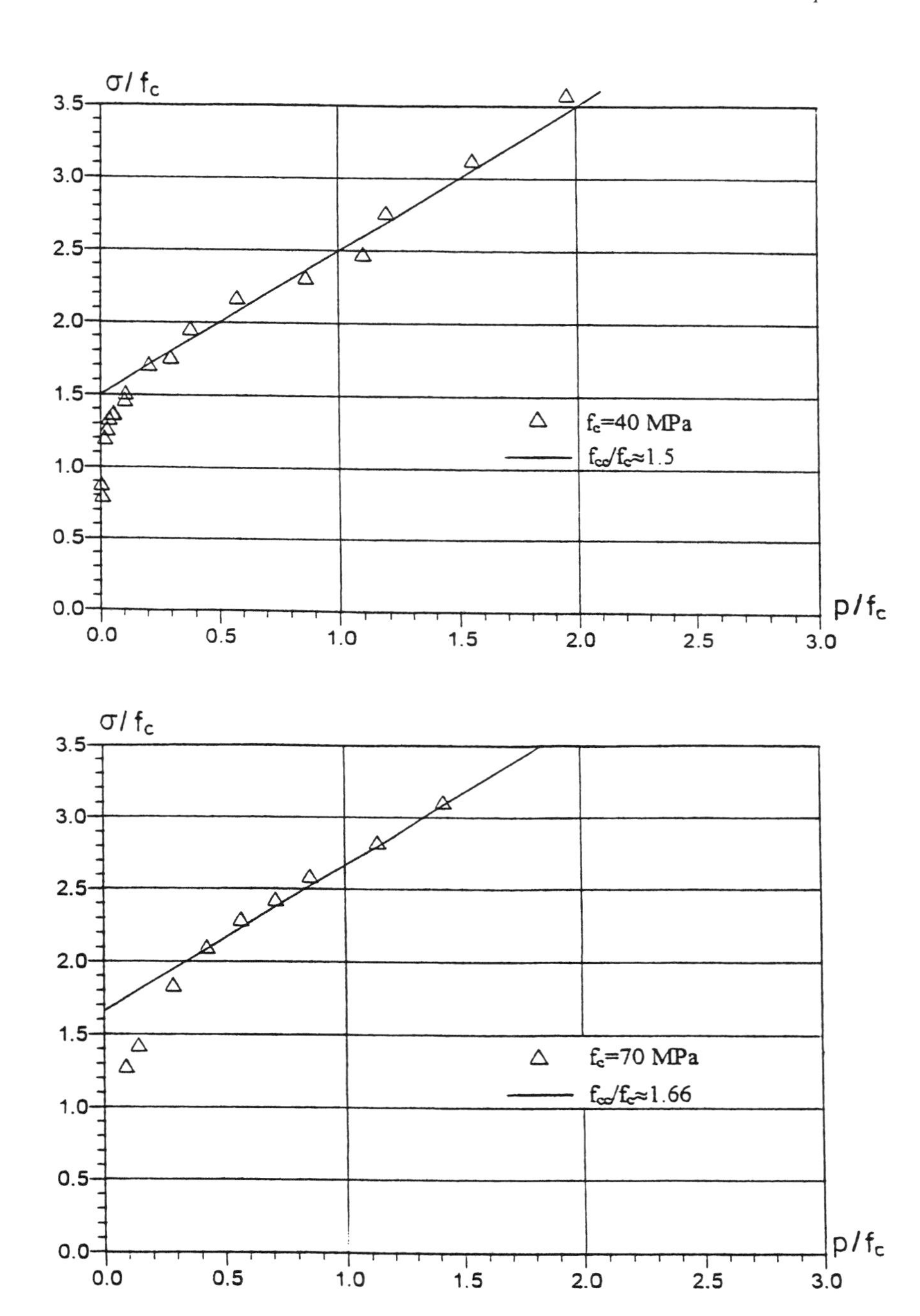

Figure 2.1.15 (continued).

We might get a rough idea about the failure condition for concrete by the following, greatly simplified, lower bound reasoning.

Suppose that the aggregate particles are so closely packed that they may act as a granular friction material carrying the load by point contact forces possibly transferred through a thin layer of cement paste.

Let the cohesion of this granular material be $c = 0$ and let the friction angle be φ.

Consider again the cylindrical specimen with the total confining pressure p and with the total stress on the end faces σ.

If the cement paste carries the confining pressure p_c, the axial stress σ_c carried by the cement paste will be $\sigma_c = p_c + 2c$. If the aggregate particles carry the confining pressure p_a, the axial stress σ_a carried by the aggregates will be $k\,p_a$, where $k > 1$ is calculated by formula (2.1.25). We assume the stress field in the cement paste to be transferred through the aggregate particles without being changed.

The total confining pressure is thus

$$p = p_c + p_a \tag{2.1.28}$$

and the total axial stress is

$$\sigma = p_c + 2c + kp_a = p_c(1-k) + 2c + kp \tag{2.1.29}$$

It appears that σ is maximum when the confining pressure carried by the cement paste is $p_c = 0$ and we have

$$\sigma = 2c + kp \tag{2.1.30}$$

Since $2c$ is the uniaxial compressive strength of the composite material, we have arrived at a Coulomb type failure condition $\sigma = f_c + kp$, the friction angle being attributed entirely to the aggregate particles and the cohesion entirely to the cement paste.

Of course, in reality the assumptions made are not likely to be even approximately fulfilled. To make a concrete workable, the volume of the cement paste must be considerably higher than the voids of the aggregate particles, making the assumption of point contact rather unrealistic.

A more realistic model may be developed on the basis of upper bound calculations as for the cement paste. In this model the role of the aggregate particles is to displace the yield lines developing in the cement paste. Consider again the cylindrical specimen with confining pressure p and axial pressure σ, shown in Fig. 2.1.16.

For sufficiently high confining pressure the yield lines in cement paste tend to develop roughly under 45° with the axial load because for cement paste $\varphi = 0$. Fig. 2.1.16 shows a yield line in cement paste meeting an aggregate particle. Since aggregate particles are normally stronger than cement paste, the yield line cannot go through the aggregate particle. It will be displaced as shown in the figure. The vertical displacement will take place along a crack making little contribution to the dissipation. The crack may be partly load induced and partly existing before loading takes place.

The total displacement of a yield line due to the aggregate particles obviously depends on the number of particles the yield line meets and the size of the aggregate particles. The total displacement can only be determined approximately. In the model a simple function depending on the aggregate content has been established.

Some important conclusions may be drawn from this model without detailed calculations.

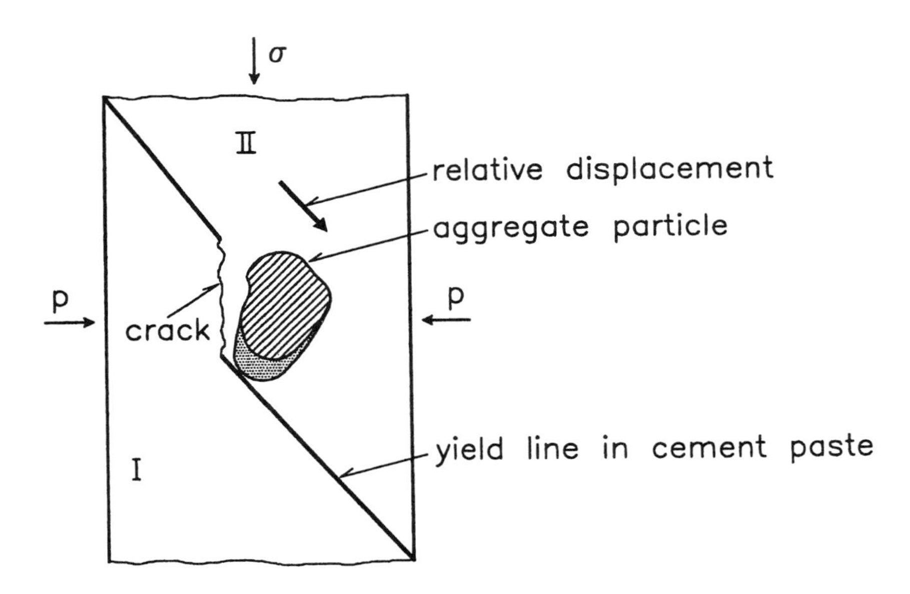

Figure 2.1.16 Yield line in cement paste displaced by an aggregate particle.

First, the uniaxial compressive strength is not affected by the aggregates, i.e., concrete has the same uniaxial compressive strength as cement paste irrespective of the volume content of aggregate particles. This conclusion is not always borne out by experiment, which may be due to differences in microcracking when the aggregate content is varied. However, the assumption of equal compressive strength of concrete and cement paste is always used in mix designs of concrete. In the Danish tests referred to, no significant difference was found between the two compressive strengths.

Second, the increase in strength above that found for cement paste by applying a confining pressure p is solely due to the displacement caused by the aggregate particles of the yield lines in the cement paste. This means that a concrete with a low volume content of aggregates exhibits a lower strength increase due to a confining pressure than a concrete with a high volume content of aggregates, i.e., mortar will behave somewhat between cement paste and normal concrete.

In Fig. 2.1.17 the Danish tests are compared with the detailed calculations by the model for a number of concretes with different strengths. The concrete strengths have the values of 10/15, 64/72/77, and 100 MPa.

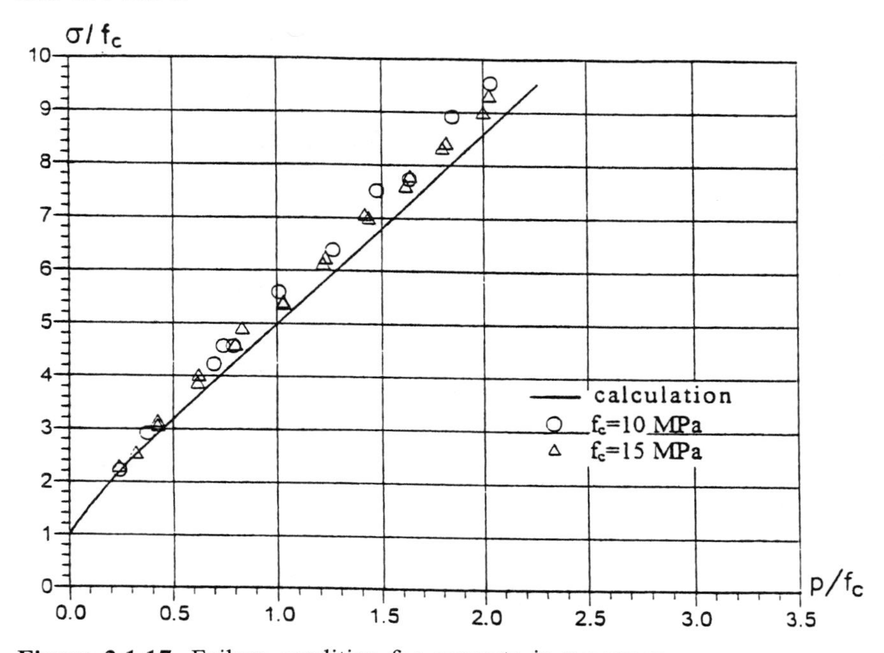

Figure 2.1.17 Failure condition for concrete in σ,p-space.

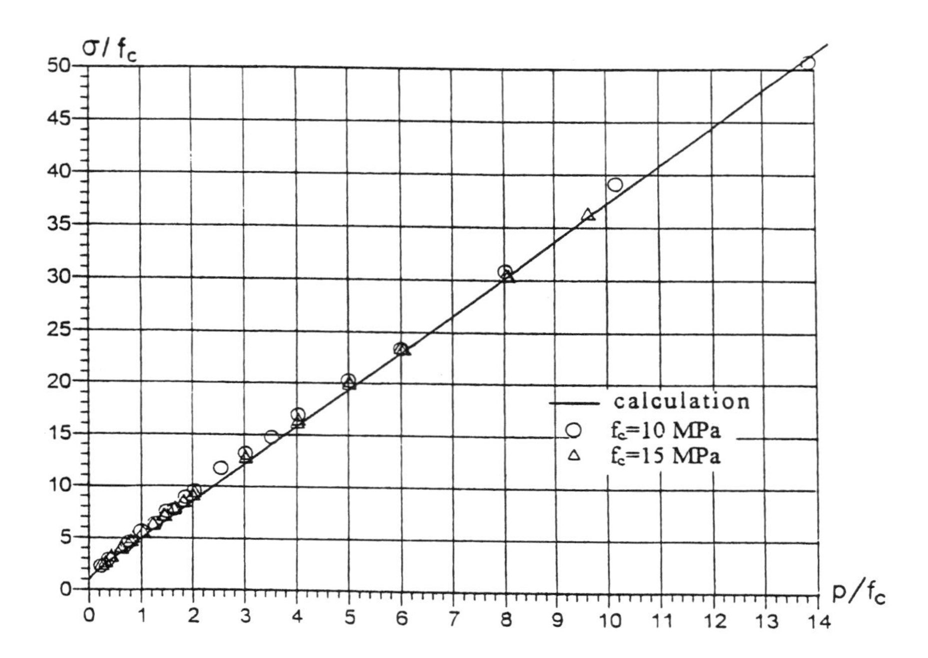

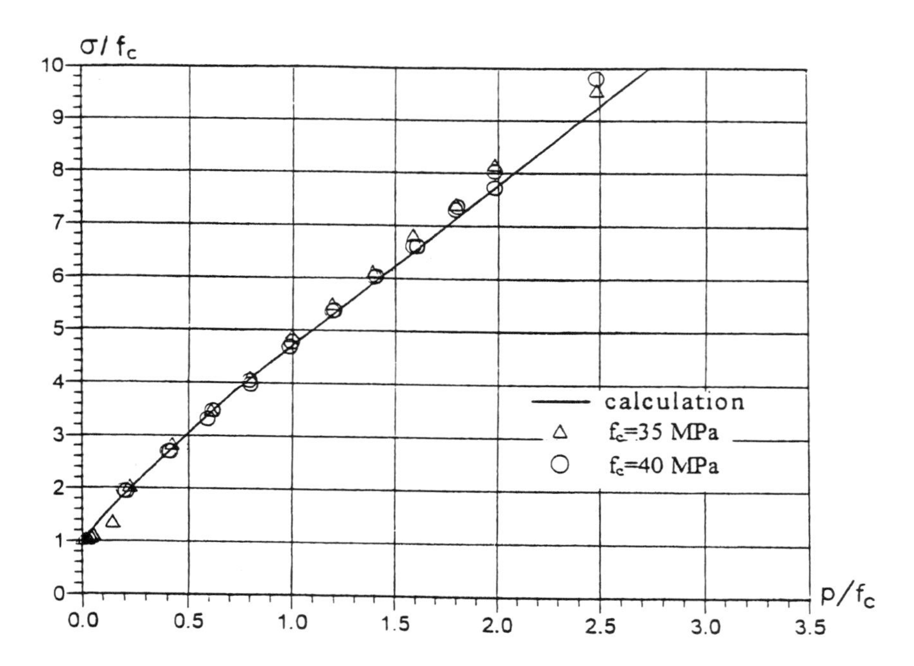

Figure 2.1.17 (continued).

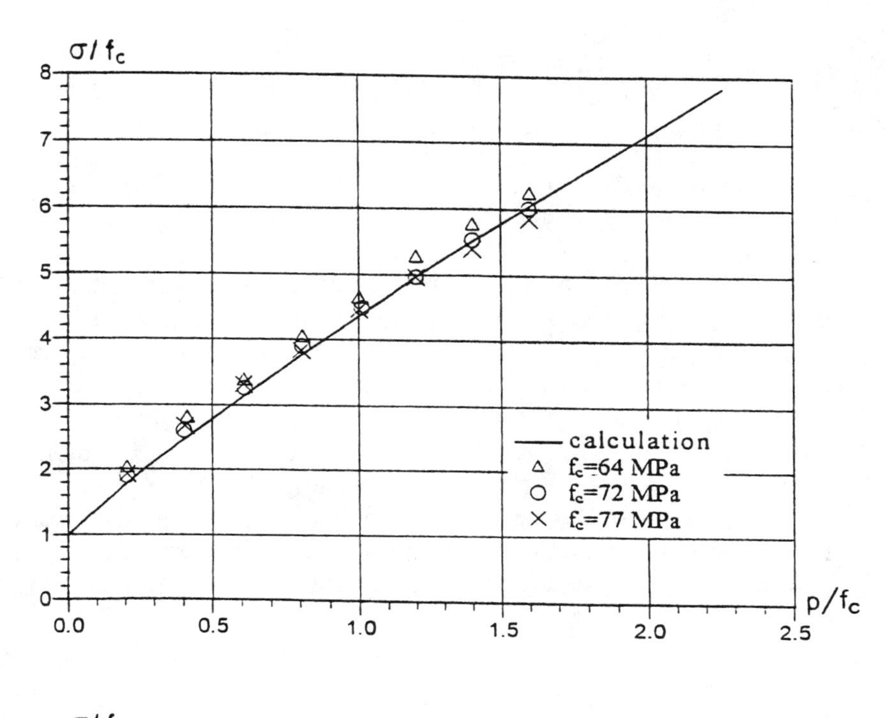

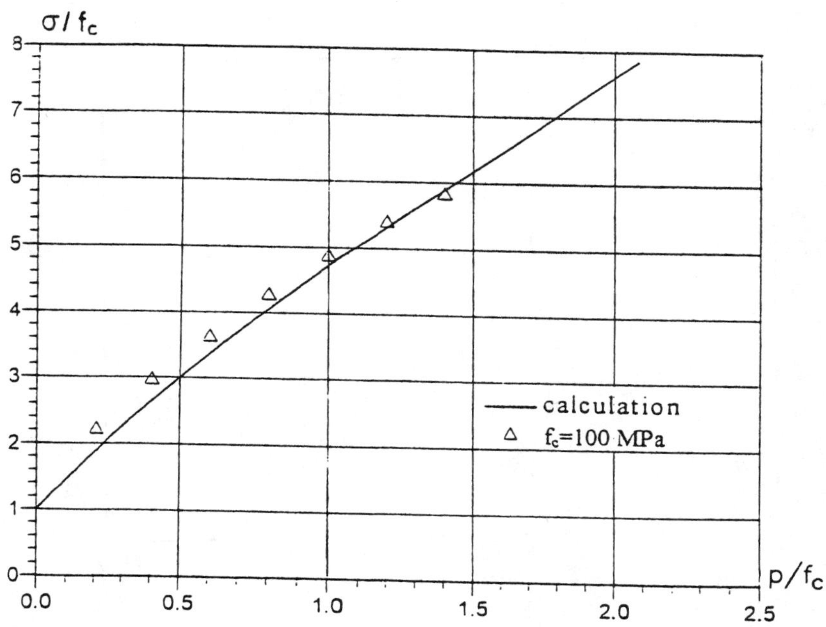

Figure 2.1.17 (continued).

It appears that for sufficiently high confining pressure, the σ-p-relation is a straight line, the equation for which may be written

$$\sigma = f_{cc} + kp \tag{2.1.31}$$

The inclination k is decreasing for increasing f_c values. This is due to the fact that higher strength normally leads to lower aggregate content.

From the inclination an apparent friction angle φ may be calculated by means of (2.1.25). The result is shown in Fig. 2.1.18. For low strength concrete the friction angle is around 37° corresponding to $k = 4$. In fact this value was used already by Coulomb (cf. Heyman [72.1]). For increasing strength the friction angle decreases almost linearly up to a concrete strength of about 65 MPa. For higher strengths it is constant at a value around 28°.

As for cement paste, the straight line (2.1.31) intersects the σ-axis at a point corresponding to an apparent uniaxial compressive strength f_{cc}, which is higher than the real uniaxial strength f_c. The reason is the same as for cement paste.

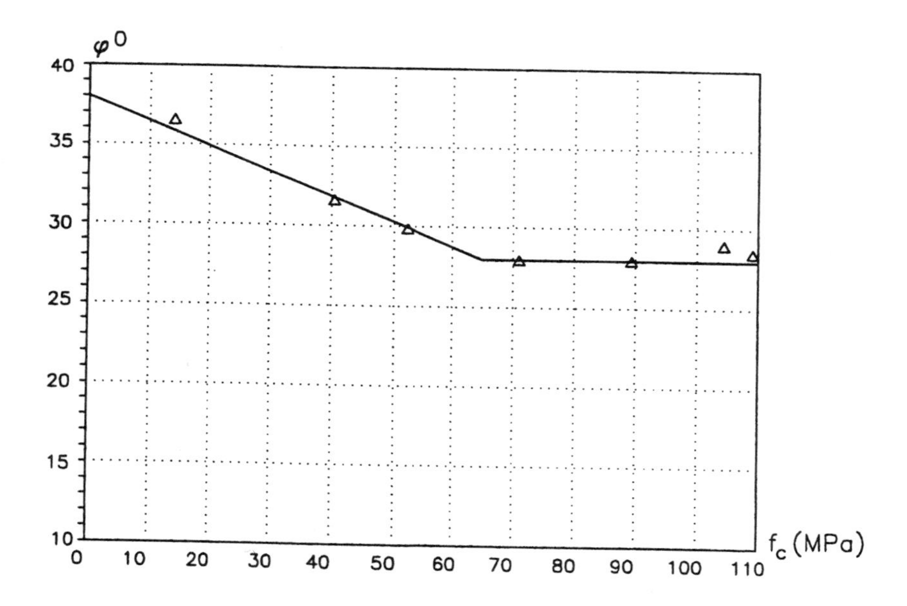

Figure 2.1.18 Friction angle for different strengths of concrete.

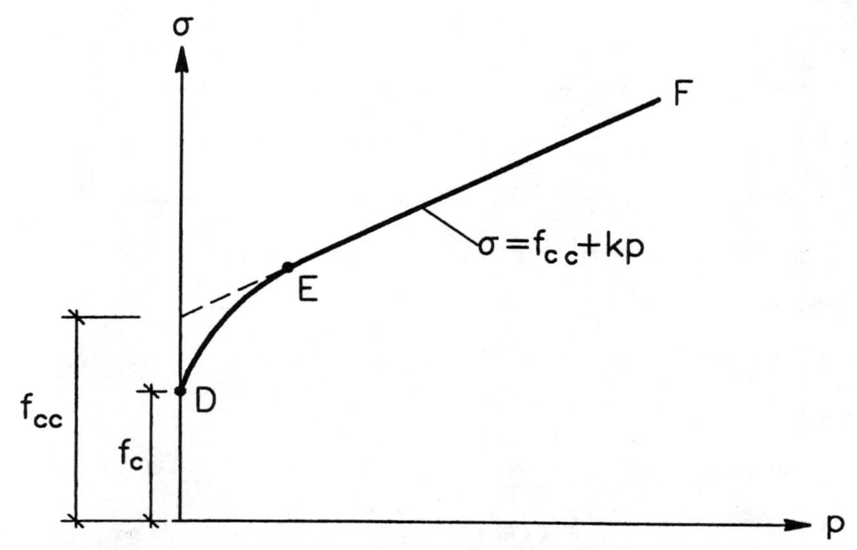

Figure 2.1.19 Failure condition for concrete in σ,p-space.

The final form of the relationship between σ and p is shown in Fig. 2.1.19.

As for cement paste there is a transition curve DE leading from the uniaxial compressive strength f_c to the point E where the transition curve intersects the straight line EF.

The ratio f_{cc}/f_c has been plotted in Fig. 2.1.20 for all Danish tests. Even for low strength concrete the ratio is as high as 1.2. For a concrete strength of f_c = 100 MPa the ratio is around 1.75.

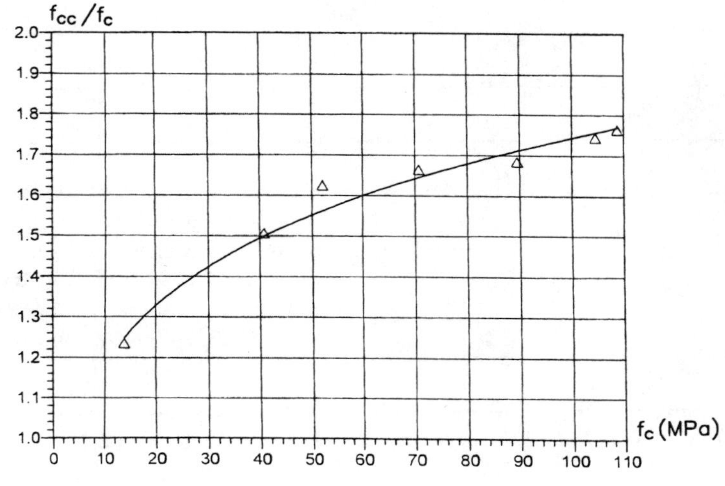

Figure 2.1.20 Ratio between f_{cc} and f_c.

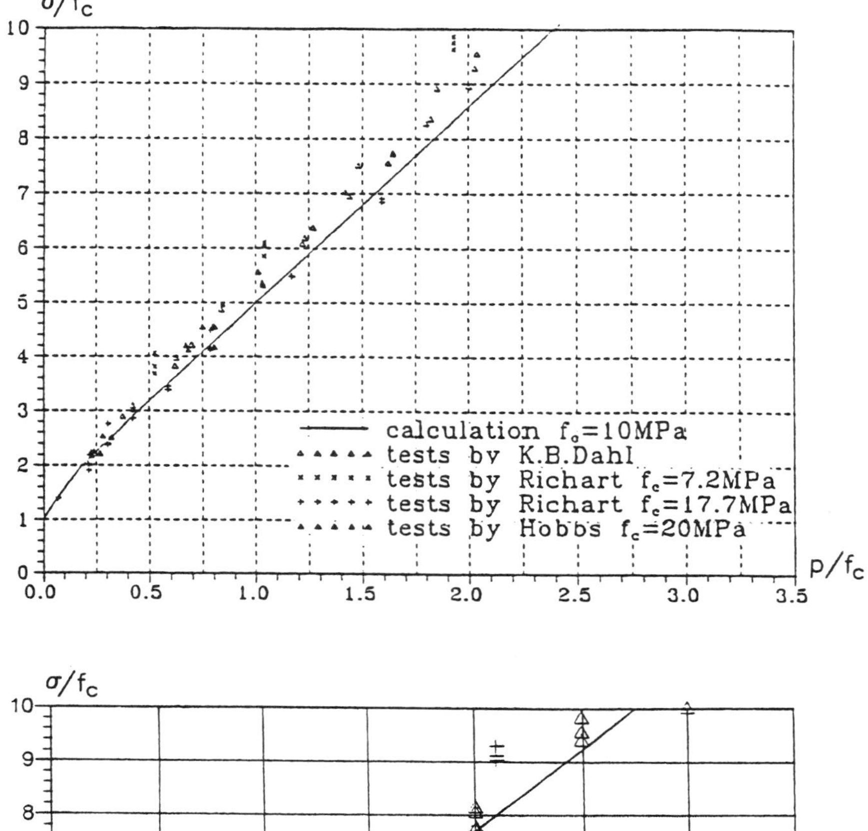

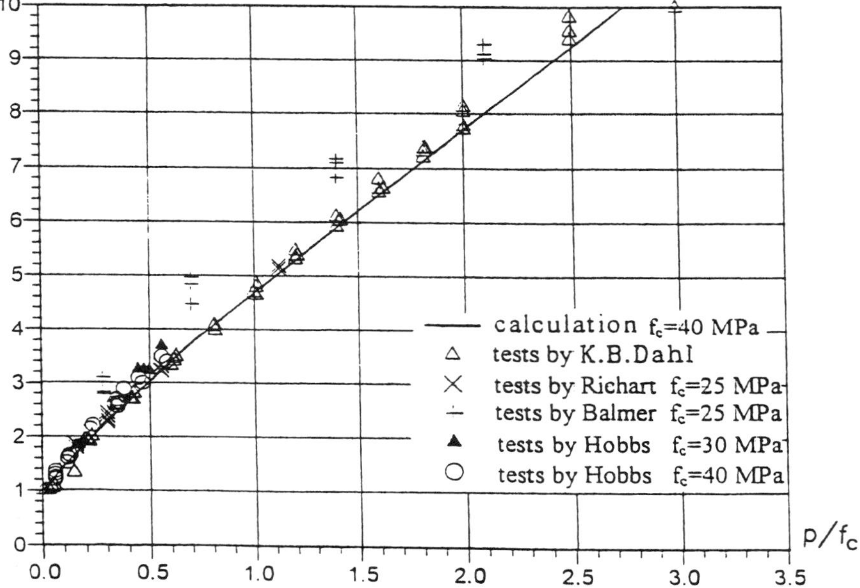

Figure 2.1.21 Failure condition for concrete. Comparison with test results.

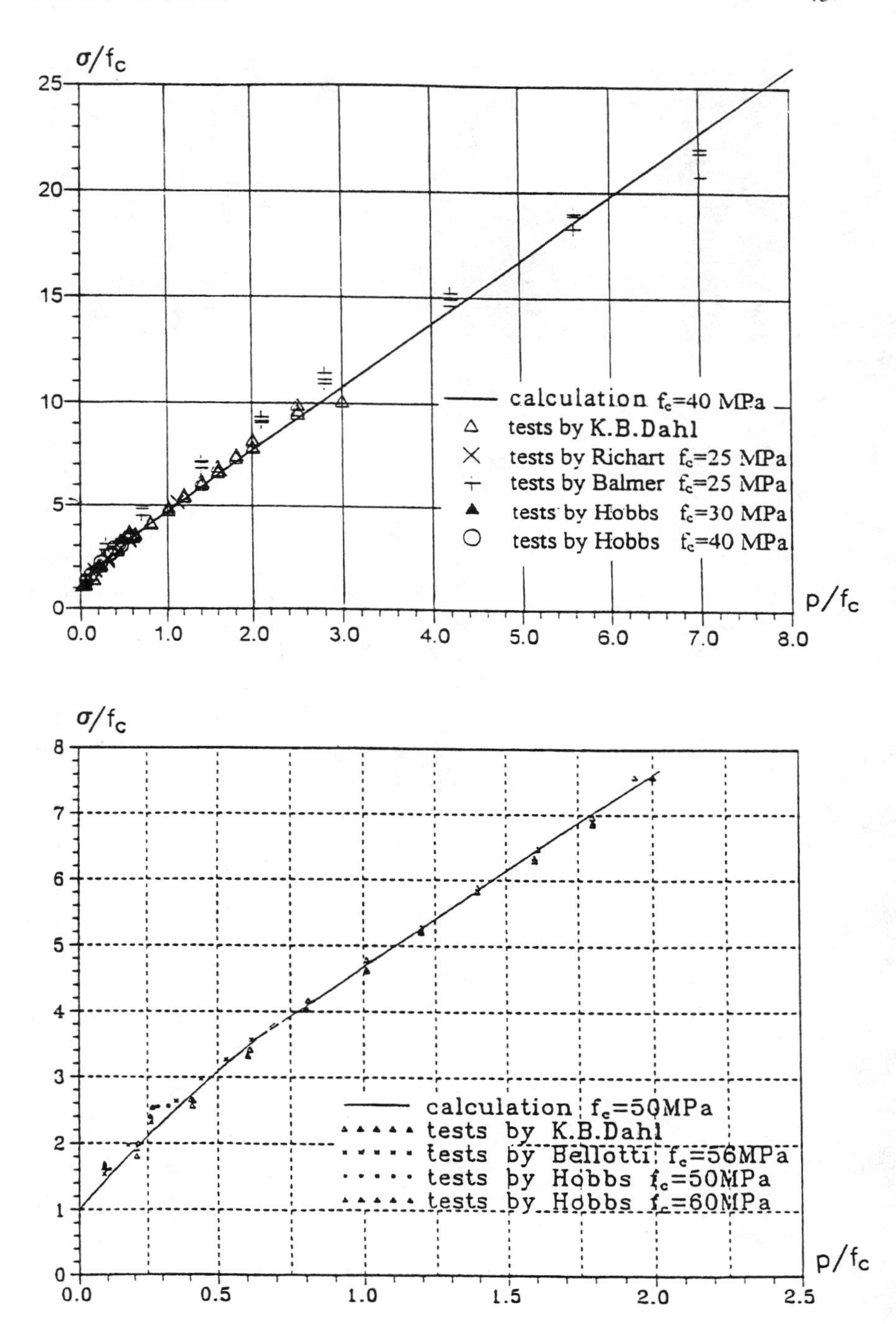

Figure 2.1.21 (continued).

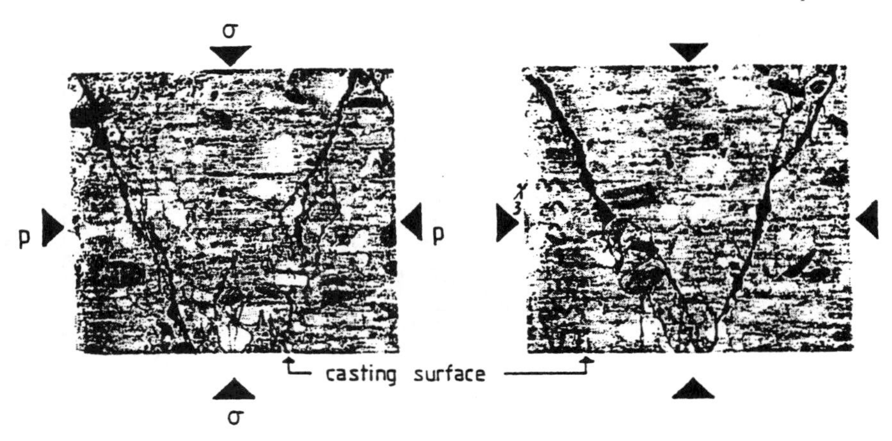

Figure 2.1.22 The location of yield lines with $p/\sigma = 0.05$ in specimens from [86.11].

Further comparisons between calculations by the model and tests from the literature have been shown in Fig. 2.1.21. The test results have been taken from [28.3], [52.5], [70.6], [74.11], and [84.10].

The predictions of the location of the yield lines have also been compared in experiments, and it seems that the predictions are rather good. Only experiments where the lateral displacements along the loaded surfaces are approximately free can be used. In Fig. 2.1.22 an example is shown. The tests were performed by van Mier [86.11]. The figure shows the observed yield lines in a case with $p/\sigma = 0.05$. The test specimens were cubes. In the transverse direction the stress was $0.33\ \sigma$.

It should be noted that the yield lines are much steeper than predicted by the Coulomb failure condition.

We shall briefly mention some other types of tests.

Sometimes, instead of using the principal stresses, it may be advantageous to depict test results using the parameters

$$\sigma_m = \frac{1}{2}(\sigma_1 + \sigma_3) \tag{2.1.32}$$

$$\tau_m = \frac{1}{2}(\sigma_1 - \sigma_3) \tag{2.1.33}$$

We may find the corresponding failure conditions by solving these expressions with regard to σ_1 and σ_3. Thus we find that

$$\sigma_1 = \sigma_m + \tau_m \tag{2.1.34}$$

$$\sigma_3 = \sigma_m - \tau_m \tag{2.1.35}$$

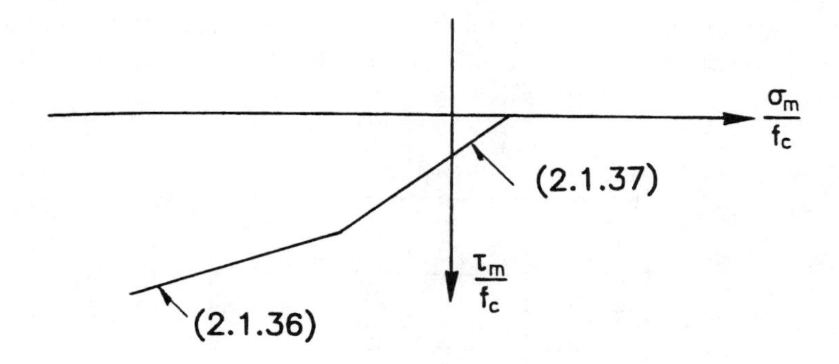

Figure 2.1.23 General outline of failure conditions for a modified Coulomb material in a σ_m, τ_m-coordinate system.

whereupon conditions (2.1.10) and (2.1.8) can be written

$$(k-1)\,\sigma_m + (k+1)\,\tau_m = f_c \qquad (2.1.36)$$

$$\sigma_m + \tau_m = f_A \qquad (2.1.37)$$

The general picture of these conditions is shown in Fig. 2.1.23.

If compressive stresses are assumed positive, the resulting curves can be found by a reflection in the τ-axis.

Tests with one principal stress equal to zero and with the other two principal stresses having opposite signs can, for instance, be carried out by subjecting a thin-walled tubular specimen to combined torsion and axial force. Fig. 2.1.24 shows some test results of this type. From the figure it appears that the tensile strength of the concrete was about $0.08f_c$ in these tests. The straight lines correspon-ding to (2.1.36) are drawn for $k = 4$.

Biaxial tests can also be performed by subjecting disk-formed or cubic specimens to compression or tension on the lateral surfaces. Many such experiments have been performed. Figure 2.1.25 illustrates the results of one such test series. A certain strengthening effect seems to occur when there are two compressive principal stresses different from zero, an effect that is not in agreement with the Coulomb frictional hypothesis.

The conclusion from this survey of our knowledge about failure conditions for concrete, obviously would be that when disregarding the load induced anisotropy and the influence of the intermediate principal stress, the failure condition must be of the form shown in

Fig. 2.1.19. An accurate failure condition must contain a transition curve leading from the uniaxial compressive strength f_c to the straight line corresponding to the Coulomb sliding failure criterion with an apparent uniaxial compressive strength $f_{cc} > f_c$.

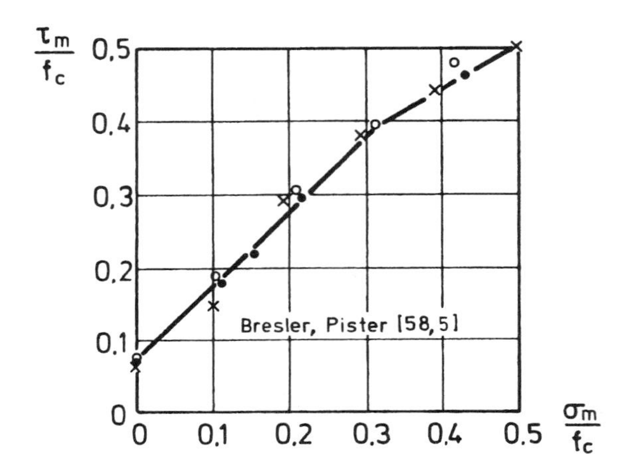

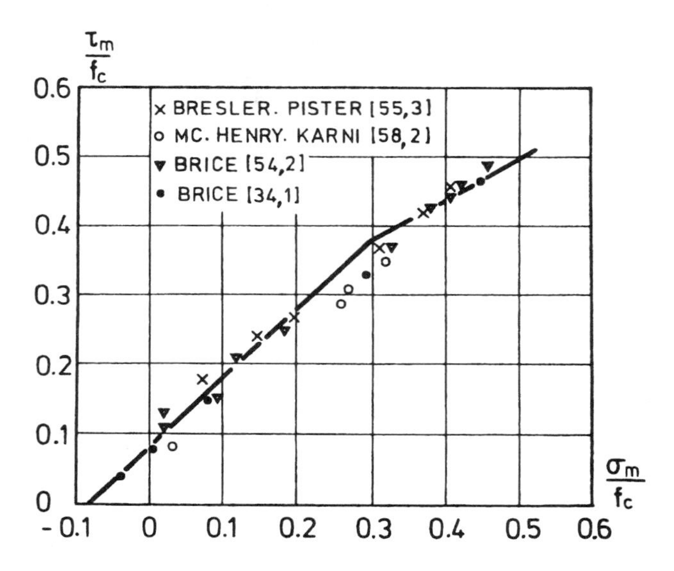

Figure 2.1.24 Test results for concrete compared to the failure criterion for a modified Coulomb material.

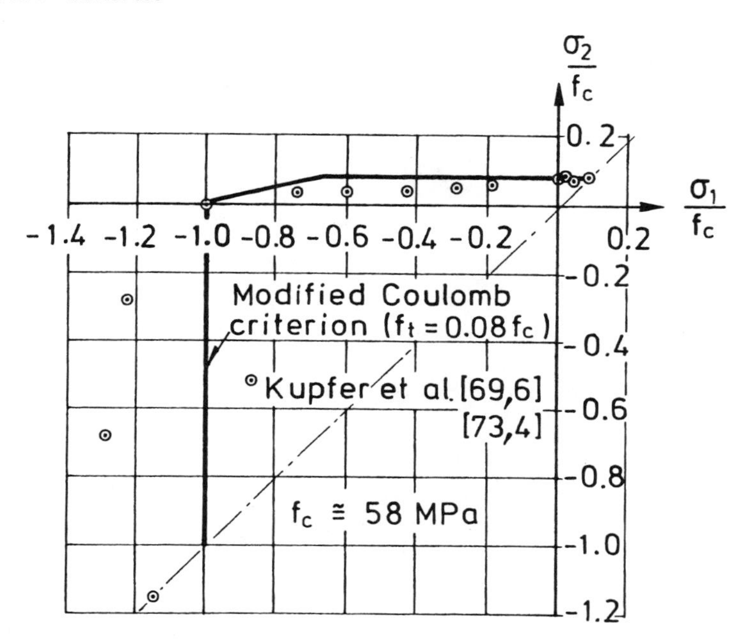

Figure 2.1.25 Test results for concrete in biaxial stresses compared to the failure criterion for a modified Coulomb material.

However, most of the calculations in this book are based on the utterly simple condition arrived at when disregarding the transition curve, and assuming $f_{cc}/f_c = 1$. This far-reaching simplification is justified by the fact that in most cases of practical interest we are more interested in the strength of concrete in a structural element than in the strength of the virgin material. The structural strength of concrete is treated in the next section.

If concrete is identified with a modified Coulomb material, we can conclude that the parameter k has a value of around 4 for low strength concrete. If this value is selected, we find from (2.1.6)

$$\mu = 0.75 \tag{2.1.38}$$

corresponding to an angle of friction

$$\varphi = 37^o \tag{2.1.39}$$

From (2.1.9) we get

$$c = \frac{f_c}{2\sqrt{k}} = \frac{1}{4} f_c \tag{2.1.40}$$

In the tests we find that both the tension failure and the shear failure are separation failures. If the tensile strength is f_t, we therefore have

$$f_v = f_t \tag{2.1.41}$$

The failure conditions (2.1.10) and (2.1.8) then get the final form

$$k\sigma_1 - \sigma_3 = f_c \qquad k = 4 \tag{2.1.42}$$

$$\sigma_1 = f_A = f_t \tag{2.1.43}$$

The failure condition in the σ,τ-coordinate system is shown in Fig. 2.1.26.

Since the structural strength of concrete is normally quite different from the strength of the virgin material (see Section 2.1.4) there are no reasons for refining the parameters for higher concrete strengths. The low strength values normally will suffice.

For the plane stress field we find in conformity with Fig. 2.1.11 the hexagon shown in Fig. 2.1.27. This hexagon is often approximated by a square in the third quadrant; that is, the tensile strength is assumed to be zero.

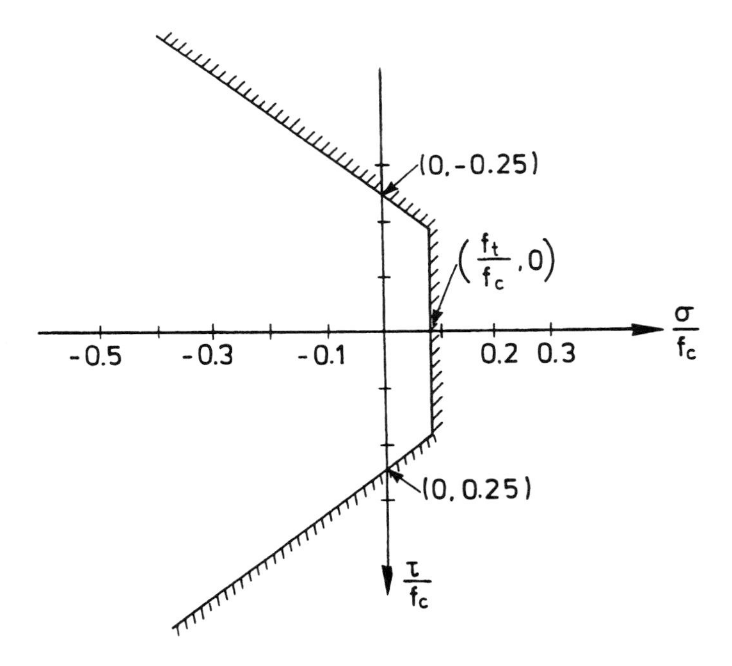

Figure 2.1.26 Failure criterion for concrete in a σ,τ-coordinate system.

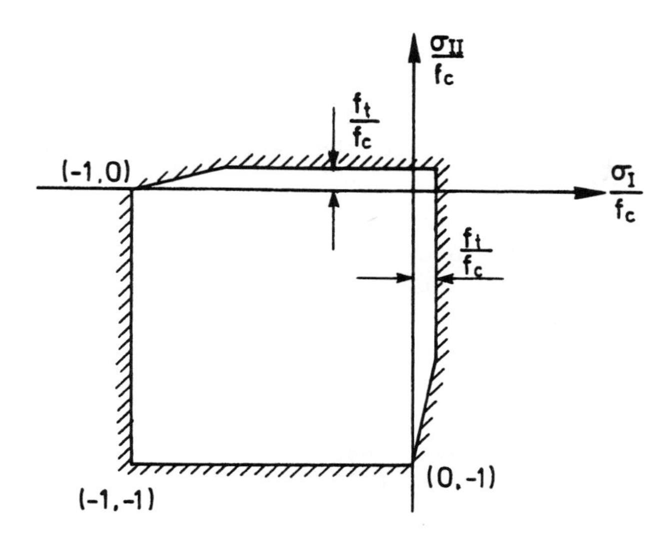

Figure 2.1.27 Failure criterion for concrete in plane stress.

Since it is difficult to determine the tensile strength of concrete by a direct tensile test, a splitting test is often used. The tensile strength determined in this way is often called the *splitting tensile strength.*

According to Danish experience, if the specimens are water cured, the direct tensile strength is approximately

$$f_{tdirect} \cong 0.26\, f_c^{2/3} \qquad (f_c \text{ in MPa}) \tag{2.1.44}$$

while the splitting tensile strength is

$$f_{tsplit} \cong 0.29\, f_c^{2/3} \qquad (f_c \text{ in MPa}) \tag{2.1.45}$$

If the specimens are cured for say 2 days in the mold, 8 days in water and 18 days in air (combined curing), we have

$$f_{tdirect} \cong 0.35\, f_c^{1/2} \qquad (f_c \text{ in MPa}) \tag{2.1.46}$$

$$f_{tsplit} \cong 0.32\, f_c^{2/3} \qquad (f_c \text{ in MPa}) \tag{2.1.47}$$

The difference in strength from the curing conditions is, of course, due to residual stresses.

Instead of (2.1.46) we will often use the formula

$$f_t = \sqrt{0.1 f_c} \qquad (f_c \text{ in MPa}) \tag{2.1.48}$$

which gives about 10% less strength than (2.1.46).

The modified Coulomb criterion with a zero tension cutoff was used by Drucker and Prager [52.4] as a yield condition for soil. For concrete, Chen and Drucker [69.3] introduced a nonzero tensile strength.

The square yield locus for concrete in plane stress was applied by the author [63.2] to slabs, and later to disks [69.2], and to shear in beams [67.2].

To account for the influence of the intermediate principal stress, more sophisticated criteria have been formulated. Surveys of proposals for concrete failure criteria have been given by Chen [78.18] and Ottosen [77.2].

2.1.4 Structural Concrete Strength

Unfortunately, the strength of concrete we observe when testing a structure usually is very different from the strength measured on standard laboratory specimens.

The main reason is that the concrete is cracked, and the cracking reduces the strength.

The most important consequence of this fact regarding the application of plastic theory is that the strength parameters, which we have to insert into the theoretical solutions, normally are lower than the standard values. We call the strength values to be inserted the *effective strengths*. The effective concrete compressive strength f_{cef} is defined by

$$f_{cef} = \nu f_c \tag{2.1.49}$$

where $\nu \leq 1$ is called the *effectiveness factor* for the compressive strength and f_c is the standard compressive strength measured on specimens of specified size, cured and tested in a specified manner. In this book f_c means the cylinder compressive strength measured on water-cured cylinders with a diameter of 150 mm and a depth of 300 mm.

Normally, cracked concrete may be assumed to have the same friction angle as the virgin material. Since the compressive strength is proportional to the cohesion c, see formula (2.1.26), we may also write

$$c_{ef} = \nu c \tag{2.1.50}$$

where c_{ef} is the effective cohesion of cracked concrete.

In a similar way the effective concrete tensile strength f_{tef} is defined by

$$f_{tef} = \nu_t f_t = \rho f_c \tag{2.1.51}$$

where $\nu_t \leq 1$ and $\rho << 1$ are effectiveness factors for the tensile strength and f_t is a standard tensile strength measured on specimens of specified size and cured and tested in a specified way. The tensile strength is not always known. In this book we will normally take it as a function of f_c given by the formula (2.1.48) of the previous section. We repeat the formula here, i.e.,

$$f_t = \sqrt{0.1 f_c} \qquad (f_c \text{ in MPa}) \tag{2.1.52}$$

There are a few cases where $\nu = 1$ may be used with good approximation. Among them is the case of a concentrated force on a concrete face where the triaxial stress field near the force prevents cracking.

The effective tensile strength is always low, ρ being only a few percent.

The strength reduction due to cracking might be subdivided into (a) strength reduction due to microcracking present even before any load is applied, (b) strength reduction due to load induced microcracking, and finally (c) strength reduction due to macrocracking. While the microcracking present before loading may be assumed to lead to an isotropic material, load induced microcracking and macrocracking will cause anisotropy, i.e., the strength parameters, for instance the compressive strength, will vary with direction. Strictly speaking, cracked concrete should be treated as an anisotropic material.

However, this cannot be done in a fully rational way with present knowledge. We must content ourselves with a more primitive approach. It consists either of considering cracked concrete to be isotropic with the effective strength parameters or by dealing with the strength parameters only in certain selected directions depending on the crack system.

Microcracks are normally defined as cracks having a crack width $\leq 10\mu m$. Macrocracks in a concrete surface are visible cracks. Microcracking is also just called internal cracking.

Taking into account the strength reduction due to cracking is not unique for the plastic theory. This strength reduction must be considered in any theory or any calculation of the strength of a concrete structure.

In plastic theory we have one more problem which is due to the softening behavior of concrete both in compression and tension, i.e., after the peak value, where the strength has been reached, the stress decreases when deformation is increased. Since the theory of perfectly plastic materials is unable to take into account the softening in a rational manner, this phenomenon must be dealt with by effectiveness factors as well, i.e., softening effects must be included in the effectiveness factors.

A decade or so ago, it was generally believed that the necessity of introducing effectiveness factors in plastic theory was almost solely due to softening effects. However, it is now clear that the effect of cracking is at least as important. In fact, in tests with approximately homogeneous stress fields cracking is the main concern.

With today's knowledge it is only possible to calculate the effectiveness factor for bending, see Chapter 5. This means that empirical formulas must be established for the different cases. The formulas known up to now will be given in the relevant sections. Here we will make an attempt to clarify the physics of strength reduction due to cracking. This cannot be done in any complete way. Much more research is needed in this important field.

As mentioned before, microcracking is already present even before any load is applied. This kind of microcracking is due to shrinkage of the cement paste, which is prevented by the unhydrated cement grains and by the aggregate particles. In the aforementioned model of Jin-Ping Zhang [97.2], a simple measure of microcracking in cement paste is introduced. This is the average length cracked relative to the average distance between unhydrated cement grains. If we again refer to Figure 2.1.12, a fully microcracked cement paste is a cement paste where the only contribution to strength stems from the yield lines formed in front of the unhydrated cement grains. On the other hand, a cement paste which has no microcracks is a cement paste having contributions to strength, not only from the yield lines in front of the hard particles, but also along the whole distance between the hard particles.

Calculations using this model show that a fully microcracked cement paste may lose a substantial part of its strength.

When concrete is subjected to external loads, the microcracking will be increased more or less.

It is very likely that load induced microcracking is a main cause of the strength reduction always found in shear problems.

Particularly if, after the formation of a macrocrack, reinforcement crossing the cracks is stressed, the transfer of shear stresses from reinforcement to concrete and the resulting bursting stresses in regions near the cracks may give rise to a substantial increase of microcracking.

Let us delve a little into this case, since it is important in almost all shear problems in beams, shear walls and panels, etc.

Fig. 2.1.28 shows two parallel macrocracks, which are crossed by a reinforcement bar perpendicular to the cracks. The reinforcement bar is stressed in tension. Let us assume that some kind of microcracking is spreading out from the macrocracks when the reinforcement stress is increased. The microcracks are shown schematically in the figure. The real microcracking is of course much more complicated.

We want to find a suitable parameter for the determination of the compressive strength of the cracked concrete, the compressive load being parallel to the macrocracks. At a stage where the microcracking has penetrated a distance $\frac{1}{2}\ell_o$ from both sides, it will be quite natural to assume that the reduced compressive strength will be a function of the parameter

$$\chi = \frac{\ell_o}{a} \tag{2.1.53}$$

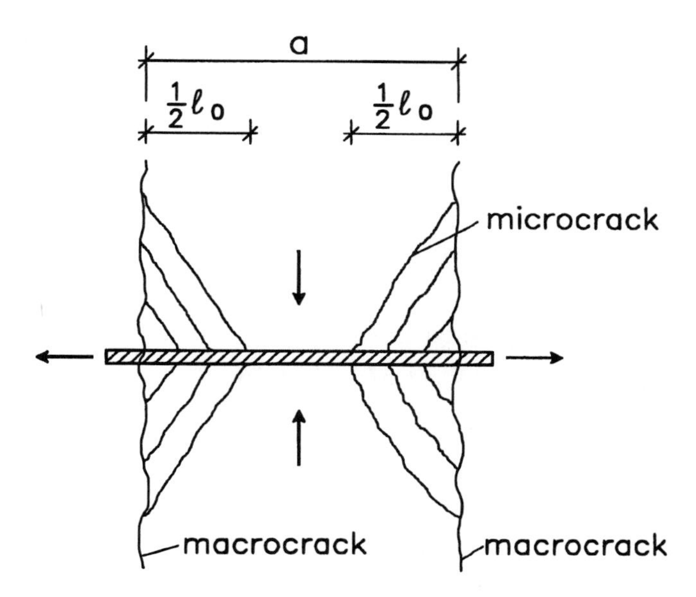

Figure 2.1.28 Formation of microcracks between two macrocracks.

A rough estimate of the crack distance for a fully developed crack system may be found using the simple formula, see [90.16],

$$a = \frac{\beta}{\lambda} \tag{2.1.54}$$

where the crack distance parameter

$$\beta = \frac{d}{4r} \tag{2.1.55}$$

d being the reinforcement diameter and r the reinforcement ratio. The parameter λ depends on the surface texture of the bars. Roughly $\lambda = 2$ for most deformed bars and $\lambda = 1$ for plain bars. Assuming further that ℓ_o is proportional to the reinforcement stress σ_s and the reinforcement diameter d and inversely proportional to the tensile strength of concrete f_t, i.e., $\ell_o = C\,\sigma_s\, d/f_t$, where C is a constant, we find

$$\chi = \frac{4C\lambda r\sigma_s}{f_t} \tag{2.1.56}$$

When, theoretically, $\chi = 1$, the microcracking has penetrated the whole crack distance and further reduction in the compressive strength is not likely to occur. However, the reasoning is not that exact. What can at best be achieved is to find a parameter, which can be used when evaluating the tests. In Section 4.6 it is shown that ν may be taken as a linear function of χ. Thus the value of C is not important for our purpose here. It has arbitrarily been chosen so as to make χ equal to 1 when $\lambda = 2$, $r\sigma_s = 10$ MPa and $f_t = \sqrt{0.1 f_c} = \sqrt{0.1 \cdot 50}$ MPa. Then for deformed bars ($\lambda = 2$)

$$\chi = \frac{r\sigma_s}{1.41\sqrt{f_c}} \qquad (f_c \text{ in MPa}) \tag{2.1.57}$$

Tests show, see Section 4.6, that the straight line representing ν as a function of χ may be taken to be

$$\nu = \nu_o = 1.52 - 0.83\chi \ngtr 1 \tag{2.1.58}$$

The straight line is shown in Fig. 2.1.29. The reason for the notation ν_o will appear below.

According to formula (2.1.58) there is no reduction in strength for $\chi < 0.63$, i.e., $r\sigma_s < 0.89\sqrt{f_c} \cong 2.8\, f_t$. It is not known if ν will be decreasing for higher values of χ or if there is a lower limit. This is not so important, since the region covered by the tests will be sufficient for most practical purposes.

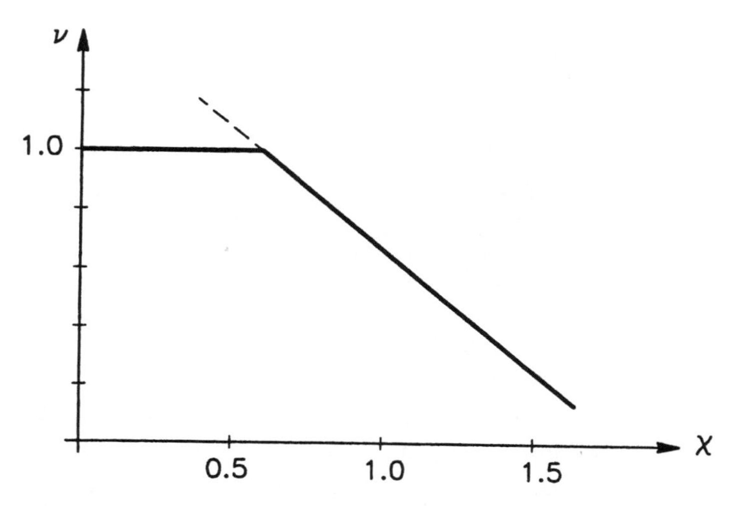

Figure 2.1.29 Effectiveness factor versus microcracking parameter χ.

The tests available are mostly with deformed bars or similar. It is known that plain bars give rise to a higher effective strength. This is quite natural, since plain bars cause less microcracking. But a formula of the type (2.1.58) cannot be established at present.

For pure shear and normal strength concrete (2.1.58) is approximately equivalent to, see Section 4.6,

$$\nu = \nu_o = 0.7 - \frac{f_c}{200} \quad (f_c \text{ in MPa}) \tag{2.1.59}$$

This formula was originally suggested for beam shear, see Chapter 5, but it has turned out to be more generally valid.

When there are more reinforcement directions making an angle different from 90° with the crack, the χ value may be put to the value belonging to the reinforcement direction with the highest value of $r\sigma_s$.

The general validity of formula (2.1.58) has been confirmed by finite element calculations for a number of different structural elements by Lars Jagd [96.2]

It appears that the effective strength is relatively lower for higher concrete strengths. It is well known that the higher the strength the higher the intensity of microcracking. It seems that higher strengths also lead to higher sensitivity to load induced microcracking.

Many tests have been performed to determine the strength reduction of cracked concrete with transverse reinforcement. They will be briefly reviewed in Chapter 4, as well as other empirical formulas for the strength reduction.

We now turn to the effects of macrocracks on strength.

Here also we have to face many complications. When a concrete structure is loaded up to failure, crack directions may change several times. First we have, in regions with high tensile stresses, the initial cracks following approximately the stress trajectories of the linear elastic uncracked concrete. When the structure is fully cracked but the concrete and the reinforcement are still in the linear elastic range, new crack directions may appear. Finally, a new change of crack directions may take place when the ultimate state is reached, characterized by reinforcement yielding and/or concrete crushing. Thus several crack systems may develop during a loading process.

The failure condition for a macrocrack may also be formulated by using the modified Coulomb failure condition. However, the cohesion is strongly reduced, to about 50% of the cohesion of the virgin material. Of course, the tensile strength is even more reduced. For crack widths exceeding say 0.1 mm, it is practically zero. Strangely, the friction angle seems to be unchanged, about 37°. All these things are described in detail in Chapter 4 and Chapter 8.

Cracks strictly parallel to the load direction will not change the strength even if the cracks have no strength at all, see Fig. 2.1.30.

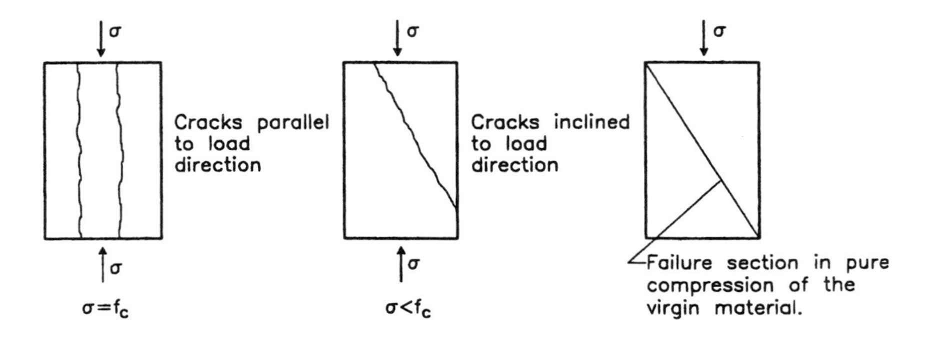

Figure 2.1.30 Influence of cracks on the compressive strength.

However, if the cracks have an inclination to the load direction strength may be reduced. When the modified Coulomb failure condition is used even the slightest inclination may cause a reduction if the tensile strength of the crack is zero. Then all inclinations between the load direction and the failure sections in pure compression of the virgin material, see Fig. 2.1.31, will give the same strength, roughly half the strength of the uncracked virgin material. Flatter cracks than these failure sections give greater strength. Fortunately, in reality concrete does not seem to be that sensitive to cracks near the load direction, see the Sections 4.6, 8.2.8 and 8.3.2.

If the cohesion of the crack is termed c' and the cohesion of the virgin material is c as before, we introduce an effectiveness factor ν_s defined by

$$c' = \nu_s c \tag{2.1.60}$$

The parameter ν_s is also called the *sliding reduction factor* and its value is around 0.5.

If the friction angle is unchanged, which as mentioned is normally the case, then $\nu_s f_c$ will be the reduced compressive strength for the most dangerous cracks.

When the macrocrack with stressed transverse reinforcement is contained in a microcracked material, experiments show that the resulting cohesion in the crack is

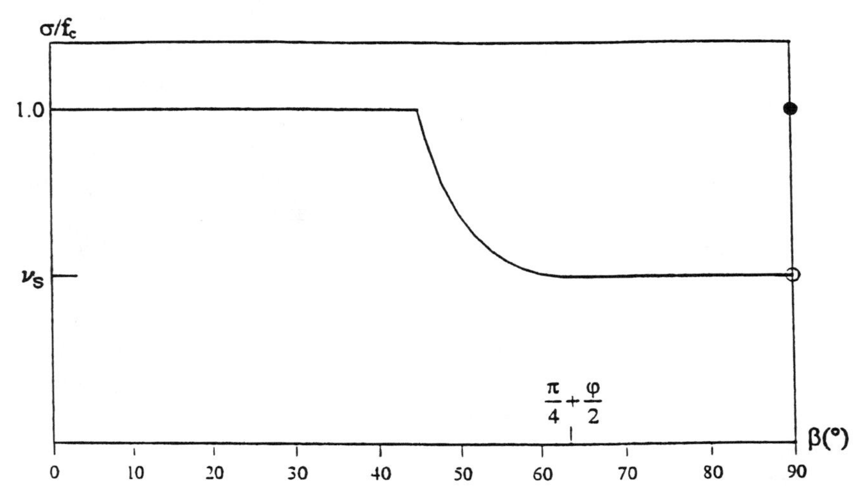

Figure 2.1.31 Compressive strength of a specimen with a crack. The angle β is the angle between the crack and the sections with pure compression. (Illustration from Section 8.2.8).

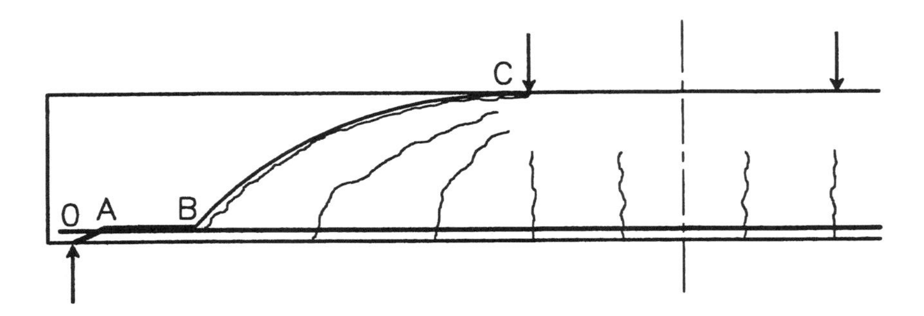

Figure 2.1.32 Failure in a nonshear reinforced beam by sliding in a crack.

$$c' = \nu_s \nu_o c \tag{2.1.61}$$

where ν_o is the effectiveness factor of the microcracked material.

This means that for the most dangerous cracks the compressive strength will be $\nu_s\ \nu_o f_c$, a value far below the standard value of the compressive strength.

The strength reduction due to macrocracks may be reduced by providing extra reinforcement traversing the cracks, see Chapter 4.

An important example, where crack sliding plays a dominating role, is the shear failure of nonshear reinforced beams. The final solution to this old problem utilizing the above ideas was given by Jin-Ping Zhang [94.10], see also Chapter 5. Briefly explaining the idea follows.

Figure 2.1.32 shows a simply supported nonshear reinforced beam loaded with two symmetrical concentrated forces. The crack picture as it has developed until failure is schematically shown in the figure. The curved cracks in the shear zone are running from the tensile face toward the nearest force when the load is increased. One of these cracks turns out to lead to failure because of the low sliding resistance along a crack. The final failure will be a sliding failure along *OA* and *BC* and a separation failure along *AB*, which is situated just above the longitudinal reinforcement. The yield line along the crack will be more dangerous than a yield line through concrete without macrocracks.

This explanation of the shear failure in nonshear reinforced beams implies that the load-carrying capacity should not be very much dependent on whether the beam is cracked from other reasons than external loads before loading or not. This conclusion is confirmed by experiments on beams heavily cracked by alkali-silica reactions [92.21]. The shear capacity turned out to be almost the same as for normal beams.

Another example where sliding in cracks plays an important role is the shear failure of orthotropic panels, i.e., panels reinforced in two perpendicular directions with different reinforcement ratios. The situation is shown in Fig. 2.1.33. The initial cracks will be formed under 45° with the sections with pure shear, and this initial crack direction will be roughly independent of the reinforcement.

However, if the reinforcement ratios in the two reinforcement directions are different the final crack direction will be different from the initial one, see Section 2.2.2. This means that the final compression direction in the concrete might be as shown in the figure. It follows that compressive failure may take place by sliding failure along the initial cracks for a very low compressive stress compared to the compressive strength of the virgin material.

Such a reduced compressive strength has been measured by Vecchio and Collins [82.4], and they also measured the strains. One of their stress-strain curves are reproduced in Fig. 2.1.34.

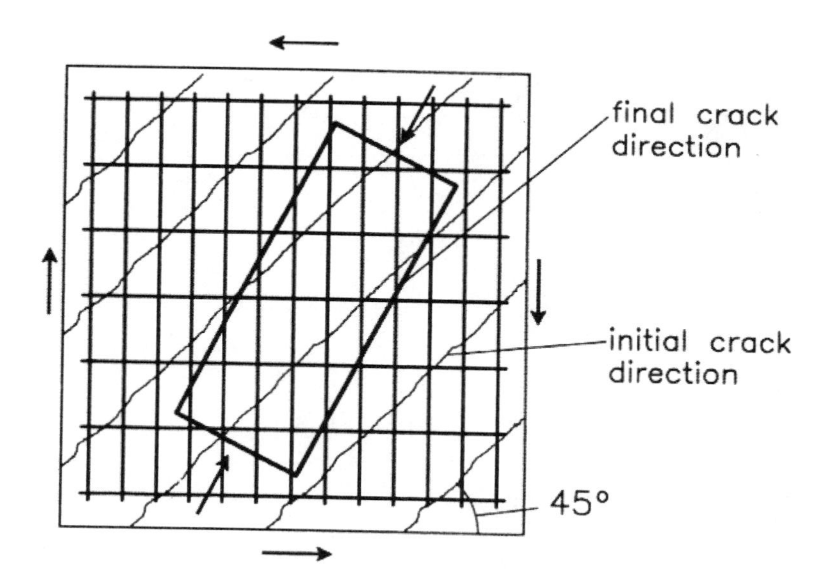

Figure 2.1.33 Orthotropic panel in pure shear.

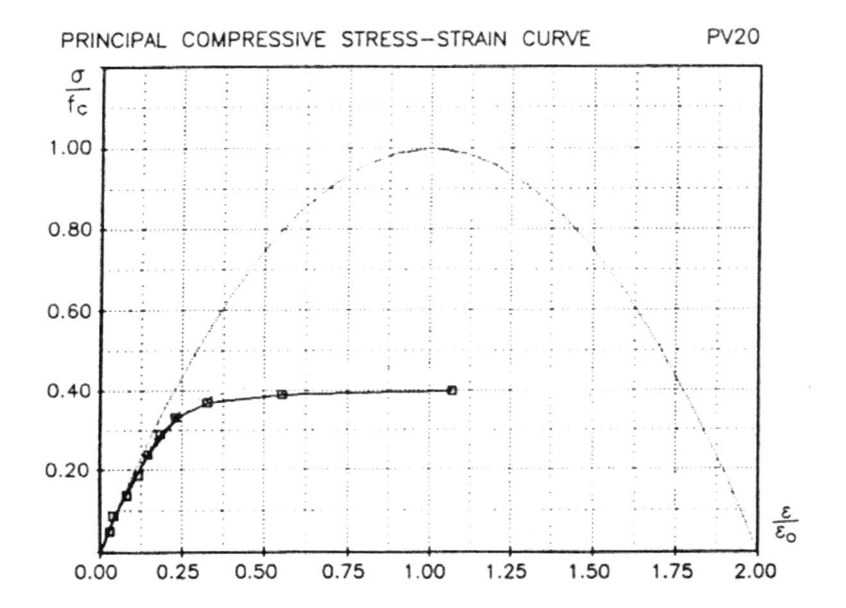

Figure 2.1.34 Stress-strain relation of an orthotropic panel, [82.4].

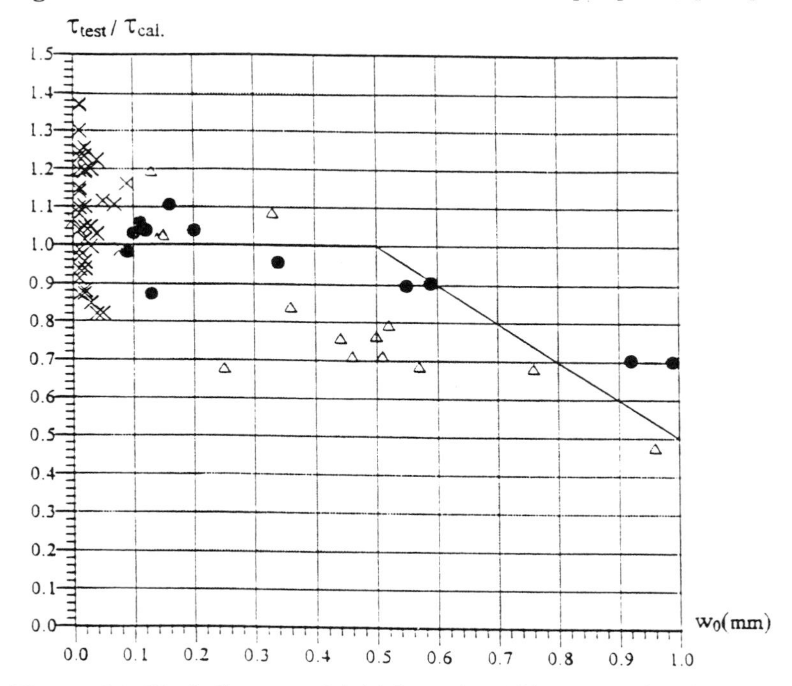

Figure 2.1.35 Influence of initial crack width w_o on the shear carrying capacity of cracks [97.2].

The sliding in initial cracks gives rise to an almost perfectly plastic behavior, but at a strongly reduced compressive stress.

The effects of crack sliding have not yet been studied to a degree that they may be taken into account in a general way. But as far as possible they are treated in the following chapters.

The sliding resistance of a crack will, of course, disappear if the crack width is sufficiently large. The large number of push-off tests published in the literature renders possible a rough estimation of the crack width making the sliding resistance disappear. Fig. 2.1.35, which has been taken from [97.2], shows the experimental value τ_{test} divided by the calculated one τ_{cal} determined without taking into account any reduction from the crack width. This ratio is given as a function of the initial crack width w_0

The scatter is considerable, but it seems reasonable to conclude that an initial crack width up to 0.5 to 0.7 mm does not seriously affect the load-carrying capacity.

Finally, we turn our attention to the effects of softening. Typical stress-strain curves for concrete in uniaxial compression are schematically shown in Fig. 2.1.36.

For a low strength concrete, say f_c = 10 MPa, the stress σ as a function of the average strain ϵ in the loading direction is characterized by a rather flat descending branch after the peak value f_c has been reached. On the other hand, for a high strength concrete, say f_c = 100 MPa, the descending branch is very steep.

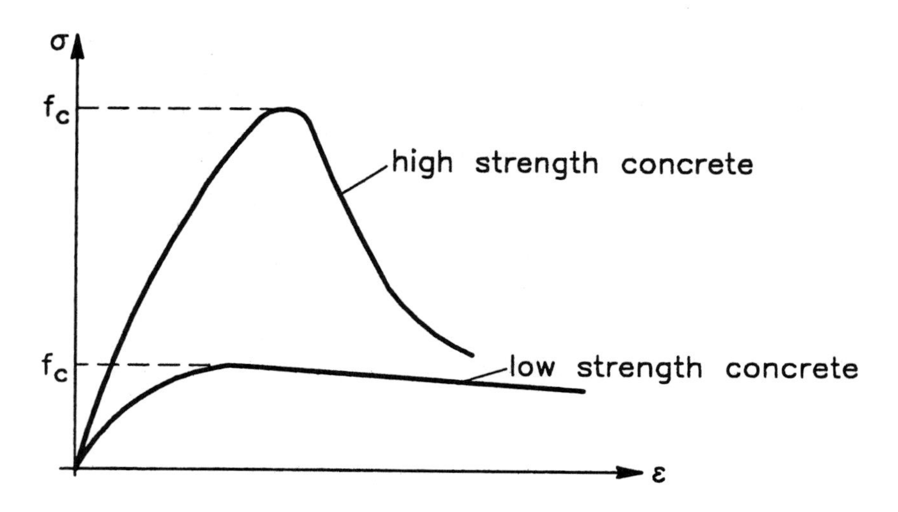

Figure 2.1.36 Typical stress-strain curves for concrete in uniaxial compression.

Unfortunately, the descending branch depends on a large number of parameters. First of all it depends on the length along which the average strain ϵ is measured, which is due to the fact that yield lines develop and the average strain will depend on whether the region with yield lines is included in the measuring length in whole or in part or not at all. It also depends on the type of test specimen. If the stress-strain relation is determined from a bending test, the descending branch will be rather flat even for a high strength concrete, which is due to the fact that in this case a larger number of yield lines may develop. Finally, the descending branch is severely dependent on the confinement supplied by friction between the loading platen at the end surfaces of the specimen and that supplied by transverse pressure or by closed stirrup reinforcement.

From these reasons it does not seem worthwhile to draw too many conclusions on the basis of one type of measurement of the descending branch.

It would seem rather impossible to use the theory of perfectly plastic materials when the stress field at the peak value of the load is strongly inhomogeneous.

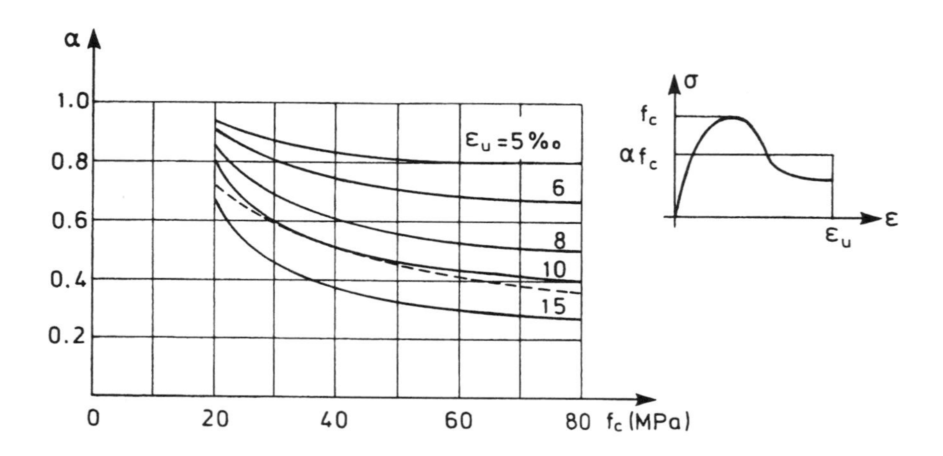

Figure 2.1.37 Diagram showing the area "under the stress-strain curve" versus the compressive strength and the ultimate strain.

Nevertheless, however strange it might seem, in such situations, the simple plastic theory has also provided useful solutions when effective strength parameters are used.

First of all, bending problems have been treated in this way, see Chapter 5.

For pure bending of a rectangular section with only tensile reinforcement, ν has been determined by Exner [83.2] using stress-strain curves measured by P. T. Wang et al. [78.21] (see Section 5.1). It turns out that ν is a function of the uniaxial compressive strength f_c, the yield stress of the reinforcement f_Y, and the reinforcement ratio r. For small values of r, it appears that ν is independent of r and decreases with increasing concrete strength. For a certain r-value, the curves have a minimum. When this is passed, ν increases with increasing r. It also appears that ν is rather close to 1 for low degrees of reinforcement and low concrete strengths.

Also in many shear problems the effectiveness factor due to softening is a decreasing function for increasing concrete strength.

It may be shown that if one compares the load-carrying capacity of two structures of elastic materials, one of a material with a concrete-like stress-strain curve and one of a material with a perfectly plastic stress-strain curve having the same area, one finds that the load-carrying capacity of the structure made of a concrete-like material is greater than or equal to the load-carrying capacity of the perfectly plastic structure, provided that the strains do not exceed the ultimate strain ϵ_u used to calibrate the yield stress of the perfectly plastic material [83.2]. In reality, unloading might of course take place in certain points (i.e., the real materials cannot be identified with elastic materials), but the theorem gives a theoretical basis for operating with effective strength parameters in softening problems.

If we compare the real stress-strain curve with a corresponding stress-strain curve for a perfectly plastic material with the yield stress αf_c and having the same area as the real stress-strain curve, we find that α will be a function of f_c and ϵ_u. Some results are shown in Fig. 2.1.37.

It is seen that α decreases for increasing f_c which is, of course, due to the fact that the area "under the stress-strain curve" $\int_o^{\epsilon_u} \sigma d\epsilon$ is relatively smaller for high concrete strengths than for low concrete strengths, cf. Fig. 2.1.36.

The calculations have been performed by Exner [79.2] using the same stress-strain curves as those mentioned above. This α-dependence on f_c is very similar to the ν-dependence on f_c observed in a number of cases, that is, approximately a square-root dependence $\nu = K/\sqrt{f_c}$. In Fig. 2.1.37 the curve drawn with the dashed line is an effectiveness factor ν determined from shear tests with beams without shear reinforcement (see Chapter 5). The curve has been taken from [79.4]. The curve has been repeated in Fig. 2.1.38, where the test results are also shown. The value of K in $\nu = K/\sqrt{f_c}$ is 3.2 when f_c is measured in MPa.

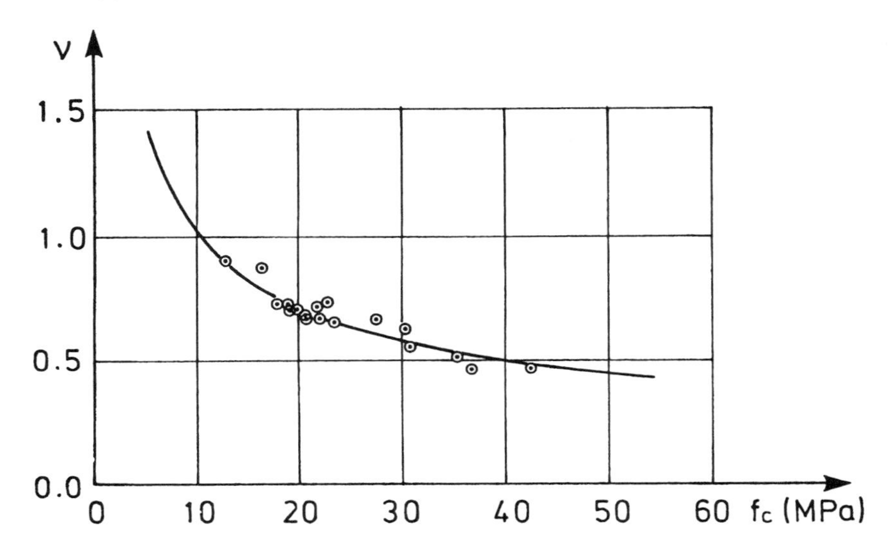

Figure 2.1.38 Effectiveness factor for beams in shear versus the compressive strength.

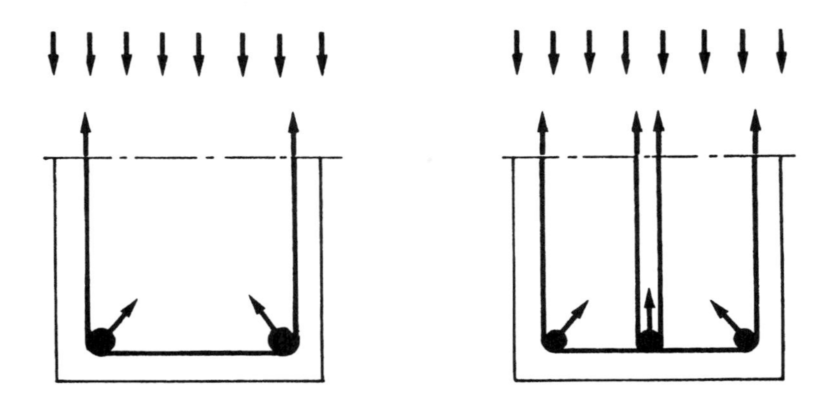

Figure 2.1.39 Stress concentration effects in beams in shear.

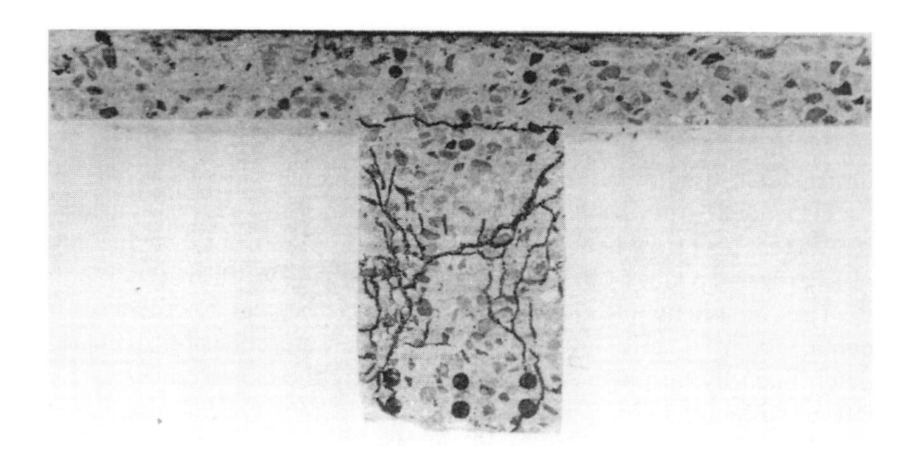

Figure 2.1.40 Cracking in a shear reinforced beam.

Since the softening is depending on both the geometry of the structural element considered and its loading, the effectiveness factors due to softening cannot be given as a function of the material parameters alone.

Geometrical effects may be of the type found in beams subjected to shear, where stress concentrations around bars play a role. This is illustrated in Fig. 2.1.39. Here the concrete stresses have to be transferred to longitudinal bars supported by stirrups. It is evident that the number of bars and their distribution along the concrete section may have an influence on the effective strength. The higher the number of supported bars,[1] the higher the effective strength that can be obtained [76.3; 80.1]. This effect, however, is rather small from a practical point of view, and may be neglected for beams as they are normally found in practice. In Chapter 5 it is shown that only the dependence on concrete strength has to be taken into account.

The stress concentrations around the supported bars, of course, also lead to a highly complicated state of microcracking. The cracking found in a normal section of a shear reinforced beam is shown in Fig. 2.1.40 (see [76.3]).

[1] In one test series described in [80.1], the stirrups were welded to a steel plate covering the whole concrete face. The result was a substantial increase in effective strength compared with the effective strength of normal beams.

In this case the effectiveness factor might as well be explained as a consequence of microcracking. Actually, formula (2.1.59) may be used in this case (see Chapter 5).

Another geometrical effect that has been found [79.4] is dependence on the absolute value of the dimensions. The larger the dimensions, the lower the effective strength.

Dependence on the loading has been demonstrated for nonshear reinforced beams, where concentrated loading and uniform loading give different values of the effective strengths (see [94.10]).

The treatment in [94.10] shows that when a macrocrack is contained in a softening region the resulting cohesion may be determined by formula (2.1.61), ν_o now being the effectiveness factor due to softening.

The foregoing considerations refer mainly to the effective concrete compressive strength. Although the ductility of concrete in tension is very limited, cases exist where plastic solutions involving the tensile strength have been shown to be meaningful. The effectiveness factor is very low, however, ρ as already mentioned being of the order of a few percent.

The cases for which it is meaningful to take into account the tensile strength in plastic solutions are characterized by very small crack widths in the failure pattern. Typical examples in this category of problems are concentrated loads on a concrete face and anchorage problems.

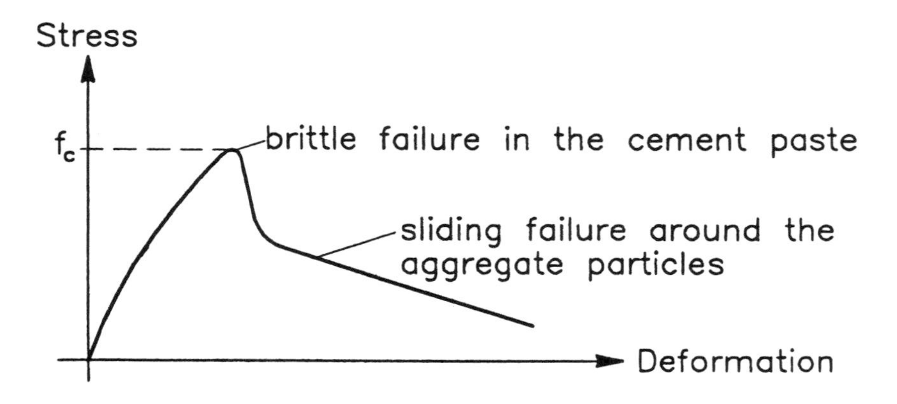

Figure 2.1.41 Stress-deformation curve for concrete in tension.

The physical reason why tensile stresses in cracks are present at the ultimate load is that after tensile failure in the cement paste, a residual strength is mobilized due to sliding failures around the aggregate particles. The initial tensile failure in the cement paste is very brittle, but the sliding failure around the aggregate particles is much more ductile. The tensile stress-deformation curve is shown schematically in Fig. 2.1.41.

For a more detailed explanation of the tensile behavior of concrete, see [97.2].

In the following chapters the available information about effective concrete strength is treated or referred to. In several important practical situations the available information will be judged to be insufficient. In such cases the effective strength has to be estimated on the conservative side on the basis of the available experimental evidence in the actual situation or in similar situations. Such crude, but conservative information is often included in design codes. Take as an example that a concrete stress σ_c according to a code rule has to be limited to, say αf_c, α being less than 1. This is just another way of saying that the effectiveness factor in this case is α.

If no experimental information at all is available, one has to design with very low concrete stresses. In fixing the effectiveness factor in these cases, one may be guided by the fact that it is very unlikely that the effectiveness factor will be lower than

$$\nu = \nu_s \nu_o \tag{2.1.62}$$

where ν_s may be set to 0.5 and ν_o is determined by (2.1.59). This value of ν corresponds to a concrete both fully microcracked and fully macrocracked.

In Eurocode 2, [91.23], the lower value has been set to

$$\nu = \frac{2}{\sqrt{f_c}} \quad (f_c \text{ in MPa}) \tag{2.1.63}$$

which is the lowest value found for shear in nonshear reinforced beams including the effect of sliding in initial cracks. This value was also suggested in the first edition of this book [84.11].

We have now finished our review of the rather incomplete knowledge about the effective strength of concrete.

It appears that an accurate description of the real behavior of concrete is not possible by simple means. In the main part of this book, we will therefore take an engineering approach to the problem.

$$f_c \rightarrow \nu f_c$$

$$f_t \rightarrow \nu_t f_t \text{ or } \rho f_c$$

Box 2.1.2 Scheme used to modify a theoretical plastic solution.

The main line will be to develop solutions using plastic theory based on the modified Coulomb failure conditions. These solutions are then modified by introduction of effective strength parameters determined on the basis of the tests available. The physics of the problems may then be well hidden, but it is believed that such an approach will be the most useful one for the engineering profession at the present stage of development.

By any measure it will be far more useful than a completely empirical approach, which is still dominating many areas of the concrete field.

The plastic solutions will be found using the notations f_c and f_t for the strength parameters. In applications f_c and f_t are then replaced by effective strengths using the scheme shown in Box 2.1.2.

2.2 YIELD CONDITIONS FOR REINFORCED DISKS

2.2.1 Assumptions

In this section we deal with yield conditions for reinforced concrete, a composite material containing concrete as well as reinforcement bars, these normally being made by steel. The yield condition for the composite material will be developed using the rigid-plastic theory (see Chapter 1), and by using the failure criteria for concrete developed in Section 2.1 as yield conditions for concrete. The shortcomings of applying the rigid-plastic model will be discussed in some detail later.

The concrete is considered to be a rigid-plastic material obeying the modified Coulomb criterion with zero tensile cutoff (see Fig. 2.2.1). Yield conditions taking into account the tensile strength of concrete were developed by Marti and Thürlimann [77.7].

The reason for setting the tensile strength of concrete equal to zero is rather obvious when using a plastic model, since the tensile failure of concrete is far from being ductile. However, as demonstrated later, plastic solutions with a nonzero cutoff are sometimes very useful.

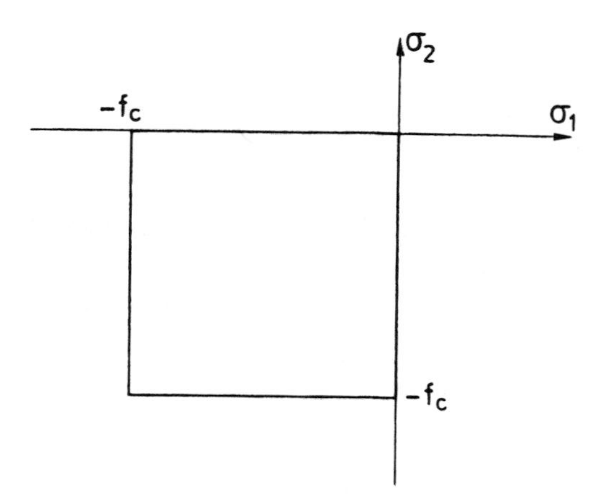

Figure 2.2.1 Yield condition for concrete in plane stress, the tensile strength being set to zero.

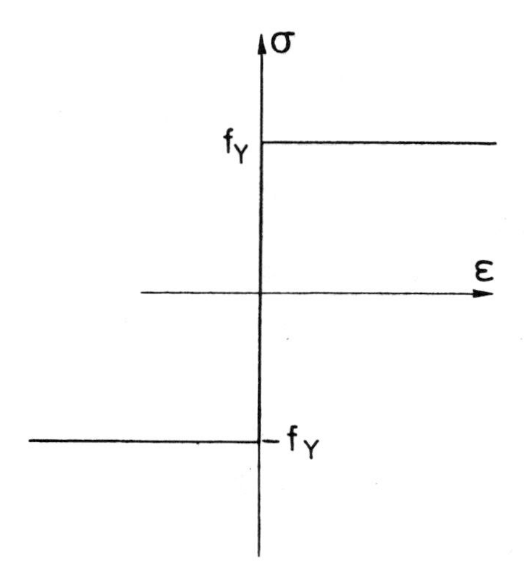

Figure 2.2.2 Uniaxial stress-strain relation for a reinforcement bar.

As for the reinforcement, we shall assume that it consists of straight bars with a stress-strain relation for tension and compression as shown in Fig. 2.2.2, in which f_Y indicates the yield stress. For steel without a definite yield point, the yield stress f_y is defined in a suitable manner (e.g., as the 0.2% offset strength).

Furthermore, we assume that the reinforcing bars are capable of carrying longitudinal tensile and compressive stresses only. This assumption seems to be natural on the basis of the lower bound theorem, proved in Chapter 1, from which it is seen that this assumption will result in stresses in the reinforcement which are statically admissible and safe if the stress is less than or equal to the yield stress. If the stresses of the concrete satisfy the same condition, the method will result in values of load-carrying capacity which, in relation to the assumptions of the strength characteristics of concrete and reinforcement, are on the safe side. This is, in fact, the most essential reason for dealing only with longitudinal tension or compression in the reinforcement.

An idea of the magnitude of the possible shear stresses in the bars may be rendered in the following way. Shear stresses in normal sections of the bars may arise if on the surface of the bar there are shear stresses or normal stresses, the latter producing varying bending moments. An idea of the magnitude of these shear stresses may be rendered by considering a couple of extreme distributions of normal and shear stresses on the surfaces of the bars. The distribution of shear stresses on the surface τ_c giving the maximum average value of the shear stresses τ_s in the normal sections are shown in Fig. 2.2.3.

This distribution corresponds to the largest possible external moment per unit of length on the surface. By a moment equation for a small segment limited by two consecutive normal sections we find, d being the diameter of the bar,

$$\frac{\pi}{4} d^2 \tau_s = \frac{\pi}{2} d \tau_c \cdot \frac{2}{\pi} d \tag{2.2.1}$$

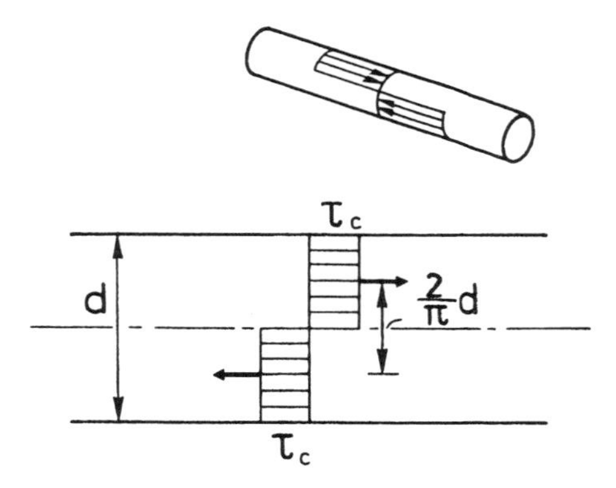

Figure 2.2.3 Shear stress distribution on the surface of a reinforcement bar giving maximum external moment per unit length.

i.e., $\tau_s \approx \tau_c$, meaning that τ_s is of the same order of magnitude as τ_c. As τ_c in a plane stress field can at most be approximately $\frac{1}{2}f_c$, it means that this is the order of magnitude of τ_s as well.

The distribution of normal stresses, giving the maximum average value of the shear stress τ_s, is shown in Fig. 2.2.4, in which for a distance of $\frac{1}{2}L$ we have normal stresses f_c directed upward and considered to be uniformly distributed over the diameter of the bar d, and for similar distance $\frac{1}{2}L$, normal stresses f_c, downward directed (dowel action).

The longer the distance L, the bigger the maximum shear stress will be in the bar. L, however, cannot possibly be longer than its value in the case where the moment of these normal stresses equals twice the yield moment at pure bending M_p of the reinforcing bar, corresponding to bending moments acting on the ends of the loaded interval as shown in the figure. As $M_p = \frac{1}{6}d^3 f_Y$ for a round bar, the following equation is found for determining the upper limit of L:

$$\frac{1}{4}dL^2 f_c = 2 \cdot \frac{1}{6}d^3 f_Y \tag{2.2.2}$$

that is,

$$L = \frac{2}{\sqrt{3}}d\sqrt{\frac{f_Y}{f_c}} \tag{2.2.3}$$

Thus the maximum value of τ_s is

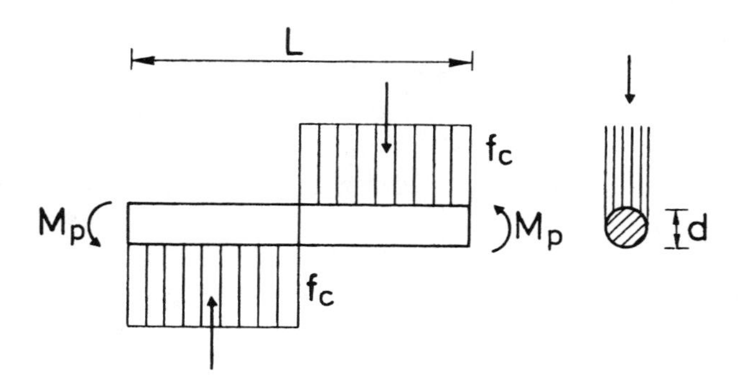

Figure 2.2.4 Normal stress distribution on the surface of a reinforcement bar giving maximum shear force in normal sections of the bar.

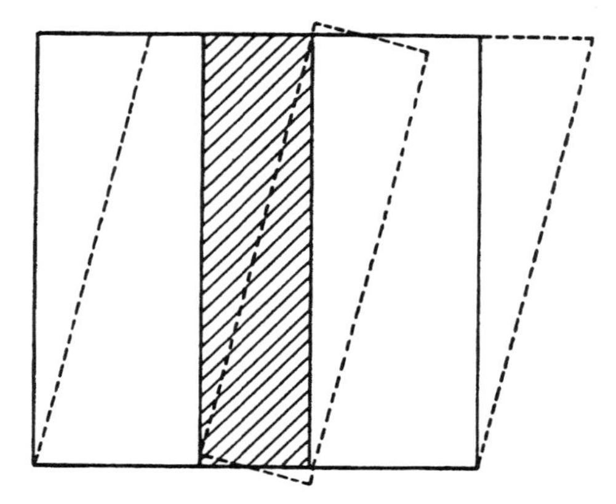

Figure 2.2.5 Rotation of a reinforcement bar in the case of pure shear strain in the surrounding concrete.

$$\tau_s = \frac{\frac{1}{2} L d f_c}{(\pi/4) d^2} = \frac{4}{\sqrt{3}\,\pi}\sqrt{f_Y f_c} \simeq 0.7\sqrt{f_Y f_c} \tag{2.2.4}$$

For instance, if $f_c \approx \frac{1}{10} f_Y$, we have

$$\tau_s \approx \frac{1}{5} f_Y \tag{2.2.5}$$

This shear stress is, of course, in no way negligible as it equals about half the shear strength of the bar. The stress distribution which gives this value of τ_s is not especially improbable, and in addition, the stresses of the concrete may locally exceed f_c several times.

Consequently, our conclusion is that the shear stresses in the reinforcing bars cannot always be considered inferior. Normally, they will possibly be so, at least when the reinforcement is placed for the purpose of carrying tensile stresses. For this and the reason mentioned above we shall assume in the following discussion that the reinforcing bars are able to carry only longitudinal tension and compression.

The normal stress in the bars is thought to be determined by the unit elongation in their axes. This assumption, of course, is a result of calculating only with tension and compression in the reinforcing bars. It would also lead to irrational results to calculate with stresses in the bars resulting from shear strains, since, according to the foregoing considerations, the shear strains of a bar will be small due to the inferior values of the shear stresses of the concrete near the bar. So we must expect that even at an early state local bond failures are taking place between concrete and bars when the concrete is subjected to shear strain in the direction of the bar. Therefore, the bar only gets a rotation of the same size as the one of concrete, but the bar itself gets no significant shear strain (see Fig. 2.2.5). At this stage we have to make clear that the measures of deformation we will use must be considered as average quantities, which do not describe the state of deformation in detail. That at a certain point we are operating with a shear strain does not indicate that bars as well as concrete have strain, but only that concrete has it.

Finally, the reinforcing bars are considered to be placed at such small intervals that the forces in them can be replaced by an equivalent continuous stress distribution in the concrete. A condition of using this approximation is that the "wave length" of the stresses is much longer than the distance between the reinforcing bars. This assumption may not be considered to be in good agreement with the real stress distribution in all cases, but it is necessary to make a mathematical treatment practical at all.

The phenomenon of crack sliding defined in Section 2.1.4 is only slightly touched upon. It is discussed more thoroughly in Chapter 4.

It is seen from this description of the assumptions that theory and reality cannot harmonize very well in detail. The detailed description of the behavior of a composite material such as reinforced concrete is so complicated that a simple theory must be inadequate in certain respects. Fortunately, it turns out that a more accurate description of the detailed behavior is not essential when the primary purpose is to determine the load-carrying capacity of a reinforced concrete structure.

The foregoing considerations concerning the magnitude of shear stresses in reinforcing bars are in agreement with the assumption of classical reinforced concrete theories. It was demonstrated by Mörsch [06.1] that the stresses of shear adhesion are unable to yield great shear stresses in the reinforcement. The magnitude of the shear stresses in bending was investigated by K. W. Johansen [28.2] in considering the conditions in joints in which the stress distribution along a bar is analogous to the distribution shown in Fig. 2.2.4.

2.2.2 Orthogonal Reinforcement

The reinforcement degree. We shall examine a disk reinforced in two directions, x and y, at right angles to each other. Areas of reinforcing bars per unit length are denoted as A_{sx} and A_{sy}, corresponding to the bars in the direction of the x-axis and the y-axis, respectively. Components of the stress tensor are termed σ_x, σ_y, and τ_{xy} as usual. Stresses of the bars are termed σ_{sx} and σ_{sy}, and the stresses of the concrete, σ_{cx}, σ_{cy}, and τ_{cxy}. Principal stresses in the concrete are indicated as σ_{c1} and σ_{c2}.

The stresses obtained when the forces in the reinforcing bars are considered to be evenly distributed over the thickness of the disk are called the *equivalent stresses* of the bars, or simply the equivalent stresses (especially the equivalent normal or shear stresses). The sum of the stresses of the concrete and the equivalent stresses of the bars in the same section are called the *total stresses*.

We introduce the term *reinforcement degree*, defined as being the ratio between the force per unit of length that the reinforcement is able to carry and the force per unit of length that the concrete is capable of carrying in pure compression. The reinforcement degree is denoted Φ, and so, for the x-axis and the y-axis, respectively, we have

$$\Phi_x = \frac{A_{sx} f_Y}{t f_c}, \qquad \Phi_y = \frac{A_{sy} f_Y}{t f_c} \tag{2.2.6}$$

in which t signifies the thickness of the disk.

If the yield stress of the bars in the x-direction, f_{Yx}, is different from the yield stress of the bars in the y-direction, f_{Yy}, here and in the following we just have to replace the term $A_{sx} f_Y$ by $A_{sx} f_{Yx}$ and the term $A_{sy} f_Y$ by $A_{sy} f_{Yy}$.

Tension and compression. The tensile strength in the direction of the x-axis of the composite material f_{tx} is immediately seen to be

$$f_{tx} = \frac{A_{sx} f_Y}{t} = \Phi_x f_c \tag{2.2.7}$$

t indicating the thickness of the disk. The stress f_{tx} is the equivalent stress in this simple case. With analogous notation the tensile strength will be

$$f_{ty} = \Phi_y f_c \tag{2.2.8}$$

in the direction of the y-axis. Correspondingly, the compressive strengths f_{cx} and f_{cy} are

$$f_{cx} = (1 + \Phi_x) f_c \tag{2.2.9}$$

$$f_{cy} = (1 + \Phi_y) f_c \tag{2.2.10}$$

In an isotropic disk[2] for which $A_{sx} = A_{sy}$ and, therefore $\Phi_x = \Phi_y = \Phi$, we obtain

$$f_{tx} = f_{ty} = \Phi f_c \tag{2.2.11}$$

$$f_{cx} = f_{cy} = (1 + \Phi) f_c \tag{2.2.12}$$

Pure shear. Next we treat pure shear in the x, y-system (see Fig. 2.2.6). Supposing that the stress field of the concrete is characterized by the principal stresses

$$\sigma_{c1} = 0 , \qquad \sigma_{c2} = -\sigma_c \tag{2.2.13}$$

and the second principal axis forms an angle α to the x-axis, we obtain

$$\sigma_{cx} = -\sigma_c \cos^2 \alpha , \qquad \sigma_{cy} = -\sigma_c \sin^2 \alpha \tag{2.2.14}$$

Defining

$$\frac{A_{sy}}{A_{sx}} = \mu \tag{2.2.15}$$

we obtain the following statical conditions for determining α, assuming the reinforcement to be stretched to yield:

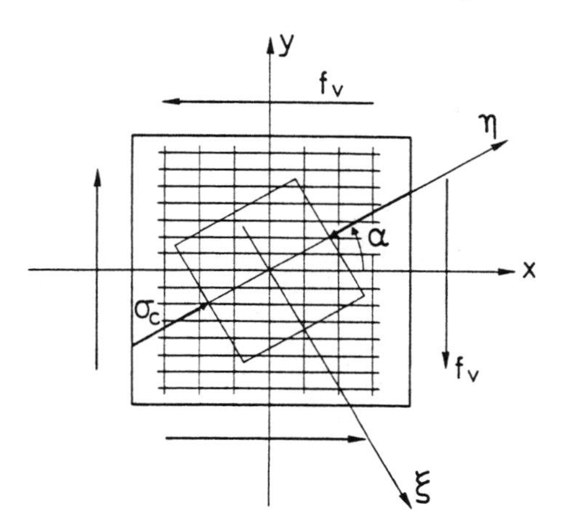

Figure 2.2.6 Disk with orthogonal reinforcement subjected to pure shear.

[2] The reason for the expression "isotropic disk" will be seen below.

$$A_{sx} f_Y = t \sigma_c \cos^2 \alpha \tag{2.2.16}$$

$$A_{sy} f_Y = t \sigma_c \sin^2 \alpha \tag{2.2.17}$$

or

$$\tan^2 \alpha = \mu \,, \qquad \sigma_c = \frac{A_{sx} f_Y}{t \cos^2 \alpha} = \frac{\Phi_x f_c}{\cos^2 \alpha} = \Phi_x (1 + \mu) f_c \tag{2.2.18}$$

Having

$$|\tau_{cxy}| = \frac{1}{2} \sigma_c \sin 2\alpha \tag{2.2.19}$$

we find the shear strength

$$f_v = \frac{1}{2} \sigma_c \sin 2\alpha = \frac{1}{2} \frac{\Phi_x f_c}{\cos^2 \alpha} \sin 2\alpha = \Phi_x f_c \tan \alpha$$

$$= \sqrt{\mu} \Phi_x f_c = \sqrt{\Phi_x \Phi_y} f_c = \frac{\sqrt{A_{sx} A_{sy}}}{t} f_Y = \sqrt{f_{tx} f_{ty}} \tag{2.2.20}$$

If $\mu = 1$ (i.e., $\Phi_x = \Phi_y = \Phi$), we obtain

$$f_v = \Phi f_c = f_{tx} = f_{ty} \tag{2.2.21}$$

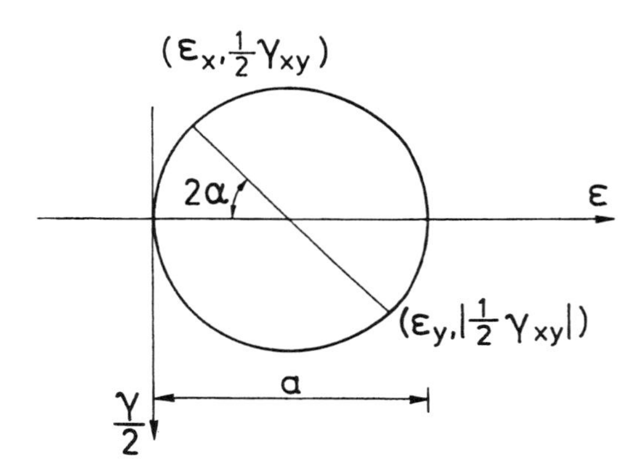

Figure 2.2.7 Mohr's circle showing the strain distribution in a disk subjected to pure shear.

In an isotropic disk the tensile strength and the shear strength in the directions of reinforcement are equal. In order to obtain yielding of the reinforcement it is seen from (2.2.18) that we must require

$$\Phi_x + \Phi_y \leq 1 \tag{2.2.22}$$

this being a condition of determining the shear strength by (2.2.20).

The solution (2.2.20) is seen to be exact according to our assumptions. The strain field at collapse is illustrated in Fig. 2.2.7 by means of Mohr's circle, the longitudinal strains being indicated by ϵ_x and ϵ_y, and the shear strains $\epsilon_{xy} = \frac{1}{2}\gamma_{xy}$, where γ_{xy} is the decrease of the right angle between the coordinate lines. Here it is assumed that $\Phi_x + \Phi_y \leq 1$.

The smallest principal strain is zero in accordance with $\sigma_c \leq f_c$. The directions of the principal strains are the ξ, η-directions in Fig. 2.2.6. According to the notation in Fig. 2.2.7, $\epsilon_\xi = a$, $\epsilon_\eta = \epsilon_{\xi\eta} = 0$. As assumed, it is seen that extensions in the directions of the bars occur, and consequently the assumption that the stress in the bars is equal to the tensile yield stress is, according to the rigid-plastic model, correct.

This example of using our assumptions clearly shows the importance of the assumption that the tensile strength is zero. We find that in all sections parallel to the η-axis the stress of the concrete is zero, which can be said to correspond to a continuous distribution of "cracks". In reality the initial cracks will be in sections under 45° with the coordinate axes x and y. Tensile stresses in these sections would appear before cracking, and the elongations in the x and y axes would be zero (i.e., the bars would not come into action). As soon as cracking appears, however, the bars get into the action and the uniaxial concrete compression stress rotates to a new direction if $\mu \neq 1$. New cracks may be formed. Thus we may get crack sliding in the initial cracks. This phenomenon is discussed shortly at the end of this section and more thoroughly in Chapter 4. It should be noted, however, that in reality the cracks will be discontinuously distributed, and therefore the strains on the bars uneven. Finally, it should be noted that the strain field corresponding to statical pure shear is not geometrical pure shear.

For higher degrees of reinforcement than those according to the limit in (2.2.22), yielding will not occur either in one direction or in both directions of reinforcement. Indicating the smaller quantity of Φ_x and Φ_y by Φ_i, we find for $\Phi_x + \Phi_y > 1$ that the shear strength is $f_v = \sqrt{\Phi_i(1 - \Phi_i)}\, f_c$ when $\Phi_i \leq 0.5$ while being $f_v = \frac{1}{2} f_c$ if $\Phi_i > 0.5$.

When $\Phi_i \leq 0.5$, yielding takes place in the direction of reinforcement corresponding to Φ_i, whereas there is no yielding in the other direction of reinforcement. When $\Phi_i > 0.5$, there is no yielding in any direction of reinforcement. The shear strength is then determined simply by the compressive strength of the concrete and is equal to the maximum shear stress $\frac{1}{2}f_c$ occurring in a concrete disk subjected to plane stress.

The yield condition in the isotropic case. We now go on to derive the yield condition in the isotropic case, that is, for $\Phi_x = \Phi_y = \Phi$ or $A_{sx} = A_{sy} = A_s$. The yield condition is depicted in a rectangular σ_x, σ_y, τ_{xy}-system of coordinates. The appearance of the yield condition in the σ_x, σ_y-plane (see Fig. 2.2.8) is evident.

We aim at determining the magnitude of the shear stresses τ_{xy} which can be carried. Starting at point B we have $\sigma_{sx} = \sigma_{sy} = f_Y$ and $\sigma_{c1} = \sigma_{c2} = 0$ (i.e., only the reinforcement is acting). Retaining these values of the stresses of the bars, it is seen that we obtain the same equivalent normal stress σ in all sections, as

$$\sigma = \frac{A_s f_Y}{t}\cos^2\theta + \frac{A_s f_Y}{t}\sin^2\theta = \frac{A_s f_Y}{t} = \Phi f_c$$

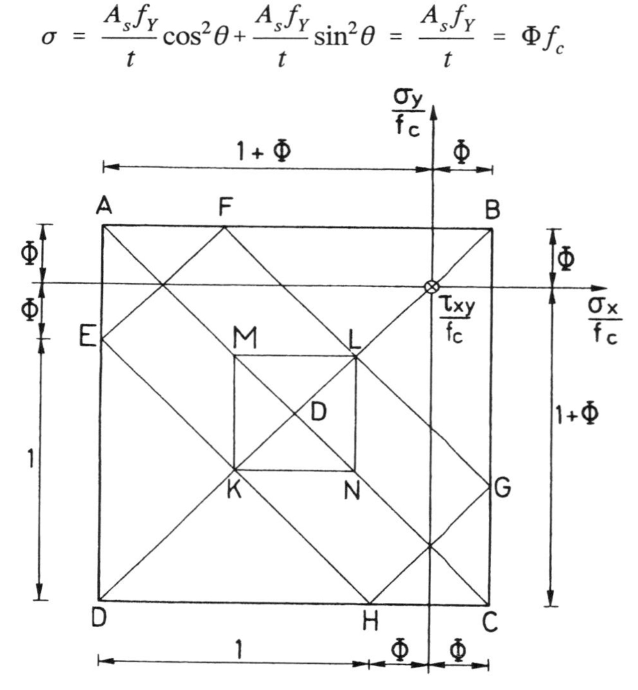

Figure 2.2.8 Yield condition for a disk with orthogonal reinforcement, the isotropic case.

and the equivalent shear stress τ in all sections is zero. From this it is obvious that in the vicinity of B there must be an area in which the yield condition is isotropic; that is, the stresses which can be carried are independent of the directions of the principal sections. In this area we are able to carry the principal tensile stress $\sigma_1 = \Phi f_c$, and by varying the stress of the concrete from zero to $-f_c$ in the other direction of principal stress, it is seen that we can carry the principal stresses $\Phi f_c \geq \sigma_2 \geq -(1-\Phi)f_c$. In this area the yield condition is determined by the equations

$$\sigma_1 = \frac{1}{2}(\sigma_x + \sigma_y) + \sqrt{\frac{1}{4}(\sigma_x - \sigma_y)^2 + \tau_{xy}^2} = \Phi f_c \tag{2.2.23}$$

$$-(1-\Phi)f_c \leq \frac{1}{2}(\sigma_x + \sigma_y) - \sqrt{\frac{1}{4}(\sigma_x - \sigma_y)^2 + \tau_{xy}^2} \leq \Phi f_c \tag{2.2.24}$$

Transforming the first equation, we get

$$-(\Phi f_c - \sigma_x)(\Phi f_c - \sigma_y) + \tau_{xy}^2 = 0 \tag{2.2.25}$$

representing a conical surface with the apex at B and the axis in the direction BD (see Fig. 2.2.8). Furthermore, from (2.2.23) and (2.2.24), we obtain

$$\sigma_x + \sigma_y \geq -(1-2\Phi)f_c \tag{2.2.26}$$

that is, (2.2.25) is valid for the area BGF in Fig. 2.2.8. Similarly, the equation of the yield condition for area EDH is found to be

$$-[(1+\Phi)f_c + \sigma_x][(1+\Phi)f_c + \sigma_y] + \tau_{xy}^2 = 0 \tag{2.2.27}$$

For the remaining area the yield condition is not that simple. We are, however, immediately able to construct statically admissible and safe stress fields within these areas. Let us look at the distance LG, at which we have the tensile yield stress f_Y in both directions of reinforcement and uniaxial compression in concrete with variable directions of the principal stress.

Moving vertically down (i.e., parallel to the σ_y-axis), we have to carry numerically larger stresses in the y-sections. This can be obtained by changing only the stresses of the bars σ_{sy}. By changing this stress from f_Y to $-f_Y$ we are moving from LG to NC. As the stresses of the concrete are unchanged, we are able to carry the same shear stresses; that is, the yield condition is a cylindrical surface with the direction of generatrix LN parallel to GC. Moving from NC to KH along lines parallel to the σ_x-axis, this can be obtained in the same way by changing σ_{sx} from f_Y to $-f_Y$ without changing the stresses of the concrete; in this case the yield condition is a cylindrical surface, the direction of generatrix being *NK* parallel to *CH*. The areas *MAFL* in which the direction of generatrix is *LM* parallel to *FA*, and *AMKE*, the direction of generatrix being *MK* parallel to *AE*, are treated in the same way. The value of $|\tau_{xy}|$ along the square *LMKN* will be the same; by (2.2.25) or (2.2.27) it is found to be $\frac{1}{2}f_c$. This shear stress can be carried in the whole area *LMKN*, the yield surface thus being a plane parallel to the σ_x, σ_y-plane. It is easily seen that the stress fields established in the different areas are identical along the limitations of the areas.

Now a safe yield surface has been constructed. By geometrical considerations analogous to the ones made in calculating the shear strength, it is clear that the stresses found are the maximum stresses that can be carried. The appearance of the yield surface is illustrated in Fig. 2.2.9, and in Fig. 2.2.10 contour lines of τ_{xy} for $\Phi = 0.25$ are recorded.

The curve of intersection of the yield surface with the τ_{xy}, σ_x-plane, identical to the curve of intersection between the yield surface and the τ_{xy}, σ_y-plane, is shown in Fig. 2.2.11. Within the interval $-(1-2\Phi)f_c \leq \sigma_x \leq \Phi f_c$, the curve of intersection is determined by the equation

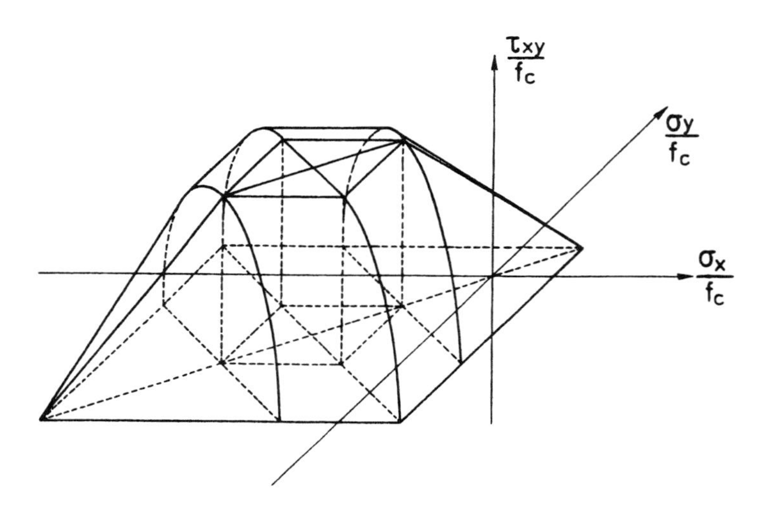

Figure 2.2.9 Yield condition for a disk with orthogonal reinforcement, the isotropic case.

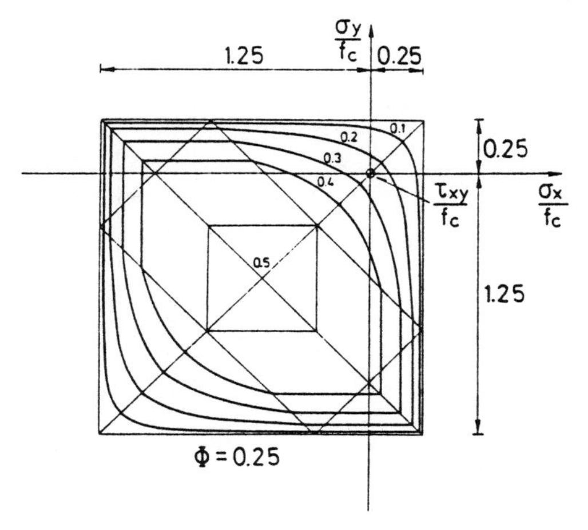

Figure 2.2.10 Shear stress contour lines for an isotropic disk with a reinforcement degree $\Phi = 0.25$.

$$\tau_{xy} = \pm\sqrt{\Phi f_c(\Phi f_c - \sigma_x)} \tag{2.2.28}$$

Within the interval $-f_c \leq \sigma_x \leq -(1 - 2\Phi)f_c$, we have a constant shear stress

$$\tau_{xy} = \pm\sqrt{\Phi(1 - \Phi)}f_c \tag{2.2.29}$$

Finally, within the interval $-(1 + \Phi)f_c \leq \sigma_x \leq -f_c$, we obtain a circle determined by the equation

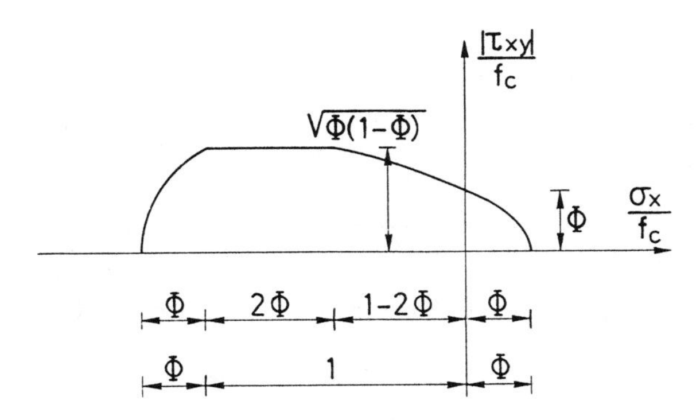

Figure 2.2.11 Yield condition for an isotropic disk in the case $\sigma_y = 0$.

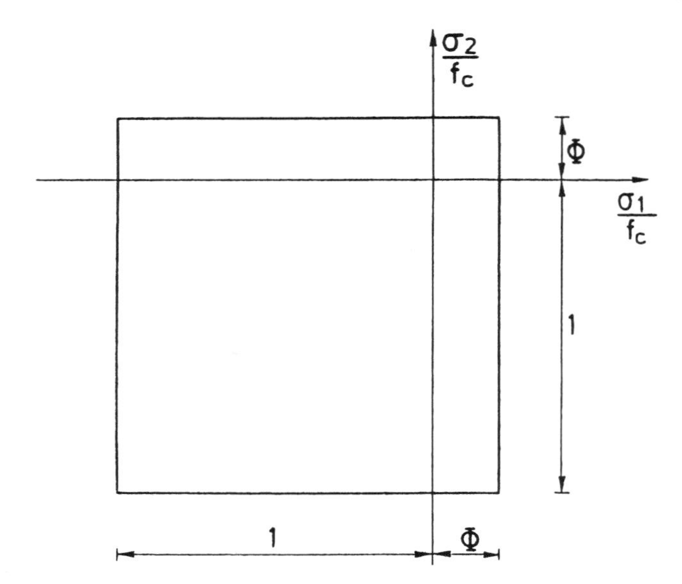

Figure 2.2.12 Approximate yield condition for an isotropic disk.

$$\tau_{xy} = \pm\sqrt{\frac{1}{4}f_c^2 - \left[\sigma_x + \left(\frac{1}{2} + \Phi\right)f_c\right]^2} \tag{2.2.30}$$

It is seen that, at small degrees of reinforcement, area *AFGCHE* in Fig. 2.2.8 is small. If we disregard the special conditions in this area and approximate the yield condition in the whole area with the two conical surfaces, valid in areas *BGF* and *EHD*, the yield surface can be expressed by the simple analytical expressions (2.2.25) and (2.2.27):

$$\sigma_x + \sigma_y \geq -f_c: \ -(\Phi f_c - \sigma_x)(\Phi f_c - \sigma_y) + \tau_{xy}^2 = 0 \tag{2.2.31}$$

$$\sigma_x + \sigma_y \leq -f_c: \ -[(1+\Phi)f_c + \sigma_x][(1+\Phi)f_c + \sigma_y] + \tau_{xy}^2 = 0 \tag{2.2.32}$$

When this approximation is made, it will also be reasonable to put $1 + \Phi \approx 1$ in (2.2.32), obtaining

$$\sigma_x + \sigma_y \geq -(1-\Phi)f_c: \ -(\Phi f_c - \sigma_x)(\Phi f_c - \sigma_y) + \tau_{xy}^2 = 0 \tag{2.2.33}$$

$$\sigma_x + \sigma_y \leq -(1-\Phi)f_c: \ -(f_c + \sigma_x)(f_c + \sigma_y) + \tau_{xy}^2 = 0 \tag{2.2.34}$$

Formula (2.2.33) must, of course, be used only when $\sigma_x \leq \Phi f_c$ and $\sigma_y \leq \Phi f_c$, and (2.2.34) only when $\sigma_x \geq -f_c$ and $\sigma_y \geq -f_c$. Similar remarks apply to (2.2.25), (2.2.27), (2.2.31), and (2.2.32).

At $\Phi < 0.1$, frequently obtained in practice, the maximum error is less than 10% of the correct maximum shear stress $\frac{1}{2}f_c$, which must be regarded as insignificant, considering that the basic assumptions will presumably give rise to errors of at least the same magnitude. The formulas (2.2.33) and (2.2.34) correspond to an isotropic yield condition, hence the designation "isotropic disk".

The isotropic yield condition is illustrated in Fig. 2.2.12, in which σ_1 and σ_2 indicate the principal stresses. In this case the curve of intersection of the yield surface with the τ_{xy}, σ_x-plane is the curve shown in Fig. 2.2.13.

Within the interval $-(1 - \Phi)f_c \leq \sigma_x \leq \Phi f_c$ we obtain the following equation for the curve:

$$\tau_{xy} = \pm\sqrt{\Phi f_c(\Phi f_c - \sigma_x)} \tag{2.2.35}$$

Within the interval $-f_c \leq \sigma_x \leq -(1 - \Phi)f_c$, we obtain

$$\tau_{xy} = \pm\sqrt{f_c(f_c + \sigma_x)} \tag{2.2.36}$$

Both curves of intersection are parabolas.

That a given disk with isotropic reinforcement is able to carry a prescribed stress field may be verified by showing that the following conditions are fulfilled:

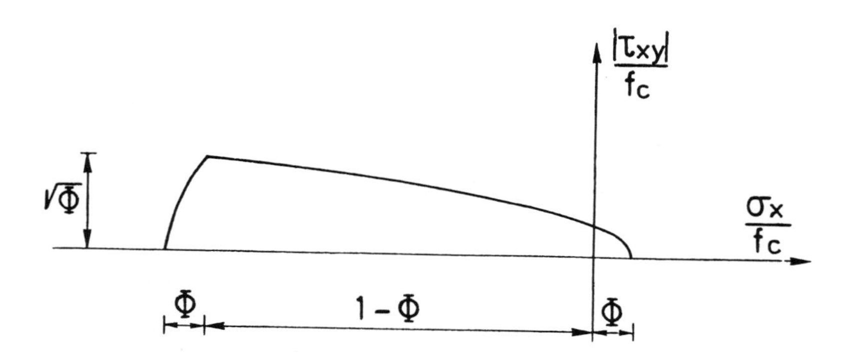

Figure 2.2.13 Approximate yield condition for an isotropic disk in the case $\sigma_y = 0$.

$$
\begin{aligned}
\sigma_x &\leq \Phi f_c \\
\sigma_y &\leq \Phi f_c \\
\sigma_x &\geq -f_c \\
\sigma_y &\geq -f_c
\end{aligned}
\qquad (2.2.37)
$$

$$-(\Phi f_c - \sigma_x)(\Phi f_c - \sigma_y) + \tau_{xy}^2 \leq 0$$

$$-(f_c + \sigma_x)(f_c + \sigma_y) + \tau_{xy}^2 \leq 0$$

How to introduce the effectiveness factor is treated in the following section.

The yield condition in the orthotropic case. Only low degrees of reinforcement are considered. When the degree of reinforcement in the two directions of reinforcement at right angles to each other differs (i.e., $\Phi_x \neq \Phi_y$), the disk is called an orthotropic disk. In this case the shear strength in the x, y-system was found earlier in this section.

Let us consider an arbitrary combination of stresses σ_x, σ_y and τ_{xy} in an isotropic disk, satisfying the yield condition (2.2.33). The isotropic disk is assumed to have the degrees of reinforcement Φ_x in both directions. When for the orthotropic disk

$$\mu = \frac{\Phi_y}{\Phi_x} = \frac{A_{sy}}{A_{sx}} = \frac{f_{ty}}{f_{tx}} \qquad (2.2.38)$$

a combination of stresses characterized by the total stresses $\sigma_x' = \sigma_x$, $\sigma_y' = \sigma_y - (1 - \mu)\Phi_x f_c$, and the same stresses in the concrete will be safe in the orthotropic disk. Since we are assuming low degrees of reinforcement, the area considered is the whole area ABC in Fig. 2.2.8.

Now using the fact that σ_x, σ_y and τ_{xy} satisfy (2.2.33), we obtain $-[\Phi_x f_c - \sigma_x'][\Phi_x f_c - (\sigma_y' + (1 - \mu)\Phi_x f_c)] + \tau_{xy}^2 = 0$ or $-(\Phi_x f_c - \sigma_x')(\Phi_y f_c - \sigma_y') + \tau_{xy}^2 = 0$; that is, the yield condition of an orthotropic disk within the area, corresponding to area ABC for an isotropic disk, reads

$$-(\Phi_x f_c - \sigma_x)(\Phi_y f_c - \sigma_y) + \tau_{xy}^2 = 0 \qquad (2.2.39)$$

when we have removed the primes in the components of the stress tensor. The yield condition in the area corresponding to CDA, in our approximation independent of the degree of reinforcement, remains unchanged (2.2.34):

$$-(f_c + \sigma_x)(f_c + \sigma_y) + \tau_{xy}^2 = 0 \qquad (2.2.40)$$

Using (2.2.7) and (2.2.8), the resulting yield conditions in the orthotropic case can also be written:

$$-(f_{tx} - \sigma_x)(f_{ty} - \sigma_y) + \tau_{xy}^2 = 0 \qquad (2.2.41)$$

$$-(f_c + \sigma_x)(f_c + \sigma_y) + \tau_{xy}^2 = 0 \qquad (2.2.42)$$

Formula (2.2.41) is valid for

$$\sigma_y \geq -\eta\,\sigma_x + \eta f_{tx} - f_c \qquad (2.2.43)$$

in which

$$\eta = \frac{f_{ty} + f_c}{f_{tx} + f_c} \qquad (2.2.44)$$

and (2.2.42) applies to

$$\sigma_y \leq -\eta\,\sigma_x + \eta f_{tx} - f_c \qquad (2.2.45)$$

With these transition conditions the two yield surfaces match. Of course, (2.2.41) must be used only for $\sigma_x \leq \Phi_x f_c = f_{tx}$ and $\sigma_y \leq \Phi_y f_c = f_{ty}$ and (2.2.42) only for $\sigma_x \geq -f_c$ and $\sigma_y \geq -f_c$.

The yield condition (2.2.41) may be derived in another way from the yield condition in the isotropic case (i.e., from two affinities). Let us consider an arbitrary combination of stresses in an isotropic disk, the stresses of the bars being σ_{sx} and σ_{sy} and the stresses of the concrete being σ_{cx}, σ_{cy}, and τ_{cxy}. Let us first assume that the largest principal stress of the concrete is zero, $\sigma_{c1} = 0$, and that the smallest principal stress satisfies the condition $-f_c \leq \sigma_{c2}$. Therefore, we have $\sigma_{cx}\sigma_{cy} = \tau_{cxy}^2$. If changing the stresses of the bars to $\sigma_{sx}' = \sigma_{sx}$, $\sigma_{sy}' = \mu\sigma_{sy}$, in which $\Phi_y/\Phi_x = \mu$, and the stresses of the concrete to $\sigma_{cx}' = \sigma_{cx}$, $\sigma_{cy}' = \mu\sigma_{cy}$, and $\tau_{cxy}' = \sqrt{\mu}\,\tau_{cxy}$, we still have $\sigma_{cx}'\sigma_{cy}' = \tau'^2_{cxy}$ (i.e., the largest principal stress σ_{c1}' is zero again). We assume that $\mu \leq 1$. Considering Mohr's circle it is seen that $-f_c \leq \sigma_{c2}'$ is also valid. If both σ_{c1} and σ_{c2} are in the interval $-f_c \leq \sigma_{c1,2} \leq 0$, it is seen, considering Mohr's circle once more, that this also applies to the principal stresses after changing the stresses.

It is seen from above that if stresses σ_{sx}, σ_{sy}, σ_{cx}, σ_{cy}, and τ_{cxy} are safe in the case $\mu = 1$, the stresses σ_{sx}, $\mu\sigma_{sy}$, σ_{cx}, $\mu\sigma_{cy}$, and $\sqrt{\mu}\,\tau_{cxy}$ will be safe when $\mu < 1$.

This signifies that if the stresses σ_x, σ_y, and τ_{xy} (the total stresses) are safe in the case $\mu = 1$, the stresses σ_x, $\mu\sigma_y$, and $\sqrt{\mu}\,\tau_{cxy}$ will be safe when $\mu < 1$. It means that a safe yield surface in the case of $\mu < 1$ may be found from the yield surface in the case of $\mu = 1$ by two affinities, one in the direction of the σ_y-axis in the scale $1/\mu$, and the other in the direction of the τ_{xy}-axis in the scale $1/\sqrt{\mu}$. These affinities lead to the same yield condition (2.2.41) as the considerations above. The latter view has proved useful in deriving the yield condition of orthotropic slabs (see Section 2.3.2).

That a given disk with orthotropic reinforcement is able to carry a prescribed stress field may be verified by showing that the following conditions are fulfilled:

$$
\begin{aligned}
&\sigma_x \le f_{tx} \\
&\sigma_y \le f_{ty} \\
&\sigma_x \ge -f_c \\
&\sigma_y \ge -f_c
\end{aligned}
\tag{2.2.46}
$$

$$-(f_{tx} - \sigma_x)(f_{ty} - \sigma_y) + \tau_{xy}^2 \le 0$$

$$-(f_c + \sigma_x)(f_c + \sigma_y) + \tau_{xy}^2 \le 0$$

As mentioned, these yield conditions are only valid for small reinforcement degrees.

For higher reinforcement degrees, one may use a simplified form of the correct yield condition. The simplification consists of introducing an upper limit of the shear stress corresponding to $|\tau_{xy}| = \frac{1}{2} f_c$. This means that (2.2.46) is replaced by

$$
\begin{aligned}
&\sigma_x \le f_{tx} \\
&\sigma_y \le f_{ty} \\
&\sigma_x \ge -f_{cx} \\
&\sigma_y \ge -f_{cy}
\end{aligned}
\tag{2.2.47}
$$

$$-(f_{tx}-\sigma_x)(f_{ty}-\sigma_y)+\tau_{xy}^2\le 0$$

$$-(f_{cx}+\sigma_x)(f_{cy}+\sigma_y)+\tau_{xy}^2\le 0$$

$$|\tau_{xy}|\le \frac{1}{2}f_c$$

Here we have reintroduced the possibility of distinguishing between the compressive strengths in the reinforcement directions by taking into account the influence of the reinforcement.

If the yield stresses are different in the x- and y-directions, the tensile strenths f_{tx} , f_{ty} and the compressive strengths f_{cx} and f_{cx} , are calculated by introducing the relevant yield strengths. The yield condition (2.2.47) is still valid. Also account is easily taken of possible differences in tensile yield strengths and compressive yield strengths.

A correct introduction of the effectiveness factor in the yield conditions is extremely difficult. For instance in pure compression we have normally $\nu = 1$ but when shear is present we have $\nu < 1$.

Thus ν varies over the σ_x , σ_y-plane. To take into account crack sliding is even more difficult, because its effect depends on the crack directions, particularly the initial ones. The crack directions again depend on the loading history. Only for proportional loading it would be a simple task to calculate the initial crack directions. In this case the effect of crack sliding is relatively unimportant if the tensile strengths f_{tx} and f_{ty} in the reinforcement directions are not too different, say one is not less than 1/4 and not larger than 4 times the other one. Due to these difficulties the effectiveness factor is, in practice, introduced in a very crude way. It is done by only changing the last condition in (2.2.47).

Then the yield conditions are

$$
\begin{aligned}
&\sigma_x \le f_{tx}\\
&\sigma_y \le f_{ty}\\
&\sigma_x \ge -f_{cx}\\
&\sigma_y \ge -f_{cy} \qquad (2.2.48)\\
&-(f_{tx}-\sigma_x)(f_{ty}-\sigma_y)+\tau_{xy}^2\le 0
\end{aligned}
$$

$$-(f_{cx}+\sigma_x)(f_{cy}+\sigma_y)+\tau_{xy}^2\le 0$$

$$|\tau_{xy}|\le \frac{1}{2}\nu f_c$$

Conservatively, ν may be set equal to ν_o determined by formula (2.1.59) in Section 2.1.4, i.e.,

$$\nu = 0.7-\frac{f_c}{200} \quad (f_c \text{ in MPa}) \tag{2.2.49}$$

The yield condition (2.2.48) and the ν-value (2.2.49) have been introduced in Eurocode 2 [91.23].

A more thorough discussion of the effectiveness factor for disks will be given in Chapter 4.

2.2.3 Skew Reinforcement

Only low degrees of reinforcement are considered. We aim at finding the yield condition of a disk reinforced in arbitrary directions and having an arbitrary number of directions of reinforcement n. Let the area of bar per unit of length (measured orthogonally to the direction of reinforcement) in the ith direction of reinforcement be A_{si} and the angle between the direction of reinforcement and the x-axis be α_i. If the stress of all the bars equals the tensile yield stress f_Y, the stresses are transformed to an arbitrary rectangular system of coordinates ξ, η (see Fig. 2.2.14), in which the ξ-direction forms the angle θ to the abscissa, by means of

$$\sigma_\xi=\frac{f_Y}{t}\sum_{i=1}^{n}A_{si}\cos^2(\theta-\alpha_i)=f_c\sum_{i=1}^{n}\Phi_i\cos^2(\theta-\alpha_i) \tag{2.2.50}$$

$$\sigma_\eta=\frac{f_Y}{t}\sum_{i=1}^{n}A_{si}\sin^2(\theta-\alpha_i)=f_c\sum_{i=1}^{n}\Phi_i\sin^2(\theta-\alpha_i) \tag{2.2.51}$$

$$\begin{aligned}\tau_{\xi\eta} &= -\frac{f_Y}{t}\sum_{i=1}^{n}A_{si}\sin(\theta-\alpha_i)\cos(\theta-\alpha_i)\\ &= -\frac{1}{2}f_c\sum_{i=1}^{n}\Phi_i\sin(2\theta-2\alpha_i)\end{aligned} \tag{2.2.52}$$

Setting $\tau_{\xi\eta}=0$ for determining θ, we obtain

$$\cos 2\theta\sum_{i=1}^{n}\Phi_i\sin 2\alpha_i-\sin 2\theta\sum_{i=1}^{n}\Phi_i\cos 2\alpha_i=0 \tag{2.2.53}$$

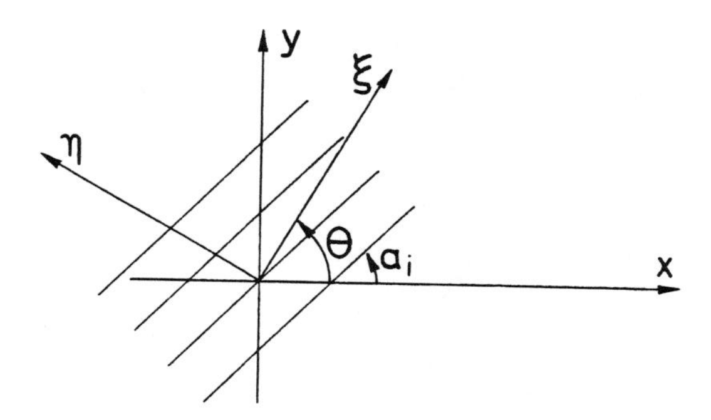

Figure 2.2.14 Skew reinforcement with an arbitrary number of reinforcement bands.

If we assume that $\Sigma_{i=1}^{n} \Phi_i \cos 2\alpha_i \neq 0$ (otherwise, $\tau_{\xi\eta} = 0$ for $\theta = \pm\, \pi/4$, or if also $\Sigma_{i=1}^{n} \Phi_i \sin 2\alpha_i = 0$ for any value of θ), we get

$$\tan 2\theta = \frac{\sum_{i=1}^{n} \Phi_i \sin 2\alpha_i}{\sum_{i=1}^{n} \Phi_i \cos 2\alpha_i} \tag{2.2.54}$$

which always has a solution θ_1 in the interval $-\pi/4 < \theta < \pi/4$.

From this it appears that a ξ-direction corresponding to $\theta = \theta_1$ always exists where $\tau_{\xi\eta} = 0$. The normal stresses σ_ξ and σ_η in these directions are found to be

$$\sigma_\xi = \frac{1}{2} f_c \sum_{i=1}^{n} \Phi_i \left(\cos 2\left(\theta_1 - \alpha_I\right) + 1\right)$$

$$= \frac{1}{2} f_c \sum_{i=1}^{n} \Phi_i \left(\cos 2\theta_1 \cos 2\alpha_i + \sin 2\,\theta_1 \sin 2\alpha_i + 1\right)$$

$$= \frac{1}{2} f_c \left[\cos 2\theta_1 \sum_{i=1}^{n} \Phi_i \cos 2\alpha_i + \sin 2\theta_1 \sum_{i=1}^{n} \Phi_i \sin 2\alpha_i + \sum_{i=1}^{n} \Phi_i \right]$$

$$= \frac{1}{2} f_c \left[\sum_{i=1}^{n} \Phi_i + \cos 2\theta_1 \left(\sum_{i=1}^{n} \Phi_i \cos 2\alpha_i + \frac{\left(\sum_{i=1}^{n} \phi_i \sin 2\alpha_i\right)^2}{\sum_{i=1}^{n} \Phi_i \cos 2\alpha_i} \right) \right]$$

$$= \frac{1}{2} f_c \left[\sum_{i=1}^{n} \Phi_i \pm \sqrt{\left(\sum_{i=1}^{n} \Phi_i \cos 2\alpha_i \right)^2 + \left(\sum_{i=1}^{n} \Phi_i \sin 2\alpha_i \right)^2} \right] \tag{2.2.55}$$

the sign + being valid when $\sum_{i=1}^{n} \Phi_i \cos 2\alpha_i > 0$, and $-$ when $\sum_{i=1}^{n} \Phi_i \cos 2\alpha_i < 0$.

Correspondingly, we obtain

$$\sigma_\eta = \frac{1}{2} f_c \left[\sum_{i=1}^{n} \Phi_i \mp \sqrt{\left(\sum_{i=1}^{n} \Phi_i \cos 2\alpha_i \right)^2 + \left(\sum_{i=1}^{n} \Phi_i \sin 2\alpha_i \right)^2} \right] \tag{2.2.56}$$

$-$ being valid for $\sum_{i=1}^{n} \Phi_i \cos 2\alpha_i > 0$ and + for $\sum_{i=1}^{n} \Phi_i \cos 2\alpha_i < 0$.

The quantities σ_ξ and σ_η, always being positive, are the principal stresses which can be carried by the reinforcing bars when fully utilized. Having low degrees of reinforcement, the yield condition may be estimated to be as for an orthotropic disk with the degrees of reinforcement Φ_ξ and Φ_η in the directions determined by (2.2.54), by which Φ_ξ and Φ_η are found using

$$\Phi_\xi = \frac{1}{2} \sum_{i=1}^{n} \Phi_i \pm \frac{1}{2} \sqrt{\left(\sum_{i=1}^{n} \Phi_i \cos 2\alpha_i \right)^2 + \left(\sum_{i=1}^{n} \Phi_i \sin 2\alpha_i \right)^2} \tag{2.2.57}$$

$$\Phi_\eta = \frac{1}{2} \sum_{i=1}^{n} \Phi_i \mp \frac{1}{2} \sqrt{\left(\sum_{i=1}^{n} \Phi_i \cos 2\alpha_i \right)^2 + \left(\sum_{i=1}^{n} \Phi_i \sin 2\alpha_i \right)^2} \tag{2.2.58}$$

This is so because the stresses corresponding to an arbitrary point of the yield condition in such an orthotropic disk may be obtained in the disk with skew reinforcement by small changes of the stresses of the concrete, being of the same order of magnitude as the degree of reinforcement times the compressive strength of the concrete.

In this way skew reinforcement becomes a special case of orthotropic reinforcement, and further treatment of skew reinforcement is unnecessary. These considerations correspond closely to those of K. W. Johansen concerning slabs with skew reinforcement [43.1; 62.1].

An analytical expression, analogous with (2.2.41), valid for any orientation of the coordinate axes, was given by Bræstrup [70.1].

2.2.4 Uniaxial Stress and Strain

In this section we illustrate an important difference in load-carrying capacity in a uniaxial stress field (uniaxial tension) and in a uniaxial strain field in an orthotropic disk. Consider an orthotropic disk (see Fig. 2.2.15). In the directions of reinforcement x and y, the tensile strengths are $\Phi_x f_c = f_{tx}$ and $\Phi_y f_c = f_{ty}$, respectively. The stress field is uniaxial tension, corresponding to the normal stress σ. The principal direction (the direction of the tension) forms an angle θ to the x-direction. We need only consider the interval $0 \le \theta \le \pi/2$.

The stresses referred to in the x and y directions are

$$\sigma_x = \sigma\cos^2\theta \ , \quad \sigma_y = \sigma\sin^2\theta \ , \quad |\tau_{xy}| = \sigma\cos\theta\sin\theta \qquad (2.2.59)$$

Inserting into the yield condition (2.2.41), we obtain the following tensile strength σ as a function of θ:

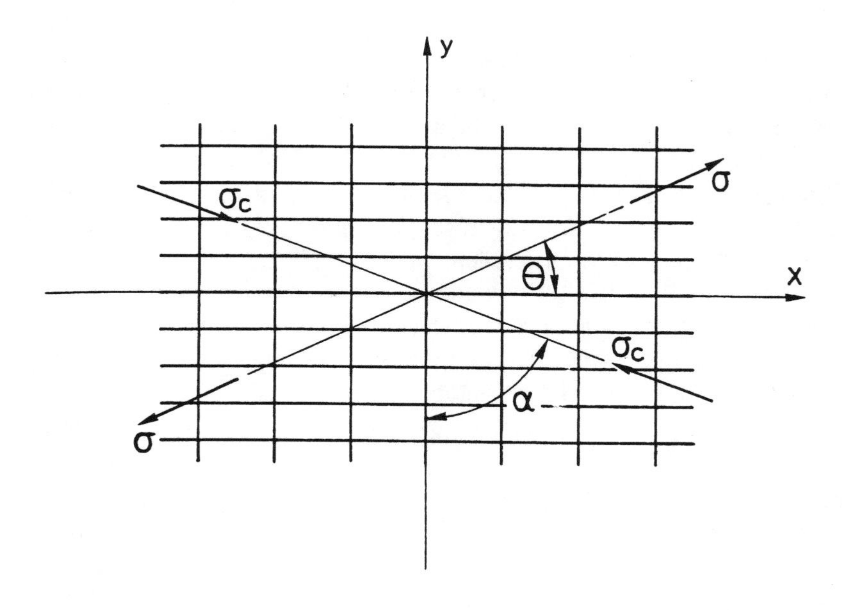

Figure 2.2.15 Uniaxial stress or strain in an orthotropic disk.

$$\sigma = \frac{f_{tx}}{\cos^2\theta + (f_{tx}/f_{ty})\sin^2\theta} \tag{2.2.60}$$

which is the well-known Hankinson's formula for the variation of the tensile or compressive strength of wood with the angle of the principal direction to the direction of the fibers, and consequently valid also for reinforced concrete disks. The direction of the compression of the concrete is found by inserting (2.2.60) in (2.4.4) of Section 2.4.1 and solving the equation with respect to $\tan\alpha$. We thus obtain

$$\tan\alpha = \frac{f_{tx}}{f_{ty}}\tan\theta \tag{2.2.61}$$

where α is the angle of the compression of the concrete to the y-axis (see Fig. 2.2.15).

The stress of the concrete σ_c is found by using (2.4.3) of Section 2.4.1. We obtain

$$\sigma_c = -f_{tx}\frac{\sin^2\theta + (f_{ty}/f_{tx})^2\cos^2\theta}{\sin^2\theta + (f_{ty}/f_{tx})\cos^2\theta} \tag{2.2.62}$$

When the disk is isotropic (i.e., $f_{tx} = f_{ty}$), $\sigma = f_{tx} = f_{ty}$ is independent of θ, and $\tan\alpha = \tan\theta$; that is, the compression of the concrete is perpendicular to the principal direction of the tension. Furthermore, $\sigma_c = -f_{tx} = -f_{ty}$. Formula (2.2.62) does not apply in the limits $\theta = 0$ and $\pi/2$, as it assumes yielding in both directions of reinforcement. In the limiting cases yielding takes place only in one of the directions of reinforcement and $\sigma_c = 0$ (see below).

The strain field has the second principal direction in the direction of the compression of the concrete (where the strain is zero when $\sigma_c < f_c$) and the first principal direction at right angles to this. If there is a strain $\epsilon > 0$ in the first principal direction, we have

$$\epsilon_x = \epsilon\cos^2\alpha, \quad \epsilon_y = \epsilon\sin^2\alpha, \quad \gamma_{xy} = 2\epsilon_{xy} = 2\epsilon\cos\alpha\sin\alpha \tag{2.2.63}$$

by which the ratio between the plastic strains is determined. These expressions are in accordance with the flow rule for orthotropic disks corresponding to (4.4.1) of Section 4.4.1. Referring the strain field to the direction of the tension and the direction at right angles to this, we obtain in the direction of the tension a strain $\epsilon\,\cos^2(\alpha - \theta)$, and in the direction perpendicular to this a strain $\epsilon\,\sin^2(\alpha - \theta)$, and the change of the right angle is $2\epsilon\,\cos(\alpha - \theta)\sin(\alpha - \theta)$.

Next consider a uniaxial strain field. Let us determine the stress field necessary for producing a uniaxial strain in the direction forming the angle θ to the x-direction. The most simple way of obtaining the answer is found by taking into account that by uniaxial strain the compression of the concrete must be perpendicular to the direction of the uniaxial strain, for which reason the stresses in sections parallel to the compression of the concrete only receive contributions from the stresses of the bars. By the usual transformation formulas for plane stress fields we find that the normal stress σ in sections perpendicular to the direction with the angle θ to the x-axis (the direction having uniaxial strain) is

$$\sigma = f_{tx}\cos^2\theta + f_{ty}\sin^2\theta \tag{2.2.64}$$

The shear stress in the same sections (and in sections perpendicular to this) is

$$\tau = -(f_{tx} - f_{ty})\cos\theta\sin\theta \tag{2.2.65}$$

The sign refers to a coordinate system having the same orientation as the x, y-system. Thus a normal stress σ as well as a shear stress τ have to be applied in order to produce a uniaxial strain, except when the disk is isotropic. If so, $\sigma = f_{tx} = f_{ty}$ and $\tau = 0$.

If it is assumed that the total stress is zero in sections parallel to the direction having uniaxial strain, the compression in the concrete is found to be

$$\sigma_c = -(f_{tx}\sin^2\theta + f_{ty}\cos^2\theta) \tag{2.2.66}$$

which, just as (2.2.62), is invalid in the limits $\theta = 0$ and $\pi/2$ (see further below).

These results may also be found by using the flow rule for disks developed in Section 4.4.1. If the strain in the direction with uniaxial strain is ϵ, we have

$$\epsilon_x = \epsilon\cos^2\theta, \quad \epsilon_y = \epsilon\sin^2\theta, \quad \gamma_{xy} = 2\epsilon_{xy} = 2\epsilon\cos\theta\sin\theta \tag{2.2.67}$$

By using the flow rule (4.4.1), we get

$$\begin{aligned} \epsilon\cos^2\theta &= \lambda(f_{ty} - \sigma_y) \\ \epsilon\sin^2\theta &= \lambda(f_{tx} - \sigma_x) \qquad (\lambda \geq 0) \\ \epsilon\cos\theta\sin\theta &= \lambda\tau_{xy} \end{aligned} \tag{2.2.68}$$

from which (assuming $\lambda > 0$)

$$\sigma_x = f_{tx} - \frac{\epsilon}{\lambda}\sin^2\theta$$

$$\sigma_y = f_{ty} - \frac{\epsilon}{\lambda}\cos^2\theta \qquad (2.2.69)$$

$$\tau_{xy} = \frac{\epsilon}{\lambda}\cos\theta\sin\theta$$

The normal stress in sections perpendicular to the direction with the angle θ to the x-axis (the direction having uniaxial strain) then is

$$\sigma = \sigma_x\cos^2\theta + \sigma_y\sin^2\theta + 2\tau_{xy}\cos\theta\sin\theta = f_{tx}\cos^2\theta + f_{ty}\sin^2\theta \qquad (2.2.70)$$

in agreement with (2.2.64).

The shear stress τ in the same sections is

$$\tau = -(f_{tx} - f_{ty})\cos\theta\sin\theta \qquad (2.2.71)$$

the same result as (2.2.65).

It should be noted that if the normal stress in sections parallel to the direction of the uniaxial strain is determined in the same way as σ, we obtain a normal stress $\sigma' = f_{tx}\sin^2\theta + f_{ty}\cos^2\theta - \epsilon/\lambda$. For an arbitrary choice of $\lambda > 0$, the stresses σ, σ', and τ will satisfy the yield condition. The normal stress σ' may consequently assume many different values at a uniaxial strain. Formula (2.2.66) corresponds to the case $\sigma' = 0$. The results above are also valid for slabs when normal stresses are replaced by bending moments, and shear stresses by twisting moments (see Chapter 6).

It has been mentioned that (2.2.62) and (2.2.66) do not apply in the limits. These limits are easily determined in the case of uniaxial strain. According to (2.2.67), we have the strains $\epsilon_x = \epsilon\cos^2\theta$ and $\epsilon_y = \epsilon\sin^2\theta$ in the direction of the reinforcement. When $\theta \leq \pi/4$ we have $\epsilon_x \geq \epsilon_y$. Setting $\epsilon_x = \epsilon_{su}$, where ϵ_{su} is the strain at failure of the reinforcement, we obtain

$$\epsilon_y = \epsilon_x\tan^2\theta = \epsilon_{su}\tan^2\theta \qquad (2.2.72)$$

In order to get yielding in both directions of reinforcement before failure occurs in the reinforcing bars in one of the directions, we have to require that $\epsilon_y \geq \epsilon_Y$, where ϵ_Y indicates the strain at the onset of yielding. By this we obtain

$$\tan^2\theta \geq \frac{\epsilon_Y}{\epsilon_{su}} \qquad (2.2.73)$$

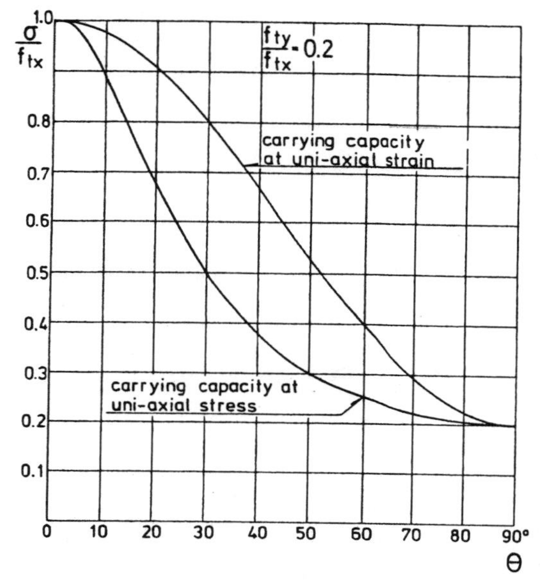

Figure 2.2.16 Normal stress σ at an angle θ to the x-direction versus θ in the cases of uniaxial stress and uniaxial strain, respectively. Reinforcement directions x and y.

For $\theta \geq \pi/4$ we have $\epsilon_x \leq \epsilon_y$, and similarly we find the limit

$$\tan^2\theta \leq \frac{\epsilon_{su}}{\epsilon_Y} \tag{2.2.74}$$

If, as an example, $\epsilon_Y = 2‰$ and $\epsilon_{su} = 50‰$, we obtain $\theta \geq 11.3°$. From this we can conclude that (2.2.66) is generally valid rather close to the limits.

In the case of uniaxial stress the limits cannot be determined without an analysis of the conditions on the way to the collapse state, as the direction of the compression of the concrete is generally not the same in the elastic and the plastic states.

It should be noted that even if the assumption of yielding in both directions of reinforcement will therefore not always be satisfied, the formulas for the load-carrying capacity, (2.2.60) and (2.2.64) can be used completely up to the limits $\theta = 0$ and $\pi/2$ because the direction of reinforcement in which there is probably no yielding contributes only moderately to the load-carrying capacity close to the limit values.

In Fig. 2.2.16 the formulas for the load-carrying capacity, (2.2.60) and (2.2.64), are recorded for the case $f_{ty}/f_{tx} = 0.2$. We see that there may be such a large difference in the load-carrying capacity in the two cases so that in the interpretation of test results it is important to realize which case is in question.

2.2.5 Experimental Verification

The yield conditions for disks loaded in their own plane have been tested in several cases [69.2]. Further the reader is referred to Chapter 4.

2.3 YIELD CONDITIONS FOR SLABS

2.3.1 Assumptions

The slabs considered are assumed to be reinforced in two layers, at the top and bottom, by straight bars. The difference in effective depth d arising from the fact that bars in any side of the slab cannot be placed at the same level is neglected in the following. The yield conditions for the concrete and the reinforcing bars are assumed to be as for the disks considered in Section 2.2. This means that the concrete is treated by the plane stress yield condition disregarding any out-of-plane deformations. We shall treat only slabs for which the reinforcement is stressed to yielding (i.e., low degrees of reinforcement are assumed).

2.3.2 Orthogonal Reinforcement

Pure bending. The area of reinforcement per unit length in a section is denoted A_s, with the index x or y. Quantities referring to the top of a slab are denoted with a prime; for example, A_{sx}' is the area of reinforcement per unit length at the top of a section at right angles to the x-axis (x-section).

The simplest possible case is considered first, pure bending in a beam (strip of slab), in which the tensile reinforcement is parallel to the axis of the beam (x-axis). It is further assumed that $A_{sx}' = A_{sy}' = A_{sy} = 0$ and that $A_{sx} = A_s$ (see Fig. 2.3.1).

In accordance with the assumptions above, the stress distribution at failure will be as shown in Fig. 2.3.1. By projection,

$$A_s f_Y = a f_c \tag{2.3.1}$$

By introducing the reinforcement degree

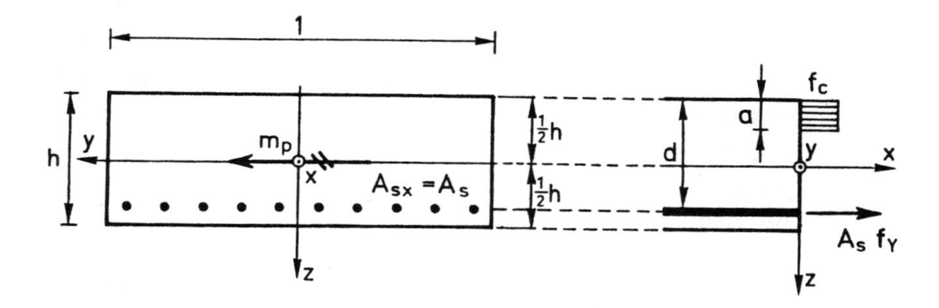

Figure 2.3.1 Slab element subjected to pure bending in the x-direction (reinforcement direction).

$$\Phi = \frac{A_s f_Y}{d f_c} \tag{2.3.2}$$

the following is obtained from (2.3.1):

$$\frac{a}{d} = \Phi \tag{2.3.3}$$

The yield moment per unit length in pure bending m_p can then be determined:

$$m_p = A_s f_Y\left(d - \frac{1}{2}a\right) = \left(1 - \frac{1}{2}\Phi\right)A_s f_Y d = \left(1 - \frac{1}{2}\Phi\right)\Phi\, d^2 f_c \tag{2.3.4}$$

The coefficient $(1 - ½\Phi)$ varies only slightly in cases of a low reinforcement degree, and the yield moment is thus mainly proportional to A_s. Scaling on the basis of a known yield moment should always be done to smaller areas of reinforcement since this is on the safe side, whereas it is unsafe to scale up to larger areas.

Solution (2.3.4) is naturally exact according to the assumptions, since a geometrically admissible strain field in the x, y-plane corresponding to the solution is constituted as follows:

$$\epsilon_x = \kappa\left[z + \left(\frac{1}{2}h - a\right)\right] \tag{2.3.5}$$

all other components of the strain tensor being zero. The parameter $\kappa > 0$ is an arbitrary constant (curvature of the slab).

If there is also compressive reinforcement, the load-carrying capacity will naturally be higher, but the increase is very slight for low degrees of reinforcement. However, for the sake of completeness, formulas will be given for a cross section with compressive reinforcement. In these formulas it is assumed that there is the same distance h_c from either side of the slab to the nearest layer of reinforcement.

The following notation is introduced:

$$\Phi_0 = \frac{A_s f_Y}{h f_c} \tag{2.3.6}$$

$$\mu = \frac{A_{sc}}{A_s} \tag{2.3.7}$$

where Φ_o is a reinforcement degree, $A_s = A_{sx}$ is the area of tensile reinforcement, and $A_{sc} = A_{sx}'$ is the area of compressive reinforcement. The following expressions are now obtained:

$$\mu \geq 1 - \frac{1}{\Phi_0}\frac{h_c}{h}$$

$$\Phi_0 \leq \frac{1}{1+\mu}\frac{h_c}{h}, \quad m_p = \left[1 - 2\frac{h_c}{h} + (1+\mu)\left(\frac{h_c}{h} - \frac{1}{2}\Phi_0(1+\mu)\right)\right]A_s f_Y h \tag{2.3.8}$$

$$\Phi_0 \geq \frac{1}{1+\mu}\frac{h_c}{h}, \quad m_p = \left(1 - 2\frac{h_c}{h} + \frac{1}{2}\frac{1}{\Phi_0}\left(\frac{h_c}{h}\right)^2\right)A_s f_Y h \tag{2.3.9}$$

$$\mu < 1 - \frac{1}{\Phi_0}\frac{h_c}{h}, \quad m_p = \left[1 - 2\frac{h_c}{h} + (1-\mu)\left(\frac{h_c}{h} - \tfrac{1}{2}\Phi_0(1-\mu)\right)\right]A_s f_Y h \tag{2.3.10}$$

Formulas (2.3.8) and (2.3.9) cover cases in which the neutral axis lies above or in the top reinforcement, and formula (2.3.10) covers cases in which it lies below the top reinforcement.

In the special case $\mu = 1$, $\Phi_0 \leq \frac{1}{2}(h_c/h)$, where both layers of reinforcement act as tensile reinforcement, (2.3.8) is rewritten as follows:

$$m_p = (1 - 2\Phi_0)A_s f_Y h = (1 - 2\Phi_0)\Phi_0 h^2 f_c \tag{2.3.11}$$

Pure torsion. A slightly more complicated case will now be considered, namely, pure torsion in the *x*, *y*-system shown in Fig.

2.3.2, a coordinate system in which the directions of the coordinates coincide with those of the reinforcing bars. The slab is assumed to be isotropic; that is, $A_{sx}' = A_{sx} = A_{sy}' = A_{sy} = A_s$ (cf. Section 2.2.2).

The principal moments are numerically equal, with opposite signs, and equal to the torsional moment in the x and y sections. The principal sections form an angle of 45° with the x-axis and the y-axis [n, t-axes in Fig. 2.3.2]. When the torsional moment acts as shown in Fig. 2.3.2, it is reasonable to expect a uniaxial state of compressive stress in the concrete, with the second principal direction in the direction of the n-axis at the top and the t-axis at the bottom. If the second principal stress at both top and bottom is made equal to f_c (i.e., $\sigma_{c1} = 0$, $\sigma_{c2} = -f_c$) and the compressive zones are extended to the depth a at the top and the bottom, the following stresses are obtained:

$$\sigma_{cx} = \sigma_{cx}' = \sigma_{cy} = \sigma_{cy}' = -\frac{1}{2}f_c \tag{2.3.12}$$

$$|\tau_{cxy}| = |\tau_{cxy}'| = \frac{1}{2}f_c \tag{2.3.13}$$

Assuming a tensile yield stress f_Y in the reinforcement, projection on the x- and y-axes gives

$$\frac{1}{2}f_c a = A_s f_Y \tag{2.3.14}$$

or, by substituting the reinforcement degree Φ_0 (see (2.3.6)),

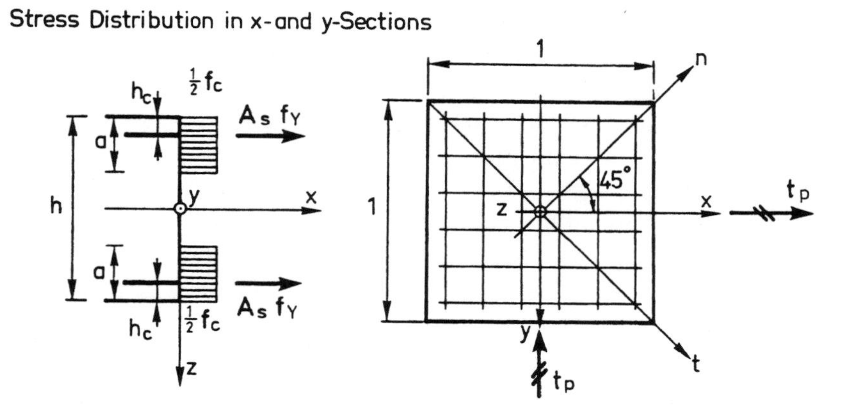

Figure 2.3.2 Slab element subjected to pure torsion in the x, y-coordinate system (reinforcement directions).

$$\frac{a}{h} = 2\Phi_0 \tag{2.3.15}$$

In the following we assume that $a \le \frac{1}{2}h$ (i.e., $\Phi_0 \le 0.25$).

A comparison between expressions (2.3.3) and (2.3.15) shows that the extent of the compressive zone in pure torsion is almost twice that in pure bending. When a is determined from (2.3.15), the stresses in the x and y sections are equivalent to a torsional moment per unit length t_p, which becomes

$$t_p = \tau_{cxy} a(h-a) = (1-2\Phi_0)\Phi_0 h^2 f_c = (1-2\Phi_0)A_s f_Y h \tag{2.3.16}$$

The solution is exact, according to the assumptions, since a geometrically admissible strain field in the x, y-plane corresponding to the assumed stress field is constituted as follows:

$$\begin{aligned} \epsilon_n &= \kappa_1\left[z + \left(\frac{1}{2}h - a\right)\right] \\ \epsilon_t &= \kappa_1\left[-z + \left(\frac{1}{2}h - a\right)\right] \end{aligned} \tag{2.3.17}$$

all other components being equal to zero.

In the above, $\kappa_1 > 0$ is an arbitrary constant equal to the principal curvatures of the slab. It will immediately be seen that this results in compressive strains in the second principal direction for the stress field at the top and the bottom of the concrete, but it will also be seen, for instance by plotting Mohr's circle, that the strain field results in extensions in the direction of the reinforcement, from which it may be concluded that the assumed stress field corresponds to geometrically admissible strains. Expression (2.3.16) therefore gives the load-carrying capacity in pure torsion, a quantity that is called the yield moment in pure torsion.

In a real reinforced concrete slab in which the concrete has a certain tensile strength, tensile stresses will naturally not arise in the reinforcement until the concrete has cracked. At the top, the cracks will run in the direction of the n-axis, and at the bottom, in the direction of the t-axis.

A comparison between the expression for t_p (2.3.16) and the abovementioned formulas for the yield moment in pure bending m_p shows that the expressions give practically the same result; indeed, for $\Phi_0 \le \frac{1}{2}(h_c/h)$ [see (2.3.11)], the expressions are identical. For $\Phi_0 \le 0.1$, which is frequently encountered in practice, the deviation is less than 6%. Hence

$$t_p \simeq m_p \tag{2.3.18}$$

If the area of reinforcement in the y-section is μ times the area of reinforcement in the x-section (i.e., $A_{sx}' = A_{sx} = A_s$, $A_{sy}' = A_{sy} = \mu A_s$) and it is assumed that the second principal direction for the stress field at the top and bottom of the concrete forms an unknown angle α with the x-direction, all other assumptions remaining unchanged, the following is obtained

$$\sigma_{cx}' = \sigma_{cx} = -f_c \cos^2\alpha,\ \ \sigma_{cy}' = \sigma_{cy} = -f_c \sin^2\alpha,\ \ |\tau_{cxy}'| = |\tau_{cxy}| = \frac{1}{2} f_c \sin 2\alpha \tag{2.3.19}$$

By substitution of

$$\frac{\sigma_{cy}'}{\sigma_{cx}'} = \mu \tag{2.3.20}$$

(i.e., $\tan^2\alpha = \mu$), the projection equations will be fulfilled when

$$a f_c \cos^2\alpha = A_s f_Y \tag{2.3.21}$$

that is [see (2.3.6)],

$$\frac{a}{h} = \frac{\Phi_0}{\cos^2\alpha} = \Phi_0(1+\mu) \tag{2.3.22}$$

As

$$|\tau_{cxy}| = |\tau_{cxy}'| = f_c \frac{\sqrt{\mu}}{1+\mu} \tag{2.3.23}$$

t_p can be expressed as follows in this case:

$$\begin{aligned} t_p = \tau_{cxy} a(h-a) &= \sqrt{\mu}\left(1 - \Phi_0(1+\mu)\right)\Phi_0 h^2 f_c \\ &= \sqrt{\mu}\left(1 - \Phi_0(1+\mu)\right) A_s f_Y h \end{aligned} \tag{2.3.24}$$

It will be seen from the way in which this solution is derived that the load-carrying capacity is greater than or equal to that found from (2.3.24) (lower bound theorem). The solution is thus a lower bound solution.

An upper bound solution can be obtained by using (2.3.17) as a geometrically admissible strain field. This gives the factor ½(1 + μ) instead of $\sqrt{\mu}$ in (2.3.24). The factor obtained from the upper bound theorem gives, for ¼ ≤ μ ≤ 4, a maximum increase in the carrying capacity of 25%, and considerably less than this over the greater part of the interval.

In this case, for low degrees of reinforcement, t_p is approximately

$$t_p \simeq \sqrt{\mu}\, m_p \tag{2.3.25}$$

m_p referring to the area of reinforcement A_s. When applying this formula, μ should be taken as less than 1.

It can now be realized that the maximum torsional moment that can be resisted depends solely on the ratio between the sum of the areas of reinforcement per unit length in the x and y sections, not on how it is distributed between top and bottom, and that this torsional moment usually acts together with bending moments in these sections. With the notation given in Fig. 2.3.3, the following is obtained by means of (2.3.24):

$$t_{\max} = \frac{1}{2}\sqrt{\frac{\mu+\mu'}{1+\kappa}}\left(1 - \frac{1}{2}(1+\kappa+\mu+\mu')\frac{A_s f_Y}{h f_c}\right)(1+\kappa)A_s f_Y h$$
$$\simeq \tfrac{1}{2}\sqrt{(1+\kappa)(\mu+\mu')}\, m_p \tag{2.3.26}$$

where m_p refers to the area of reinforcement A_s.

At the same time, the torsional moment t_{max} is found to act together with the bending moments:

$$m_x = \tfrac{1}{2}(1-\kappa)(h-2h_c)A_s f_Y \simeq \tfrac{1}{2}(1-\kappa)\, m_p \tag{2.3.27}$$

$$m_y = \tfrac{1}{2}(\mu-\mu')(h-2h_c)A_s f_Y \simeq \tfrac{1}{2}(\mu-\mu')\, m_p \tag{2.3.28}$$

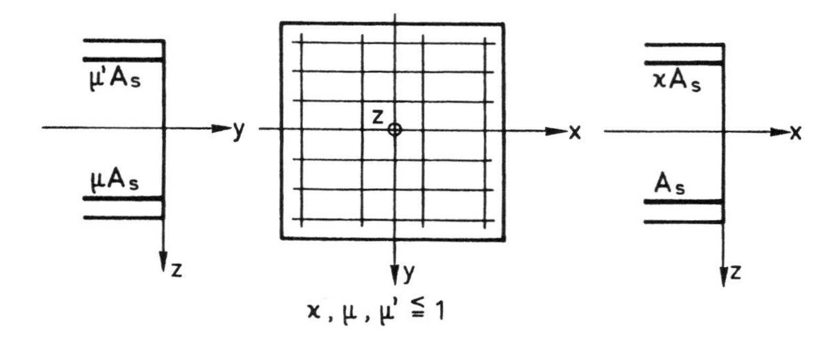

Figure 2.3.3 Reinforcement areas at top and bottom of an orthotropic slab element.

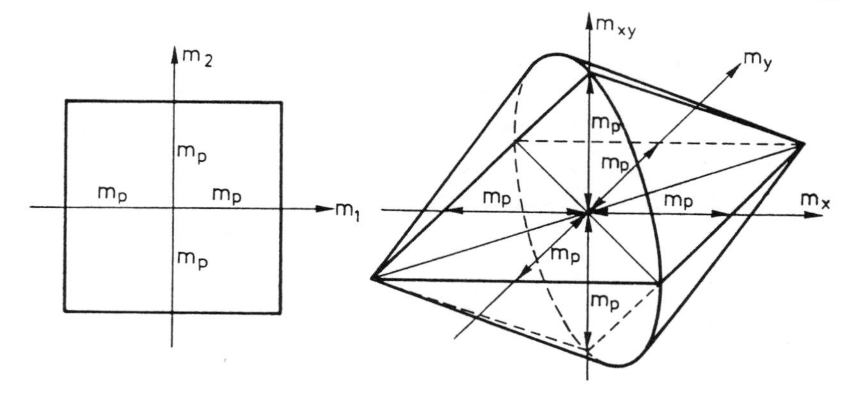

Figure 2.3.4 Yield condition for an isotropic slab element having the same reinforcement areas in all four reinforcement directions.

Thus, in the x-section, there is a bending moment equal to half the difference between the yield moments in pure bending corresponding to the area of reinforcement A_s and κA_s and in the y-section a bending moment equal to half the difference between the yield moments corresponding to the areas of reinforcement μA_s and $\mu' A_s$.

Combined bending and torsion. The yield conditions for reinforced concrete slabs can now be dealt with, that is, determination of the bending moments m_x and m_y and the torsional moment m_{xy}, which can be resisted by slabs reinforced in different ways.

The simplest case is an isotropic slab in which there is the same amount of reinforcement at the top and bottom of the slab (i.e., $A_{sx}' = A_{sx} = A_{sy}' = A_{sy} = A_s$). In this case the yield condition shown in Fig. 2.3.4 and hereafter called the *square yield condition* was proposed intuitively by Johansen [43.1; 62.1].

According to this, such combinations of m_x, m_y, and m_{xy} that result in one or both of the principal moments being numerically equal to the positive and the negative yield moment in pure bending can be resisted. This yield criterion is isotropic because, according to the criterion, it does not matter in which section the principal moments occur. In a Cartesian coordinate system, with axes m_x, m_y, and m_{xy}, the equation for the yield surface can be obtained from

$$\frac{1}{2}(m_x + m_y) \pm \sqrt{\frac{1}{4}(m_x - m_y)^2 + m_{xy}^2} = \pm m_p \tag{2.3.29}$$

the left-hand side of which expresses the principal moments m_1 and m_2. Formula (2.3.29) represents two cones (see Fig. 2.3.4). The curve of intersection between this surface and the m_x, m_y-plane is naturally a square with the sides $2m_p$; the curve of intersection between the plane $m_x = -m_y$ and the cones is an ellipse with the principal axes $2\sqrt{2}m_p$ and $2m_p$, and the curve of intersection between the plane $m_x = m_y$ and the cones is a rhomboid with the sides $\sqrt{3}m_p$. It will be seen that the yield criterion enables the slab to resist a torsional moment $m_{xy} = \pm m_p$, which is in agreement with the result expressed in (2.3.18).

The question now is how this yield criterion corresponds to the assumptions made regarding concrete and reinforcement. It will be shown that the assumptions lead very near, and indeed in some cases right to, the square yield criterion. For the sake of simplicity, only the important cases $m_x = -m_y$ and $m_x = m_y$ will be dealt with.

Consider first the case $m_x = -m_y$. As a statically admissible stress field, a state of stress in the concrete is selected that corresponds to the smallest principal stress of $-f_c$ and the greatest principal stress equal to zero, and that is characterized by the fact that the second principal direction forms an angle α with the y-axis at the top, and an angle α with the x-axis at the bottom (see Fig. 2.3.5). The stresses are:

$$\sigma_{cy}' = -f_c\cos^2\alpha, \quad \sigma_{cx}' = -f_c\sin^2\alpha, \quad |\tau_{cxy}'| = \tfrac{1}{2}f_c\sin2\alpha \tag{2.3.30}$$

$$\sigma_{cy} = -f_c\sin^2\alpha, \quad \sigma_{cx} = -f_c\cos^2\alpha, \quad |\tau_{cxy}| = \tfrac{1}{2}f_c\sin2\alpha \tag{2.3.31}$$

For reasons of symmetry it is necessary only to consider the cases $m_y \geq 0$ and $m_{xy} \geq 0$, and the following stresses in the reinforcement are therefore selected:

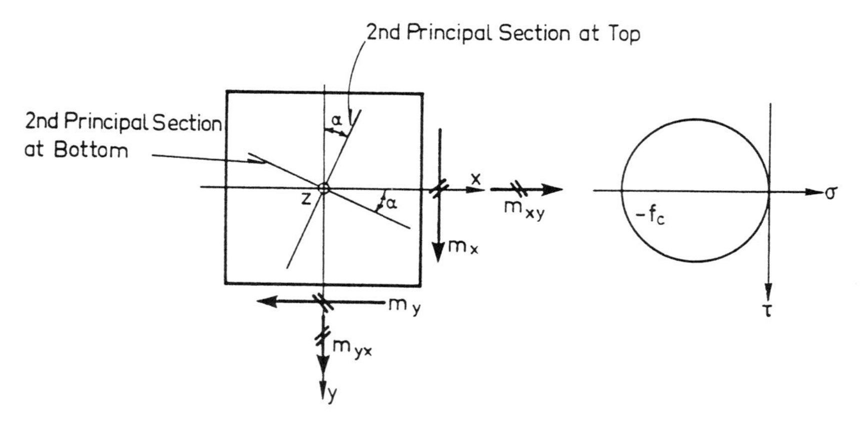

Figure 2.3.5 Slab element subjected to combinations of bending moments $m_x = -m_y$ and torsional moments m_{xy}.

$$\sigma_{sy}' = \beta f_Y, \quad \sigma_{sx}' = f_Y \tag{2.3.32}$$

$$(-1 \leq \beta \leq 1)$$

$$\sigma_{sy} = f_Y, \quad \sigma_{sx} = \beta f_Y \tag{2.3.33}$$

By projection of the normal stresses on the x and y axes, the stresses extending over a depth a at both top and bottom, the following formula is obtained:

$$af_c\cos^2\alpha + af_c\sin^2\alpha = A_s f_Y + \beta A_s f_Y \tag{2.3.34}$$

Hence [see (2.3.6)]

$$\frac{a}{h} = \Phi_0(1 + \beta) \tag{2.3.35}$$

It will be seen that the projection equations for the shear stresses are satisfied. The bending moment $m = m_y$ then becomes

$$\begin{aligned} m = m_y = -m_x &= \frac{1}{2}(h-a)af_c\cos^2\alpha - \frac{1}{2}(h-a)af_c\sin^2\alpha \\ &+ A_s f_Y\left(\frac{1}{2}h - h_c\right) - \beta A_s f_Y\left(\frac{1}{2}h - h_c\right) \end{aligned} \tag{2.3.36}$$

where h_c is the distance from the surface of the slab to the nearest layer of reinforcement. The torsional moment $t = m_{xy}$ can be found in a similar way:

$$t = m_{xy} = \frac{1}{2}f_c a(h-a)\sin 2\alpha \tag{2.3.37}$$

These expressions can be written as follows:

$$M = \frac{m}{\Phi_0 h^2 f_c} = \frac{m}{A_s f_Y h} = \frac{1}{2}(1+\beta)\big(1 - \Phi_0(1+\beta)\big)\cos 2\alpha + \omega(1-\beta) \tag{2.3.38}$$

$$T = \frac{t}{\Phi_0 h^2 f_c} = \frac{t}{A_s f_Y h} = \frac{1}{2}(1+\beta)\big(1 - \Phi_0(1+\beta)\big)\sin 2\alpha \tag{2.3.39}$$

In (2.3.38) ω is defined as follows:

$$\omega = \frac{1}{2}\left(1 - 2\frac{h_c}{h}\right) \tag{2.3.40}$$

Thus, if

$$T = \frac{t}{A_s f_Y h} = \eta(1 - 2\Phi_0) \tag{2.3.41}$$

then $\eta = 1$ for $t = t_p$, in accordance with (2.3.16), is obtained, and $\eta = 0$ for pure bending.

From (2.3.38) and (2.3.39),

$$\sin 2\alpha = \frac{\eta(1-2\Phi_0)}{\frac{1}{2}(1+\beta)\left[1-\Phi_0(1+\beta)\right]} \tag{2.3.42}$$

$$M = (1-2\Phi_0)\sqrt{\left[\frac{\frac{1}{2}(1+\beta)\left[1-\Phi_0(1+\beta)\right]}{1-2\Phi_0}\right]^2 - \eta^2} + \omega(1-\beta) \tag{2.3.43}$$

By substituting $\beta = 1$ the following is obtained:

$$M = \frac{m}{A_s f_Y h} = (1-2\Phi_0)\sqrt{1-\eta^2} \tag{2.3.44}$$

or

$$m = \sqrt{1-\eta^2}\,(1-2\Phi_0)A_s f_Y h \tag{2.3.45}$$

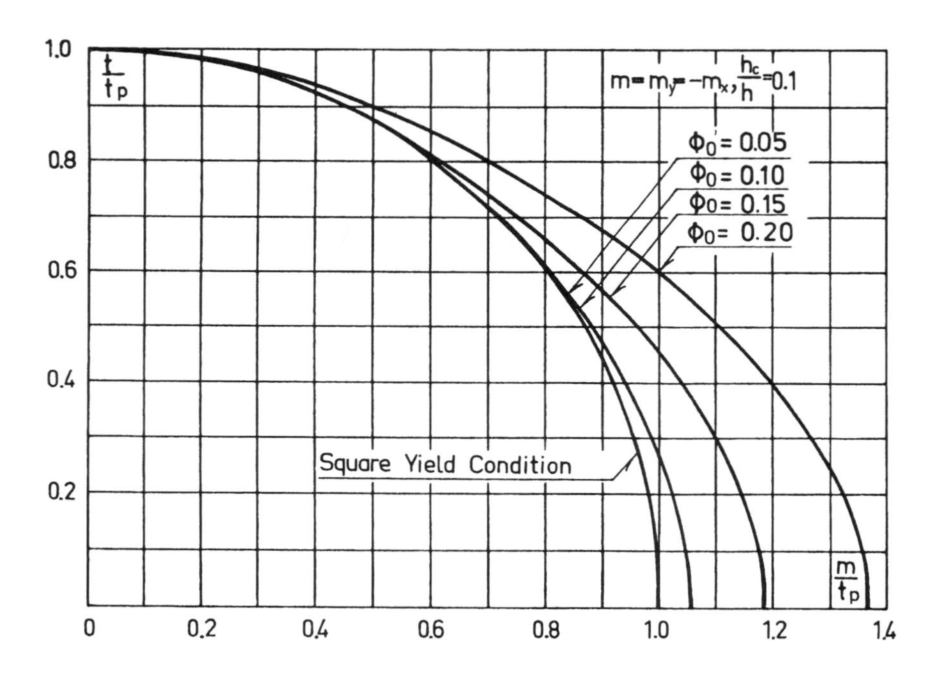

Figure 2.3.6 Yield condition for a slab element compared to the square yield condition in the case $m_x = -m_y$.

Thus, when $\eta = 0$ (pure bending) and $\beta = 1$, that is, both layers of reinforcement are stressed to their tensile yield point, (2.3.45) can be rewritten

$$m_{\eta=0} = (1 - 2\Phi_0)A_s f_Y h \tag{2.3.46}$$

This is identical to (2.3.11), which is valid for $\Phi_0 \le \frac{1}{2}(h_c/h)$. In this case, (2.3.45) represents precisely the aforementioned elliptical intersection curve.

In Fig. 2.3.6, the function (2.3.43) is depicted for various values of Φ_0. These curves are drawn so as to give the maximum value of m/t_p for given $\eta = t/t_p$ that is obtainable by varying β.

It will be seen from Fig. 2.3.6 that for low degrees of reinforcement (i.e., $\Phi_0 \le 0.1$) there is good agreement between the yield condition (2.3.29), the square yield condition, and expression (2.3.43), which represents a lower bound solution in relation to our assumptions concerning concrete and reinforcement.

The case $m_x = m_y$ can be dealt with in a similar way. For reasons of symmetry it is necessary to consider only the cases $m_x \ge 0$ and $m_{xy} \ge 0$. At the top of the slab, a state of stress characterized by the smallest principal stress being $\sigma_{c2} = -f_c$ and $\sigma_{cx} = \sigma_{cy}$, and at the bottom, a state of stress characterized by the greatest principal stress being zero and $\sigma_{cx} = \sigma_{cy}$ is assumed.

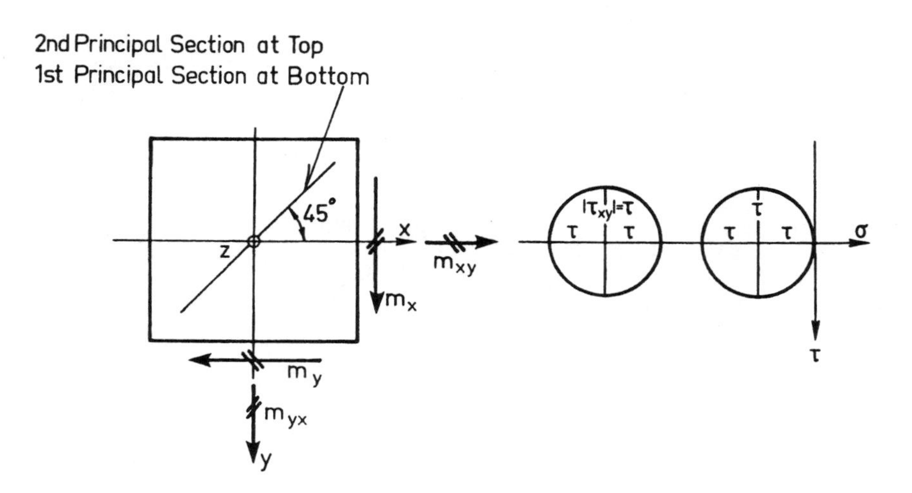

Figure 2.3.7 Slab element subjected to combinations of bending moments $m_x = m_y$ and torsional moments m_{xy}.

The same shear stress τ is assumed at top and bottom in the x and y sections, whereby the two Mohr's circles become as shown in Fig. 2.3.7. The zones of compression in the concrete are extended to the depth a at top and bottom. The stresses in the reinforcement are fixed at

$$\sigma_{sx} = \sigma_{sy} = f_Y, \quad \sigma_{sx}' = \sigma_{sy}' = \beta f_Y, \quad -1 \le \beta \le 1 \tag{2.3.47}$$

whereby the following projection equation is obtained:

$$(f_c - \tau)a + \tau a = A_s f_Y (1 + \beta) \tag{2.3.48}$$

from which [see (2.3.6)]

$$\frac{a}{h} = \Phi_0 (1 + \beta) \tag{2.3.49}$$

By application of the notation from the preceding case,

$$M = \frac{m_x}{A_s f_Y h} = \frac{m}{A_s f_Y h} = \frac{1}{2}(1+\beta)\left[1 - \Phi_0(1+\beta)\right]\left(1 - 2\frac{\tau}{f_c}\right) + \omega\,(1-\beta) \tag{2.3.50}$$

$$T = \frac{m_{xy}}{A_s f_Y h} = \frac{t}{A_s f_Y h} = \frac{\tau}{f_c}(1 + \beta)\left[1 - \Phi_0(1 + \beta)\right] \tag{2.3.51}$$

If, as previously, $T = \eta(1 - 2\Phi_0)$, then

$$\frac{\tau}{f_c} = \frac{\eta(1 - 2\Phi_0)}{(1 + \beta)\left[1 - \Phi_0(1 + \beta)\right]} \tag{2.3.52}$$

and

$$M = \frac{1}{2}(1+\beta)\left[1 - \Phi_0(1+\beta)\right]\left[1 - 2\frac{\eta(1 - 2\Phi_0)}{(1+\beta)\left[1 - \Phi_0(1+\beta)\right]}\right] + \omega(1 - \beta) \tag{2.3.53}$$

For $\beta = 1$, this formula gives

$$M = \frac{m}{A_s f_Y h} = (1 - 2\Phi_0)(1 - \eta) \tag{2.3.54}$$

or

$$m = (1 - \eta)(1 - 2\Phi_0)A_s f_Y h \tag{2.3.55}$$

As previously, it is found that $m_{\eta=0} = (1 - 2\Phi_0)A_s f_Y h$. Thus, for $\Phi_0 \le ½(h_c/h)$, the curve of intersection mentioned earlier is obtained, being a straight line between the conical surface and the plane $m_x = m_y$. For other values of Φ_0 see Fig. 2.3.8 which is analogous to Fig. 2.3.6.

Correspondingly, it is possible to construct a safe, statically admissible stress field for all possible combinations of m_x and m_y, but the cases already treated should suffice to show clearly that the square yield condition for isotropic slabs corresponds well to the behavior assumed here for slabs, with low degrees of reinforcement.

Let us now consider a slab with the same reinforcement in each direction but a different area of reinforcement per unit length at the top and bottom (i.e., $A_{sx} = A_{sy} = A_s$, $A_{sx}' = A_{sy}' = A_s'$). In this case it is to be expected that the square yield condition shown in Fig. 2.3.9 can be used. In the figure the reinforcement at the bottom, A_s, corresponds to the yield moment m_p, and the reinforcement at the top, A_s', corresponds to the yield moment m_p'.

In the m_x , m_y , m_{xy}-coordinate system, this condition is equivalent to two cones with the axis of the cones in the direction *AC* (see Fig. 2.3.9). The common base lies in a plane perpendicular to the m_x, m_y-plane through *BD*. The yield surface is identical to the yield surface for an isotropic slab with positive and negative yield moments corresponding to the average value of the moments m_p and m_p', although with a different location in the coordinate system.

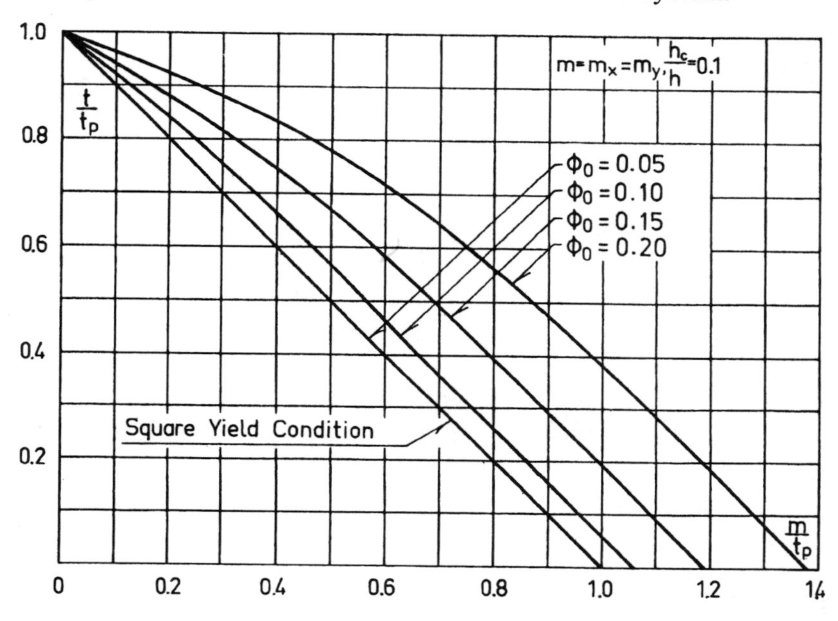

Figure 2.3.8 Yield condition for a slab element compared to the square yield condition in the case $m_x = m_y$.

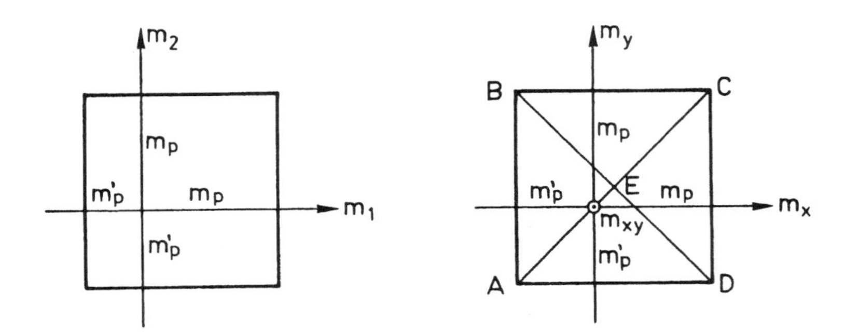

Figure 2.3.9 Yield condition for a slab element reinforced isotropically but differently at top and bottom.

The maximum torsional moment arises for $m_x = m_y = ½(m_p - m_p')$, corresponding to the point of intersection E of the diagonals AC and BD, and this has a value of $½(m_p + m_p')$. This yield criterion is also justifiable on the basis of the assumptions made.

By means of (2.3.26)-(2.3.28), it may be shown that the maximum torsional moment is equal to $½(m_p + m_p')$, and that this moment arises for $m_x = m_y = ½(m_p - m_p')$.

By making $\mu = 1$, $\kappa = \mu' \simeq m_p'/m_p$ in (2.3.26), the following is obtained:

$$t_{\max} = \frac{1}{2}\sqrt{\left(1 + \frac{m_p'}{m_p}\right)\left(1 + \frac{m_p'}{m_p}\right)}\, m_p = \frac{1}{2}\left(m_p + m_p'\right) \tag{2.3.56}$$

Then, from (2.3.27) and (2.3.28),

$$m_x = m_y \simeq \frac{1}{2}\left(m_p - m_p'\right) \tag{2.3.57}$$

The yield surface as a whole can be obtained by imagining the stress field corresponding to an arbitrary point of the yield surface of an isotropic slab in which $A_{sx}' = A_{sx} = A_{sy}' = A_{sy} = ½(A_s + A_s')$, superposed by the forces $½(A_s - A_s')f_Y$ in the reinforcement at the top and bottom, acting as shown in Fig. 2.3.10. These result in the bending moments $m_x = m_y \simeq ½(m_p - m_p')$. (Note that the parameter β [see (2.3.32) and (2.3.33)] must be given such a value here that the numerical value of the resultant stress in the reinforcement does not exceed f_Y.) This superposition results in precisely the yield criterion shown in Fig. 2.3.9.

We can now deal with a more general case of an orthotropic slab. The reinforcement is characterized by

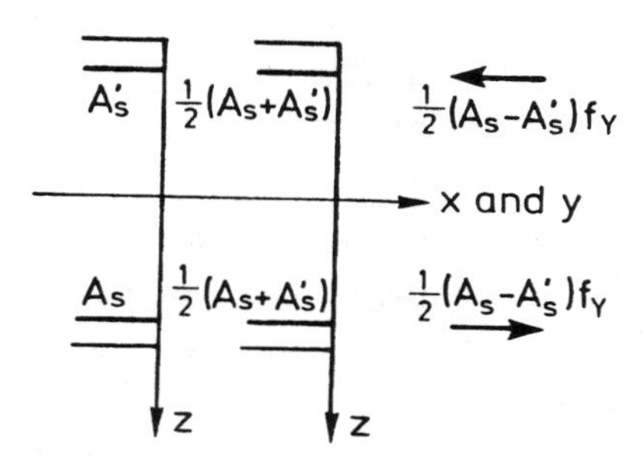

Figure 2.3.10 Superposition rule used to develop the yield condition for a slab element reinforced isotropically but differently at top and bottom.

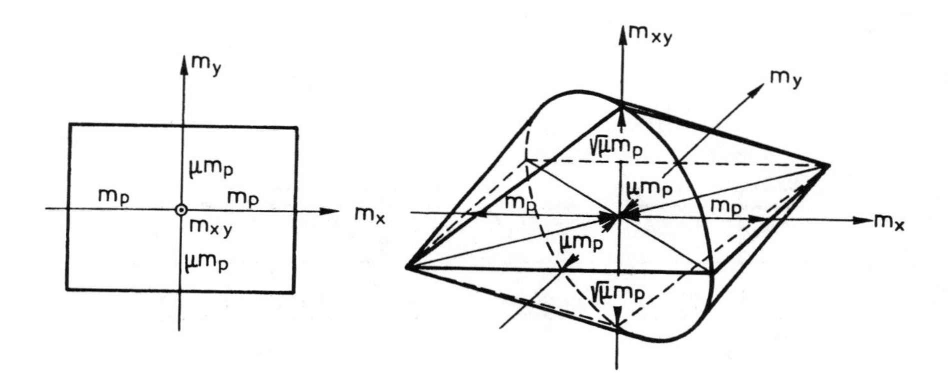

Figure 2.3.11 Yield condition for a slab element reinforced orthotropically and identically at top and bottom.

$$A_{sx}' = A_{sx} = A_s$$
$$A_{sy}' = A_{sy} = \mu A_s \quad (\mu < 1) \tag{2.3.58}$$

In this case $t_p = \sqrt{\mu}\, m_p$ [see (2.3.25)], where m_p corresponds to the areas of reinforcement A_s.

In Section 2.2.2 it was shown that if the state of stress σ_{sx}, σ_{sy}, σ_{cx}, σ_{cy}, and τ_{cxy} is safe and statically admissible for $\mu = 1$, then the state of stress σ_{sx}, $\mu\sigma_{sy}$, σ_{cx}, $\mu\sigma_{cy}$, and $\sqrt{\mu}\,\tau_{cxy}$ is safe in the case $\mu \leq 1$. The latter state of stress naturally results in the same bending moment in the x-section, but μ times as great a bending moment in the y-section and $\sqrt{\mu}$ as great a torsional moment.

From this it can be concluded that if the moments m_x, m_y, and m_{xy} are safe in the case $\mu = 1$, then the moments m_x, μm_y and $\sqrt{\mu}\, m_{xy}$ are safe in the case $\mu \leq 1$. This means that an applicable yield criterion (a lower bound solution) in the case $\mu < 1$ can be constructed on the basis of the yield criterion in the case $\mu = 1$ by two affinities, one in the direction of the m_y-axis in the scale $1/\mu$, and one in the direction of the m_{xy}-axis in the scale $1/\sqrt{\mu}$. This still results in two cones (see Fig. 2.3.11).

Finally, the most general case of an orthotropic slab can be dealt with. The reinforcement is characterized by

$$\begin{aligned} A_{sx} &= A_s, \qquad & A_{sx}' &= \kappa A_s \\ A_{sy} &= \mu A_s, \qquad & A_{sy}' &= \mu' A_s \end{aligned} \tag{2.3.59}$$

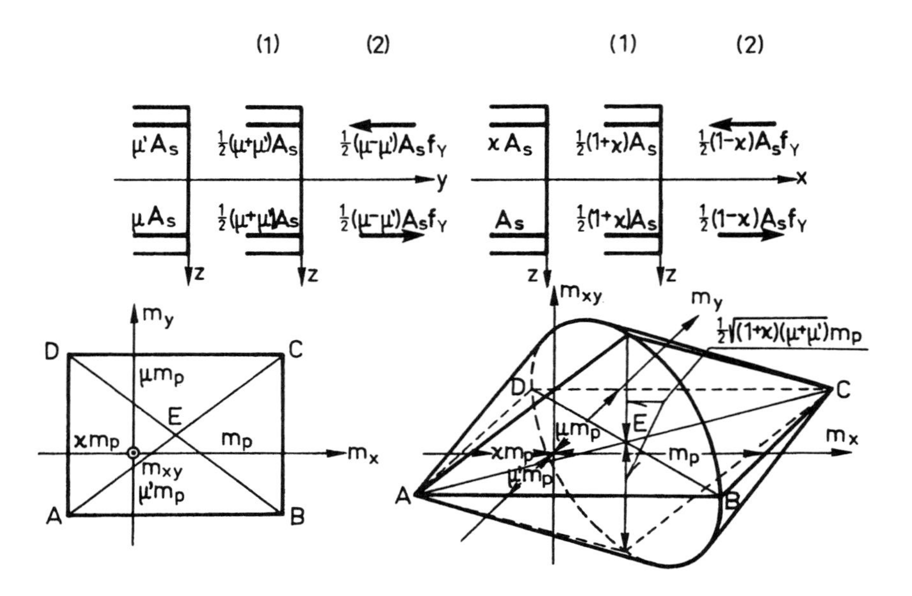

Figure 2.3.12 Yield condition for a slab element in the general orthotropic case.

It is assumed that κ, μ, and μ' are less than 1.

The appearance of the yield condition in the m_x, m_y-plane is obvious (see Fig. 2.3.12). The maximum torsional moment is determined from (2.3.26) and the simultaneous bending moments from (2.3.27) and (2.3.28). The values for the bending moments correspond to the point of intersection E of the diagonals AC and BD.

The yield surface as a whole can be found by superposition of the following states of stress:

1. Stress fields corresponding to arbitrary points on the yield surface for a slab with the following reinforcements:

$$A_{sx}' = A_{sx} = \frac{1}{2}(1+\kappa)A_s$$
$$A_{sy}' = A_{sy} = \frac{1}{2}(\mu+\mu')A_s \qquad (2.3.60)$$

   that is, a reinforcement of type (2.3.58), with equal reinforcement per unit length at top and bottom in the x-section and correspondingly in the y-section.
2. The forces ½$(1 - \kappa)A_s f_Y$ in the reinforcement in the x-section and the forces ½$(\mu - \mu')A_s f_Y$ in the reinforcement in the y-section, acting as shown in Fig. 2.3.12. These forces result in the following bending moments:

$$m_x \simeq \frac{1}{2}(1-\kappa)m_p$$
$$m_y \simeq \frac{1}{2}(\mu-\mu')m_p \qquad (2.3.61)$$

The state of stress 1 corresponds to a yield surface similar to that shown in Fig. 2.3.11. Superposition of the two states of stress therefore results in the yield surface shown in Fig. 2.3.12. This consists of two cones, similar to the yield surface shown in Fig. 2.3.11, except for a different positioning of the surface in relation to the coordinate axes.

Analytical expressions for the yield conditions. Analytical expressions can now be given for the yield conditions derived in the foregoing. In the most general case of an orthotropic slab, the following analytical expression is obtained, introducing the notation $m_{px} = m_p$, $m_{px}' = \kappa m_p$, $m_{py} = \mu m_p$, and $m_{py}' = \mu' m_p$, which is more general, also permitting m_p to be zero.

$$m_y \geq -\eta m_x + \eta m_{px} - m_{py}': \; -(m_{px}-m_x)(m_{py}-m_y) + m_{xy}^2 = 0 \qquad (2.3.62)$$

$$m_y \leq -\eta m_x + \eta m_{px} - m_{py}' : -(m_{px}' + m_x)(m_{py}' + m_y) + m_{xy}^2 = 0 \qquad (2.3.63)$$

where

$$\eta = \frac{m_{py} + m_{py}'}{m_{px} + m_{px}'} \qquad (2.3.64)$$

Of course, (2.3.62) will be used only for $m_x \leq m_{px}$ and $m_y \leq m_{py}$. Similarly, (2.3.63) will be used only for $m_x \geq -m_{px}$ and $m_y \geq -m_{py}'$.

If it has to be verified that a slab can carry a prescribed moment field, one must show that the following conditions are fulfilled:

$$\begin{aligned} m_x &\leq m_{px} \\ m_y &\leq m_{py} \\ m_x &\geq -m_{px}' \\ m_y &\geq -m_{py}' \end{aligned}$$

$$\begin{aligned} -(m_{px} - m_x)(m_{py} - m_y) + m_{xy}^2 \leq 0 \\ -(m_{px}' + m_x)(m_{py}' + m_y) + m_{xy}^2 \leq 0 \end{aligned} \qquad (2.3.65)$$

For the sake of clearness, the meaning of the quantities involved in the yield condition will be repeated:

- m_{px} is the numerical value of the positive yield moment in pure bending in an x-section (i.e., a section perpendicular to the x-axis).
- m_{px}' is the numerical value of the negative yield moment in pure bending in an x-section.
- m_{py} is the numerical value of the positive yield moment in pure bending in a y-section.
- m_{py}' is the numerical value of the negative yield moment in pure bending in a y-section.

In the derivation it was assumed that κ, μ, and μ' were less than 1 in order to point out that scaling from a known value of the yield moment in bending and the corresponding area of reinforcement should be done to smaller areas of reinforcement. This restriction in the coefficients vanishes when the formulations (2.3.62), (2.3.63) and (2.3.65) are used.

If

$$\begin{aligned} m_{px} = m_{py} = m_p \\ m_{px}' = m_{py}' = m_p' \end{aligned} \qquad (2.3.66)$$

we are in the isotropic case, and the expressions (2.3.62) and (2.3.63) are equivalent to the isotropic yield condition shown in Fig. 2.3.9. The yield conditions above were derived by the author [63.2]. They have been adopted in Eurocode 2 [91.23].

It should be mentioned that Massonnet and Save [63.1], Wolfensberger [64.1], and Kemp [65.2] have obtained the same yield conditions by using K. W. Johansen's method of calculating the bending moments in a yield line [43.1]. This method, however, is not satisfactory, as both the yield condition and the formulas of stress resultants in a yield line should be derived on the basis of the properties of the basic materials, concrete and reinforcement. The yield conditions have also been studied by Morley [66.2] along lines similar to those of the author.

Effectiveness factors. Effectiveness factors for pure bending will be given in Chapter 5. These effectiveness factors may be used when calculating m_{px}, m_{py}, m_{px}', and m_{py}'. For small reinforcement degrees these effectiveness factors may be used when calculating the bending yield moments.

Since for pure torsion we have compressive stresses in the concrete together with tensile stresses in the reinforcement crossing the cracks parallel to the compressive stresses, the effectiveness factors will be reduced (cf. Section 2.1.4). For higher reinforcement degrees the yield moment in pure torsion should be calculated using a reduced value of the effectiveness factor. The reduced value of the effectiveness factor may be taken to be 70% of the value valid for bending (cf. Chapter 5 and 7).

If the value determined for the yield moment in pure torsion in this way is still designated t_p, an approximate yield condition for higher reinforcement degrees may be taken to be

$$
\begin{aligned}
& m_x \le m_{px} \\
& m_y \le m_{py} \\
& m_x \ge -m_{px}' \\
& m_y \ge -m_{py}' \\
& -(m_{px} - m_x)(m_{py} - m_y) + m_{xy}^2 \le 0 \\
& -(m_{px}' + m_x)(m_{py}' + m_y) + m_{xy}^2 \le 0
\end{aligned}
\qquad (2.3.67)
$$

$$|m_{xy}| \le t_p$$

The modification introduced by the last condition is equivalent to the modification introduced in Section 2.2 in the yield conditions for disks.

2.3.3 An Alternative Derivation of the Yield Conditions for Slabs

The yield conditions for slabs with low degrees of reinforcement may be derived in a similar way utilizing the yield conditions for disks developed in Section 2.2.

When regarding a slab reinforced both at the top and bottom in two directions x and y at right angles to each other as two parallel disks of equal thickness, we are immediately able to derive yield conditions for slabs by using the yield conditions (2.2.41) and (2.2.42). Yielding may occur if one of the disks or both the disks are yielding. Denoting the imaginary thickness of the disks t, the distance between the center planes of the disks h_i, and finally, the bending moments per unit length m_x and m_y and the torsional moment per unit length m_{xy}, we obtain the stresses (see Fig. 2.3.13)

$$\sigma_x = \frac{m_x}{h_i t}, \quad \sigma_y = \frac{m_y}{h_i t}, \quad \tau_{xy} = \frac{m_{xy}}{h_i t}$$

$$\sigma_x' = -\frac{m_x}{h_i t}, \quad \sigma_y' = -\frac{m_y}{h_i t}, \quad \tau_{xy}' = -\frac{m_{xy}}{h_i t} \tag{2.3.68}$$

where the stresses of the upper disk are primed.

As the yield condition (2.2.41) has to be taken into consideration only at low degreees of reinforcement, we obtain the following yield conditions of the two disks:

$$-\left(\Phi_x f_c - \frac{m_x}{h_i t}\right)\left(\Phi_y f_c - \frac{m_y}{h_i t}\right) + \left(\frac{m_{xy}}{h_i t}\right)^2 = 0 \tag{2.3.69}$$

$$-\left(\Phi_x' f_c + \frac{m_x}{h_i t}\right)\left(\Phi_y' f_c + \frac{m_y}{h_i t}\right) + \left(\frac{m_{xy}}{h_i t}\right)^2 = 0 \tag{2.3.70}$$

the degrees of reinforcement of the upper disk being primed.

If we multiply by $(h_i\, t)^2$ and insert

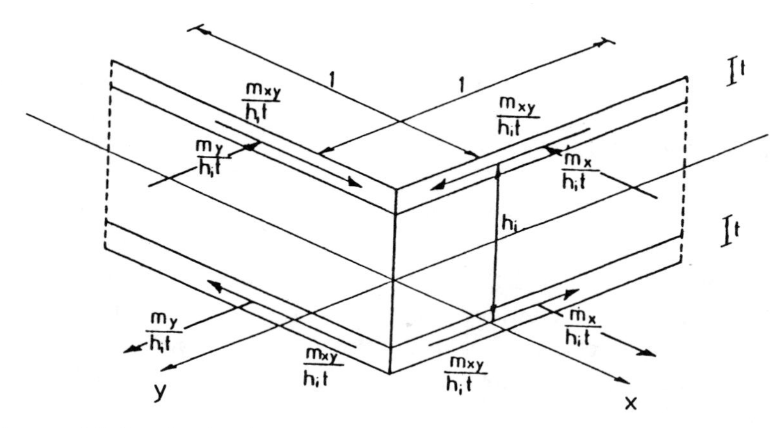

Figure 2.3.13 Derivation of the yield condition for slabs by means of a disk analogy.

$$m_{px} = h_i t \Phi_x f_c \quad , \quad m_{py} = h_i t \Phi_y f_c$$
$$m_{px}' = h_i t \Phi_x' f_c \quad , \quad m_{py}' = h_i t \Phi_y' f_c \tag{2.3.71}$$

these being the positive and the negative yield moments per unit length at pure bending in the directions of reinforcement, we obtain

$$-(m_{px} - m_x)(m_{py} - m_y) + m_{xy}^2 = 0 \tag{2.3.72}$$

$$-(m_{px}' + m_x)(m_{py}' + m_y) + m_{xy}^2 = 0 \tag{2.3.73}$$

These formulas are in agreement with (2.3.62) and (2.3.63).

2.3.4 Arbitrarily Reinforced Slabs

For a disk reinforced in several directions forming any angles, it was shown in Section 2.2.3 that the yield condition corresponds to an equivalent orthotropically reinforced disk. For a slab with the same lines of symmetry at the top and at the bottom, the yield condition therefore corresponds to that of an equivalent orthotropic slab.

If the lines of symmetry are not the same at the top and bottom, yield conditions can be derived by means of the yield conditions for disks, properly transformed to a common coordinate system. Bræstrup [70.1] showed that the yield condition may also be formulated in moments referred to axes x, y arbitrarily oriented with respect to any reinforcement direction. The yield surface is biconical, as shown in Fig. 2.3.12, but the vertices A and C no longer lie in the plane $m_{xy} = 0$. We shall, however, not pursue this matter further here.

2.3.5 Experimental Verification

The yield conditions have been confirmed experimentally by tests on slabs in pure torsion, which gave very good agreement between theory and tests [69.2]. Other tests have also been carried out [63.8; 67.3; 67.4], but the confidence in the yield conditions derived lies mainly in the agreement between numerous tests on slabs and the load-carrying capacity determined on the basis of the yield conditions.

2.3.6 Yield Conditions for Shells

If a slab element is subjected to in-plane forces as well as bending and torsional moments, we have a shell element with six generalized stresses. It has not yet been possible to derive a simple yield condition in this case in closed analytical form. A parametrical description may be obtained by calculating the forces and moments as functions of a proper set of parameters determining the six generalized strains in a shell element [66.5; 79.22; 79.23].

Rajendran and Morley [74.10] have developed a numerical procedure for the determination of points on the exact yield surface. Using the method introduced by Wolfensberger and Kemp for slabs (Section 2.3.2; see Bræstrup [70.3]) has also been attempted.

2.4 REINFORCEMENT DESIGN

2.4.1 Disks with Orthogonal Reinforcement

In Section 2.2 we have solved the problem of determining the stresses that can be carried by a disk with a given reinforcement. We are now turning to the inverse problem: to investigate the reinforcement that is necessary to carry given stresses. The most important case in practice is orthogonal reinforcement; therefore, this case is treated separately.

We assume that the concrete is able to carry the stresses at points where both principal stresses are negative. At points where one or both principal stresses are tensile stresses, reinforcement is added. The directions of reinforcement are the x and y directions in Fig.

2.4.1. Denoting the second principal stress σ_{c2} in the concrete as $-\sigma_c$, the angle between the second principal direction and the y-axis as α, and the areas of bar per unit of length in the two directions of reinforcement as A_{sx} and A_{sy}, respectively, we find that the stresses which can be carried when $\sigma_{sx} = \sigma_{sy} = f_Y$ and the stress field in the concrete is uniaxial are

$$\sigma_x = -\sigma_c \sin^2\alpha + \frac{A_{sx} f_Y}{t} \tag{2.4.1}$$

$$\sigma_y = -\sigma_c \cos^2\alpha + \frac{A_{sy} f_Y}{t} \tag{2.4.2}$$

$$\tau_{xy} = \sigma_c \sin\alpha \cos\alpha \tag{2.4.3}$$

From these equations the reinforcement may be determined if σ_c or α are known. However, it is natural to derive the formulas corresponding to the smallest amount of reinforcement (minimum reinforcement). We express σ_c by τ_{xy} using (2.4.3) and insert it into (2.4.1) and (2.4.2), and so obtain

$$\sigma_x = -\tau_{xy} \tan\alpha + \frac{A_{sx} f_Y}{t} \tag{2.4.4}$$

$$\sigma_y = -\tau_{xy} \cot\alpha + \frac{A_{sy} f_Y}{t} \tag{2.4.5}$$

or

$$\frac{A_{sx} f_Y}{t} = \sigma_x + \tau_{xy} \tan\alpha \tag{2.4.6}$$

$$\frac{A_{sy} f_Y}{t} = \sigma_y + \tau_{xy} \cot\alpha \tag{2.4.7}$$

As a measure of the amount of reinforcement per unit area we now have

$$R = |\sigma_x + \tau_{xy} \tan\alpha| + |\sigma_y + \tau_{xy} \cot\alpha| \tag{2.4.8}$$

As $\tau_{xy} \tan\alpha$ must always be positive, we content ourselves with investigating the case in which τ_{xy} is positive and $0 \leq \alpha \leq \pi/2$.

For $\sigma_x \geq -\tau_{xy} \tan\alpha$ and $\sigma_y \geq -\tau_{xy} \cot\alpha$, both terms within the numerical signs in (2.4.8) are positive, and we get

$$R = \sigma_x + \sigma_y + \tau_{xy}(\tan\alpha + \cot\alpha) \tag{2.4.9}$$

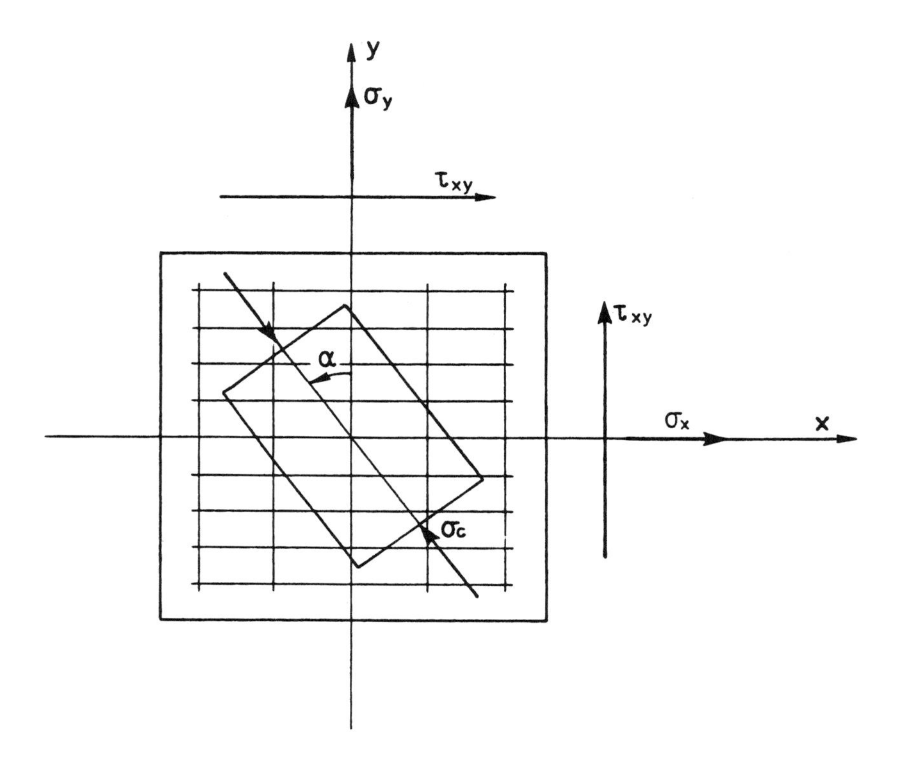

Figure 2.4.1 Disk element with uniaxial stress in the concrete and subjected to arbitrary stress combinations.

The minimum of R is then found for $\alpha = \pi/4$ and (2.4.6) and (2.4.7) read

$$\frac{A_{sx} f_Y}{t} = \sigma_x + \tau_{xy} \tag{2.4.10}$$

$$\frac{A_{sy} f_Y}{t} = \sigma_y + \tau_{xy} \tag{2.4.11}$$

If one or both terms are negative within the numerical signs in (2.4.8), this result is invalid; that is, the investigation only comprises the following interval of α:

$$\tan\alpha \geq -\frac{\sigma_x}{\tau_{xy}} , \qquad \cot\alpha \geq -\frac{\sigma_y}{\tau_{xy}} \tag{2.4.12}$$

If we now introduce the condition $\sigma_x \leq \sigma_y$, which may be done without rendering the formulas less general, $\alpha = \pi/4$ is true only within the permissible interval of α given by (2.4.12), if

$$\sigma_x \geq -\tau_{xy} \tag{2.4.13}$$

If $\sigma_x < -\tau_{xy}$, the term $\sigma_x + \tau_{xy}\tan\alpha$ in (2.4.8) and possibly also the term $\sigma_y + \tau_{xy}\cot\alpha$ may be negative for $\alpha = \pi/4$. For $\tan\alpha = -\sigma_x/\tau_{xy}$, that is, corresponding to the minimum of α according to (2.4.12), the former term in (2.4.8) is nought. If α exceeds this value, the term $\tau_{xy}\tan\alpha$ increases, while the term $\tau_{xy}\cot\alpha$ decreases.

However, as $\alpha > \pi/4$ the increase of $\tau_{xy}\tan\alpha$ is bigger than the decrease of $\tau_{xy}\cot\alpha$, for which reason the total amount of reinforcement is increased. If α decreases, the amount of reinforcement is increasing (compressive reinforcement in the direction of the x-axis). The minimum is then found for $\tan\alpha = -\sigma_x/\tau_{xy}$. If the second term $\sigma_y + \tau_{xy}\cot\alpha$ is zero or negative for $\tan\alpha = -\sigma_x/\tau_{xy}$, no reinforcement is necessary, as then $\sigma_y - \tau^2_{xy}/\sigma_x \leq 0$ or $\sigma_x\sigma_y \geq \tau^2_{xy}$, signifying that both principal stresses are negative or, especially when the equal sign is valid, that the largest principal stress is zero. By this we have obtained the following set of formulas for determining the minimum reinforcement:

$$\sigma_x \leq \sigma_y$$

Case 1:

$$\sigma_x \geq -|\tau_{xy}|$$

$$f_{tx} = \frac{A_{sx}f_Y}{t} = \sigma_x + |\tau_{xy}| \tag{2.4.14}$$

$$f_{ty} = \frac{A_{sy}f_Y}{t} = \sigma_y + |\tau_{xy}| \tag{2.4.15}$$

$$\sigma_c = 2|\tau_{xy}| \tag{2.4.16}$$

Case 2:

$$\sigma_x < -|\tau_{xy}|$$

If $\sigma_y < 0$, reinforcement is required for

$$\sigma_x\sigma_y \leq \tau_{xy}^2 \tag{2.4.17}$$

and the reinforcement is determined by

$$A_{sx} = 0 \tag{2.4.18}$$

$$f_{ty} = \frac{A_{sy} f_Y}{t} = \sigma_y + \frac{\tau_{xy}^2}{|\sigma_x|} \tag{2.4.19}$$

$$\sigma_c = |\sigma_x| \left[1 + \left(\frac{\tau_{xy}}{\sigma_x} \right)^2 \right] \tag{2.4.20}$$

Formulas (2.4.16) and (2.4.20) make it possible to determine the stress in the concrete σ_c. These are immediately derived from (2.4.3). The stress of the concrete must not exceed νf_c, ν being the effectiveness factor (see below).

The directions of reinforcement may often be chosen freely. From the formulas derived, the natural result appears to be that the minimum reinforcement is obtained if the directions of reinforcement are identical to the principal directions. Therefore, this ought to be the goal. In cases where it is a matter of demonstrating that a given reinforcement is able to carry given stresses, the yield conditions derived in Section 2.2 may, of course, be used.

However, an important case should be mentioned, that in which the reinforcement in one direction is known. If we insert $\gamma = \tan\alpha$ in (2.4.3), (2.4.6), and (2.4.7), we get

$$f_{tx} = \frac{A_{sx} f_Y}{t} = \sigma_x + \gamma |\tau_{xy}| \tag{2.4.21}$$

$$f_{ty} = \frac{A_{sy} f_Y}{t} = \sigma_y + \frac{1}{\gamma} |\tau_{xy}| \tag{2.4.22}$$

$$\sigma_c = |\tau_{xy}| \left(\gamma + \frac{1}{\gamma} \right) \tag{2.4.23}$$

γ being a positive quantity which may theoretically be arbitrarily chosen only if $\sigma_c \leq \nu f_c$, or which may be determined when the reinforcement in one direction is known.

Formulas (2.4.21)-(2.4.23) can also be used instead of (2.4.14)-(2.4.20) if minimum reinforcement is not sought. In this case γ can theoretically be freely chosen.

In order not to get too high reinforcement stresses in the serviceability limit state, it is necessary to put limits on the choice of γ.

To find reasonable limits for γ, we first remark that if the direction of the compressive stress in the concrete is known in the serviceability limit state, then the reinforcement formulas may be used to determine the reinforcement stresses, since the reinforcement formulas are nothing else than equilibrium equations. In the formulas we just have to replace f_Y with σ_{sx} and σ_{sy}, respectively.

If we regard the minimum reinforcement formulas to represent the conditions in the serviceability limit state as an approximation (cf. Section 1.4), and if we require σ_{sx} and σ_{sy} not to exceed f_Y in the serviceability limit state, then we may determine limits for γ if the ratio between the ultimate load and the serviceability limit state load is known. As an example taking this ratio to be 2, we must choose γ within such limits that the formulas (2.4.21)-(2.4.23) never give more than twice and never less than half the reinforcement provided by the minimum reinforcement formulas.

If we denote the reinforcement areas corresponding to minimum reinforcement $A_{sx,\min}$ and $A_{sy,\min}$, respectively, then the reinforcement areas determined by (2.4.21)-(2.4.23), A_{sx} and A_{sy}, must satisfy the conditions

$$\begin{aligned} \frac{1}{2}A_{sx,\min} \le A_{sx} \le 2A_{sx,\min} \\ \frac{1}{2}A_{sy,\min} \le A_{sy} \le 2A_{sy,\min} \end{aligned} \tag{2.4.24}$$

These conditions fix the limits for γ.

It should be mentioned that when, for example, σ_y is a compressive stress it is always possible to determine a value of $\gamma > 0$ in such a way that $A_{sy} = 0$. Then we have

$$f_{tx} = \frac{A_{sx}f_Y}{t} = \sigma_x + \frac{\tau_{xy}^2}{|\sigma_y|} \tag{2.4.25}$$

in which case, as appears from the formulas (2.4.14)-(2.4.19), we do not always obtain minimum reinforcement.

Finally, we shall write down the formulas for the case where the yield stresses are different in the two reinforcement directions. The yield stresses are denoted f_{Yx} and f_{Yy}. The formulas may be developed by the same procedure as that used above.

Case 1:

$$\sigma_x \ge -|\tau_{xy}|\sqrt{\lambda}\,, \qquad \sigma_y \ge -\frac{|\tau_{xy}|}{\sqrt{\lambda}}$$

$$f_{tx} = \frac{A_{sx}f_{Yx}}{t} = \sigma_x + |\tau_{xy}|\sqrt{\lambda} \tag{2.4.26}$$

$$f_{ty} = \frac{A_{sy} f_{Yy}}{t} = \sigma_y + \frac{|\tau_{xy}|}{\sqrt{\lambda}} \tag{2.4.27}$$

$$\sigma_c = |\tau_{xy}| \left(\sqrt{\lambda} + \frac{1}{\sqrt{\lambda}} \right) \tag{2.4.28}$$

Case 2:

$$\sigma_x \le \sigma_y \lambda \, , \qquad \sigma_x < - |\tau_{xy}| \sqrt{\lambda}$$

Reinforcement is necessary only if $\sigma_x \sigma_y \le \tau^2_{xy}$.

$$f_{tx} = 0 \tag{2.4.29}$$

$$f_{ty} = \sigma_y + \frac{\tau_{xy}^2}{|\sigma_x|} \tag{2.4.30}$$

$$\sigma_c = |\sigma_x| \left[1 + \left(\frac{\tau_{xy}}{\sigma_x} \right)^2 \right] \tag{2.4.31}$$

Case 3:

$$\sigma_x \ge \sigma_y \lambda \, , \qquad \sigma_y < - \frac{|\tau_{xy}|}{\sqrt{\lambda}}$$

Reinforcement is necessary only if $\sigma_x \sigma_y \le \tau^2_{xy}$.

$$f_{tx} = \sigma_x + \frac{\tau_{xy}^2}{|\sigma_y|} \tag{2.4.32}$$

$$f_{ty} = 0 \tag{2.4.33}$$

$$\sigma_c = |\sigma_y| \left[1 + \left(\frac{\tau_{xy}}{\sigma_y} \right)^2 \right] \tag{2.4.34}$$

The parameter λ is given by

$$\lambda = \frac{f_{Yx}}{f_{Yy}} \tag{2.4.35}$$

Instead of using the last mentioned formulas one may, without any significant error, replace in the foregoing formulas f_Y on the left-hand sides with f_{Yx} and f_{Yy}, respectively.

The concrete stress σ_c must satisfy the condition

$$\sigma_c \le \nu f_c \tag{2.4.36}$$

On the safe side ν may be taken to be

$$\nu = 0.7 - \frac{f_c}{200} \qquad (f_c \text{ in MPa}) \tag{2.4.37}$$

A more accurate value may be determined by (2.1.58). The limits (2.4.24) in a crude way also take into account crack sliding. This is discussed in more detail in Chapter 4.

Having determined the necessary reinforcement by means of the formulas above, it is often found advantageous to use another, perhaps more economical or practical distribution of the reinforcement. In the case of homogeneous stress fields, it should be noted that the reinforcement theoretically might be distributed in any other way that results in the same statical equivalence of the reinforcement forces in sections perpendicular to the reinforcement directions. Sometimes, for instance in slabs and shells, such a transformation changes the compression forces in the concrete and, if so, the concrete stresses have to be calculated to take into account these extra forces. Also, the complete equilibrium of the whole transformed stress field at the boundaries must be considered. Examples of reinforcement transformations are given in Chapter 5.

Formulas (2.4.21) and (2.4.22) were derived by Leitz [23.2; 25.1; 25.2; 26.1; 30.1] ($\gamma = 1$) and Marcus [26.2] by using truss analogies, and by Hillerborg [53.1] and K. W. Johansen [57.1]. Hillerborg's starting point was the formulas for the moments in a yield line of an orthotropic slab, given by K. W. Johansen [43.1]. This method is not satisfactory, as both the yield conditions and the design formulas should be derived from considerations on the strength of the concrete and the reinforcement. The theory presented here has been extended to cover compression reinforcement by Clark [76.9]. It should be noted that if (2.4.21) and (2.4.22) are solved regarding the last term on the right side, the yield condition (2.2.41) is obtained by multiplication.

The method of determining the reinforcement in the plane stress field given here has been adopted in Eurocode 2 [91.23].

2.4.2 Examples

To illustrate the use of the reinforcement formulas, we show their application to some simple stress fields.

Pure tension. In pure tension $\sigma_x = \sigma$, $\sigma_y = \tau_{xy} = 0$, it is, of course, most obvious to reinforce in the direction of the x-axis using reinforcement determined by

$$f_{tx} = \frac{A_{sx} f_Y}{t} = \sigma \tag{2.4.38}$$

The stresses can be carried by the reinforcement without assistance from the concrete.

If we reinforce in the directions ξ and η instead (see Fig. 2.4.2 in which the ξ-direction forms the angle θ to the x-axis), the stresses for which to reinforce are

$$\sigma_\xi = \sigma\cos^2\theta\ , \qquad \sigma_\eta = \sigma\sin^2\theta\ , \qquad |\tau_{\xi\eta}| = \frac{1}{2}\sigma\sin2\theta \tag{2.4.39}$$

and the reinforcement necessary is, from (2.4.14) and (2.4.15), determined to be

$$\begin{aligned} f_{t\xi} &= \frac{A_{s\xi} f_Y}{t} = \sigma\cos^2\theta + \frac{1}{2}\sigma\sin2\theta \\ f_{t\eta} &= \frac{A_{s\eta} f_Y}{t} = \sigma\sin^2\theta + \frac{1}{2}\sigma\sin2\theta \end{aligned} \tag{2.4.40}$$

The volume of the reinforcement per unit of area of the disk V is

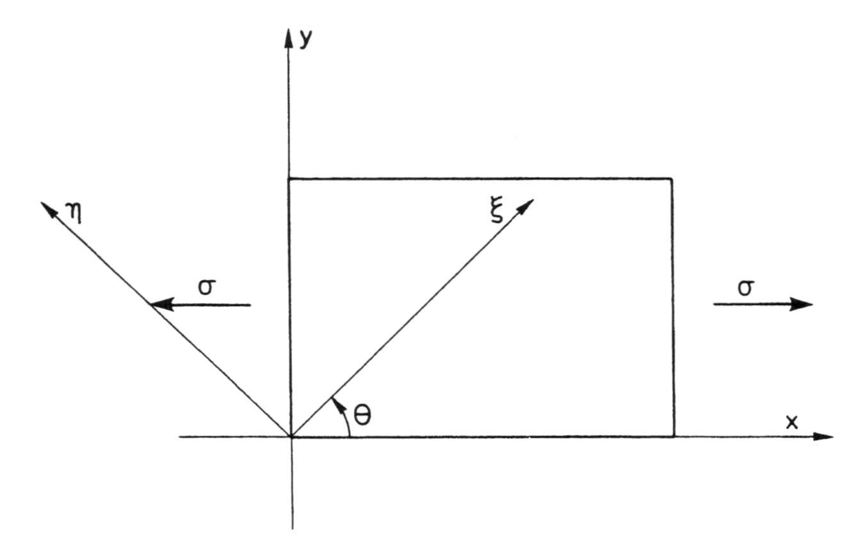

Figure 2.4.2 Disk subjected to uniaxial stress in the x-direction and reinforced orthogonally in the ξ- and η-directions.

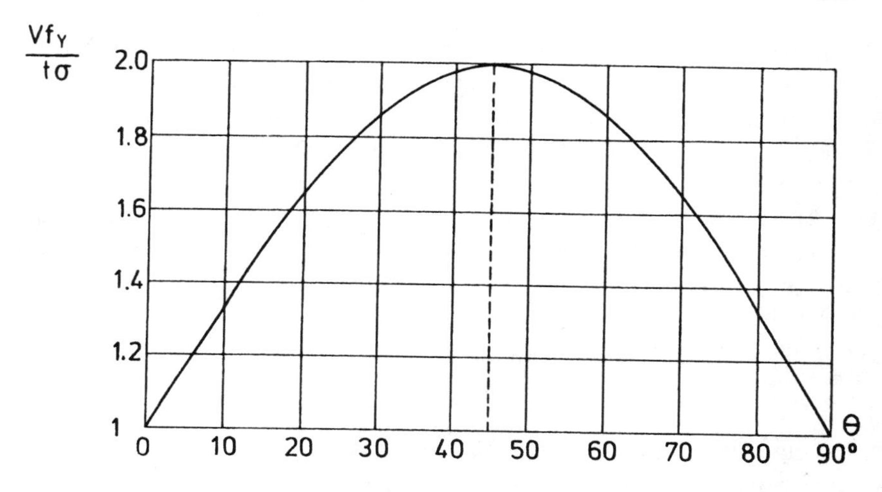

Figure 2.4.3 Reinforcement volume for a disk subjected to uniaxial stress and reinforced orthogonally at an angle with the uniaxial stress.

$$\frac{Vf_Y}{t} = \frac{A_{s\xi}f_Y}{t} + \frac{A_{s\eta}f_Y}{t} = \sigma(1 + \sin 2\theta) \tag{2.4.41}$$

In Fig. 2.4.3 the ratio $Vf_Y/t\sigma$ is plotted as a function of θ. The maximum amount of reinforcement is obtained for $\theta = 45°$, being twice the amount when we reinforce in the first principal direction.

Let us examine the behavior for $\theta = 45°$ a little further. In this case the stresses (2.4.39) are

$$\sigma_\xi = \frac{1}{2}\sigma\ , \qquad \sigma_\eta = \frac{1}{2}\sigma\ , \qquad |\tau_{\xi\eta}| = \frac{1}{2}\sigma \tag{2.4.42}$$

From (2.4.40) we obtain

$$f_{t\xi} = \frac{A_{s\xi}f_Y}{t} = \sigma\ , \qquad f_{t\eta} = \frac{A_{s\eta}f_Y}{t} = \sigma \tag{2.4.43}$$

and for the stresses of the concrete according to (2.4.16),

$$\sigma_c = \sigma \tag{2.4.44}$$

Thus the reinforcement is unable to carry the stresses without assistance of the concrete. The uniaxial compression of the concrete has the direction of the y-axis; that is, the boundary conditions of the stresses of the concrete are not satisfied if there are free boundaries parallel to the x-axis. This is evident in advance. When the reinforcing bars intersect the boundary, these being considered subjected to the yield stress, it is impossible to satisfy the boundary conditions. It follows that another stress field must develop along the boundaries in connection with the anchorage of the reinforcement. A simple model showing how the concrete stress may be developed in the case considered is shown in Fig. 2.4.4. Let us assume that the bars in the two directions intersect as shown in Fig. 2.4.4.

If we assume that the area A is subjected to uniaxial compression in the direction of the boundary by a stress equal to the compressive stress σ_c', the necessary anchor length ℓ measured perpendicular to the boundary is determined by the equilibrium condition

$$\frac{\pi}{4} d^2 f_Y = 2 \cdot \frac{1}{2} \sigma_c' d \ell \sqrt{2} \tag{2.4.45}$$

having for simplicity assumed the stresses of the concrete to be uniformly distributed over a thickness equal to the diameter of the bar d. From (2.4.45) we find that

$$\ell = \frac{\pi}{4\sqrt{2}} d \frac{f_Y}{\sigma_c'} \simeq \frac{1}{2} d \frac{f_Y}{\sigma_c'} \tag{2.4.46}$$

The equilibrium also requires the stresses of the concrete $-\sigma_c'$ along the distance 2ℓ in the direction perpendicular to the boundary, as illustrated in the figure.

Since σ_c' can at most be of the order f_c, it appears that the upper limit of the concrete stress σ_c calculated according to (2.4.44) would be only a fraction of the compressive strength f_c.

Of course, the calculation is rather theoretical. The bars are never anchored in this simple way in practice. A better way is shown in Fig. 2.4.5, where the bars are formed as closed stirrups running around longitudinal bars along the boundary.

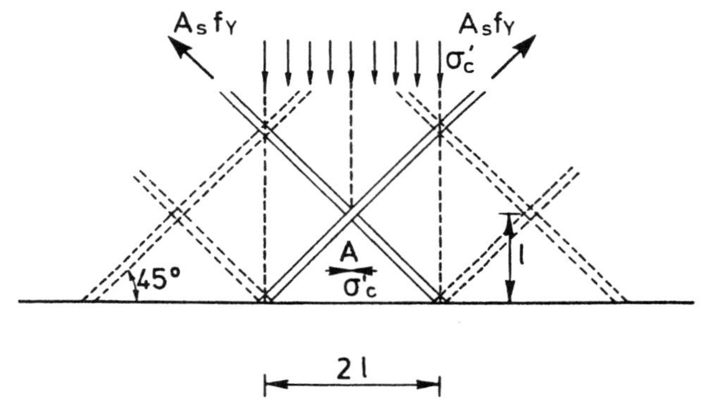

Figure 2.4.4 Stress field at boundary.

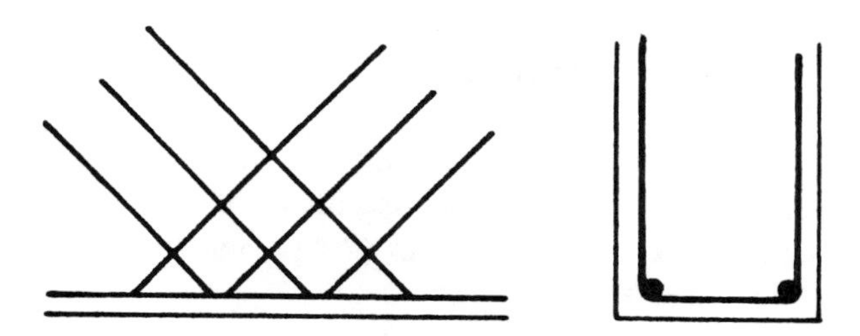

Figure 2.4.5 Proper reinforcement anchorage at boundaries.

Then the upper limit determined by (2.4.37) may be used.

Suenson [19.1; 22.1] did some experiments with slab strips reinforced as described here. The experiments showed that regarding strength, the behavior of the reinforcement was even a little more favorable than found here, which is due to a slight change in the direction of force in the bars after the cracking of the concrete.

Shear. For the stress field pure shear (i.e., $\sigma_x = \sigma_y = 0$, $\tau_{xy} = \tau$), from (2.4.14) and (2.4.15) for reinforcement in the x and y directions we find the necessary reinforcements to be

$$f_{tx} = \frac{A_{sx} f_Y}{t} = f_{ty} = \frac{A_{sy} f_Y}{t} = \tau \tag{2.4.47}$$

that is, equal amounts of reinforcement per unit of length must be used in both directions. The stress of the concrete according to (2.4.16) will be

$$\sigma_c = 2\tau \tag{2.4.48}$$

If we reinforce in the direction of the principal stresses ξ, η instead, these forming an angle of 45° to the x and y directions (see Fig. 2.4.2), we obtain the stresses

$$\sigma_\xi = -\sigma_\eta = \tau \ , \quad \tau_{\xi\eta} = 0 \tag{2.4.49}$$

That is, according to (2.4.18) and (2.4.19),

$$f_{t\xi} = \frac{A_{s\xi} f_Y}{t} = \tau \ , \quad f_{t\eta} = \frac{A_{s\eta} f_Y}{t} = 0 \tag{2.4.50}$$

and according to (2.4.20), the stress of the concrete

$$\sigma_c = \tau \tag{2.4.51}$$

The uniaxial compression of the concrete is in the η-direction. In this case the stress of the concrete and the amount of reinforcement are only half the size.

Finally, let us examine the case in which we want μ times the reinforcement per unit of length in one direction, for example, the y-direction, as in the x-direction. From (2.4.6) and (2.4.7) we obtain

$$f_{tx} = \frac{A_{sx} f_Y}{t} = \tau \tan\alpha \tag{2.4.52}$$

$$f_{ty} = \frac{A_{sy} f_Y}{t} = \frac{\mu A_{sx} f_Y}{t} = \tau \cot\alpha \tag{2.4.53}$$

from which

$$\tan^2\alpha = \frac{1}{\mu} \tag{2.4.54}$$

leading to

$$\frac{A_{sx} f_Y}{t} = \frac{\tau}{\sqrt{\mu}}, \quad A_{sy} = \mu A_{sx} \tag{2.4.55}$$

The stress of the concrete σ_c is determined from (2.4.3) to be

$$\sigma_c = \tau\left(\sqrt{\mu} + \frac{1}{\sqrt{\mu}}\right) \tag{2.4.56}$$

These results could, of course, also be found directly by means of (2.4.21)-(2.4.23).

2.4.3 Disks with Skew Reinforcement

We shall confine ourselves to treating the case in which we have two directions of reinforcement, forming an angle ω to each other. When, exceptionally, skew reinforcement is actual, we shall as a rule refer the stresses to a skew system of coordinates. Therefore, it is assumed that the stresses $\sigma_{x'}$, $\sigma_{y'}$, and $\tau_{x'y'}$ refer to a skew system of coordinates x' , y' in which the axes coincide with the direction of reinforcement (see Fig. 2.4.6).

Let the second principal direction of the stresses of the concrete form an angle α to the y'-axis. We thus find that the following stresses can be carried:

$$\sigma_{x'} = -\sigma_c \sin^2\alpha \frac{1}{\sin\omega} + \frac{A_{sx'} f_Y}{t} \sin\omega \tag{2.4.57}$$

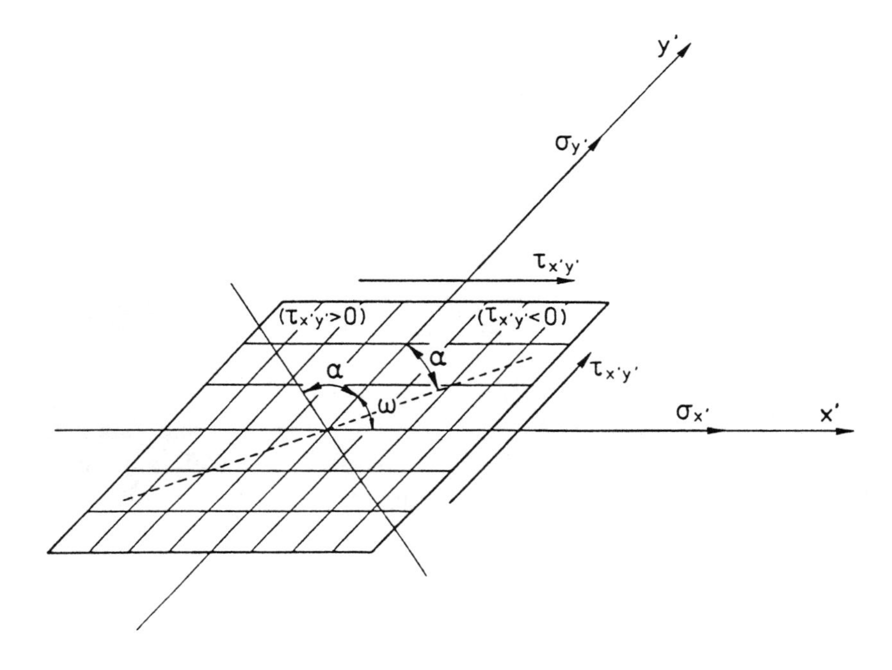

Figure 2.4.6 Notation used for disks with two reinforcement directions forming an arbitrary angle to each other.

$$\sigma_{y'} = -\sigma_c \sin^2(\omega \pm \alpha)\frac{1}{\sin\omega} + \frac{A_{sy'} f_Y}{t}\sin\omega \tag{2.4.58}$$

$$|\tau_{x'y'}| = \sigma_c \sin\alpha \sin(\omega \pm \alpha)\frac{1}{\sin\omega} \tag{2.4.59}$$

using + for $\tau_{x'y'} > 0$ and − for $\tau_{x'y'} < 0$. In these equations $A_{sx'}$ and $A_{sy'}$ still mean areas of bar per unit of length measured in the direction perpendicular to the x' and y' directions, respectively.

Continuing as in the case of orthogonal reinforcement, that is, determining σ_c from (2.4.59) and inserting into (2.4.57) and (2.4.58), we obtain

$$\frac{A_{sx'} f_Y}{t}\sin\omega = \sigma_{x'} + |\tau_{x'y'}|\frac{\sin\alpha}{\sin(\omega \pm \alpha)} \tag{2.4.60}$$

$$\frac{A_{sy'} f_Y}{t}\sin\omega = \sigma_{y'} + |\tau_{x'y'}|\frac{\sin(\omega \pm \alpha)}{\sin\alpha} \tag{2.4.61}$$

If we put

$$\gamma = \frac{\sin\alpha}{\sin(\omega \pm \alpha)} \tag{2.4.62}$$

(2.4.60) and (2.4.61) may be transformed to

$$\frac{A_{sx'} f_Y}{t} \sin\omega = \sigma_{x'} + \gamma |\tau_{x'y'}| \tag{2.4.63}$$

$$\frac{A_{sy'} f_Y}{t} \sin\omega = \sigma_{y'} + \frac{1}{\gamma} |\tau_{x'y'}| \tag{2.4.64}$$

We now see that the formulas for determining $(A_{sx'} f_Y / t)\sin\omega$ and $(A_{sy'} f_Y / t)\sin\omega$ are analogous to (2.4.6) and (2.4.7), which apply to orthogonal reinforcement. Therefore, we are immediately able to put down the expressions giving the minimum reinforcement,

$$\sigma_{x'} \leq \sigma_{y'}$$

Case 1:

$$\sigma_{x'} \geq -|\tau_{x'y'}|$$

$$\frac{A_{sx'} f_Y}{t} \sin\omega = \sigma_{x'} + |\tau_{x'y'}| \tag{2.4.65}$$

$$\frac{A_{sy'} f_Y}{t} \sin\omega = \sigma_{y'} + |\tau_{x'y'}| \tag{2.4.66}$$

Case 2:

$$\sigma_{x'} < -|\tau_{x'y'}|$$

$$A_{sx'} = 0 \tag{2.4.67}$$

$$\frac{A_{sy'} f_Y}{t} \sin\omega = \sigma_{y'} + \frac{\tau_{x'y'}^2}{|\sigma_{x'}|} \tag{2.4.68}$$

The stresses of the concrete are determined by (2.4.59). In case 1 we have $\gamma = 1/\gamma$ (i.e., $\gamma^2 = 1$), whereby α is determined by the equation

$$\sin\alpha = \sin(\omega \pm \alpha) \tag{2.4.69}$$

that is, the second principal direction bisects the angle between the direction of the coordinates.

In case 2[3] we have $\gamma = -\sigma_{x'} / |\tau_{x'y'}|$, from which α is determined by

$$\frac{\sin\alpha}{\sin(\omega \pm \alpha)} = -\frac{\sigma_{x'}}{|\tau_{x'y'}|} \tag{2.4.70}$$

Compression reinforcement in connection with skew reinforcement has been considered by Clark [76.9].

2.4.4 Slabs

The foregoing formulas for the reinforcement necessary in disks may be used directly to determine the reinforcement in slabs if we regard a slab as consisting of two parallel disks, as in Section 2.3.3. We thereby directly obtain the following formulas, in which m_{px} signifies the necessary positive yield moment per unit of length in pure bending for sections perpendicular to the x-axis, m_{px}' the corresponding negative yield moment, and a similar notation for sections perpendicular to the y-axis. Finally, m_x, m_y, and m_{xy} signify the bending moment per unit of length in sections perpendicular to the x-axis, the bending moment per unit of length within sections perpendicular to the y-axis, and the torsional moment per unit of length within both sections, respectively (i.e., the moments the slab has to carry). The formulas read

$$m_x \leq m_y$$

Case 1:

$$m_x \geq -|m_{xy}|$$

$$m_{px} = m_x + |m_{xy}|, \qquad m_{py} = m_y + |m_{xy}| \tag{2.4.71}$$

(a) If $m_y \leq |m_{xy}|$, then

$$m_{px}' = -m_x + |m_{xy}|, \qquad m_{py}' = -m_y + |m_{xy}| \tag{2.4.72}$$

(b) If $m_y > |m_{xy}|$, then

$$m_{px}' = -m_x + \frac{m_{xy}^2}{|m_y|}, \qquad m_{py}' = 0 \tag{2.4.73}$$

[3] In this case it is also valid that $\sigma_{x'}\sigma_{y'} = \tau_{x'y'}^2$ when one of the principal stresses is zero. $A_{sy'} = 0$ is therefore taking place when the stresses correspond to the negative principal stresses.

Case 2:

$$m_x < -|m_{xy}|$$

$$m_{px} = 0 \ , \quad m_{py} = m_y + \frac{m_{xy}^2}{|m_x|} \tag{2.4.74}$$

(a) If $m_y \leq |m_{xy}|$: as in condition (a) of case 1 (2.4.75)
(b) If $m_y > |m_{xy}|$: as in condition (b) of case 1 (2.4.76)

For slabs a set of formulas of the type (2.4.21)-(2.4.23) can be written down immediately:

$$\begin{aligned} m_{px} &= m_x + \gamma|m_{xy}| \quad , \quad m_{py} = m_y + \frac{1}{\gamma}|m_{xy}| \\ m_{px}' &= -m_x + \gamma'|m_{xy}| \ , \quad m_{py}' = -m_y + \frac{1}{\gamma'}|m_{xy}| \end{aligned} \tag{2.4.77}$$

The parameters γ and γ' are positive numbers, which can be chosen freely. They should, however, not deviate too much from the values that give a minimum amount of reinforcement. The same guidelines as given for disks in Section 2.4.1 may be used, i.e., γ and γ' are chosen in such a way that the reinforcement found is between half the value and twice the value determined by the formulas giving minimum reinforcement, formulas (2.4.71)-(2.4.76).

These formulas are used in cases where a statically permissible distribution of moments is known, just as in the preceding formulas for disks.

By using (2.4.65)-(2.4.68) it is immediately possible to put down the formulas for reinforcement of slabs with skew reinforcement.

A conservative set of formulas in relation to (2.4.71)-(2.4.76), which was first given by the author [64.3], is

$$\begin{aligned} m_{px} &= m_x + |m_{xy}| \quad , \quad m_{py} = m_y + |m_{xy}| \\ m_{px}' &= -m_x + |m_{xy}| \ , \quad m_{py}' = -m_y + |m_{xy}| \end{aligned} \tag{2.4.78}$$

If one or more of the terms m_{px} , m_{py} , m_{px}' , and m_{py}' are less than zero according to the formulas above, the terms in question are set at zero.

Formulas (2.4.78) lead to paradoxical results in certain cases; for example, for pure bending in only one direction we find that the compressive zone has to be reinforced in certain cases, which, of course, is unnecessary.

When the bending yield moments m_{px}, m_{py}, m_{py}', and m_{py}' have been determined, the necessary reinforcement may be found by a simple bending analysis. For bending, the ν-values are given in Section 5.1. These values may be used for small degrees of reinforcement. For higher degrees of reinforcement, particularly when the torsional moment is dominating, the ν-value may be taken at 70% of the value for bending, as already mentioned in Section 2.3.2. Of course, for practical purposes one or another kind of interpolation formula between the ν-values for bending and for torsion might be established. We shall, however, not dwell on this, since the procedure described will be sufficient for most practical purposes.

2.4.5 Shells

For shells (i.e., structural elements having both in-plane forces and bending and torsional moments), simple reinforcement formulas have not yet been developed except for cylindrical shells [71.2]. Safe, statically admissible solutions can be derived by considering the shell as consisting of two disks, one at the top and one at the bottom. Then the forces and moments acting on these two disks can easily be determined. Sometimes more than two disks are appropriate. The necessary amount of reinforcement is found using the formulas of Section 2.4.1 or 2.4.3 [74.7]. The reinforcement found is placed in the middle planes of the disks. If this is not convenient from a practical point of view, the reinforcement may be moved to other positions, provided that the reinforcement forces belonging to the new positions are changed in such a way that renders them statically equivalent to the reinforcement forces found in the middle planes of the disks. If the reinforcement transformation changes a reinforcement force from a tensile force to a compression force, either compression reinforcement is used or the compression force must be taken into account when calculating the concrete compression stresses. More accurate methods have been developed by Morley and Gulvanessian [77.5] and by Karl Erik Hansen et al. [78.13].

2.4.6 Three-Dimensional Stress Fields

In the case of a three-dimensional stress field, the formulas for the necessary reinforcement are, of course, more complicated. Three perpendicular reinforcement directions have been treated in [85.4].

2.4.7 Reinforcement Design According to the Elastic Theory

Since the linear elastic stress distribution is an equilibrium solution, there is fundamentally no objection to the use of the stress distribution according to the linear elastic theory[4] for determining the necessary reinforcement. The most obvious reason for doing so is that we obtain a strong reinforcement in places where formation of cracks first occurs. Knowledge of the solution of the elastic theory further provides the possibility of calculating the cracking load and of estimating the stiffness of the structure up to the cracking load.

A completely consistent design of reinforcement according to the elastic theory is, however, very rare. If the stress distribution of the elastic theory has to be correct after the formation of cracks, a closely meshed reinforcement which is everywhere in line with the first principal direction (trajectorial reinforcement) has to be provided, and this reinforcement must be of the same stiffness as the uncracked concrete. The necessary amount of reinforcement is very large in that case, and the arrangement of reinforcement is frequently very expensive to carry out, as shown by the following considerations. If a principal tensile stress of the concrete is σ, it leads to a strain $\varepsilon_c = \sigma/E_c$ if we neglect Poisson's ratio. E_c is Young's modulus for the concrete. If the concrete cracks, and if the reinforcement ratio corresponding to a reinforcement in the direction perpendicular to the section considered is equal to r, the reinforcement stress σ_s corresponding to the stress σ of the section will be

$$\sigma_s = \frac{\sigma}{r} \tag{2.4.79}$$

and the strain of the bars will be

$$\varepsilon_s = \frac{\sigma}{E_s r} \tag{2.4.80}$$

E_s being Young's modulus for the steel. Requiring that $\varepsilon_c = \varepsilon_s$, we obtain

$$r = \frac{E_c}{E_s}, \quad \sigma_s = \sigma \frac{E_s}{E_c} \tag{2.4.81}$$

[4] By linear elastic theory we mean the classical linear-elastic theory applied to the uncracked concrete structure.

This is a very large reinforcement ratio and a very small stress of the bars.

If, instead, the reinforcement is designed on the assumption of the usual utilization of the bars up to the design values, a plastic method is in fact applied, and a redistribution of stresses, necessary in order to ensure compatibility, is accepted. Therefore, there is no guarantee that the actual stresses of the bars are equal to the design stresses assumed. As it is a plastic method which forms the basis of the design of the reinforcement, the formulas in previous sections can be used in such cases as well.

A method of reinforcing frequently used in the case $\sigma_1 > 0$, $\sigma_2 < 0$, when the directions of reinforcement and the principal directions are not identical, is to provide equal amounts of reinforcement in two perpendicular, arbitrarily chosen directions, and to provide such an amount of reinforcement that a stress equal to the principal tensile stress σ_1 can be carried in both directions.

If we denote the directions of reinforcement by x and y, the necessary reinforcement must consequently be determined by

$$\frac{A_{sx} f_Y}{t} = \frac{A_{sy} f_Y}{t} = \sigma_I \tag{2.4.82}$$

This is, of course, in complete accordance with the considerations in previous sections, but we obtain a larger amount of reinforcement. The method is irrational when we are reinforcing in a direction near the direction of the principal tensile stress. The amount of reinforcement is then up to twice the amount necessary.

When this method is used, no change in the direction of the compression of the concrete is required; that is, the total stresses have the same principal directions as the stresses of the concrete. The stress of the concrete σ_c, however, [cf. (2.4.20)], is larger than $|\sigma_2|$:

$$\sigma_c = |\sigma_2| + \sigma_I \tag{2.4.83}$$

Long experience clearly shows, however, that we can also reinforce according to a stress field, which requires a change in the direction of the compression of the concrete. However, the change requires attention to crack sliding (cf. Chapter 4).

Maximum stiffness (minimum crack widths) and a minimum amount of reinforcement are obtained if the reinforcement is placed in the principal directions. As mentioned, this is practically feasible only in exceptional cases. With given directions of reinforcement the minimum reinforcement is obtained by using the formulas derived in previous sections.

Further, even with fixed directions of reinforcement, the elastic theory will often give a very inconvenient arrangement of reinforcement. Therefore, it is customary to examine only certain sections, and for these ensure that the resultant of the tensile stresses are equal to the resultant of the forces of reinforcement when the stresses of the bars equal the design values. This is, of course, a quite unsatisfactory method.

In some cases plastic solutions have remarkable properties concerning reinforcement stresses in the linear-elastic fully cracked state. For beam structures, see Chapter 1.

2.4.8 Stiffness in the Cracked State

We have not yet dealt with the condition in the elastic state. Here we are going to examine the elastic properties of an isotropically reinforced disk element in pure shear (see Fig. 2.2.6). We assume that the concrete is linear-elastic, Poisson's ratio and the tensile strength being zero. Young's modulus for the concrete is indicated as E_c. The reinforcing bars are considered to be linear-elastic and - as previously - subjected only to tension or compression. Young's modulus for the reinforcement is indicated as E_s. The strains in the ξ and η directions are indicated as ε_ξ and ε_η, respectively. The equations of projection corresponding to (2.2.16) and (2.2.19) will be, the ξ and η directions forming an angle of 45° to the directions of reinforcement:

$$-\frac{1}{2}E_c\varepsilon_\eta t = \frac{1}{2}(\varepsilon_\xi + \varepsilon_\eta)A_s E_s \tag{2.4.84}$$

$$\frac{1}{2}E_c\varepsilon_\eta = -\tau \tag{2.4.85}$$

from which we obtain

$$\varepsilon_\eta = -\frac{2\tau}{E_c} \tag{2.4.86}$$

$$\varepsilon_\xi = -\varepsilon_\eta\frac{1+\psi}{\psi} = \frac{2\tau}{E_c}\frac{1+\psi}{\psi} \tag{2.4.87}$$

where the parameter ψ is defined as

$$\psi = \frac{A_s E_s}{tE_c} = nr \tag{2.4.88}$$

Here n signifies the ratio E_s/E_c and r is the reinforcement ratio.

In the uncracked state we would obtain $\varepsilon_\xi = \tau/E_c$ and $\varepsilon_\eta = -\tau/E_c$, so the stiffness in the cracked state is considerably smaller. As ψ is of the order 0.1, ε_ξ will be of the order 20 times the value in the uncracked state. Of course, it should be noted that the strain field will actually be discontinuous, with the greater part of the gradients of displacement concentrated in the cracks.

By uniaxial tension $\sigma_x = \sigma$ with the bars in the x-direction we obtain in the uncracked state $\varepsilon_x = \sigma/E_c$, and in the cracked state in which the bars carry all the forces $\varepsilon_x = (1/\psi)(\sigma/E_c)$. In this case the decrease in the stiffness is only about half as big as the decrease by pure shear.

A completely general analysis of the stiffness by all possible combinations of stresses is difficult. Furthermore, as the practical use of such investigations is doubtful, we shall not go into this further.

If we compare the stresses of the concrete in the case above and the case in which the shear reinforcement is in the direction of the tensile principal stress (ξ-direction), it is seen that in the case above it is twice as big. This is equivalent to the conditions in slabs by pure torsion, treated first by N. J. Nielsen [20.1].

2.4.9 Concluding Remarks

Formulas for determining the necessary reinforcement have, of course, always occupied engineers. This is indeed one of the fundamental problems of reinforced concrete theory. Here we have based our calculation on the plastic theory; we have contented ourselves with finding an arrangement of reinforcement which is statically able to carry the stresses, whereas we have not dealt very much with conditions in the elastic state or in the real collapse state.

The design formulas give the smallest amount of reinforcement necessary corresponding to the given strength of materials. The most essential assumption is that the tensile strength of the concrete be assumed to be zero. Thus, when a structure of reinforced concrete is given less reinforcement than stated by these formulas, we are relying on the tensile strength of the concrete. Since this may vanish for many reasons; for example, the load has earlier exceeded the cracking load or residual stresses from shrinkage have arisen with subsequent formation of cracks, in the author's opinion the design principles given here also represent the minimum reinforcement in question.

Furthermore, there may be many reasons for providing more reinforcement than that corresponding to the design formulas given. For example, in areas where theoretically no reinforcement is necessary, a slight, uniformly distributed reinforcement should normally be placed. This problem is discussed in more detail in Chapter 4.

Chapter 3

THE THEORY OF PLAIN CONCRETE

3.1 STATICAL CONDITIONS

If the stress tensor $\boldsymbol{S}$ in Cartesian coordinates x, y, and z is denoted

$$\boldsymbol{S} = \begin{bmatrix} \sigma_x & \tau_{xy} & \tau_{xz} \\ \tau_{xy} & \sigma_y & \tau_{yz} \\ \tau_{xz} & \tau_{yz} & \sigma_z \end{bmatrix} \tag{3.1.1}$$

we have the equilibrium equations

$$\begin{aligned} &\frac{\partial \sigma_x}{\partial x} + \frac{\partial \tau_{xy}}{\partial y} + \frac{\partial \tau_{xz}}{\partial z} + \rho f_x = 0 \\ &\frac{\partial \tau_{xy}}{\partial x} + \frac{\partial \sigma_y}{\partial y} + \frac{\partial \tau_{yz}}{\partial z} + \rho f_y = 0 \\ &\frac{\partial \tau_{xz}}{\partial x} + \frac{\partial \tau_{yz}}{\partial y} + \frac{\partial \sigma_z}{\partial z} + \rho f_z = 0 \end{aligned} \tag{3.1.2}$$

where ρ is the mass density and f_x , f_y , and f_z are the components of the force per unit mass.

The statical boundary conditions are

$$\boldsymbol{p} = \boldsymbol{S}\boldsymbol{n} \tag{3.1.3}$$

where $\boldsymbol{p}$ is the stress vector along the boundary and $\boldsymbol{n}$ is a unit outwardly normal to the boundary. We also have to deal with discontinuity conditions of the type mentioned in Section 1.3. In cylinder coordinates r, θ, and z the equilibrium equations are

$$\begin{aligned} &\frac{\partial \sigma_r}{\partial r} + \frac{1}{r}\frac{\partial \tau_{r\theta}}{\partial \theta} + \frac{\partial \tau_{rz}}{\partial z} + \frac{1}{r}(\sigma_r - \sigma_\theta) + \rho f_r = 0 \\ &\frac{\partial \tau_{r\theta}}{\partial r} + \frac{1}{r}\frac{\partial \sigma_\theta}{\partial \theta} + \frac{\partial \tau_{\theta z}}{\partial z} + 2\frac{\tau_{r\theta}}{r} + \rho f_\theta = 0 \end{aligned} \tag{3.1.4}$$

$$\frac{\partial \tau_{rz}}{\partial r} + \frac{1}{r}\frac{\partial \tau_{\theta z}}{\partial \theta} + \frac{\partial \sigma_z}{\partial z} + \frac{1}{r}\tau_{rz} + \rho f_z = 0$$

3.2 GEOMETRICAL CONDITIONS

If the strain tensor in Cartesian coordinates x, y, and z is denoted

$$\boldsymbol{\varepsilon} = \begin{bmatrix} \varepsilon_x & \varepsilon_{xy} & \varepsilon_{xz} \\ \varepsilon_{xy} & \varepsilon_y & \varepsilon_{yz} \\ \varepsilon_{xz} & \varepsilon_{yz} & \varepsilon_z \end{bmatrix} \tag{3.2.1}$$

and if the components of the displacement vector are u_x , u_y , and u_z, we have the geometrical conditions

$$\varepsilon_x = \frac{\partial u_x}{\partial x}, \quad \varepsilon_y = \frac{\partial u_y}{\partial y}, \quad \varepsilon_z = \frac{\partial u_z}{\partial z}$$

$$\begin{aligned} \gamma_{xy} &= 2\varepsilon_{xy} = \frac{\partial u_x}{\partial y} + \frac{\partial u_y}{\partial x} \\ \gamma_{xz} &= 2\varepsilon_{xz} = \frac{\partial u_x}{\partial z} + \frac{\partial u_z}{\partial x} \\ \gamma_{yz} &= 2\varepsilon_{yz} = \frac{\partial u_y}{\partial z} + \frac{\partial u_z}{\partial y} \end{aligned} \tag{3.2.2}$$

γ_{xy} ,... are the changes of angles between lines parallel to the coordinate axes.

In cylinder coordinates r, θ, and z, the corresponding equations are

$$\varepsilon_r = \frac{\partial u_r}{\partial r}, \quad \varepsilon_\theta = \frac{1}{r}\left(\frac{\partial u_\theta}{\partial \theta} + u_r\right), \quad \varepsilon_z = \frac{\partial u_z}{\partial z}$$

$$\begin{aligned} \gamma_{r\theta} &= 2\varepsilon_{r\theta} = \frac{\partial u_\theta}{\partial r} + \frac{1}{r}\left(\frac{\partial u_r}{\partial \theta} - u_\theta\right) \\ \gamma_{rz} &= 2\varepsilon_{rz} = \frac{\partial u_r}{\partial z} + \frac{\partial u_z}{\partial r} \\ \gamma_{\theta z} &= 2\varepsilon_{\theta z} = \frac{1}{r}\frac{\partial u_z}{\partial \theta} + \frac{\partial u_\theta}{\partial z} \end{aligned} \tag{3.2.3}$$

3.3 VIRTUAL WORK

The virtual work equation reads

$$\iiint (\sigma_x \varepsilon_x + \sigma_y \varepsilon_y + \sigma_z \varepsilon_z + \tau_{xy}\gamma_{xy} + \tau_{xz}\gamma_{xz} + \tau_{yz}\gamma_{yz})\,dx\,dy\,dz$$
$$= \iiint (\rho f_x u_x + \rho f_y u_y + \rho f_z u_z)\,dx\,dy\,dz + \int (p_x u_x + p_y u_y + p_z u_z)\,dA \tag{3.3.1}$$

Here p_x , p_y , and p_z are the components of the stress vector along the boundary A.

The integral on the left-hand side and the first integral on the right-hand side are extended over the whole volume V of the body and the second integral on the right-hand side along the boundary A.

The equation is an identity if the statical conditions are fulfilled and is then valid for any displacement field satisfying the geometrical conditions. If, on the contrary, the equation is known to be valid for any displacement field, the statical conditions are fulfilled.

3.4 CONSTITUTIVE EQUATIONS

3.4.1 Plastic Strains in Coulomb Materials

In this chapter concrete is identified as a rigid-plastic Coulomb material or a rigid-plastic, modified Coulomb material. As mentioned previously, a *Coulomb material* is a material satisfying the failure condition of the frictional hypothesis (2.1.10),

$$k\sigma_1 - \sigma_3 = f_c \tag{3.4.1}$$

where σ_1 is the greatest of the principal stresses and σ_3 the smallest. The parameter k is defined by formula (2.1.6). We recall the alternative expressions (2.1.25) for later use:

$$k = \left(\frac{1+\sin\varphi}{\cos\varphi}\right)^2 = \left(\frac{\cos\varphi}{1-\sin\varphi}\right)^2 = \frac{1+\sin\varphi}{1-\sin\varphi} = \tan^2\left(\frac{\pi}{4}+\frac{\varphi}{2}\right) \tag{3.4.2}$$

If σ_1, σ_2, and σ_3 can mean any of the principal stresses, we have to work with the six yield conditions (see Fig. 3.4.1)

$$\begin{aligned}
&1.\quad k\sigma_1 - \sigma_3 - f_c = 0\ , \qquad \sigma_1 \ge \sigma_2 \ge \sigma_3\\
&2.\quad k\sigma_3 - \sigma_1 - f_c = 0\ , \qquad \sigma_3 \ge \sigma_2 \ge \sigma_1\\
&3.\quad k\sigma_1 - \sigma_2 - f_c = 0\ , \qquad \sigma_1 \ge \sigma_3 \ge \sigma_2
\end{aligned} \tag{3.4.3}$$

$$4. \quad k\sigma_2 - \sigma_1 - f_c = 0 \, , \qquad \sigma_2 \geq \sigma_3 \geq \sigma_1$$

$$5. \quad k\sigma_2 - \sigma_3 - f_c = 0 \, , \qquad \sigma_2 \geq \sigma_1 \geq \sigma_3$$

$$6. \quad k\sigma_3 - \sigma_2 - f_c = 0 \, , \qquad \sigma_3 \geq \sigma_1 \geq \sigma_2$$

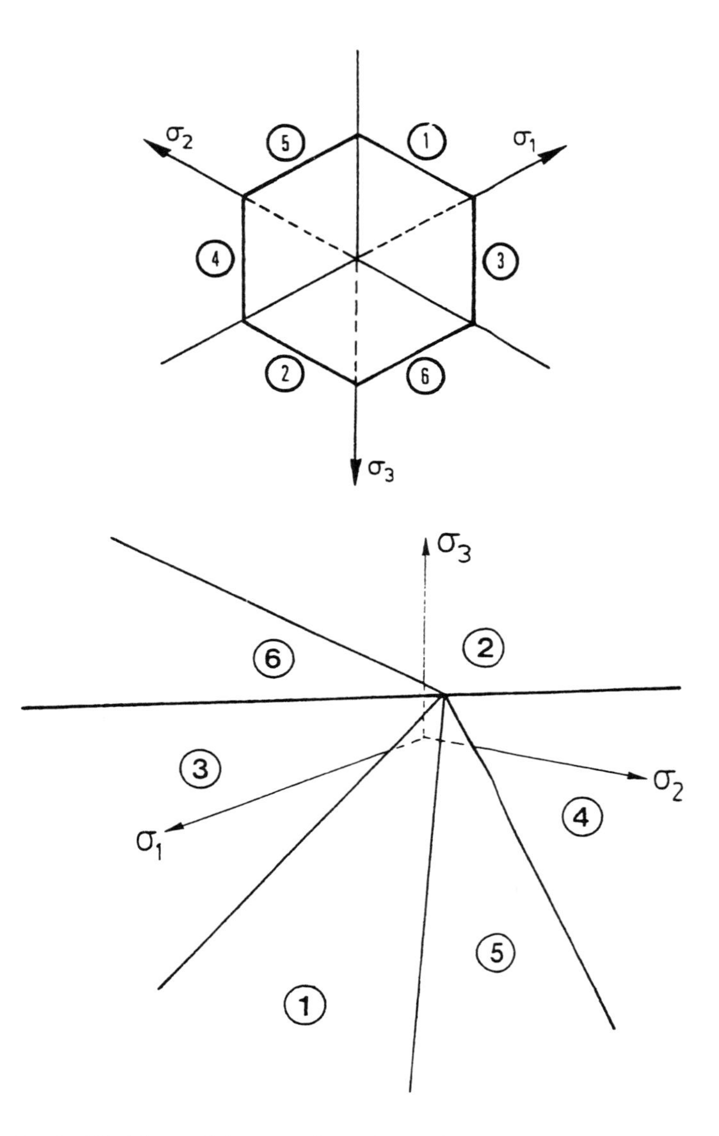

Figure 3.4.1 Yield conditions in the principal stress space for a Coulomb material.

The intersections of the individual planes with the coordinate planes are shown in Fig. 3.4.1, where the lines are marked with the number of relevant surfaces. The appearance of the surface in the σ_1, σ_2, σ_3-space is also illustrated in the figure.

According to (1.1.9.), we have along the various planes the following strains:

$$\begin{aligned}
&1.\quad \varepsilon_1 = \lambda k, \quad \varepsilon_2 = 0, \quad \varepsilon_3 = -\lambda \\
&2.\quad \varepsilon_1 = -\lambda, \quad \varepsilon_2 = 0, \quad \varepsilon_3 = \lambda k \\
&3.\quad \varepsilon_1 = \lambda k, \quad \varepsilon_2 = -\lambda, \quad \varepsilon_3 = 0 \\
&4.\quad \varepsilon_1 = -\lambda, \quad \varepsilon_2 = \lambda k, \quad \varepsilon_3 = 0 \qquad \lambda \geq 0 \\
&5.\quad \varepsilon_1 = 0, \quad \varepsilon_2 = \lambda k, \quad \varepsilon_3 = -\lambda \\
&6.\quad \varepsilon_1 = 0, \quad \varepsilon_2 = -\lambda, \quad \varepsilon_3 = \lambda k
\end{aligned} \tag{3.4.4}$$

Notice that

$$\varepsilon_1 + \varepsilon_2 + \varepsilon_3 = \lambda (k-1) \tag{3.4.5}$$

As $k \geq 1$, it is seen that a Coulomb material dilatates when yielding except in the special case $k = 1$.

The yield surface has edges where the following planes intersect:

1/5 4/5 2/4 2/6 3/6 1/3

The strains along an edge or an apex are found as positive linear combinations of the strains belonging to the adjacent planes.

If we consider, for example, the edge along 1 and 5, 1 has the normal

$$\varepsilon_1 = \lambda_1 k, \quad \varepsilon_2 = 0, \quad \varepsilon_3 = -\lambda_1, \quad \lambda_1 \geq 0 \tag{3.4.6}$$

and 5 the normal

$$\varepsilon_1 = 0, \quad \varepsilon_2 = \lambda_2 k, \quad \varepsilon_3 = -\lambda_2, \quad \lambda_2 \geq 0 \tag{3.4.7}$$

Along edge 1/5 we therefore have

$$\begin{aligned}
\varepsilon_1 &= \lambda_1 k \\
\varepsilon_2 &= \lambda_2 k \\
\varepsilon_3 &= -(\lambda_1 + \lambda_2) \quad (\lambda_1 \geq 0, \quad \lambda_2 \geq 0)
\end{aligned} \tag{3.4.8}$$

Analogously, along the other edges we get

$$
\begin{array}{llll}
4/5: & \varepsilon_1 = -\lambda_1 & \varepsilon_2 = (\lambda_1 + \lambda_2)k & \varepsilon_3 = -\lambda_2 \\
2/4: & \varepsilon_1 = -(\lambda_1 + \lambda_2) & \varepsilon_2 = \lambda_2 k & \varepsilon_3 = \lambda_1 k \\
2/6: & \varepsilon_1 = -\lambda_1 & \varepsilon_2 = -\lambda_2 & \varepsilon_3 = (\lambda_1 + \lambda_2)k \\
3/6: & \varepsilon_1 = \lambda_1 k & \varepsilon_2 = -(\lambda_1 + \lambda_2) & \varepsilon_3 = \lambda_2 k \\
1/3: & \varepsilon_1 = (\lambda_1 + \lambda_2)k & \varepsilon_2 = -\lambda_2 & \varepsilon_3 = -\lambda_1
\end{array}
\tag{3.4.9}
$$

The yield surface has an apex in the point corresponding to

$$\sigma_1 = \sigma_2 = \sigma_3 = \frac{f_c}{k-1} = c \cot \varphi \tag{3.4.10}$$

since, according to (2.1.9) we have $f_c = 2c\sqrt{k}$. Here the strain vector may be an arbitrary positive linear combination of all the strain vectors determined according to (3.4.6) and similar equations. We get

$$
\begin{aligned}
\varepsilon_1 &= (\lambda_1 + \lambda_3)k - (\lambda_2 + \lambda_4) \\
\varepsilon_2 &= (\lambda_4 + \lambda_5)k - (\lambda_3 + \lambda_6) \qquad (\lambda_i \geq 0) \\
\varepsilon_3 &= (\lambda_2 + \lambda_6)k - (\lambda_1 + \lambda_5)
\end{aligned}
\tag{3.4.11}
$$

By going through the cases dealt with we find, disregarding the apex, that the permissible combinations of the strains satisfy the condition

$$\frac{\Sigma \varepsilon^+}{\Sigma |\varepsilon^-|} = k \tag{3.4.12}$$

where $\Sigma \varepsilon^+$ is the sum of the positive principal strains, and $\Sigma |\varepsilon^-|$ is the sum of the numerical values of the negative principal strains.

In the apex we get

$$\frac{\Sigma \varepsilon^+}{\Sigma |\varepsilon^-|} \geq k \tag{3.4.13}$$

Strain fields where

$$\frac{\Sigma \varepsilon^+}{\Sigma |\varepsilon^-|} < k \tag{3.4.14}$$

cannot occur. It is seen that the change of volume (3.4.5) can be written

$$\varepsilon_1 + \varepsilon_2 + \varepsilon_3 = \Sigma \varepsilon^+ - \Sigma |\varepsilon^-| \tag{3.4.15}$$

in all cases.

If we disregard the apex, we get from (3.4.2) and (3.4.12)

$$\varepsilon_1 + \varepsilon_2 + \varepsilon_3 = \Sigma|\varepsilon^-|(k-1) = \Sigma|\varepsilon^-|\frac{2\sin\varphi}{1-\sin\varphi} \tag{3.4.16}$$

By this formula we can calculate the angle of friction for a Coulomb material from strain measurements. We find

$$\sin\varphi = \frac{\varepsilon_1 + \varepsilon_2 + \varepsilon_3}{|\varepsilon_1| + |\varepsilon_2| + |\varepsilon_3|} \tag{3.4.17}$$

3.4.2 Dissipation Formulas for Coulomb Materials

Now the stresses belonging to a given strain field can be determined. There is no unique correspondence between strains and stresses.

The dissipation per unit volume W corresponding to a given strain field is for stresses along plane 1:

$$W = \sigma_1\varepsilon_1 + \sigma_2\varepsilon_2 + \sigma_3\varepsilon_3 = \sigma_1\lambda k - \sigma_3\lambda = \lambda(k\sigma_1 - \sigma_3) = \lambda f_c \tag{3.4.18}$$

The same expression is found along the other planes.

Along the edge 1/5 we have

$$k\sigma_1 - \sigma_3 - f_c = k\sigma_2 - \sigma_3 - f_c = 0 \tag{3.4.19}$$

The dissipation is

$$W = \sigma_1\lambda_1 k + \sigma_2\lambda_2 k - \sigma_3(\lambda_1 + \lambda_2) = (\lambda_1 + \lambda_2)f_c \tag{3.4.20}$$

The same expression is obtained along the other edges.

At the apex we have

$$\begin{aligned} W = {} & \sigma_1[(\lambda_1 + \lambda_3)k - (\lambda_2 + \lambda_4)] \\ & + \sigma_2[(\lambda_4 + \lambda_5)k - (\lambda_3 + \lambda_6)] \\ & + \sigma_3[(\lambda_2 + \lambda_6)k - (\lambda_1 + \lambda_5)] \\ = {} & (\lambda_1 + \lambda_2 + \lambda_3 + \lambda_4 + \lambda_5 + \lambda_6)f_c \end{aligned} \tag{3.4.21}$$

using the condition at the apex (3.4.10). It is seen that for $k > 1$, in all cases we may write

$$W = \frac{1}{k-1}(\varepsilon_1 + \varepsilon_2 + \varepsilon_3)f_c = c\cot\varphi(\varepsilon_1 + \varepsilon_2 + \varepsilon_3) \tag{3.4.22}$$

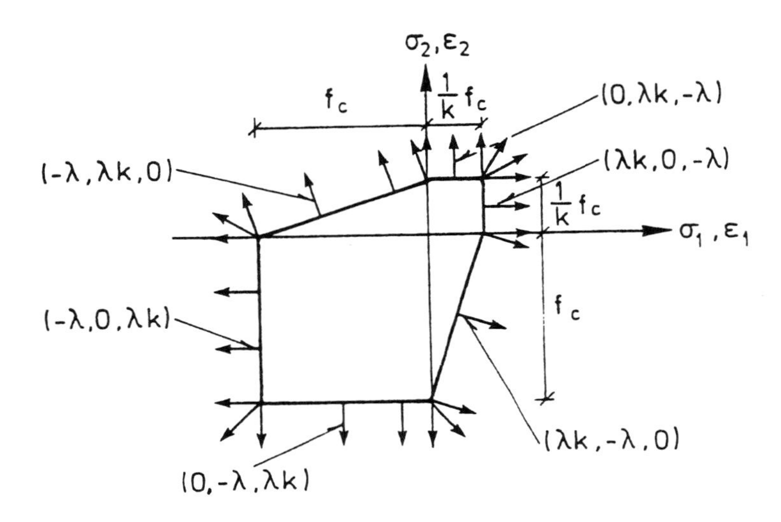

Figure 3.4.2 Yield condition and flow rule for a Coulomb material in plane stress.

By using (3.4.12) we find, when the apex is disregarded, that W can be written

$$W = \frac{f_c}{k} \Sigma \varepsilon^+ \tag{3.4.23}$$

or

$$W = f_c \Sigma |\varepsilon^-| \tag{3.4.24}$$

By plane stress fields we previously found the yield condition shown in Fig. 3.4.2. The strains that might occur are also illustrated, by indicating the coordinates of the strain vector in the various areas.

In plane strain, for example $\varepsilon_3 = 0$, it is seen that we are at either plane 3 or plane 4, where the greatest principal strain is λk and the smallest principal strain is $-\lambda$. From the yield conditions it is seen that σ_3 is lying between the principal stresses in the plane. The yield condition is shown in Fig. 3.4.3. Along planes 3 and 4 we thus have

$$\frac{\Sigma \varepsilon^+}{\Sigma |\varepsilon^-|} = \frac{\varepsilon^+}{|\varepsilon^-|} = k \tag{3.4.25}$$

in conformity with (3.4.12). In the apex we have

$$\frac{\varepsilon^{+}}{|\varepsilon^{-}|} \geq k \tag{3.4.26}$$

According to (3.4.22), the dissipation will be

$$W = \frac{1}{k-1}(\varepsilon_1 + \varepsilon_2) f_c = c\cot\varphi\,(\varepsilon_1 + \varepsilon_2) \tag{3.4.27}$$

Disregarding the apex, the dissipation can, according to (3.4.23), be written

$$W = \frac{f_c}{k}\varepsilon^{+} \tag{3.4.28}$$

or according to (3.4.24),

$$W = f_c|\varepsilon^{-}| \tag{3.4.29}$$

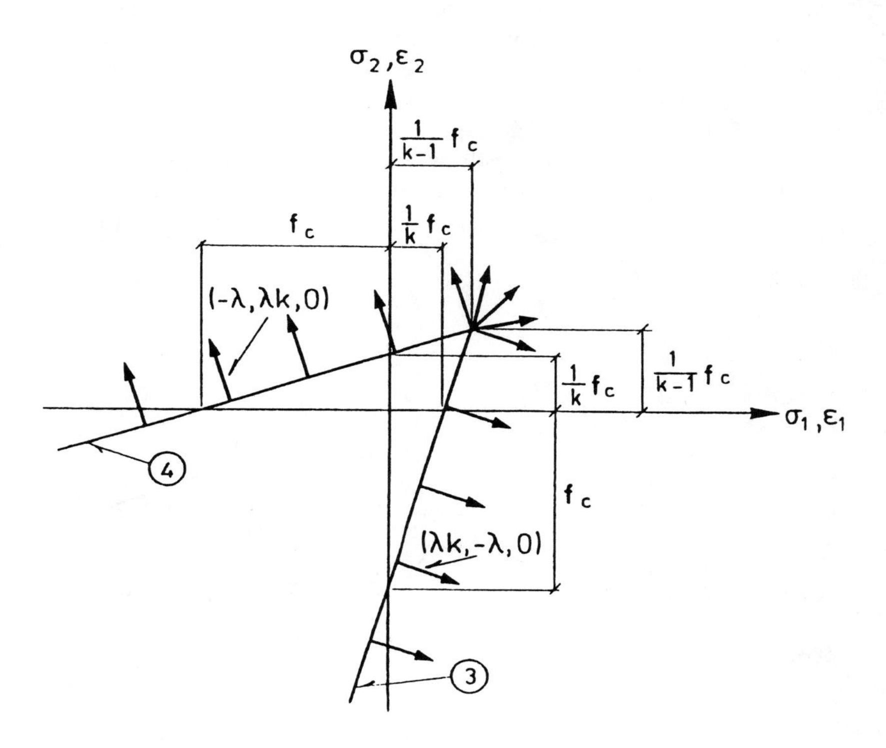

Figure 3.4.3 Yield condition and flow rule for a Coulomb material in plane strain.

The flow rule and so on for a Coulomb material can also be derived by referring to a coordinate system which has as one of the coordinate planes one of the sections in which Coulomb's failure condition is satisfied. Since this is instructive in several ways, this alternative procedure will be illustrated briefly.

As previously, let the stresses in the failure sections be denoted σ and τ. The plane determined by σ and τ is, of course, coinciding with the principal section with the intermediate principal stress σ_2. We assume that $\sigma_1 > \sigma_2 > \sigma_3$. In this principal section we insert an n, t-coordinate system as shown in Fig. 3.4.4. Perpendicular to the n, t-plane we have the second principal direction.

In this coordinate system the dissipation per unit volume is

$$W = \sigma\varepsilon + \tau\gamma + \sigma'\varepsilon' + \sigma_2\varepsilon_2 \tag{3.4.30}$$

Here ε and γ are the longitudinal strain in the n-direction and the change of angle between the n- and t-directions, respectively. The meaning of σ' and the corresponding strain ε' is seen from Fig. 3.4.4. By using von Mises's principle of maximum plastic work (i.e., by requiring that $\delta W = 0$ when the stresses vary on the yield surface), we find the natural result that the vector (ε, γ) in the σ, τ-coordinate system is a normal to the yield condition, as shown in Fig. 3.4.4, and that ε' and ε_2 are both zero. Thus the strain field is a plane strain field in the n, t-plane, in conformity with the expressions (3.4.4), which apply when the principal stresses differ. If δ is the length of the strain vector (ε, γ), it is seen that

$$\varepsilon = \delta \sin\varphi \ , \quad \gamma = \delta\cos\varphi \tag{3.4.31}$$

From this the significant result

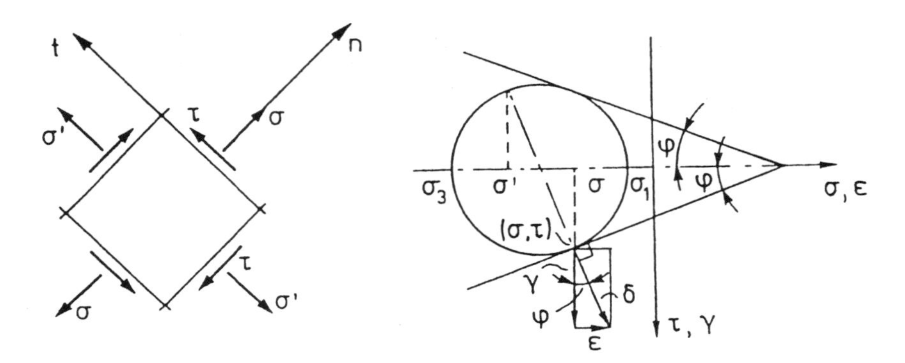

Figure 3.4.4 The flow rule for a Coulomb material developed in a coordinate system having a coordinate plane parallel to a failure section.

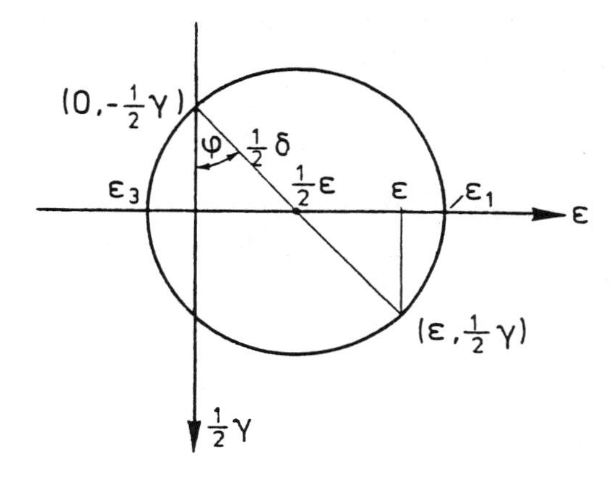

Figure 3.4.5 Mohr's circle for a strain field in a coordinate system having a coordinate plane parallel to a failure section.

$$\varepsilon = \gamma \tan \varphi = \gamma \mu \tag{3.4.32}$$

is derived.

Mohr's circle for the strains in the n, t-plane can then easily be constructed, as shown in Fig. 3.4.5. The quantity δ and the angle of friction φ have simple geometric meanings, as shown in the figure. The principal strains are

$$\begin{aligned} \varepsilon_1 &= \frac{1}{2}\varepsilon + \frac{1}{2}\delta = \frac{1}{2}\delta(1 + \sin\varphi) \\ \varepsilon_3 &= \frac{1}{2}\varepsilon - \frac{1}{2}\delta = -\frac{1}{2}\delta(1 - \sin\varphi) \end{aligned} \tag{3.4.33}$$

It is seen that the ratio between the principal strains is [see (3.4.2)]

$$\frac{\varepsilon_1}{\varepsilon_3} = -\frac{1 + \sin\varphi}{1 - \sin\varphi} = -k \tag{3.4.34}$$

in conformity with (3.4.4). From (3.4.34) we get

$$\sin\varphi = \frac{\varepsilon_1 + \varepsilon_3}{\varepsilon_1 - \varepsilon_3} \tag{3.4.35}$$

complying with (3.4.17), as for the plane strain field considered we have $\varepsilon_2 = 0$ and $|\varepsilon_3| = -\varepsilon_3$.

3.4.3 Plastic Strains in Modified Coulomb Materials

As mentioned previously a *modified Coulomb material* is a material which, apart from the failure condition (3.4.1), satisfies the condition (separation failure criterion)

$$\sigma_1 - f_t = 0 \tag{3.4.36}$$

Here σ_1 is the greatest principal stress and f_t is the tensile strength. Of course, it must be assumed that

$$f_t \leq \frac{1}{k} f_c \tag{3.4.37}$$

[cf. (2.1.12)].

If σ_1, σ_2, and σ_3 can mean any of the principal stresses we have to use, apart from the six conditions (3.4.3),

$$\begin{array}{llll} 7. & \sigma_1 - f_t = 0\,, & \sigma_1 \geq \sigma_2\,, & \sigma_1 \geq \sigma_3 \\ 8. & \sigma_2 - f_t = 0\,, & \sigma_2 \geq \sigma_1 & \sigma_2 \geq \sigma_3 \\ 9. & \sigma_3 - f_t = 0\,, & \sigma_3 \geq \sigma_1\,, & \sigma_3 \geq \sigma_2 \end{array} \tag{3.4.38}$$

Along these surfaces we get the strains

$$\begin{array}{llll} 7. & \varepsilon_1 = \lambda\,, & \varepsilon_2 = 0\,, & \varepsilon_3 = 0 \\ 8. & \varepsilon_1 = 0\,, & \varepsilon_2 = \lambda\,, & \varepsilon_3 = 0 \\ 9. & \varepsilon_1 = 0\,, & \varepsilon_2 = 0\,, & \varepsilon_3 = \lambda \end{array} \tag{3.4.39}$$

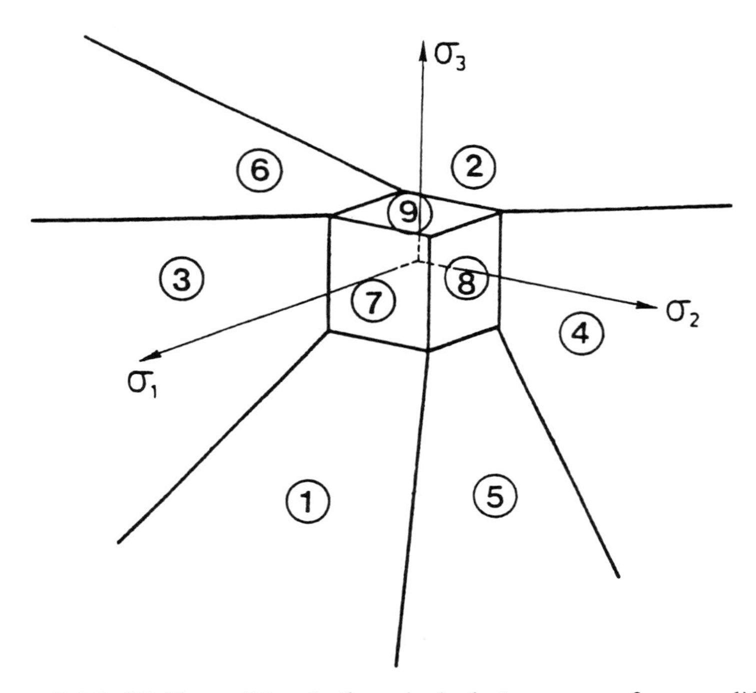

Figure 3.4.6 Yield condition in the principal stress space for a modified Coulomb material.

The yield surface thus appearing has, in addition to the edges dealt with in the preceding section, the following edges:

1/7 3/7 6/9 2/9 4/8 5/8 8/7 7/9 8/9

and apexes where the following yield surfaces intersect:

4/5/8 7/8/5/1 1/3/7 3/7/6/9 2/6/9 8/9/2/4 7/8/9

The appearance of the yield surface in the σ_1, σ_2, σ_3-space is illustrated in Fig. 3.4.6. As before, the strains along the edges and the apexes are found as positive linear combinations of the strains belonging to the adjacent planes.

One finds that if

$$\frac{\Sigma\varepsilon^+}{\Sigma|\varepsilon^-|} = k \tag{3.4.40}$$

the corresponding point lies on one of surfaces 1 to 6 or one of the edges in between.

If

$$\frac{\Sigma\varepsilon^+}{\Sigma|\varepsilon^-|} > k \tag{3.4.41}$$

the corresponding point lies on one of the surfaces, edges, or apexes of this section.

In uniaxial compression it appears from the above formulas that the ratio of the sum of incremental transversal strains and the absolute value of the incremental strain in the force direction (axial strain) is k both for a Coulomb material and a modified Coulomb material. This is not in particularly good agreement with measured values.

Measurements are, of course, difficult due to the confinement by the loading platen. This influence may be reduced by applying the load through a large number of slender steel brushes or by using layers of teflon.

In Fig. 3.4.7 some measurements of this kind reported by van Mier [86.11] have been plotted. The test specimens were cubes. In one case the force direction was in the casting direction and in another case the force direction was perpendicular to the casting direction.

In the last mentioned case the scatter is considerable. In both cases it appears that the abovementioned ratio of incremental strains in the softening range generally is considerably larger than k which is 4 or less (cf. Section 2.1). Even the individual transversal strain increments far exceeds 4 times the absolute value of axial strain increment as illustrated in the figure.

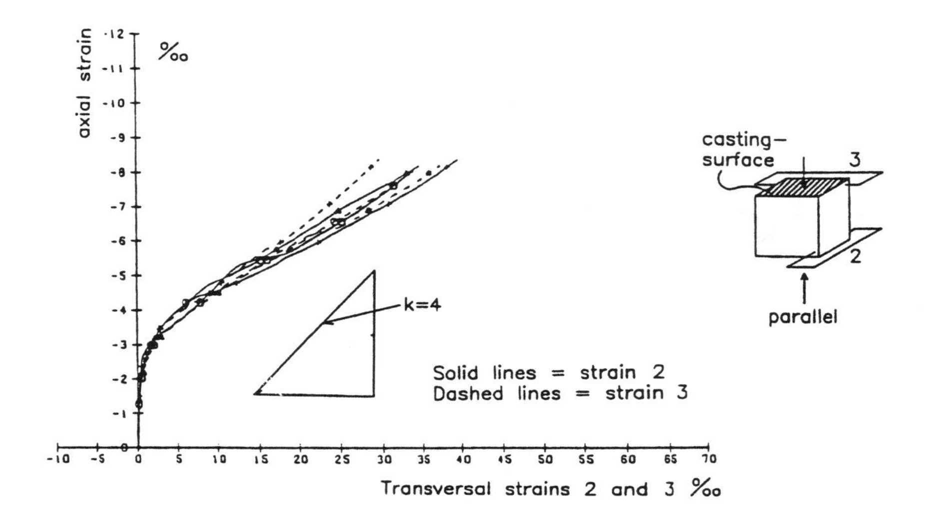

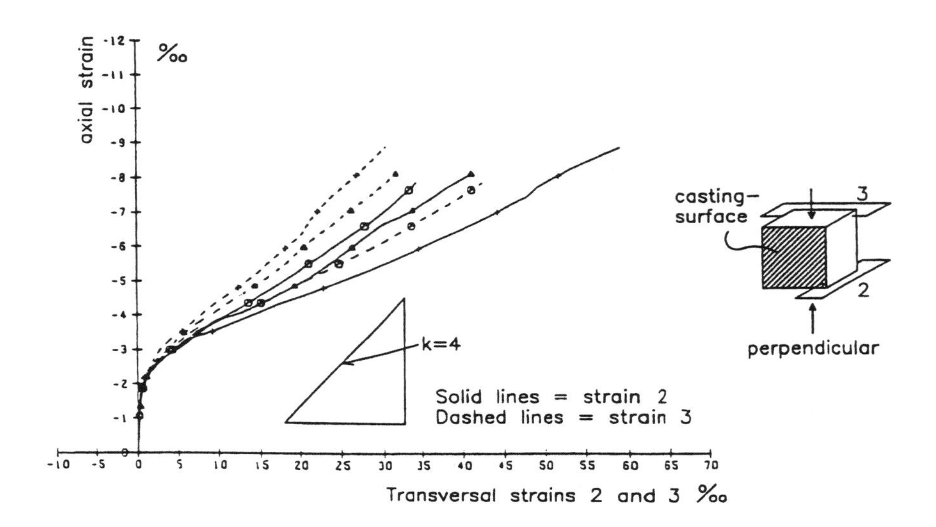

Figure 3.4.7 Axial and transversal strains in uniaxial compression [86.11].

According to the model of Jin-Ping Zhang (cf. Section 2.1), this discrepancy may be traced back to the fact that the yield lines in reality are steeper than the yield lines calculated by plastic theory for a Coulomb material.

At the peak value of the compression stress the agreement might be better.

Thus, we may conclude that the plastic strain increments determined on the basis of a Coulomb type yield condition may be far from reality. The justification of using such a simplified material model lies solely in the practical usefulness of the solutions obtained for the load-carrying capacity. The predictions regarding the plastic strain increments may be poor.

3.4.4 Dissipation Formulas for Modified Coulomb Materials

If (3.4.40) is satisfied, the dissipation is determined by (3.4.22) or one of the alternative expressions (3.4.23) or (3.4.24).

Along edge 1/7 we have

$$k\sigma_1 - \sigma_3 - f_c = \sigma_1 - f_t = 0 \tag{3.4.42}$$

and the dissipation is

$$W = \sigma_1(\lambda_1 k + \lambda_2) + \sigma_2 \cdot 0 + \sigma_3(-\lambda_1) = \lambda_1 f_c + \lambda_2 f_t \tag{3.4.43}$$

The same expression is found along edges 3/7, 6/9, 2/9, 4/8, and 5/8. Along edge 8/7 we have

$$\sigma_1 - f_t = \sigma_2 - f_t = 0 \tag{3.4.44}$$

The dissipation is

$$W = \sigma_1\lambda_1 + \sigma_2\lambda_2 = (\lambda_1 + \lambda_2) f_t \tag{3.4.45}$$

The same expression is obtained along edges 7/9 and 8/9.

In apex 4/5/8 we have

$$k\sigma_2 - \sigma_1 - f_c = k\sigma_2 - \sigma_3 - f_c = \sigma_2 - f_t = 0 \tag{3.4.46}$$

The dissipation is

$$W = \sigma_1(-\lambda_1) + \sigma_2[(\lambda_1 + \lambda_2)k + \lambda_3] + \sigma_3(-\lambda_2)$$
$$= (\lambda_1 + \lambda_2) f_c + \lambda_3 f_t \tag{3.4.47}$$

The same expression is obtained in apexes 1/3/7 and 2/6/9.

In apex 7/8/5/1 we have

$$k\sigma_1 - \sigma_3 - f_c = k\sigma_2 - \sigma_3 - f_c = \sigma_1 - f_t = \sigma_2 - f_t = 0 \tag{3.4.48}$$

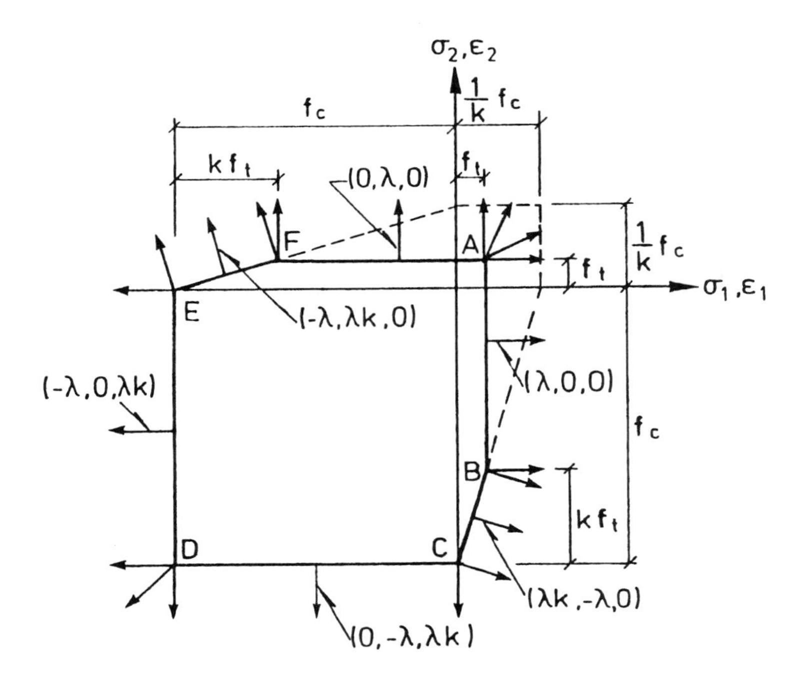

Figure 3.4.8 Yield condition and flow rule for a modified Coulomb material in plane stress.

The dissipation is

$$
\begin{aligned}
W &= \sigma_1(\lambda_1 k + \lambda_3) + \sigma_2(\lambda_2 k + \lambda_4) + \sigma_3[-(\lambda_1 + \lambda_2)] \\
&= (\lambda_1 + \lambda_2) f_c + (\lambda_3 + \lambda_4) f_t
\end{aligned}
\tag{3.4.49}
$$

The same expression is obtained in apexes 3/7/6/9 and 8/9/2/4.

Finally, in apex 7/8/9 we have

$$\sigma_1 - f_t = \sigma_2 - f_t = \sigma_3 - f_t = 0 \tag{3.4.50}$$

and the dissipation is

$$W = \sigma_1\lambda_1 + \sigma_2\lambda_2 + \sigma_3\lambda_3 = (\lambda_1 + \lambda_2 + \lambda_3) f_t \tag{3.4.51}$$

By studying the various cases it is seen that in a modified Coulomb material the dissipation is determined as follows. When

$$\frac{\Sigma\varepsilon^+}{\Sigma|\varepsilon^-|} = k \tag{3.4.52}$$

we have [corresponding to (3.4.29)]

$$W = f_c \Sigma |\varepsilon^-| \tag{3.4.53}$$

When

$$\frac{\Sigma \varepsilon^+}{\Sigma |\varepsilon^-|} > k \tag{3.4.54}$$

we have

$$W = f_c \Sigma |\varepsilon^-| + f_t (\Sigma \varepsilon^+ - k \Sigma |\varepsilon^-|) \tag{3.4.55}$$

It is seen that (3.4.55) applies in all cases, as the last term in (3.4.55) vanishes when (3.4.52) is satisfied. Strain fields where

$$\frac{\Sigma \varepsilon^+}{\Sigma |\varepsilon^-|} < k \tag{3.4.56}$$

cannot occur. In the important special case $f_t = 0$ we have simply

$$W = f_c \Sigma |\varepsilon^-| \tag{3.4.57}$$

The yield condition for plane stress fields has been calculated in Section 2.1.2. It is illustrated in Fig. 3.4.8, where the strain vectors in the various areas are also shown.

It appears that in the general case of plane stress the formula for W expressed as a function of ε_1 and ε_2 is rather complicated and therefore not given. In the special case $f_t = 0$ we find that

$$W = \frac{1}{2} f_c [|\varepsilon_1| + |\varepsilon_2| - (\varepsilon_1 + \varepsilon_2)] \tag{3.4.58}$$

At plane strain fields, where for example $\varepsilon_3 = 0$, it is seen that the strain vector must be placed on plane 3, 4, 7, or 8. The yield condition is shown in Fig. 3.4.9.

Since there can only be one negative principal strain, the expressions for the dissipation W are reduced as follows. When

$$\frac{\Sigma \varepsilon^+}{|\varepsilon^-|} = k \tag{3.4.59}$$

we have

$$W = f_c |\varepsilon^-| \tag{3.4.60}$$

When

$$\frac{\Sigma \varepsilon^+}{|\varepsilon^-|} > k \tag{3.4.61}$$

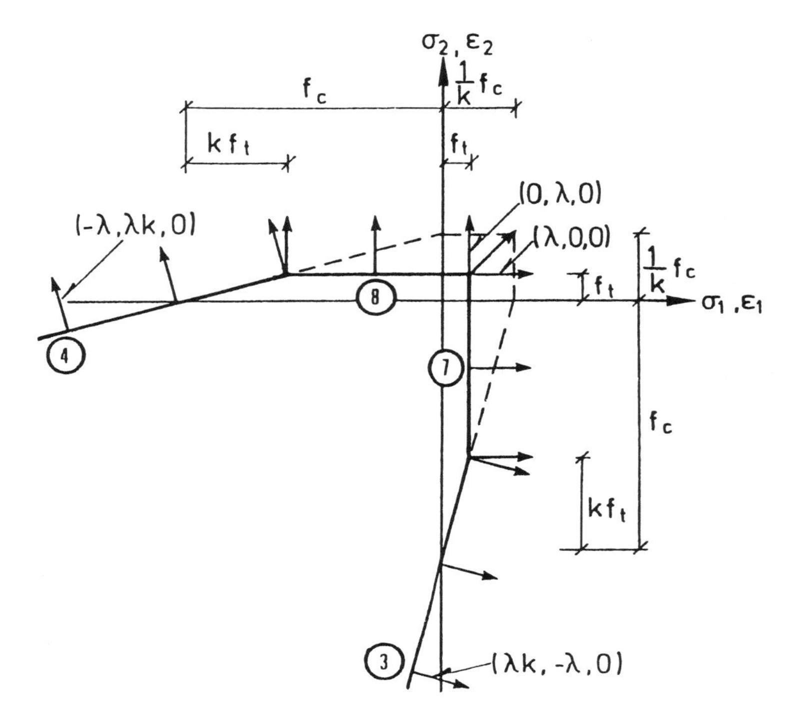

Figure 3.4.9 Yield condition and flow rule for a modified Coulomb material in plane strain.

we have

$$W = f_c\,|\varepsilon^-| + f_t(\Sigma\varepsilon^+ - k\,|\varepsilon^-|) \tag{3.4.62}$$

It is seen that (3.4.62) can be used in all cases, as the last term in (3.4.62) vanishes when (3.4.59) is satisfied. General dissipation formulas for Coulomb materials and modified Coulomb materials were derived by Nielsen et al. [78.14].

3.4.5 Planes and Lines of Discontinuity

Strains in a plane of discontinuity. In the plastic theory it is necessary, as mentioned previously, to operate with *planes and lines of discontinuity* along which are jumps in the displacements. Let us consider a volume bounded by two parallel planes with the distance δ. Let us assume that there is a plane, homogeneous strain field in the volume and that parts I and II outside the volume move as rigid bodies in a n, t-plane. The coordinate system shown in Fig. 3.4.10 is inserted. Part I is assumed to be not moving, and part II is assumed to have the displacements u_n and u_t. Therefore, in the volume we have the strains

$$\varepsilon_n = \frac{u_n}{\delta}\ , \qquad \varepsilon_t = 0\ , \qquad \gamma_{nt} = \frac{u_t}{\delta} \tag{3.4.63}$$

The strains can also be expressed by the numerical value of the displacement u of part II and the angle α which the displacement vector forms with the t-axis (see Fig. 3.4.10). We find that

$$u_n = u \sin\alpha\ , \qquad u_t = u\cos\alpha \tag{3.4.64}$$

and thus

$$\varepsilon_n = \frac{u\sin\alpha}{\delta}\ , \qquad \varepsilon_t = 0\ , \qquad \gamma_{nt} = \frac{u\cos\alpha}{\delta} \tag{3.4.65}$$

The principal strains are

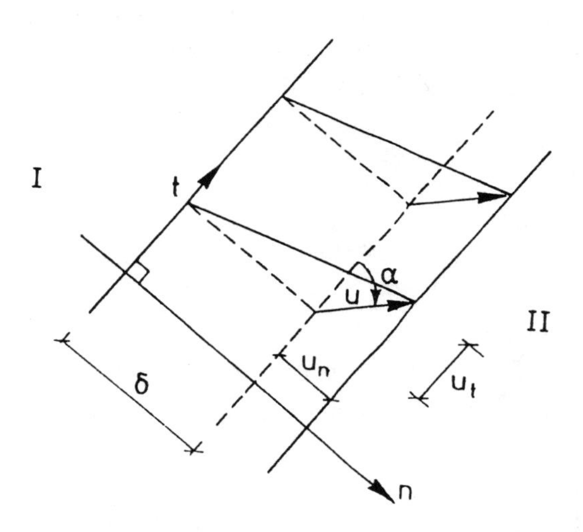

Figure 3.4.10 Plane of displacement discontinuity or yield line defined by a narrow strip with homogeneous strain.

$$\left.\begin{matrix}\varepsilon_1 \\ \varepsilon_2\end{matrix}\right\} = \frac{1}{2}\frac{u\sin\alpha}{\delta} \pm \frac{1}{2}\sqrt{\frac{u^2\sin^2\alpha}{\delta^2} + \frac{u^2\cos^2\alpha}{\delta^2}} \tag{3.4.66}$$

$$= \frac{1}{2}\frac{u}{\delta}(\sin\alpha \pm 1)$$

The maximum principal strain in the plane is

$$\varepsilon_{max} = \varepsilon^{+} = \frac{1}{2}\frac{u}{\delta}(\sin\alpha + 1) \tag{3.4.67}$$

and the smallest principal strain is

$$\varepsilon_{min} = \varepsilon^{-} = \frac{1}{2}\frac{u}{\delta}(\sin\alpha - 1) \tag{3.4.68}$$

The first principal direction of strain is found to bisect the angle between the displacement vector and the n-axis.

Plane strain. First we consider the case of plane strain. For a Coulomb material, (3.4.12) applies when we disregard the apex. Using (3.4.67) and (3.4.68), we find that

$$\sin\alpha + 1 = k(1 - \sin\alpha) \tag{3.4.69}$$

or

$$\sin\alpha = \frac{k-1}{k+1} \tag{3.4.70}$$

Thus it is seen that $0 \le \alpha \le \pi$.

Using (3.4.2), (3.4.70) is seen to be equivalent with the condition

$$\tan\alpha = \pm\tan\varphi \tag{3.4.71}$$

or

$$\alpha = \begin{cases}\varphi \\ \pi - \varphi\end{cases} \tag{3.4.72}$$

Thus by (3.4.23) and (3.4.24),

$$W = f_c \cdot \frac{1}{2}\frac{u}{\delta}(1 - \sin\varphi) = \frac{f_c}{k} \cdot \frac{1}{2}\frac{u}{\delta}(1 + \sin\varphi) \tag{3.4.73}$$

The dissipation per unit length W_ℓ measured in the direction of the t-axis is

$$W_\ell = Wb\delta \tag{3.4.74}$$

where b is the dimension of the body perpendicular to the n, t-plane.

By this we have

$$W_\ell = ub \cdot \frac{1}{2} f_c (1 - \sin\varphi)$$

$$= ub \cdot \frac{1}{2} \frac{f_c}{k} (1 + \sin\varphi) \tag{3.4.75}$$

The dissipation is thus independent of δ. Therefore, we need to deal only with the relative motion of the two rigid bodies, and thus for $\delta \to 0$ we get a *discontinuity plane* or, in the n, t-plane, a *discontinuity line* in the displacements. In the following a discontinuity line is also called a *yield line*. The formula for W_ℓ also applies to a curved surface of discontinuity and a curved yield line.

Introducing the cohesion $c = f_c/2\sqrt{k}$ [see (2.1.9)], we get the simple result

$$W = \frac{u}{\delta} c \cos\varphi \tag{3.4.76}$$

or

$$W_\ell = ubc \cos\varphi \tag{3.4.77}$$

When $\varphi < \alpha < \pi - \varphi$ we are at the apex. By the expression (3.4.22) for W we have

$$W_\ell = ubc \cot\varphi \sin\alpha \tag{3.4.78}$$

This result is, however, of minor practical interest.

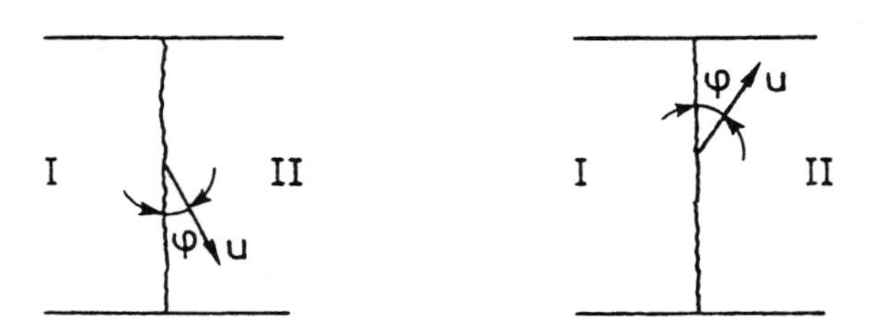

Figure 3.4.11 Displacement direction in a yield line in a Coulomb material in plane strain.

The foregoing result implies that the displacement vector forms the angle φ with the discontinuity line when the apex is disregarded (see Fig. 3.4.11).

For a modified Coulomb material we also find that

$$\varphi \le \alpha \le \pi - \varphi \tag{3.4.79}$$

The dissipation is found by means of (3.4.55) as

$$W_\ell = \frac{1}{2} f_c u b \left[1 - \sin\alpha + \frac{f_t}{f_c} \{ -(k-1) + (k+1)\sin\alpha \} \right] \tag{3.4.80}$$

Introducing

$$k - 1 = \frac{2\sin\varphi}{1 - \sin\varphi} \tag{3.4.81}$$

$$k + 1 = \frac{2}{1 - \sin\varphi} \tag{3.4.82}$$

(3.4.80) can be written

$$W_\ell = \frac{1}{2} f_c u b \left[1 - \sin\alpha + 2\frac{f_t}{f_c} \frac{\sin\alpha - \sin\varphi}{1 - \sin\varphi} \right] \tag{3.4.83}$$

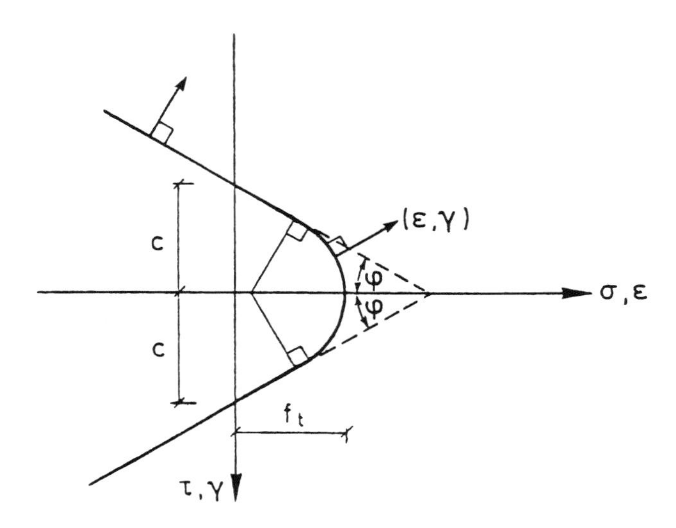

Figure 3.4.12 Dissipation in a yield line developed in a coordinate system having a coordinate plane parallel to a failure section.

Introducing the parameters

$$\ell = 1 - \frac{f_t}{f_c}(k-1) = 1 - 2\frac{f_t}{f_c}\frac{\sin\varphi}{1-\sin\varphi} \qquad (3.4.84)$$

$$m = 1 - \frac{f_t}{f_c}(k+1) = 1 - 2\frac{f_t}{f_c}\frac{1}{1-\sin\varphi} \qquad (3.4.85)$$

the brief version

$$W_\ell = \tfrac{1}{2} f_c u b(\ell - m\sin\alpha) \qquad (3.4.86)$$

is obtained.

The expressions above for dissipation in a discontinuity line can be found in a different way, and even more simply, by using the yield condition in a σ, τ-coordinate system as a basis and using the normality condition in this system (cf. Fig. 3.4.12). It is left to the reader as an exercise to show this for some of the cases described above.

Plane stress. We now consider plane stress occurring in thin disks. The disk thickness is denoted as b in the following. For a modified Coulomb material the yield condition in plane stress was shown in Fig. 3.4.8. The areas not covered by the formulas for plane strain are *CD* and *ED*. Thus it is seen that in (3.4.55) we should insert

$$\Sigma|\varepsilon^-| = \frac{1}{2}\frac{u}{\delta}(1-\sin\alpha) \qquad \text{for} \begin{cases} \alpha \le \varphi \\ \alpha \ge \pi - \varphi \end{cases} \qquad (3.4.87)$$

In this α-interval we then have

$$W_\ell = \tfrac{1}{2} f_c u b(1-\sin\alpha) \qquad (3.4.88)$$

For a modified Coulomb material the following formulas apply:

$$W_\ell = \frac{1}{2} f_c ub\left[1 - \sin\alpha + \frac{f_t}{f_c}\{-(k-1)+(k+1)\sin\alpha\}\right] \quad \text{for } \varphi \le \alpha \le \pi - \varphi \qquad (3.4.89)$$

$$W_\ell = \tfrac{1}{2} f_c ub(1-\sin\alpha) \quad \text{for } \alpha \le \varphi \text{ and } \alpha \ge \pi - \varphi \qquad (3.4.90)$$

Instead of (3.4.89) the alternative formulas (3.4.83) or (3.4.86) can be used.

The expression is strongly simplified if $f_t = 0$, as then, in the whole α-interval, we get

Dissipation Formulas

General formulas

Coulomb material

$$W = f_c \Sigma |\epsilon^-| \tag{3.4.24}$$

Modified Coulomb material

$$W = f_c \Sigma |\epsilon^-| + f_t(\Sigma \epsilon^+ - k \Sigma |\epsilon^-|) \tag{3.4.55}$$

$$(\Sigma \epsilon^+ / \Sigma |\epsilon^-| \geq k)$$

Yield lines

Plane strain, Coulomb material

$$W_\ell = ubc \cos\varphi \tag{3.4.77}$$

Plane strain, Modified Coulomb material

$$W_\ell = \tfrac{1}{2} f_c ub(\ell - m \sin\alpha) \tag{3.4.86}$$

ℓ and m are defined by equations (3.4.84) and (3.4.85)

Plane stress, Coulomb material

$$W_\ell = \frac{1}{2}\frac{f_c}{k} ub(1 + \sin\alpha) \qquad \varphi \leq \alpha \leq \pi - \varphi \tag{3.4.92}$$

$$W_\ell = \tfrac{1}{2} f_c ub(1 - \sin\alpha) \qquad \alpha \leq \varphi \qquad \alpha \geq \pi - \varphi \tag{3.4.93}$$

Plane stress, Modified Coulomb material

$$W_\ell = \tfrac{1}{2} f_c ub(\ell - m \sin\alpha) \qquad \varphi \leq \alpha \leq \pi - \varphi \tag{3.4.86}$$

$$W_\ell = \tfrac{1}{2} f_c ub(1 - \sin\alpha) \qquad \alpha \leq \varphi \qquad \alpha \geq \pi - \varphi \tag{3.4.90}$$

ℓ and m are defined by equations (3.4.84) and (3.4.85)

If $f_t = 0$

$$W_\ell = \tfrac{1}{2} f_c ub(1 - \sin\alpha) \tag{3.4.91}$$

Box 3.4.1 Summary of dissipation formulas.

$$W_\ell = \frac{1}{2} f_c u b (1 - \sin\alpha) \tag{3.4.91}$$

The Coulomb material can be treated as a special case of a modified Coulomb material, since in the expression above we only have to put $f_t = (1/k) f_c$, as seen from Figs. 3.4.2 and 3.4.8. We then get

$$W_\ell = \begin{cases} \frac{1}{2} \frac{f_c}{k} u b (1 + \sin\alpha) & \text{for } \varphi \le \alpha \le \pi - \varphi \quad (3.4.92) \\ \frac{1}{2} f_c u b (1 - \sin\alpha) & \text{for } \alpha \le \varphi \text{ and } \alpha \ge \pi - \varphi \quad (3.4.93) \end{cases}$$

As a control we have that these two expressions must be identical for $\alpha = \varphi$ and $\alpha = \pi - \varphi$, which from (3.4.2) is seen to be the case.

The most important dissipation formulas for Coulomb materials and modified Coulomb materials have been collected in Box 3.4.1.

Dissipation formulas for yield lines in Coulomb materials and modified Coulomb materials were developed by B. C. Jensen [76.1] and Nielsen et al. [78.14]. Yield lines with rotations as well as in-plane displacements, *generalized yield lines*, have been treated by Janas [63.9] and Janas and Sawczuk [66.5].

Due to the conditions concerning the strains perpendicular to the stress plane, there will also be a discontinuity in the displacement perpendicular to the disk at the transition from the rigid parts to the strip δ. We could then assume a narrow transition area that counteracts this discontinuity, and by letting $\delta \to 0$ we will still find the dissipation W_ℓ. It is more natural, however, to consider the two-dimensional theory as self-contained with its individual two-dimensional yield condition and so on. Thus it is seen that the procedure used leads to valid upper bound solutions, the derivation of which often is greatly simplified when using discontinuous displacement fields.

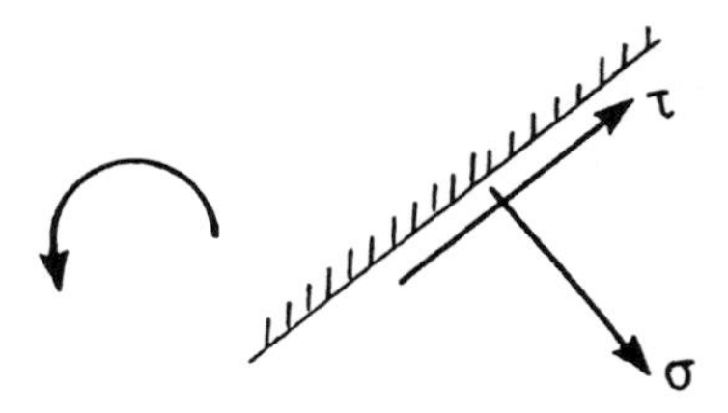

Figure 3.5.1 Positive sign of the shear stress.

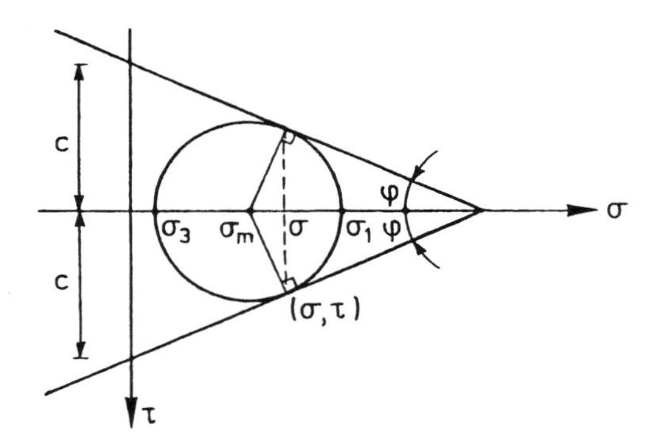

Figure 3.5.2 Mohr's circle for a stress field causing failure.

3.5 THE THEORY OF PLANE STRAIN FOR COULOMB MATERIALS

3.5.1 Introduction

In the following some plain strain solutions for Coulomb materials are treated. Such solutions are of great importance in soil mechanics, but it is believed that their usefulness in concrete and rock mechanics will also become evident in the future.

Concerning the sign convention for normal and shear stresses in an arbitrary section, the following should be noted. A positive direction of rotation, an orientation of the plane, normally counterclockwise, is introduced. In a section the normal stress σ is assumed positive as a tensile stress, and the shear stress τ is assumed positive when a tensile stress and shear stress, in this order, describe the positive direction of rotation of the plane (see Fig. 3.5.1). When the positive orientation is counterclockwise, the σ and τ axes in Mohr's circle have to be oriented as shown in Fig. 3.5.2.

3.5.2 The Stress Field

Mohr's circle for a stress field causing failure in a Coulomb material is shown in Fig. 3.5.2. The *failure sections*, the sections in which the shear stress and the normal stress satisfy the Coulomb failure criterion, form an acute angle $90° - \varphi$ with each other, where

φ is the angle of friction. The failure sections are often termed *slip lines* in the literature. The normal stress in the failure sections is σ, and the numerical value of the shear stress is τ. These magnitudes satisfy Coulomb's criterion (2.1.2),

$$\tau = c - \sigma \tan\varphi \tag{3.5.1}$$

where c is the cohesion.

The stress field is illustrated in Fig. 3.5.3. Note that the shear stress in the failure sections, forming the acute angle $90° - \varphi$, points off the mutual edge.

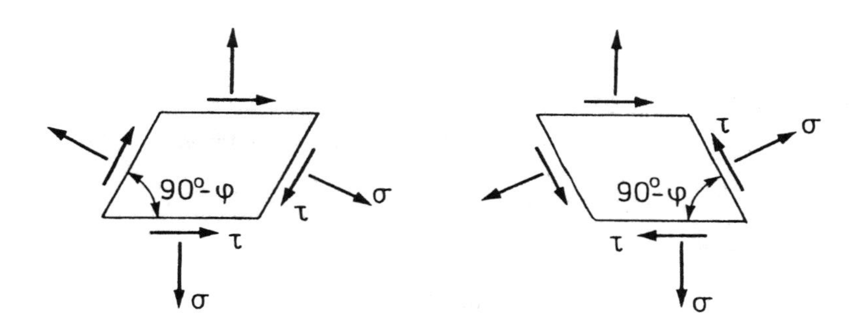

Figure 3.5.3 Failure sections with corresponding normal and shear stress.

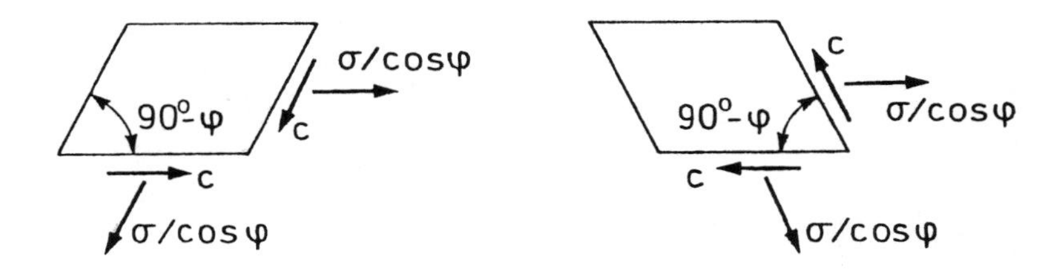

Figure 3.5.4 Stresses in failure sections resolved along the directions of the failure sections.

The normal stress σ_2 in a section parallel to the plane satisfies the condition $\sigma_3 \leq \sigma_2 \leq \sigma_1$, where σ_1 and σ_3 are the greatest and smallest principal stresses, respectively.

When we resolve the stresses in the failure sections along the directions of the failure sections, we obtain the results shown in Fig. 3.5.4. When these components are used, the component along the direction of the failure section is always equal to the cohesion c.

If we want to refer the stress field in the plane to a rectangular n, t-coordinate system, where, for example, the t-axis is identical with a failure section, we obtain the conditions shown in Fig. 3.5.5. If the conditions are as shown in Fig. 3.5.5 to the left, we have

$$\sigma_n = \sigma \tag{3.5.2}$$

$$\tau_{nt} = \tau = c - \sigma \tan\varphi \tag{3.5.3}$$

By Mohr's circle it is seen that

$$\sigma_t = \sigma - 2\tau\tan\varphi = \left(1 + 2\tan^2\varphi\right)\sigma - 2c\tan\varphi \tag{3.5.4}$$

Thus we get the stress field

$$\begin{aligned} \sigma_n &= \sigma \\ \tau_{nt} &= c - \sigma\tan\varphi \\ \sigma_t &= \left(1 + 2\tan^2\varphi\right)\sigma - 2c\tan\varphi \end{aligned} \tag{3.5.5}$$

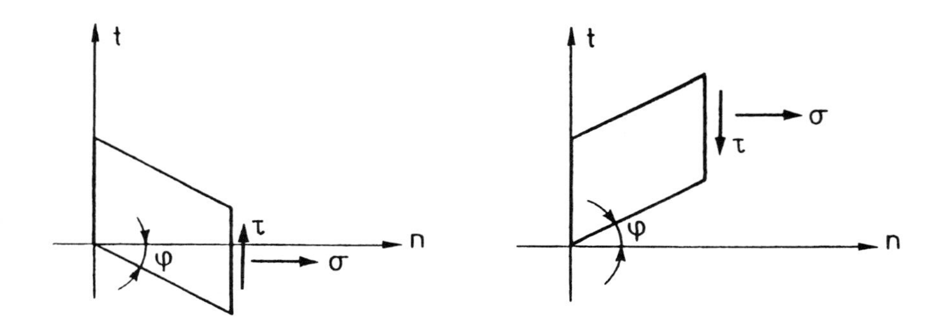

Figure 3.5.5 Coordinate system with one coordinate plane parallel to a failure section.

Here the stresses are expressed by σ alone.

If the conditions are as shown in Fig. 3.5.5 to the right, we have

$$\sigma_n = \sigma$$

$$\tau_{nt} = -(c - \sigma \tan\varphi) \tag{3.5.6}$$

$$\sigma_t = (1 + 2\tan^2\varphi)\,\sigma - 2c\tan\varphi$$

The principal stresses can also be expressed by σ. From Mohr's circle we have

$$\left.\begin{matrix}\sigma_1\\ \sigma_3\end{matrix}\right\} = \sigma - \tau\tan\varphi \pm \frac{\tau}{\cos\varphi} \tag{3.5.7}$$

giving

$$\left.\begin{matrix}\sigma_1\\ \sigma_3\end{matrix}\right\} = \sigma\frac{1 \mp \sin\varphi}{\cos^2\varphi} \pm c\frac{1 \mp \sin\varphi}{\cos\varphi} \tag{3.5.8}$$

Instead of using the normal stress σ as a parameter, we can use the mean stress σ_m in the plane:

$$\sigma_m = \tfrac{1}{2}(\sigma_1 + \sigma_3) \tag{3.5.9}$$

It is seen that

$$\sigma_m = \sigma - \tau\tan\varphi = \left(1 + \tan^2\varphi\right)\sigma - c\tan\varphi \tag{3.5.10}$$

If we want to calculate the stresses in an arbitrary section, it can be done by noticing that the radius in a Mohr's circle corresponding to failure is $c\cos\varphi - \sigma_m \sin\varphi$. Introducing the angle θ between the failure section with positive shear stress and the section with the sought stresses σ_x, τ_{xy} (see Fig. 3.5.6), we get by Mohr's circle

$$\sigma_x = \sigma_m + \left(c\cos\varphi - \sigma_m\sin\varphi\right)\cos\left(\frac{\pi}{2} - \varphi - 2\theta\right) \tag{3.5.11}$$

Correspondingly, the stress σ_y in the section perpendicular to the section with the stresses σ_x and τ_{xy} can be determined. The result is

$$\sigma_x = c\cot\varphi + \left(\sigma_m - c\cot\varphi\right)[1 - \sin\varphi\sin(2\theta + \varphi)]$$

$$\sigma_y = c\cot\varphi + \left(\sigma_m - c\cot\varphi\right)[1 + \sin\varphi\sin(2\theta + \varphi)] \tag{3.5.12}$$

$$\tau_{xy} = -\left(\sigma_m - c\cot\varphi\right)\sin\varphi\cos(2\theta + \varphi)$$

For $c = 0$ we find

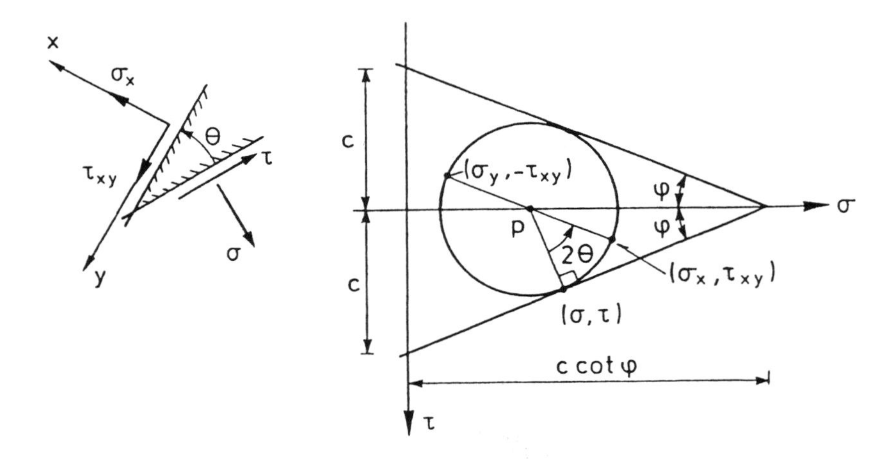

Figure 3.5.6 Stresses in arbitrary sections determined by Mohr's circle.

$$\begin{aligned} \sigma_x &= \sigma_m[1 - \sin\varphi \sin(2\theta + \varphi)] \\ \sigma_y &= \sigma_m[1 + \sin\varphi \sin(2\theta + \varphi)] \\ \tau_{xy} &= -\sigma_m \sin\varphi \cos(2\theta + \varphi) \end{aligned} \qquad (3.5.13)$$

Formulas (3.5.13) for $c = 0$ can be derived from formulas (3.5.12) for $c \neq 0$ when in the last-mentioned formulas c cotφ is subtracted from all normal stresses, which is also obvious when the geometrical meaning of the quantity $c \cot\varphi$ is noted (see Fig. 3.5.6).

Setting $\theta = 0$ in (3.5.12) we get (3.5.5) when (n, t) is replaced by (x, y). Setting $2\theta = 90° - \varphi$, we get the following formulas for the principal stresses:

$$\left.\begin{matrix} \sigma_1 \\ \sigma_3 \end{matrix}\right\} = \sigma_m(1 \mp \sin\varphi) \pm c\cos\varphi \qquad (3.5.14)$$

complying with (3.5.8).

If the stresses in an arbitrary section σ_x , τ_{xy} are given, and if we want to determine the failure sections and their stresses, this can be done by solving the equations

$$\sigma_x = c\cot\varphi + (\sigma_m - c\cot\varphi)[1 - \sin\varphi\sin(2\theta + \varphi)] \qquad (3.5.15)$$

$$\tau_{xy} = -(\sigma_m - c\cot\varphi)\sin\varphi\cos(2\theta + \varphi) \tag{3.5.16}$$

with regard to θ and σ_m.

Introducing an angle β defined by

$$\tan\beta = -\frac{\tau_{xy}}{\sigma_x - c\cot\varphi} \tag{3.5.17}$$

we get

$$\frac{\sin\varphi\cos(2\theta + \varphi)}{1 - \sin\varphi\sin(2\theta + \varphi)} = \tan\beta \tag{3.5.18}$$

which can be rewritten as

$$\cos(2\theta + \varphi - \beta) = \frac{\sin\beta}{\sin\varphi} \tag{3.5.19}$$

from which θ can be determined. Note that θ is assumed positive in the positive direction of the plane, as shown in Fig. 3.5.6.

When θ is determined we have

$$\sigma_m = c\cot\varphi + \frac{\sigma_x - c\cot\varphi}{1 - \sin\varphi\sin(2\theta + \varphi)} \tag{3.5.20}$$

There will normally be two solutions for θ and σ_m.

When σ_m is determined, the normal stress in the failure sections σ can be determined by (3.5.10), giving

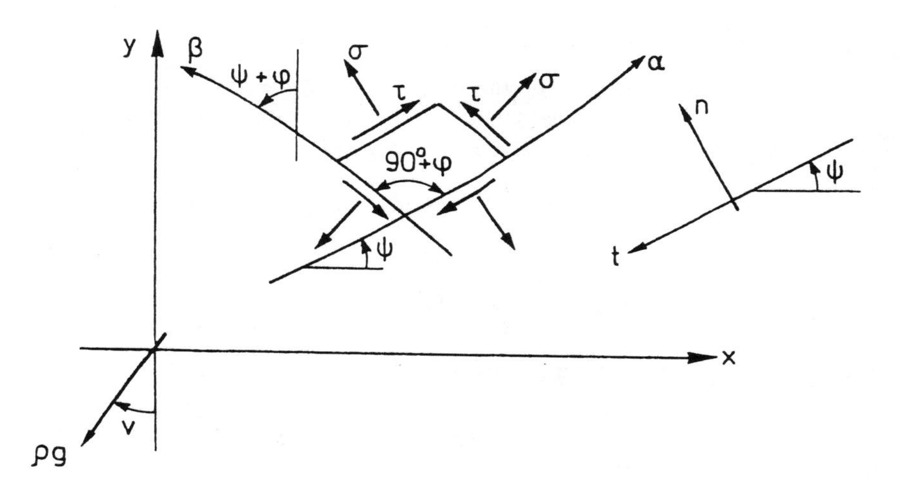

Figure 3.5.7 Definition of α- and β-lines.

$$\sigma = \left(\sigma_m + c\tan\varphi\right)\cos^2\varphi \tag{3.5.21}$$

If we want to state the failure criterion in an arbitrary, rectangular x, y-coordinate system, it can be done by noting that the radius in a Mohr circle, corresponding to failure, is $c\cos\varphi - \frac{1}{2}\left(\sigma_x + \sigma_y\right)\sin\varphi$. Expressing the radius by the stresses σ_x, σ_y, and τ_{xy}, we get the condition

$$\sqrt{\frac{1}{4}\left(\sigma_x - \sigma_y\right)^2 + \tau_{xy}^2} + \frac{1}{2}\left(\sigma_x + \sigma_y\right)\sin\varphi = c\cos\varphi \tag{3.5.22}$$

In a failure zone, only those zones that satisfy the equilibrium conditions are possible.

We consider an arbitrary failure zone referred to a rectangular x, y-coordinate system (see Fig. 3.5.7). The failure sections are called α- and β-lines and these are oriented in such a way that they have the same orientation as the x, y-system. The lines are designated in such a way that the convex angle between the α- and β-lines is $90^\circ + \varphi$. The angle between the x-axis and the tangent to an α-line is denoted ψ and is assumed positive in the positive direction of rotation of the x, y-plane.

Using the transformation formulas for stresses and formulas (3.5.6), we find that

$$\begin{aligned}
\sigma_x &= \sigma\sin^2\psi + \left[\left(1 + 2\tan^2\varphi\right)\sigma - 2c\tan\varphi\right]\cos^2\psi - (c - \sigma\tan\varphi)\sin 2\psi \\
\sigma_y &= \sigma\cos^2\psi + \left[\left(1 + 2\tan^2\varphi\right)\sigma - 2c\tan\varphi\right]\sin^2\psi + (c - \sigma\tan\varphi)\sin 2\psi \\
\tau_{xy} &= \frac{1}{2}\left[\sigma - \left(\left(1 + 2\tan^2\varphi\right)\sigma - 2c\tan\varphi\right)\right]\sin 2\psi + (c - \sigma\tan\varphi)\cos 2\psi
\end{aligned} \tag{3.5.23}$$

We assume that gravity acts on the body at an angle v with the y-axis, as shown in Fig. 3.5.7. The density is ρ and the acceleration due to gravity is g. The equilibrium conditions are in this case

$$\begin{aligned}
\frac{\partial\sigma_x}{\partial x} + \frac{\partial\tau_{xy}}{\partial y} - \rho g\sin v &= 0 \\
\frac{\partial\sigma_y}{\partial y} + \frac{\partial\tau_{xy}}{\partial x} - \rho g\cos v &= 0
\end{aligned} \tag{3.5.24}$$

Having made the differentiations indicated, we obtain for $\psi = 0$

$$\frac{\partial\sigma}{\partial x} + 2\tan^2\varphi\frac{\partial\sigma}{\partial x} - 2(c - \sigma\tan\varphi)\frac{\partial\psi}{\partial x} + 2(\sigma\tan^2\varphi - c\tan\varphi)\frac{\partial\psi}{\partial y}$$

$$-\tan\varphi \frac{\partial \sigma}{\partial y} - \rho g \sin\nu = 0$$

$$\frac{\partial \sigma}{\partial y} + 2(c - \sigma\tan\varphi)\frac{\partial \psi}{\partial y} + 2\left(\sigma\tan^2\varphi - c\tan\varphi\right)\frac{\partial \psi}{\partial x}$$

$$-\tan\varphi \frac{\partial \sigma}{\partial x} - \rho g \cos\nu = 0 \tag{3.5.25}$$

Multiplying the last equation by $\tan\varphi$ and adding the equations, we get

$$\frac{\partial \sigma}{\partial x} - 2(c - \sigma\tan\varphi)\frac{\partial \psi}{\partial x} = \rho g \frac{\sin\nu + \cos\nu\tan\varphi}{1 + \tan^2\varphi} \tag{3.5.26}$$

Since the x, y-system can be placed arbitrarily and for $\psi = 0$ we have $\partial(\cdot)/\partial x = \partial(\cdot)/\partial s_\alpha$, where s_α is the curve length along an α-line, the equation applies to any point of an α-line, when x is replaced by s_α, and ν is the angle with the α-line of the force of gravity. If we again consider the x, y-system as fixed and if it is assumed that the force of gravity goes in the negative direction of the y-axis, it is seen that we must put $\nu = \psi$, and then we get

$$\frac{\partial \sigma}{\partial s_\alpha} - 2(c - \sigma\tan\varphi)\frac{\partial \psi}{\partial s_\alpha} = \rho g \cos\varphi \sin(\psi + \varphi) \tag{3.5.27}$$

A corresponding equation can be derived for a β-line. The result is

$$\frac{\partial \sigma}{\partial s_\beta} + 2(c - \sigma\tan\varphi)\frac{\partial \psi}{\partial s_\beta} = \rho g \cos\varphi \cos\psi \tag{3.5.28}$$

These results can also be expressed by using σ_m or τ as parameters instead of σ.

Using τ, for example, we get

$$\frac{\partial \tau}{\partial s_\alpha} + 2\tau\tan\varphi \frac{\partial \psi}{\partial s_\alpha} + \rho g \sin\varphi \sin(\psi + \varphi) = 0 \tag{3.5.29}$$

$$\frac{\partial \tau}{\partial s_\beta} - 2\tau\tan\varphi \frac{\partial \psi}{\partial s_\beta} + \rho g \sin\varphi \cos\psi = 0 \tag{3.5.30}$$

which do not contain the cohesion c. These equations are called *Kötter's equations* [03.1] and they play an important role in soil mechanics [53.5].

3.5.3 Simple, Statically Admissible Failure Zones

The simplest conceivable failure zone for a Coulomb material is made up by two sets of parallel lines forming the angle $90° - \varphi$ with each other. Placing the y-axis along one set of lines, we get the conditions shown in Fig. 3.5.8. The stresses are found by (3.5.5) and (3.5.6) as

$$\sigma_x = \sigma$$

$$\sigma_y = (1 + 2\tan^2\varphi)\,\sigma - 2c\tan\varphi \tag{3.5.31}$$

$$\tau_{xy} = \pm(c - \sigma\tan\varphi)$$

where the upper sign applies to the conditions in Fig. 3.5.8 to the left, and the lower sign applies to the conditions to the right.

Assuming a weightless material, we have from the equilibrium equations,

$$\frac{\partial\sigma}{\partial x} \pm \frac{\partial(c - \sigma\tan\varphi)}{\partial y} = 0$$

$$\frac{\partial\left[(1 + 2\tan^2\varphi)\,\sigma - 2c\tan\varphi\right]}{\partial y} \pm \frac{\partial(c - \sigma\tan\varphi)}{\partial x} = 0 \tag{3.5.32}$$

Determining $\partial\sigma/\partial x$ and $\partial\sigma/\partial y$ from these equations, we have

$$\frac{\partial\sigma}{\partial x} = \frac{\partial\sigma}{\partial y} = 0 \tag{3.5.33}$$

meaning that

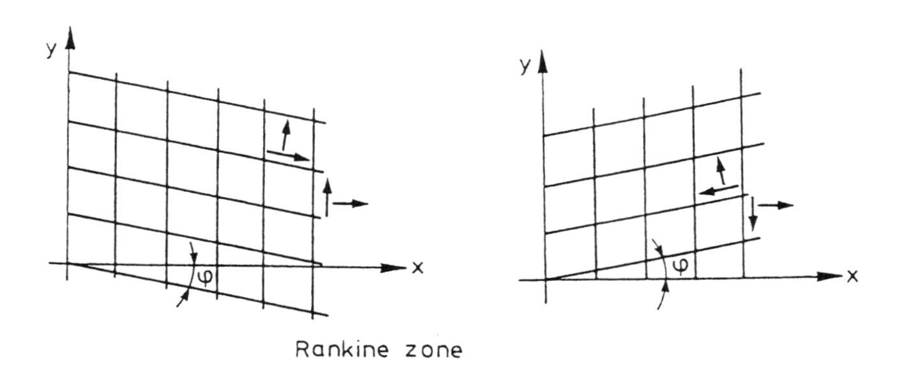

Figure 3.5.8 Failure lines in a Rankine zone.

$$\sigma = \text{constant} \tag{3.5.34}$$

The field is called a *Rankine field* or a *Rankine zone.*

It is easy to take into account the weight of the material and we find simple formulas for the derivatives $\partial\sigma/\partial x$ and $\partial\sigma/\partial y$. However, we omit the calculations here.

If, instead, we let the two sets of curves be logarithmic spirals, with the angle of inclination φ and straight lines from the pole, respectively, we also get curves that form the angle $90° - \varphi$ (see Fig. 3.5.9).

It is natural here to use polar coordinates. The stresses are

$$\begin{aligned} \sigma_r &= (1 + 2\tan^2\varphi)\,\sigma - 2c\tan\varphi \\ \sigma_\theta &= \sigma \\ \tau_{r\theta} &= \pm(c - \sigma\tan\varphi) \end{aligned} \tag{3.5.35}$$

where the upper sign applies to the conditions in Fig. 3.5.9 to the left, and the lower sign applies to the conditions to the right. If these stresses are inserted into the equilibrium equations for a weightless body, we find that

$$\begin{aligned} \frac{\partial[(1 + 2\tan^2\varphi)\,\sigma - 2c\tan\varphi]}{\partial r} \pm \frac{1}{r}\frac{\partial(c - \sigma\tan\varphi)}{\partial\theta} + 2\,\frac{\sigma\tan^2\varphi - c\tan\varphi}{r} &= 0 \\ \frac{1}{r}\frac{\partial\sigma}{\partial\theta} \pm \frac{\partial(c - \sigma\tan\varphi)}{\partial r} \pm 2\,\frac{c - \sigma\tan\varphi}{r} &= 0 \end{aligned} \tag{3.5.36}$$

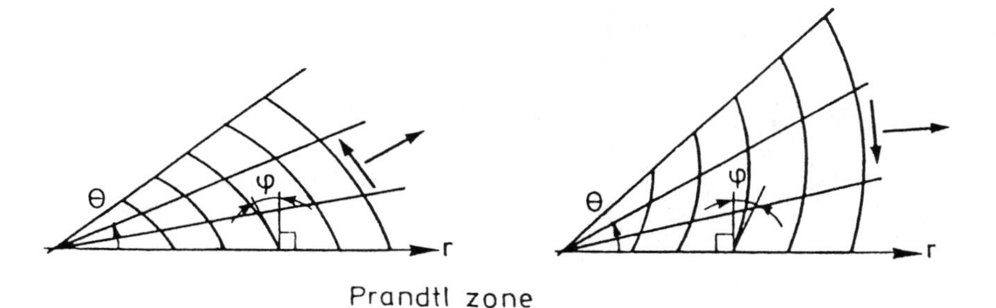

Figure 3.5.9 Failure lines in a Prandtl zone.

Multiplying the last equation by tanφ and adding or subtracting the equations, we see that

$$\frac{\partial \sigma}{\partial r} = 0 \tag{3.5.37}$$

Thus σ is a constant along the straight failure lines. Then the last equation gives

$$\frac{\partial \sigma}{\partial \theta} \pm 2(c - \sigma \tan\varphi) = 0 \tag{3.5.38}$$

with the solution

$$\sigma = -Ce^{\pm 2\theta \tan\varphi} + c\cot\varphi \tag{3.5.39}$$

where C is a constant.

It is seen that the shear stresses are

$$\tau_{r\theta} = \pm C \tan\varphi \, e^{\pm 2\theta \tan\varphi} \tag{3.5.40}$$

Thus C must be > 0 in order that the assumption concerning the sign of the shear stresses is correct. The failure zone dealt with is called a *Prandtl field* or a *Prandtl zone.*

3.5.4 The Strain Field

It is seen from the formulas for strains in a Coulomb material in Section 3.4.1 that for plane strain conditions the ratios between the principal strains in the plane are given by (the apex is disregarded)

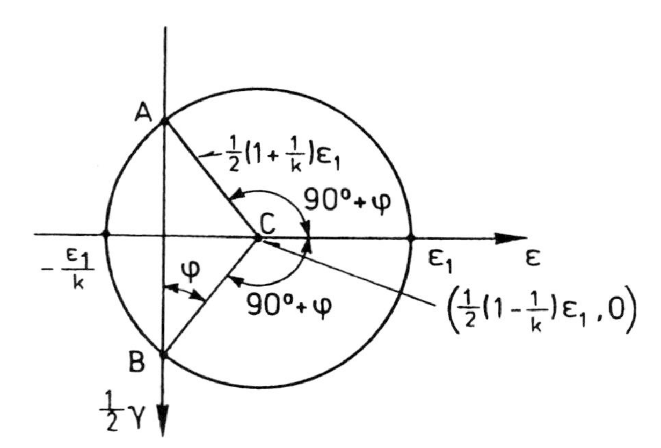

Figure 3.5.10 Mohr's circle for the strain field in a Coulomb material in plane strain.

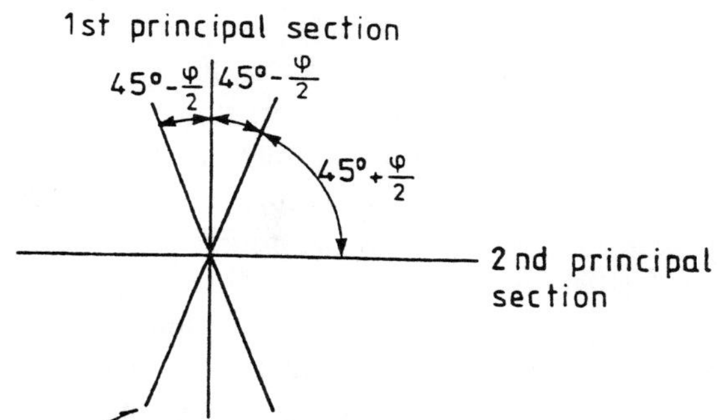

Figure 3.5.11 Principal directions and failure sections.

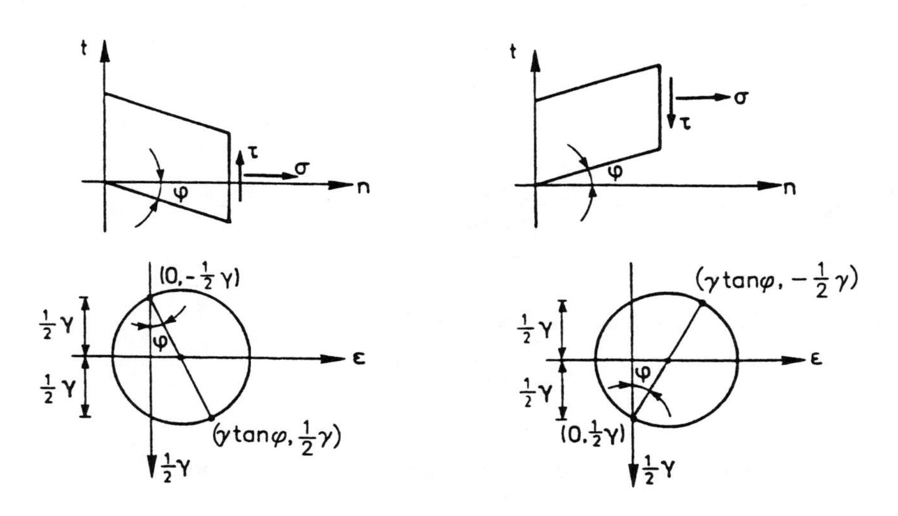

Figure 3.5.12 Strains referred to a coordinate system with one coordinate plane parallel to a failure section.

$$\frac{\varepsilon_1}{\varepsilon_3} = -k = -\frac{1+\sin\varphi}{1-\sin\varphi} \tag{3.5.41}$$

where ε_1 is the greatest and ε_3 the smallest principal strain. Therefore, Mohr's circle is as shown in Fig. 3.5.10. From the figure it is seen that angle *ABC* is φ. Further, it is seen that there are always two directions where the longitudinal strains are zero, and that these directions in fact are the failure sections. *That is, there are no longitudinal strains along the failure sections.*[1] The conditions are illustrated in Fig. 3.5.11.

Referring the strain field to a rectangular *n, t*-coordinate system where, for example, the *t*-axis is identical with one of the failure sections, we have the conditions shown in Fig. 3.5.12. From the left of the figure we have

$$\varepsilon_n = \gamma\tan\varphi$$

$$\varepsilon_t = 0 \tag{3.5.42}$$

$$\gamma_{nt} = \gamma$$

where γ is the numerical value of the change of angle between the *n*- and *t*-direction. If the conditions are as shown to the right in the figure, we have

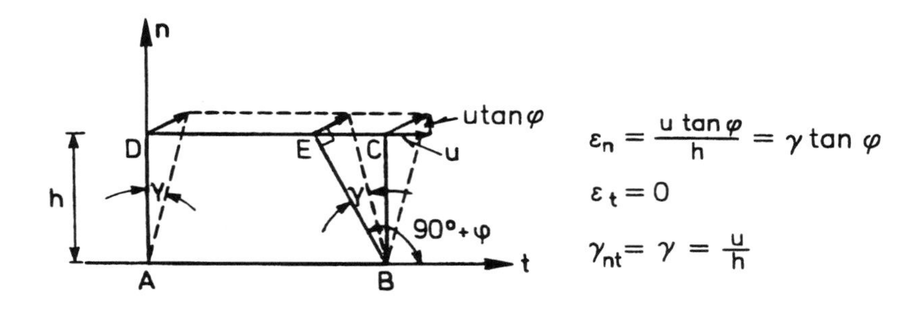

Figure 3.5.13 Homogeneous strain field in a Rankine zone.

[1] So it is possible to get a qualitative impression of the deformation of a failure zone by considering the zone as a truss system with bars in the directions of the failure sections and hinges in the nodal points.

$$\varepsilon_n = \gamma \tan\varphi$$

$$\varepsilon_t = 0 \tag{3.5.43}$$

$$\gamma_{nt} = -\gamma < 0$$

We have already found this result in another way in Section 3.4.2. Bearing in mind that the change of an angle between two arbitrary directions can be determined from the shear strains in the actual directions, because the rotation is the same for both directions, we see that γ *also means the reduction of the obtuse angle* $90° + \varphi$ *between the failure sections*. The distance between the failure sections is thus increased by the deformation. A change in volume per unit volume $\gamma \tan\varphi$ takes place.

3.5.5 Simple Geometrically Admissible Strain Fields

The simplest strain field having strains of the described type is a homogeneous strain field as illustrated in Fig. 3.5.13. The displacements are characterized by the fact that one side *DC* of a rectangle *ABCD* gets a displacement u in the direction of *DC* relative to another side, *AB*, parallel with *DC*, and a displacement $u \tan\varphi$ perpendicular to and away from *DC*. It is seen that if we draw a line *BE* at the angle of friction φ with side *BC*, the resulting displacement vector is perpendicular to *BE*, and the change of angle γ might just as well be calculated as the rotation of *BE* or of *AD* or *BC*. This has been applied in Fig. 3.5.14, where the deformation described is assumed to take place in triangular areas. The deformation is seen to correspond to a Rankine field. The failure lines of the Rankine field and the sign of the shear stresses are shown in the figure.

The dissipation in a Rankine field is determined by (3.4.27). Since $\varepsilon_1 + \varepsilon_3 = \gamma \tan\varphi$, we have the following dissipation per unit volume:

$$W = c \cot\varphi (\varepsilon_1 + \varepsilon_3) = c\gamma \tag{3.5.44}$$

The justification of the result is easily seen when we use the normality condition in connection with the failure condition in the form $\tau = c - \sigma \tan\varphi$ (see, e.g., Fig. 3.4.4).

Let us then consider a simple displacement field in polar coordinates. Tentatively setting

$$u_r = 0$$
$$u_\theta = \mp u(\theta) \tag{3.5.45}$$

where $u(\theta) > 0$ and where the upper sign applies to the conditions in Fig. 3.5.15 to the left and the lower applies to the conditions to the right, we obtain [by (3.2.3)]

$$\varepsilon_r = 0$$

$$\varepsilon_\theta = \mp \frac{1}{r}\frac{du}{d\theta} \tag{3.5.46}$$

$$\gamma_{r\theta} = \pm \frac{u}{r}$$

If, in conformity with (3.5.42) and (3.5.43), we put $\varepsilon_\theta = \pm \gamma_{r\theta} \tan\varphi$, we get

$$\frac{1}{r}\frac{du}{d\theta} = \mp \frac{u}{r}\tan\varphi \tag{3.5.47}$$

with the solution

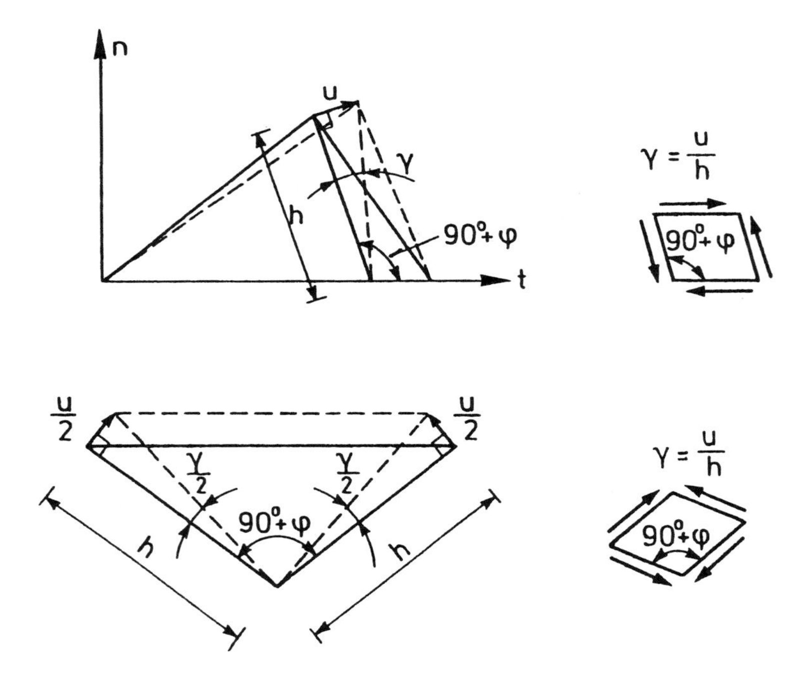

Figure 3.5.14 Homogeneous strain field in a Rankine zone.

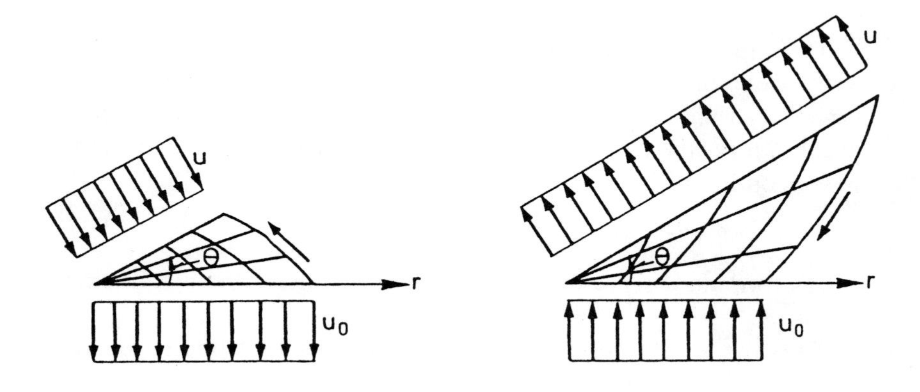

Figure 3.5.15 Strain field in a Prandtl zone.

$$u = u_0 e^{\mp \theta \tan\varphi} \tag{3.5.48}$$

Here u_0 is the u-value for $\theta = 0$.

From this we derive the strains

$$\varepsilon_r = 0$$

$$\varepsilon_\theta = \frac{1}{r} u_0 \tan\varphi \, e^{\mp \theta \tan\varphi} \tag{3.5.49}$$

$$\gamma_{r\theta} = \pm \frac{1}{r} u_0 e^{\mp \theta \tan\varphi}$$

The dissipation W per unit volume is, according to (3.5.44),

$$W = c \cot\varphi \cdot \frac{1}{r} u_0 \tan\varphi \, e^{\mp \theta \tan\varphi}$$

$$= \frac{c}{r} u_0 e^{\mp \theta \tan\varphi} \tag{3.5.50}$$

The displacement field is seen to correspond to the Prandtl field.

Considering especially a Prandtl field limited by two radii $\theta = 0$ and $\theta = \Theta$ and a logarithmic spiral with the equation $r = R_0 e^{\mp \theta \tan\varphi}$, we find the total dissipation as

$$D_1 = bc \cot\varphi \int_0^\Theta \int_0^{R_0 e^{\mp \theta \tan\varphi}} \frac{1}{r} u_0 \tan\varphi \, e^{\mp \theta \tan\varphi} r \, dr \, d\theta$$

$$= \mp \frac{1}{2} bc \cot\varphi R_0 u_0 \left[e^{\mp 2\Theta \tan\varphi} - 1 \right] \tag{3.5.51}$$

The quantity b is the extension of the body transverse to the plane.

The limiting line in the form of the logarithmic spiral will be a permissible discontinuity line if the body outside the zone is not moving, as the displacement vector forms the angle φ with the discontinuity line (see Fig. 3.4.11). The dissipation per unit length of the discontinuity line W_ℓ is found by (3.4.77):

$$W_\ell = bcu\cos\varphi \qquad (3.5.52)$$

In this formula u represents the numerical value of the displacement vector.

Thus the total dissipation along the discontinuity line is found as

$$D_2 = bc\cos\varphi \int_L u\,ds = bc\cos\varphi \int_0^\Theta u_0 e^{\mp\theta\tan\varphi} \frac{\gamma d\theta}{\cos\varphi} \qquad (3.5.53)$$

where ds is the curve length along the discontinuity line and L is its length. Inserting $r = R_0 e^{\mp\theta\tan\varphi}$, we get the simple result

$$D_2 = D_1 \qquad (3.5.54)$$

so the total dissipation is

$$D = D_1 + D_2 = \mp bc\cot\varphi R_0 u_0 [e^{\mp 2\Theta\tan\varphi} - 1] \qquad (3.5.55)$$

The result is summarized in Fig. 3.5.16.

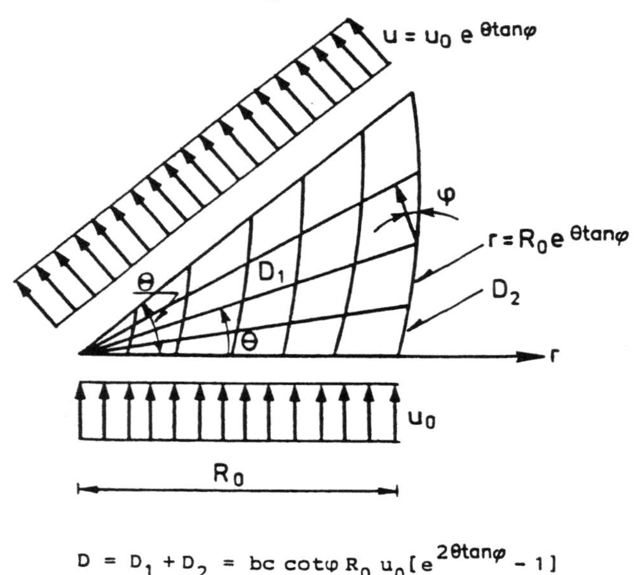

Figure 3.5.16 Total dissipation in a Prandtl zone.

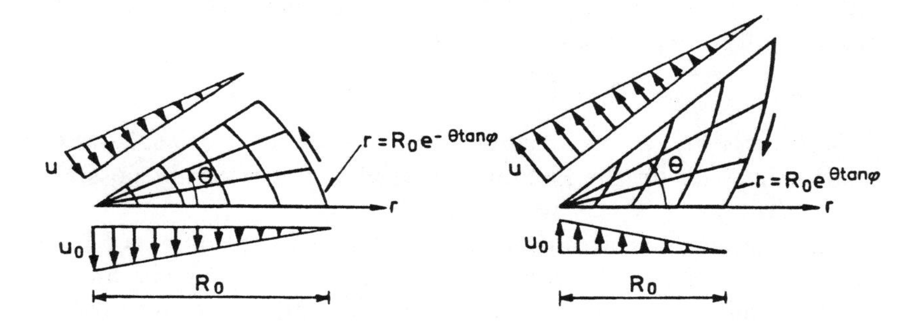

Figure 3.5.17 Strain field in a Prandtl zone.

There is another simple displacement field in a Prandtl zone. If we tentatively set

$$u_r = 0$$
$$u_\theta = \mp u(\theta)\left(1 - e^{\pm\theta\tan\varphi}\frac{r}{R_0}\right) \tag{3.5.56}$$

corresponding to $u_\theta = 0$ along the limiting line in the form of the logarithmic spiral $r = R_0 e^{\mp\theta\tan\varphi}$ (see Fig. 3.5.17), we have

$$\varepsilon_r = 0$$
$$\varepsilon_\theta = \mp\frac{1}{r}\frac{du}{d\theta}\left(1 - e^{\pm\theta\tan\varphi}\frac{r}{R_0}\right) + \frac{u}{r}\tan\varphi\, e^{\pm\theta\tan\varphi}\frac{r}{R_0} \tag{3.5.57}$$
$$\gamma_{r\theta} = \pm\frac{u}{R_0}e^{\pm\theta\tan\varphi} \pm \frac{u}{r}\left(1 - e^{\pm\theta\tan\varphi}\frac{r}{R_0}\right)$$

Setting, as before, $\varepsilon_\theta = \pm\gamma_{r\theta}\tan\varphi$, we have the same differential equation for determination of u, meaning that

$$u = u_0 e^{\mp\theta\tan\varphi} \tag{3.5.58}$$

The strains are found as

$$\varepsilon_r = 0$$
$$\varepsilon_\theta = \frac{1}{r}u_0\tan\varphi\, e^{\mp\theta\tan\varphi} \tag{3.5.59}$$
$$\gamma_{r\theta} = \pm\frac{1}{r}u_0 e^{\mp\theta\tan\varphi}$$

the same values as above. The dissipation D_1 in the zone is therefore the same as before, and since there is no contribution from the discontinuity line, the total dissipation is

$$D = D_1 = \mp \frac{1}{2} b c \cot\varphi R_0 u_0 \left[e^{\mp 2\Theta \tan\varphi} - 1 \right] \tag{3.5.60}$$

It is obvious that an arbitrary combination of the two displacement fields dealt with for the Prandtl field is also geometrically possible.

For a geometrically possible failure zone, the conditions that the displacements have to satisfy in order that the longitudinal strains are zero along the failure lines can be found in the following way. If, for example, the displacement field is characterized by the components of the displacement vector in directions perpendicular to the α- and β-lines, as shown in Fig. 3.5.18 (cf. the conditions in Figs. 3.5.13 and 3.5.14), we find by projection on the x- and y-axes, respectively,

$$\begin{aligned} u_x &= u_\alpha \cos(\psi + \varphi) - u_\beta \sin\psi \\ u_y &= u_\alpha \sin(\psi + \varphi) + u_\beta \cos\psi \end{aligned} \tag{3.5.61}$$

By calculating $\varepsilon_x = \partial u_x / \partial x$ and $\varepsilon_y = \partial u_y / \partial y$ and putting $\varepsilon_x = 0$ for $\psi = 0$ and $\varepsilon_y = 0$ for $\psi + \varphi = 0$, we get by using the fact that $\partial(\cdot)/\partial x = \partial(\cdot)/\partial s_\alpha$ for $\psi = 0$ and $\partial(\cdot)/\partial y = \partial(\cdot)/\partial s_\beta$ for $\psi + \varphi = 0$, the following formulas:

$$\begin{aligned} \cos\varphi \frac{\partial u_\alpha}{\partial s_\alpha} - \left(u_\beta + u_\alpha \sin\varphi \right) \frac{\partial \psi}{\partial s_\alpha} &= 0 \\ \cos\varphi \frac{\partial u_\beta}{\partial s_\beta} + \left(u_\alpha + u_\beta \sin\varphi \right) \frac{\partial \psi}{\partial s_\beta} &= 0 \end{aligned} \tag{3.5.62}$$

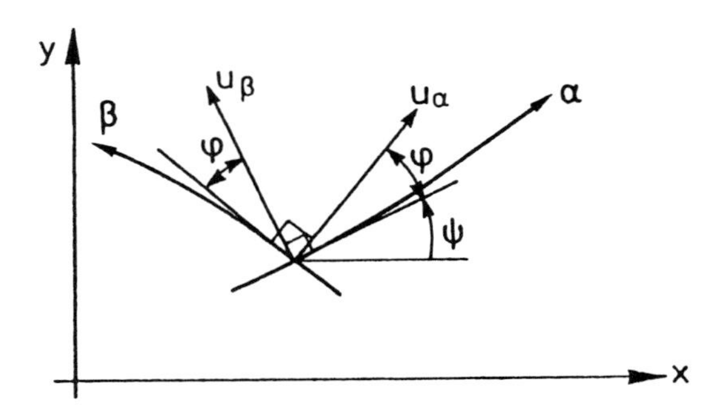

Figure 3.5.18 Displacement field for a Coulomb material in plane strain.

Since the x, y-system can be placed arbitrarily, these formulas apply to an arbitrary point of a geometrically possible failure zone. The conditions are, however, only necessary conditions. To ensure that the displacement field is admissible it is necessary that the angle between the α- and β-lines decreases or, what is the same, that the change in volume is positive.

3.6 APPLICATIONS

3.6.1 Pure Compression of a Prismatic Body

Consider a prismatic body of a Coulomb material with cohesion c and angle of friction φ. The body has plane end surfaces perpendicular to the axis. At the same ends the body is subjected to a uniformly distributed compression stress p (see Fig. 3.6.1).

Two failure mechanisms with yield planes as shown in Fig. 3.6.1 are considered. The failure mechanism to the left has one yield plane at the angle β with the end surfaces. Part II is assumed to be fixed while part I moves the distance u at an angle equal to the angle of friction φ with the yield plane.

The mechanism to the right has two yield planes, *AB* and *DC*. The upper and lower part I move the distance u_1 in the direction of the force while part II, *AEC*, and *BED* move outward perpendicular to the direction of the force. The displacements u_1 and u_2 are chosen in such a way that the relative displacement between parts I and II is a displacement at angle φ with the yield planes. In the figure the relative displacement u_{12} has been constructed for the yield plane *DE*, so that this condition is satisfied.

If the work equation for determination of an upper bound for p is set up, and if p is minimized with regard to β, we find that

$$\beta = \frac{\pi}{4} + \frac{\varphi}{2} \tag{3.6.1}$$

The corresponding value of p is

$$p = f_c = \frac{2c\cos\varphi}{1 - \sin\varphi} = 2c\tan\left(\frac{\pi}{4} + \frac{\varphi}{2}\right) \tag{3.6.2}$$

where f_c is the compressive strength. It is left to the reader to carry out the necessary calculations as the dissipation is determined from (3.4.75) or the equivalent expression (3.4.77). The result implies that the yield planes are identical with the sections where Coulomb's failure condition is satisfied (see, e.g., Fig. 2.1.6).

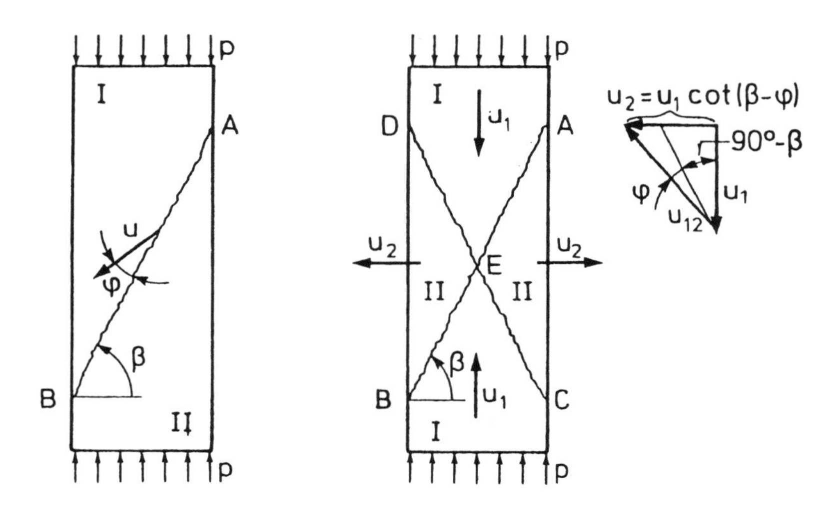

Figure 3.6.1 Failure mechanisms in pure compression of a prismatic body.

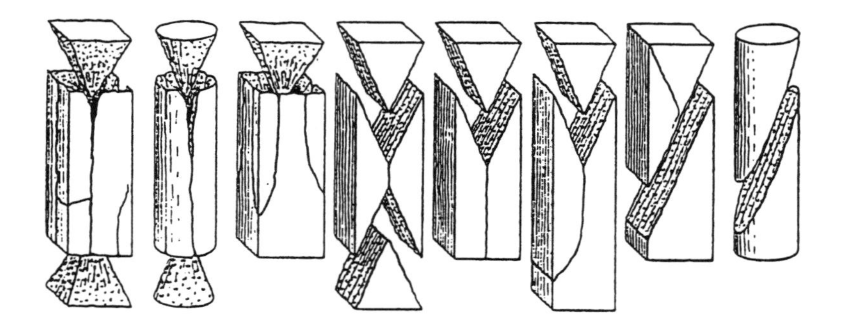

Figure 3.6.2 Failure patterns observed in tests (Suenson [42.1]).

In Fig. 3.6.2 some drawings made by the late professor E. Suenson reproducing the failure patterns, which he observed in the laboratory, are shown. The picture is from his book [42.1].

Instead, let the body now be a circular cylinder. Consider an axisymmetrical failure mechanism of the same type as that shown on the right of Fig. 3.6.1, and let the material be a modified Coulomb material. The load-carrying capacity will now be dependent on the tensile strength since the failure mechanism will give rise to radial cracks, which on the surface will appear as rectilinear cracks parallel to the generatrix. The upper bound is now found to be

$$p = f_c + k f_t \tag{3.6.3}$$

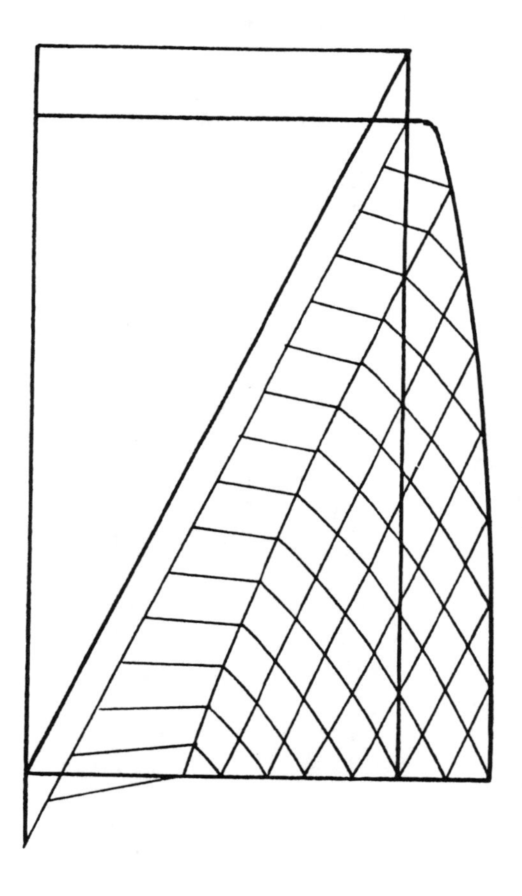

Figure 3.6.3 Axisymmetrical displacement field in pure compression of a cylindrical body.

Since the exact solution is, of course, $p = f_c$, the result is correct only for $f_t = 0$.

An axisymmetrical displacement field giving the correct solution has been developed by Exner [83.2]. His solution is sketched in Fig. 3.6.3.

3.6.2 Pure Compression of a Rectangular Disk

We consider a rectangular disk of a modfied Coulomb material with the tensile strength equal to zero. The disk has thickness b and along the two opposite sides it is loaded with a uniformly distributed compressive stress p (see Fig. 3.6.4).

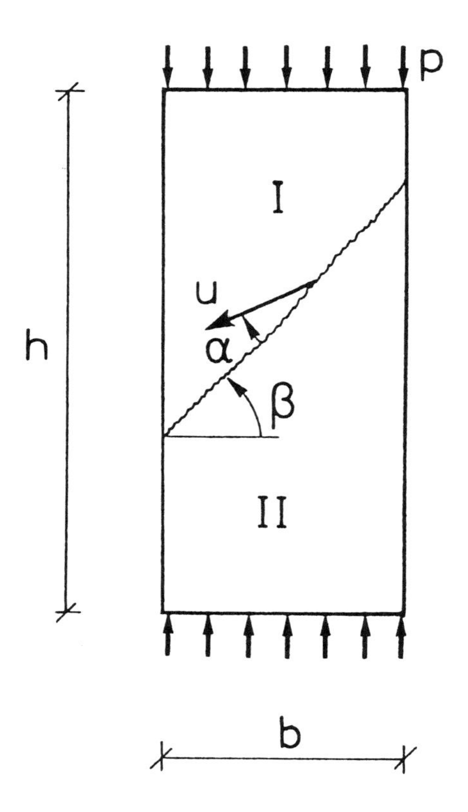

Figure 3.6.4 Failure mechanism in pure compression of a rectangular disk.

If the work equation is set up for a failure mechanism with a straight yield line at the angle β with the sides b, corresponding to the assumption that disk part II is fixed and part I moves the distance u at the angle α with the yield line, we have

$$pbu\cos\left[\frac{\pi}{2}-(\beta-\alpha)\right]=\frac{1}{2}f_c u(1-\sin\alpha)\frac{b}{\cos\beta} \tag{3.6.4}$$

where f_c is the compressive strength and where the dissipation is determined by (3.4.91).

When the upper bound solution for p is minimized with regard to α and β, we obtain the natural result

$$p=f_c \tag{3.6.5}$$

The angle β is determined by

$$\beta=\frac{\pi}{4}+\frac{\alpha}{2} \tag{3.5.6}$$

The calculations are left to the reader, and it is also left to the reader to find the direction of the displacement relative to the yield line in some characteristic cases (e.g., $\beta = 0$, $\beta = \pi/4$, and $\beta = \pi/4 + \varphi/2$, where φ is the angle of friction). Compare the result for the last-mentioned case with the result of Section 3.6.1.

3.6.3 A Semi-infinite Body

A semi-infinite body of a Coulomb material is considered. Over a strip of arbitrary width the body is subjected to a compressive stress p (see Fig. 3.6.5). We want to determine the load-carrying capacity. We try to combine Rankine fields with Prandtl fields, as shown in Fig. 3.6.5. In the outermost Rankine fields the horizontal boundary line is a principal section. The failure sections will then have to form either the angle $45° + \varphi/2$ or $45° - \varphi/2$ with the horizontal. Since τ must have the direction shown, it is seen, for example by Mohr's circle, that the angle must be $45° - \varphi/2$. Correspondingly, it is seen that in the Rankine field below the loaded part the failure sections must form the angle $45° + \varphi/2$ with the horizontal. The opening angle of the Prandtl zone is thus 90°. It is seen that the slope of the α-and β-lines of the failure zones is continuous.

Furthermore, in the outermost Rankine fields, it is seen, for example by Mohr's circle, that the principal stress (equal to zero) on the surface of the body is the greatest principal stress, (i.e., $\sigma_1 = 0$). Formulas (3.5.8) for the principal stresses then give the condition

$$\sigma \frac{1 - \sin\varphi}{\cos^2\varphi} + c\frac{1 - \sin\varphi}{\cos\varphi} = 0 \tag{3.6.7}$$

from which

$$\sigma = -c\cos\varphi \tag{3.6.8}$$

Therefore, we must have this σ-value along the outermost boundary lines along the Prandtl fields. To determine the constant C in (3.5.39) we therefore obtain, since it is seen by comparison with Fig. 3.5.9 that the uppermost sign is valid:

$$-Ce^{2\tan\varphi \cdot 0} + c\cot\varphi = -c\cos\varphi \tag{3.6.9}$$

from which

$$C = c\cot\varphi(1 + \sin\varphi) \tag{3.6.10}$$

Along the other boundary line of the Prandtl fields we therefore have

$$\sigma = -c\cot\varphi(1 + \sin\varphi)e^{\pi\tan\varphi} + c\cot\varphi \tag{3.6.11}$$

where we have used the fact that the opening angle of the Prandtl fields is $\Theta = \pi/2$.

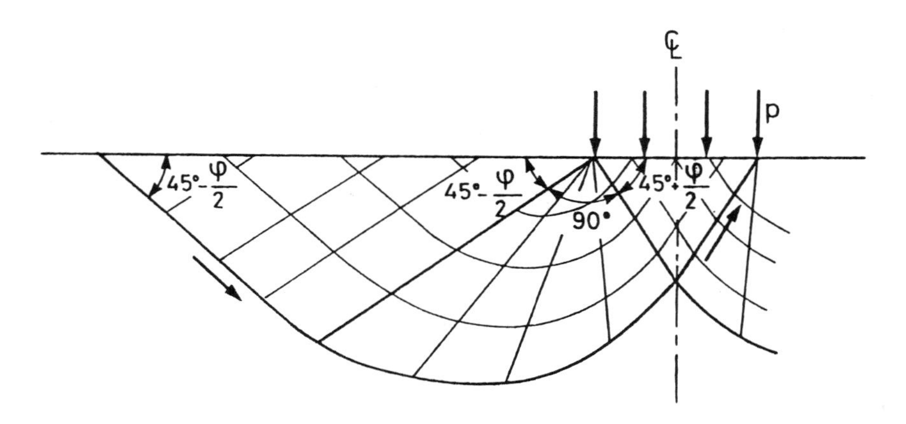

Figure 3.6.5 Failure zone in a semi-infinite body loaded by a compressive stress along a strip of arbitrary width.

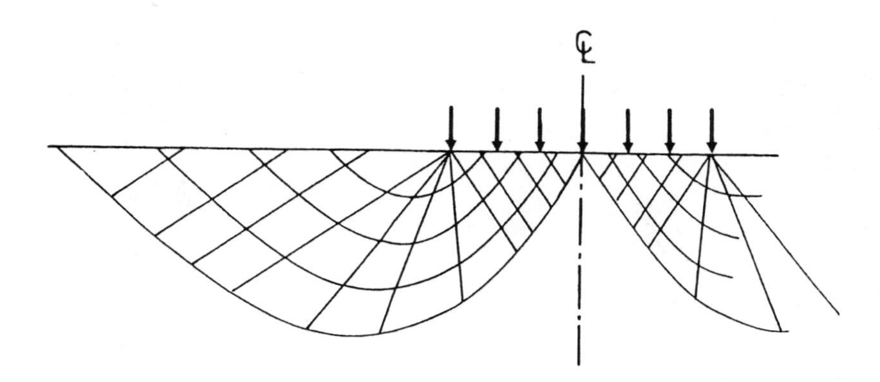

Figure 3.6.6 Alternative failure zone in a semi-infinite body loaded by a compressive stress along a strip of arbitrary width.

In the Rankine field below the loaded area it is seen by Mohr's circle that the load p corresponds to the smallest principal stress. Formulas (3.5.8) therefore give

$$\sigma_3 = -p = (-c\cot\varphi(1+\sin\varphi)e^{\pi\tan\varphi} + c\cot\varphi)\frac{1+\sin\varphi}{\cos^2\varphi} - c\frac{1+\sin\varphi}{\cos\varphi} \tag{3.6.12}$$

that is,

$$p = c\cot\varphi\left[e^{\pi\tan\varphi}\tan^2\left(\frac{\pi}{4}+\frac{\varphi}{2}\right)-1\right] \tag{3.6.13}$$

It can be shown [53.4] that one can find a statically admissible stress field corresponding to points within or on the yield surface outside the failure zones. The solution is due to Prandtl [20.2].

In Fig. 3.6.6 a combination of Rankine and Prandtl fields is shown which is immediately seen to give the same load-carrying capacity. In the special case $\varphi = 0$, this solution was given by Hill [50.2].

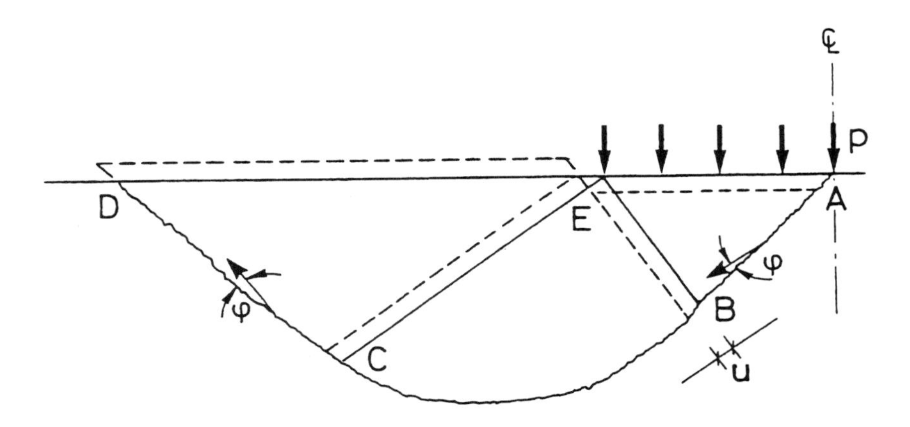

Figure 3.6.7 Displacement field in semi-infinite body loaded by a compressive stress along a strip of arbitrary width.

The plastic theory cannot determine which of the solutions is correct. As shown in Chapter 1, there is a unique solution only for the load-carrying capacity. Only a description of the extension of the yielding, step by step as the load grows from zero to the maximum value, can show which solution is correct.

To verify that the load-carrying capacity found is correct we have to show that there is a displacement field which corresponds to the stress field. We assume that the Rankine zone below the loaded area moves the distance u, perpendicular to the direction BE, as a rigid body (see Fig. 3.6.7). AB is thus an admissible yield line. The displacement field in the Prandtl zone is therefore of a type dealt with in Section 3.5.5. Thus BC is also a yield line. The displacement along CE can be determined by (3.5.48). This displacement is perpendicular to CE, so CED moves as a rigid body in the direction perpendicular to CE (i.e., CD becomes a yield line). It is seen that the displacement field corresponds to the stress field, so the solution (3.6.13) is an exact solution.

There are other geometrically possible displacement fields corresponding to the solution found. It is left to the reader to consider a displacement field corresponding to Prandtl's solution, where the Rankine zone below the loading moves vertically downward as a rigid body, while the displacement field in the Prandtl zone is as it was previously. (*Hint:* Show that the displacement fields in the

Rankine and Prandtl zones can be adjusted to each other in such a way that the relative displacements in lines *AB* and *BE* shown in Fig. 3.6.7 will take place at the correct angle, so that the lines will be admissible yield lines.)

If the load is transferred to the semi-infinite body through another sufficiently strong body, Hill's solution together with the first displacement field treated can be geometrically possible only if the body through which the load is transferred is completely smooth. The other solution, however, will be possible in any case, especially if the surface of the body is sufficiently rough to prevent sliding. From this it can be concluded that the load-carrying capacity in all cases is determined by (3.6.13) irrespective of the magnitude of the coefficient of friction between the two bodies.

The solutions presented in this section are important special cases in soil mechanics, since they serve as a basis for calculation of the load-carrying capacity of foundations. A number of problems concerning the load-carrying capacity for foundations on soil have been treated according to plastic theories by Chen [75.2].

3.6.4 A Slope with Uniform Load

Consider a semi-infinite, weightless body of a Coulomb material in the form of a slope with uniform load p (see Fig. 3.6.8). The load-carrying capacity will be determined under plane strain conditions. A work equation for the determination of an upper bound solution is set up for a failure mechanism in the following way. The triangle *ABE* moves as a rigid body the distance u_0 at the angle of friction φ with *AB*, which in this way becomes a yield line. *EBC* is a Prandtl zone with constant displacement u_0 and $u_0 e^{\Theta\tan\varphi}$ [see (3.5.48)] along the radial boundary lines of lengths R_0 and $R_0 e^{\Theta\tan\varphi}$, respectively. The triangle *ECD* moves as a rigid body the distance $u_0 e^{\Theta\tan\varphi}$ at the angle φ with *CD*, which thus becomes a yield line. The angle of the yield lines with the boundary lines of the body has been chosen with an eye to the statical conditions for Rankine zones (see Fig. 3.6.5).

The work equation has the form

$$p \cdot 2R_0 \cos\left(\frac{\pi}{4}+\frac{\varphi}{2}\right) u_0 \cos\left[\frac{\pi}{2}-\left(\frac{\pi}{4}+\frac{\varphi}{2}\right)+\varphi\right]$$

$$= cR_0 u_0 \cos\varphi + cR_0 u_0 \cot\varphi\left[e^{2\Theta\tan\varphi}-1\right] + cR_0 e^{\Theta\tan\varphi} u_0 e^{\Theta\tan\varphi}\cos\varphi \tag{3.6.14}$$

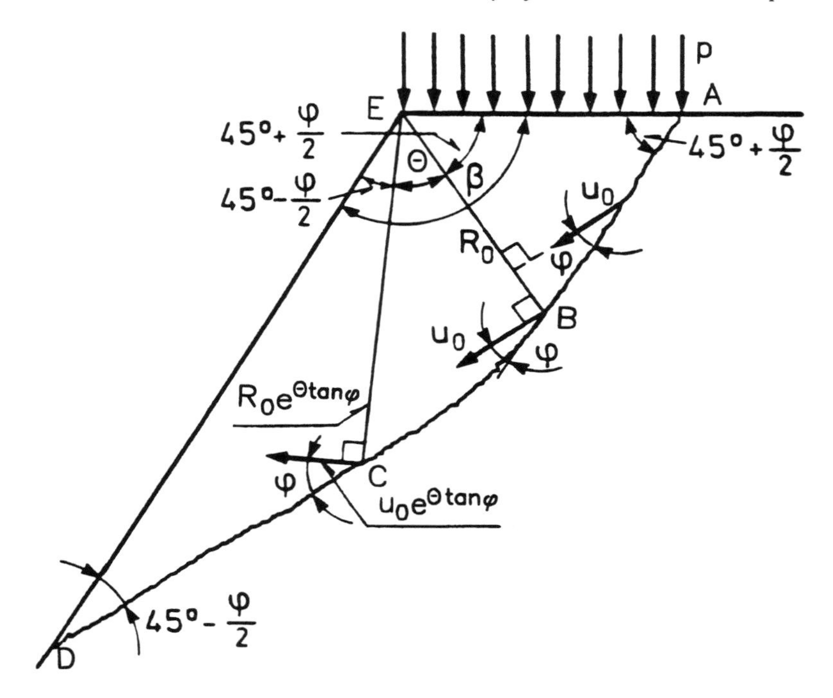

Figure 3.6.8 Failure mechanism for a slope with uniform load, $\beta \geq \pi/2$.

where the left-hand side is the external work and the right-hand side is the dissipation.

The solution as regards p can be written

$$p = c\cot\varphi\left[\tan^2\left(\frac{\pi}{4} + \frac{\varphi}{2}\right)e^{(2\beta - \pi)\tan\varphi} - 1\right] \tag{3.6.15}$$

This is in agreement with (3.6.13) for $\beta = \pi$.

By calculations analogous to those derived in Section 3.6.3, it may be shown that the equilibrium conditions are satisfied when *ABE* and *ECD* are considered as Rankine zones and *EBC*, as in the upper bound solution, is considered as a Prandtl zone. In this way it may be shown that the solution is exact for $\beta \geq \pi/2$.

For $\beta < \pi/2$ a statically admissible stress field can be found by combining two Rankine zones separated by a stress discontinuity line ℓ (see Fig. 3.6.9).

The corresponding lower bound solution is found to be

$$p = c\cot\varphi\left[\frac{\sin(\beta - \mu)}{\sin(\beta + \mu)}\tan^2\left(\frac{\pi}{4} + \frac{\varphi}{2}\right) - 1\right] \tag{3.6.16}$$

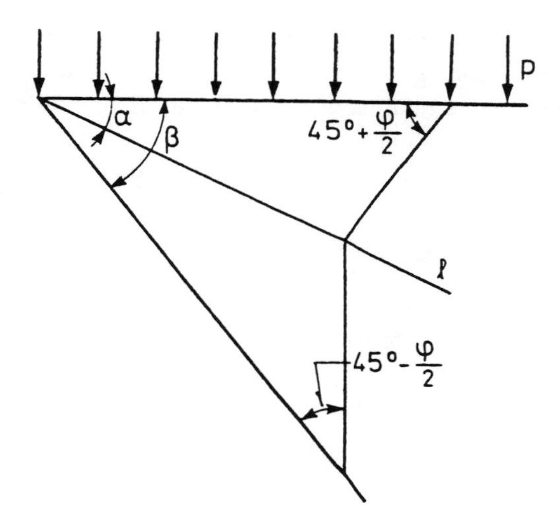

Figure 3.6.9 Discontinuous stress field for $\beta < \pi/2$.

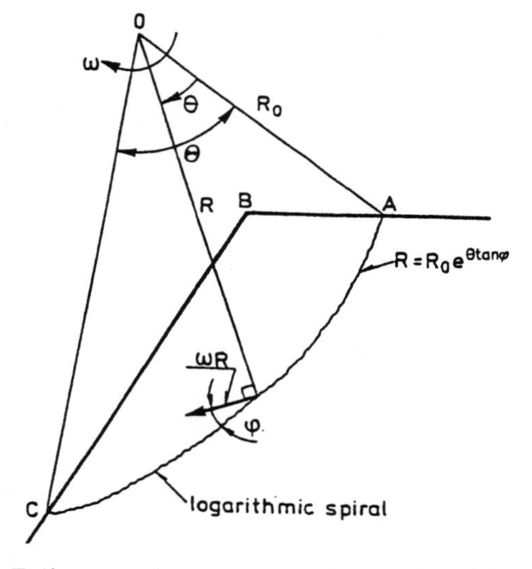

Figure 3.6.10 Failure mechanism for a slope with arbitrary loads.

The angle μ is determined by

$$\sin\mu = \sin\varphi \sin\beta \, , \qquad 0 \le \mu \le \frac{\pi}{2} \tag{3.6.17}$$

and angle α shown in the figure is equal to $(\beta + \mu)/2$. This solution may be shown to be exact for $\beta \le \pi/2$ (see [50.4; 54.1]).

Upper bound solutions for arbitrary loads can be found by considering yield lines in the form of logarithmic spirals AC with the angle of inclination φ (see Fig. 3.6.10). A failure mechanism is formed by the rigid body ABC rotating a small angle ω about the pole of the spiral. Thus it is seen that the displacement vector everywhere forms the angle φ with the tangent to the logarithmic spiral, which thus becomes a yield line.

With the conditions of the figure the dissipation in the yield line is found to be

$$D = \frac{1}{2} b c \cot\varphi \, \omega \, R_0^2 \left[e^{2\Theta \tan\varphi} - 1 \right] \tag{3.6.18}$$

where b is the thickness of the body transverse to the plane (cf. Section 3.5.5). Failure mechanisms of this kind are often used in soil mechanics for analyzing the stability of slopes (see also Section 3.6.5).

3.6.5 Strip Load on a Concrete Block

Consider a rectangular concrete block resting on a smooth surface as shown in Fig. 3.6.11. The block is loaded on the surface along a relatively narrow strip through a strong body. The load per unit thickness of the block is P. We shall determine the load-carrying capacity under plane strain conditions, when the material is a modified Coulomb material.

Loading far from the edge. Consider a failure mechanism with yield lines as shown in Fig. 3.6.12. The corresponding upper bound solution was developed by Chen and Drucker [69.3].

The body is separated into three rigid parts. Part I is translating vertically downward, and parts II to III horizontally apart. The relative displacement of I to II is a separation and it makes an angle φ to the yield line (see Fig. 3.6.12). The relative displacement is u. The displacement of I is u_1, and the displacement of II and III is u_2, where

$$u_1 = u\cos(\beta + \varphi) \ , \qquad u_2 = u\sin(\beta + \varphi) \tag{3.6.19}$$

The dissipation is, according to (3.4.83),

$$D = f_t(h - a\cot\beta)\cdot 2u\sin(\beta + \varphi) + \frac{1 - \sin\varphi}{2} f_c \frac{2a}{\sin\beta} \cdot u \tag{3.6.20}$$

The external work is

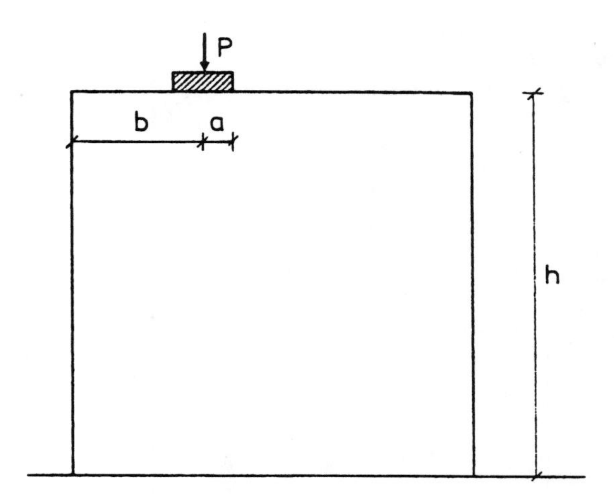

Figure 3.6.11 Rectangular block loaded by a concentrated strip load.

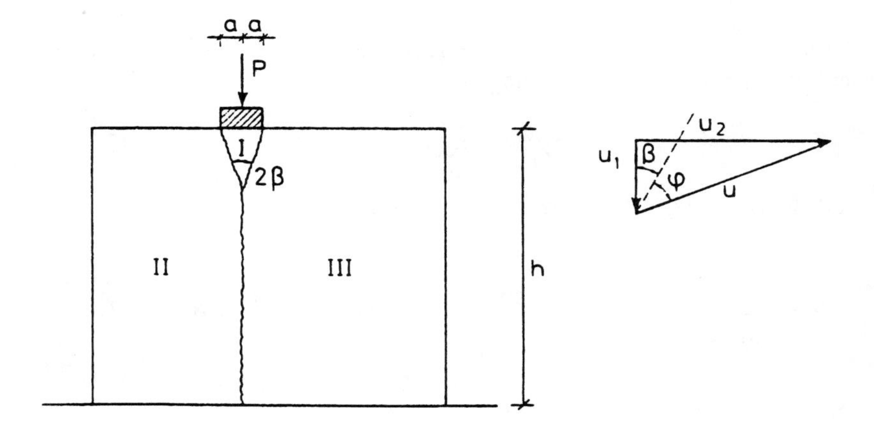

Figure 3.6.12 Failure mechanism for a rectangular block loaded by a concentrated strip load far from the edge.

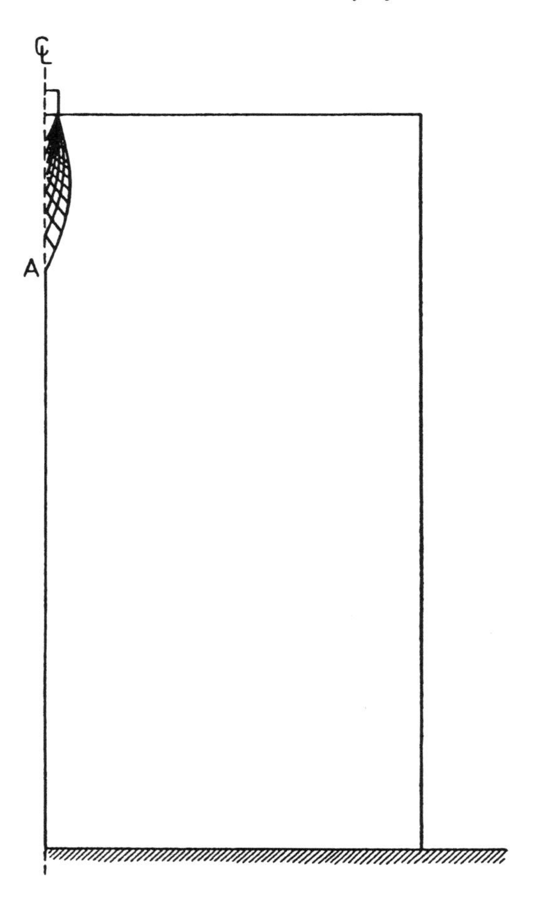

Figure 3.6.13 Failure zone for a rectangular block loaded by a concentrated strip load far from the edge.

$$W_E = Pu\cos(\beta + \varphi) \tag{3.6.21}$$

Setting $W_E = D$, we get the upper bound solution

$$P = \frac{2a}{\sin\beta\cos(\beta + \varphi)}\left[\frac{1 - \sin\varphi}{2}\, f_c + \sin(\beta + \varphi)\left(\frac{h}{a}\sin\beta - \cos\beta\right) f_t\right] \tag{3.6.22}$$

This expression should be minimized with respect to β. If the tensile strength is zero, we get the solution

$$P = 2af_c \tag{3.6.23}$$

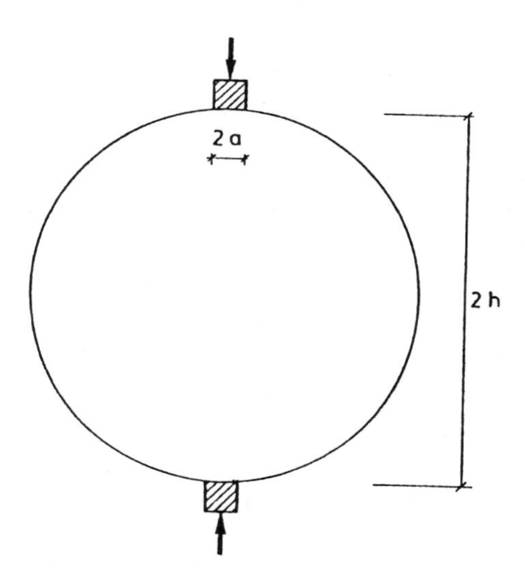

Figure 3.6.14 Splitting test of a concrete cylinder.

for $\beta = \pi/4 - \varphi/2$, which is evidently an exact one. In the general case β can be found from the equation

$$\cot\beta = \tan\varphi + \frac{1}{\cos\varphi}\sqrt{1 + \frac{\frac{h}{a}\cos\varphi}{\frac{f_c}{f_t}\frac{1-\sin\varphi}{2} - \sin\varphi}} \tag{3.6.24}$$

and if $\cot\beta \leq h/a$, the load-carrying capacity can be written

$$P = 2af_t\left[\frac{h}{a}\tan(2\beta + \varphi) - 1\right] \tag{3.6.25}$$

The exact solution in the case considered, where the loading is sufficiently far from the edge, is known. It was given by Izbicki [72.8]. The solution involves a slip line field as shown in Fig. 3.6.13. Following the line of symmetry from above, the normal stress in the line of symmetry starts being negative and then increases until it reaches the tensile strength f_t at A. From this point a yield line is developed with the normal stress equal to the tensile strength. This solution is also exact in the case where a concrete cylinder is compressed between two narrow plates (splitting test), provided that the width of the plate is sufficiently small (see Fig. 3.6.14).

The solution has been illustrated in Fig. 3.6.15. The calculations have been carried out by H. Exner [83.2]. The curves show $P/2hf_c$ as a function of f_t/f_c for different values of a/h. It is left to the reader to compare the exact solution with the upper bound solution (3.6.25) and to show that for realistic tensile strengths the deviation is less than 11%.

The elastic solution, under the assumption that the loading plates are small compared with the diameter, says that the tensile stress along the splitting surface is uniform and has the magnitude $P/\pi h$. If this stress is set equal to the tensile strength, it gives a straight line in Fig. 3.6.15, starting at (0,0) and having the inclination $\pi/2$.

Loading near the edge. If the loading is near the edge, other types of failure mechanisms have to be considered. A simple one is illustrated in Fig. 3.6.16. A rigid body *ABC*, containing the loaded surface, moves relative to the remaining part of the block, which is at rest. If the yield line *AB* is straight, the mechanism involves two degrees of freedom, the inclination of the yield line and the direction of the translation. It is left to the reader to verify that the best solution is obtained by putting the angle *CAB* equal to $\pi/4 + \varphi/2$ and letting the displacement vector form an angle φ with *AB*. This upper bound solution gives the load-carrying capacity

$$P = (a + b)f_c \tag{3.6.26}$$

corresponding to a compressive stress equal to the uniaxial compressive strength along *CA*. This solution is obviously correct if $b = a$. For $b > a$ the solution is not correct. A number of possible solutions have been studied by H. Exner [83.2].

In Fig. 3.6.17a, a mechanism involving rotation about a point *O* and a yield line *AB* in the form of a logarithmic spiral is shown (see Section 3.6.4). This mechanism also has two degrees of freedom. In the figure, these are chosen to be the two angles v_1 and v_2.

A better solution, of course, might be obtained if all possible yield lines starting at *A* and ending at a point on the edge of the block could be examined. This is possible without great effort, since it may be shown (see J. F. Jensen [81.2] and H. Exner [83.2]) that an optimal yield line separating rigid bodies of a weightless, modified

Coulomb material in plane strain is composed of a logarithmic spiral, as considered earlier, or part of a hyperbola. Yield lines in the form of hyperbolas were also discovered in a special case by Müller [78.9] and utilized by Marti [78.15].

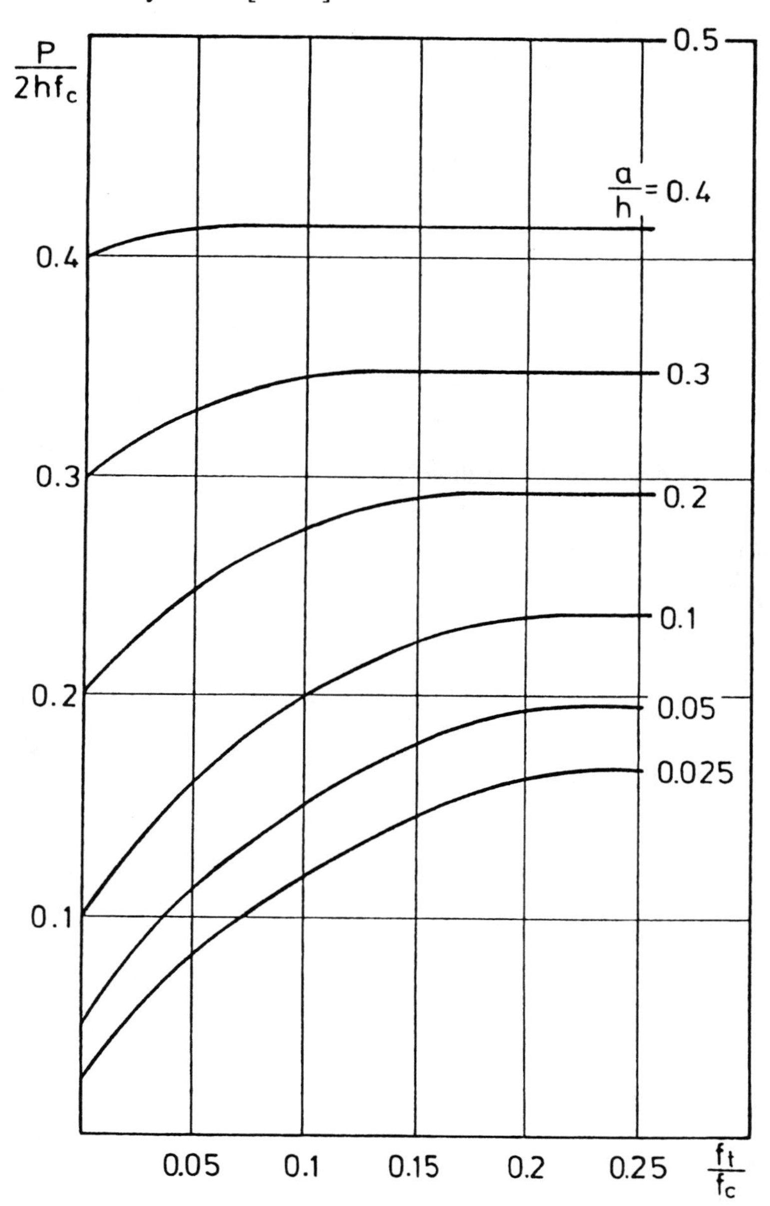

Figure 3.6.15 Splitting strength of a concrete cylinder.

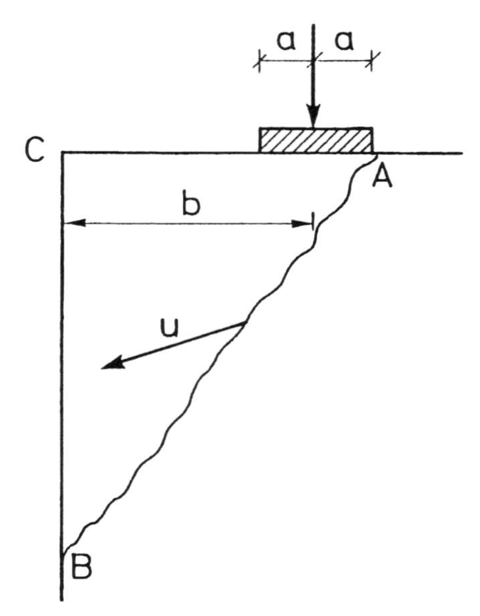

Figure 3.6.16 Failure mechanism for a rectangular block loaded by a concentrated strip load near the edge.

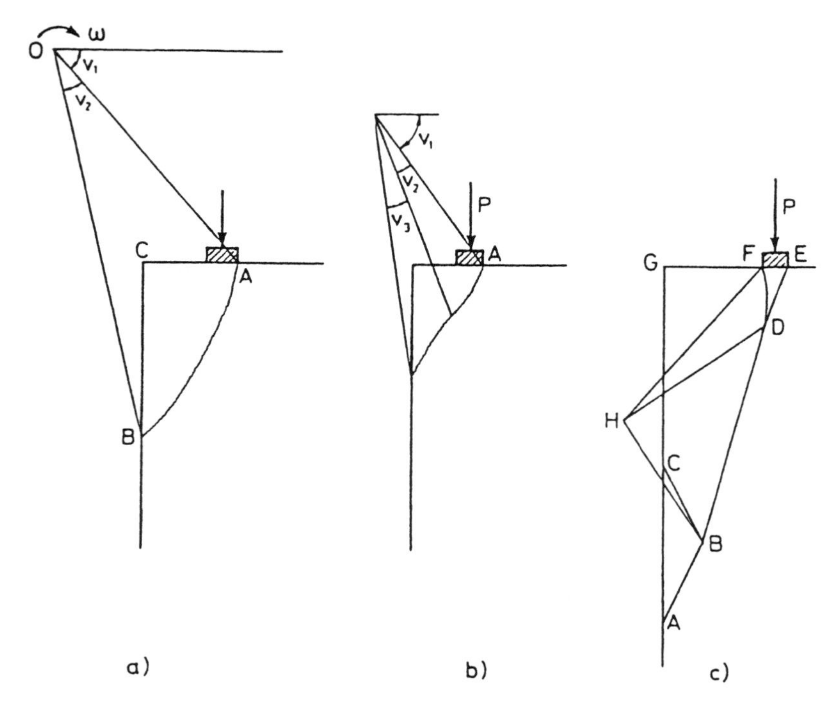

Figure 3.6.17 Failure mechanism for a rectangular block loaded by a concentrated strip load near the edge.

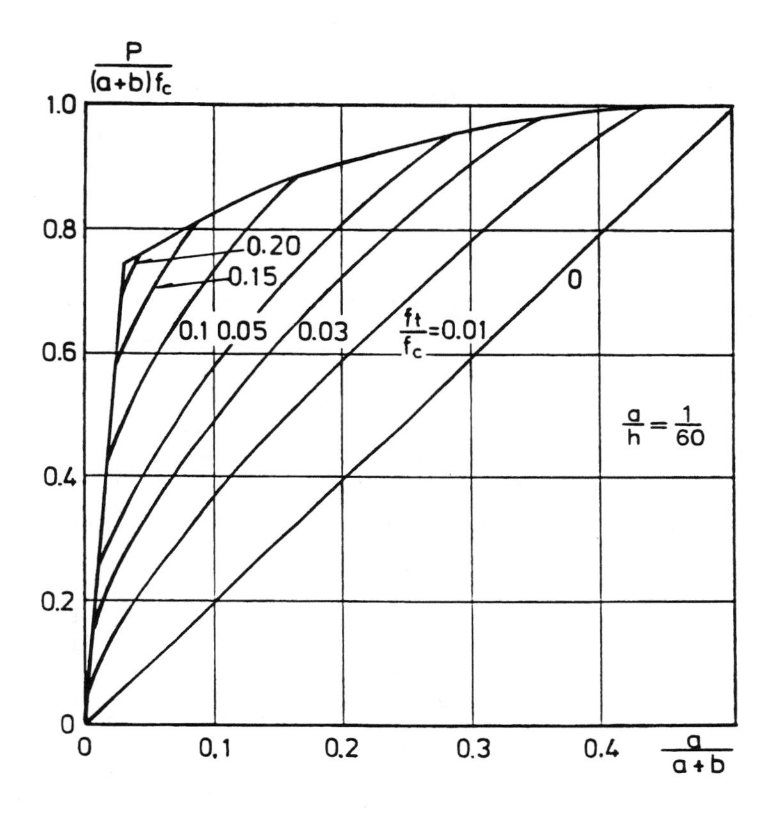

Figure 3.6.18 Splitting strength of a rectangular block.

The pole of the spiral is at the center of relative rotation and the asymptotes of the hyperbola are orthogonal and intersect at the center of relative rotation. The curve is smooth at the transition between the spiral and the hyperbola. In this case, therefore, the optimal upper bound solution with just one yield line is included in the mechanism shown in Fig. 3.6.17b. The mechanism has three degrees of freedom chosen as the angles v_1, v_2, and v_3. The load-carrying capacity will now be dependent on the tensile strength.

The mechanism in Fig. 3.6.17c contains a compressive zone *ABC* with angles $BCA = BAC = \pi/4 - \varphi/2$ with a homogeneous strain, body *BDFGC* rotating counterclockwise about *B*, and body *FDE* translating at an angle φ to *DE*. The relative rotation point of the two rigid, moving bodies is *H*, and therefore the curve *FD* is a logarithmic spiral with its pole at *H*. The acute angle between *HB* and *DE* is $\pi/2 - \varphi$. This mechanism has four degrees of freedom. It satisfies the natural requirement that there must be compressive stresses at the region where the yield line meets the edge of the block.

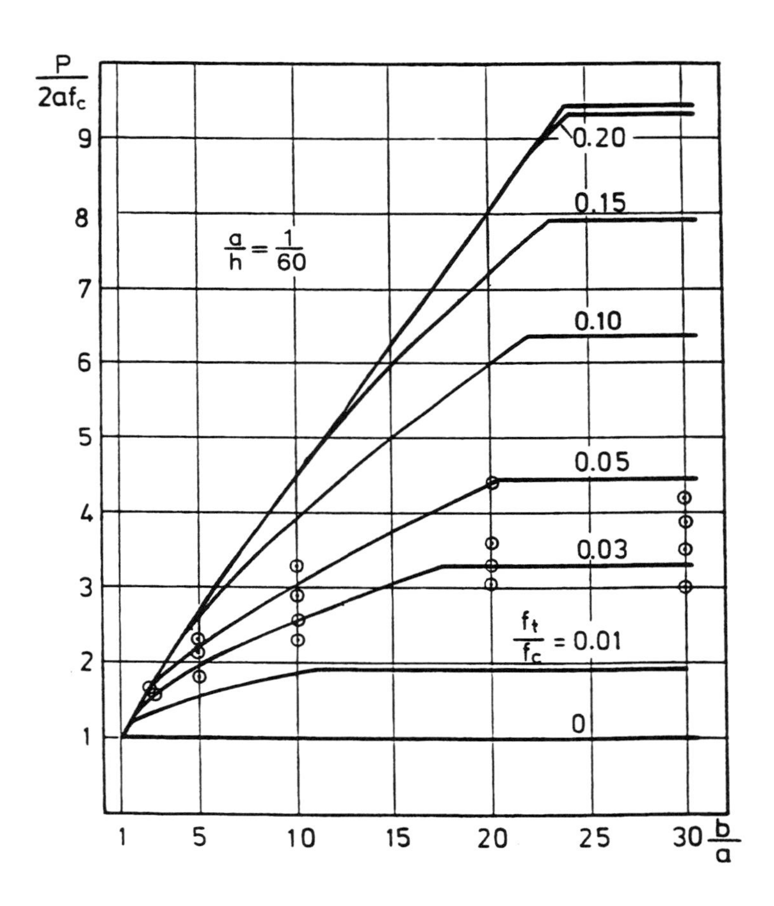

Figure 3.6.19 Splitting strength of a rectangular block compared with some test results.

Of course, the load-carrying capacity cannot be higher than that obtained by the Prandtl solution of Section 3.6.3, giving $P/2a = 13.73f_c$ for $\tan\varphi = 0.75$ ($\varphi = 37°$).

By combining the solutions mentioned, one finds the load-carrying capacity shown in Fig. 3.6.18. In Fig. 3.6.19 the same results are given using another set of parameters. In this figure some test results due to Rathkjen and treated in [76.1] are also plotted. It turns out that in this case an effective tensile strength of about 3% of the compressive strength f_c gives a good correlation with the test results.

3.6.6 Point Load on a Cylinder or Prism

Consider a circular cylinder with radius b and depth h loaded by a point load in the centre of the end face. The point load is distributed over a circular area with radius a (see Fig. 3.6.20).

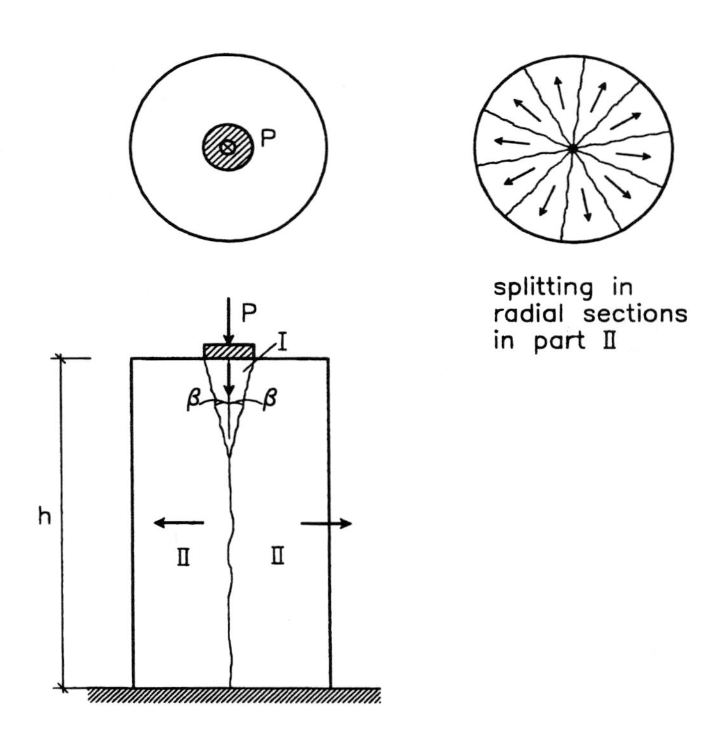

Figure 3.6.20 Point load on a circular cylinder.

In the figure a failure mechanism is shown, where a conical part I is pressed into the cylinder thereby splitting part II of the cylinder. The splitting is in this case taking place along radial sections in part II, in which we have tensile strains due to the outward directed displacements caused by the downward moving conical part I.

The calculations are analogous to those of the corresponding plane case treated in the previous section. The result is the following.

The angle β shown in the figure is determined by

$$\cot\beta = \tan\beta + \frac{1}{\cos\varphi}\sqrt{1 + \frac{\frac{2bh}{a^2}\cos\varphi}{\frac{f_c}{f_t}\frac{1-\sin\varphi}{2} - \sin\varphi}} \tag{3.6.27}$$

and the load-carrying capacity is determined by

$$\sigma = \frac{P}{\pi a^2} = f_t\left(\frac{2bh}{a^2}\tan(2\beta + \varphi) - 1\right) \tag{3.6.28}$$

The formulas are completely analogous to (3.6.24) and (3.6.25) if h/a is replaced by $2bh/a^2$.

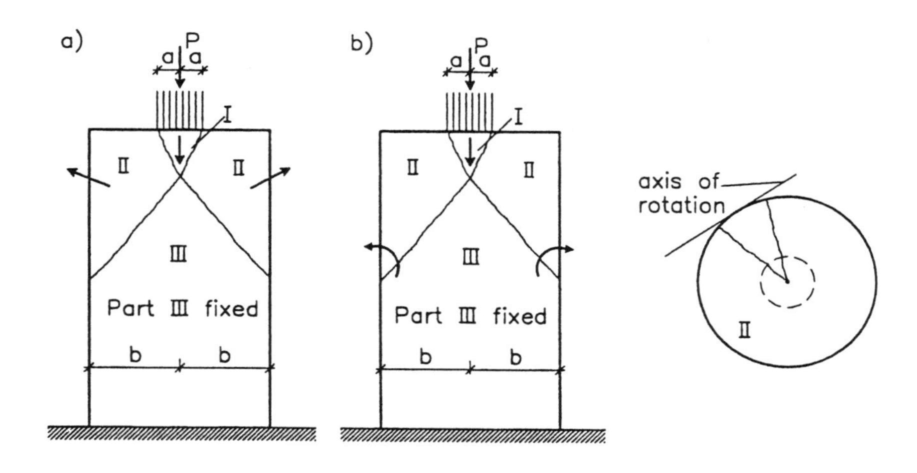

Figure 3.6.21 Local failure produced by a point load.

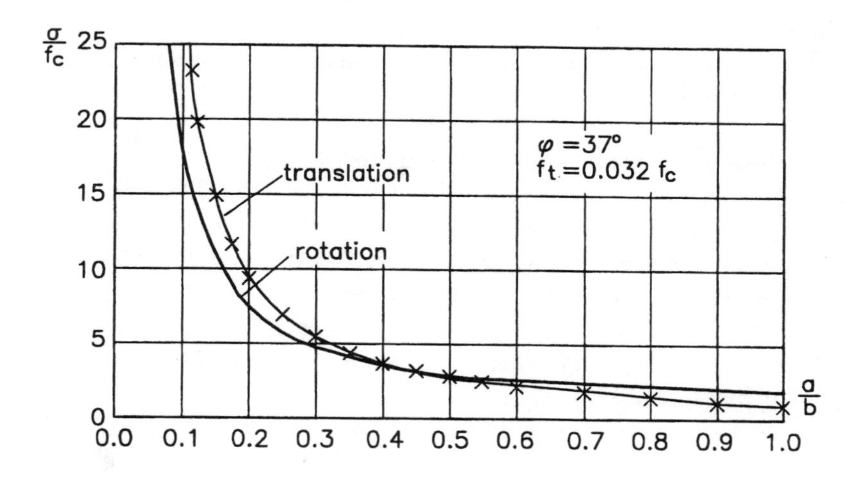

Figure 3.6.22 Load-carrying capacity, point load on a cylinder.

The solution is due to Chen and Drucker [69.3].

The same solution may be applied to a prism with a square cross section with side lengths $2b$ and loaded in the center over a square with sides parallel to the sides of the cross section and with side lengths $2a$.

If the cylinder is high enough the load is not able to split the cylinder along the whole depth. The failure mechanism will be local. Two local failure mechanisms are shown in Fig. 3.6.21. In case a), part III is fixed, part II is moving apart by translation and part I is as before moving downwards in the force direction. As before, in part II, we get radial yield lines.

In case b), part II is rotating along infinitely many rotation axes situated where the yield lines intersect the surface of the cylinder. The rotation axes are tangents to the cylindrical surface and lying in a plane perpendicular to the axis of the cylinder. As before, in part II we have radial yield lines.

Case a) was treated by Chen and Drucker [69.3] and case b) by Karen Grøndahl Lorenzen [95.9] [97.15] in a master thesis. We will not deal with the calculations in detail since they are rather lengthy and have to be done numerically. To illustrate the results, Fig. 3.6.22 shows σ/f_c as a function of a/b in the case $\varphi = 37°$. The tensile strength has been put to 3.2% of f_c, which gave good results in the strip load case (see Fig. 3.6.19). Actually this value is too high in this case if one compares it with test results. We will return to this problem in the next section.

It appears that the rotation mechanism is the best one for a/b less than about 0.5.

3.6.7 Design Formulas for Concentrated Loading

Approximate formulas. We shall conclude this chapter by showing how plastic theory may be used to develop simple design formulas for concentrated loading on unreinforced concrete.

The theoretical solutions are too complicated to use in practice and they also give very little feeling about the importance of the different parameters.

It turns out that practically all cases may be treated by the same type of approximation formula. We may arrive at such a type of formula by considering again the case of a strip load on a concrete block (cf. Section 3.6.5).

If in formula (3.6.22) we put $\beta = \pi/4 - \varphi/2$, which is the β-value for uniaxial compression, and if we further put $\varphi = 37°$, we find the simple formula

$$\sigma = \frac{P}{2at} = f_c + 4\left(\frac{h}{2a} - 1\right) f_t \qquad (3.6.29)$$

or

$$\frac{\sigma}{f_c} = 1 + 4\left(\frac{h}{2a} - 1\right)\frac{f_t}{f_c} \qquad (3.6.30)$$

This formula cannot be used directly as an approximation formula. It turns out that the factor 4 in the second term is too high. It also turns out that if this factor is lowered somewhat, very good agreement with the upper bound solution is obtained. If it is required that the upper bound solution coincides with the formula (3.6.30) for, say $f_t \cong 0.03 f_c$ and for $h/2a = 50$, it is found that the factor 4 has to be reduced to 2.48. Then the deviation between the upper bound solution and the formula

$$\frac{\sigma}{f_c} = 1 + 2.48\left(\frac{h}{2a} - 1\right)\frac{f_t}{f_c} \qquad (3.6.31)$$

will normally be less than 10%, a deviation which is of no practical significance.

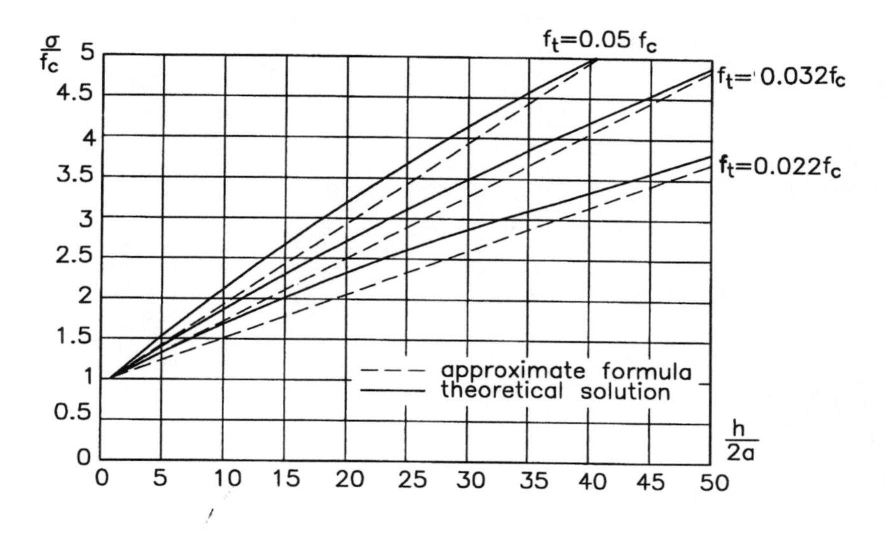

Figure 3.6.23 Comparison between the theoretical solution and the approximation formula (3.6.31) (strip load).

In Fig. 3.6.23 the formula (3.6.31) is compared with the theoretical solution for some values of f_t.

For a strip load near the edge we may be guided by Fig. 3.6.19 in suggesting approximate formulas. In this figure the horizontal lines for large *b/a*-values correspond to splitting failure along the whole depth. The remaining *b/a*-interval may be approximated by two straight lines. For very small *b/a*-values we may use (3.6.26) which may be written

$$\frac{\sigma}{f_c} = \frac{1}{2}\left(1 + \frac{b}{a}\right) \tag{3.6.32}$$

The remaining *b/a*-interval may be approximated by a straight line. It may be verified that

$$\frac{\sigma}{f_c} = 1.5 + 3.6\left(\frac{b}{a} - 1\right)\frac{f_t}{f_c} \tag{3.6.33}$$

is in good agreement with the curves in Fig. 3.6.19 for the relevant f_t / f_c-values, i.e., f_t / f_c-values of a few percent.

The three formulas covering the strip load case are summarized in Fig. 3.6.24 also showing the type of yield line pattern connected to the formulas. In the figure the constants K_1 and K_2 are defined and the theoretical values indicated. The values may later be modified when compared with test results.

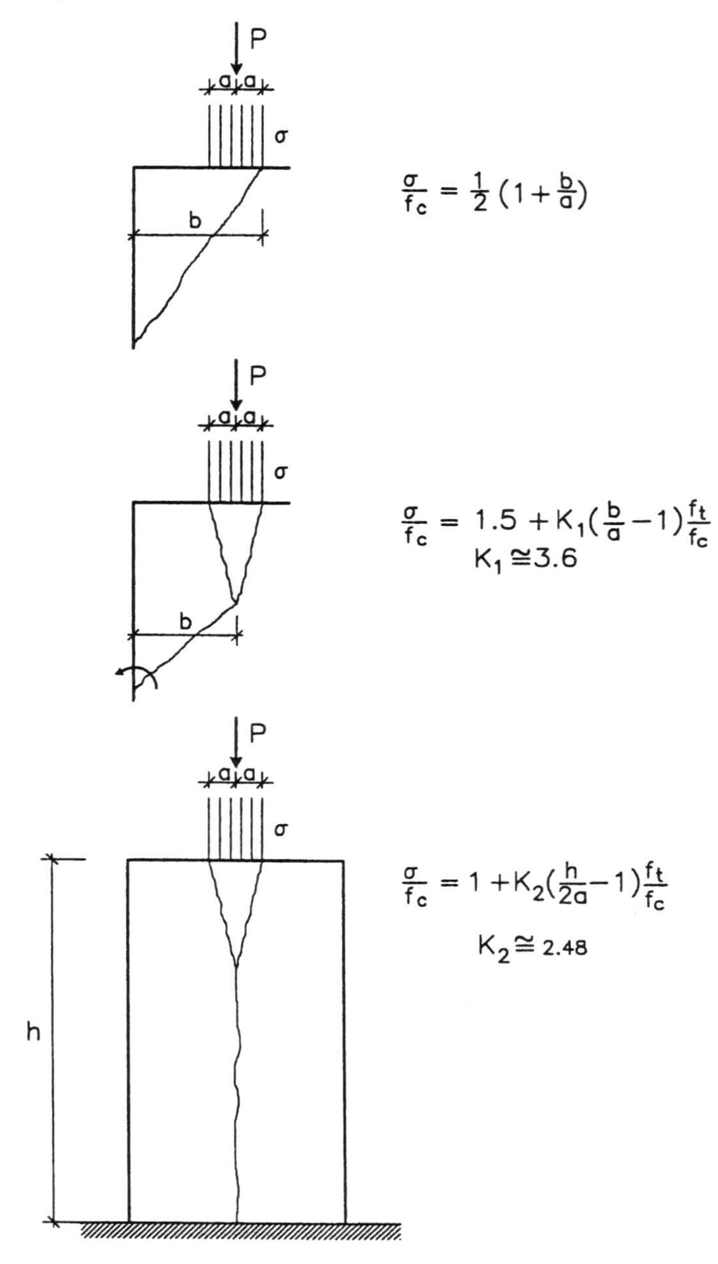

Figure 3.6.24 Yield line patterns and approximate formulas for strip load.

Turning now to the case of point loads, it is easy to write down a formula for the splitting strength of a cylinder or a square prism. This is so because the theoretical solution (3.6.27) and (3.6.28) may be derived from the strip load case by exchanging h/a with $2bh/a^2$ as already mentioned.

This means that a good approximation formula will be

$$\frac{\sigma}{f_c} = 1 + 2.48\left(\frac{bh}{a^2} - 1\right)\frac{f_t}{f_c} \tag{3.6.34}$$

As already mentioned this formula is also valid for a prism with a square cross section with side length $2b$ loaded in the center along a square with side length $2a$.

In the case $b = a$, $h = 2a = 2b$, which are the usual dimensions in a standard compression test, the formula should, strictly speaking, give $\sigma/f_c = 1$. The reason why formula (3.6.34) gives a higher value is the contribution from the radial yield lines. The formula could be corrected by replacing $bh/a^2 - 1$ with $bh/a^2 - 2$. This change is, however, of no importance in the usual region of application, so we will stay with (3.6.34).

When depth h is large, the cylinder or prism will not split along the whole depth, so we need a formula for large depths. The theoretical solution is shown in Fig. 3.6.22. An approximation formula of the type (3.6.33) may cover this case too, i.e., we tentatively write

$$\frac{\sigma}{f_c} = 1 + K_4\left(\frac{b}{a} - 1\right)\frac{f_t}{f_c} \tag{3.6.35}$$

To reproduce the lowest curves in Fig. 3.6.22 the constant K_4 must be around 50. However, this value is too high according to the test results which must be due to a lower effective tensile strength than that assumed in Fig. 3.6.22, which was about 3% of f_c.

In fact in this case the effective tensile strength seems to be only about half that obtained for a strip load and for the splitting case of a point load. If f_t is fixed at the same value as for a strip load and for the splitting case of a point load, which we will do in the final formulas (see below), K_4 will be around 25 to 30. The reason for the low value of the effective tensile strength in this case probably is the following: The first cracks to develop are the radial cracks. Only when these have been formed, can the other yield lines with tension be formed. Therefore the tensile stresses in the radial cracks might have almost disappeared when the final failure takes place, causing a low average value of the tensile stresses at failure.

The approximation formulas in the case of a point load are summarized in Fig. 3.6.25.

This is as far as we can get using theoretical considerations. In the final formulas, f_c must be replaced by νf_c and f_t must be replaced by the effective tensile strength $\nu_t f_t$ or ρf_c (cf. Section 2.1.4).

Regarding the effective compressive strength it will be seen that, at least for normal strength concrete, i.e., $f_c \leq 50$ MPa, ν may be chosen to be 1. The reason for this is, of course, that the triaxial stress field near the concentrated load will prevent cracking before failure, i.e. there will be no reduction in strength because of cracking. Also, it seems that there will be no reduction because of softening. However, there are some test results giving better agreement if ν is chosen around 0.6-0.7. In these cases a crack has probably developed

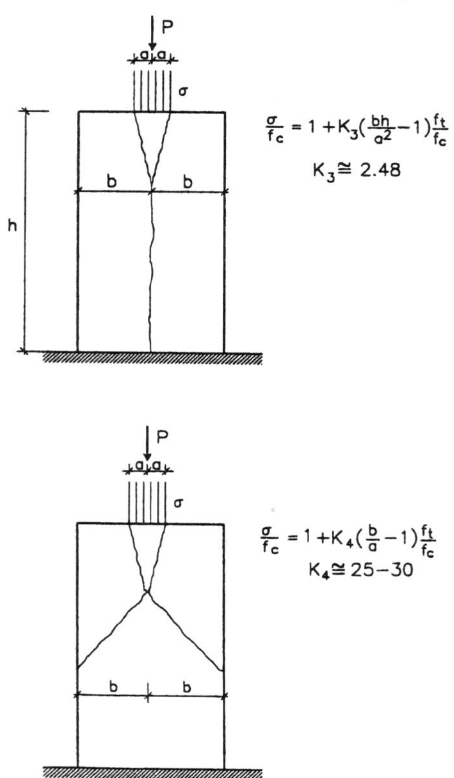

Figure 3.6.25 Yield line patterns and approximate formulas for a point load.

along some or the whole of the boundary of the conical part of the failure mechanism leading to a reduced strength. However, as we shall see, rather good results are obtained with $\nu = 1$, so we will maintain this value in doing the calibration with test results.

The most natural way of determining the effective tensile strength after having chosen $\nu = 1$ would be to use the theoretical values of our constants K_1, ..., K_4 and determine an optimized value of ν_t by comparison with tests. However, our approximate formulas are such, that we may just as well choose a reasonable value of the effective tensile strength and then determine optimized values of the constants.

We are coming close to the theoretical constants for a strip load and the splitting case of a point load by setting

$$f_{tef} = 0.5\sqrt{0.1\, f_c} \qquad (f_c \text{ in MPa}) \tag{3.6.36}$$

Since $\sqrt{0.1\, f_c}$ is a lower value of the tensile strength (cf. Section 2.1.3), it means that the effectiveness factor ν_t has been set at 0.5.

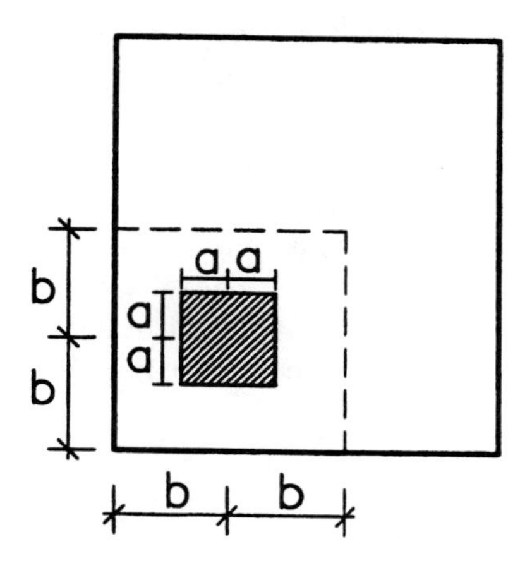

Figure 3.6.26 Eccentrical load on a rectangular cross section. Loaded area in the form of a square.

As mentioned this value seems to be too high in the point load case when local failure is governing. However, as already mentioned, since a lower value of ν_t in this case may be counteracted by selecting a lower value of the constant K_4, we may also use $\nu_t = 0.5$ in this case.

Semi-empirical formulas. In practice one has to solve a large variety of problems not covered by the theoretical solutions. Fortunately it seems possible to develop a set of semi-empirical formulas, which will cover a large part of the cases encountered in practice.

Consider first the case where a prism with square section is loaded eccentrically over a square with side lengths $2a$ (see Fig. 3.6.26). The distances from the center of the square to the edges of the square section of the prism are b. Further we assume that the sides of the loaded square are parallel to the sides of the section. Let us imagine that we know (which we do not) a lower bound solution for the point load case. Then this lower bound solution could be used in the eccentrical case considered, too, taking the stress field outside the area $2b \cdot 2b$ to be zero. This means that a safe value for the load-carrying capacity may be determined by our formula for the point load case inserting a and b into the formula.

Now what to do if the loaded area is rectangular $2e \cdot 2f$ and the distances to the edges of the section of the prism are different, say c and d, respectively, as shown in Fig. 3.6.27. A semi-empirical formula may be found by assuming that if the loaded area is not too far from a square (or a circle) and the effective surrounding area, the rectangle $2c \cdot 2d$ is also not too far from a square (or a circle), then we may determine the load-carrying capacity by transforming to the square (or circle) case keeping the loaded area and the effective surrounding area constant. This transformation may be written

$$\begin{aligned} 4a^2 &= 2e \cdot 2f = A_o \quad \Rightarrow \quad a = \tfrac{1}{2}\sqrt{A_o} \\ 4b^2 &= 2c \cdot 2d = A \quad \Rightarrow \quad b = \tfrac{1}{2}\sqrt{A} \end{aligned} \tag{3.6.37}$$

having denoted the loaded area by A_o and the surrounding effective area by A. If this is right, we must insert in our point load formula

$$\frac{b}{a} = \sqrt{\frac{A}{A_o}} \tag{3.6.38}$$

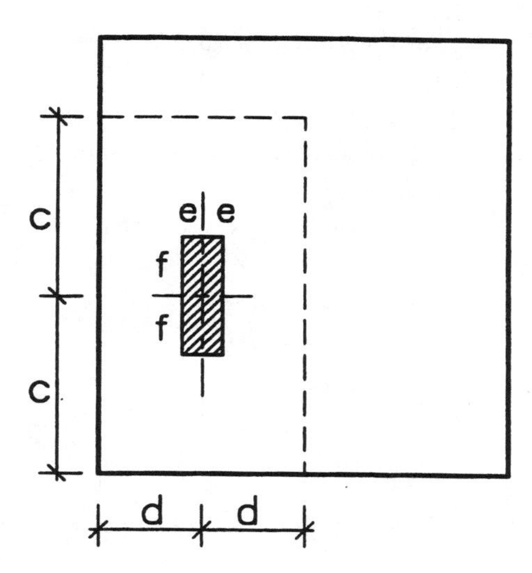

Figure 3.6.27 Eccentrical load on a rectangular section. Loaded area in the form of a rectangle.

A transformation of this kind has turned out to be very successful in the case of deep prisms or cylinders, i.e., specimens which do not split along the whole depth but have a local failure. The use of the area ratio A/A_o is not an invention in plastic theory. It was used previously by Bauschinger in 1876, although his formula was quite different. (He dealt with rock and used the third root of the area ratio. He did not consider the influence of tensile strength.)

If we insert (3.6.38) into (3.6.35) we find

$$c_{\text{local}} = \frac{\sigma}{f_c} = 1 + K_4 \left(\sqrt{\frac{A}{A_o}} - 1 \right) \frac{f_t}{f_c} \qquad (3.6.39)$$

Here we have introduced an enhancement factor $c_{\text{local}} = \sigma/f_c$, i.e., the ratio of the average strength over the loaded area divided by the uniaxial compressive strength f_c.

In practice the transformation rule introduced is used for arbitrary loaded areas not too far from the square or the circle. The effective surrounding area A is chosen as the largest area which may be placed in the section, having the same center of gravity as the loaded area. The rule is illustrated in Fig. 3.6.28. In this figure the cross section with the concentrated load is drawn as a rectangle although the section may be of any form.

One would imagine that formula (3.6.39) cannot be valid in the case of a strip load, differing fundamentally from a point load. However, it turns out that (3.6.39) may also be used for a strip load when local failure takes place. This may be shown in the following way.

Local failure for a strip load is governed by the formulas (3.6.32) and (3.6.33), i.e,

$$c_{\text{local}} = \frac{\sigma}{f_c} = \min \begin{cases} \dfrac{1}{2}\left(1 + \dfrac{b}{a}\right) \\ 1.5 + 3.6\left(\dfrac{b}{a} - 1\right)\dfrac{f_t}{f_c} \end{cases} \tag{3.6.40}$$

the smallest value being valid.

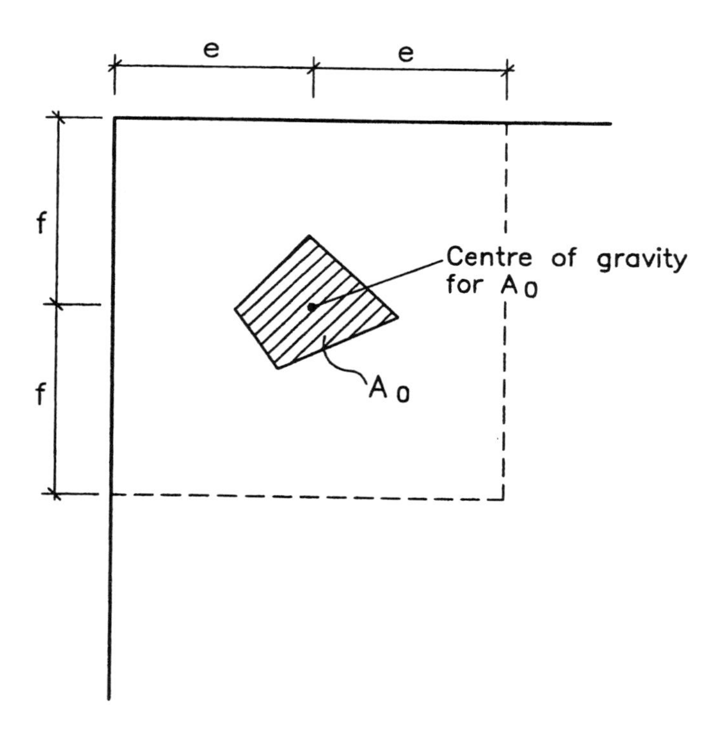

Figure 3.6.28 Rule to determine the effective area A for an arbitrary concentrated load.

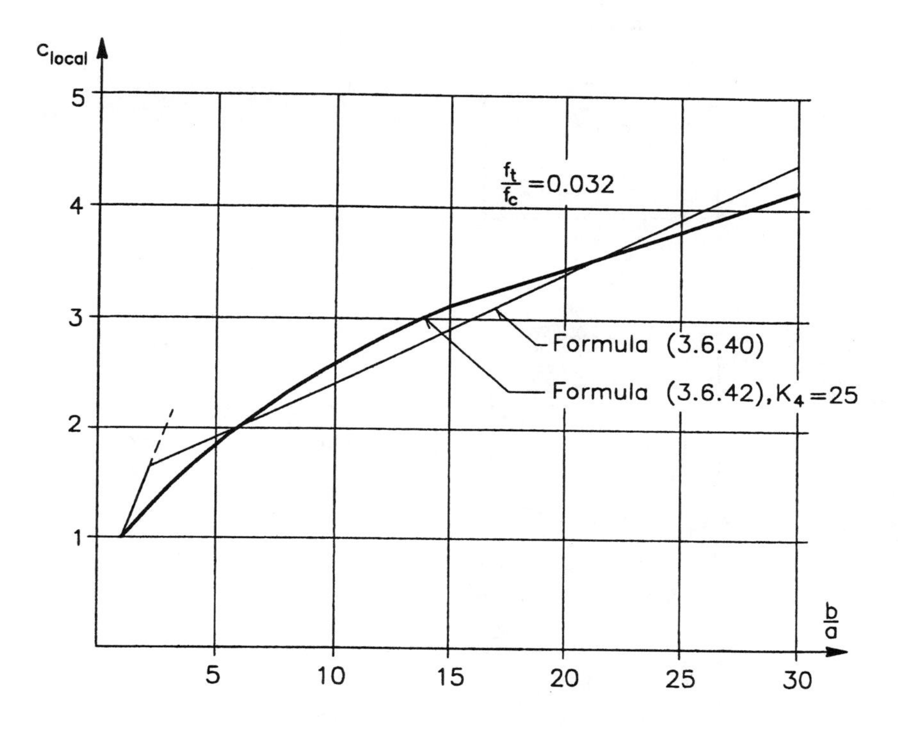

Figure 3.6.29 Local failure, comparison between formulas (3.6.40) and (3.6.42).

For a strip load we have

$$\sqrt{\frac{A}{A_o}} = \sqrt{\frac{2bt}{2at}} = \sqrt{\frac{b}{a}} \tag{3.6.41}$$

whereby (3.6.39) reads

$$c_{\text{local}} = \frac{\sigma}{f_c} = 1 + K_4\left(\sqrt{\frac{b}{a}} - 1\right)\frac{f_t}{f_c} \tag{3.6.42}$$

In Figure 3.6.29 the two formulas have been compared using $K_4 = 25$ and $f_t/f_c = 0.032$. It appears that the results of the two formulas are almost identical, taking into account the rather large scatter we will see when considering the test results.

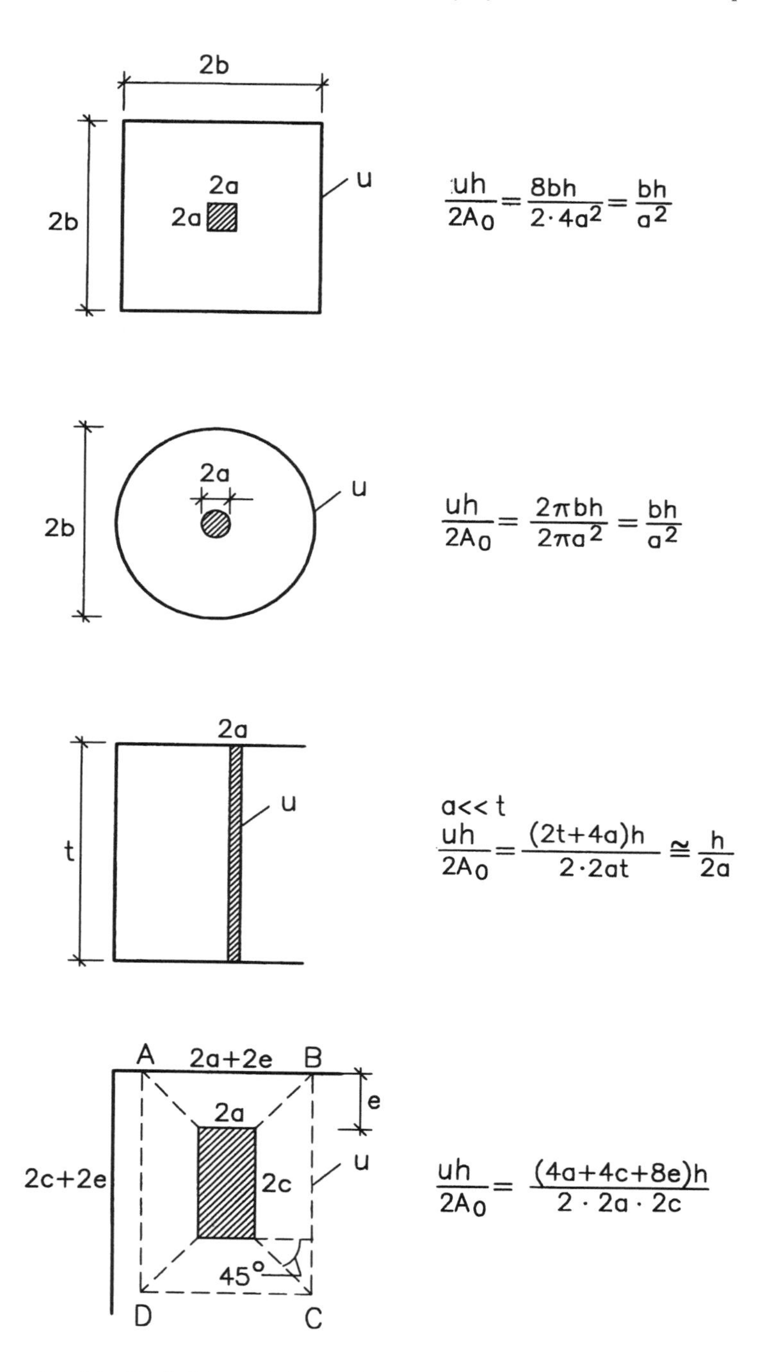

Figure 3.6.30 Definition of the perimeter u.

We conclude that local failure may be dealt with using only one semi-empirical formula, namely (3.6.39), a rather amazing simplification.

The local failure is by far the most important in design. This is rather fortunate because it seems more difficult to simplify the splitting failure into a single formula.

We have shown that splitting in the case of a strip load is governed by the parameter $h/2a$ and in the case of a point load by the parameter bh/a^2.

One possibility, which has been discussed in [95.9] [97.15], is to introduce a parameter $uh/2A_o$, u being the perimeter of a curve to be defined in a moment, h the depth, and A_o the loaded area. The curve for which to calculate the perimeter u is determined by the smallest distance from the loaded area to the edge of the cross section loaded by concentrated force. In the case of a square cross section or a circular cross section loaded in the center, u is simply the perimeter of the cross section. In the case of a strip load, where the smallest distance to the edge of the cross section is zero, u is the perimeter of the loaded area. In Fig. 3.6.30 it is shown that $uh/2A_o$ is the right parameter to use in the case of a strip load, when the width of the strip $2a$ is small compared with the thickness t. In the figure it is also shown how to define u when the loaded area is a rectangle.

It would be easy to extend the definition of u to a more general case analogous to the one shown in Fig. 3.6.28 for the local failure. However, it is a little speculative at the present stage of development, since experimental evidence is lacking except for rectangular loaded areas on rectangular cross sections. So the generalization must at the moment be left to the authors of codes of practice.

If we accept that such a generalization is possible, then the splitting case may also be treated by a single formula, which may be written

$$c_{\text{split}} = \frac{\sigma}{f_c} = 1 + K_3\left(\frac{uh}{2A_o} - 1\right)\frac{f_t}{f_c} \tag{3.6.43}$$

Here we have introduced the enhancement factor c_{split}, which is analogous to c_{local}.

We are left with only two constants K_3 and K_4 to calibrate against experiments. The reader will probably share our wonder that such a complicated problem like the load-carrying capacity at concentrated loads may be reduced to two such simple formulas.

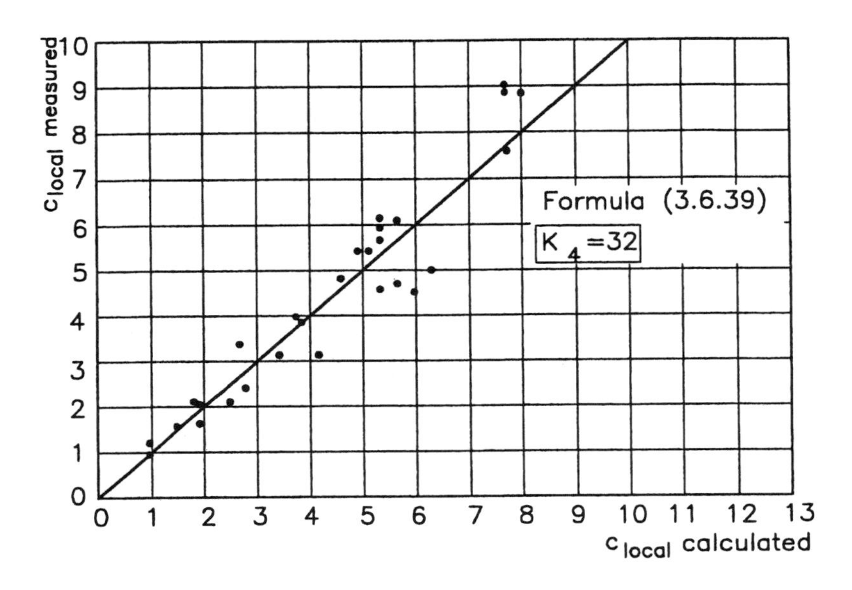

Figure 3.6.31 Comparison with test results, local failure, point load.

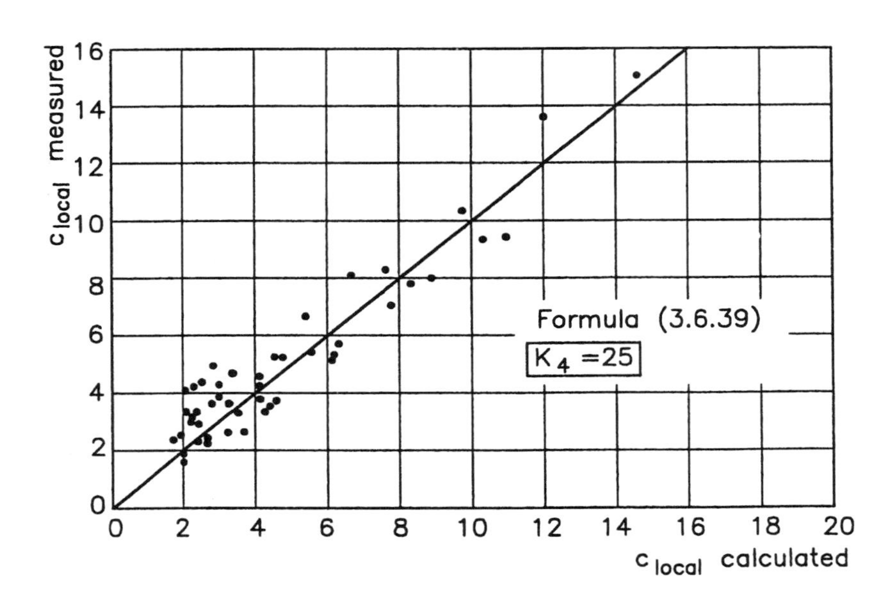

Figure 3.6.32 Comparison with test results, local failure, rectangular load.

Of course, in practice the formulas should only be used in cases comparable to the point of departure for developing them. Loaded areas for instance in the form of a combination of several strip loads are not covered. This case is met if, say, a steel column with an I-shape is supported on an unreinforced concrete block. Due to the limited ductility in tension of the concrete one may not be sure that such more complicated cases may be treated by extrapolation. In the case mentioned it is, of course, easy to change the design to be covered by the above formulas by supporting the column on a sufficiently stiff and strong transverse plate welded to the column.

Comparison with tests. The available test results have been compared with the formulas developed above in [95.9] [97.15], to which the reader is referred for a more detailed description of the test specimens, the strength parameters, test procedure, etc.

Test results with local failure in the case of a central point load are shown in Fig. 3.6.31. Only tests which according to the final formulas would lead to local failure have been included.

The figure shows the measured value of c_{local} versus the calculated value of c_{local} using formula (3.6.39). The points should lie on a straight line through origin with the inclination 45°. The best value of the constant K_4 is, in this case, 32. It has been assumed that the effective tensile strength may be calculated according to formula (3.6.36), i.e., $\nu_t = 0.5$. The agreement is astonishingly good, bearing in mind that we are dealing with a case where a prime contribution to the load-carrying capacity stems from the tensile strength of concrete.

In Fig. 3.6.32 tests with loaded areas in the form of a rectangle placed in different positions in a rectangular cross section are shown.

The best value of K_4 is, in this case, 25. The effective tensile strength is, of course, calculated as before. The agreement is also rather good in this case except in a region of small values of c_{local}. In these particular tests the loaded area was placed rather close to the edge of the cross section, giving rise to a failure pattern not really covered by the formulas. We will return to this point at the end of this section.

Finally, Fig. 3.6.33 shows some test results with strip load. They have been treated using the original formulas (3.6.40) as well as the general formula (3.6.39). The agreement is rather good when the original formulas are used, less good when the general formula with $K_4 = 25$ is used.

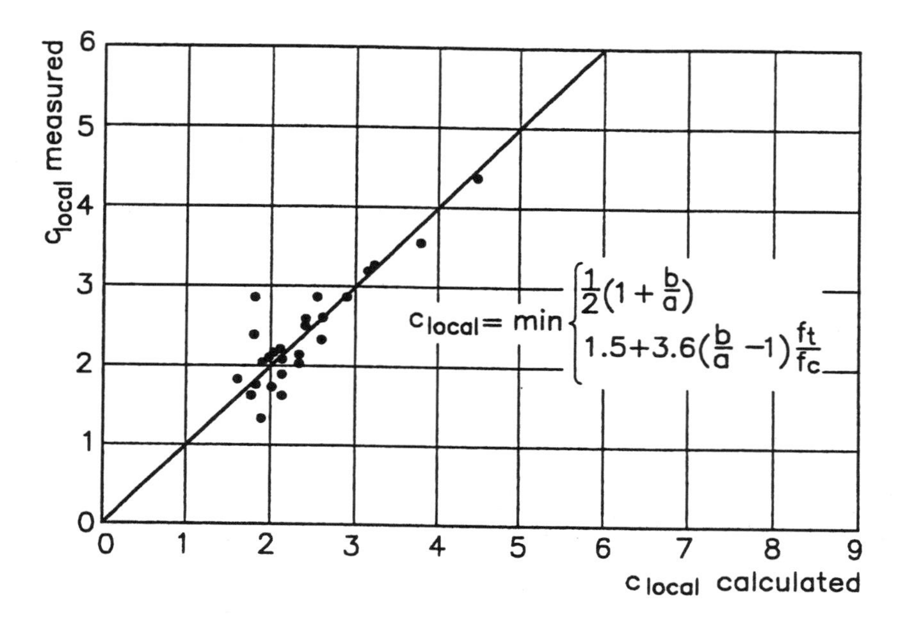

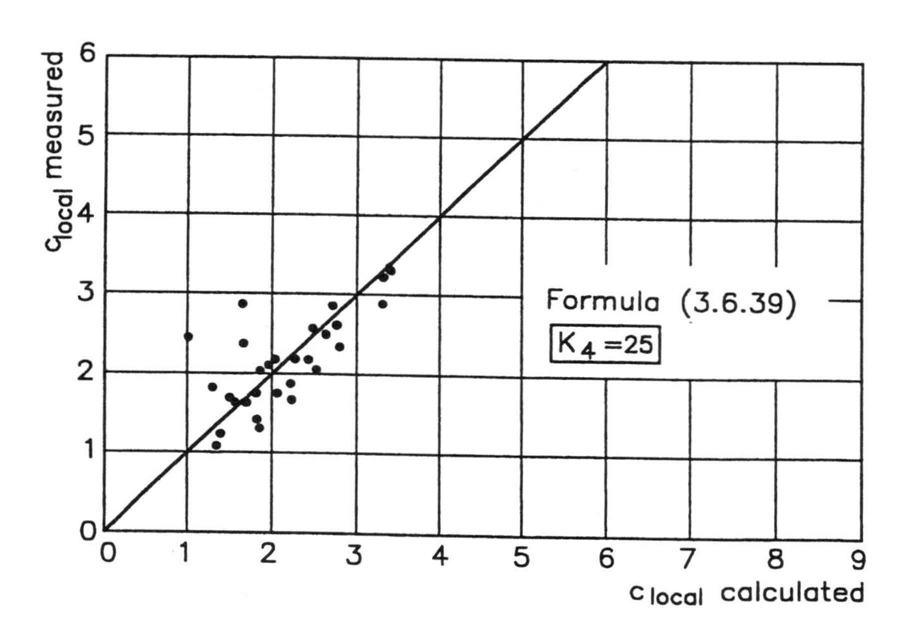

Figure 3.6.33 Comparison with test results, local failure, strip load.

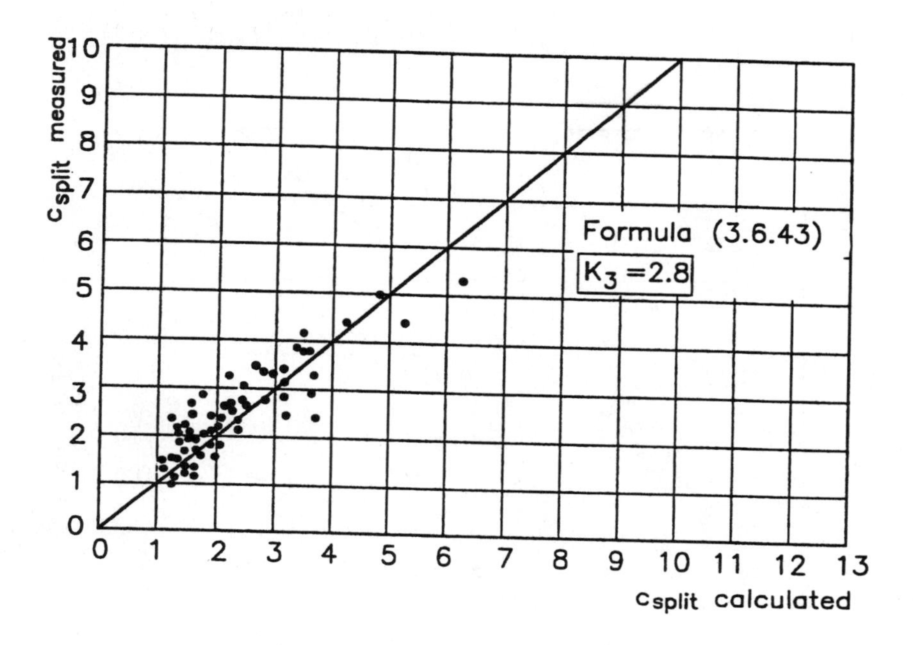

Figure 3.6.34 Comparison with test results, splitting failure, point load.

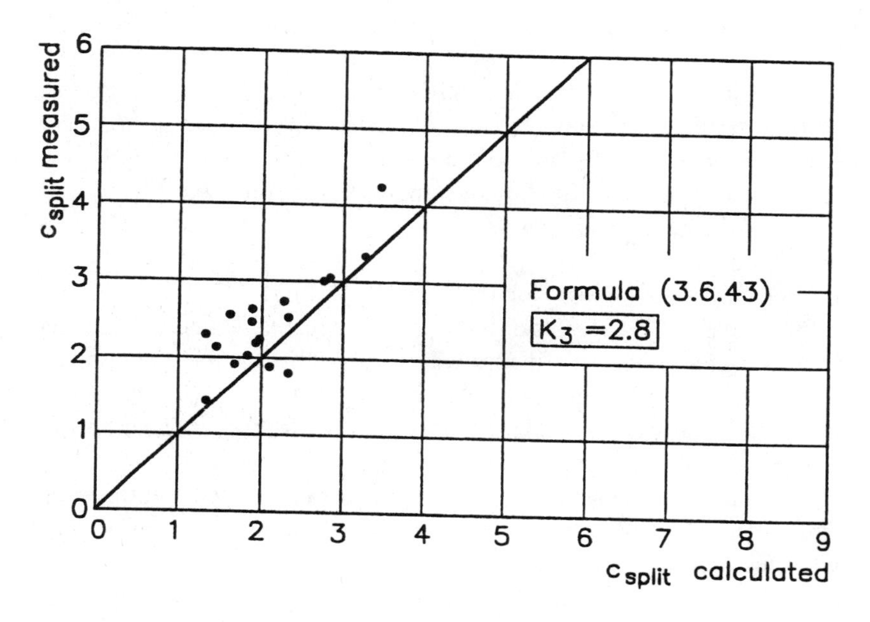

Figure 3.6.35 Comparison with test results, splitting failure, rectangular load.

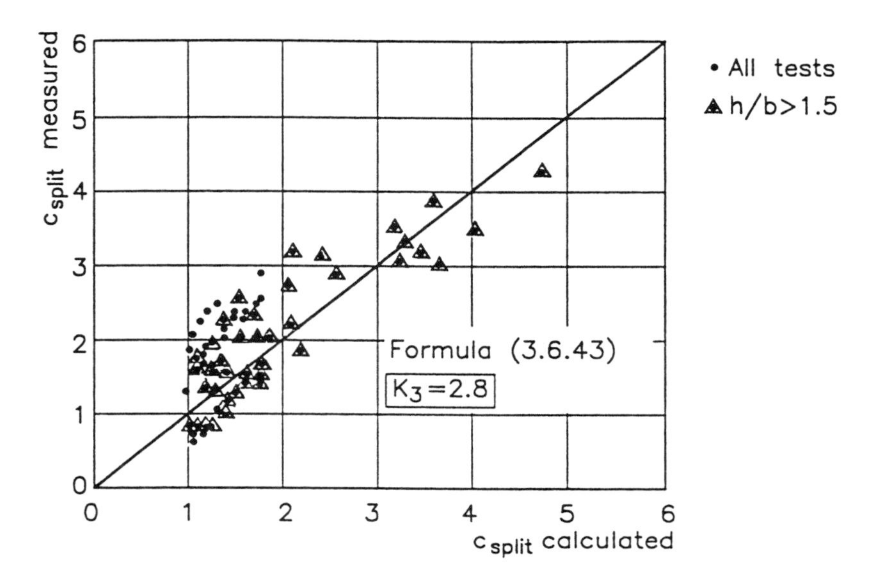

Figure 3.6.36 Comparison with test results, splitting failure, strip load.

In Figures 3.6.34 to 3.6.36 the corresponding test results for splitting failure are shown. The calculated value of c_{split} has been determined by formula (3.6.43) and the effective tensile strength as before. The best value of the constant K_3 is found to be 2.8.

The agreement is considerably poorer than what we found for the local failure.

There are two reasons for that.

First, many of these tests have been performed using a specimen supported in a rather indefinable way. The friction along a supporting surface will act as a reinforcement, but this action is difficult to take into account in the calculation because the friction properties are largely unknown. This explains why many test results are much higher than the theoretical values. The theoretical assumptions are better fulfilled if the specimen is made with double depth and loaded by concentrated forces in both ends. In this way the symmetry section acts as a perfectly frictionless support. This is illustrated in Fig. 3.6.37. In such tests the h-value to insert in our formulas is, of course, half the total depth of the specimen as indicated in the figure.

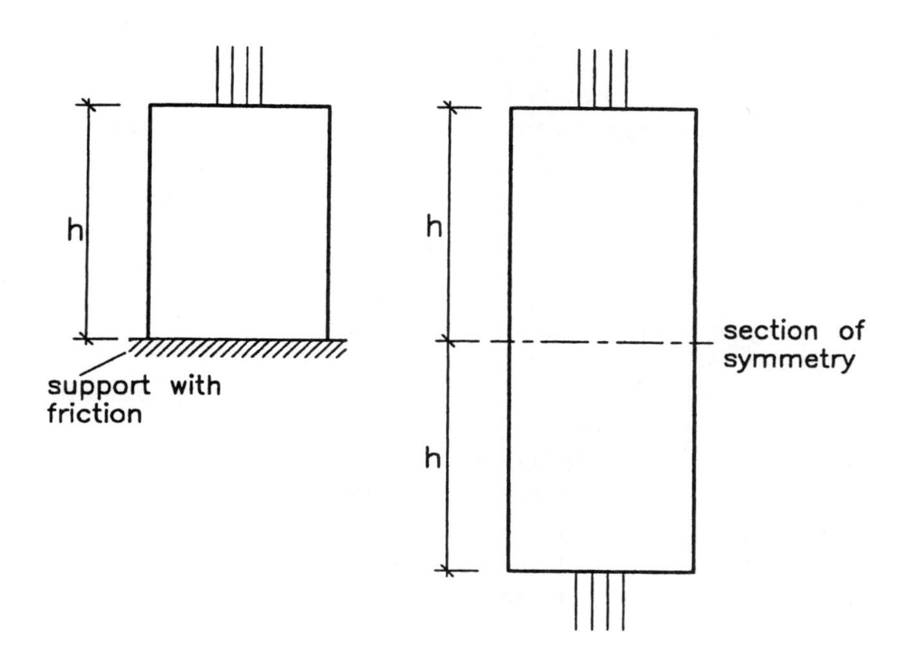

Figure 3.6.37 Eliminating the effect of friction along support by using double depth.

Another reason for the discrepancy is that some tests in the low value region of c_{split} have been done using very small depths, $h/2a$ being of the order 1. For $h = 2a$ we have exactly the same conditions as in a standard compression test and, as mentioned earlier, the formulas developed cover this case in a poor manner.

Some of the tests with $h/2a$ around 1 and a little higher sometimes give smaller values of σ/f_c than 1. This is, as mentioned before, probably due to crack formation in the compression zone near the concentrated force. For $h/2a < 1$ the load-carrying capacity σ/f_c is normally greater than 1. The reason is the same as why the cube strength is normally higher than the cylinder strength. Due to the friction between the loading platen and the test specimen, the usual yield lines cannot develop.

In some test series the load-carrying capacity is constantly lower than the one we get by taking the numerical value 1 as the first term on the right-hand side of our formulas for σ/f_c. These test results are typically such that the number 1 should be replaced by, say 0.7. This is probably also due to crack formation in the compression zone. This problem could be dealt with using an effectiveness factor < 1 on the compressive strength f_c . If ν equals, say 0.7, one would get 0.7 as the first term in our formulas, letting f_c still mean the uniaxial compressive strength. Having chosen to use $\nu = 1$, we must, as mentioned earlier, take the reduction in load-carrying due to crack formation into account by reducing the values of the constants K_1, ..., K_4.

If all tests with small depths are taken out, say tests with $h / b < 1.5$, the agreement is much better as appears from Fig. 3.6.36. Strictly speaking, all tests which are not really splitting tests should be taken out and only tests of the type shown in Fig. 3.6.37 to the right should be retained.

Tests with double depth were carried out at the Technical University of Denmark in 1995. They are reported in [95.9] [97.15]. The test specimens and the loading arrangement are shown schematically in Fig. 3.6.38. The total depth was varied from 200 to 600 mm and the length of the strip load ℓ from 60 to 200 mm. The width of the strip load was constant at 30 mm. The distance b was constant at 50 mm. The compressive strength was in the range 20 to 30 MPa.

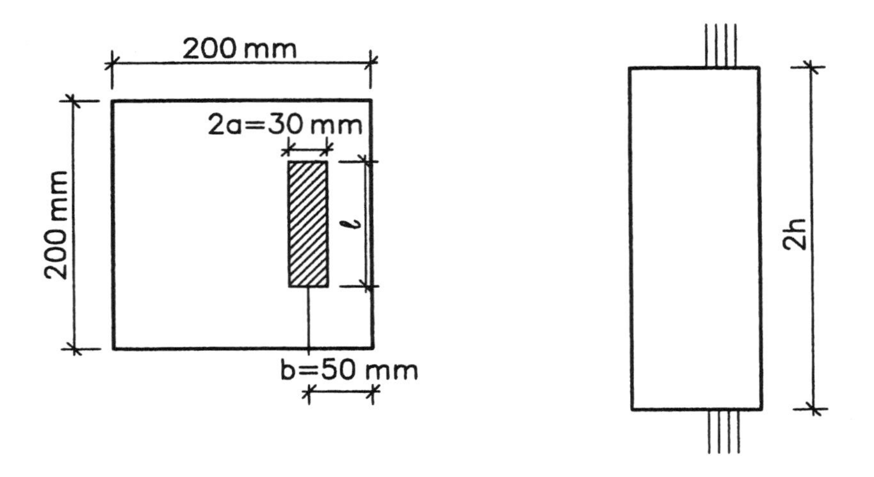

Figure 3.6.38 Tests with double depth to eliminate effect of friction along support [95.9] [97.15].

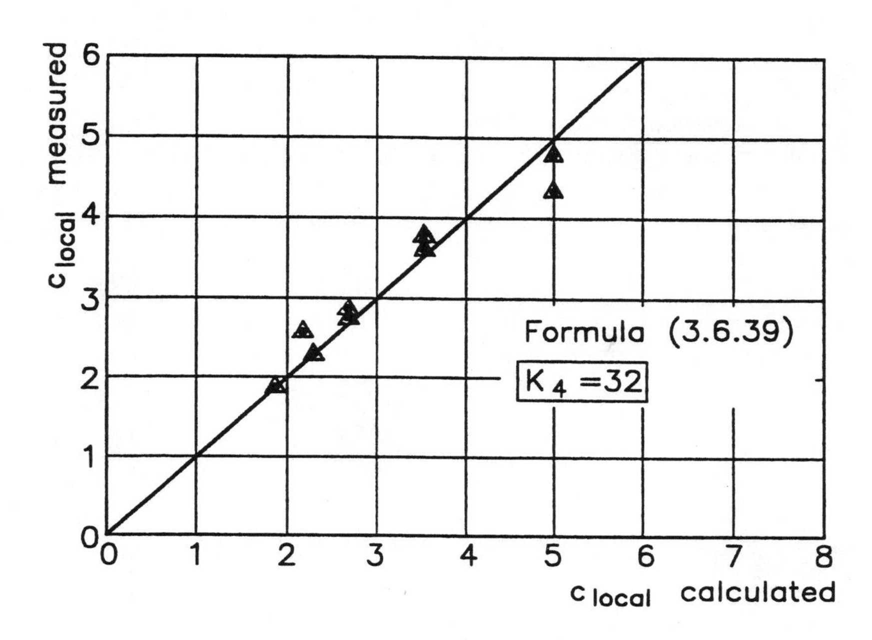

Figure 3.6.39 Comparison with test results, local failure, rectangular load.

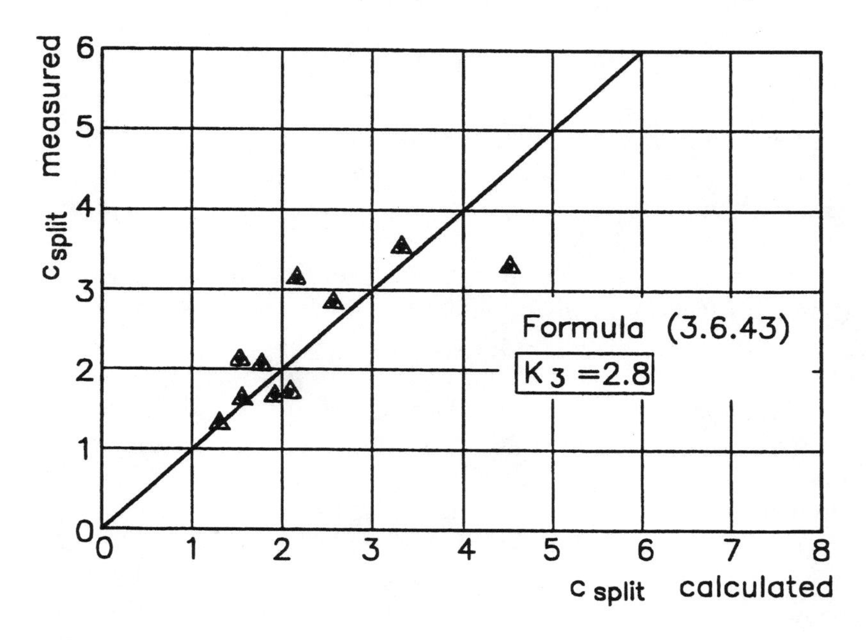

Figure 3.6.40 Comparison with test results, splitting failure, rectangular load.

The test results are shown in Figs. 3.6.39 and 3.6.40. When calculating c_{local}, formula (3.6.39) has been used with $K_4 = 32$. In formula (3.6.43) for c_{split} the value $K_3 = 2.8$ has been selected.

The agreement is rather perfect.

Our formulas were even able to predict the failure mode, i.e.,whether local failure or splitting failure would occur depending on which formula gave the lowest value of the enhancement factor.

Conclusion. We may conclude that the load-carrying capacity at concentrated loads on unreinforced concrete may be calculated using two simple, semi-empirical formulas.

Local failure is governed by the formula

$$c_{\text{local}} = \frac{\sigma}{f_c} = 1 + 25\left(\sqrt{\frac{A}{A_o}} - 1\right)\frac{f_t}{f_c} \tag{3.6.44}$$

having set the constant $K_4 = 25$, which, according to the test results, will be on the safe side.

Regarding the definition of A, see Fig. 3.6.28. A_o is the loaded area. The stress σ is the average stress over the loaded area at failure.

Splitting failure is governed by the formula

$$c_{\text{split}} = \frac{\sigma}{f_c} = 1 + 2.8\left(\frac{uh}{2A_o} - 1\right)\frac{f_t}{f_c} \tag{3.6.45}$$

Regarding the definition of the perimeter u, see Fig. 3.6.30.

When applying the formulas, the effectiveness factor on the compression strength may be set to 1 at least for normal strength concrete, i.e., $f_c \leq 50$ MPa.

The effective tensile strength may be calculated by formula (3.6.36), which means that in the two formulas above we may insert

$$f_t = 0.5\sqrt{0.1 f_c} \qquad (f_c \text{ in MPa}) \tag{3.6.46}$$

The allowable c-values are normally restricted by code rules to limit in a simple way the local deformations around the concentrated force. Popular limits on c are in the range 2-4. However, from a pure strength point of view, the limits are, as already mentioned, in the strip load case the Prandtl solution given in Section 3.6.3 limiting σ to 13.73 f_c. The corresponding solution in the point load case is not known. It is often assumed to be around 25% higher, limiting σ to about 17 f_c. Of course, these extreme values of σ are of little practical interest and in this range only a very limited number of test results are available.

The reader may have noticed that almost nothing has been said about lower bound solutions. The whole development has been based on upper bound solutions. In fact very little has been done up to now regarding lower bound solutions.

Effect of reinforcement. There is an important point which should be commented upon. That is the influence of reinforcement on the load-carrying capacity. Some preliminary studies have been made by B. C. Jensen [76.1] and Jens Christoffersen [96.1] [97.10], but a lot remains to be done in this field. Theoretically it is easy to include reinforcement in upper bound solutions. For instance in a yield line crossing a reinforcement bar, the contribution to the dissipation from the bar is the displacement discontinuity in the bar direction times the bar area times the yield strength.

Probably reinforcement influences the load-carrying capacity in very much the same way as for anchor bars and lap splices. In Chapter 9 there is a thorough discussion on this point. For small amounts of reinforcement the reinforcement will not yield when maximum load is reached and the load-carrying capacity is pretty much the same as for unreinforced concrete. The reinforcement will slightly increase the effective tensile strength but this is of rather small practical significance for the concentrated loads problem and may be disregarded. When the reinforcement is sufficiently large, maximum load is reached when the reinforcement is yielding and the contribution from the tensile strength of concrete has vanished.

When a reinforcement corresponding to the reinforcement ratio r is supplied in the sections suffering tensile failure, a simple estimate of the load-carrying capacity may be obtained by replacing f_t/f_c in the above formulas with $rf_Y/f_c = \Phi$, Φ being as usual the reinforcement degree. If the load-carrying capacity obtained in this way is higher than that determined for unreinforced concrete, this higher value will be an estimate of the load-carrying capacity for the reinforced case. Otherwise the load-carrying capacity is as for unreinforced concrete.

In cases with tensile failure in many directions, reinforcement in three perpendicular directions, each direction with a reinforcement ratio r, may be used. Probably, in most cases, reinforcement in two directions perpendicular to the loading direction will suffice. In the strip load case only reinforcement in one direction is needed.

The load-carrying capacity for reinforced concrete obtained in this way will probably be rather accurate in most cases. However, no complete comparison with test results has yet been carried out. The estimate obtained will be conservative in the point load case. The reader may recall that in this case, the theoretical value of the constant K_4 was found to be higher than the value we have used in the final formula due to a lower effective tensile strength of the concrete. It would probably be allowable to use the theoretical value when dealing with reinforcement. However, it is recommended not to take this into account at the present stage of development.

From the previous analysis we may draw an important conclusion regarding minimum reinforcement for concentrated loading. Since the effective tensile strength of the concrete has been found to be about half the standard tensile strength, the necessary reinforcement ratio r to make the failure at concentrated loading highly ductile must be of the order

$$r = 0.5 \frac{f_t}{f_Y} \qquad (3.6.47)$$

f_t being the tensile strength of the concrete and f_Y the yield stress of the reinforcement. This reinforcement is provided in perpendicular directions according to the remarks above. To make sure there is proper anchorage of this reinforcement, it should be carried out as closely spaced, closed stirrups.

If a reinforcement equal to or higher than this is supplied, the load-carrying capacity does not depend on the tensile strength of concrete, which is always to be preferred in any stuctural element of importance.

Edge and corner loads. Although the theory of the load-carrying capacity at concentrated loads on unreinforced concrete is now rather complete, there are still some points in the theory which need further work. One point is related to edge and corner loads.

Consider a load at or near an edge of the cross section. If the distances to the other edges of the cross section are large, the load-carrying capacity derived from our formulas may be too high. In Fig. 3.6.41 (left) such a case is shown. Also indicated are the two areas A and A_o involved when using our formula for c_{local} (in such a case, of course, splitting may be left out). The area A may in this case attain a value much higher than the area really involved in the failure, which will be a kind of sliding failure involving the near surroundings of the loaded area.

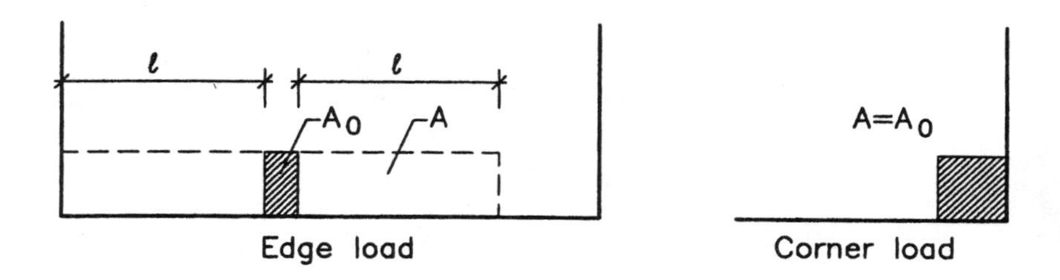

Figure 3.6.41 Edge and corner loads.

On the other hand, if we have a corner load, shown in Fig. 3.6.41 (right), the area A equals A_o and we will find $\sigma/f_c = 1$. However, it is evident that normally a part of the surrounding concrete will be involved in the failure, so σ/f_c will be larger than one.

These problems have been treated in a preliminary investigation by L. Hauge [86.9] in a masters thesis and by K. Madsen[2] [86.10] [90.19]. But it seems that much work remains to be done. Theoretical work is lacking. So these problems still have to be dealt with in a fully empirical manner. At the present time, strictly speaking, this is a problem for authors of codes of practice. Let us, however, try to play this game, which may be useful once in a while even in a book that is supposed to be scientifically oriented. A code-like proposal is shown in Fig. 3.6.42. If we want to apply the formula for c_{local} we would like to find a suitable area A to insert. According to the figure, the area sought is found by drawing lines from the corners of the loaded area to the edges of the cross section. The inclination of these lines is 2:1. In this way areas like $ABCD$ or $ABCDEF$ appear, which are the areas to be inserted as A in the formula for c_{local}. The lines with the inclination 2:1 do not indicate the concrete volume involved in the failure. The calculation rule is quite formal (as many other code rules).

A calculation based on this A-value replaces the former calculation, when the areas mentioned, $ABCD$ or $ABCDEF$, are lying within the actual cross section. If parts of the areas A are outside the cross section, then the former calculation is valid. The proposal may lead to both higher and lower load-carrying capacity than hitherto found.

[2] K. Madsen [86.10], [88.10], [90.19], formerly at the Danish Engineering Academy, has put a lot of effort into solving concentrated load problems and he has inspired most of the work reported here.

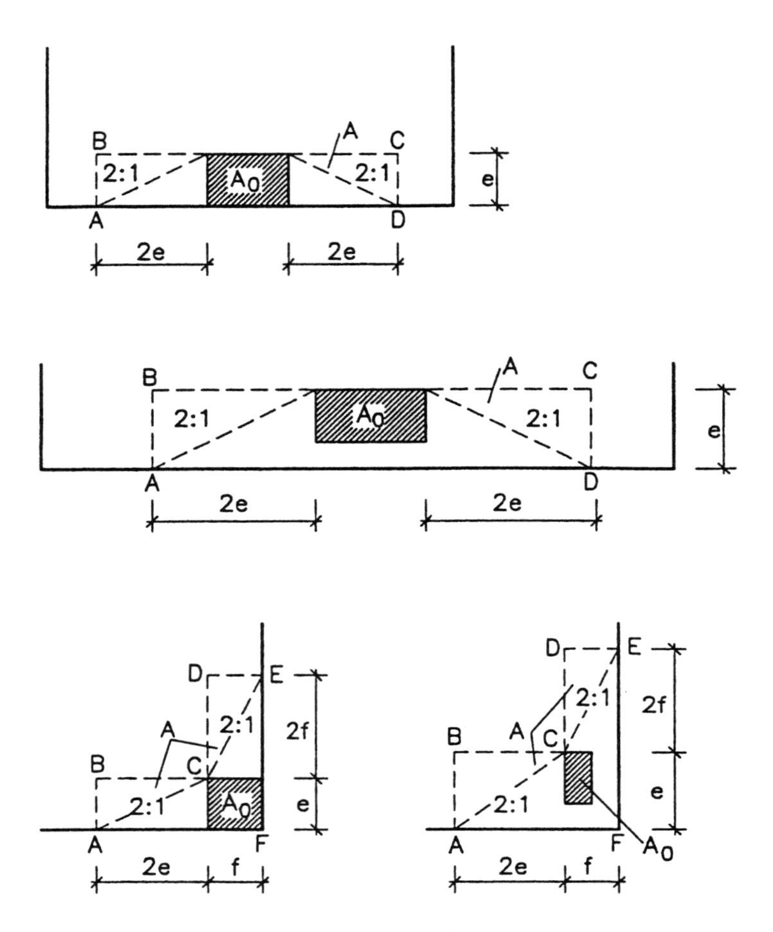

Figure 3.6.42 Edge and corner loads. Definition of A.

In the case shown at the top of Fig. 3.6.42 the sides AB and CD would be placed along the edges of the cross section giving rise to a much higher A. In the corner case the empirical rule often leads to higher load-carrying capacity.

For loaded areas which are not rectangular or with sides not parallel to the edges of the cross section, the rules must be modified. We will not deal with such cases here.

Fig. 3.6.43 shows the results of a test series with edge failures reported by K. Madsen [86.10] [90.19]. Also shown is the result of a calculation using formula (3.6.39) with $K_4 = 32$ and 25, respectively. It appears that $K_4 = 32$ gives very good results.

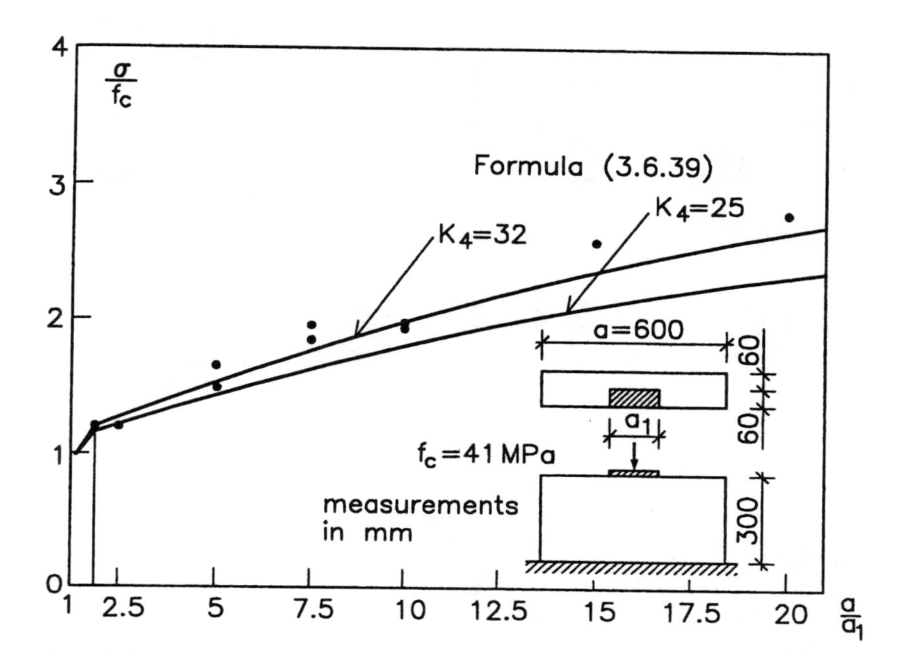

Figure 3.6.43 Comparison with test results, edge load.

It should be emphasized that the edge and corner rules are based on very few tests, and the rules should be applied in a conservative manner. The best way is, of course, to avoid loading near the edge and the corners, and if necessary, to supply some reinforcement which may help carry the load, and render the failure more ductile.

Group action. The next problem to be touched upon is the problem of group action. A group action problem arises when a number of concentrated loads are so close as to interfere with each other in the failure mechanism. In relation to plastic theory this problem is completely untouched.

In Fig. 3.6.44 a simple problem of 4 concentrated forces, which are rather close, is shown. Let us assume that the depth of the concrete body is large, so that we have a local failure problem. Let us also assume for simplicity that all loads are equal. The present way of dealing with such a problem in practice is to assign to each load a surrounding effective area A. Each area A has the same center of gravity as the loaded area and the areas A are chosen as large as possible so as not to interfere with each other. In Fig. 3.6.44 all the areas A are equal due to the simplicity of the loading arrangement.

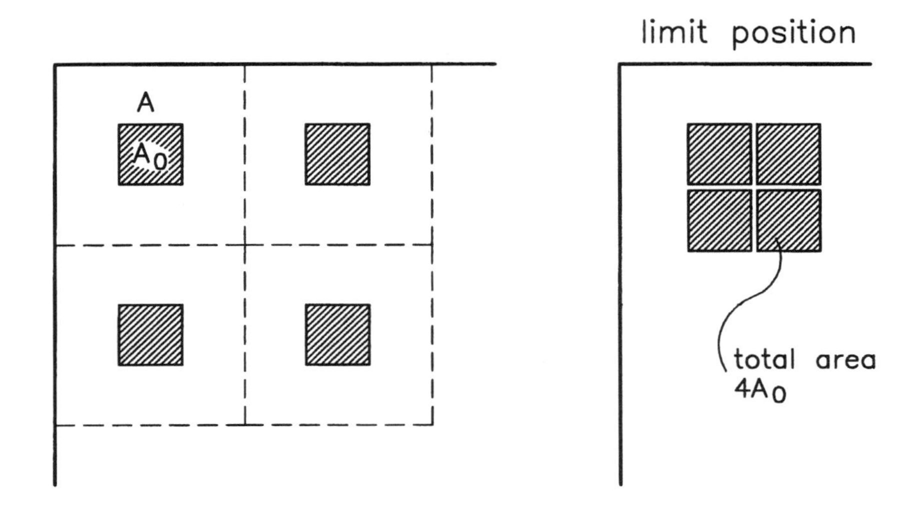

Figure 3.6.44 Group action, four loads near a corner.

After having assigned an area A to each load, the load-carrying capacity corresponding to each loaded area may be calculated. Such a procedure is, of course, in complete agreement with the lower bound theorem.

However, if the loads are very close the procedure may give unrealistic results. In this respect a theorem of plastic theory may be helpful. It deals with combining a number of loads P_1, P_2, ..., P_n. For each load acting alone, the load-carrying capacity is supposed to be known. Let us denote them P_{1p} , P_{2p} , ..., P_{np} . Then it may be shown that when the loads are acting simultaneously, the combined loading case is safe if

$$\frac{P_1}{P_{1p}} + \frac{P_2}{P_{2p}} + ... + \frac{P_n}{P_{np}} \leq 1 \tag{3.6.48}$$

The theorem is proved in Chapter 6 for slabs where a more accurate description of the assumptions is also given.

It would be tempting to use this formula to calculate the load-carrying capacity in combined loading cases. However, this does not work out, the load-carrying capacity obtained in this way is normally too conservative.

However, the theorem is useful since it means that if a set of concentrated loads is moved to a position where the load-carrying capacity corresponding to the individual loads is higher, then, in the combined loading case, the load-carrying capacity will also be higher after the loads have been moved to the new position.

Consider again Fig. 3.6.44 and imagine now that the four loads are moved to a position where they are as close as possible keeping the position of the load nearest to the corner fixed. This situation is shown in the figure to the right.

The load-carrying capacity in this situation may be calculated inserting into our formula for c_{local} an A_o-value equal to four times the A_o-value of the individual loads and an area A determined in the usual way. The σ/f_c value found may then be used to calculate the load-carrying capacity in the original loading case.

It follows from the theorem mentioned that the load-carrying capacity, when the loads are close to each other, does not need to be assessed lower than the load-carrying capacity found in the limit position in Fig. 3.6.44 to the right, where they are as close as possible.

If the individual loads have different magnitudes, the procedure must be modified.

This way of thinking may be used to prevent the traditional calculation procedure to render completely unrealistic results. But it is clear that group action must be researched much more before a satisfactory state is reached. Also test results are missing.

Size effects. The last problem to be mentioned is the size effect. Normally when the load-carrying capacity is highly dependent on the tensile strength of concrete, there is a substantial size effect.

Often size effects may be roughly described by a so-called Weibull root, defined in the following way: If a strength f_1 belongs to dimensions of a structural element characterized by the volume V_1, and a strength f_2 belongs to a volume V_2, then

$$\frac{f_1}{f_2} = \left(\frac{V_2}{V_1}\right)^{\frac{1}{m}} \tag{3.6.49}$$

m being the Weibull root. If all dimensions are changed in the same ratio, then

$$\frac{f_1}{f_2} = \left(\frac{L_2}{L_1}\right)^{\frac{3}{m}} \tag{3.6.50}$$

L_1 and L_2 characterizing the linear dimensions. Very little is known about the Weibull roots for concrete. For the uniaxial compressive strength the order of m is 30 and for the tensile strength the order of m is 10 [97.12]. This means that the size effect on the tensile strength is substantially more serious than on the compressive strength.

A root $m = 10$ means that if the linear dimensions are increased by a factor of 10, the strength will be decreased by a factor of 2.

No systematic study of the precise size effect rule to be used in connection with our formulas for concentrated loading has been carried out due to the lack of test data.

However, the size effect on the tensile strength is so important that a warning should be given about using the formulas for dimensions substantially larger than the size of normal laboratory specimens without reducing the tensile strength.

The tensile strength to be used for large elements may be roughly estimated using for L_1 and L_2 in formula (3.6.50) the respective depths in the case of splitting failure and a typical cross section dimension, say $\sqrt{A}$, in the case of local failure. The tensile strength given in this chapter may then be assumed to be valid for a depth or for a typical cross section dimension within the range of 100-200 mm.

The size effect on the compressive strength may be neglected.

Chapter 4

DISKS

Disks and slabs with constant thickness are structural members bounded by two parallel planes having a distance, the thickness, which is small compared with the other dimensions. Furthermore, disks are loaded by forces statically equivalent to a plane force system in the middle plane of the disk, while slabs are loaded perpendicular to the middle plane. The stress field in disks is approximately a plane stress field.

In this chapter the plastic theory for reinforced concrete disks is described. The basic theory has been worked out primarily by the author [69.2].

4.1 STATICAL CONDITIONS

In rectangular coordinates x and y, the equilibrium conditions read (see Section 3.1)

$$\frac{\partial \sigma_x}{\partial x} + \frac{\partial \tau_{xy}}{\partial y} + \rho f_x = 0 \tag{4.1.1}$$

$$\frac{\partial \sigma_y}{\partial y} + \frac{\partial \tau_{xy}}{\partial x} + \rho f_y = 0 \tag{4.1.2}$$

The statical boundary conditions are

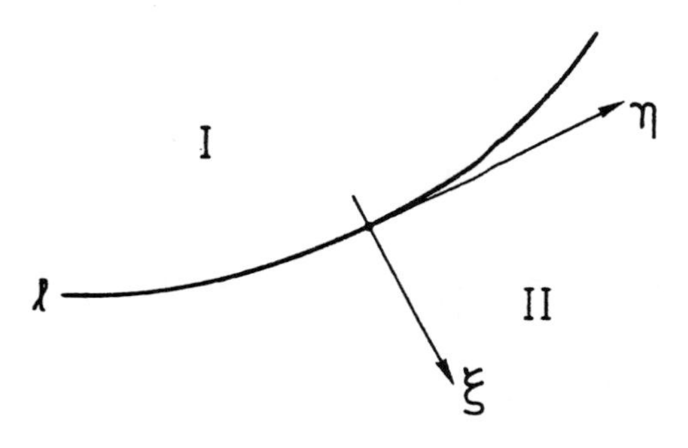

Figure 4.1.1 Local coordinate system at a boundary point.

$$p_x = \sigma_x \cos\alpha + \tau_{xy} \sin\alpha \tag{4.1.3}$$

$$p_y = \sigma_y \sin\alpha + \tau_{xy} \cos\alpha \tag{4.1.4}$$

in which p_x and p_y are the components of the stresses on the boundary and α is the angle between the x-axis and the outward directed normal to the boundary, calculated positive in the positive direction of the x, y-plane.

In polar coordinates r, θ, the corresponding formulas are

$$\frac{\partial \sigma_r}{\partial r} + \frac{1}{r} \frac{\partial \tau_{r\theta}}{\partial \theta} + \frac{\sigma_r - \sigma_\theta}{r} + \rho f_r = 0 \tag{4.1.5}$$

$$\frac{1}{r} \frac{\partial \sigma_\theta}{\partial \theta} + \frac{\partial \tau_{r\theta}}{\partial r} + 2 \frac{\tau_{r\theta}}{r} + \rho f_\theta = 0 \tag{4.1.6}$$

As the stresses calculated are the total stresses appearing as the sum of stresses of the concrete and the statical equivalence of the stresses of the bars, the meaning of the boundary conditions formulated in the total stresses demands a comment.

The boundary conditions (4.1.3) and (4.1.4) will be satisfied provided that the total stresses are equal to the stresses applied at the boundaries. Consequently, the boundary conditions will be satisfied along a free boundary, for example, provided that the statical equivalence of the stresses of the bars and the concrete vanishes (only in special cases will the stresses of the bars vanish on the boundary). There is, however, no other possibility of formulating the boundary conditions when the theory is formulated with the total stresses as generalized stresses, as by this formulation of the boundary conditions there is a unique solution (cf. Chapter 1).

Along a line of discontinuity ℓ (see Fig. 4.1.1), we have

$$\sigma_\xi^{\mathrm{I}} = \sigma_\xi^{\mathrm{II}} \tag{4.1.7}$$

$$\tau_{\xi\eta}^{\mathrm{I}} = \tau_{\xi\eta}^{\mathrm{II}} \tag{4.1.8}$$

where the ξ, η-coordinate system is a local rectangular system having the η-axis in the direction of the tangent at an arbitrary point of the line of discontinuity and where the disk has been divided into parts I and II by the line of discontinuity.

4.2 GEOMETRICAL CONDITIONS

The relationship between the displacements u_x and u_y and the strains ε_x, ε_y, and γ_{xy} referred to as rectangular coordinates is (see Section 3.2)

$$\varepsilon_x = \frac{\partial u_x}{\partial x}, \quad \varepsilon_y = \frac{\partial u_y}{\partial y}, \quad \gamma_{xy} = \frac{\partial u_x}{\partial y} + \frac{\partial u_y}{\partial x} \tag{4.2.1}$$

The strains have to satisfy the compatibility equation

$$\frac{\partial^2 \varepsilon_x}{\partial y^2} + \frac{\partial^2 \varepsilon_y}{\partial x^2} = \frac{\partial^2 \gamma_{xy}}{\partial x \partial y} \tag{4.2.2}$$

The fulfillment of (4.2.2) verifies (for simply connected regions) that continuous displacements u_x and u_y, corresponding to the strains, exist.

In polar coordinates the corresponding formulas are:

Strain-displacement relations:

$$\varepsilon_r = \frac{\partial u_r}{\partial r} \tag{4.2.3}$$

$$\varepsilon_\theta = \frac{1}{r}\frac{\partial u_\theta}{\partial \theta} + \frac{u_r}{r} \tag{4.2.4}$$

$$\gamma_{r\theta} = \frac{1}{r}\frac{\partial u_r}{\partial \theta} + \frac{\partial u_\theta}{\partial r} - \frac{u_\theta}{r} \tag{4.2.5}$$

Compatibility equation:

$$\frac{\partial^2 \varepsilon_\theta}{\partial r^2} + \frac{1}{r^2}\frac{\partial^2 \varepsilon_r}{\partial \theta^2} - \frac{1}{r}\frac{\partial \varepsilon_r}{\partial r} + \frac{2}{r}\frac{\partial \varepsilon_\theta}{\partial r} = \frac{1}{r}\frac{\partial^2 \gamma_{r\theta}}{\partial r \partial \theta} + \frac{1}{r^2}\frac{\partial \gamma_{r\theta}}{\partial \theta} \tag{4.2.6}$$

4.3 VIRTUAL WORK

The virtual work equation reads (see Section 3.3)

$$\iint (\sigma_x \varepsilon_x + \sigma_y \varepsilon_y + \tau_{xy}\gamma_{xy})\,dx\,dy$$

$$= \iint (\rho f_x u_x + \rho f_y u_y)\,dx\,dy + \int (p_x u_x + p_y u_y)\,ds \tag{4.3.1}$$

The equation is an identity if the statical conditions are fulfilled and is then valid for any displacement field. If, on the contrary, the equation is valid for any displacement field, the statical conditions are satisfied.

4.4 CONSTITUTIVE EQUATIONS

4.4.1 Plastic Strains in Disks

We shall restrict ourselves to the case of small degrees of reinforcement, which means that the yield conditions can be considered to be described by (2.2.41) and (2.2.42). According to the flow rule (1.1.9), the plastic strains are found to be

$$\sigma_y \geq -\eta\sigma_x + \eta\Phi_x f_c - f_c$$

$$\begin{aligned} \varepsilon_x &= \lambda(\Phi_y f_c - \sigma_y) \\ \varepsilon_y &= \lambda(\Phi_x f_c - \sigma_x) \qquad (\lambda \geq 0) \\ \gamma_{xy} &= 2\varepsilon_{xy} = 2\lambda\tau_{xy} \end{aligned} \tag{4.4.1}$$

$$\sigma_y < -\eta\sigma_x + \eta\Phi_x f_c - f_c$$

$$\begin{aligned} \varepsilon_x &= -\lambda(f_c + \sigma_y) \\ \varepsilon_y &= -\lambda(f_c + \sigma_x) \qquad (\lambda \geq 0) \\ \gamma_{xy} &= 2\varepsilon_{xy} = 2\lambda\tau_{xy} \end{aligned} \tag{4.4.2}$$

By means of the yield conditions it can easily be verified that

$$\varepsilon_x\varepsilon_y = \varepsilon_{xy}^2 \tag{4.4.3}$$

which means that one principal strain is zero. The expression above is not valid in the apexes *A* and *C* (see Fig. 4.4.1).

Further, the formulas are not valid along the edge, the projection of which on the σ_x, σ_y-plane is *BD*. Here any positive linear combination of the strain vectors corresponding to the limiting positions for the normals of the two conical surfaces that intersect along the edge can be used.

The following therefore applies along *BD*:

$$\begin{aligned} \varepsilon_x &= \lambda_1(\Phi_y f_c - \sigma_y) + \lambda_2(-f_c - \sigma_y) \\ \varepsilon_y &= \lambda_1(\Phi_x f_c - \sigma_x) + \lambda_2(-f_c - \sigma_x) \\ \gamma_{xy} &= 2\lambda_1\tau_{xy} + 2\lambda_2\tau_{xy} \end{aligned} \tag{4.4.4}$$

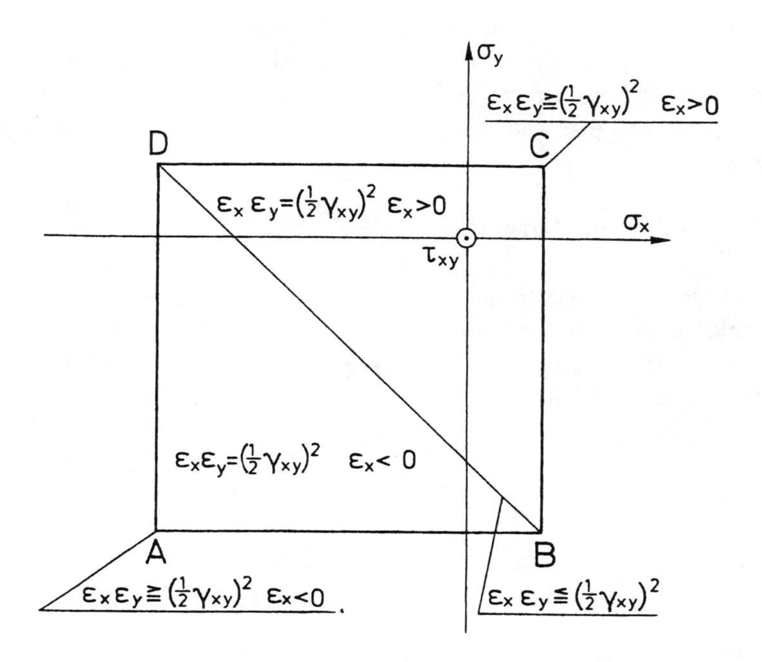

Figure 4.4.1 Plastic strains in an orthotropic disk.

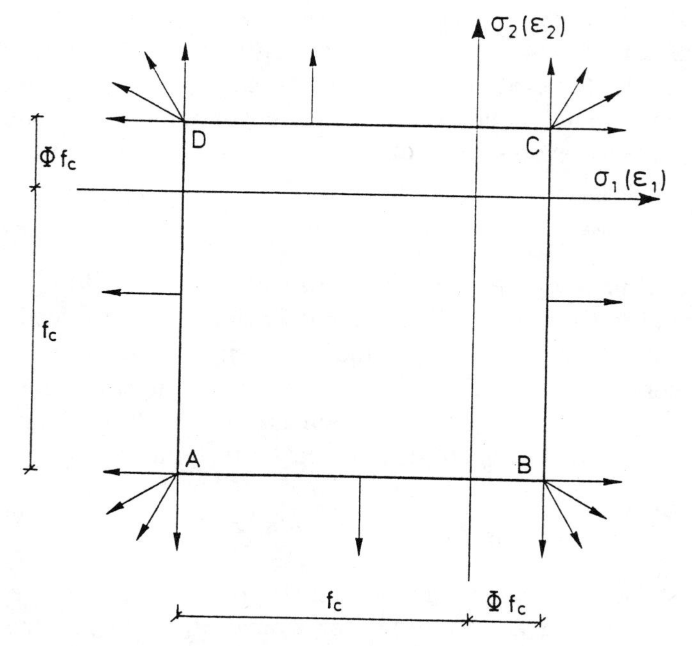

Figure 4.4.2 Plastic strains in an isotropic disk.

where λ_1 and λ_2 are arbitrary nonnegative factors.

In Fig. 4.4.1 the properties of the plastic strains characteristic for the different parts of the yield condition are shown. A survey of the plastic strains that can occur in an isotropic disk is given in Fig. 4.4.2.

4.4.2 Dissipation Formulas

In the region where one principal strain is zero there exists a very simple formula for the dissipation per unit volume. Consider, for example, the region where (2.2.41) is valid. Here we get the dissipation per unit volume

$$\begin{aligned} W &= \sigma_x \varepsilon_x + \sigma_y \varepsilon_y + \tau_{xy}\gamma_{xy} \\ &= \lambda \sigma_x(\Phi_y f_c - \sigma_y) + \lambda \sigma_y(\Phi_x f_c - \sigma_x) + 2\lambda \tau_{xy}^2 \end{aligned} \tag{4.4.5}$$

Invoking the yield condition, we find that

$$W = \Phi_x f_c \varepsilon_x + \Phi_y f_c \varepsilon_y = \Phi_x f_c |\varepsilon_x| + \Phi_y f_c |\varepsilon_y| \tag{4.4.6}$$

In the region where (2.2.42) is valid we find that

$$W = f_c(|\varepsilon_x| + |\varepsilon_y|) \tag{4.4.7}$$

The expressions (4.4.6) and (4.4.7) are also seen to be valid in the apexes A and C, respectively. Along the edge a more complicated expression is found, which is not written down here. For an isotropic disk, a general expression valid for all cases can be given. The formula is

$$W = \tfrac{1}{2} f_c[(1 + \Phi)(|\varepsilon_1| + |\varepsilon_2|) - (1 - \Phi)(\varepsilon_1 + \varepsilon_2)] \tag{4.4.8}$$

We deal next with the important special case of a yield line in an isotropic disk. Consider a narrow region having width δ, in which the strain field is homogeneous with the increment u_ξ in displacement in the direction of the ξ-axis and the increment u_η in displacement in the direction of the η-axis. The region is assumed to be limited by rigid bodies. The strains are determined by (see Fig. 4.4.3)

$$\varepsilon_\xi = \frac{u_\xi}{\delta}, \quad \varepsilon_\eta = 0, \quad \gamma_{\xi\eta} = \frac{u_\eta}{\delta} \tag{4.4.9}$$

The stress field is homogeneous and may be found by using the flow rule, as u_ξ and u_η are inserted for ε_ξ and $\gamma_{\xi\eta}$. Now, letting $\delta \to 0$ but keeping u_ξ and u_η constant, we obtain a yield line.

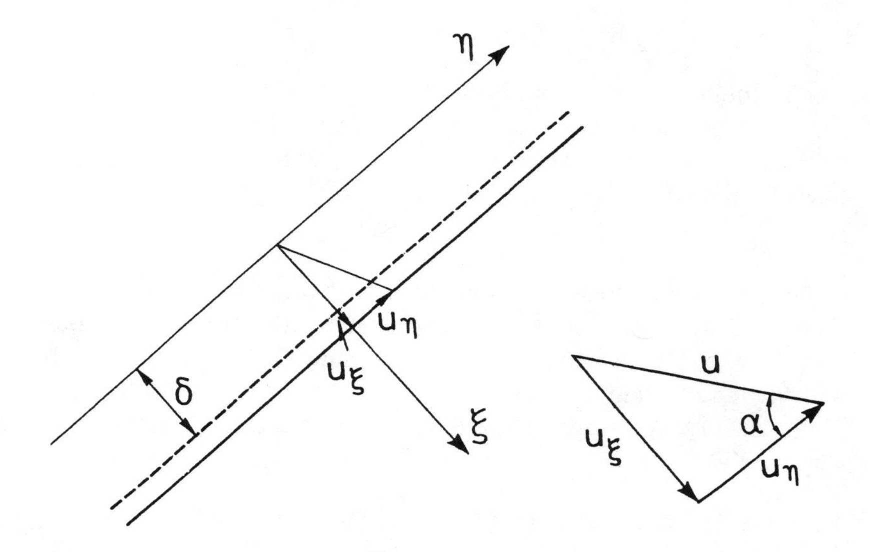

Figure 4.4.3 Yield line defined by a narrow strip with homogeneous strain.

The principal strains are

$$\left.\begin{matrix}\varepsilon_1 \\ \\ \varepsilon_2\end{matrix}\right\} = \frac{1}{2}(\varepsilon_\xi + \varepsilon_\eta) \pm \frac{1}{2}\sqrt{(\varepsilon_\xi - \varepsilon_\eta)^2 + \gamma_{\xi\eta}^2} \tag{4.4.10}$$

Since ε_1 and ε_2 have opposite signs, we have

$$|\varepsilon_1| + |\varepsilon_2| = \sqrt{(\varepsilon_\xi - \varepsilon_\eta)^2 + \gamma_{\xi\eta}^2} \tag{4.4.11}$$

Further,

$$\varepsilon_1 + \varepsilon_2 = \varepsilon_\xi + \varepsilon_\eta \tag{4.4.12}$$

In the deforming region we therefore get

$$W = \frac{1}{2} f_c \left[(1+\Phi)\sqrt{\left(\frac{u_\xi}{\delta}\right)^2 + \left(\frac{u_\eta}{\delta}\right)^2} - (1-\Phi)\frac{u_\xi}{\delta} \right] \tag{4.4.13}$$

which means that the dissipation per unit length of the yield line $W_\ell = W\delta t$, t being the thickness, is

$$W_\ell = \frac{1}{2} f_c t \left[(1+\Phi)\sqrt{u_\xi^2 + u_\eta^2} - (1-\Phi) u_\xi \right] \tag{4.4.14}$$

Introducing the angle α shown in Fig. 4.4.3, we have $u_\xi = u \sin\alpha$, $u_\xi = u \cos\alpha$. Hence

$$W_\ell = \frac{1}{2} f_c t u [(1+\Phi) - (1-\Phi)\sin\alpha] \tag{4.4.15}$$

The formula is identical with (3.4.91) for $\Phi = 0$, as it should be.

The large plastic work dissipated in a yield line of pure shear ($u_\xi = 0$) or a zone of pure shear strain, is remarkable. In a straight yield line of length ℓ with $u_\xi = 0$, we have

$$D = \frac{1}{2} f_c t (1+\Phi) \, |u_\eta| \, \ell \tag{4.4.16}$$

that is, the shear stress corresponding to this deformation is not the shear strength Φf_c, but $\frac{1}{2}(1+\Phi) f_c$, substantially greater. Of course this is due to the fact that the strain field pure shear is not compatible with the stress field pure shear due to the dilatancy of the concrete. The great resistance against pure shear surely indicates that it occurs very rarely.

4.5 EXACT SOLUTIONS FOR ISOTROPIC DISKS

4.5.1 Various Types of Yield Zones

Let us examine the possibilities of yield zones in isotropic disks. Consider first the case in which the stresses in the zone correspond to one of the corners of the yield condition (see Fig. 4.4.2). The lines along which the principal stresses correspond to the corner of the yield condition form an orthogonal net which we shall use as lines of coordinates α and β (see Fig. 4.5.1). The equilibrium conditions in arbitrary orthogonal coordinates for a weightless disk (see e.g., [64.4]) read

$$\frac{\partial B \sigma_\alpha}{\partial \alpha} + \frac{\partial A \tau_{\alpha\beta}}{\partial \beta} + \frac{\partial A}{\partial \beta} \tau_{\alpha\beta} - \frac{\partial B}{\partial \alpha} \sigma_\beta = 0 \tag{4.5.1}$$

$$\frac{\partial B \sigma_\beta}{\partial \beta} + \frac{\partial A \tau_{\alpha\beta}}{\partial \alpha} + \frac{\partial B}{\partial \alpha} \tau_{\alpha\beta} - \frac{\partial A}{\partial \beta} \sigma_\alpha = 0 \tag{4.5.2}$$

In these expressions A and B are functions of α and β which determine the square of the arc differential ds corresponding to the differentials of the curved coordinates $d\alpha$ and $d\beta$, that is

$$ds^2 = A^2 d\alpha^2 + B^2 d\beta^2 \tag{4.5.3}$$

Now setting $\sigma_\alpha = a$, $\sigma_\beta = b$, $\tau_{\alpha\beta} = 0$, a and b being constants, we obtain as equilibrium conditions

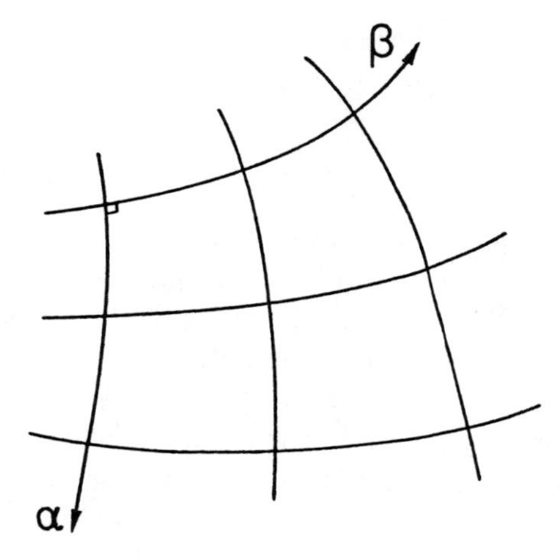

Figure 4.5.1 Yield zone in an isotropic disk.

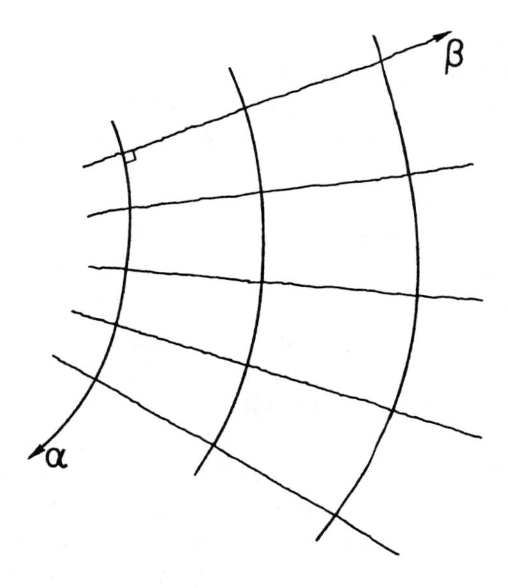

Figure 4.5.2 Yield zone in an isotropic disk with one set of coordinate lines being straight lines.

$$(a - b)\frac{\partial B}{\partial \alpha} = 0 \tag{4.5.4}$$

$$(b - a)\frac{\partial A}{\partial \beta} = 0 \tag{4.5.5}$$

When $a = b$ the equilibrium equations are identically satisfied, as known in advance, but when $a \neq b$ we must demand that

$$\frac{\partial B}{\partial \alpha} = \frac{\partial A}{\partial \beta} = 0 \tag{4.5.6}$$

These equations are satisfied only when the coordinate lines are straight, as the derivatives in (4.5.6) are proportional to the curvature of the coordinate lines. In yield zones with stresses corresponding to the corners of the yield condition, the principal stress trajectories therefore form a rectilinear, orthogonal net, a result that is different from the corresponding result in the plastic theory for slabs in which the lines of coordinates form a Hencky net (see Section 6.5.1).

Generally, therefore, we expect to obtain yielding in only one principal direction in a yield zone. The stress trajectories will then also be straight lines (cf. Fig. 4.5.2), which may be seen from equilibrium equations (4.5.1) and (4.5.2) as well.

In such a yield zone in which we have for example, $\sigma_\alpha = a =$ constant and $\tau_{\alpha\beta} = 0$, σ_β may be determined from (4.5.2) by integration. An example of such a yield zone will be given in Example 1 of Section 4.5.3, in which one set of stress trajectories has a common point of intersection.

4.5.2 A Survey of Known Solutions

A number of exact solutions exist for isotropic disks. They have been derived by the author [69.2]. A survey of the solutions is given in Fig. 4.5.3. In all cases we are concerned with homogeneous, isotropically reinforced disks with a reinforcement degree Φ. The quantities p and q are stresses at loaded areas, and t is the thickness.

In order that the solution h may correspond to a geometrically admissible failure mechanism, we must, strictly speaking, assume the support is being divided into two parts, where the two parts are allowed to rotate freely. If the support consists of an undivided, sufficiently strong, and wide abutment platen, the carrying capacity will correspond to the case where the bearing pressure is equal to f_c on the extreme two parts of the abutment platen, each having the length $y_0 \tan\alpha$, where the angle α is shown in Fig. 4.5.4, and where y_0 is the depth of the compression zone in the region of maximum moment, being of length $c - 2y_0 \tan\alpha$ (see Fig. 4.5.4). The shaded parts will then be subjected to hydrostatic, biaxial compression and the failure mechanism will be similar to the one illustrated in Fig.

4.5.9. Similar remarks can be made regarding some of the other examples, for which, however, we shall not go into detail.

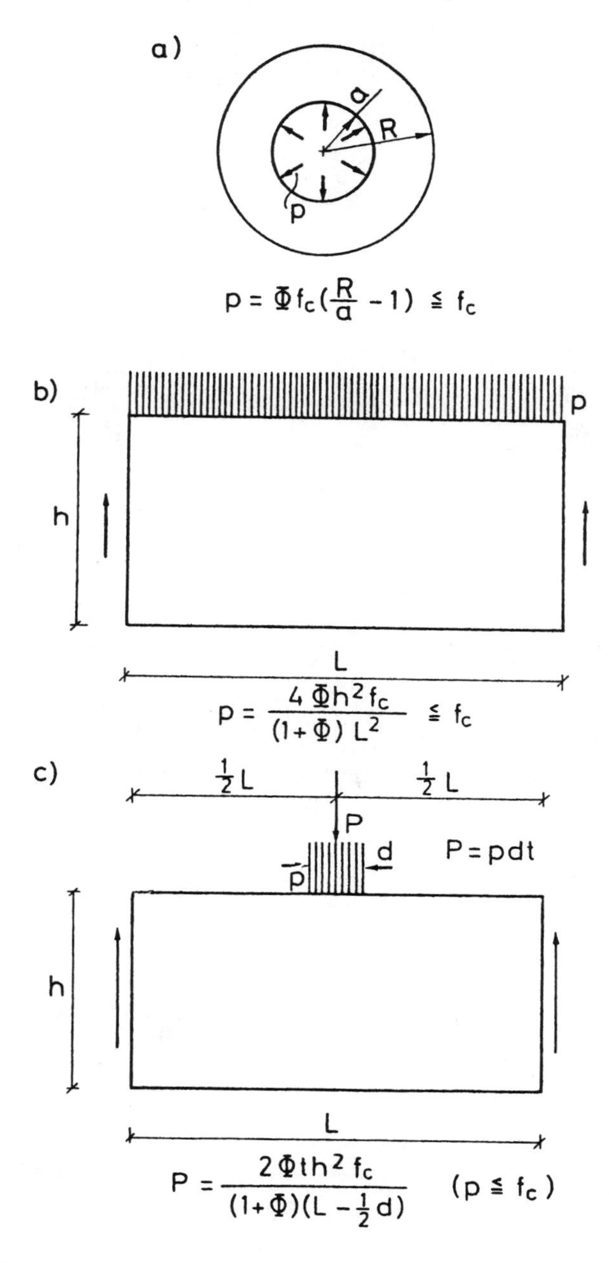

Figure 4.5.3 Survey of exact solutions for isotropic disks.

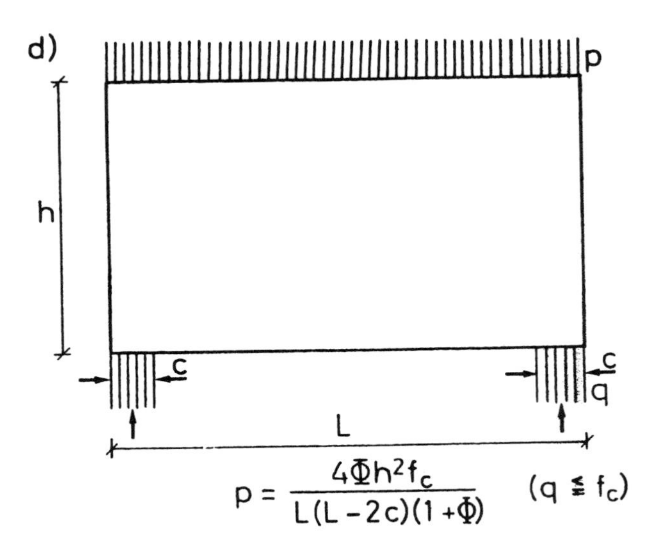

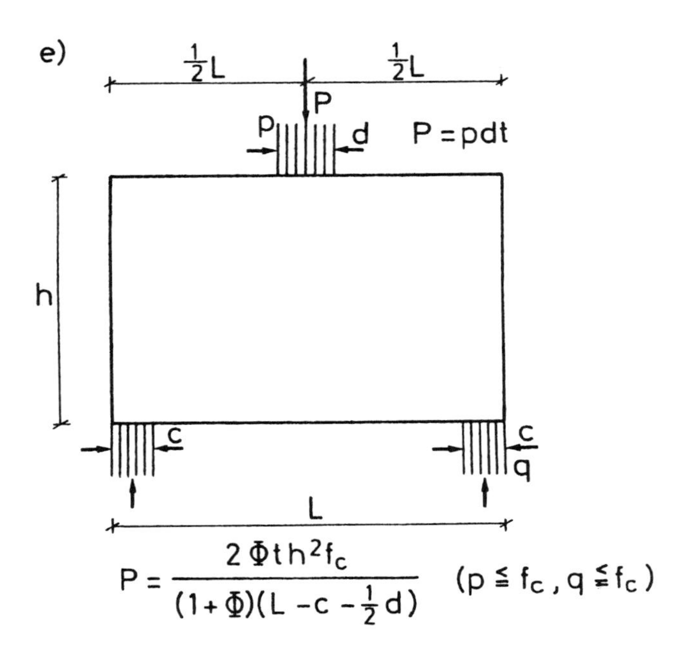

Figure 4.5.3 (continued)

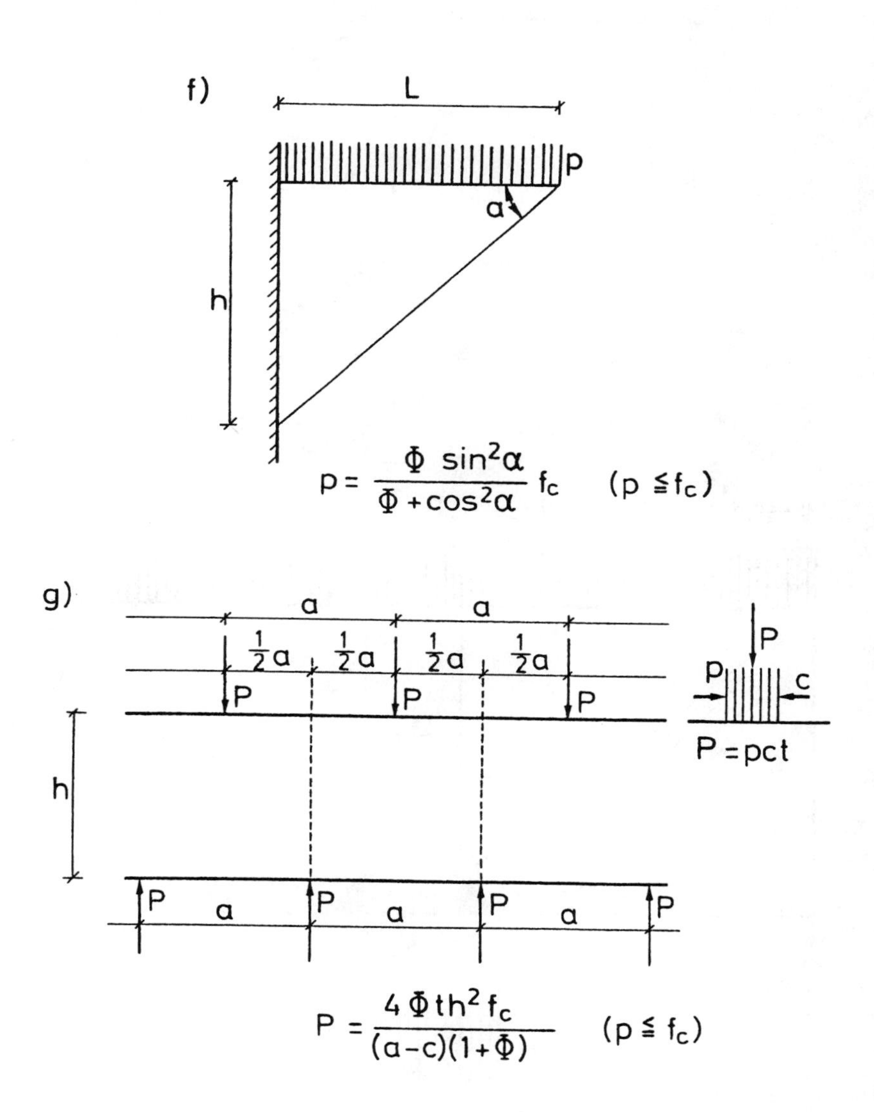

Figure 4.5.3 (continued)

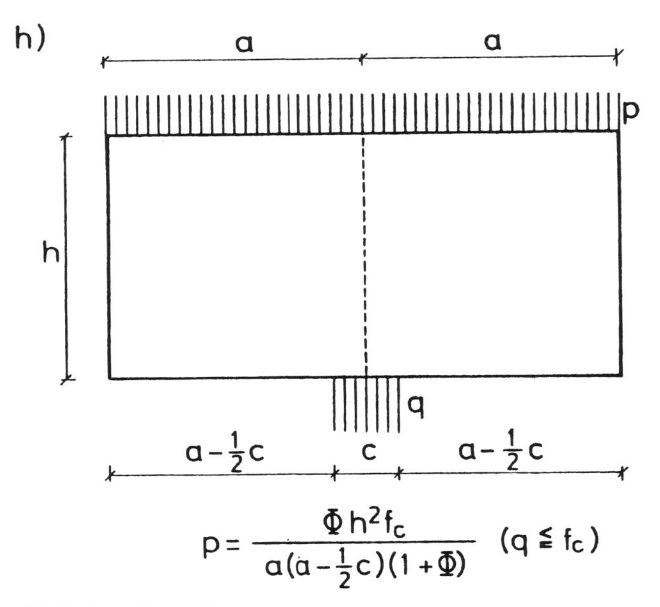

Figure 4.5.3 (continued)

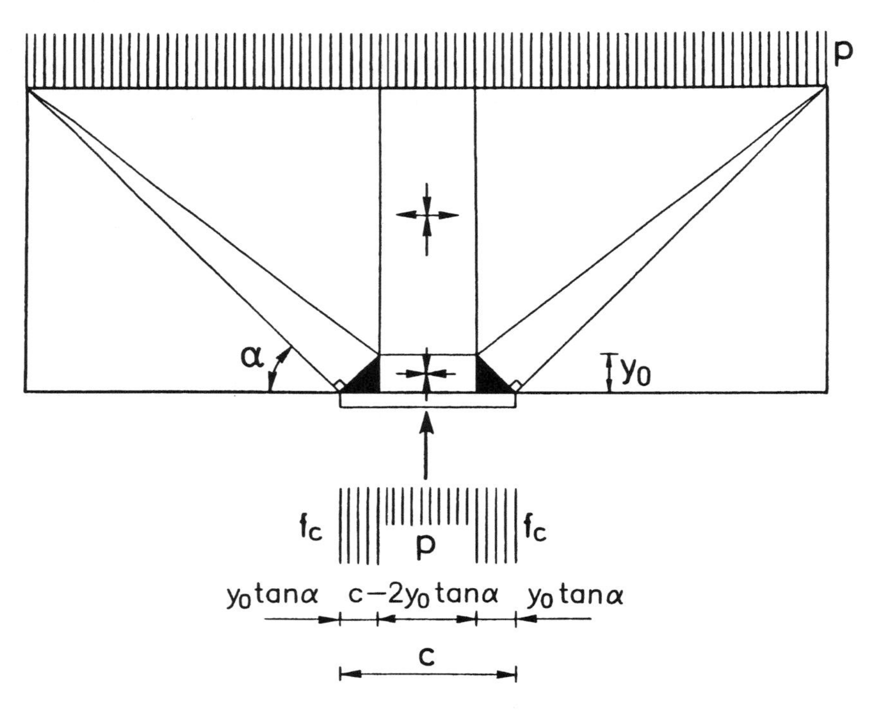

Figure 4.5.4 Modification of an exact solution in the case of a rigid abutment platen.

4.5.3 Illustrative Examples

Example 1

In this example we are considering an isotropic, circular disk with a circular hole. The boundaries of the disk are concentric circles. On the boundary of the hole, the disk is loaded by a radial pressure p. A polar system of coordinates is used as shown in Fig. 4.5.5.

The stress σ_θ at collapse is estimated to be equal to the tensile yield stress Φf_c. The equilibrium condition is, according to (4.1.5);

$$\frac{d\sigma_r}{dr} + \frac{\sigma_r - \sigma_\theta}{r} = 0$$

Setting $\sigma_\theta = \Phi f_c$, we obtain

$$\frac{d\sigma_r}{\sigma_r - \Phi f_c} = -\frac{dr}{r}$$

or, as $\sigma_r = -p$ for $r = a$,

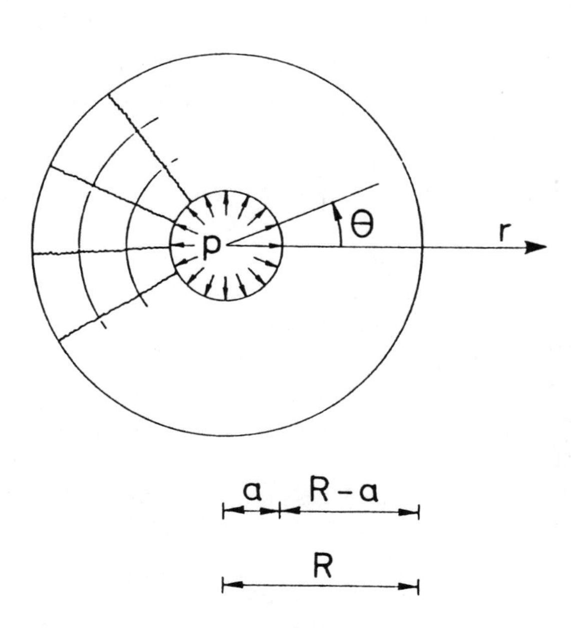

Figure 4.5.5 Circular disk with a concentric circular hole loaded by a radial pressure.

$$\sigma_r = \frac{\Phi f_c r - a(p + \Phi f_c)}{r}$$

Setting $\sigma_r = 0$ for $r = R$, we obtain

$$p = \Phi f_c \left(\frac{R}{a} - 1 \right)$$

σ_r has the numerical largest value for $r = a$. Therefore, if we require

$$p = \Phi f_c \left(\frac{R}{a} - 1 \right) \leq f_c$$

that is,

$$\frac{R}{a} \leq \frac{1}{\Phi} + 1$$

the stresses will correspond to points within or on the yield surface.

The principal stress trajectories are, of course, the coordinate lines. There is yielding in radial sections only.

A geometrically admissible displacement field corresponding to these stresses exists. In polar coordinates we have, according to (4.2.3) and (4.2.4) in rotationally symmetrical cases,

$$\varepsilon_r = \frac{du_r}{dr}, \qquad \varepsilon_\theta = \frac{u_r}{r}$$

u_r being the displacement in the direction of the r-axis. Now, inserting u_r = constant = $c > 0$, we obtain $\varepsilon_r = 0$ and $\varepsilon_\theta = c/r$, that is, we may put

$$\varepsilon_r = 0, \qquad \varepsilon_\theta = c/r$$

by which a geometrically admissible displacement field with strains, corresponding to the stresses, is indicated.

For $R/a > 1/\Phi + 1$ the carrying capacity is $p = f_c$. It is obvious that it is possible to find a safe, statically admissible stress field corresponding to this load, as for the part of the disk corresponding to $a \leq r \leq a(1 + 1/\Phi)$ we can use the stress field above, and for $a(1 + 1/\Phi) \leq r \leq R$ we can assume all stresses to be zero.

As the carrying capacity, on the other hand, cannot exceed f_c, it is obvious that this is the carrying capacity. We shall, however, go a little further into the corresponding displacement field.

Consider the following displacements:

$$u_r = u_0 \frac{\rho - r}{\rho - a} \qquad (u_0 > 0,\ \rho > a,\ a \le r \le \rho)$$

which render the displacement u_0 for $r = a$ and the displacement 0 for $r = \rho$. By using (4.2.3) and (4.2.4), we obtain

$$\varepsilon_r = -\frac{u_0}{\rho - a}, \qquad \varepsilon_\theta = \frac{u_0}{r}\frac{\rho - r}{\rho - a}$$

which by comparison with the flow rule is seen to correspond to the stresses

$$\sigma_r = -f_c, \qquad \sigma_\theta = \Phi f_c$$

By this we obtain, using the work equation,

$$2\pi a t p u_0 = t\int_a^\rho \int_0^{2\pi} \left(f_c \frac{u_0}{\rho - a} + \Phi f_c \frac{u_0}{r}\frac{\rho - r}{\rho - a} \right) r\, dr\, d\theta$$

$$= \pi t f_c u_0 (\rho + a) + \pi t \Phi f_c u_0 (\rho - a)$$

For any value of ρ we obtain an upper bound of the load-carrying capacity. Setting $\rho \to a$, we obtain

$$p \to f_c$$

This displacement field is thus giving a load-carrying capacity equal to the lower bound solution.

This deformation is of the same kind as the deformation along the lines of discontinuity mentioned previously, but there is the difference that here the line of discontinuity is the loaded boundary itself and as the two parts meeting in the line of discontinuity are not moving relative to each other as rigid bodies, the line of discontinuity gets longitudinal strains here. The displacement field may be regarded as a simple plastic indentation u_0.

Example 2

In this example we consider a rectangular, isotropic disk, having depth h and length L, loaded with a uniformly distributed load corresponding to the compression stress p on one face of the disk (the top face) (see Fig. 4.5.6). The disk is assumed to be simply supported at the points (0,0) and (L,0). In this case, the reactions are supposed to be transferred along the sides $x = 0$ and $x = L$.

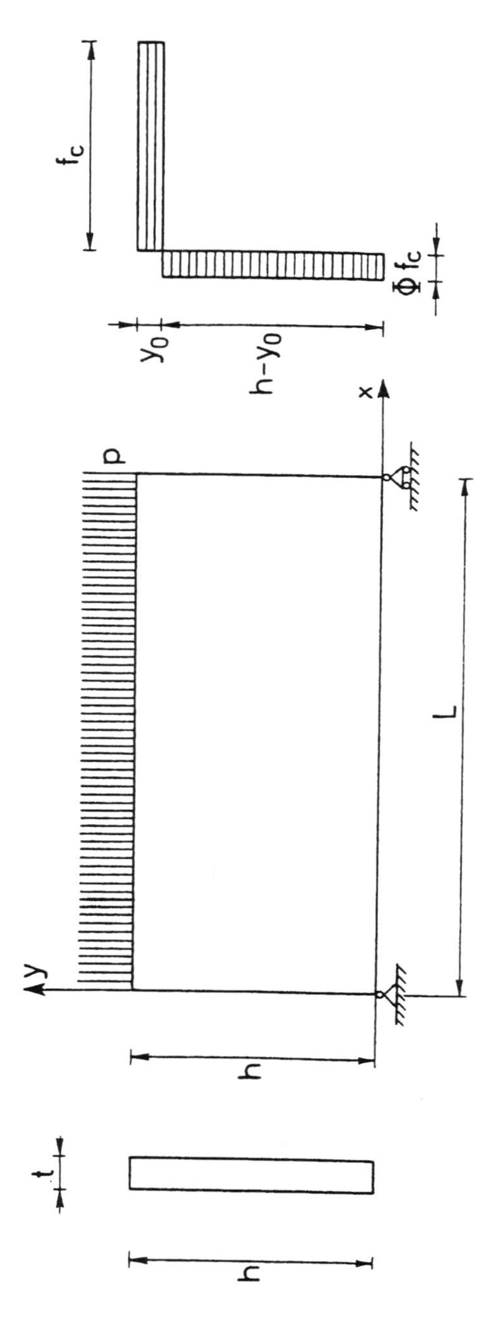

Figure 4.5.6 Rectangular disk supported along the lateral faces and loaded by a uniform load.

First, the cross-sectional yield moment in pure bending is determined. As the neutral axis is situated at a distance y_0 from the top face and the degree of reinforcement of the disk is Φ, we obtain by projection

$$f_c y_0 = (h - y_0)\Phi f_c$$

from which

$$y_0 = \frac{\Phi}{1+\Phi} h$$

By this we obtain the yield moment M_p:

$$M_p = \frac{1}{2} f_c y_0 th = \frac{1}{2}(h - y_0)\Phi f_c th = \frac{1}{2}\frac{\Phi}{1+\Phi} th^2 f_c$$

t being the thickness of the disk.

If, on the other hand, the problem is to determine the degree of reinforcement Φ corresponding to a given moment M_p, we find that

$$\Phi = \frac{\mu}{1-\mu}$$

where

$$\mu = \frac{M_p}{\frac{1}{2} th^2 f_c}$$

The uniformly distributed tensile stress $\sigma_t = \Phi f_c$ corresponding to a given moment M_p and a given compressive strength f_c is therefore

$$\sigma_t = \frac{\mu f_c}{1-\mu} = \frac{1}{1-\mu}\frac{M_p}{\frac{1}{2} th^2}$$

When Φ is small we obtain a close approximation

$$\sigma_t \approx \frac{M_p}{\frac{1}{2} th^2}$$

In order to reach the yield moment M_p in the center cross section, the load must be equal to

$$p = \frac{4\Phi h^2 f_c}{(1+\Phi)L^2} = \frac{4 y_0 h f_c}{L^2} = \frac{4(h - y_0)\Phi h f_c}{L^2}$$

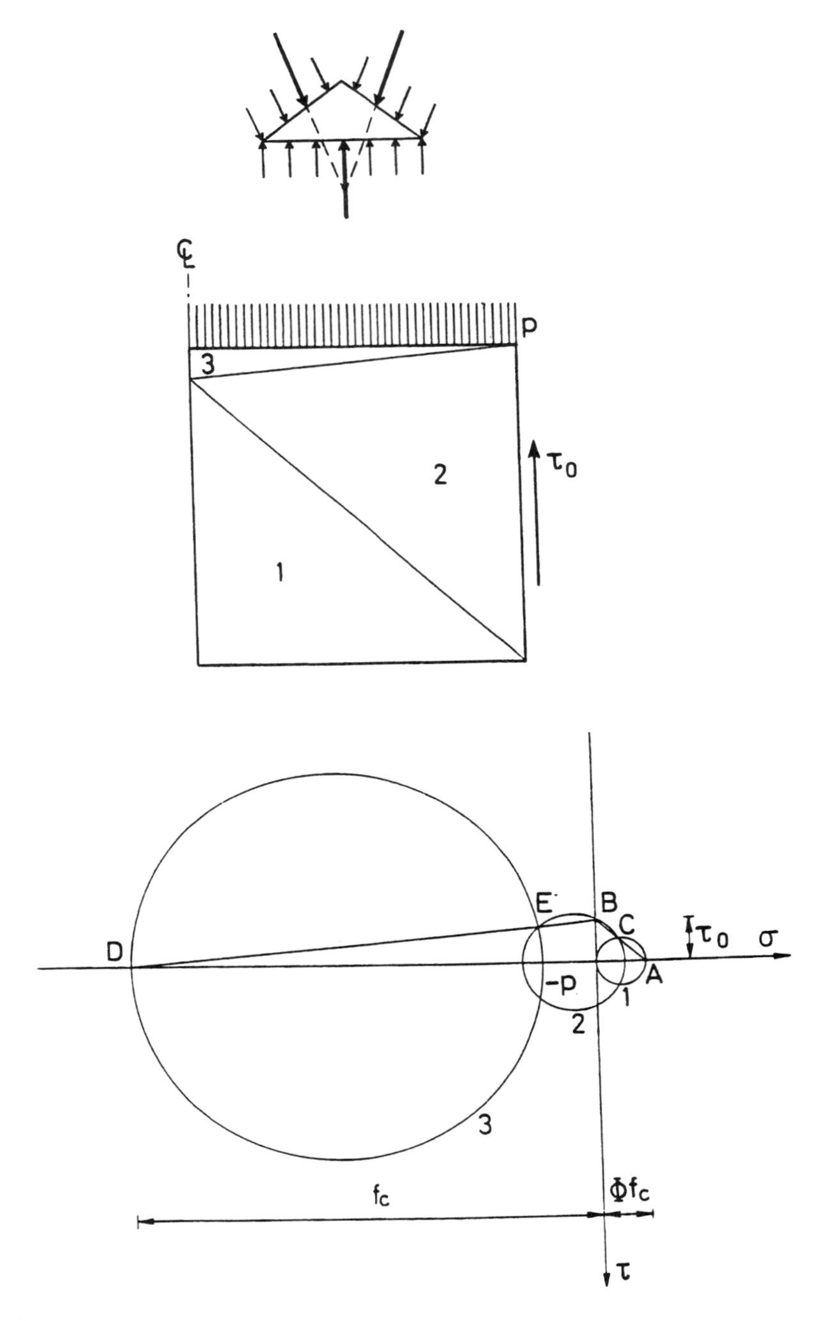

Figure 4.5.7 Discontinuous stress field in a rectangular disk supported along the lateral faces and loaded by a uniform load.

It may now be shown that this is an exact solution if $p \leq f_c$. We shall demonstrate this by constructing a stress field with lines of discontinuity as illustrated in Fig. 4.5.7 and homogeneous stress fields within the three parts 1, 2, and 3 formed by the lines of discontinuity. We use the general theorem that if a weightless, triangular area is loaded by stresses along the boundaries, uniformly distributed, then if the resultant forces are in equilibrium a homogeneous stress field within the area exists. Notice that the stresses along the boundaries may have any direction, i.e., they may be equivalent to both normal stresses and shear stress along the boundaries.

The theorem is illustrated in Fig. 4.5.7.

The homogeneous stress field is easily determined by projection equations for parts of the area containing the sections in which we look for the stresses.

The center cross section is considered fully utilized.

If the stress field in part 1 is homogeneous, it must be uniaxial, as one side is stress free. Mohr's circle for part 1 may, therefore, be drawn immediately, as point A is situated at a distance of Φf_c from the origin. Through A a line is then drawn parallel with the line of discontinuity between parts 1 and 2 to intersect the τ-axis at point B. Together with C, which is the point of intersection of AB with circle 1, the point of intersection B determines Mohr's circle for part 2, which can therefore be drawn. If marking out a point D at a distance f_c from the origin, DB, as seen by a simple geometrical consideration, will be parallel with the line of discontinuity between parts 2 and 3. The point of intersection of DB with circle 2, called E, determines, together with D, Mohr's circle for part 3, which can finally be drawn.[1] By checking it is now easily seen that Mohr's circles thus drawn correspond to stress fields which, in the lines of discontinuity, have similar shear and normal stresses.[2]

Furthermore, it is seen that circle 2 will always correspond to a smaller principal tensile stress than circle 1. When $p \rightarrow f_c$, circle 3 will, in the limit, be a point D, and the maximum principal compressive stress in part 2 will then approach f_c, and thus the solution is

[1] This way of constructing Mohr's circles was used by Szmodits [64.11].

[2] It should be noted that the shear stress is constant and equal to τ_0 along the entire vertical lateral face. That this is possible is due, of course, to the emerging of lines of discontinuity from the corners of the disk.

safe when $p \le f_c$. When $p = f_c$ there will be a local compression failure at the top face.

The curve of the carrying capacity for $\Phi = 0.1$ is shown in Fig. 4.5.8, in which the moment in the center cross section $M = \frac{1}{8} p t L^2$ is plotted as a function of L/h.

The stress field, referred to a system of coordinates with a horizontal x-axis as shown in Fig. 4.5.6, can immediately be calculated. In part 1 we obtain

$$\sigma_x = \Phi f_c \ , \qquad \sigma_y = 0 \ , \qquad \tau_{xy} = 0$$

In part 2, σ_y is determined by considering the stresses along a horizontal line through the neutral axis and projecting on the vertical. We thus obtain

$$\sigma_y = -\frac{2\tau_0}{L}\left(h - y_0\right)$$

As

$$\tau_0 = \frac{1}{2} p \frac{L}{h} = \frac{2\Phi h f_c}{(1+\Phi) L}$$

we obtain in part 2

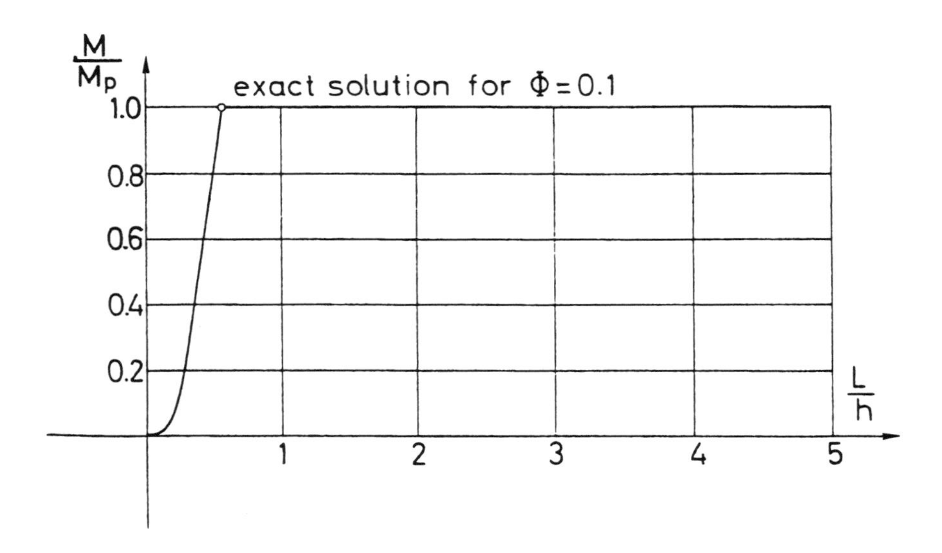

Figure 4.5.8 Exact solution for a rectangular disk supported along the lateral faces and loaded by a uniform load.

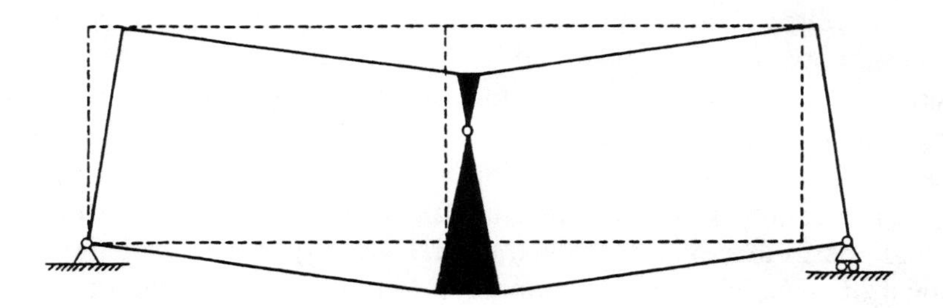

Figure 4.5.9 Failure mechanism for a rectangular disk supported along the lateral faces and loaded by a uniform load.

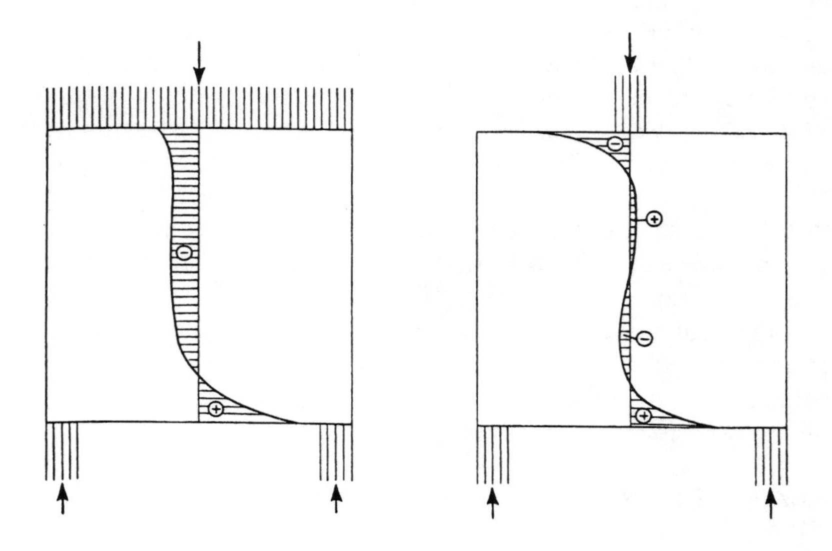

Figure 4.5.10 Examples of normal stress distributions in deep beams according to the elastic theory.

$$\sigma_x = 0\ , \quad \sigma_y = -\frac{4\Phi h^2 f_c}{(1+\Phi)^2 L^2}\ , \quad \tau_{xy} = \frac{2\Phi h f_c}{(1+\Phi) L}$$

Finally, we obtain in part 3

$$\sigma_x = -f_c\ , \quad \sigma_y = -p = -\frac{4\Phi h^2 f_c}{(1+\Phi) L^2}\ , \quad \tau_{xy} = 0 \quad (p \le f_c)$$

An upper bound solution corresponding to the lower bound solution above is easily derived by considering the displacement field sketched in Fig. 4.5.9. When the two halves of the disk rotate as rigid bodies, discontinuities in u_x, corresponding to the stresses assumed, are produced. The solution is therefore exact.

In addition to the lower and upper bound technique, the calculations in this example illustrate the strongly intuitive method that must often be used in determining solutions in plastic theory.

4.5.4 Comparison with the Elastic Theory

The stress distributions found according to the plastic theory differ considerably from the stress distributions found according to the theory of elasticity assuming uncracked concrete. In Fig. 4.5.10 the stress distributions in the center section of two disks having a uniformly distributed load on the top face and a concentrated force applied in the center of the top face, respectively, are outlined. The depth of the disks is assumed to be of the same order of magnitude as the span. The maximum compressive stress below the concentrated force is highly dependent on the area over which the force is distributed, so that the bigger the area, the smaller the maximum compressive stress. Therefore, on the way to the collapse load, there is a considerable redistribution of the stresses.

Some solutions based on the elastic theory of rectangular disks may be found in [31.2, 33.1, 56.4, 58.3]. Nowadays linear elastic solutions for uncracked concrete may be obtained almost as a routine by the use of finite element programs.

4.6 THE EFFECTIVE COMPRESSIVE STRENGTH OF REINFORCED DISKS

As already mentioned in Chapter 2, there are three main reasons for the necessity of working with reduced strength parameters, the so-called effective strengths, in concrete. They are softening, strength reductions due to internal cracking due to stressed reinforcement crossing the cracks, and strength reduction due to sliding in initial cracks. For shear in disks with a certain amount of minimum reinforcement everywhere, it seems that softening is rather unimportant. In what follows we describe what is known at present about the two remaining effects.

4.6.1 Strength Reduction due to Internal Cracking

The effective compressive strength of disks, or panels as this structural element is often named, has been the subject of many experimental investigations in recent years. A lot has been learned[3] but the subject is definitely far from being closed.

The first ones to investigate the reduction of the compressive strength of cracked concrete due to stressed transverse reinforcement were, to the author's knowledge, Demorieux [69.8] and Robinson and Demorieux [77.8]. The test arrangement in the 1977 tests is shown schematically in Fig. 4.6.1. A rectangular disk 450 x 420 mm^2 with thickness 100 mm was subjected to uniaxial compression while, at the same time, transverse reinforcement could be subjected to tensile stresses. Even reinforcement under an angle ω (see fig. 4.6.1) different from 90° was dealt with.

The main results for $\omega = 90°$ were the following:

> Compressive strength of unreinforced disk: $0.82\,f_c$.
>
> Compressive strength of reinforced disk with zero stress in transverse reinforcement: $0.87\,f_c$.
>
> Compressive strength when transverse reinforcement was stressed to the yield point: $0.75\,f_c$.

The uniaxial compressive strength of the disk was 18% lower than the cylinder strength. This may be due to the disk being rather slender, compared to the standard cylinder, making it more sensitive to small eccentricities. There are possibly also other reasons like the curing conditions which may differ from those of the standard cylinder, different strengths at top and bottom depending on casting direction, etc.

What is important for us here is that the stress in the transverse reinforcement caused a reduction of the compressive strength from $0.87\,f_c$ to $0.75\,f_c$, i.e., 14% when the transverse reinforcement was stressed to the yield point. Tests with $\omega \neq 90°$ gave similar results.

Since then many tests of this kind have been done. A review is undertaken in [97.3]. The conclusion from all these efforts is clear.

[3] In the first edition of this book [84.11], this section was nonexistent.

Transverse stressed reinforcement reduces the compressive strength of a disk.

The reduction must be caused by the internal cracking between the primary cracks, an aspect which was already discussed in Section 2.1.4. This view is confirmed by some tests with plain bars. They cause less internal cracking and give rise to less strength reduction.

It is clear that the internal cracking depends on a large number of parameters, the most important ones being the reinforcement stress, the concrete strength, the cover, the reinforcement ratio, the diameter of the bars and the distribution of the reinforcement in the sections. Probably the most important aspect of strength reduction is the bursting stresses around the reinforcement bars, which might possibly lead to more or less spalling off the cover. No comprehensive theory has yet been developed to take all these parameters into account.

An attempt to model strength reduction due to internal cracking has been made in [97.3]. The conclusions are by no means final, but they show that the parameter χ introduced in Section 2.1.4 seems to be a reasonable one to work with. More on that later.

This discussion illustrates the difficulties we are facing when we want to determine the compressive strength of a cracked, reinforced disk. And there is more. As mentioned in Section 2.1.4 and further discussed in Chapter 5, it is difficult to anchor reinforcement bars along the boundaries without causing additional severe internal cracking, unless the reinforcement bars are attached to, say a sufficiently strong and stiff steel plate, which is seldom done. So anchorage effects must be added to the number of causes of strength reduction. This has been clearly demonstrated by some of the test series made, where several test specimens suffered anchorage failure or other kinds of edge failures.

At the present stage of development it is impossible to give more than a qualitative explanation of the strength reduction due to internal cracking. So we must rely on empirical formulas derived from tests.

For pure shear in disks with normal strength concrete, i.e., $f_c < 50$ MPa, it seems that the simple formula for the effectiveness factor

$$\nu = \nu_0 = 0.7 - \frac{f_c}{200} \qquad (f_c \text{ in MPa}) \tag{4.6.1}$$

gives reasonable agreement with tests. The formula was originally suggested for shear in beams (cf. Chapter 5) but it has turned out to be more generally applicable. The reason for the double notation ν and ν_0 will appear later.

To illustrate the validity of (4.6.1) for disks we may refer to an extensive test series carried out by Vecchio/Collins [82.4]. The test specimens were square with side lengths 890 mm and thickness 70 mm. In the tests considered first, the disks were subjected to pure shear and the reinforcement was either isotropic or orthotropic with directions parallel to the sides. The bars were plain bars but they were welded in the cross points and probably acted more or less as deformed bars. The compressive strength of the concrete varied between 11.5 MPa and 34.5 MPa. The yield strength of the bars was around 250 MPa.

The shear strength may be calculated by the yield conditions of Section 2.2, i.e.,

$$
\begin{aligned}
&-f_{cx} \le \sigma_x \le f_{tx} \\
&-f_{cy} \le \sigma_y \le f_{ty} \\
&-(f_{tx} - \sigma_x)(f_{ty} - \sigma_y) + \tau_{xy}^2 \le 0 \\
&-(f_{cx} + \sigma_x)(f_{cy} + \sigma_y) + \tau_{xy}^2 \le 0 \\
&|\tau_{xy}| \le \tfrac{1}{2}\, v f_c
\end{aligned}
\tag{4.6.2}
$$

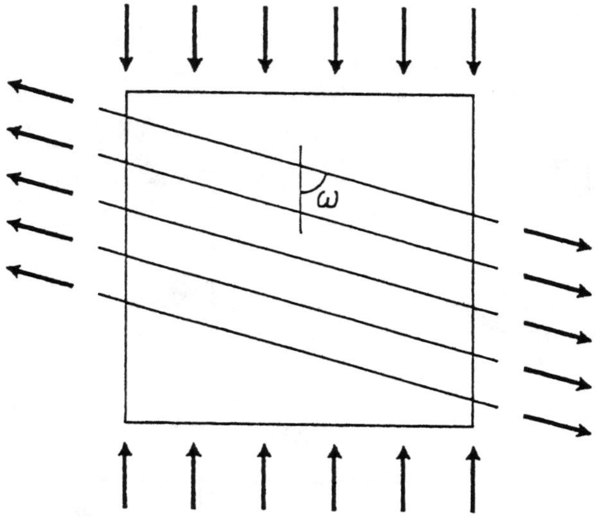

Figure 4.6.1 Schematic test arrangement used by Robinson and Demorieux [77.8].

Here f_{tx} and f_{ty} are the tensile strengths in the reinforcement directions (due to the reinforcement alone) and f_{cx} and f_{cy} are the compressive strengths in the reinforcement directions (possibly calculated by taking into account the reinforcement contribution). The normal stresses σ_x and σ_y are positive as tension.

The shear strength f_v may be determined setting $\sigma_x = \sigma_y = 0$ and $|\tau_{xy}| = f_v$.

We get

$$f_v = \min \begin{cases} \sqrt{f_{tx} f_{ty}} \\ \frac{1}{2} \nu f_c \end{cases} \tag{4.6.3}$$

The results of such a calculation are shown in Fig. 4.6.2. The effectiveness factor has been calculated using formula (4.6.1).

In some cases the disks failed because of anchorage failure and in one case there was only reinforcement in one direction, in which case (4.6.3) leads to $f_v = 0$. The shear strength measured in this case is, of course, due to the tensile strength of concrete. These special tests have been marked in the figure.

The tests covered failure by yielding of the reinforcement and failure by concrete crushing.

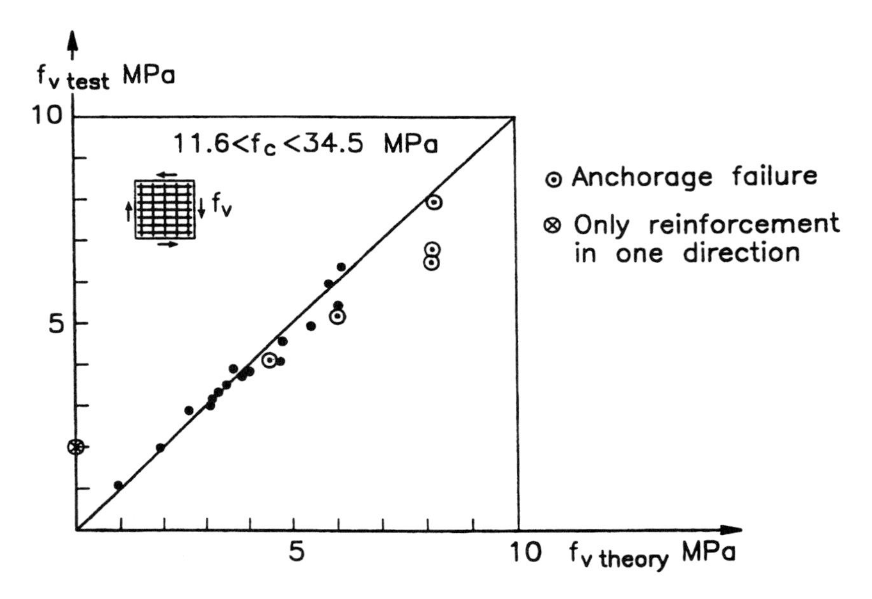

Figure 4.6.2 Comparison between yield conditions and pure shear tests by Vecchio and Collins [82.4].

It appears that the agreement with the yield conditions is excellent.

For the highly orthotropic disks the agreement is probably accidental. For large differences in reinforcement degrees in the two reinforcement directions, it is to be expected (see Section 4.6.2), that the load-carrying capacity is governed by crack sliding. Calculations show that the expected reduction is of an order up to about 25%. Such a reduction was not found. This is probably due to the test arrangement. The way of supporting the disks did not allow a continuous yield line along the initial cracks under 45° with the reinforcement directions. Further, a strong band of higher strength concrete was supplied along the boundary. These two effects are likely to have caused a higher load-carrying capacity.

The ν-formula (4.6.1) is not very accurate for high strength concrete. It has been suggested (see Eurocode 2, [91.23]) to improve it by introducing a lower limit of 0.5, i.e.,

$$\nu = \nu_0 = 0.7 - \frac{f_c}{200} \not< 0.5 \qquad (f_c \text{ in MPa}) \tag{4.6.4}$$

However, a better formula might be derived from one of the test series performed by Takeda, Yamaguchi and Naganuma [91.8]. Their test specimens were square with side lengths of 1200 mm and thickness of 200 mm. The compressive strengths were varied from about 20 MPa to about 80 MPa. The reinforcement was, in the tests we are dealing with, parallel to the sides of the disk. It was ribbed and had a yield strength of a little more than 400 MPa.

The test results leading to compression failure in the concrete (overreinforced disks) are depicted in Fig. 4.6.3, which shows the shear strength f_v as a function of f_c. There are a number of tests with pure shear in isotropic disks and a few tests with shear plus biaxial hydrostatic compression in isotropic disks.

The pure shear test results are lying rather close to the following curve suggested by Takeda, Yamaguchi and Naganuma

$$f_v = 0.95 f_c^{0.66} \qquad (f_c \text{ in MPa}) \tag{4.6.5}$$

which, using $f_v = \frac{1}{2}\nu f_c$, corresponds to

$$\nu = \nu_0 = \frac{1.9}{f_c^{0.34}} \not> 1 \qquad (f_c \text{ in MPa}) \tag{4.6.6}$$

This formula may be used for high strength concrete, say $50 < f_c < 100$ MPa, instead of (4.6.1) and (4.6.4). It appears from Fig. 4.6.3

that the overreinforced disks with shear plus biaxial hydrostatic compression showed higher load-carrying capacity than the pure shear disks. The reason is, that the stress in the reinforcement is diminished for increasing biaxial compression, keeping everything else constant. When the reinforcement stress is decreased the amount of internal cracking is also decreased and then the strength is increased.

To take this into account we may introduce the parameter

$$\chi = \frac{r\sigma_s}{1.41\sqrt{f_c}} \qquad (f_c \text{ in MPa}) \tag{4.6.7}$$

(cf. Section 2.1.4). Here r is the reinforcement ratio and σ_s the stress in the reinforcement.

For a shear stress τ plus a biaxial hydrostatic compressive stress σ we have

$$\chi = \frac{\tau - \sigma}{1.41\sqrt{f_c}} \qquad (f_c \text{ in MPa}) \tag{4.6.8}$$

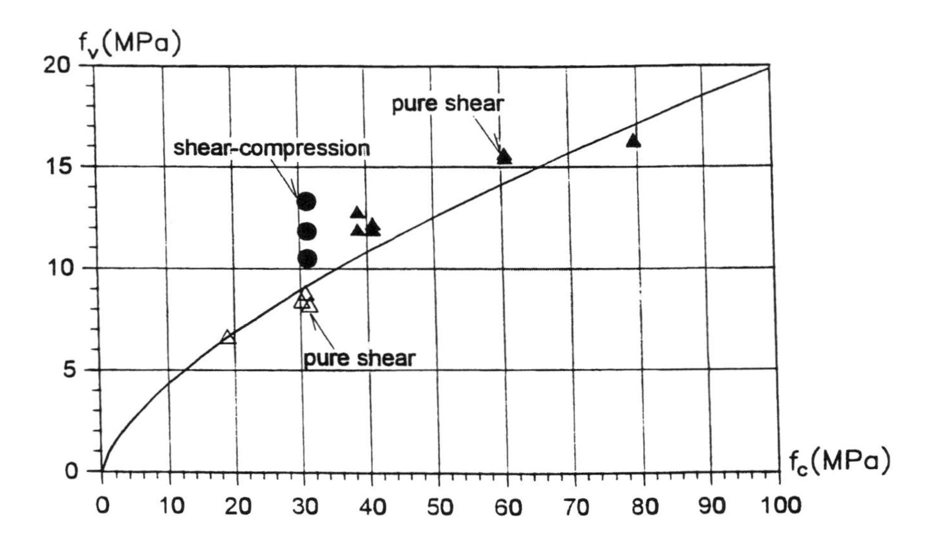

Figure 4.6.3 Tests with overreinforced disks by Takeda, Yamaguchi and Naganuma [91.8].

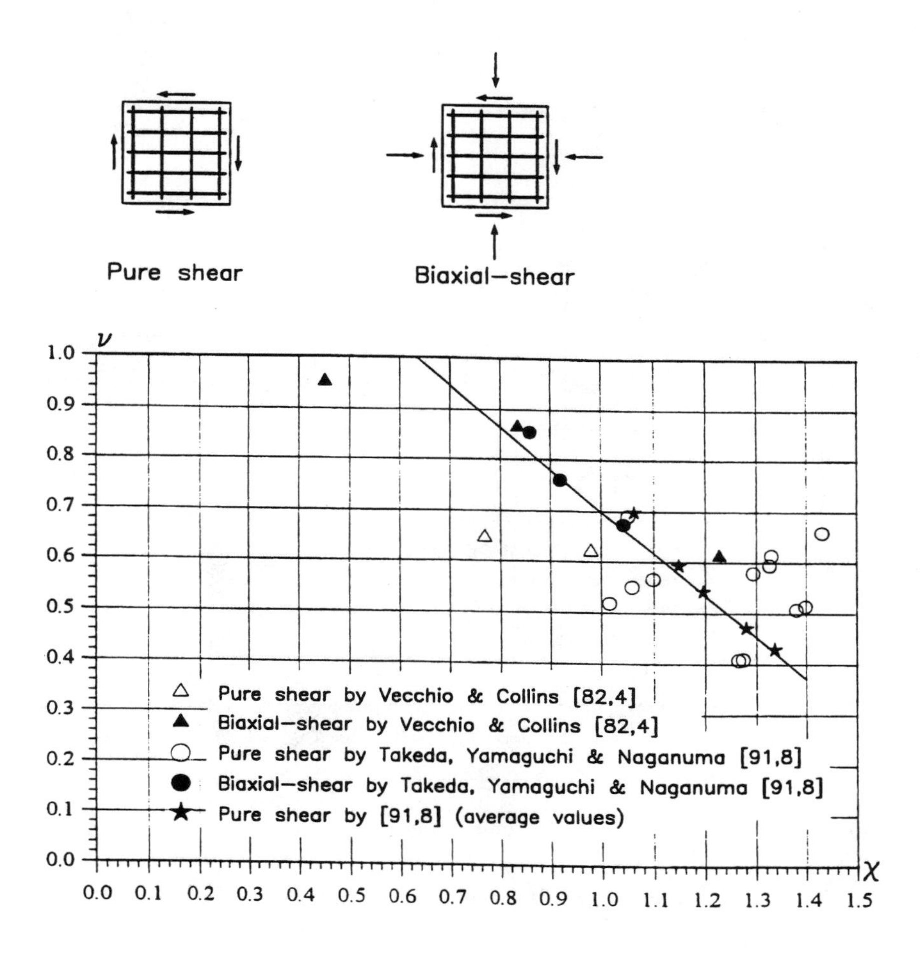

Figure 4.6.4 Effectiveness factor ν versus the parameter $\chi = r\,\sigma_s/1.41\sqrt{f_c}$.

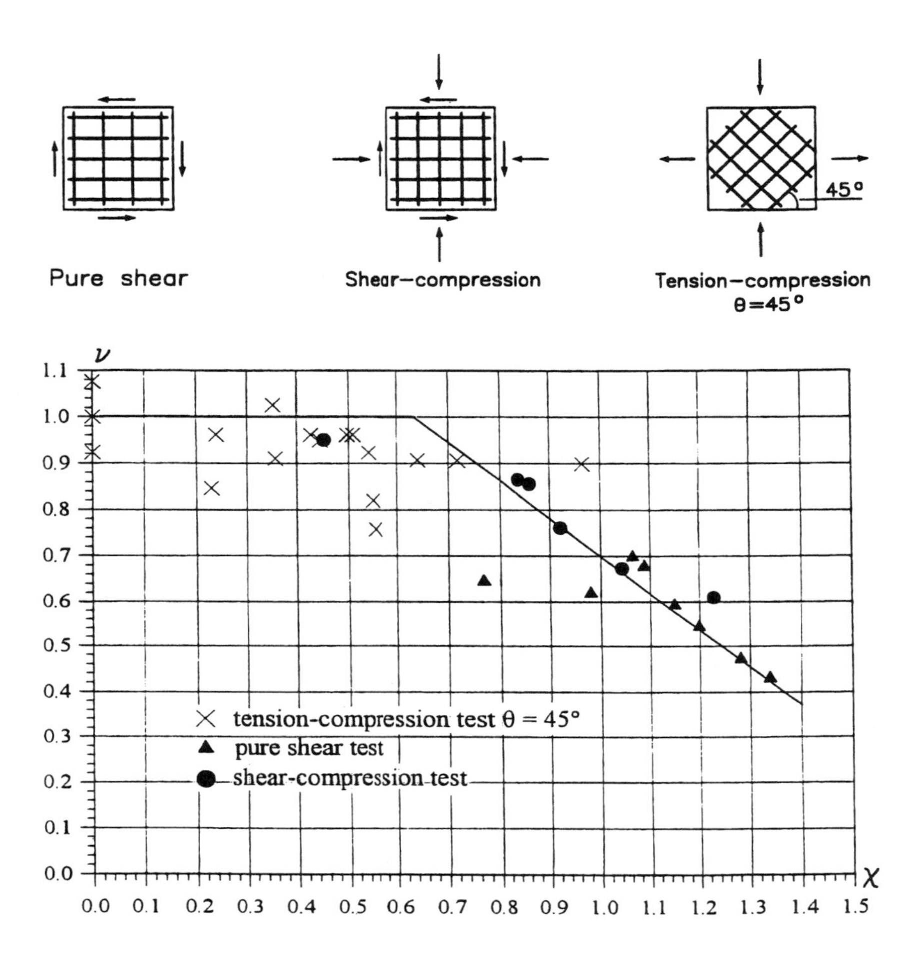

Figure 4.6.5 Effectiveness factor ν versus χ.

If the pure shear results, using the average curve (4.6.5) as representing the test results, and the shear plus biaxial compression results are plotted in a diagram showing $\nu = \tau / \frac{1}{2} f_c$ as a function of χ, we get the straight line shown in Fig. 4.6.4. The points to focus on are the five stars (pure shear) and the three black circles (shear plus biaxial compression). These points are with good approximation lying on the straight line

$$\nu = 1.52 - 0.83\chi \tag{4.6.9}$$

If the true test results are plotted instead of the interpolated values, the scatter is considerably larger. In Fig. 4.6.4 those tests of Vecchio/Collins with shear and biaxial compression and those with isotropic reinforcement suffering compression failure have also been plotted.

Further, in Fig. 4.6.5 tests by different authors for low values of χ have been included. They have been carried out by applying compression and tension to specimens with reinforcement under an angle $\theta = 45°$ with the direction of the external forces.

Finally, in Fig. 4.6.6 tension-compression tests made on specimens with reinforcement in the direction of the external forces ($\theta = 0$) have been shown together with the other types of tests mentioned. Some of the tests with $\theta = 0$ have given rise to less reduction in compressive strength than the other tests. The scatter is quite large and it is clear that more refined methods are required to describe the situation accurately. Such methods will include more parameters than those combined in the χ-value, the most important ones being the cover, the reinforcement diameter, and parameters describing the distribution of the reinforcement bars in the sections.

Regarding the detailed treatment of the tests the reader is referred to [97.3].

Although it might seem a little premature, we conclude that to cover shear plus biaxial compression the ν-value to be used is

$$\nu = 1.52 - 0.83\chi \ngtr 1 \tag{4.6.10}$$

This will include (4.6.1) for normal strength concrete and (4.6.6) for high strength concrete as special cases.

The reader may find it useful to demonstrate this by himself. For pure shear $\chi = \tau/1.41\sqrt{f_c}$. Inserting this into (4.6.10), setting $\tau = \frac{1}{2}\nu f_c$ and solving the equation for τ, we will find a resulting ν-value $(\tau/\frac{1}{2}f_c)$ very close to (4.6.6). In the case of shear and biaxial compression we have, as mentioned, $\chi = (\tau - \sigma)/1.41\sqrt{f_c}$. Inserting this into (4.6.10) and solving for τ, we find the shear strength as a function of the biaxial compressive stress σ.

Schematically the shear-carrying capacity as a function of σ will be the straight line a) shown in Fig. 4.6.7. This line is, of course, only valid in the overreinforced case where we have compression failure in the concrete. If for low values of σ the reinforcement is yielding, then according to the yield conditions (4.6.2)

$$\tau = f_v = f_t + \sigma \tag{4.6.11}$$

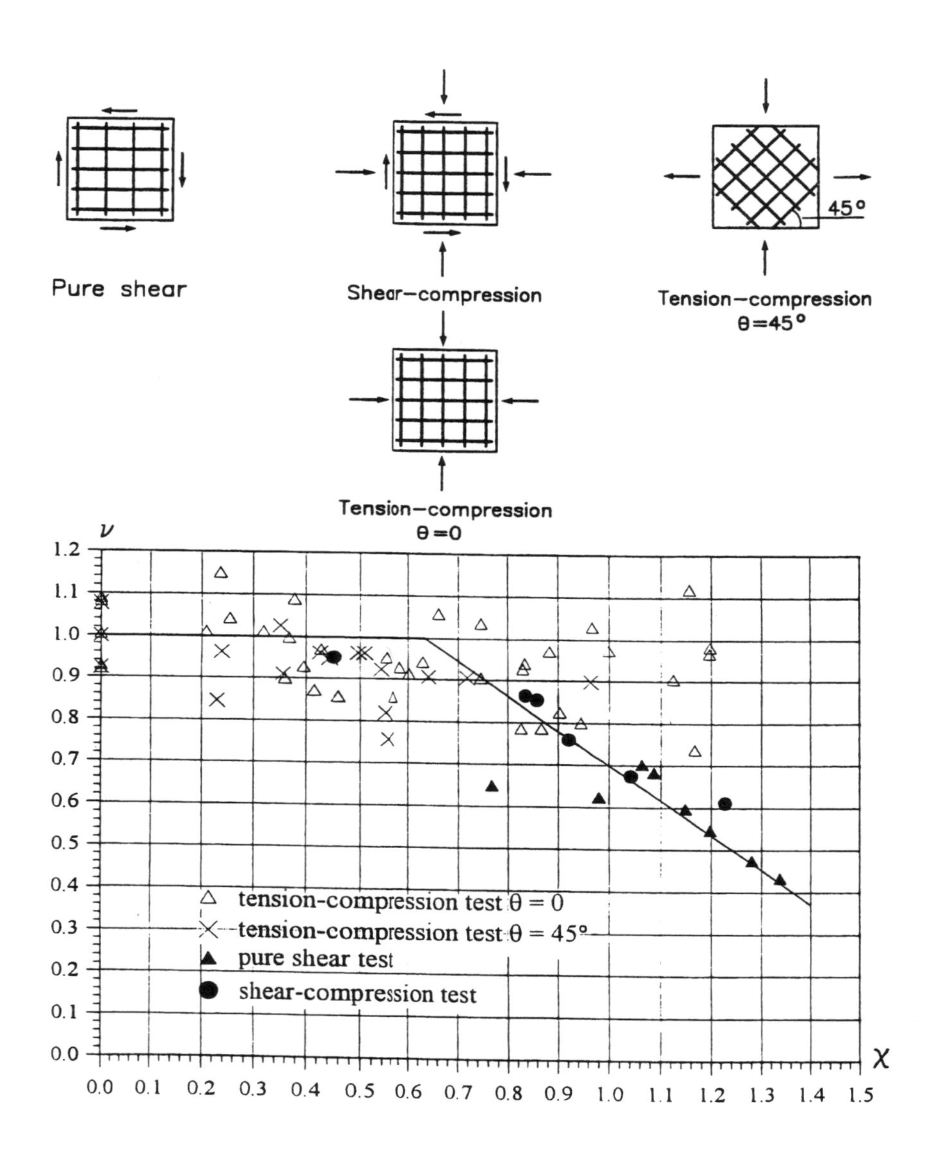

Figure 4.6.6 Effectiveness factor ν versus χ.

still considering the isotropic case, i.e., $f_{tx} = f_{ty} = f_t$ and counting σ positive as compression, i.e., $\sigma_x = \sigma_y = -\sigma$.

In Fig. 4.6.7 the formula (4.6.11) also appears as a straight line shown as b). The lowest value of a) and b) is, of course, valid. Finally, the absolute upper limit of τ equal to $\frac{1}{2}f_c$ must also be respected.

The curve shown in Figure 4.6.7 is only valid for compressive stresses σ up to a certain limit. For sufficiently high values of σ the compression condition of the yield conditions is taking over (condition number four of (4.6.2)). This region has not been considered in the figure.

The inclination of a) is typically only around 60% of the inclination of b), as illustrated schematically in the figure.

If no account was taken of the increase of ν with compressive stresses the upper limit would have been $\frac{1}{2}\nu f_c$ independent of σ, ν being determined by (4.6.1) or (4.6.6). If no account was taken of the reduction of the compressive strength, the shear capacity would be determined by the line b) and the horizontal line corresponding to the upper limit $\frac{1}{2}f_c$.

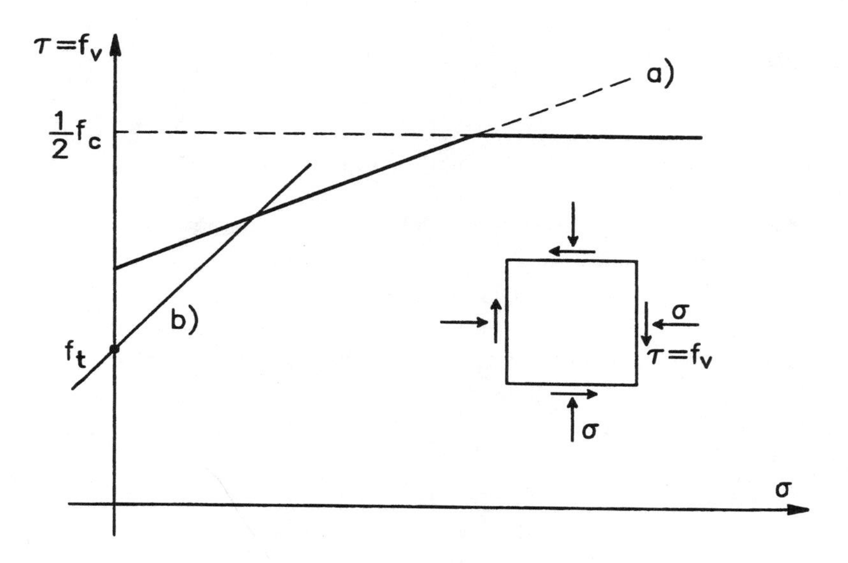

Figure 4.6.7 Shear capacity versus biaxial compressive stress σ.

The increase of shear capacity due to compression stresses is sometimes important in design. For instance, it serves to diminish the dimensions of compression flanges in beams carrying both normal stresses and shear (see Chapter 5).

Vecchio/Collins suggested on the basis of their tests a reduction of the compressive strength, which is equivalent to a ν-formula, ν being a function of the average strain transverse to the compression direction. For constant reinforcement ratio and constant compressive strength this is equivalent to a ν-formula rendering ν as a function of χ.

However, it is unlikely that a transverse strain in a panel with a small amount of reinforcement causes the same internal damage as the same strain in a panel with a larger amount of reinforcement. Indeed, a final decision on this point must await further research.

Here we are rather content not having to calculate the transverse strain, since it is extremely difficult or even impossible using the simple plastic theory.

In practical design we, of course, meet many cases other than isotropic reinforcement and cracks under 45° or 90° with the reinforcement directions. Fortunately it seems that the formula (4.6.10) may be extended to much more general cases taking the value of $r\ \sigma_s$ to be the largest one for the reinforcement transverse (but not necessarily perpendicular) to the direction of the compression. This has been confirmed, first of all by a large number of shear tests on beams where we have normally only one reinforcement direction (the stirrups) transverse to the compression direction. All these tests lead to the same formula for ν as (4.6.1) (cf. Chapter 5). Most of these tests were with normal strength concrete, but it seems likely that (4.6.6) may be used for high strength concrete beams as well as for high strength concrete disks.

4.6.2 Strength Reduction due to Sliding in Initial Cracks

The theory of sliding in initial cracks is in its infancy. It should have been initiated long ago. Preliminary studies for disks are reported in [97.3]. For beams the level is more advanced (see Chapter 5). When our knowledge is more mature we will know the yield conditions taking into account crack sliding. The implications on reinforcement design are treated below. A particularly interesting limiting case is when the concrete may be assumed to be cracked in all directions. Such a concrete may have suffered alkali-silica reactions. Work is in progress but results cannot yet be reported.

In the Vecchio/Collins tests the stress-strain relation of cracked concrete was measured. It appears from the measurements that compressive failure sometimes took place for a value of the stress in the concrete much less than the value calculated by using the ν-formulas (4.6.1) or (4.6.6). We will show that this is due to sliding in initial cracks.

Consider Fig. 4.6.8. This figure illustrates that if the initial crack direction is different from the final uniaxial compression direction, there is an evident possibility that compression failure may take place by sliding along the initial cracks instead of sliding along the usual failure lines. In the figure the angle β is indicating the initial crack direction and the angle α is indicating the final direction of the uniaxial compression. Cracks may be formed in the new direction of the compression but not always.

Now we assume that the failure condition in the crack is of the Coulomb type, i.e.,

$$|\tau| = c' + \mu' \sigma \tag{4.6.12}$$

Here σ and τ are the normal stress, positive as compression, and the shear stress, respectively, in the crack. The parameters c' and μ' are the cohesion and the coefficient of friction, respectively, in the crack. These values are always less than or equal to the values valid for uncracked concrete.

It will be shown in Chapter 8 that μ' may be taken as the value valid for uncracked concrete, i.e., $\mu' = \mu = 0.75$ having assumed the friction angle to be $\varphi = 37^{\circ}$. However, c' is strongly reduced compared to the value for uncracked concrete. We put

$$c' = \nu_s \cdot \frac{1}{4} \nu_0 f_c \tag{4.6.13}$$

where ν_s is the so-called sliding reduction factor (cf. Section 2.1.4), and $\frac{1}{4}\nu_0 f_c$ is the cohesion for concrete with a compressive strength $\nu_0 f_c$. The formula means that c' is measured relative to the already reduced compressive strength $\nu_0 f_c$. If we express the normal stress and the shear stress along the cracks by the compressive stress f_{cs} corresponding to sliding in the initial cracks and insert these stresses into the failure condition (4.6.12), we get by solving the equation with respect to f_{cs},

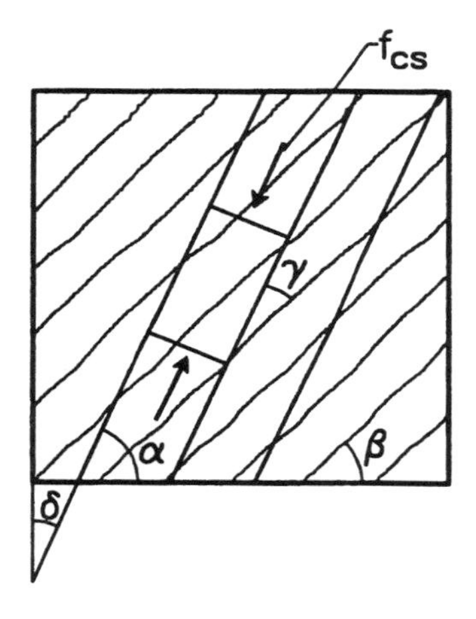

Figure 4.6.8 Disk with initial crack inclination different from final compression direction (crack direction).

$$\frac{f_{cs}}{\nu_0 f_c} = \frac{\frac{1}{4}\nu_s}{|\sin(\alpha-\beta)\cos(\alpha-\beta)| - 0.75\sin^2(\alpha-\beta)} \ngtr 1 \qquad (4.6.14)$$

By means of this formula the compressive strength f_{cs} for sliding failure in the initial cracks may be calculated, when ν_s and the angles β and α are known.

The calculation is, of course, only meaningful if any reinforcement present is utilized for carrying an external load. Otherwise the reinforcement will increase the strength by resisting the sliding (see below).

Sometimes it is more convenient to express f_{cs} by the angle $\gamma = \alpha - \beta$ between the initial cracks and the final cracks (see Fig. 4.6.8). We find

$$\frac{f_{cs}}{\nu_0 f_c} = \frac{\frac{1}{4}\nu_s}{|\sin\gamma\cos\gamma| - 0.75\sin^2\gamma} \ngtr 1 \qquad (4.6.15)$$

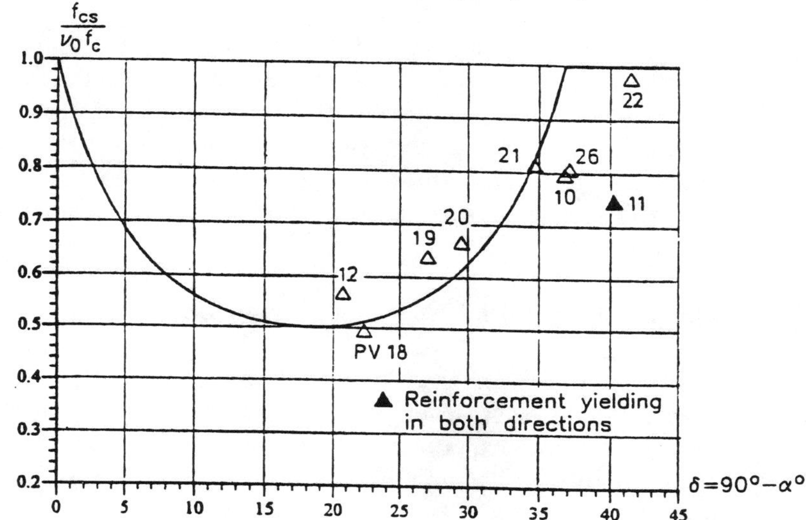

Figure 4.6.9 Tests compared with theory. Orthotropic disks by Vecchio/Collins [82.4].

The orthotropic disks in the test series of Vecchio/Collins were subjected to pure shear which means that the initial crack angle $\beta = 45^\circ$. The angle α may be approximately calculated by assuming that both reinforcement directions are yielding. The necessary formula may be found in Section 2.2.

In Fig. 4.6.9 f_{cs} has been drawn as a function of the angle $\delta = \pi/2 - \alpha$ (see Fig. 4.6.8). The effectiveness factor $\nu = \nu_0$ has been calculated by (4.6.1) and ν_s has been set at 0.5. In the figure the measured compressive strengths have also been plotted. At the individual points the names of the specimens adopted by Vecchio/Collins are also shown.

The scatter is considerable, but it appears that reasonable agreement is found when using $\nu_s = 0.5$. This means that the apparent cohesion in the cracks has been reduced to half the value for concrete with internal cracking only. This value of the sliding reduction factor will be rediscovered in Chapter 8 on joints.

The consequences of these experimental facts are rather dramatic. They mean that a change in crack direction may lead to a substantial reduction of the compressive strength beyond that caused by softening and by internal cracking.

Take as an example a typical value of $\nu_0 = 0.6$ and $\nu_s = 0.5$. Then the resulting compressive strength is in the worst case only $0.6 \cdot 0.5 = 0.3 = 30\%$ of the standard cylinder strength. Because of internal cracking and sliding in initial cracks, in this example only 30% of the uncracked compressive strength is left for the benefit of the designer.

The only alleviation is, as mentioned in Section 2.1.4, that the sliding in initial cracks seems to result in a very ductile behavior. The reader may once again study Fig. 2.1.34 which illustrates an almost perfectly plastic stress-strain curve for one of the orthotropic Vecchio/Collins disks.

4.6.3 Implications of Initial Crack Sliding on Design

In the past a large number of empirical formulas for the effectiveness factors have been developed on the basis of experiments. The majority of these experiments have been with proportional loading. For proportional loading these formulas may, of course, still be used.

Probably crack sliding is only important at proportional loading, from a practical point of view, when very low values of the effectiveness factors have been found in experiments. The most prominent example is the shear failure of nonshear reinforced beams. Here the ν-formulas contained the shear span as a parameter. By taking crack sliding into account, more reasonable ν-formulas have been established (cf. Section 4.8.4 and Chapter 5). Crack sliding is unimportant at proportional loading if the reinforcement is designed for a stress field with compression directions not deviating too much from the initial ones.

Now we describe how to take crack sliding into account. This must be done when a ν-formula has not been established or when we are not dealing with proportional loading.

If the crack directions of the structure are known and the stress field to design for has been selected, it is easy by means of the reinforcement formulas to design a safe structure. If a uniaxial compression direction equal to the crack direction is used, then the upper limit of σ_c may be put to $\nu_0 f_c$. If another direction of the uniaxial compression is chosen, the upper limit of σ_c is f_{cs} determined by (4.6.15).

Of course, in any case it will be on the safe side to use a ν-value equal to $\nu = \nu_s \nu_0$.

Any direction of the uniaxial compression stress may be used if the upper limit of σ_c is taken to be f_{cs}. Especially, the direction corresponding to local minimum reinforcement may be chosen.

If there are two crack directions at a point, the direction of the uniaxial compression may be taken as the direction bisecting the angle between the cracks. The upper limit is f_{cs} calculated by (4.6.15) with γ equal to half the angle between the cracks.

Unfortunately, in practice the crack directions are seldomly known with any high accuracy.

Only when dealing with homogeneous stress fields can the initial crack directions be easily calculated.

In other cases the crack system may be rather complicated. In fact it may consist of several crack systems crossing each other. First we have the crack directions corresponding to stress trajectories of uncracked concrete. They might be calculated rather easily using a standard linear elastic computer program. After cracking in some regions with high stresses, the stiffness in these regions will be dramatically reduced changing, sometimes completely, the subsequent cracking. The crack directions for a fully cracked structure may be determined by minimizing the complementary elastic energy. Such calculations have been reported in [78.6] and [97.17].

If the reinforcement starts yielding, a third crack system may develop and then a final one is developed in the ultimate state.

If the structure is loaded by cyclic loading a larger number of crack systems may be developed.

The conclusion is that even for a simple structural system with inhomogeneous stress fields the crack directions can only be predicted by rather advanced methods, and only when knowing the loading history. Of course, the designer may, as a first approximation, use the linear elastic, uncracked stress distribution as a guide for how to select the crack directions to take into account in the design.

Often it is convenient to evaluate the effects of crack sliding by upper bound methods (see the next section).

To maintain a sliding resistance in cracks, it is important to keep the crack widths small (cf. Chapter 8).

A necessary condition is that the reinforcement is able to carry a stress in any section which is larger than the effective tensile strength of concrete. If this condition is not met the reinforcement is not able to furnish a well-distributed crack system and only one or a few large cracks may develop.

The effective tensile strength depends a lot on the circumstances. When analyzing test data for specimens with homogeneous stress fields it is normally found that the effective tensile strength, i.e., the stress causing cracking, is substantially lower than the tensile strength measured on standard test specimens. This is due to stress concentrations, including stress concentrations around the reinforcement bars induced by the ribs, shrinkage stresses, etc. On the other hand, if the stress field is inhomogeneous, like in bending, the nominal cracking stress based on a linear elastic calculation might be substantially higher than the standard tensile strength. This is due to the development of a plastic tensile zone. The average tensile stress in this zone is, however, also lower than the standard tensile strength. In such cases there is a pronounced size effect. The nominal cracking stress is decreasing when the size is increasing.

Based on tests and practical experience it seems reasonable to assume that, as a general rule, the effective tensile strength may be put at half the value measured on standard test specimens.

The necessary reinforcement ratio r to transfer the effective tensile strength of concrete f_{tef}, when the reinforcement is stressed to the yield point, is determined by

$$r f_Y = f_{tef} \tag{4.6.16}$$

Setting $f_{tef} = 0.5 f_t = 0.5\sqrt{0.1 f_c} = 0.16\sqrt{f_c}$ (f_c in MPa) we get

$$r f_Y = 0.16\sqrt{f_c} \quad (f_c \text{ in MPa}) \tag{4.6.17}$$

It means that the reinforcement degree Φ required is

$$\Phi = \Phi_{\min} = \frac{0.16}{\sqrt{f_c}} \quad (f_c \text{ in MPa}) \tag{4.6.18}$$

A reinforcement degree of at least this magnitude should be present in any part of the element.

If the crack direction is known, a reinforcement satisfying (4.6.18) in the direction perpendicular to the cracks will suffice. Normally for disks, reinforcement in two perpendicular directions with the same reinforcement degree must be used.

When applying (4.6.18) in practice we should, strictly speaking, use an upper fractile value of the tensile strength. However, the usual characteristic values are often used. We will not elaborate further on this point here.

Changes in compression direction are believed to be small when the stress field used for reinforcement design is relatively close to the global minimum reinforcement solution. Therefore the designer should pay attention to optimizing the reinforcement not only for saving reinforcement, but also for helping the structure to easily reach a high load-carrying capacity. Minimum reinforcement structures were discussed in Chapter 1. It was mentioned that it is believed, although it has not been generally proved, that the redistribution of stresses on the way to the ultimate limit state is kept to a minimum for minimum reinforcement structures.

Regarding minimum reinforcement solutions we note in passing that when the stress field to reinforce for has been selected, the reinforcement formulas of Section 2.4, giving local minimum reinforcement, will also give global minimum reinforcement for this particular stress field.

The designer may, however, be rather relaxed about obtaining minimum reinforcement or not. Experience shows that large deviations from minimum reinforcement are allowable. The structure generally will adapt itself to the reinforcement chosen.

When considering cracking it should be borne in mind that a structure might crack from many reasons other than those covered by the design. Cracks may be due to unforeseen loadings or temperature, shrinkage or creep deformations. Finally cracking may be due to chemical reactions like alkali-silica reactions.

4.6.4 Plastic Solutions Taking into Account Initial Crack Sliding

The designer who wants to avoid any speculations on the effect of crack sliding may provide extra reinforcement in all parts of the structure. How to determine this is treated below.

How to design the reinforcement taking into account crack sliding was treated above. Otherwise lower bound solutions taking into account sliding in cracks are an almost untouched subject. Work is, however, in progress.

Upper bound solutions for sliding in initial cracks may be determined in the usual way. Sliding in cracks is always treated as a plane strain problem.

If the dissipation formula (3.4.86) is used with $f_t = 0$ and f_c replaced by $\nu_s \nu_0 f_c$, it must be emphasized that this is equivalent to the failure condition shown in Fig. 4.6.10 (left).

In the region corresponding to small compressive stresses the Coulomb lines are effectively replaced by a circle touching the Coulomb lines and the τ-axis. However, the procedure used to develop formula (4.6.15) is equivalent to a failure condition shown in Fig. 4.6.10 (right) or, alternatively to use (3.4.86) with a small tensile strength rendering the Coulomb lines to be valid for all compressive stress states. Although the difference between the two conditions may seem small, the results they give may be rather different. This is examined in Chapter 8 on joints. The failure condition in Fig. 4.6.10 (left) has rendered very good results for non-shear reinforced beams (see [94.10]), while for reinforced disks we found reasonable results developing formula (4.6.15) in a way equivalent to using the failure condition in Fig. 4.6.10 (right).

The analysis carried out in Chapter 8 shows that the truth lies somewhere between the two conditions. So at present both may be used, whichever is convenient.

If the failure condition of Fig. 4.6.10 (right) is used in upper bound calculations, we need a dissipation formula. It is easily shown that for a crack with a relative displacement vector of length u making an angle α with the crack, we have

$$W_\ell = c' ub \cos\alpha \qquad \varphi \le \alpha \le \pi - \varphi \tag{4.6.19}$$

c' being the cohesion of the crack. We still have assumed the friction angle to be the same as for uncracked concrete. The formula (4.6.19) may be written

$$W_\ell = \frac{1}{4}\nu_s \nu_0 f_c ub \cos\alpha \qquad \varphi \le \alpha \le \pi - \varphi \tag{4.6.20}$$

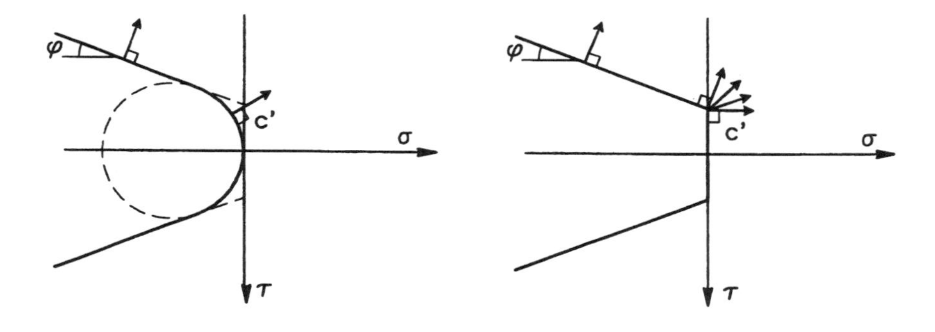

Figure 4.6.10 Possible failure conditions for sliding in cracks.

Besides investigating failure mechanisms with sliding in cracks, failure mechanisms in uncracked, or rather concrete with only internal cracking, must be dealt with since they might govern the load-carrying capacity.

Upper bound solutions for sliding in cracks have been examined in [94.10] [97.4] [97.7] [97.8] for beams and slabs without shear reinforcement. It turns out that sliding in initial cracks explains the rather low shear capacity obtained for these stuctural elements. The theory is described in Chapter 5. In [97.9] shear reinforced beams have been treated.

In [97.3] upper bound solutions for orthotropic disks with sliding in initial cracks have been determined. They confirm the results given in Fig. 4.6.9.

It is possible to supply an additional reinforcement to avoid sliding in initial cracks, the additional reinforcement resisting the sliding (confinement effect).

Let us conclude this section by calculating a safe value for this reinforcement.

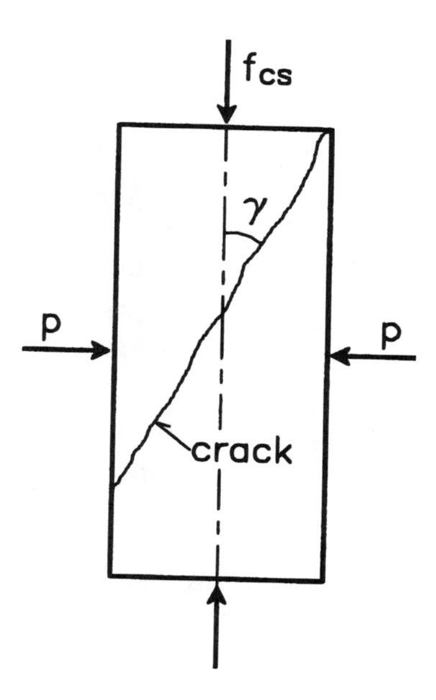

Figure 4.6.11 Confined, cracked element under compression.

An estimate may be found by first considering the effect on f_{cs} of a transversal pressure p (see Fig. 4.6.11). By expressing the stresses in the cracks forming an angle γ with the direction of f_{cs} and inserting these stresses into the failure condition of the crack, (4.6.12), we find

$$f_{cs} = \frac{c' + p\left(|\sin\gamma\cos\gamma| + \mu'\cos^2\gamma\right)}{|\cos\gamma\,\sin\gamma| - \mu'\sin^2\gamma} \tag{4.6.21}$$

On the other hand, if a specific value of f_{cs} is required, the necessary value of the side pressure p is

$$p = \frac{f_{cs}\left(|\cos\gamma\,\sin\gamma| - \mu'\sin^2\gamma\right) - c'}{|\sin\gamma\,\cos\gamma| + \mu'\cos^2\gamma} \tag{4.6.22}$$

The most dangerous crack direction is, of course, the direction of the yield line in an uncracked material. The corresponding angle is $\gamma = \pi/4 - \varphi/2$ (cf. Section 3.6.1). For $\varphi = 37°$ we have $\gamma = 27°$. Inserting into (4.6.22) $\mu = \mu' = 0.75$ and $c' = \nu_s \cdot \frac{1}{4}\nu_0 f_c$, we find

$$4p = f_{cs} - \nu_s \nu_0 f_c \tag{4.6.23}$$

Any additonal reduction in compressive strength by sliding in initial cracks is avoided by requiring f_{cs} to be $\nu_0 f_c$ leading to

$$p = \frac{1}{4}\nu_0 f_c(1 - \nu_s) \tag{4.6.24}$$

For $\nu_s = 0.5$

$$p = \frac{1}{8}\nu_0 f_c \tag{4.6.25}$$

This means that, on the safe side, sliding in initial cracks may be avoided by supplying additional reinforcement perpendicular to the compression direction corresponding to a reinforcement ratio

$$r = \frac{1}{8}\,\frac{\nu_0 f_c}{f_Y} \tag{4.6.26}$$

f_Y being the yield stress of the additional reinforcement. This is the same as requiring a reinforcement degree

$$\Phi = \frac{1}{8}\nu_0 \tag{4.6.27}$$

The additional reinforcement may also be composed of an isotropic reinforcement in two perpendicular directions, the reinforcement ratios of which are determined by (4.6.26).

Taking as an example $f_c = 20$ MPa, $f_Y = 500$ MPa and $\nu_0 = 0.6$, we get $r = 0.3\%$ which is about twice the minimum reinforcement ratio given by

(4.6.17). Of course, the additional reinforcement need only be supplied in areas with concrete stresses exceeding $\nu_s \nu_0 f_c$.

If the required compressive strength f_{cs} is somewhere in between $\nu_s \nu_0 f_c$ and $\nu_0 f_c$, formula (4.6.23) may in a similar way be used to find the necessary confinement reinforcement.

4.6.5 Concluding Remarks

Unfortunately when writing this book, the theory of cracked concrete is far from finished. Since cracking seems to be a very important aspect of the strength reduction of concrete in compression, hopefully much more research will be devoted to the subject in the future.

For disks the effective concrete strength of a uniaxial stress field in an unreinforced region is also of great interest. The strength reduction in this case is attributed to softening as well as, for certain shear span/depth ratios, to crack sliding. The problem is examined in Section 4.8.

4.7 GENERAL THEORY OF LOWER BOUND SOLUTIONS

4.7.1 Statically Admissible Stress Fields

A statically admissible stress distribution for a weightless disk satisfies equilibrium conditions (4.1.1) and (4.1.2) with $\rho = 0$ and boundary conditions (4.1.3) and (4.1.4). As first demonstrated by Airy in 1862, the equilibrium conditions are identically satisfied if the stresses are derived by a function Ψ of x and y in the following way:

$$\sigma_x = \frac{\partial^2 \Psi}{\partial y^2} \tag{4.7.1}$$

$$\sigma_y = \frac{\partial^2 \Psi}{\partial x^2} \tag{4.7.2}$$

$$\tau_{xy} = -\frac{\partial^2 \Psi}{\partial x \partial y} \tag{4.7.3}$$

The function Ψ is called *Airy's stress function.*

All stress distributions derived in this way from a function Ψ consequently satisfy the equilibrium conditions. However, Ψ cannot be chosen completely arbitrarily, as the boundary conditions must also be satisfied. In fact, the stresses on the boundary determine Ψ and its first derivatives along the boundary, as shown below.

Consider a stress field that can be derived from a certain stress function Ψ and which satisfies the boundary conditions. As linear functions and constants do not change the stresses, Ψ can always be adjusted so that $\partial\Psi/\partial x = \partial\Psi/\partial y = \Psi = 0$ in an arbitrary point P_0.

In order to express $\partial\Psi/\partial x$ and $\partial\Psi/\partial y$ by the stresses, we write down the curve integrals

$$\frac{\partial\Psi}{\partial x} = \int_{P_0}^{P} \frac{\partial}{\partial x}\left(\frac{\partial\Psi}{\partial x}\right)dx + \frac{\partial}{\partial y}\left(\frac{\partial\Psi}{\partial x}\right)dy = \int_{P_0}^{P} \sigma_y dx - \tau_{xy} dy \tag{4.7.4}$$

$$\frac{\partial\Psi}{\partial y} = \int_{P_0}^{P} \frac{\partial}{\partial x}\left(\frac{\partial\Psi}{\partial y}\right)dx + \frac{\partial}{\partial y}\left(\frac{\partial\Psi}{\partial y}\right)dy = \int_{P_0}^{P} -\tau_{xy} dx + \sigma_x dy \tag{4.7.5}$$

in which we integrate along the boundary from a fixed, arbitrarily chosen point P_0, where we have $\partial\Psi/\partial x = \partial\Psi/\partial y = 0$, to the arbitrary point P having the coordinates (x, y). If we introduce the length of arc s along the curve of integration, (4.7.4) and (4.7.5) may be written

$$\frac{\partial\Psi}{\partial x} = \int_{P_0}^{P}\left(\sigma_y \frac{dx}{ds} - \tau_{xy}\frac{dy}{ds}\right)ds = -\int_{P_0}^{P} p_y ds \tag{4.7.6}$$

$$\frac{\partial\Psi}{\partial y} = \int_{P_0}^{P}\left(-\tau_{xy} \frac{dx}{ds} + \sigma_x\frac{dy}{ds}\right)ds = \int_{P_0}^{P} p_x ds \tag{4.7.7}$$

as the boundary conditions are used. Here we have assumed that s is positive in such a direction that the outward-directed normal of the boundary n and the tangent in the s-direction form a system of coordinates having the same orientation as the x, y-system.

Consequently, the first derivatives of Ψ have the simple statical interpretation of the projection sums on the axes of coordinates of the forces per unit thickness along the curve of integration. For these projection sums we introduce the terms

$$P_y = \int_{P_0}^{P} p_y ds = -\frac{\partial\Psi}{\partial x} \tag{4.7.8}$$

$$P_x = \int_{P_0}^{P} p_x ds = \frac{\partial\Psi}{\partial y} \tag{4.7.9}$$

Finally, in order to determine Ψ along the boundary, we integrate once more, using integration by parts:

$$\Psi = \int_{P_0}^{P}\left(\frac{\partial\Psi}{\partial x}\frac{dx}{ds}+\frac{\partial\Psi}{\partial y}\frac{dy}{ds}\right)ds = \int_{P_0}^{P}\left(-P_y\frac{dx}{ds}+P_x\frac{dy}{ds}\right)ds$$

$$= \int_{P_0}^{P}\left[-\frac{dP_y}{ds}(x_P-x)+\frac{dP_x}{ds}(y_P-y)\right]ds$$

$$= \int_{P_0}^{P}\left[-p_y(x_P-x)+p_x(y_P-y)\right]ds \tag{4.7.10}$$

The coordinates of point P, (x_P, y_P) in the last integral, are to be considered as constants. It is seen from (4.7.10) that the stress function is equal to the moment of the forces per unit thickness (considered to be positive in the positive direction of rotation of the x, y-plane) along the curve of integration.

As the curve of integration in the previous integrals does not necessarily have to be the boundary, but may be an arbitrary curve in the disk, the statical interpretation of Ψ and its derivates can be applied for every point within the body (see [37.1]). That an integration along an arbitrary, closed curve equals zero for both Ψ and its first derivatives indicates simply that the stresses satisfy the equilibrium conditions.

Equations (4.7.4), (4.7.5), and (4.7.10) may also be used to demonstrate the existence of a stress function satisfying (4.7.1)-(4.7.3), for any statically admissible stress field. If, conversely, Ψ and its first derivatives are calculated along the boundary of a disk with given stresses p_x and p_y at the boundary, then according to the statical interpretation above, the boundary conditions will be satisfied for every Ψ-function, assuming the values found at the boundary. The equilibrium conditions will also be satisfied.

Every Ψ-function assuming the values and derivatives found at the boundary may thus be used as a stress function. Therefore, we could in principle carry out the calculation of the necessary reinforcement by choosing Ψ in a certain number of points, which, for example, may be done graphically by drawing the curves of Ψ in the directions of the coordinates, since the values of the first derivatives found determine the initial and final slopes of these curves. Having determined Ψ in an adequate number of points, we may find the stresses by approximate difference expressions corresponding to (4.7.1)-(4.7.3). If we assume Ψ to be chosen at points in a rectangular net having the mesh sizes Δx and Δy (see Fig. 4.7.1), we find at point 0

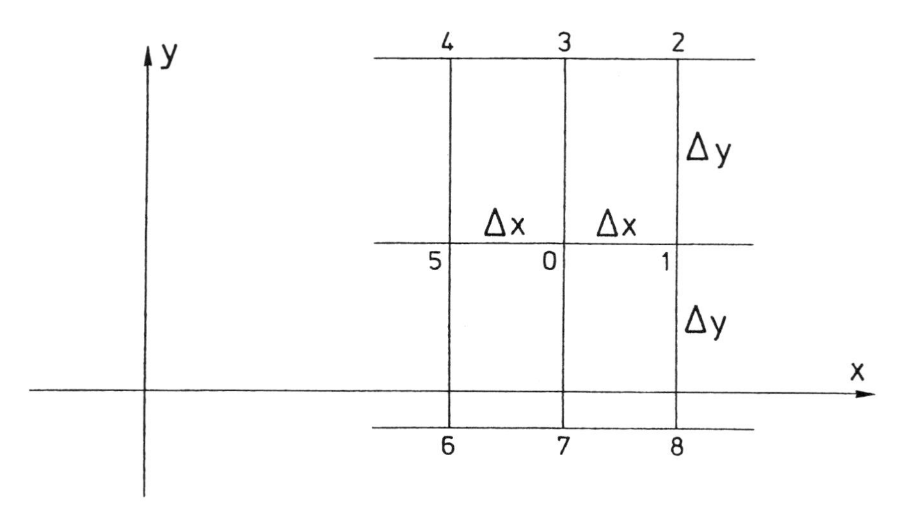

Figure 4.7.1 Net points used in difference expressions for stresses at a point.

$$\sigma_x = \frac{\Psi_3 - 2\Psi_0 + \Psi_7}{(\Delta y)^2} \tag{4.7.11}$$

$$\sigma_y = \frac{\Psi_1 - 2\Psi_0 + \Psi_5}{(\Delta x)^2} \tag{4.7.12}$$

$$\tau_{xy} = \frac{(\Psi_4 + \Psi_8) - (\Psi_2 + \Psi_6)}{4\Delta x \Delta y} \tag{4.7.13}$$

When the stresses have been determined, the reinforcement may be determined by using the formulas in Section 2.4, and the Ψ-values may, if necessary, be changed so that a more convenient arrangement of reinforcement is obtained.

In the equations above we have assumed the volume forces to be zero. If the disk is subjected to a force per unit volume having the components $\rho f_x = g_x$ and $\rho f_y = g_y$, the equilibrium conditions (4.1.1) and (4.1.2) are satisfied by putting

$$\sigma_x = \frac{\partial^2 \Psi}{\partial y^2} - \int g_x dx \tag{4.7.14}$$

$$\sigma_y = \frac{\partial^2 \Psi}{\partial x^2} - \int g_y dy \tag{4.7.15}$$

$$\tau_{xy} = -\frac{\partial^2 \Psi}{\partial x \partial y} \tag{4.7.16}$$

in which $\int g_x dx$ and $\int g_y dy$ are arbitrary integrals to g_x and g_y, respectively.

The previous reflections on the statical interpretation of the stress function Ψ are still valid when σ_x is replaced by the formal stress $\sigma_x + \int g_x dx$ and σ_y by $\sigma_y + \int g_y dy$. Along the boundary, Ψ is determined in the same way by means of these "stresses", which will be known in stress boundary value problems.

To be able to determine the stresses, it will sometimes be necessary to know Ψ-values at points of the net outside the boundary. These are calculated by using the first derivatives of Ψ determined. The same method may be used if the boundary itself cannot be fit into a rectangular net.

4.7.2 A Theorem of Affinity

Let us assume that we know a statically admissible stress distribution in a weightless disk referred to a rectangular *x, y*-system; that is, the stress distribution satisfies the equilibrium equations and the boundary conditions. Consider a weightless disk affine to the disk given. The affine disk is referred to an *x**, *y**-system and is determined by

$$x^* = x \,, \qquad y^* = \sqrt{\mu}\, y \tag{4.7.17}$$

Consider the stress distribution

$$\sigma_{x^*} = \sigma_x \,, \qquad \sigma_{y^*} = \mu \sigma_y \,, \qquad \tau_{x^*y^*} = \sqrt{\mu}\, \tau_{xy} \tag{4.7.18}$$

valid at points corresponding to each other by the affinity. The quantities σ_x, σ_y, and τ_{xy} are the stress components of the original disk. We then find

$$\frac{\partial \sigma_{x^*}}{\partial x^*} = \frac{\partial \sigma_{x^*}}{\partial x} \frac{\partial x}{\partial x^*} = \frac{\partial \sigma_x}{\partial x} \tag{4.7.19}$$

$$\frac{\partial \tau_{x^*y^*}}{\partial x^*} = \frac{\partial \tau_{x^*y^*}}{\partial x} \frac{\partial x}{\partial x^*} = \sqrt{\mu}\, \frac{\partial \tau_{xy}}{\partial x} \tag{4.7.20}$$

and similarly,

$$\frac{\partial \tau_{x^*y^*}}{\partial y^*} = \frac{\partial \tau_{xy}}{\partial y} \tag{4.7.21}$$

$$\frac{\partial \sigma_{y*}}{\partial y^*} = \sqrt{\mu}\,\frac{\partial \sigma_y}{\partial y} \tag{4.7.22}$$

By this it appears that the equilibrium conditions are also satisfied in the affine disk when the stress field is given by (4.7.18).

For the stresses at the boundary, the following rules of transformation are found:

$$p_{x*} = \sqrt{\frac{\mu}{\mu \cos^2 \alpha + \sin^2 \alpha}}\; p_x \tag{4.7.23}$$

$$p_{y*} = \mu \sqrt{\frac{1}{\mu \cos^2 \alpha + \sin^2 \alpha}}\; p_y \tag{4.7.24}$$

in which p_x and p_y are the components of the stress at the boundary of the original disk, and p_{x*} and p_{y*} are the components of the stress at the boundary at the corresponding point of the affine disk. The angle α is the angle between the x-axis and the outward directed normal of the original disk.

If the original disk is isotropic, it is seen that if the stress field σ_x, σ_y, and τ_{xy} is safe, the stress field in the affine disk is also safe, provided that $A_{sx*} = A_{sx}$ and $A_{sy*} = \mu A_{sy}$, according to the theorem of affinity mentioned in Section 2.2.2.

The orthotropic, affine disk, however, is not always fully utilized if $\mu < 1$, because compressive stresses - if any - will be reduced, and if $\mu > 1$, the compressive stresses are possibly too large in the affine disk, if the original isotropic disk is fully utilized in compression. This theorem is analogous to a theorem in the slab theory (see Section 6.8.1).

4.7.3 The Stringer Method

Using the difference equations described in Section 4.7.1 the equilibrium equations are not exactly satisfied. In this section we describe a slightly different elaboration of the method in which the equilibrium equations are found to be satisfied exactly.

We assume the external load per unit thickness to be perpendicular to the boundary and to be replaced by a statically equivalent system of concentrated forces applied at the net points. If ψ and its derivatives are calculated as before, we obtain a polygonal variation, as the derivatives are discontinuous. Then the ψ-values of the external net points closest to the boundary are determined by means of the derivatives of ψ found along the boundary. Now we choose the ψ-curves in the direction of the x- and y-axis as polygons (the ψ-surface will then consist of hyperbolic paraboloids, one for each mesh rectangle), and just as at the boundary, we obtain discontinuities in the derivatives corresponding to the normal stresses in the net points being infinite, but statically equivalent to a concentrated force equal to the differences of the slopes of the ψ-curve. The shear stress will be constant within each mesh rectangle.

Regarding a system of net points as in Fig. 4.7.2, we obtain at point 0 a force per unit thickness in the direction of the x-axis,

$$\Sigma_x = \frac{\Psi_3 - 2\Psi_0 + \Psi_7}{\Delta y} \tag{4.7.25}$$

which we can imagine as being applied in section 7-3.

In the direction of the y-axis, we obtain a force per unit thickness

$$\Sigma_y = \frac{\Psi_1 - 2\Psi_0 + \Psi_5}{\Delta x} \tag{4.7.26}$$

applied in section 5-1.

Finally, we obtain the shear force per unit thickness shown in the figure, those in part A for example, being of the magnitude

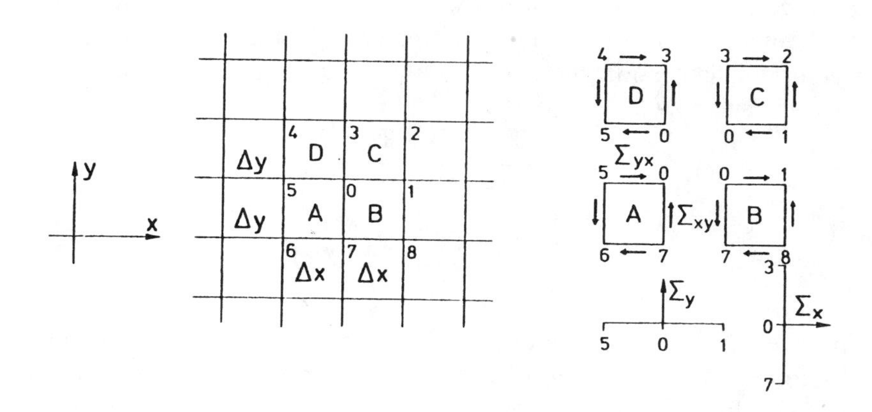

Figure 4.7.2 Net points used in difference expressions for stringer forces and shear forces in the rectangular areas.

$$\Sigma_{xy} = \frac{\Psi_5 + \Psi_7 - (\Psi_0 + \Psi_6)}{\Delta x} \tag{4.7.27}$$

and

$$\Sigma_{yx} = \frac{\Psi_5 + \Psi_7 - (\Psi_0 + \Psi_6)}{\Delta y} \tag{4.7.28}$$

Corresponding formulas are valid for the shear forces in parts *B, C,* and *D*. The forces found in this way will satisfy the equilibrium conditions exactly.

Distributing Σ_x over the distance Δy, Σ_y over Δx, and Σ_{xy} and Σ_{yx} over Δy and Δx, respectively, we obtain the stresses

$$\sigma_x = \frac{\Psi_3 - 2\Psi_0 + \Psi_7}{(\Delta y)^2} \tag{4.7.29}$$

$$\sigma_y = \frac{\Psi_1 - 2\Psi_0 + \Psi_5}{(\Delta x)^2} \tag{4.7.30}$$

$$\tau_{xy} = \tau_{yx} = \frac{\Psi_5 + \Psi_7 - (\Psi_0 + \Psi_6)}{\Delta x \Delta y} \tag{4.7.31}$$

These formulas are identical to (4.7.11)-(4.7.13).

If the disk is subjected to forces parallel to the boundaries or to forces applied at internal points, for instance, volume forces (i.e., in practice by the dead weight of the disk), these are assumed to apply at the net points. We term the components of these external forces per unit thickness in the net points Q_x and Q_y, and analogous to (4.7.14) and (4.7.15) we obtain

$$\Sigma_x = \frac{\Psi_3 - 2\Psi_0 + \Psi_7}{\Delta y} - \Sigma Q_x \tag{4.7.32}$$

$$\Sigma_y = \frac{\Psi_1 - 2\Psi_0 + \Psi_5}{\Delta x} - \Sigma Q_y \tag{4.7.33}$$

in which ΣQ_x, being equivalent to $\int g_x dx$ in (4.7.14), is the sum of the forces in the direction of the x-axis along a net line. The values of ΣQ_x along the y-axis or another axis parallel to the y-axis may be chosen arbitrarily. This can also be applied to ΣQ_y. The formulas for the shear forces remain unchanged. By calculation of the Ψ-values along the boundary, forces Σ_x and Σ_y are replaced by $\Sigma_x + \Sigma Q_x$ and $\Sigma_y + \Sigma Q_y$, respectively. We have thus obtained a system of forces which is in equilibrium with the external load and consists of concentrated forces in the net lines and constant shear forces in every mesh rectangle.[4]

If the reinforcement is placed along the net lines, the forces in the reinforcing bars correspond exactly to the concentrated forces in the net lines. In addition to these forces, however, we also have to reinforce for the shear stresses in the rectangles, a reinforcement that in Section 4.7.1 we have also concentrated in the net lines. When the reinforcement at point 0 in Fig. 4.7.2 is designed for these concentrated forces and for the average value of the shear stresses in the adjoining rectangles, we are back to the method previously described. We see, however, that the present variant of the method, which might be called the *stringer method*, also implies the possibility of inserting concentrated reinforcement inside the disk - not only at the boundaries. At the same time, it is seen that a secondary net of reinforcement can be provided between the reinforcing bars in the net lines for carrying the shear stresses in the rectangles. The mesh size used may be as large as desired provided that the external forces can be regarded as acting at the net points. Notice, however, that the method described in the following, make it easy to include shear stresses along the boundaries when modelling the load.

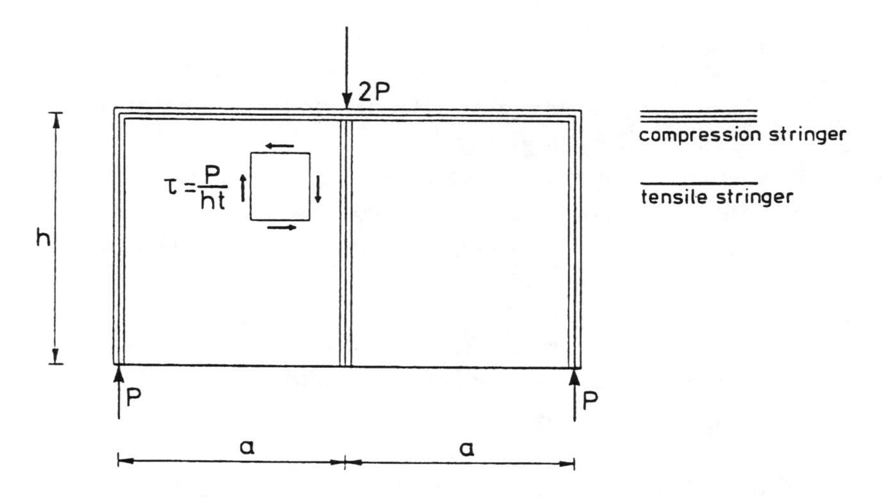

Figure 4.7.3 Simple example of a stringer system.

[4] An analogous use of Airy's stress function for the purpose of finding forces in space trusses has been made by Jørgen Nielsen [64.5].

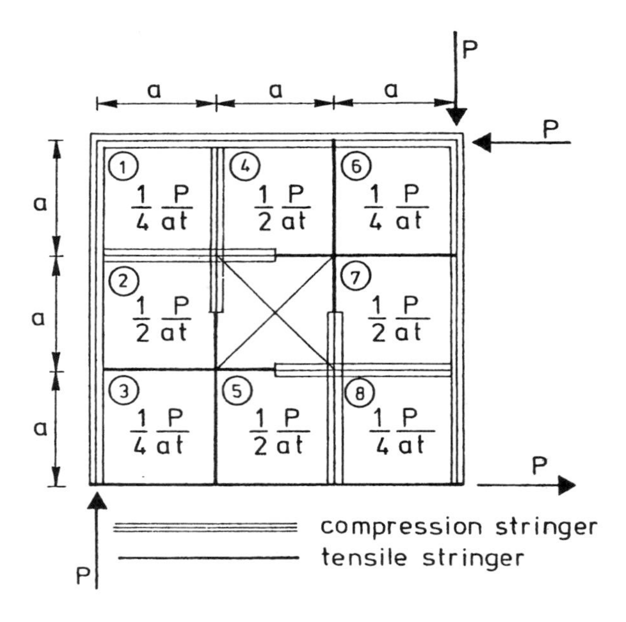

Figure 4.7.4 Stringer system in a disk with a hole.

Instead of using the stress function, we may also begin by choosing the distribution of the shear stresses in every mesh rectangle, so that the total shear force in sections parallel to the net lines is correct. Then the stringer forces (i.e., the forces in the net lines) can be easily determined. This method is normally used in practice because of its simplicity.

As a simple, theoretical example the disk in Fig. 4.7.3, which is simply supported and subjected to a concentrated force $2P$ in the center is considered. If we use the mesh size $a \cdot h$, we obtain a shear stress in every mesh rectangle of $\tau = P/ht$, where t is the thickness of the disk, while the force in the horizontal stringers varies linearly from Pa/h in the center to zero at the lateral surfaces. The top stringer is a compression stringer, and the bottom stringer is a tension stringer. The force in the vertical stringers varies from P to zero. They are all compression stringers. Consequently, the solution requires a concentrated tensile reinforcement at the bottom face and a uniformly distributed shear reinforcement in the entire disk. For deep beams, however, this solution is of minor interest. In Section 4.8 it will be shown that, theoretically, no shear reinforcement is necessary.

In Figs. 4.7.4 to 4.7.7, four additional examples of stringer solutions are shown. In Fig. 4.7.4, showing one half of a disk with two holes, we may choose the τ-value in areas 1 and 2 freely, whereby the τ-value in area 3 can be determined by vertical projection. Similarly, if the τ-value in area 4 is chosen, the τ-value in area 5 can be determined by vertical projection. The τ-values in areas 6 and 7 are then determined by horizontal projection. Finally, the τ-value in area 8 can be determined by either horizontal or vertical projection, leaving one equation as a control equation. One set of statically admissible τ-values is shown in the figure. It is left to the reader to find the stringer forces. The sign of the stringer forces is evident from the figure.

The second example, Fig. 4.7.5, is calculated in a similar way. The τ-values in areas 1 and 2 are statically determinate. The τ-values in areas 3, 7, 4, 8, 5, and 9 can be chosen freely; while the τ-values in areas 11, 12, and 13 are found by vertical projection. The τ-values in areas 6 and 10 are found by horizontal projection. Finally, the τ-value in area 14 can, as in the preceding example, be found by either horizontal or vertical projection, leaving one equation as a control equation. One set of statically admissible τ-values is shown in the figure. It is left to the reader to find the stringer forces. The sign of these forces is evident from the figure. An important reinforcement detail is the suspension reinforcement in the vertical stringer adjoining the areas 2, 3, 7, and 11.

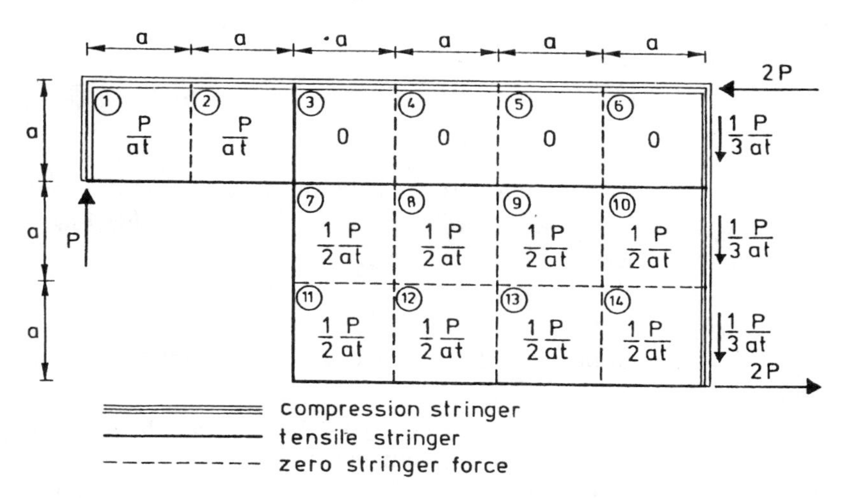

Figure 4.7.5 Stringer system in a disk to which the load is transferred through a corbel.

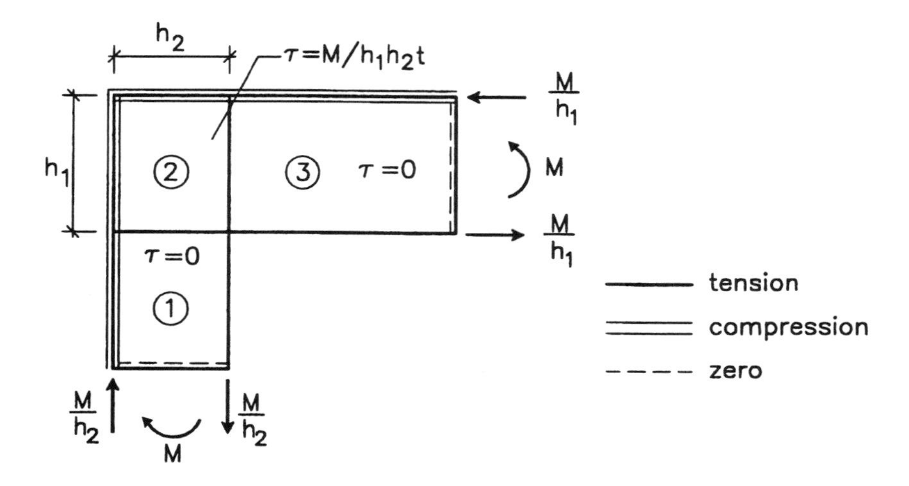

Figure 4.7.6 Frame corner.

In Fig. 4.7.6 a frame corner is shown. It has to transfer a bending moment M. This is transformed to a statically equivalent system of forces acting in the net points as shown in the figure. In this case the shear stresses are statically determinate. It is evident that the shear stress in the areas 1 and 3 is zero. By horizontal or vertical projection through area 2 we find

$$\tau = \frac{M}{h_1 h_2 t} \tag{4.7.34}$$

The sign of the stringer forces is shown in the figure. The stringer forces are decreasing linearly from maximum absolute value to zero in the regions where they adjoin area 2.

The solution shown may be used for either sign of M. The sign of the stringer forces of course changes when the sign of M is changed. The solution is easily extended to more complicated frame corners.

In Fig. 4.7.7 the problem is to reinforce a region around a hole in a shear zone. At some distance from the hole the shear stress is τ. The hole is rectangular $a \cdot b$ with sides parallel to the sections with pure shear. We imagine the zone with additional reinforcement to be determined by the lengths x and y shown in the figure. The shear stresses are easily determined by vertical and horizontal projection requiring the total shear force carried in a section through the zone with additional reinforcement to be that of a section with the same length in a zone with pure shear. In this case we may obtain equal values of shear stress in the four areas A. Further, areas B may have equal shear stress and similarly areas C may have equal shear stress.

The values of the shear stresses and the stringer forces are given in the figure. Also the lengths x and y rendering minimum reinforcement have been indicated. The lengths x and y, of course, are in practice often chosen so as to match the surrounding part of the structure in a suitable way.

It is easy to generalize the solution to the case where the zone with additional reinforcement has arbitrary lengths.

The problem of reinforcing around holes using the stringer method was treated by J. Kærn [79.7].

If the hole is not rectangular, the solution is still applicable if a rectangular zone circumscribing the hole is used in the calculation. The area between the hole and the rectangular zone should of course be supplied with some extra reinforcement.

In the above examples the sign of the shear stresses is evident. In more complicated cases a sign convention for the shear stresses should be adopted and the calculation carried through systematically using the adopted sign convention.

When the shear stress distribution is statically indeterminate the shear stresses should be chosen to make the total reinforcement a minimum. In hand calculations, the experienced designer normally is able to arrive at a good solution using only a few steps. He will start with a solution satisfying equilibrium and change this in steps to obtain better economy. Nowadays computer programs exist by which minimum reinforcement solutions may be determined [94.15].

Reinforcement in the areas with pure shear is easily determined by the reinforcement formulas. The effectiveness factors may be determined as explained in Section 4.6. Minimum cracking reinforcement must be present everywhere.

Reinforcement in the tensile stringers is easily determined. Often the reinforcement is carried through the whole stringer system without being curtailed. In this case shear reinforcement in the areas with pure shear is sometimes not supplied if the shear stresses are low, for instance lower than the shear capacity of nonshear reinforced slender beams, see Chapter 5.

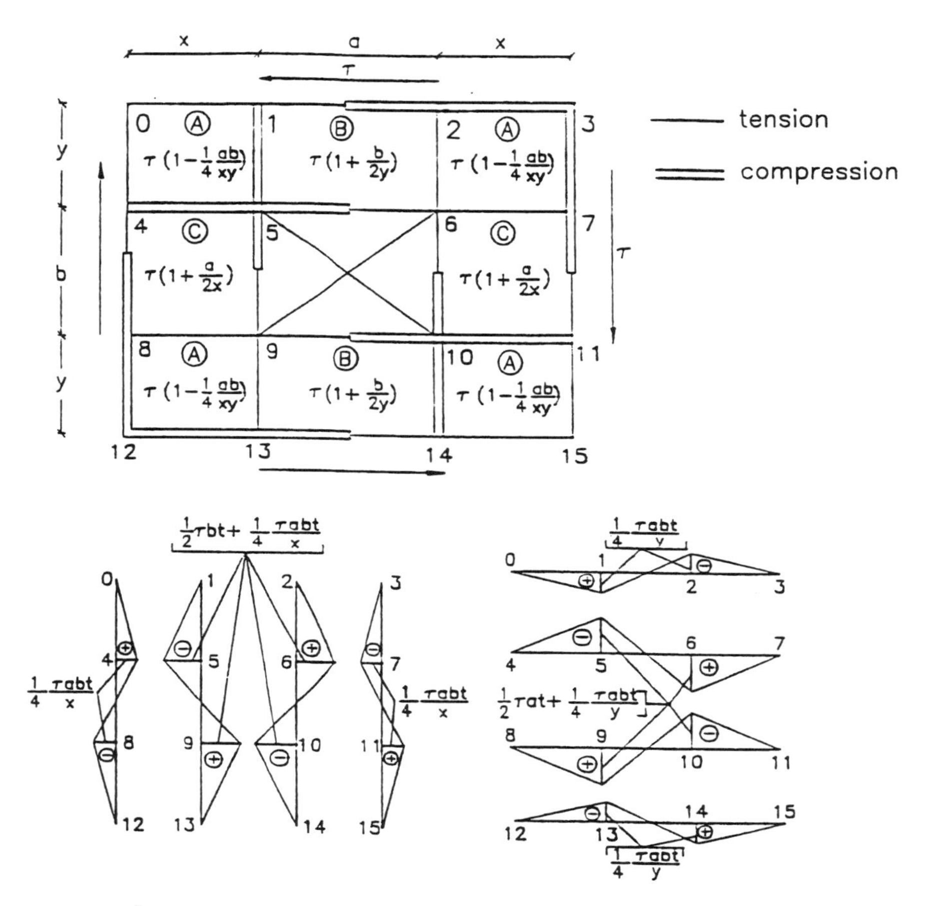

$x \cong \frac{1}{2}\sqrt[3]{b^2 a}$ however not less than the anchorage length

$y \cong \frac{1}{2}\sqrt[3]{a^2 b}$ however not less than the anchorage length

Figure 4.7.7 Reinforcement around a hole.

The rate of change of a stringer force along its length is, of course, not allowed to be higher than that corresponding to the anchorage length of the bars in question. If the anchorage length is ℓ and the number of bars in a stringer is n the rate of change dT/ds must be less than $nA_s f_Y/\ell$, T being the stringer force, s a length measured along the stringer, A_s the area of one bar and f_Y the yield stress. We have assumed a uniform shear stress distribution around a bar along the anchorage length. When this rule is applied to a stringer force increasing linearly from zero to a maximum value along a length ℓ', ℓ' must be larger than the anchorage length ℓ. If more anchorage strength is needed the bar number may be increased. Bars may be looped to increase the anchorage capacity by the effect of the concentrated compressive stresses around the circular bend.

When the mesh size is relatively small, and the areas with pure shear are reinforced in directions parallel to the stringers, sometimes the reinforcement in the areas with pure shear is transferred to the nearby stringers, i.e., the reinforcement in a nearby stringer is increased by half the value of the total reinforcement in the area with pure shear in the direction of the nearby stringer. If a nearby stringer is in compression the force in the transferred reinforcement reduces the force in the stringer, i.e., if the force in the transferred reinforcement is in absolute value less than the compression force in the stringer the transferred reinforcement does not have to be supplied. Of course such a reinforcement transformation is not allowed for areas in pure shear near a free boundary. Here the shear reinforcement is necessary to receive the skew compression stresses in the concrete as illustrated in Fig. 4.7.8. It follows that shear reinforcement near free boundaries should always be carried out as closed stirrups.

If compression stringers materialize, for instance, in the form of flanges it will probably be reasonable to use the ν-values for bending (cf. Chapter 5), when designing the compression stringers. If the stringers have the same thickness as the disk, normally rather small ν-values are used ($\nu \sim 0.5$) and it is required that the necessary thickness of the stringers is small compared with the mesh size.

We conclude that although the problem of finding a statically admissible stress distribution is simple in principle, it is, however, complicated enough to make it desirable for frequently occurring types of disks and loads to have statically admissible stress distributions available which are easier to determine. This is particularly true if global minimum reinforcement solutions are sought. In the following some simple solutions are given for rectangular disks with different loads and conditions of support.

4.7.4 Shear Zone Solutions for Rectangular Disks[5]

Distributed load on the top face. A rectangular disk with a vertical, continuous load corresponding to the compressive stress p on the top face is considered. We examine the distance a from the maximum moment to the point with zero moment (see Fig. 4.7.9).

The solution presented in this section requires a concentrated tensile reinforcement along the bottom. Along the top face we have a compression zone. Between the compression zone and the tensile zone there is a "shear zone", as in the classical beam theory.

The cross section with maximum moment is examined in the usual way. The σ_x-distribution shown in Fig. 4.7.9 is chosen for convenience. The compressive stresses f_c, being equal to the compressive strength of the concrete, are uniformly distributed over a distance y_0, and the forces in the tensile reinforcement are $A_s f_Y$, A_s being the area of the bars and f_Y the yield stress of the reinforcement. By horizontal projection we obtain

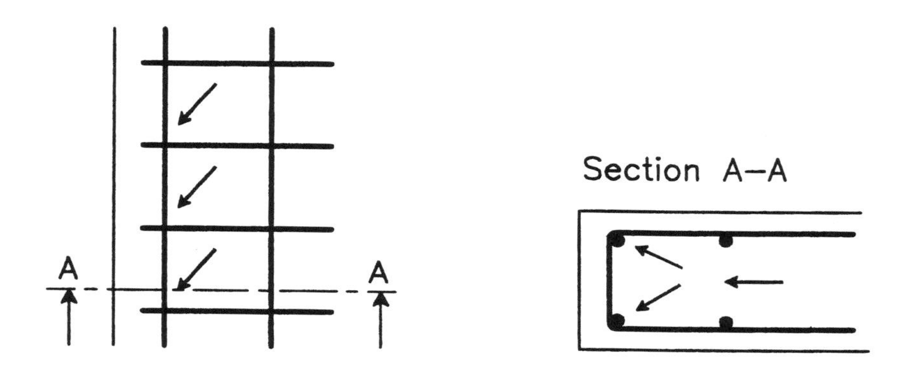

Figure 4.7.8 Skew compressive stresses near a free boundary.

[5] The following solutions apply to all combinations of depth and span, but their field of application is especially deep beams. Beams in which the depth is small compared with the span are calculated as normal beams (see Chapter 5).

$$A_s f_Y = y_0 t f_c \tag{4.7.35}$$

in which t indicates the thickness of the disk.

With d as the effective depth, we define the degree of reinforcement

$$\Phi = \frac{A_s f_Y}{t d f_c} \tag{4.7.36}$$

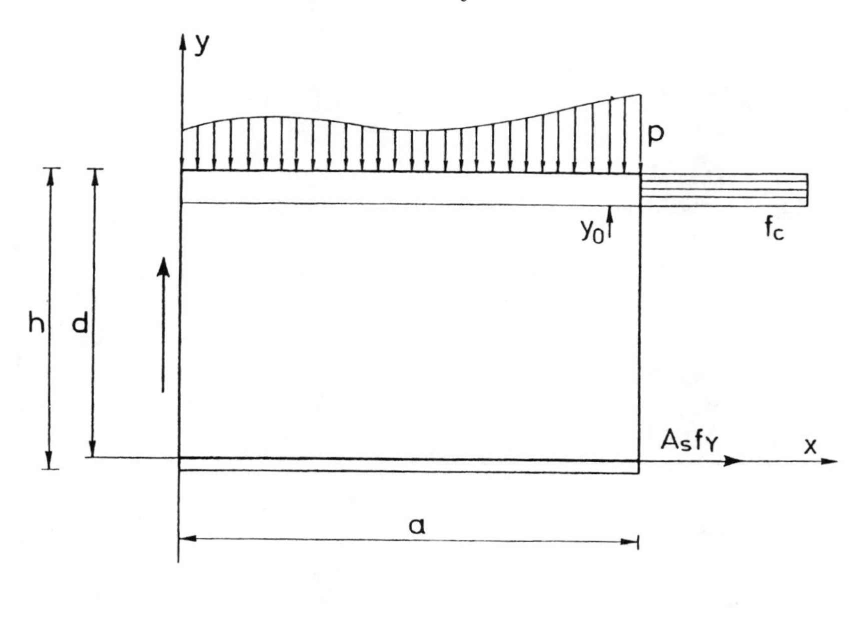

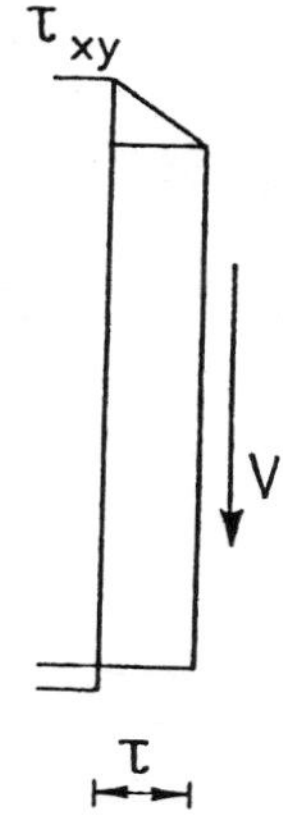

Figure 4.7.9 Normal stress and shear stress distribution in a rectangular disk.

We then obtain from (4.7.35)

$$y_0 = \Phi d \tag{4.7.37}$$

and the yield moment M_p is

$$M_p = \left(1 - \frac{1}{2}\Phi\right)\Phi t d^2 f_c = \left(1 - \frac{1}{2}\Phi\right) A_s f_Y d \tag{4.7.38}$$

From this A_s may be determined. Setting

$$\mu = \left(1 - \frac{1}{2}\Phi\right)\Phi = \frac{M_p}{t d^2 f_c} \tag{4.7.39}$$

and solving the first equation in (4.7.39) for Φ, we obtain

$$\Phi = 1 - \sqrt{1 - 2\mu} \tag{4.7.40}$$

which, when μ is small, will approximately render $\Phi \approx \mu$. From this the necessary reinforcement can be determined when M_p , f_c , and the cross-sectional dimensions are known.

Next we look for a statically admissible stress field in the entire disk. If the reinforcement is designed according to this stress field, the solution will be a lower bound solution. If $h_i = d - \frac{1}{2}y_0$, the maximum moment $M_{\max} = T_{\max}\, h_i = A_s f_Y\, h_i = t\, y_0 f_c\, h_i$. Now, consider the stress field

$$\sigma_x = \begin{cases} -\dfrac{M}{M_{\max}} f_c & \text{for } d - y_0 \le y \le d \\ 0 & \text{for } 0 \le y \le d - y_0 \end{cases} \tag{4.7.41}$$

and the force of reinforcement

$$T = \frac{M}{M_{\max}} A_s f_Y = \frac{M}{h_i} \tag{4.7.42}$$

From the equilibrium condition (4.1.1) we then obtain for a weightless disk

$$\frac{\partial \tau_{xy}}{\partial y} = -\frac{\partial \sigma_x}{\partial x} = \begin{cases} \dfrac{dM}{dx}\dfrac{f_c}{M_{\max}} = \dfrac{V f_c}{M_{\max}} = \dfrac{V}{t y_0 h_i} & \text{for } d - y_0 \le y \le d \\ 0 & \text{for } 0 \le y \le d - y_0 \end{cases} \tag{4.7.43}$$

The boundary conditions for τ_{xy} are

$$\tau_{xy} = \begin{cases} 0 & \text{for } y = d \\ -\dfrac{1}{t}\dfrac{dT}{dx} = -\dfrac{1}{t}\dfrac{VA_s f_Y}{M_{\max}} = -\dfrac{V}{th_i} & \text{for } y = 0 \end{cases} \tag{4.7.44}$$

The equilibrium condition and the boundary conditions are satisfied for

$$\tau_{xy} = \begin{cases} -\dfrac{V}{ty_0 h_i}(d-y) & \text{for } d - y_0 \le y \le d \\ -\dfrac{V}{th_i} = -\tau & \text{for } 0 \le y \le d - y_0 \end{cases} \tag{4.7.45}$$

which also satisfy the condition $t\int_0^d \tau_{xy} dy = -V$.

Finally, the equilibrium condition (4.1.2) renders

$$\frac{\partial \sigma_y}{\partial y} = -\frac{\partial \tau_{xy}}{\partial x} = \begin{cases} \dfrac{dV}{dx}\dfrac{d-y}{ty_0 h_i} & \text{for } d - y_0 \le y \le d \\ \dfrac{dV}{dx}\dfrac{1}{th_i} & \text{for } 0 \le y \le d - y_0 \end{cases} \tag{4.7.46}$$

As $dV/dx = -pt$, we obtain by using the boundary conditions

$$\sigma_y = \begin{cases} -p & \text{for } y = d \\ 0 & \text{for } y = 0 \end{cases} \tag{4.7.47}$$

the following expressions for σ_y:

$$\sigma_y = \begin{cases} -\dfrac{p}{y_0 h_i}\left(dy - \frac{1}{2}y^2 - \frac{1}{2}d^2\right) - p & \text{for } d - y_0 \le y \le d \\ -\dfrac{p}{h_i} y & \text{for } 0 \le y \le d - y_0 \end{cases} \tag{4.7.48}$$

which also satisfy the jump condition along $y = d - y_0$.

Consequently, in the "shear zone" $0 \le y \le d - y_0$ we have, at fixed x, a constant shear stress, while the shear stresses vary linearly in the compression zone. The normal stresses σ_y vary linearly in the shear zone and parabolically in the compression zone. When the stresses in the entire disk are determined in this way, the necessary reinforcement may be found by using the formulas in Section 2.4.

Concerning the reinforcement of the shear zone, the conservative approach is to disregard the normal stresses σ_y in this case, as they are negative and therefore will reduce the necessary reinforcement. Then the shear zone is reinforced only for pure shear, which can be done as described in Section 2.4.2. In deep beams, however, the stresses σ_y will often be important. If we use a constant, vertical reinforcement, the horizontal reinforcement may be varied according to the variation of σ_y.

Distributed load at the bottom face. If the load is applied to the disk along the bottom face, a statically admissible stress distribution can immediately be found by adding $q(x)$ to σ_y everywhere, where $q(x)$ is equal to the σ_y-value along the bottom face, and where $q(x)$ substitutes for $p(x)$ in the formulas for σ_y.

Distributed load along a horizontal line. If the load consists of a horizontal line load $\bar{p}$ acting along $y = r$ (see Fig. 4.7.10), the expressions for σ_x and τ_{xy} are still valid. If the σ_y-values, found by inserting $p = \bar{p}/t$ into (4.7.48), are termed σ_y', we obtain in the case illustrated in Fig. 4.7.10:

$$\sigma_y = \begin{cases} \sigma_y' & \text{for } 0 \le y < r \\ \sigma_y' + \dfrac{\bar{p}}{t} & \text{for } r < y \le d \end{cases} \tag{4.7.49}$$

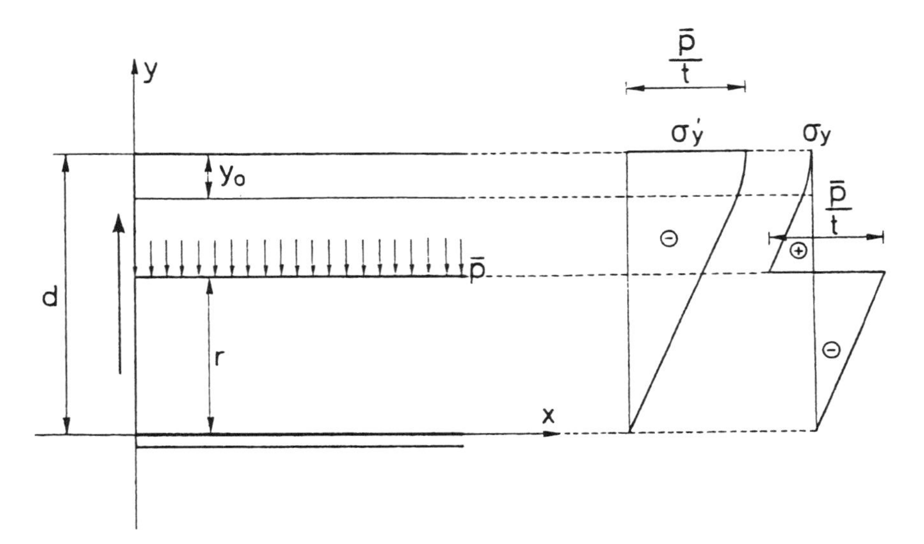

Figure 4.7.10 Normal stress (σ_y) distribution in a disk loaded along a line between the top and bottom face.

By a similar consideration the stresses for the dead load can be found. If, for example, this is constant, equal to g per unit volume of the disk and directed in the negative direction of the y-axis, the stresses are first found by using (4.7.48), as we substitute p with gd. If we term the stresses found σ_y', the correct normal stresses are $\sigma_y' + gy$. [Here we have neglected the weight of the concrete cover, but if it were important (which is rare), it can be included as a uniformly distributed load along the bottom face.]

Arbitrary loads. The stress distribution (4.7.41), (4.7.45), and (4.7.48) can be used for arbitrary combinations of stresses applied on the top and bottom faces of the disk. The distribution of σ_y is determined along an arbitrary line parallel to the y-axis as above, as the boundary conditions can always be satisfied.

In the case of a disk loaded with a "concentrated force" (see Fig. 4.7.11), we obtain stresses σ_y only in the part of the disk vertically below the loading. Here one might expect a more economical distribution of the stresses to exist, that is a distribution where the force is "spread" over a bigger area, by which the reducing effect of the stresses σ_y on the reinforcement is taken into account. This problem is treated in the following sections.

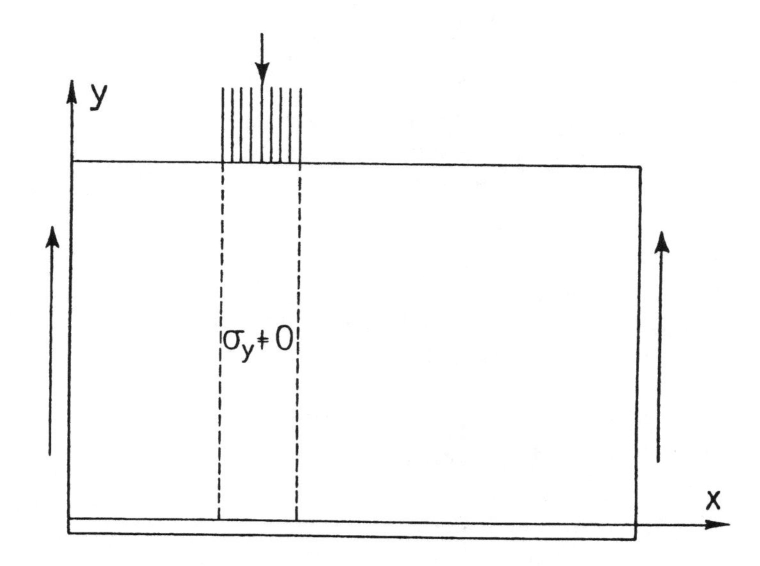

Figure 4.7.11 Area with $\sigma_y \neq 0$ for the equilibrium solution considered.

Effectiveness factors. When, as in this case, the type of stress field to reinforce for has been selected, the formulas in Section 2.4 for local minimum reinforcement, will also provide global minimum reinforcement for this particular stress field. This is, of course, only true if the concentrated reinforcement along the faces is curtailed according to the tensile force at the individual points. When crack sliding may be neglected, cf. Section 4.6, the effectiveness factor $\nu = \nu_o$ may be calculated using either formula (4.6.1) or (4.6.6). When the reinforcement stress is small, formula (4.6.10) may be advantageous to use instead. Minimum reinforcement should be provided in the web according to (4.6.18) in two perpendicular directions. If crack sliding cannot be ruled out the safe value $\nu = \nu_s \nu_o$ with $\nu_s = 0.5$ may be used. Alternatively additional uniform reinforcement is provided in areas where the concrete stress exceeds $\nu_s \nu_o f_c$, cf. Section 4.6.4.

4.8 STRUT AND TIE MODELS

4.8.1 Introduction

In strut and tie models the load is carried by "concrete bars" subjected to compression (the struts) and tensile bars made by the reinforcement bars.

This type of model is as old as reinforced concrete theory. In the old days the model was normally called a truss model and used under that name by Mörsch [22.2] and Johansen [28.2] among others. Through the years many solutions have been proposed based on such models and they have been extensively used in practical design. The model also played an essential role when the modern theory of reinforced concrete started to develop [63.2] [63.3] [67.2] [69.2]. In these works by the author the compression bar forces were spread out to furnish a continuous uniaxial stress field.

The simple strut and tie models have had a rebirth thanks to the works of Schlaich and his group in Stuttgart (see, for instance, [87.5] [91.28]).

The simple truss models are excellent for preliminary designs, but they may also be used in final designs in many cases as is evident from the following.

4.8.2 The Single Strut

Let us begin by studying the single strut.

The strut in Fig. 4.8.1 carries the uniaxial compression stress f_c, i.e., the uniaxial compressive strength of concrete. It is convenient for the calculations to imagine the strut to be loaded by the shaded areas shown in the figure. They carry a plane hydrostatic pressure f_c. These areas do not necessarily need to be present. Sometimes only the part along the so-called *shear span*, a (see the figure), is utilized. The vertical depth of the strut is h.

If the strut inclination with a horizontal axis is θ, we have

$$\tan\theta = \frac{x_0}{y_0} = \frac{h - y_0}{a + x_0} \tag{4.8.1}$$

If the strut has to carry the vertical load P, the shear force along the shear span a, we must require

$$P = x_0 t f_c \tag{4.8.2}$$

t being the thickness.

To maintain equilibrium the horizontal forces C shown in the figure must be applied. They are

$$C = y_0 t f_c \tag{4.8.3}$$

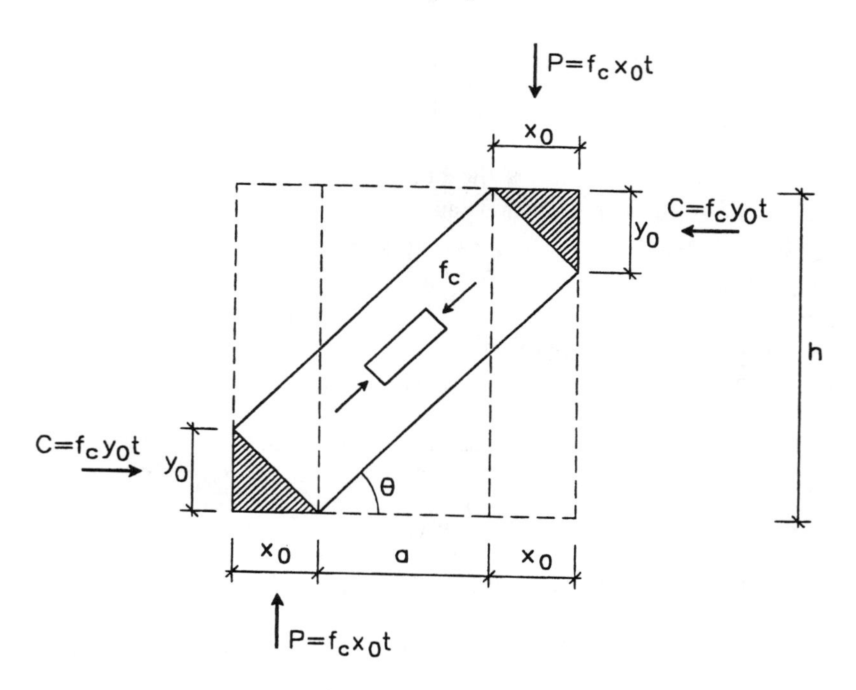

Figure 4.8.1 The single strut.

If the shear force P to be carried is given, we find

$$x_0 = \frac{P}{tf_c} \quad \Rightarrow \quad \frac{x_0}{h} = \frac{P}{thf_c} = \frac{\tau}{f_c} \tag{4.8.4}$$

τ being the average shear stress along the depth h. Then y_0 may be determined by (4.8.1) giving

$$\frac{y_0}{h} = \frac{1}{2} - \sqrt{\frac{1}{4} - \frac{x_0}{h}\left(\frac{a}{h} + \frac{x_0}{h}\right)} \tag{4.8.5}$$

There is also a solution with a plus sign in front of the square root. This solution is, however, not interesting because it leads to larger reinforcement consumption.

The maximum value of y_0/h is 1/2. This is obtained when the term under the square root in (4.8.5) is zero, which takes place when

$$\frac{x_0}{h} = \frac{1}{2}\left[\sqrt{1 + \left(\frac{a}{h}\right)^2} - \frac{a}{h}\right] \tag{4.8.6}$$

This means that the maximum value of the average shear stress is

$$\frac{\tau}{f_c} = \frac{1}{2}\left[\sqrt{1 + \left(\frac{a}{h}\right)^2} - \frac{a}{h}\right] \tag{4.8.7}$$

The formula determines the highest load which can be carried by a strut of depth h and shear span a.

In Chapter 5 we will derive this important formula in another way. The reader may verify that the strut has a maximum load-carrying capacity when $y_0 = h/2$ by a moment equation for the whole system including the hydrostatic areas.

If, on the other hand, y_0 is given, we get by solving (4.8.1) with respect to x_0

$$\frac{x_0}{h} = \frac{1}{2}\left[\sqrt{4\frac{y_0}{h}\left(1 - \frac{y_0}{h}\right) + \left(\frac{a}{h}\right)^2} - \frac{a}{h}\right] \tag{4.8.8}$$

If $y_0 = h/2$, the term $4(y_0/h)(1 - y_0/h) = 1$ and we are back to (4.8.6).

The single strut stress field has numerous applications. It will suffice to consider a few.

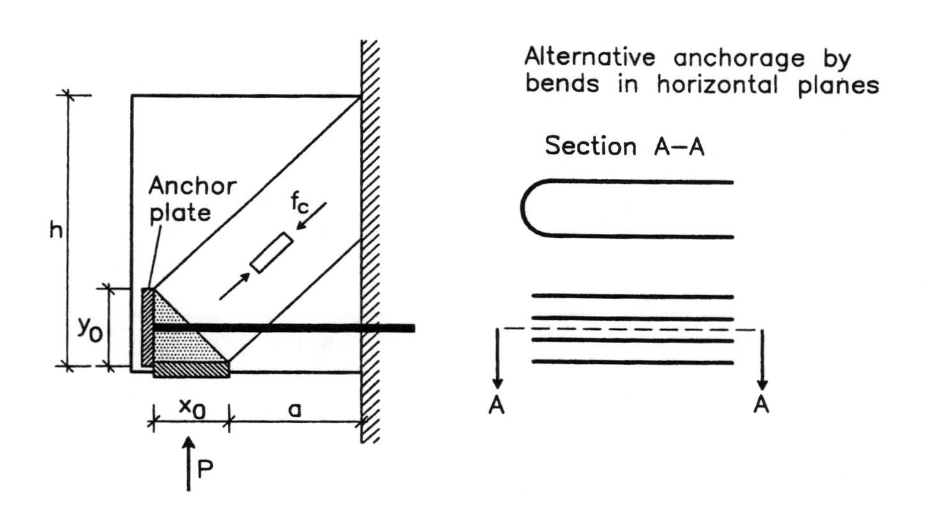

Figure 4.8.2 Strut action in a corbel.

Fig. 4.8.2 shows a corbel which has to carry the vertical force P. The horizontal extension x_0 of the hydrostatic area is calculated by formula (4.8.4). Then the shear span a may be determined.

By formula (4.8.7) it is verified whether or not the corbel is able to carry the load by strut action. If yes, the vertical extension of the hydrostatic area y_0 is calculated by (4.8.5). Then the necessary area of the reinforcement A_s is determined by

$$A_s f_Y = y_0 t f_c \tag{4.8.9}$$

f_Y being the yield stress. The necessary reinforcement degree Φ is thus

$$\Phi = \frac{A_s f_Y}{t h f_c} = \frac{y_0}{h} \leq \frac{1}{2} \tag{4.8.10}$$

The maximum value of Φ is seen to be 1/2.

In the example of Fig. 4.8.2 we imagine the reinforcement to be anchored by an anchor plate of depth y_0 and thickness t. Many other possibilities are, of course, open. Often it will be preferred to anchor the reinforcement by looped bars having bends in a horizontal plane as illustrated in the figure.

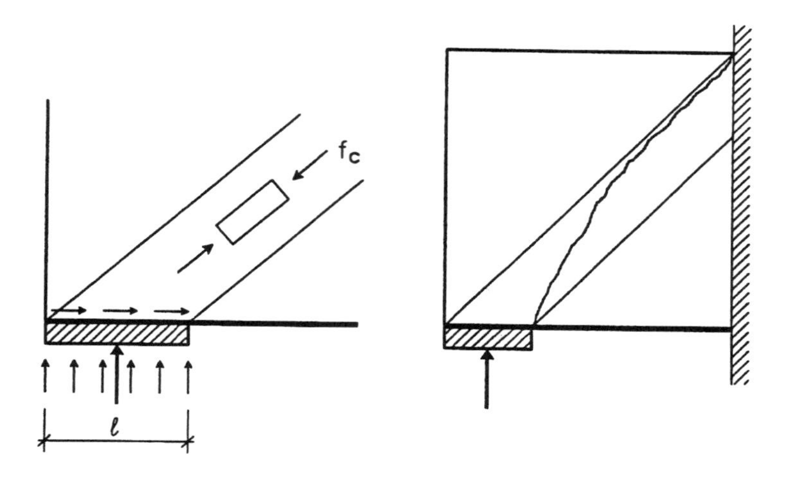

Figure 4.8.3 Effect of support length.

If the reinforcement degree for one or another reason is limited to some value $\Phi < 1/2$, then x_0 may be calculated by (4.8.8) inserting $y_0/h = \Phi$ which means that the load-carrying capacity is given by

$$\frac{\tau}{f_c} = \frac{P}{thf_c} = \frac{x_0}{h} = \frac{1}{2}\left[\sqrt{4\,\Phi\,(1-\Phi) + \left(\frac{a}{h}\right)^2} - \frac{a}{h}\right] \qquad (4.8.11)$$

The anchorage of the reinforcement and the whole design of the support regions and the regions around concentrated forces are extremely important and may be decisive for the load-carrying capacity as pointed out by J. F. Jensen [81.2]. He investigated the load-carrying capacity as a function of the length of the support and the position of the reinforcement.

Consider as an example the theoretical case where the reinforcement is placed along the bottom face and the strut force is equilibrated by a shear stress delivered by the reinforcement and a normal stress delivered by an abutment platen of length ℓ (see Fig. 4.8.3). If the length $\ell \rightarrow 0$ it will be impossible to stress the strut and the load-carrying capacity will approach zero. J. F. Jensen showed that when the reinforcement is strong enough not to yield, the exact solution will be the strut corresponding to the length of the abutment platen as long as the load-carrying capacity does not exceed the maximum value given by (4.8.7). The yield line will be curved. One possibility is illustrated in the figure. The relative displacement is a rotation around some point on the reinforcement line to the left of the abutment platen.

Yield line patterns corresponding to the maximum shear capacity and to yielding of the reinforcement will be dealt with in Chapter 5.

4.8.3 Strut and Tie Systems

Often it is easy to construct the strut and tie system graphically. This is illustrated in Fig. 4.8.4.

We assume that the compression stresses at the load and the reactions are equal to their maximum value, the compressive strength f_c. Let us assume the external force P to be given, by which the area necessary to carry the force can be determined. Knowing the thickness of the disk, we are then able to determine the distance AB. Drawing the lines AD and BG as limiting lines of the struts, we can find the necessary dimensions of the struts AC and BC by drawing AC perpendicular to AD, and BC perpendicular to BG. Assuming the triangle ABC to be subjected to a biaxial hydrostatic compression (and therefore fully utilized), this triangle transfers the compression stress f_c to the struts, which in this way will be fully utilized. Along DE and FG, the compressions of the struts are transferred to the supports and to the reinforcement $DEFG$, thus obtaining a constant force outside DE and FG. The compressions at the supports are seen to be less than the compressive strength f_c, although the transfer of force to the reinforcing bars fully utilizes the strength of the concrete. It is immediately seen that the stress field described is statically admissible, and as it only requires a constant tensile reinforcement, it is the optimal way of carrying the force P when the tensile reinforcement is not curtailed. Calculation of the necessary tensile reinforcement and the compressions at the supports is completed without difficulties.

It may, of course, be questioned whether it will be possible to transfer the compressions of the struts to the tensile reinforcement in the way described. The average shear stress in sections DE and FG, together with the normal stress in these sections, will certainly not exceed the yield condition, but as we are dealing here with average values, the local stresses around the tensile bars may be considerably greater. Therefore, in practice, this solution requires special investigation of the anchorage of the tensile reinforcement, cf. Chapter 9.

In Fig. 4.8.5, a strut and tie system for the case of two concentrated forces at a simply supported beam is shown.

The design may be carried out by calculating in the usual way the area of tensile reinforcement required in the section subjected to the largest bending moment, by controlling the magnitude of the compressions at the supports and the compressions below the concentrated forces and, finally, to ensure proper anchorage of the tensile reinforcement.

As seen above, the compression stress σ_ℓ at the supports should at most be put at

$$\sigma_{\ell max} = f_c \sin^2 \theta \tag{4.8.12}$$

the meaning of θ being clear from Fig. 4.8.5.

If the reaction is termed R and the constant force of the reinforcement $T = A_s f_Y$, we obtain

$$\sin^2 \theta = \frac{1}{1 + (T/R)^2} \tag{4.8.13}$$

by which the condition $\sigma_\ell \le \sigma_{\ell\,\max} = f_c\, sin^2\theta$ becomes

$$\sigma_\ell \le \frac{f_c}{1 + (T/R)^2} \tag{4.8.14}$$

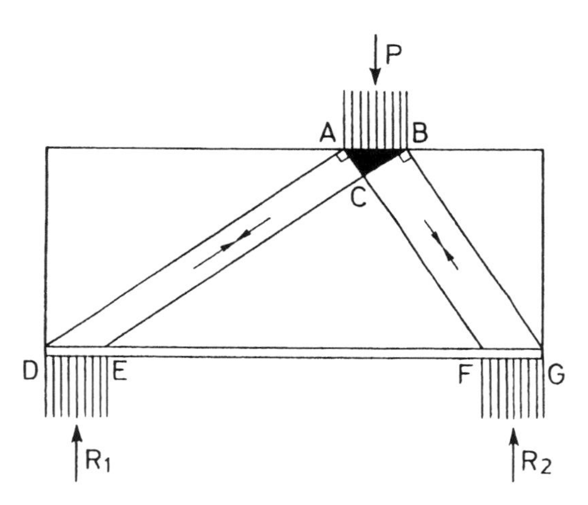

Figure 4.8.4 Strut and tie solution for a disk with one concentrated load.

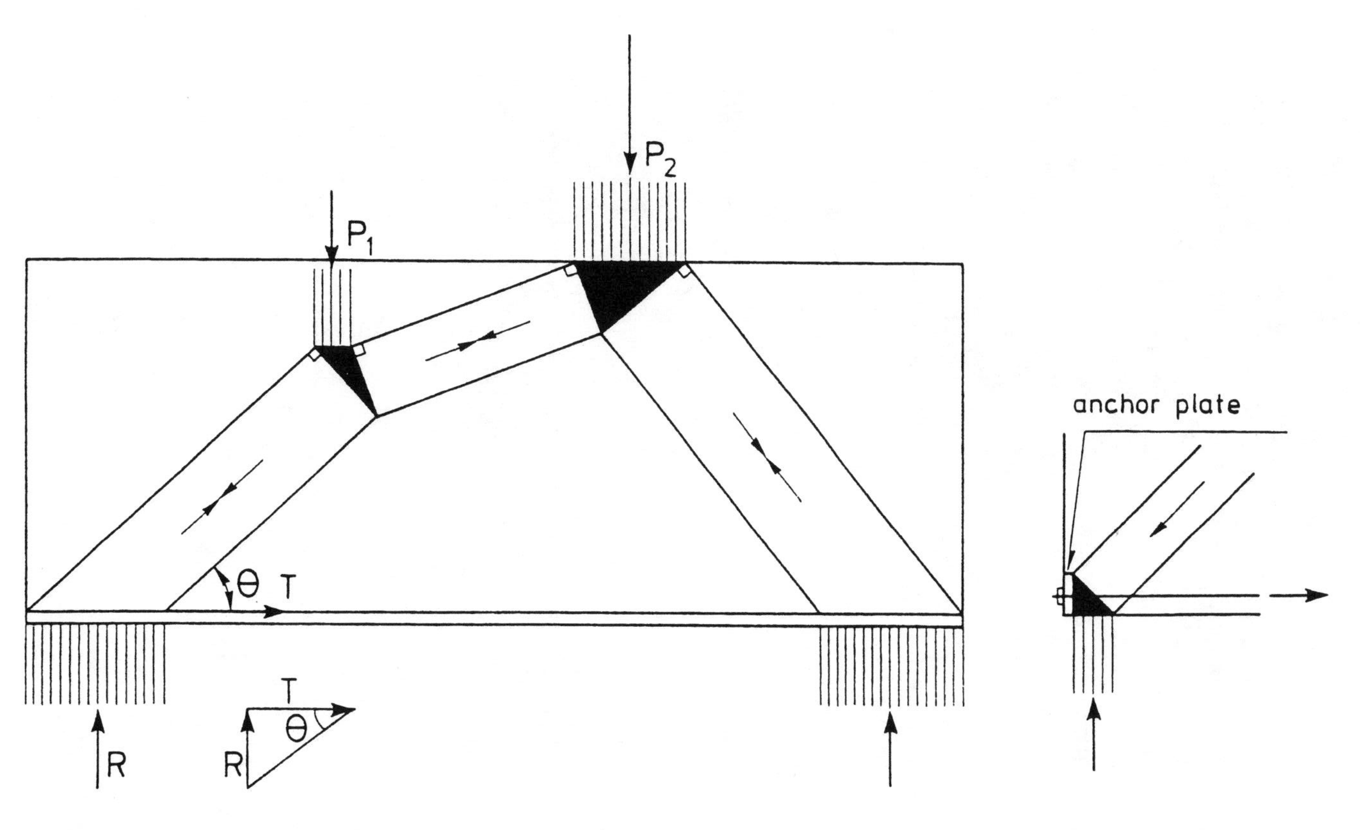

Figure 4.8.5 Strut and tie solution for a disk with two concentrated loads.

If the anchorage of the tensile bars is established by using anchor plates at the ends of the bars, we need only require a compression stress at the supports $\sigma_\ell \leq f_c$, just as the area of the anchor plates is determined by putting the maximum compression on the plate at f_c. The action is illustrated in Fig. 4.8.5, where the shaded part is considered to be subjected to biaxial hydrostatic compression.

As pointed out by J. F. Jensen [81.2], the formula (4.8.14) neglects the effect of the cover and might, therefore, be too conservative for large covers. Further, sometimes the reinforcement may be anchored beyond the abutment platen. Let us assume, as an example, that the anchorage beyond the support may have the same effect as an anchor plate with depth $2c$, see Fig. 4.8.6. In this case the requirement for the support stress σ in the intermediate part of the support length is

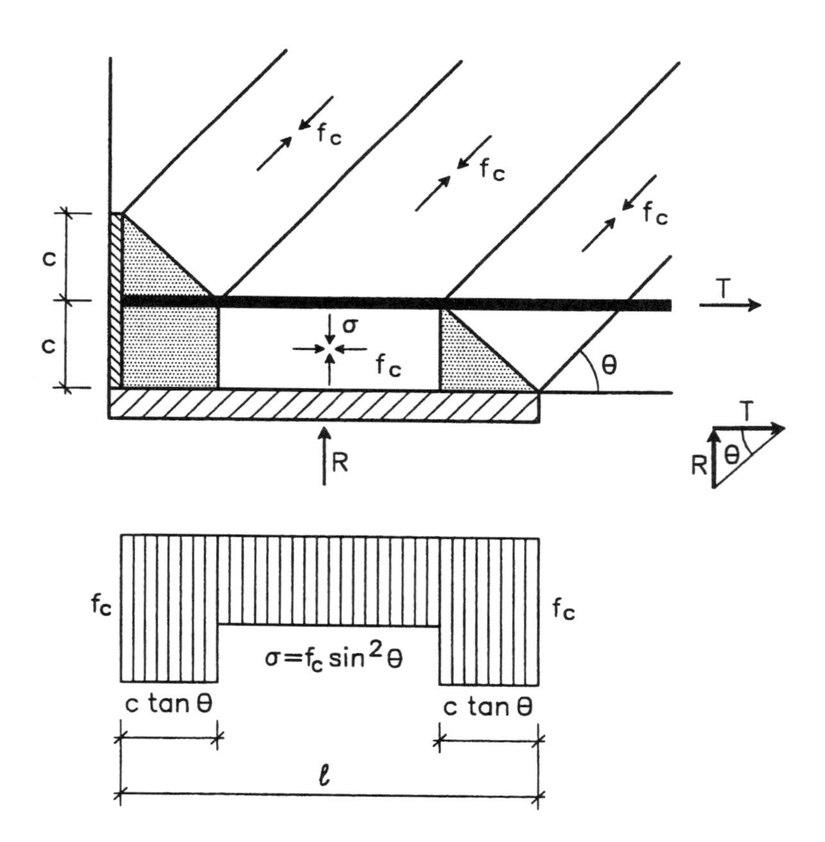

Figure 4.8.6 Stress field at support.

$$\sigma = \frac{R - 2f_c tc\,\dfrac{R}{T}}{\ell t} \le \frac{f_c}{1 + \left(\dfrac{T}{R}\right)^2} \tag{4.8.15}$$

If the numerator of the second term is zero we are back at the simple anchor plate solution.

For arbitrary, vertical loads acting at the top side of the disk the strut and tie solution may also be utilized. The section with the maximum bending moment is designed in the usual way according to (4.7.39) and (4.7.40) and the compression stresses of the supports are controlled by using (4.8.14) or (4.8.15).

For continuous disks, the stress distribution derived in Section 4.7.4 as well as the strut and tie model can be used. Consider a rectangular, horizontally spanning disk of constant thickness t, supported at the bottom face and loaded with vertical forces at the top or bottom face. In Fig. 4.8.7 the strut system is drawn within an end span. The force T_0 in the reinforcement of the top face is assumed to "load the strut" at the maximum moment.

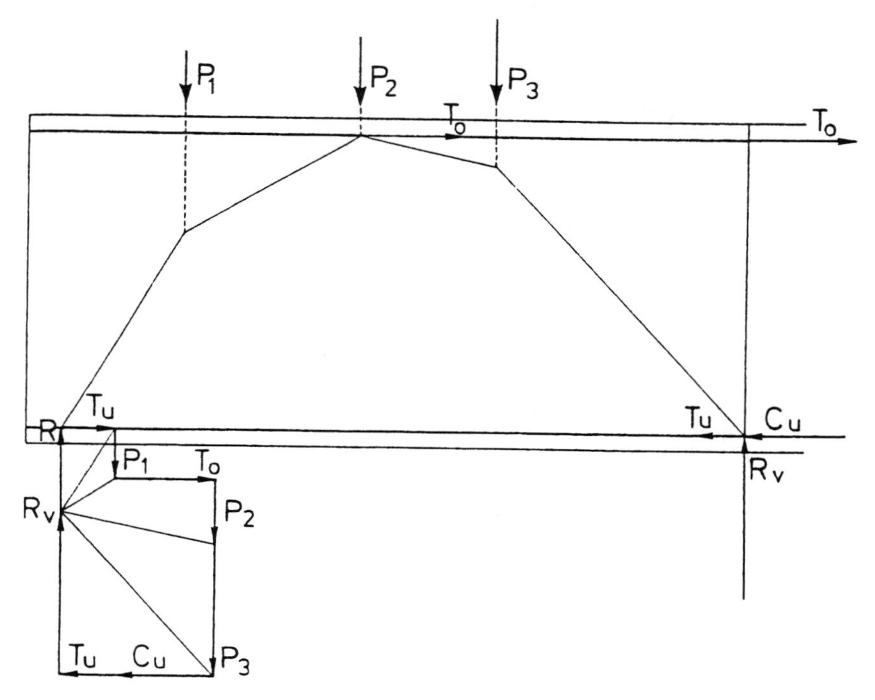

Figure 4.8.7 Strut and tie solution for a disk with loads at the top face.

The conditions at an intermediate support are illustrated in Fig. 4.8.8 where it is assumed that the entire reinforcement at the bottom face, having the force T_{uv}, is to be anchored, which is considered to take place along length c' of the support by means of a part of the reaction and a part of the compression of the strut. The remaining part of the compression of the strut is kept in balance by the compressive force C_u from the bending moment at the support and the reaction at the remaining part of the support c''. The compression of the strut is assumed to form the angle θ with the horizontal.

The smallest support length ℓ^v_{min} at the considered side of the section with maximum negative bending moment can be calculated by considering the strut fully utilized to the compressive strength f_c. If the compressive stress from C_u, distributed within the depth y_0, is also f_c, the shaded section of Fig. 4.8.8 is subjected to biaxial, hydrostatic compression, and the compression of the support along length c'' is f_c.

By this we obtain

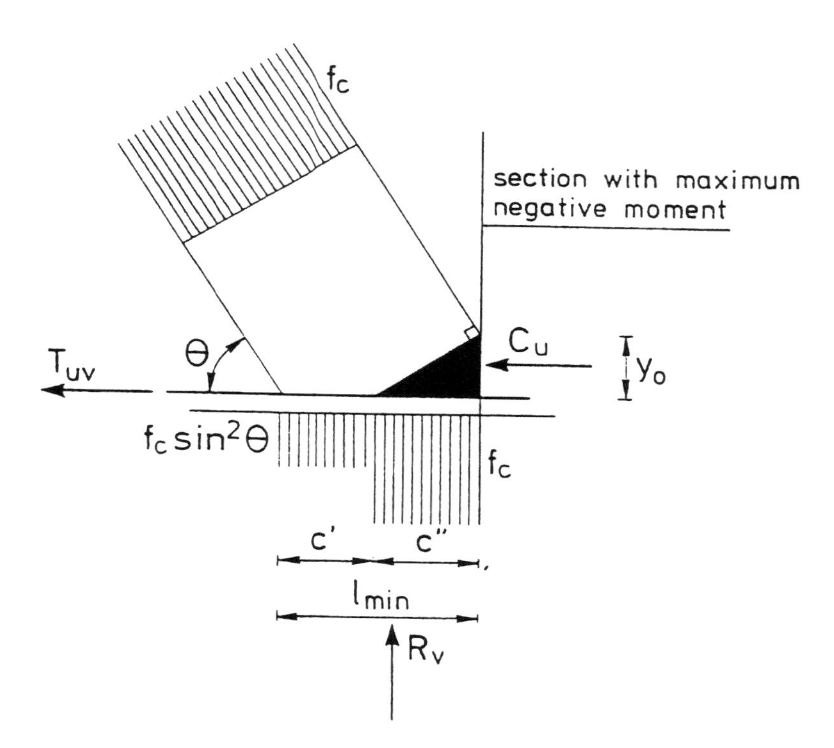

Figure 4.8.8 Stress field at an intermediate support.

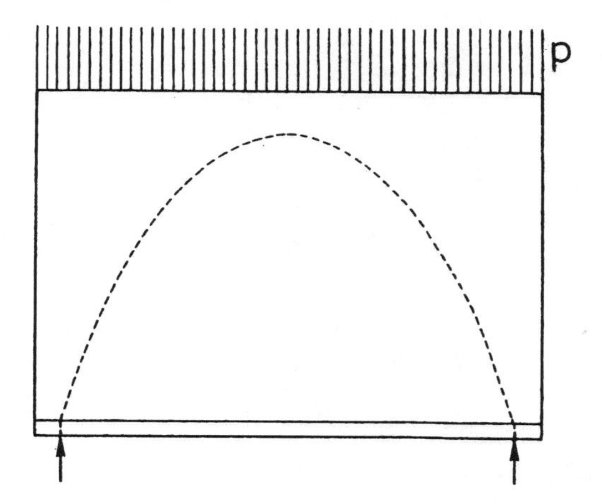

Figure 4.8.9 Arch solution for a disk with uniform load.

$$\ell^{v}_{\min} = y_0 \tan\theta + \frac{T_{uv}}{f_c t \cos\theta \sin\theta} \tag{4.8.16}$$

in which t is the thickness of the disk and $f_c\ t \cos\theta \sin\theta$ is, therefore, the maximum shear force per unit length which can be carried in a horizontal section. If R_v denotes the part of the reaction transferred to the side of the section in question, we have

$$\tan\theta = \frac{R_v}{C_u + T_{uv}} \tag{4.8.17}$$

and

$$\cos\theta = \frac{C_u + T_{uv}}{\sqrt{(C_u + T_{uv})^2 + R_v^2}} \tag{4.8.18}$$

$$\sin\theta = \frac{R_v}{\sqrt{(C_u + T_{uv})^2 + R_v^2}} \tag{4.8.19}$$

If the force in the reinforcement at the bottom face of the adjoining span is T_{uh}, and if, for example, $T_{uv} > T_{uh}$, it will often be natural to carry the reinforcement corresponding to the force T_{uh} through at the support. If so, the length of support can be reduced. The formulas can be used when y_0 is calculated as the depth necessary for transmission of the force C_u+T_{uh} [i.e., $y_0 = (C_u+T_{uh})/f_c\ t$], and also when we insert $T_{uv} - T_{uh}$ in the formula instead of T_{uv}.[6] When $T_{uv} = T_{uh}$, $\ell^{v}_{\min}$ is simply $R_v / f_c\, t$. It should be noted that the total length is found as a contribution of the form (4.8.16) from each side of the section with maximum negative moment.

The bending moments above the supports are statically indeterminate. Concerning the selection of these moments, there are, according to the theory of plasticity, no limitations. Just as with slender continuous beams, we should, however, probably not deviate too much from the elastic stress distribution. For continuous deep beams we may as a basis of evaluation use the elastic moment distribution in a slender continuous beam of the same ratios of span without significant errors.

A large number of concentrated forces may be used to model a continuous load. In the limit the strut for a uniform load, or rather the arch, will be parabolic as shown in Fig. 4.8.9.

Another statically admissible, more convenient, stress field is shown in Fig. 4.8.10. Among the various, statically admissible stress distributions, one with lines of discontinuity is shown. The distance from the lateral face to the maximum bending moment is a. We suppose that in the section with the maximum moment there is the usual stress distribution with a rectangular stress block, corresponding to a normal stress equal to the compressive strength f_c at the top, and in the reinforcement at the bottom face a tensile force equal to the compression force, assumed to be constant from the maximum moment to the supports. The disk is divided into parts 1, 2, 3, 4, and 5, separated by lines of discontinuity. The stress fields are homogeneous within the individual areas. As we shall consider part 5 to be stress free, we have a uniaxial stress field within part 3. This means that the position of the point K, dividing the top face into two parts, cannot be arbitrarily chosen.

If we imagine that the supporting area is given, the position of the line of discontinuity between parts 3 and 5 is fixed. Therefore, point K must be determined so that the resultant of the forces in part 4 has the same direction as the line of discontinuity between parts 3 and 5. This is easily done graphically, as shown in Fig. 4.8.10. Then Mohr's circles can be constructed, and from the construction it is seen that if the supporting area is of sufficient size so that the smallest

[6] That is, the force needed to anchor.

principal stress within part 2 is numerically less than the compressive strength, the disk will be able to carry the load exclusively by compressive stresses in the concrete. In practice, (4.8.14) or (4.8.15) may be used to find the maximum value of the compression stress at the support.

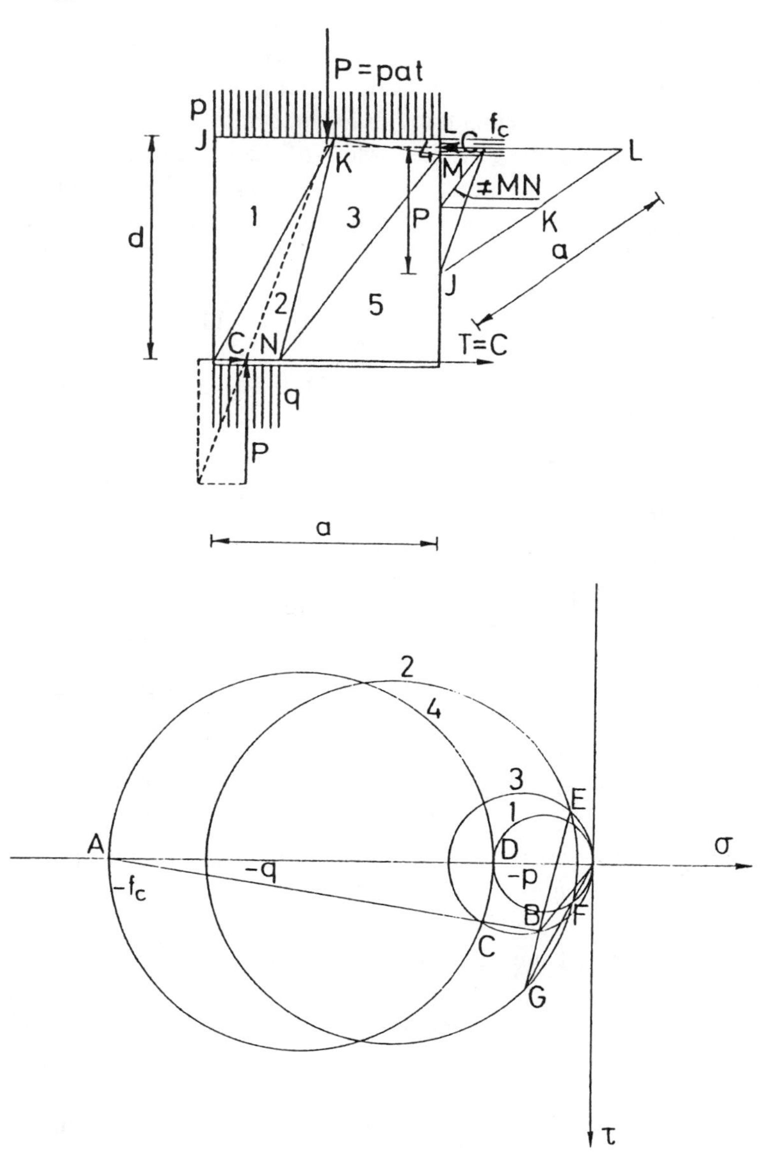

Figure 4.8.10 Discontinuous stress field for a disk with uniform load.

4.8.4 Effectiveness Factors

The effective strength of a single strut is rather well known.

The ν-value must preferably be determined for overreinforced beams and disks (reinforcement not yielding), since for normally reinforced elements (reinforcement yielding), the load-carrying capacity is rather insensitive to the compressive strength f_c.

Although the formulas (4.8.7) and (4.8.11) are, strictly speaking, only valid for a reinforcement lay-out which may provide a force in the position corresponding to the middle of the compression zone y_0, it has normally been assumed that the beams and disks have been properly designed to carry the maximum shear capacity. Undoubtedly this has not always been the case which means that the ν-values determined are too low, i.e., on the safe side.

It should be mentioned that (4.8.7) and (4.8.11) are always valid upper bound solutions (cf. Chapter 5).

Replacing f_c by νf_c in formulas (4.8.7) and (4.8.11), they may be written

$$\frac{\tau}{f_c} = \begin{cases} \frac{1}{2}\nu\left[\sqrt{4\frac{\Phi}{\nu}\left(1-\frac{\Phi}{\nu}\right)+\left(\frac{a}{h}\right)^2}-\frac{a}{h}\right] & \text{for } \Phi \le \frac{1}{2}\nu \\ \frac{1}{2}\nu\left[\sqrt{1+\left(\frac{a}{h}\right)^2}-\frac{a}{h}\right] & \text{for } \Phi > \frac{1}{2}\nu \end{cases} \tag{4.8.20}$$

Φ still being defined by (4.8.10). These formulas may be used for determining ν. Of course, bending failures must be excluded.

In Fig. 4.8.11 some test results for almost constant Φ and a constant ν-value of 0.58 have been plotted together with the theoretical best fit curve. The figure has been taken from [78.1].

It appears that when using a constant ν-value, the theory has a tendency of giving too low values of τ/f_c for the smaller and higher values of a/h and too high values for intermediate values of a/h. This means that ν must be dependent on a/h, which is, of course, extremely inconvenient. Many attempts have been made to explain this phenomenon, including the study of dowel action and the effect of the tensile strength of concrete to mention only a few. The riddle was only recently solved. We return to the solution in a moment.

By using equations (4.8.20) Roikjær, in a master thesis, found that

$$\nu = f_1(f_c)\, f_2(h)\, f_3(r)\, f_4\left(\frac{a}{h}\right) \tag{4.8.21}$$

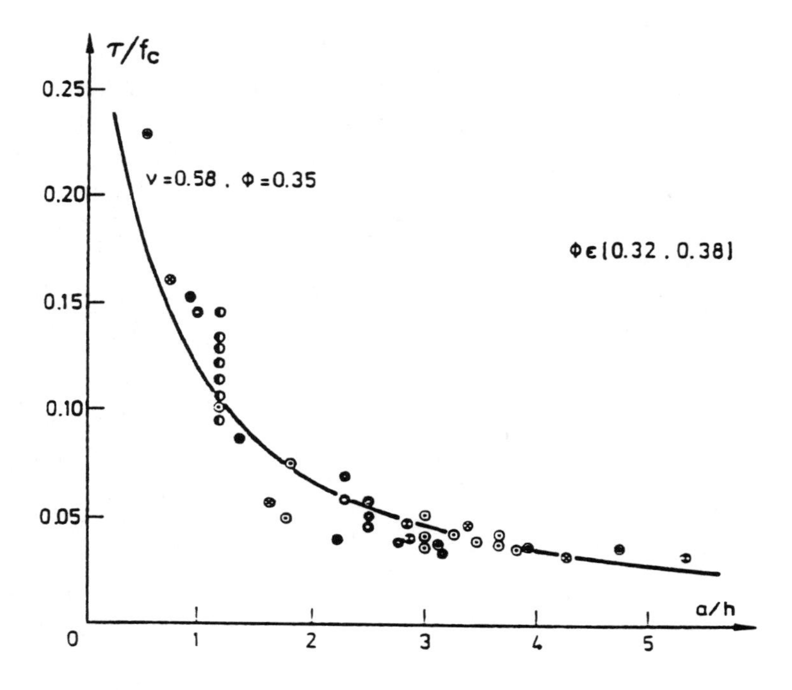

Figure 4.8.11 Test results for a single strut.

f_1, f_2, f_3, and f_4 are functions of one variable as indicated. Roikjær's results are summarized in [78.1] and [79.4].

The functions are

$$f_1 = \frac{3.5}{\sqrt{f_c}} \qquad (5 < f_c < 60 \text{ MPa})$$

$$f_2(h) = 0.27\left(1 + \frac{1}{\sqrt{h}}\right) \qquad (0.08 < h < 0.7) \quad (h \text{ in m}) \tag{4.8.22}$$

$$f_3(r) = 0.15r + 0.58 \qquad (r < 4.5\%) \quad (r \text{ in } \%)$$

$$f_4\left(\frac{a}{h}\right) = 1.0 + 0.17\left(\frac{a}{h} - 2.6\right)^2 \quad \left(\frac{a}{h} < 5.5\right)$$

The function f_1 gives the dependence on f_c and is of the form found in many other cases, i.e., ν is decreasing when f_c increases.

The function f_2 gives a size effect leading to a decreasing ν-value for increasing depth h.

The function f_3 describes the dependence of the reinforcement ratio $r = A_s / th$. Notice that r must be inserted in %.

The effects described by the functions f_2 and f_3 must be attributed to softening; the softening effect being increased with increasing depth and decreased for increasing reinforcement ratio. In f_2 there might, of course, also be a contribution from the crack widths which are higher for a structure being scaled up, keeping everything else the same, and f_3 may contain a contribution from dowel effects of the reinforcement. Whatever is the truth, these functions are quite acceptable.[7]

The function f_4 describes the above-mentioned dependency on a/h. It has a minimum for $a/h = 2.6$.

Before giving the physical explanation of this function, it should be mentioned that Chen Ganwei [88.12] by treating almost 700 tests including both ordinary beams, deep beams and corbels found a similar formula, namely

$$\nu = \frac{0.60\left(2 - 0.4\dfrac{a}{h}\right)(r+2)\,(1-0.25\,h)}{\sqrt{f_c}} \leq 1 \qquad (4.8.23)$$

$$\frac{a}{h} \ngtr 2.5 \qquad r \ngtr 2\% \qquad h \ngtr 1.0\ m$$

The units are the same as before. This formula was verified by tests with $f_c < 100$ MPa, $a/h < 10$ and $h < 1.20$ m.

According to this formula a/h should not be set larger than 2.5, neither in the formula (4.8.23) nor in (4.8.20). This means that for larger a/h-values than 2.5 the load-carrying capacity, according to Chen's formula, is a constant equal to the value determined for $a/h = 2.5$.

There are not many tests with $h > 1.20$ m. In practice disks with $h > 1.20$ m occur frequently. Until further knowledge is gained it seems reasonable to use the ν-values found by inserting the limits of h stated for the formulas, i.e., no further decrease in ν above the limits given is considered.

[7] In [97.9] it has been demonstrated that the f_3-function is entirely due to dowel action.

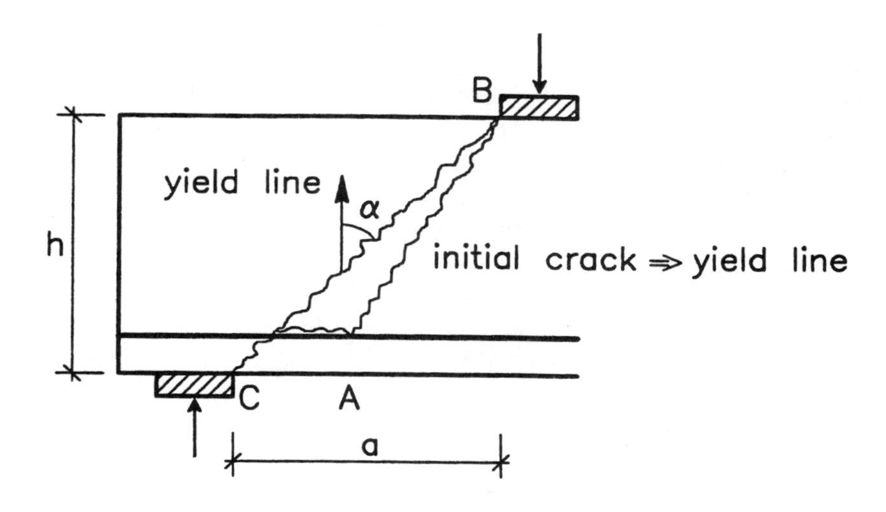

Figure 4.8.12 Sliding in initial cracks.

The coefficient of variation on the ratio test result/theory is rather large, around 16%. The mean value is, of course, one. That is how ν is determined.

The question of why it is necessary to include an *a/h*-function when the whole range of *a/h* values has to be covered was explored and finally solved by Jin-Ping Zhang [94.10] [97.14]. The low values of ν for *a/h* around 2.5 are due to sliding in initial cracks (cf. Section 4.6.2). Because of the dramatically reduced sliding resistance in a crack, sliding in a crack *AB* in Fig. 4.8.12 which gives rise to a yield line *CAB* may be more dangerous than sliding along the theoretical yield line *CB*, which for a single strut goes from support to force as shown in the figure. If the initial crack follows the theoretical yield line, the load-carrying capacity will be reduced by a factor $\nu_s = 0.5$, where ν_s is the sliding reduction factor.

Jin-Ping Zhang found that up to $a/h \approx 0.75$ the original plastic solution may be applied with

$$\nu = \nu_0 = 1.6\, f_1\, f_2\, f_3 \tag{4.8.24}$$

the functions being defined in (4.8.22).

The limit $a/h = 0.75$ may be explained by the fact that, at least for overreinforced elements (tensile reinforcement not yielding), the vertical relative displacement along a yield line running from support to force as in Fig. 4.8.12 for this *a/h*-value forms an angle α with the yield line determined by $\tan\alpha = 0.75 = \tan\varphi$. For $a/h < 0.75$ this relative displacement is not possible due to the conditions imposed by the yield condition for the crack, i.e., sliding in the crack is not possible.

For $0.75 < a/h < 2$ the test points fill up almost the whole interval between the original plastic solution with ν found by (4.8.24) and the solution obtained by determining the most dangerous crack and applying the sliding reduction factor, i.e., applying a ν-value $\nu = \nu_s \nu_0 = 0.5\ \nu_0$, ν_0 being again determined by (4.8.24). In this interval the lowest solution corresponding to sliding in initial cracks in the overreinforced case is half the original plastic solution using ν according to (4.8.24) because the most dangerous crack in this interval is a crack running from support to force, i.e., the same as the theoretical yield line. However, due to the finite distances between the cracks, the most dangerous crack is not always formed, which explains why the test points are filling up the interval mentioned.

For $a/h > 2$ a new solution had to be worked out.

The crack sliding solutions will be examined more closely in Chapter 5.

Since the strut solution is most relevant for $a/h < 2$, we may conclude that for $a/h < 0.75$ the effectiveness factor $\nu = \nu_0$ may be calculated by formula (4.8.24). For $0.75 < a/h < 2$ the effectiveness factor may as an approximation be set at $\nu = \nu_s\ \nu_0$, ν_0 being determined by (4.8.24) and $\nu_s = 0.5$.

Alternatively, for any *a/h*-value, the ν-value may be calculated using (4.8.21) or (4.8.23).

This is what is known at present for the single strut. For more complicated strut and tie systems very little has been done. Experimental studies on nodes in strut and tie systems were carried out by Jirsa et al. [91.29]. A general design procedure is outlined in Chapter 5.

It should be emphasized that the whole strut and tie system, including the support regions and the regions with concentrated forces, should be calculated using the ν-value determined in the above way.

The effect of prestressing is most relevant for beams and is examined in Chapter 5. Normal forces are treated in Section 4.9 and in Chapter 5. Structural elements with normal forces were examined by Chen Ganwei [88.12]. Compression normal forces increase the ν-value. A formula may be found in the reference cited.

4.8.5 More Refined Models

The simple strut and tie systems considered previously will not satisfy the advanced designer. He will normally, while his professional experience is progressing, make his own stock of equilibrium solutions suitable for his specific kind of problems.

The circular fan will be an inevitable part of such a library.

Let us consider two examples of circular fan solutions. This type of solution was already used in Section 4.5.3, Example 1, to solve an axisymmetrical problem. In this example we had tensile stresses in radial directions. In what follows the stresses in radial sections are assumed to be zero and there are no shear stresses in radial and circumferential sections. Thus in a polar coordinate system r, θ we have

$$\sigma_\theta = \tau_{r\theta} = 0 \tag{4.8.25}$$

i.e., there are only stresses σ_r. They will, of course, be compressive.

In Fig. 4.8.13 the simplest circular fan solution is shown. We have radial stresses of the form $\sigma_r = c/r$, c being a negative constant. It may immediately be shown that the equilibrium equations are satisfied for a weightless body.

If the radial stresses are known, say for $r = e$ and $r = e + m$, respectively, we have (see the figure)

$$pe = q(e + m) \tag{4.8.26}$$

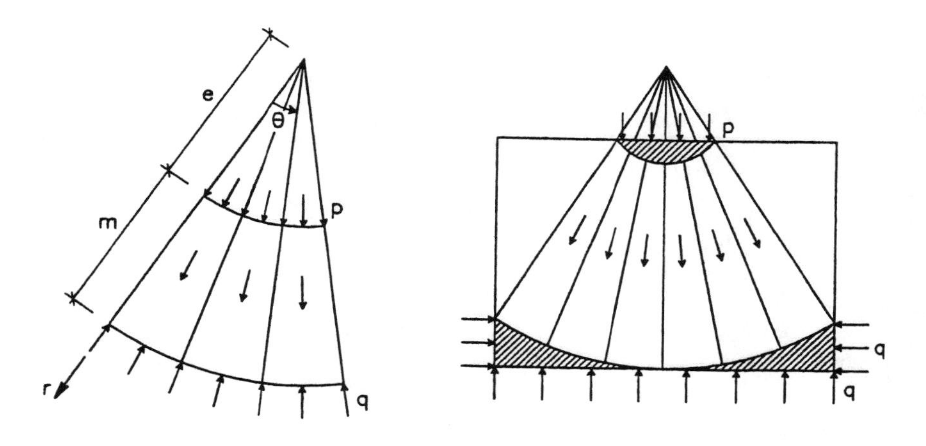

Figure 4.8.13 Simplest circular fan solution.

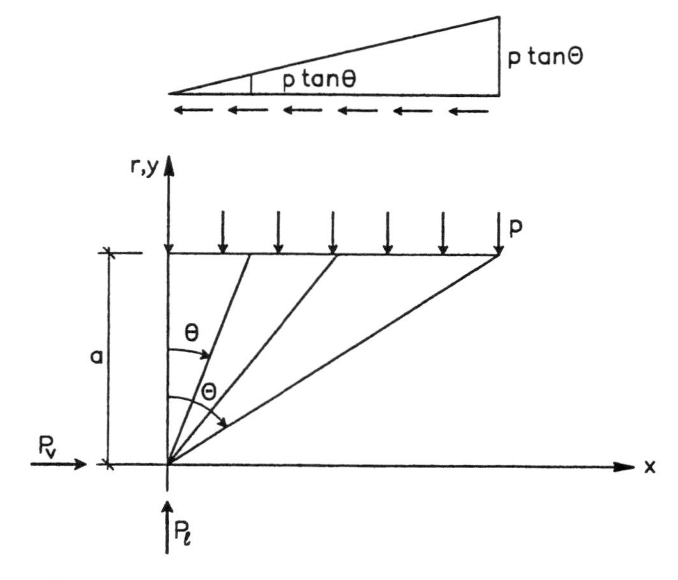

Figure 4.8.14 Circular fan solution.

An elementary application of this stress field is also shown. A concrete foundation block is loaded in the middle by a uniform compression stress p. The reaction from the ground is a uniform compression stress q. In this case the circular fan is stressed by the shaded areas which are in biaxial hydrostatic compression. If reinforcement is necessary, i.e., if the horizontal reactions cannot be delivered by the soil or by neighboring structures, the necessary reinforcement is easily calculated by determining the resultant force of the horizontal stresses q.

The reader may easily invent a number of similar solutions using the circular fan with constant σ_r along circumferential sections.

Now consider the case where c in $\sigma_r = c/r$ is a function of θ. The equilibrium equations are still satisfied. We require, see Fig. 4.8.14, the normal stress σ_y to be constant along horizontal lines y = const. If $\sigma_y = -p$ for $y = a$, i.e., for $r = a/\cos\theta$ we find, since $\sigma_y = \sigma_r \cos^2\theta$, that

$$\sigma_r = -\frac{pa}{\cos^3\theta}\,\frac{1}{r} \tag{4.8.27}$$

As before, $\sigma_\theta = \tau_{r\theta} = 0$. Along any other horizontal line, say $y = b$, we have

$$\sigma_y = -\frac{pa}{b} \tag{4.8.28}$$

Since $\tau_{xy} = \sigma_r \cos\theta \sin\theta$ the shear stress along the horizontal line $y = a$ is

$$\tau_{xy} = -p\tan\theta \tag{4.8.29}$$

The shear stress in any other horizontal line, say $y = b$, is

$$\tau_{xy} = -\frac{pa}{b}\tan\theta \tag{4.8.30}$$

Thus τ_{xy} is a linear function of x along any horizontal line (cf. Fig. 4.8.14).

The vertical force P_ℓ carried by a circular fan $0 \le \theta \le \Theta$, $0 \le y \le a$, is

$$P_\ell = pa\tan\Theta \tag{4.8.31}$$

and the horizontal force is

$$P_v = \frac{1}{2}pa\tan^2\Theta \tag{4.8.32}$$

Since $\sigma_r \to \infty$ for $r \to 0$, the pole can, of course, not belong to the region considered in any real case. The largest value of the concrete stress $\sigma_c = |\sigma_r|$ is found when θ is as large as possible and r as small as possible. We find that at any point $\sigma_c = |\tau_{xy}|(\tan\theta + \cot\theta)$.

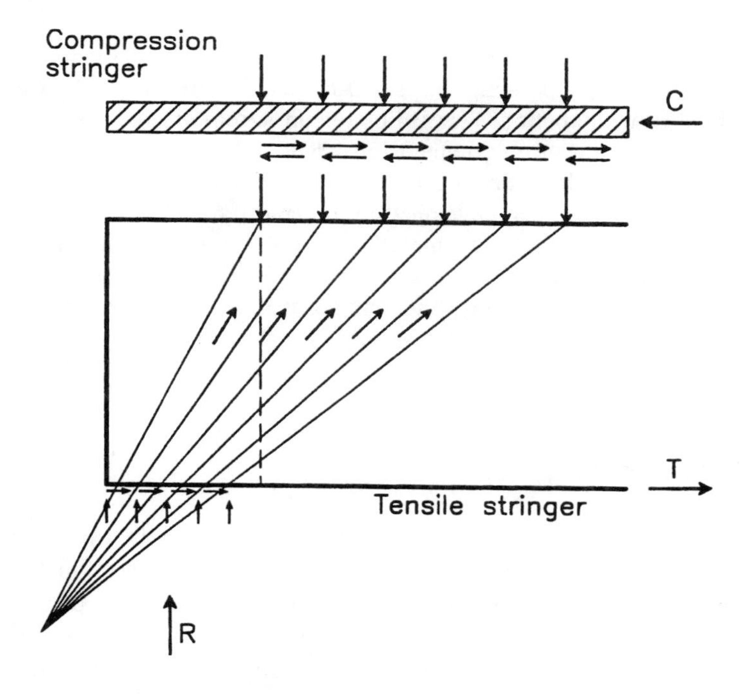

Figure 4.8.15 Beam end modelled by a circular fan.

The solution has numerous applications. In Fig. 4.8.15 a beam is modelled by means of a circular fan. The uniform normal stress at the support is delivered by the reaction R and the shear stress is delivered by a tensile stringer. The uniform stress on the top face of the beam may be an applied load or compressive stresses from the anchorage of a uniform stirrup reinforcement. The shear stresses on the top face may be delivered by a compression stringer.

Experimental evidence on effectiveness factors for solutions of the type considered here will normally be lacking. The designer must make a qualified engineering judgement based on what is known from similar cases. In each situation he must evaluate the effect of the three main causes for the reduction of the compressive strength of concrete, i.e., softening, stressed reinforcement crossing uniaxial compression fields and sliding in initial cracks.

Circular fans are often used to model stress fields for which the initial crack directions may be assumed to be equal to the final crack directions; then sliding in initial cracks may be excluded. In unreinforced regions, only softening effects are left. If so, the ν-value may be taken from Section 4.6 setting $\nu = \nu_0$ determined by formula (4.6.1) or (4.6.6). The same values may normally be used in reinforced regions if minimum reinforcement is provided.

Sometimes higher ν-values may be justified.

In the strip foundation case in Fig. 4.8.13 the softening effects may be estimated to be unimportant, i.e., it will be justified to use $\nu = 1$ except in the region near the bottom face if this region is reinforced.

In the future advanced computer programs will be available. They will take into account the reduction of the compressive strength of concrete. Such a program has been developed by L. Jagd [96.2] [97.13].

The program takes into account all the main causes of strength reduction and the solutions obtained have been compared with test results for a large number of structural elements. Such programs may in the future be used to determine effectiveness factors instead of using experiments.

4.9 SHEAR WALLS

4.9.1 Introduction

In practice the strut and tie solutions discussed in the previous section are often and always, for important structures, combined with a uniformly distributed reinforcement in two perpendicular directions. This is necessary to avoid undesired crack width behavior for the serviceability load. The concentrated reinforcement obtained in simple strut and tie solutions is only able to govern the crack width distribution in the vicinity of the reinforcement, say in a distance of 10 times the reinforcement diameter.

We will demonstrate how to combine the single strut solution with uniform reinforcement. The structural element arrived at in this way is often termed a shear wall. Due to their importance in seismic design, shear walls have been the subject of many experimental and theoretical investigations in recent years.

4.9.2 Strut Solution Combined with Web Reinforcement

Since the shear wall is seldomly an isolated structural element, we will pose the design task as illustrated in Fig. 4.9.1. The vertical load P is transferred to the wall by means of an end beam and the wall transfers the force to a neighboring structure. The shear wall may have flanges at top and bottom. These are treated as stringers. The end beam might be subjected to normal stresses along the vertical face. For simplicity we assume these normal stresses to be statically equivalent to a compression normal force N acting in the middle point of the end beam. It will be easy to generalize to a normal force with an arbitrary point of application. Further it will be easy to generalize to tensile normal forces.

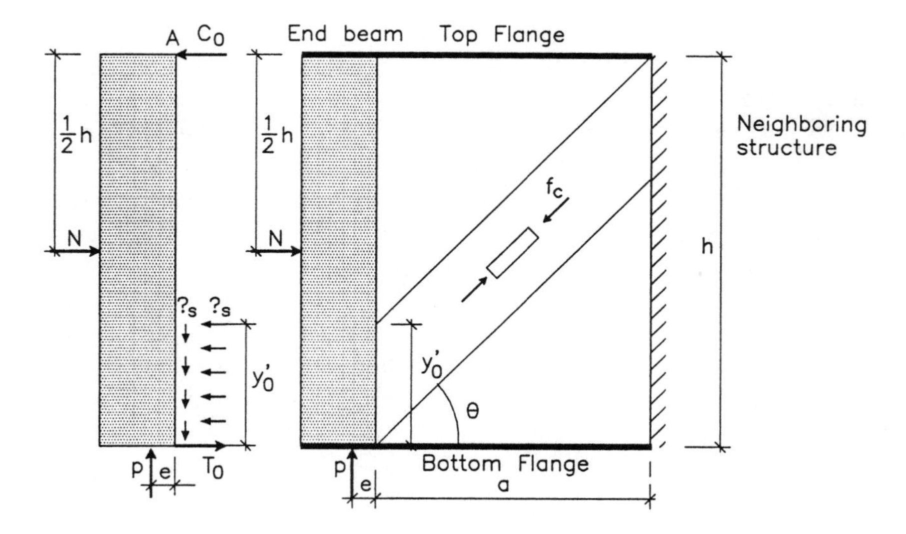

Figure 4.9.1 Shear wall with a strut.

By means of the formulas of Section 4.8 we may start finding out whether a strut alone is able to carry the load. If so we have $y_0 \leq \frac{1}{2}h$ (see Fig. 4.8.1), and we may calculate θ and thereafter the length y_0' shown in Fig. 4.9.1 to be

$$y_0' = \frac{y_0}{\cos^2\theta} \tag{4.9.1}$$

The strut delivers to the neighboring structure and to the end beams a shear stress τ_s and a compressive stress σ_s, both uniformly distributed along y_0'. These stresses are

$$\tau_s = f_c \cos\theta \sin\theta \tag{4.9.2}$$

$$\sigma_s = f_c \cos^2\theta \tag{4.9.3}$$

If the reinforcement is placed in the flanges we may find the forces T_0 and C_0 (see Fig. 4.9.1) by equilibrium equations for the end beam.

By taking moment about point A we get

$$T_0 = \frac{Pe}{h} + \frac{y_0' t}{h}\left(h - \tfrac{1}{2}y_0'\right) f_c \cos^2\theta - \tfrac{1}{2}N \tag{4.9.4}$$

and by projection

$$C_0 = T_0 - y_0' t f_c \cos^2\theta + N \tag{4.9.5}$$

The parameter e determines the position of P on the end beam and t is the thickness of the wall.

Notice that $T_0 > 0$ means a tensile force in the bottom flange at the end beam and $C_0 > 0$ means a compressive force in the top flange at the end beam. C_0 might well turn out to be negative which means that the top flange must be reinforced.

The end beam must, of course, be designed to carry the forces acting on it.

As mentioned in Section 4.6.3 a uniform minimum reinforcement in both vertical and horizontal direction should be supplied (cf. formula (4.6.18)). This reinforcement may be taken into account by using the following equations.

If P exceeds the value which may be carried by strut action, the triangular areas outside the strut may be utilized by assuming a diagonal compression field with the uniaxial concrete stress σ_c everywhere (see Fig. 4.9.2). The angle θ corresponds to $y_0 = \frac{1}{2}h$. The value of x_0 may be calculated by (4.8.6) and then θ by (4.8.1).

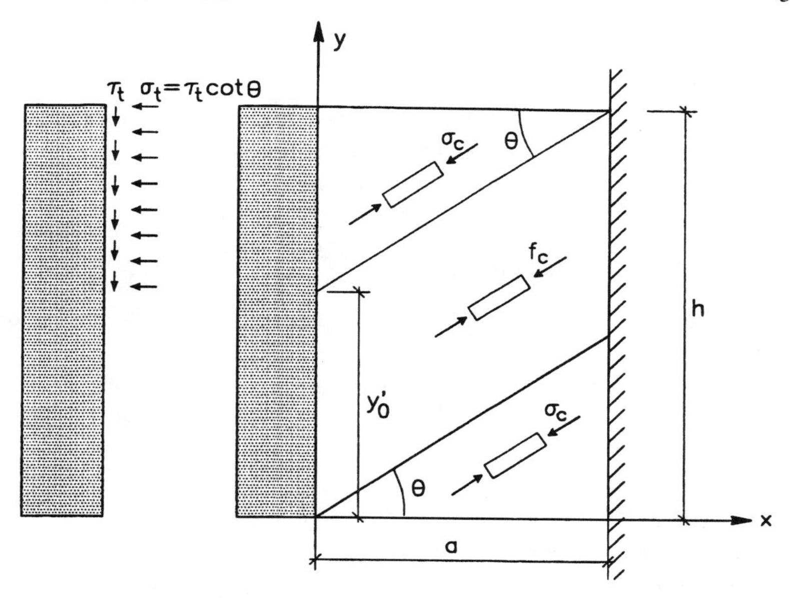

Figure 4.9.2 Shear wall with strut and diagonal compression field in triangular areas.

The concrete stresses in the triangular parts, referred to in the *x*, *y*-system shown in Fig. 4.9.2, are

$$\begin{aligned} \sigma_x &= -\sigma_c \cos^2\theta = -\tau_t \cot\theta \\ \sigma_y &= -\sigma_c \sin^2\theta = -\tau_t \tan\theta \\ \tau_t &= |\tau_{xy}| = \sigma_c \cos\theta \sin\theta \end{aligned} \tag{4.9.6}$$

having expressed σ_x and σ_y by the shear stress τ_t.

The force P_s which can be carried by the strut may be calculated by formula (4.8.7), i.e.,

$$P_s = \frac{1}{2} t h f_c \left[\sqrt{1 + \left(\frac{a}{h}\right)^2} - \frac{a}{h} \right] \tag{4.9.7}$$

The shear stress τ_t which has to be carried by the triangular areas is calculated by

$$\tau_t = \frac{P - P_s}{t(h - y_0')} \tag{4.9.8}$$

y_0' being determined by (4.9.1).

The vertical reinforcement required is

$$r_y f_{Yy} = \tau_t \tan\theta \tag{4.9.9}$$

where r_y is the reinforcement ratio in the vertical direction and f_{Yy} is the yield stress of the vertical reinforcement.

The concrete stress in the triangular areas is

$$\sigma_c = \frac{\tau_t}{\sin\theta\cos\theta} = \tau_t(\tan\theta + \cot\theta) \tag{4.9.10}$$

The maximum force which can be carried by this statical model is found by setting $\sigma_c = f_c$,

$$\begin{aligned} P &= (h - y_0')tf_c\cos\theta\sin\theta + P_s \\ &= atf_c\sin^2\theta + P_s = at\tau_t\tan\theta + P_s \end{aligned} \tag{4.9.11}$$

having utilized that $\tan\theta = (h - y_0')/a$ (see Fig. 4.9.2).

The stresses delivered to the end beam from the upper triangular area are shown in Fig. 4.9.2 (left). The equations for the flange forces T_0 and C_0 now read

$$\begin{aligned} T_0 = &\frac{Pe}{h} + \frac{y_0' t}{h}\left(h - \frac{1}{2}y_0'\right)f_c\cos^2\theta \\ &+ \frac{1}{2}\frac{(h - y_0')^2 t}{h}\tau_t\cot\theta - \frac{1}{2}N \end{aligned} \tag{4.9.12}$$

$$C_0 = T_0 - y_0' tf_c\cos^2\theta - (h - y_0')t\,\tau_t\cot\theta + N \tag{4.9.13}$$

Finally, if a uniform horizontal reinforcement is present we might imagine this to be stressed to the yield point everywhere without changing the stress field in the concrete. However, it will, of course, influence the flange forces. We find

$$\begin{aligned} T_0 = &\frac{Pe}{h} + \frac{y_0' t}{h}\left(h - \frac{1}{2}y_0'\right)f_c\cos^2\theta \\ &+ \frac{1}{2}\frac{(h - y_0')^2 t}{h}\tau_t\cot\theta - \frac{1}{2}r_x f_{Yx} h t - \frac{1}{2}N \end{aligned} \tag{4.9.14}$$

$$\begin{aligned} C_0 = &T_0 - y_0' tf_c\cos^2\theta - (h - y_0')t\,\tau_t\cot\theta \\ &+ r_x f_{Yx} h t + N \end{aligned} \tag{4.9.15}$$

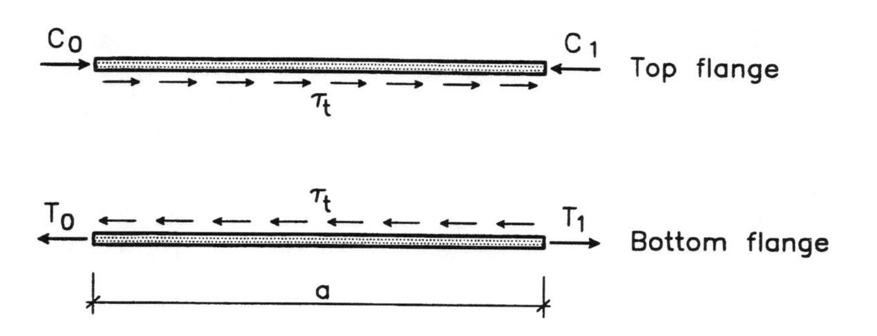

Figure 4.9.3 Forces in flanges.

r_x being the reinforcement ratio and f_{Yx} the yield stress of the horizontal reinforcement, respectively.

When the triangular areas are active the flange forces will vary. Since the shear stress τ_t is constant the forces vary linearly and their values at the point where the flanges meet the neighboring structure (see Fig. 4.9.3) will be

$$C_1 = C_0 + \tau_t at \tag{4.9.16}$$

$$T_1 = T_0 + \tau_t at \tag{4.9.17}$$

The flanges have to be designed to carry these forces, Normally, constant reinforcement should be used. It must be remembered that the force in the top flange might be a tensile force in the whole or in part of the flange.

If a prescribed uniform minimum reinforcement is utilized to carry the load, the formulas cannot be applied directly since the resulting strut geometry is not known. By estimating the strut angle θ, an iteration procedure normally will solve the problem in a few steps.

When the load P equals the value given by the right-hand side of (4.9.11) we have a diagonal compression field $\sigma_c = f_c$ corresponding to the angle θ in the whole wall. It may be shown that θ for the strut with maximum load-carrying capacity is always less than 45°. Additional load-carrying capacity may be obtained by changing the angle θ and letting it approach 45°. This is so because the maximum load-carrying capacity is obtained when $\sigma_c = f_c$ and $\theta = 45°$, i.e.,

$$P = P_{max} = \frac{1}{2} f_c ht \tag{4.9.18}$$

The diagonal compression field with an arbitrary angle θ and a uniaxial compression stress σ_c is shown in Fig. 4.9.4.

When the load to be carried is between the values (4.9.11) and (4.9.18) we may determine θ by the equations (4.9.6) of the diagonal compression field.

If the shear stress to be carried is $\tau = P/ht$, the angle θ of the uniaxial compression may be determined by the last formula in (4.9.6). For $\sigma_c = f_c$, $\tau = f_c \cos\theta \sin\theta = \frac{1}{2} f_c \sin 2\theta$, and thus

$$\sin 2\theta = \frac{2\tau}{f_c} \tag{4.9.19}$$

The necessary vertical reinforcement ratio is determined by (4.9.9), i.e.,

$$r_y f_{Yy} = \tau \tan\theta \tag{4.9.20}$$

The force T_0 in the tensile flange now becomes

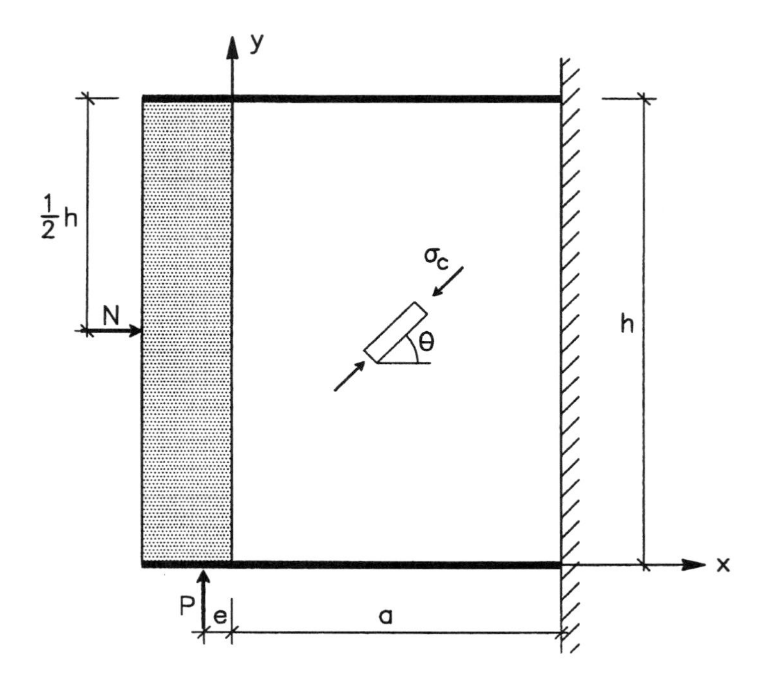

Figure 4.9.4 Diagonal compression field.

$$T_0 = \frac{Pe}{h} + \frac{1}{2} ht\,\tau \cot\theta - \frac{1}{2} r_x f_{Yx} ht - \frac{1}{2} N \qquad (4.9.21)$$

while the formula for the force C_0 in the compression flange reads

$$C_0 = T_0 - ht\,\tau\cot\theta + r_x f_{Yx}\, ht + N \qquad (4.9.22)$$

The variation of the flange forces along the shear span is determined by (4.9.16) and (4.9.17), replacing τ_t by τ.

If for $N = 0$ and $e = 0$ the horizontal reinforcement ratio r_x is chosen to be

$$r_x f_{Yx} = \tau \cot\theta \qquad (4.9.23)$$

then $T_0 = C_0 = 0$ because in this case the horizontal reinforcement will equilibrate the stresses in vertical sections from the diagonal compressive field.

It might be instructive to find the average shear stress $\tau = P/ht$ as a function of the vertical reinforcement ratio or, equivalently, the vertical reinforcement degree

$$\psi = \Phi_y = r_y \frac{f_{Yy}}{f_c} \qquad (4.9.24)$$

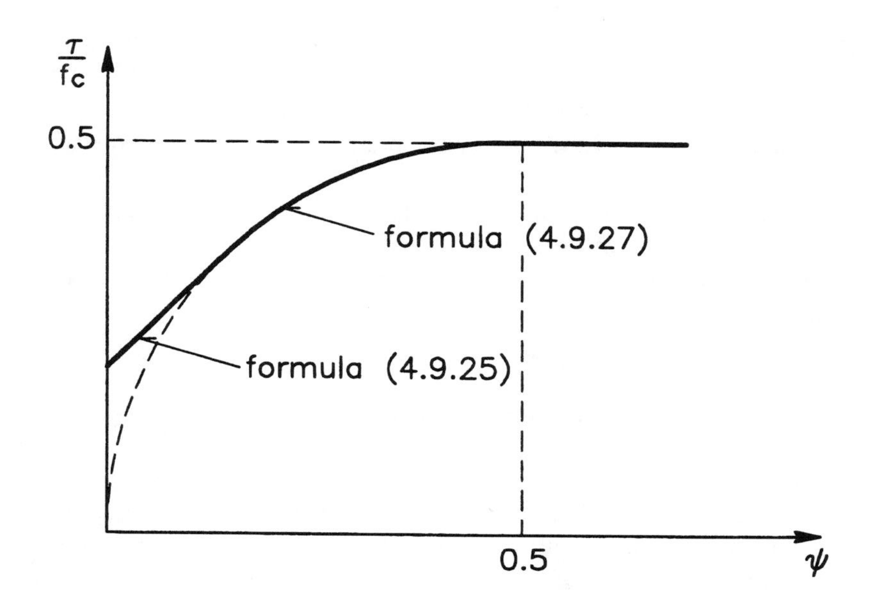

Figure 4.9.5 Maximum shear strength versus shear reinforcement degree.

By means of (4.9.9), (4.9.7) and (4.9.11) we find that the maximum average shear stress which can be carried is

$$\frac{\tau}{f_c} = \frac{1}{2}\left[\sqrt{1+\left(\frac{a}{h}\right)^2} - \frac{a}{h}\right] + \psi\,\frac{a}{h} \tag{4.9.25}$$

When the diagonal compression field with $\sigma_c = f_c$ is active in the whole web we have $r_y f_{Yy} = f_c \sin^2\theta$, i.e.,

$$\sin\theta = \sqrt{\psi} \tag{4.9.26}$$

Since $\tau = f_c \cos\theta \sin\theta$, we arrive at the formula

$$\frac{\tau}{f_c} = \sqrt{\psi(1-\psi)} \tag{4.9.27}$$

For $\psi = 0.5$ the shear capacity is $0.5 f_c$ and $\theta = 45°$. For higher values of ψ, the vertical reinforcement does not yield. It is clear that the shear capacity will be constant at $0.5 f_c$, the angle θ being constant at 45°.

This means that the shear capacity as a function of ψ is determined by (4.9.25), (4.9.27) and the formula for the maximum shear capacity $\tau = 0.5\, f_c$.

The resulting curve is depicted in Fig. 4.9.5. The straight line (4.9.25), the position of which is depending on a/h, is tangent to the circle (4.9.27), which continues in the straight line $\tau = 0.5\, f_c$. The solution derived is clearly a lower bound solution.

In Chapter 5 it is shown that corresponding upper bound solutions exist, which means that the solution is exact. Of course, it must be assumed that the end beam and the flanges are able to carry the forces acting on them.

Notice that the solution is also an exact solution for any normal force N; of course, assuming again that the end beam and the flanges are sufficiently strong.

4.9.3 Diagonal Compression Field Solution

The solution derived in the previous section is the one supplying the smallest amount of vertical reinforcement. However, often a more economical solution may be found using a diagonal compression field with a steeper inclination of the uniaxial compression. Then the vertical reinforcement will be increased and the horizontal reinforcement due to the shear will be decreased. This solution is already interesting for a shear span equal to the depth. A minimum reinforcement solution for the case considered has been developed in Chapter 5, Section 5.2.6. It has been assumed that $e = 0$ which we will assume here, too. In Section 5.2.6 the yield stresses may be different in web and flanges. Here it is assumed that all yield stresses are equal.

We may still refer to Fig. 4.9.4. The solution in Section 5.2.6 has been developed on the assumption that no horizontal web reinforcement is provided and that the flange reinforcement is constant. The solution is, however, the same, when we allow for horizontal reinforcement, except that in the solution in Section 5.2.6 the horizontal reinforcement is transferred to the flanges. This is natural for a slender beam but not for a shear wall. So we will look at the solution here from the shear wall point of view.

In a homogeneous stress field solution assuming no horizontal web reinforcement, there is only one redundant which might be taken as the normal stress in the web along vertical sections. We term this σ_0.

One part of the minimum reinforcement solution from Section 5.2.6 with no horizontal web reinforcement is

$$\begin{aligned} N &\le \tau h t & \sigma_0 &= \tau \\ \tau h t < N &\le \sqrt{2}\,\tau h t & \sigma_0 &= \frac{N}{ht} \\ \sqrt{2}\,\tau h t < N &\le N_\ell = 2\tau a t + \sqrt{2}\,\tau h t & \sigma_0 &= \sqrt{2}\,\tau \end{aligned} \tag{4.9.28}$$

For $N \le \tau ht$, $\sigma_0 = \tau$, which is more than or equal to what the normal force can supply. Therefore tensile reinforcement is necessary in the flanges to balance the end beam forces in the horizontal direction. In the limit when $N = \tau ht$, the forces balance out and no horizontal reinforcement in the flanges is required at the end beam.

In the intermediate case $\tau ht < N \le \sqrt{2}\,\tau ht$, $\sigma_0 = N/ht$, i.e., the normal force is carried by the web and no horizontal reinforcement in the flanges is necessary at the end beam. The stresses to reinforce for in the web are σ_0 and τ. When no horizontal reinforcement is provided in the web we have, according to the reinforcement formulas, $\tan\theta = \tau/\sigma_0$. In this N-interval the direction of the uniaxial compression changes from 45° to 35°.3 with horizontal when N is increasing.

In the third case, σ_0 is kept constant for increasing N. This means that the normal force is used to reduce the tensile reinforcement in the bottom flange, which is more economical than using it for decreasing the shear reinforcement. In the limit $N = N_\ell = 2\tau at + \sqrt{2}\,\tau ht$ the bottom flange will be in compression along the whole length.

For $N > N_\ell$, no flange reinforcement is necessary at all and we only have to supply vertical reinforcement in the web. Now the part of the normal force exceeding N_ℓ may be used to reduce the shear reinforcement.

In Section 5.2.6 it is shown that for

$$N_\ell < N \le \sigma_{0\max} ht + 2\tau at \qquad \sigma_0 = \sqrt{2}\,\tau + \frac{N - N_\ell}{ht} \tag{4.9.29}$$

where

$$\sigma_{0\max} = f_c\left[\frac{1}{2} + \sqrt{\frac{1}{4} - \left(\frac{\tau}{f_c}\right)^2}\right] \tag{4.9.30}$$

Higher normal forces than $\sigma_{0\max}\, ht + 2\tau at$ must be carried by the flanges, because when $\sigma_0 = \sigma_{0\max}$ the compressive strength of the web is fully utilized. In the calculation it has been tacitly assumed that the shear stress τ alone does not fully utilize the web crushing strength.

Now we consider horizontal reinforcement. In the first case, $N \le \tau ht$. When $\sigma_0 = \tau$ the total tensile force necessary for balancing the end beam forces is $\tau ht - N$, which corresponds to a uniform horizontal reinforcement in the web determined by

$$r_x f_Y = \tau - \frac{N}{ht} \tag{4.9.31}$$

The same result would, according to the reinforcement formulas, be found if the web is subjected to a normal stress $\sigma_x = -N/ht$, a normal stress $\sigma_y = 0$ and a shear stress τ. The vertical web reinforcement would also be the same, namely

$$r_y f_Y = \tau \tag{4.9.32}$$

and we would also find the same concrete stress

$$\sigma_c = 2\tau \tag{4.9.33}$$

The angle θ is 45°. Therefore the normal stress in the concrete in vertical sections would be τ.

The difference between the two solutions with or without horizontal reinforcement in the web is most clearly seen by considering the forces on the end beam. They are shown in Fig. 4.9.6.

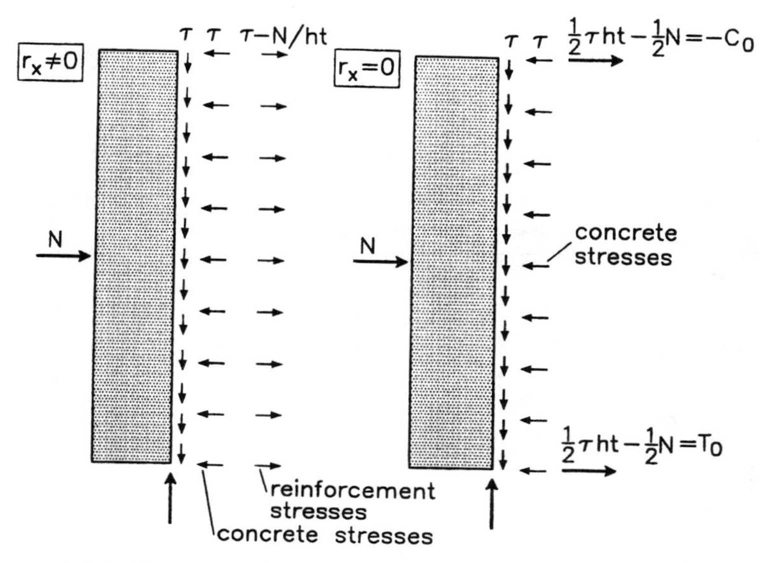

Figure 4.9.6 Forces on the end beam in a shear wall with and without horizontal web reinforcement.

In the other cases only vertical reinforcement is necessary. We have

$$\tan\theta = \frac{\tau}{\sigma_0} \tag{4.9.34}$$

The vertical reinforcement in the web is, according to the reinforcement formulas, determined by

$$r_y f_Y = \frac{\tau^2}{\sigma_0} \tag{4.9.35}$$

The concrete stress is

$$\sigma_c = \sigma_0\left(1 + \left(\frac{\tau}{\sigma_0}\right)^2\right) \tag{4.9.36}$$

Finally the flange forces at the end beam are

$$T_0 = \frac{1}{2}\sigma_0 ht - \frac{1}{2}N \tag{4.9.37}$$

$$C_0 = -\frac{1}{2}\sigma_0 ht + \frac{1}{2}N \tag{4.9.38}$$

The flange forces at the neighboring structure are determined as before.

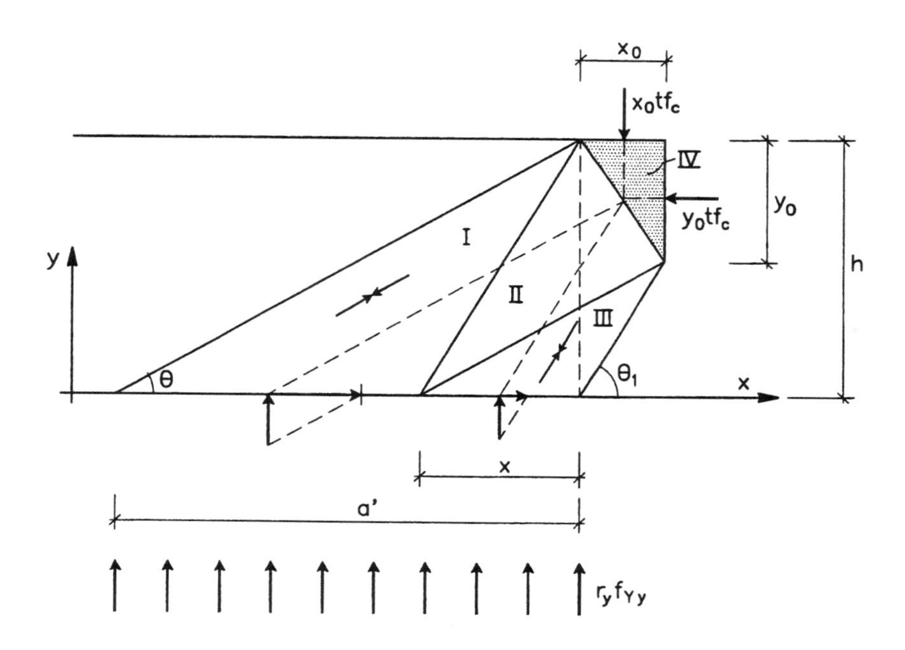

Figure 4.9.7 Modification of diagonal compression field.

It is possible to reduce the reinforcement in the stringers by a modification of the diagonal compression field when the concrete stress is less than f_c. The modification may be introduced both at maximum bending moment and at the support. The modification was suggested by J. F. Jensen [81.2] and introduced in the shear wall theory by Junying Liu [97.5]. The statics of the modified stress field are shown in Fig. 4.9.7. Along the horizontal bottom face the concrete is subjected to a uniform compressive stress $r_y f_{Yy}$ from the vertical stirrups, r_y being the reinforcement ratio and f_{Yy} the yield stress. To the bottom face shear stresses are also transferred from the stringer. The stresses at the bottom face are carried by uniaxial stress fields in areas I and III under the angles θ and θ_1 with horizontal, respectively. Areas I and III deliver forces to area II, which also receives forces from area IV. We imagine this area to be in biaxial hydrostatic compression with the stress f_c. Area II will also be in biaxial compression, but with two different principal stresses.

For the determination of x_o we have

$$x_o t f_c = r_y f_{Yy} a' t \tag{4.9.39}$$

Since $\tan\theta = h/a'$ we may write

$$\frac{x_o}{h} = \psi \cot\theta \tag{4.9.40}$$

ψ, as before, being the reinforcement degree $\psi = r_y f_{Yy}/f_c$. The length y_o may be determined by a moment equation about the middle point of a'. This reads

$$\frac{1}{2}(a' + x_o)x_o t f_c = y_o t f_c\left(h - \frac{1}{2}y_o\right) \tag{4.9.41}$$

Solving for y_o we get

$$\frac{y_o}{h} = 1 - \sqrt{1 - (1 + \psi)\,\psi \cot^2\theta} \tag{4.9.42}$$

Now

$$\tan\theta_1 = \frac{h - y_o}{x_o} = \frac{1 - \dfrac{y_o}{h}}{\dfrac{x_o}{h}} \tag{4.9.43}$$

Finally the length x may be determined by a projection equation on the horizontal. Noticing that the shear stress along x is $r_y f_{Yy} \cot\theta_1$, and along a'-x it is $r_y f_{Yy} \cot\theta$, we find

$$r_y f_{Yy} \cot\theta (a' - x) + r_y f_{Yy} \cot\theta_1 x = f_c y_o \tag{4.9.44}$$

or

$$\frac{x}{h} = \frac{\psi \cot^2\theta - \dfrac{y_o}{h}}{\psi(\cot\theta - \cot\theta_1)} \tag{4.9.45}$$

It was shown by J. F. Jensen that if the concrete stresses σ_c in areas I and III, determined by $\sigma_c/f_c = \psi(1 + \cot^2\theta)$ and $\sigma_c/f_c = \psi(1 + \cot^2\theta_1)$, respectively, are less than f_c (the first one decisive because $\cot\theta_1 < \cot\theta$) then the principal compressive stresses of area II will be less than f_c.

Fig. 4.9.7 is drawn approximately to scale for the case $\theta = 30°$ and $\psi = 0.2$.

When the stringer forces are determined on the basis of the modified stress field, the maximum tensile force will be smaller, because the shear stress along x is smaller than along $a' - x$. It should be noticed that if the modification of the stress field is carried out near the support it will influence the equilibrium of the end beam.

In the case where a strut is combined with diagonal compression fields in triangular areas a similar modification may be carried out (see [97.5]).

4.9.4 Effectiveness Factors

A large number of shear wall tests have been treated according to the plastic theory by Junying Liu [97.5].

The results are given in more detail in the next section. They show that when the shear walls have a shear span/depth ratio less than about 2 and are provided with web reinforcement in two directions, the effect of crack sliding seems to be unimportant and simple formulas for the effectiveness factor may be provided. This means that the structures have developed cracks in such a way that no failure mechanisms involving crack sliding have been decisive.

When pure strut solutions are used and web minimum reinforcement is not provided we must, of course, use the results of Section 4.8. This means that for $a/h < 0.75$, $\nu = \nu_0$ may be found, for instance, by using formula (4.8.24). For $0.75 \leq a/h \leq 2$, $\nu = \nu_s \, \nu_0$ with $\nu_s = 0.5$ and ν_0 determined by formula (4.8.24). For $a/h > 2$ the shear wall is treated as a beam (cf. Chapter 5). Alternatively, for any a/h-value, ν may be determined by either formula (4.8.21) or (4.8.23).

For webs with at least minimum reinforcement according to (4.6.18) in two perpendicular directions the following formulas due to Junying Liu [97.5] may be used.

$$\nu = 0.8 - \frac{f_c}{200} + 0.73 \frac{N}{A f_c} \qquad (f_c \leq 70 \text{ MPa}) \tag{4.9.46}$$

$$\nu = \frac{1.9}{f_c^{0.34}} + 0.73 \frac{N}{A f_c} \qquad (f_c > 70 \text{ MPa}) \tag{4.9.47}$$

In these formulas the normal force N is, as before, positive as compression. A is the total area of cross section of the wall. The formulas should not be used for tensile normal forces.

As usual, in the formulas given, f_c must everywhere be replaced by νf_c.

We conclude this section with a remark on minimum reinforcement. A normal force may, of course, to a certain extent replace reinforcement. Therefore, in the case of normal compression force on a shear wall the minimum reinforcement requirement may be relaxed in the direction of the normal force. It may be taken to be, say 40 to

50% of that given by (4.6.18), if the average stress from the normal force exceeds the effective tensile strength of concrete f_{tef}, which as before may be set at half the tensile strength measured on standard test specimens, i.e., $f_{tef} = 0.5 f_t = 0.5\sqrt{0.1 f_c}$ (f_c in MPa).

4.9.5 Test Results

It appears from the above that the design of shear walls using lower bound solutions is a rather simple task. To determine the load-carrying capacity of a shear wall with specified geometry and reinforcement is more complicated using hand calculations. However, this task may nowadays be accomplished using standard computer optimization routines. In this way a large number of shear wall tests have been treated by Junying Liu [97.5].

We shall not go into the details about reformulating the equations. Only the assumptions and the results will be dealt with.

In determining the shear capacity, the value of the depth h has been set to the total depth minus the cover. Bending capacity has been determined along the lines explained in Chapter 5.

It is assumed that the web reinforcement is uniform and that the flange reinforcement is constant. Since the compression flange may be acted upon by tensile forces, a condition regarding its tensile capacity may be included.

Similarly, conditions regarding the compression capacity of flanges may be included. Conditions regarding the strength of the end beam have not been considered. All diagonal compression field solutions have been modified along the lines described at the end of Section 4.9.3.

In Fig. 4.9.8 the results of an optimization are shown in the case $N = 0$ and $r_x = 0$. It has been assumed that the compression flange has sufficient tensile reinforcement. It has also been assumed that it has sufficient compression capacity. Further, the parameter e (see Fig. 4.9.1) is set at zero. In the figure, Φ means the longitudinal reinforcement degree defined by

$$\Phi = \frac{r_\ell f_{Y\ell}}{f_c} \tag{4.9.48}$$

Here r_ℓ is the longitudinal reinforcement ratio

$$r_\ell = \frac{A_\ell}{ht} \tag{4.9.49}$$

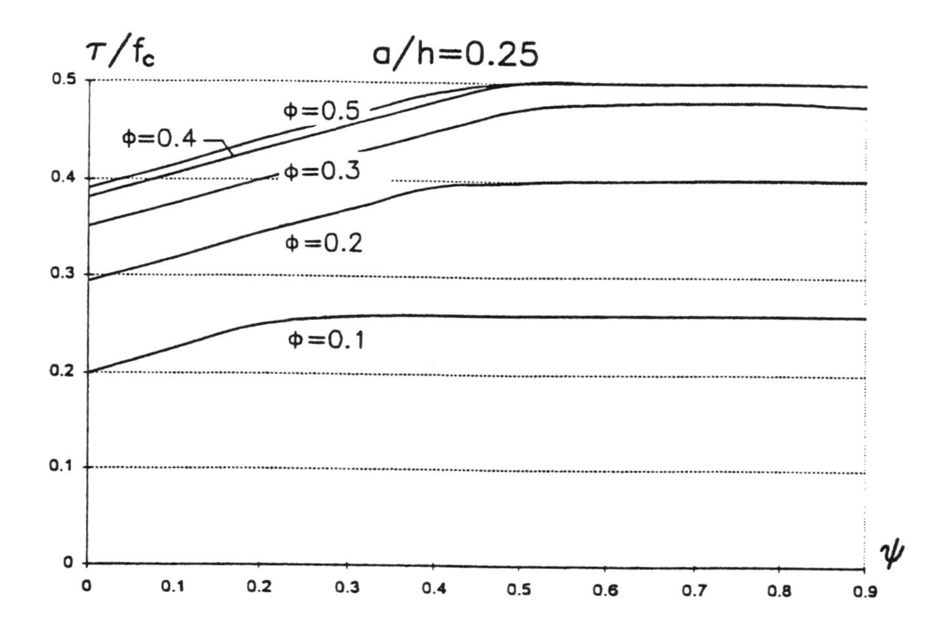

Figure 4.9.8 Shear capacity as a function of longitudinal reinforcement degree Φ and vertical web reinforcement degree ψ.

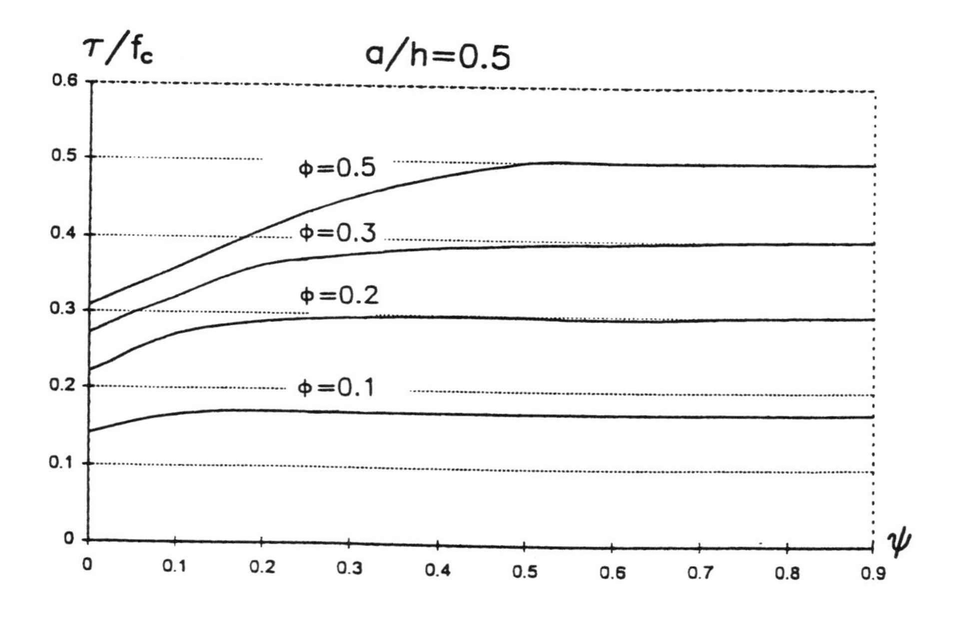

Figure 4.9.8 (continued).

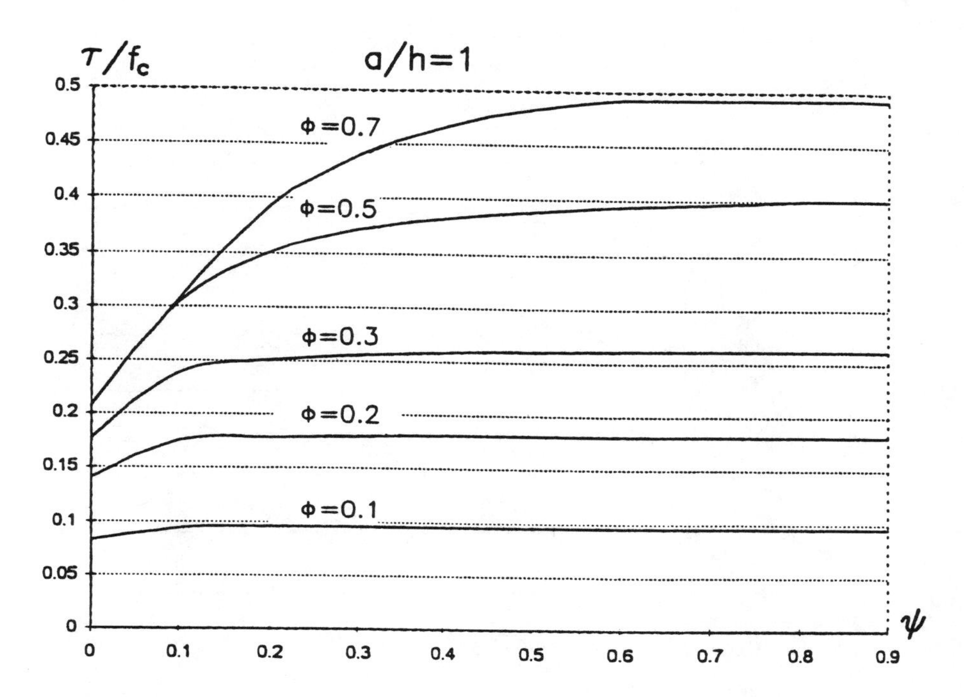

Figure 4.9.8 (continued).

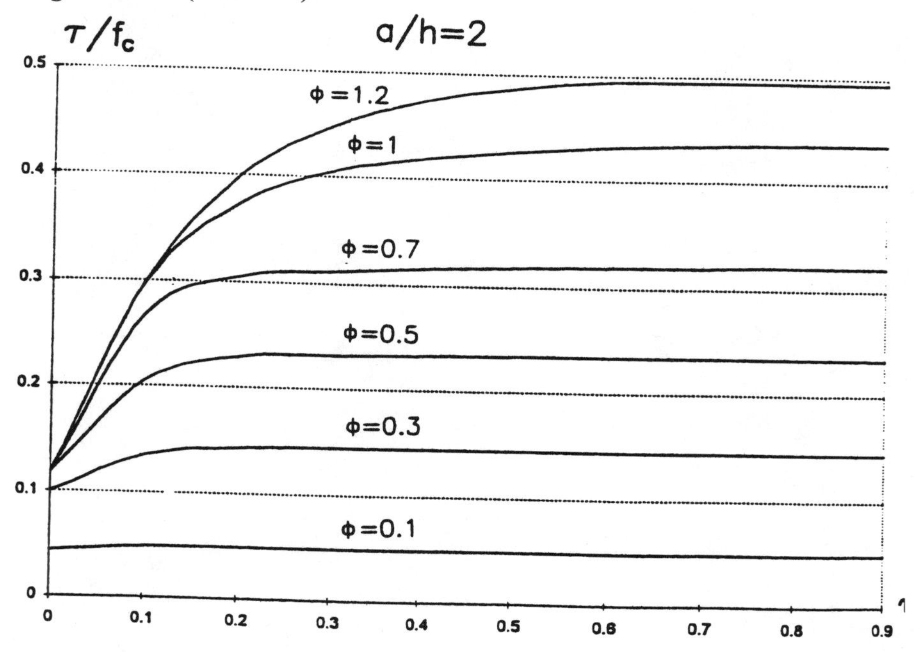

Figure 4.9.8 (continued).

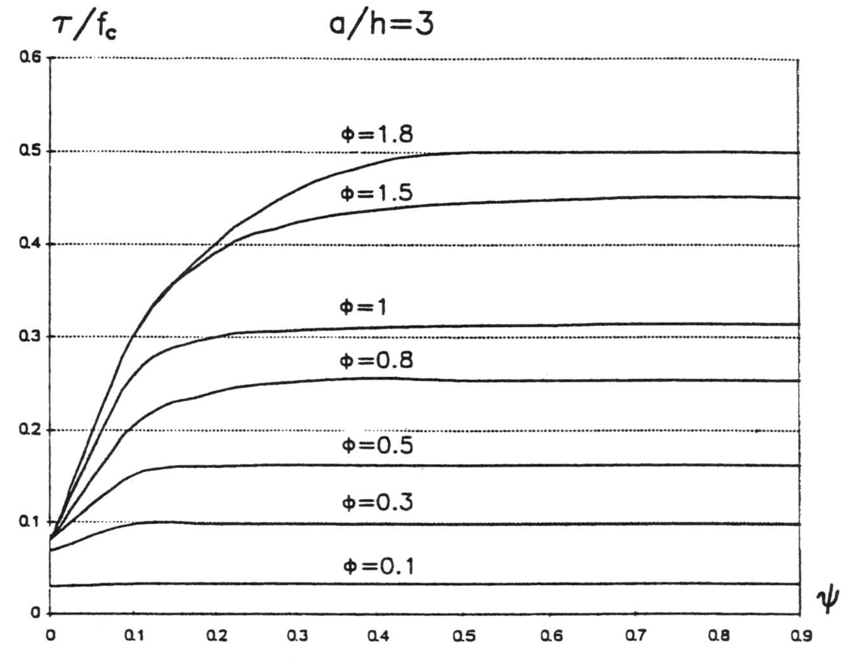

Figure 4.9.8 (continued).

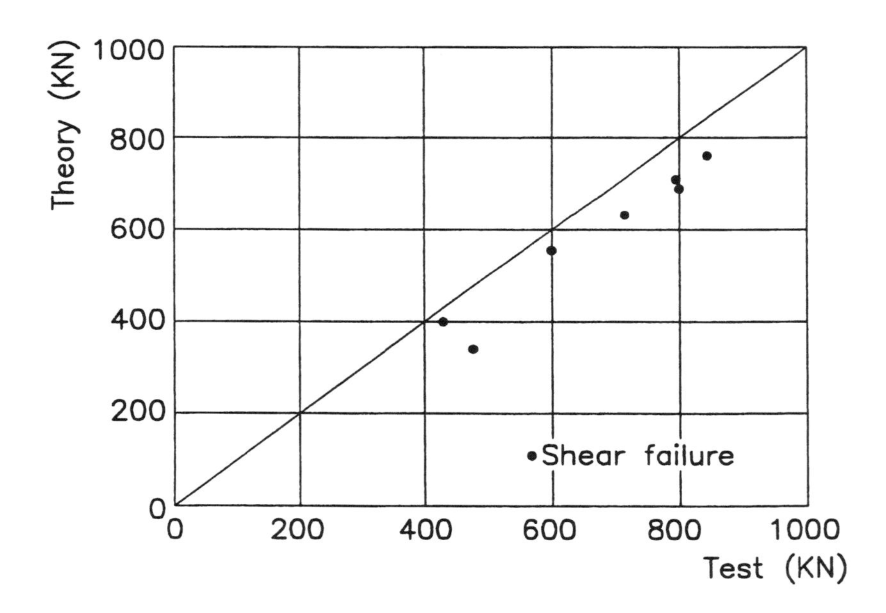

Figure 4.9.9 Shear wall tests by Gupta and Rangan [93.14] [94.14] [96.3].

A_ℓ being the tensile flange reinforcement area.

The curves show that when the bottom flange reinforcement is too small to obtain maximum shear capacity the effect of adding a vertical web reinforcement is sometimes small, sometimes even non-existent.

In Fig. 4.9.9 and Fig. 4.9.10 two test series are compared with optimized lower bound solutions. The first one was carried out at Curtin by Gupta and Rangan [93.14] [94.14] [96.3]. These tests were governed by shear failures. The second one was carried out by Maier and Thürlimann [85.7]. The agreement between theory and experiment is reasonable.

Fig. 4.9.11 shows a comparison between test results and theory for shear walls with rectangular sections. The coefficient of variation is about 12%. The range of the various parameters is

$$\begin{aligned}
&0.31 < \frac{a}{h} < 2.4 \\
&13 < f_c < 66 \text{ MPa} \\
&300 < f_{Y\ell} < 690 \text{ MPa} \\
&300 < f_{Yx} < 670 \text{ MPa} \\
&380 < f_{Yy} < 670 \text{ MPa} \\
&0.08 < r_\ell < 1.16\% \\
&0.22 < r_x < 2.9\% \quad (\text{some with } r_x = 0) \\
&0.25 < r_y < 1.6\% \quad (\text{some with } r_y = 0) \\
&0.0062 < \Phi_\ell = r_e \frac{f_{Y\ell}}{f_c} < 0.375 \\
&0.058 < \Phi_x = r_x \frac{f_{Yx}}{f_c} < 0.662 \\
&0.044 < \Phi_y = \psi = r_y \frac{f_{Yy}}{f_c} < 0.34
\end{aligned} \tag{4.9.50}$$

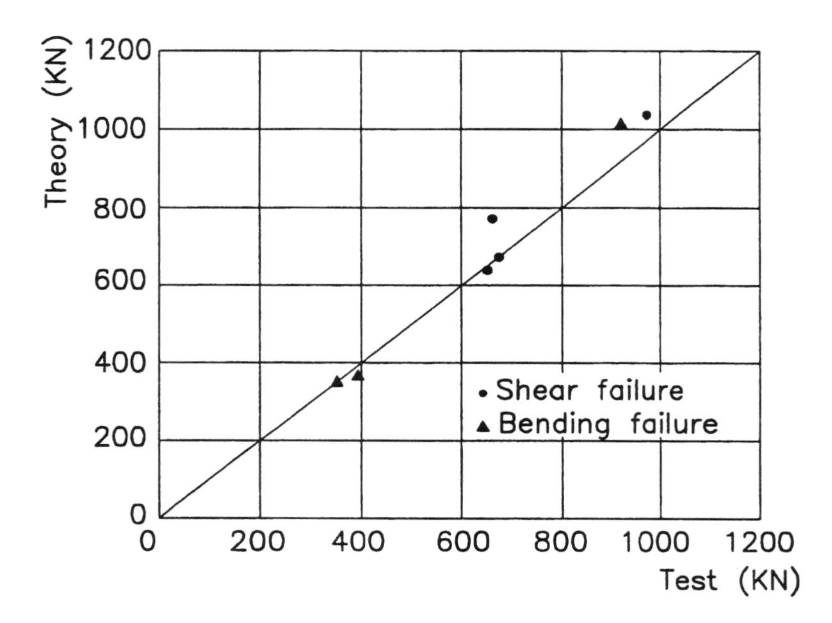

Figure 4.9.10 Shear wall tests by Maier and Thürlimann [85.7].

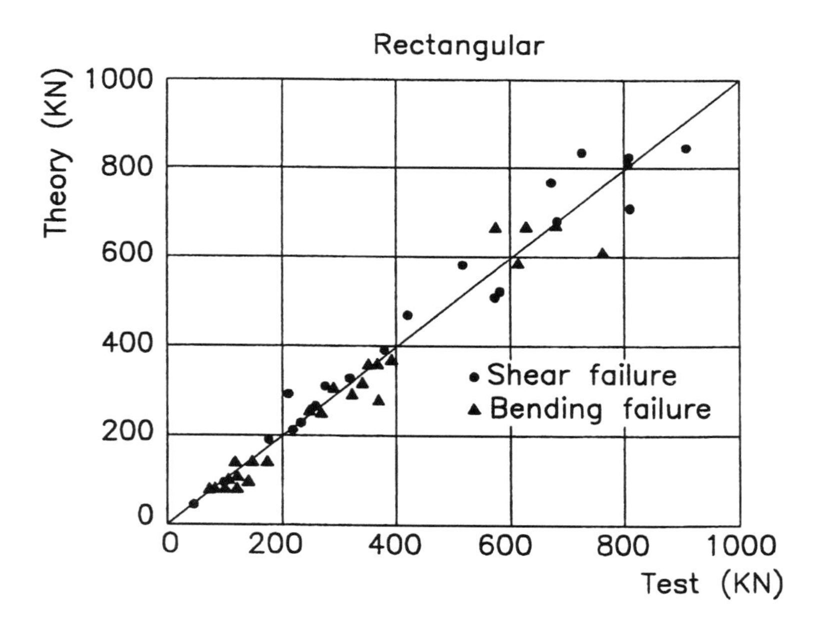

Figure 4.9.11 Comparison between test results and theory for shear walls with rectangular sections [97.5].

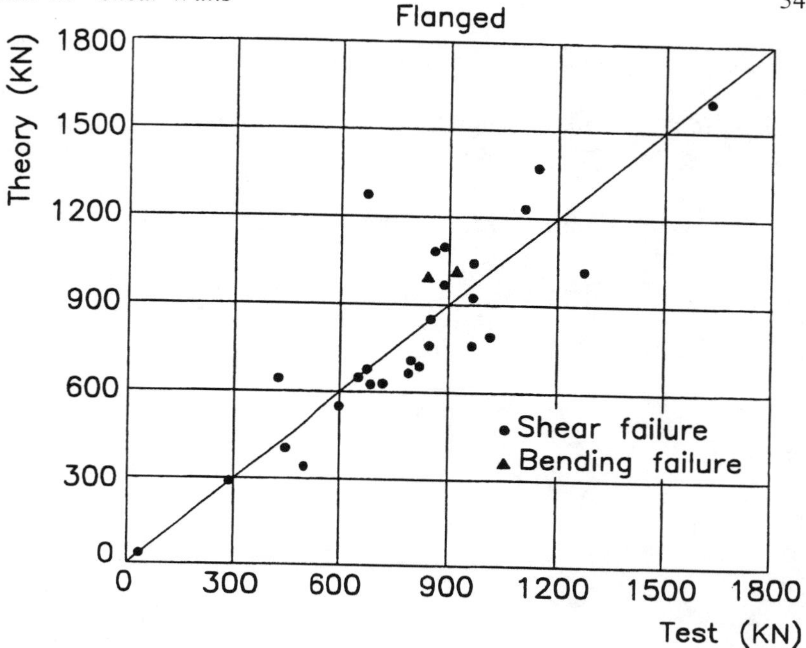

Figure 4.9.12 Comparison between test results and theory for shear walls with flanges [97.5].

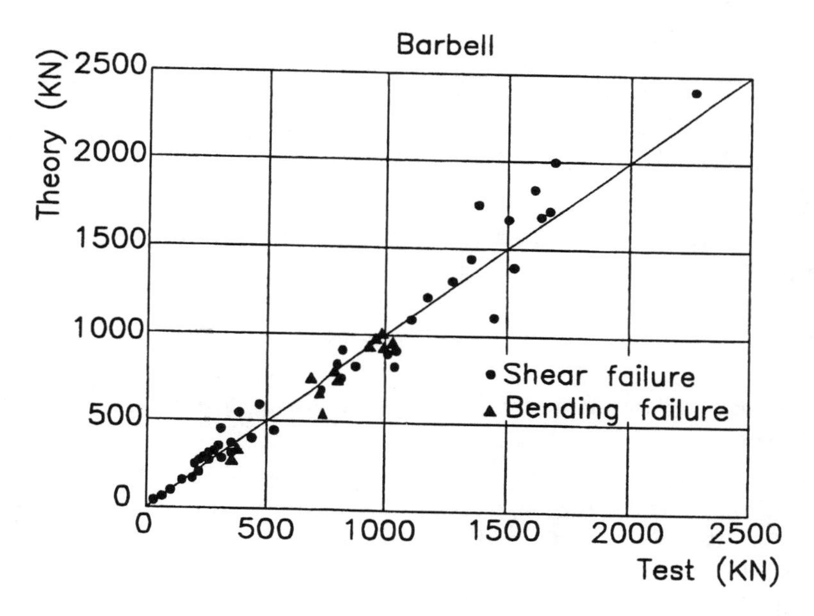

Figure 4.9.13 Comparison between test results and theory for shear walls with a barbell section [97.5].

Fig. 4.9.12 and Fig. 4.9.13 show the comparison between test results and theory for shear walls with flanges and with a barbell section (column like flanges), respectively. The coefficient of variation is about 20% for walls with flanges and 15% for walls with a barbell section.

For all types of shear walls the range of the parameters is

$$
\begin{aligned}
&0.23 < \frac{a}{h} < 2.4 \\
&13 < f_c < 137 \text{ MPa} \\
&208 < f_{Y\ell} < 1009 \text{ MPa} \\
&284 < f_{Yx} < 1420 \text{ MPa} \\
&284 < f_{Yy} < 1420 \text{ MPa} \\
&0.02 < r_\ell < 3.0\% \\
&0.07 < r_x < 2.9\% \qquad \text{(some with } r_x = 0) \\
&0.07 < r_y < 1.85\% \qquad \text{(some with } r_y = 0) \\
&0.0062 < \Phi_\ell < 0.85 \\
&0.018 < \Phi_x < 0.662 \\
&0.018 < \Phi_y = \psi < 0.36
\end{aligned}
\qquad (4.9.51)
$$

In cases of a large difference between yield strength and ultimate strength a mean value was used.

Not all the test specimens were provided with minimum cracking reinforcement in two directions. However, no clear difference in load-carrying capacity was found between elements satisfying the minimum requirement and those which did not. Thus it might be possible to reduce the minimum reinforcement when only load-carrying capacity is considered. However, minimum reinforcement is also provided to reduce crack widths. It is not possible on the basis of the available tests to show that reduction is possible in this respect.

It may be concluded that the theory works quite well.

4.10 HOMOGENEOUS REINFORCEMENT SOLUTIONS

4.10.1 Loads at the Top Face

For a number of practical cases, simple solutions which can be used to design the necessary amount of homogeneous reinforcement have been developed. The reason for the practical significance of these solutions is, as already mentioned, that disks usually have to be reinforced by combining a homogeneous reinforcement solution with a concentrated longitudinal reinforcement solution, because concentrated reinforcement alone may lead to undesirable crack distributions.

Further, cases exist where homogeneous reinforcement might even compare favorably with solutions involving concentrated reinforcement. Consider as an example a continuous disk. If we assume the same moment M above the support as at the maximum positive moment, the amount of reinforcement per span of length L, when concentrated reinforcement is provided at the top and bottom faces, is proportional to $2ML/h$, where h is the distance between the center of tension and compression, which here we may put equal to the depth of the disk. It is assumed that the reinforcement is not curtailed, which will be natural, as bending up of bars from the bottom face to the top face is precluded due to the depth. To this must be added a possible shear reinforcement or a uniformly distributed reinforcement, which should always be provided. If, instead, the horizontal reinforcement is uniformly distributed through the entire depth of the disk, the amount of reinforcement is proportional to $ML/\frac{1}{2}h = 2ML/h$, as the distance between the center of tension and compression in sections subjected to the largest moment is then $\frac{1}{2}h$. To this must be added a possible vertical reinforcement. It is seen that the same horizontal reinforcement will be necessary for resisting the bending moments.

The method that can be used for deriving homogeneous reinforcement solutions are illustrated by the case shown in Fig. 4.10.1. Parts 1, 2, and 3 are separated by lines of discontinuity. The stresses of part 2 are decisive for the vertical reinforcement, these being (see Example 2 in Section 4.5.3)

$$\sigma_x = 0 \qquad |\tau_{xy}| = \frac{1}{2}p\frac{L}{h}, \qquad \sigma_y = -p\left(1 - \frac{y_0}{h}\right) \tag{4.10.1}$$

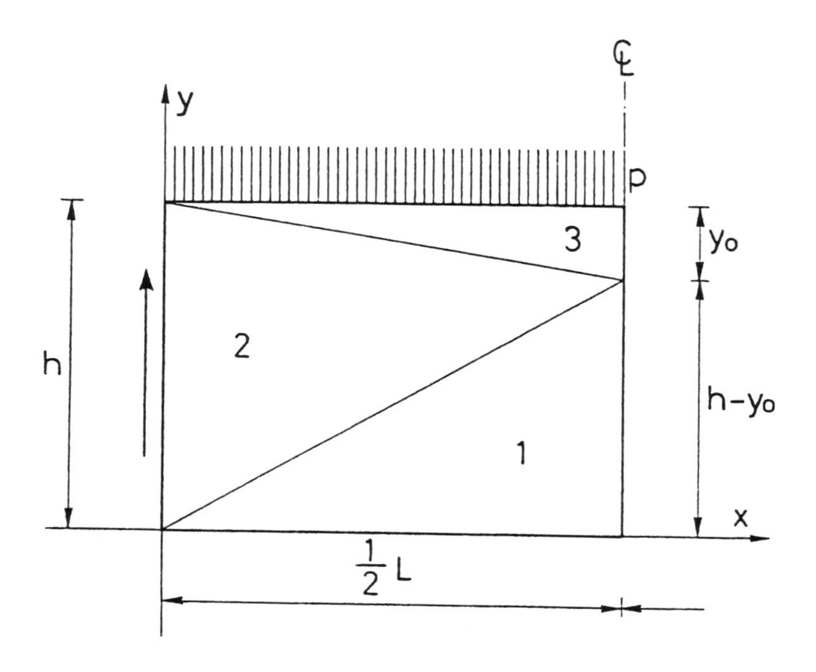

Figure 4.10.1 Discontinuous stress field for a disk supported along the lateral faces and loaded by a uniform load.

Inserting these values in the reinforcement formulas of Section 2.4.1, we get

$$\frac{A_{sx}f_Y}{t} = \sigma_x + \frac{\tau_{xy}^2}{|\sigma_y|} = \frac{\frac{1}{4}p(L/h)^2}{1 - y_0/h} \tag{4.10.2}$$

$$\frac{A_{sy}f_Y}{t} = 0 \tag{4.10.3}$$

In Example 2 of Section 4.5.3 we found that the necessary reinforcement degree Φ in the section with maximum moment could be determined by the formula (solving for Φ):

$$p = \frac{4(h - y_0)\,\Phi\, h f_c}{L^2} \tag{4.10.4}$$

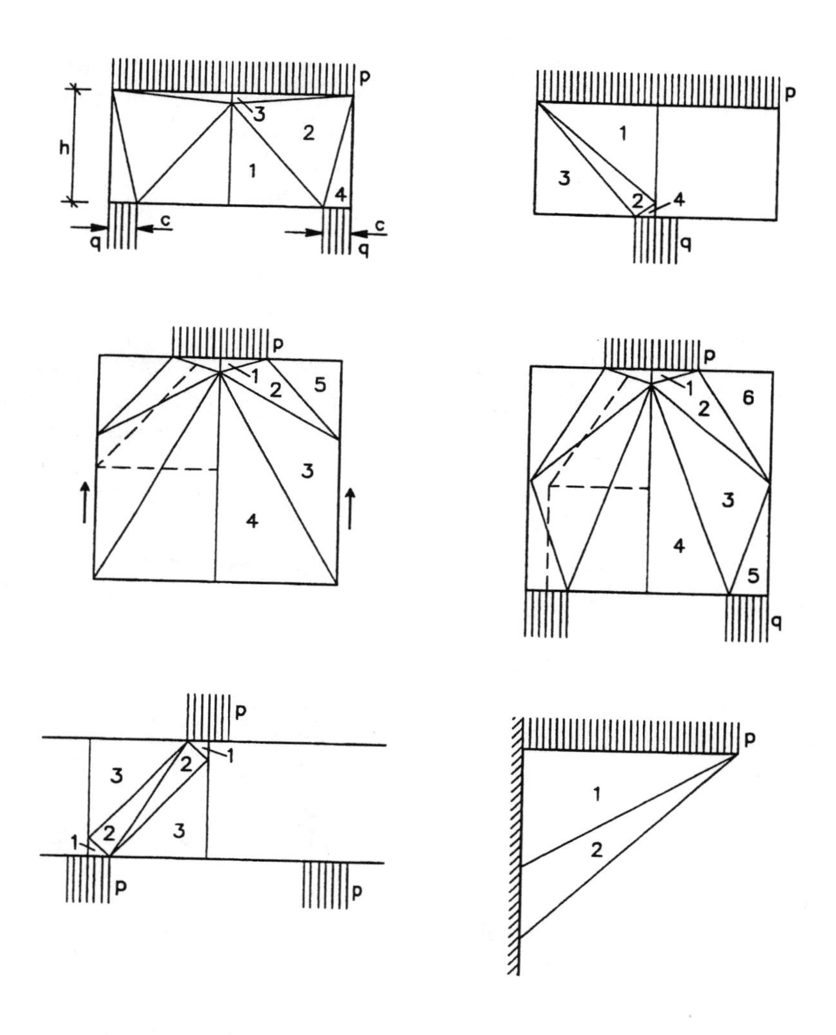

Figure 4.10.2 Homogeneous reinforcement solutions.

By comparing (4.10.2) with (4.10.4) we arrive at the interesting conclusion that in the x-direction the same amount of reinforcement suffices as that necessary to resist the bending stresses in the section with maximum bending moment. In the y-direction no reinforcement is necessary.

In [69.2] a large number of lower bound solutions of the type considered has been developed and it was shown that similar conclusions hold in all these cases, some of which are shown in Fig. 4.10.2. In the figure the subdivision into triangular areas with homogeneous stress fields has been indicated.

It seems reasonable to conclude that vertical reinforcement is theoretically unnecessary in the case of arbitrary loading on the top face and that the horizontal reinforcement may be determined so that the maximum bending moment can be resisted.

Regarding the concrete stresses it was shown in [69.2] that for isotropic reinforcement the maximum concrete stresses were always found in the section with maximum moment, at the supports and at the load application points. The concrete stresses therefore only have to be checked at these points.

Of course, the anchorage of the tensile reinforcement has to be checked as usual.

Minimum reinforcement is supplied according to formula (4.6.18).

These theoretical investigations cannot, however, exclude shear failures because of the relatively low ν-values for shear and because of the danger of crack sliding. For disks with low degrees of reinforcement shear failure is not likely to occur, neither with nor without crack sliding. An upper limit may be estimated on the basis of tests and practical experience. It may be assumed that if the reinforcement degree $\Phi \leq 0.6/\sqrt{f_c}$ (f_c in MPa), the above method may be used. For continuous disks the upper limit should be half as big. In these cases the ν-value is unimportant and may be set to 1.

For disks with higher reinforcement degrees a complete lower bound analysis may be carried out. The entire analysis is done using the ν-value $\nu = \nu_0$, where ν_0 may be found by means of formula (4.6.1) or (4.6.6). In areas where the concrete stress σ_c exceeds the value $\nu_s \nu_0 f_c$, ν_s as before being set at 0.5, additional reinforcement is supplied (cf. Section 4.6.4).

Alternatively, the shear capacity, taking into account crack sliding, may be checked using the upper bound technique developed in Chapter 5.

4.10.2 Loads at the Bottom Face

When the disk is subjected to a vertical load acting at the bottom face, the most obvious method of calculation is to transfer this load to the top face of the disk by a well-anchored tensile reinforcement (suspension reinforcement), and then calculate the remaining reinforcement as for disks loaded at the top face. The method may be utilized for arbitrary loads along the bottom face.

Similarly, if the load is acting somewhere between the top face and the bottom face a suspension reinforcement is supplied, which is able to transfer the load to the top face.

4.10.3 A Combination of Homogeneous and Concentrated Reinforcement

A disk will, as mentioned, in practice normally have to be reinforced by a uniformly distributed reinforcement, because a concentrated reinforcement often leads to excessive crack widths at some distance from the concentrated reinforcement. Therefore, it is natural to use a combination of homogeneous and concentrated reinforcement. Such solutions may be developed by superimposing the foregoing equilibrium solutions for homogeneous reinforcement and concentrated reinforcement, making use of the fact that the sum of two equilibrium solutions is again an equilibrium solution.

Since for both homogeneous reinforcement solutions and for concentrated reinforcement solutions we have shown that the necessary horizontal reinforcement may be found by analyzing the section with maximum bending moment, it seems reasonable to conclude that this also holds for the superimposed solutions.

The necessary horizontal reinforcement may therefore be calculated by first selecting the uniform reinforcement to be provided in both the horizontal and vertical directions. The value must be larger than or equal to that required by formula (4.6.18). Then by estimating the depth of the bending compression zone in the section or sections of maximum moment, a moment equation about the midpoint of the compression zone immediately furnishes the necessary amount of concentrated reinforcement. Then the estimate of the compression zone may be checked and if necessary the calculation is repeated.

When calculating the bending moment capacity of the uniform reinforcement, this reinforcement is, of course, assumed to be stressed to the yield point along the whole depth of the tensile zone.

As before, loads on the bottom face or between the bottom face and the top face are transferred to the top face by an extra suspension reinforcement, preferably in the form of closed stirrups.

A suspension reinforcement should also be supplied when disks supported along the end face are provided with concentrated reinforcement. The suspension reinforcement must be designed to transfer the part of the reaction which belongs to the concentrated reinforcement solution to the top face.

Strict control of the concrete stresses requires a complete lower bound analysis. We cannot be sure that a check of only the support stresses and the stresses at load application points is sufficient to show that the concrete stresses everywhere in the superimposed solutions may be resisted, even if this, as demonstrated above, is valid for the individual equilibrium solutions.

Fortunately, in practice, one often meets disks and deep beams for which only small reinforcement degrees are required. These structural elements are not likely to suffer a shear failure if they have been provided with minimum reinforcement. The upper limit of the reinforcement degree below which shear failure is unlikely may be estimated on the basis of tests and practical experience. It seems reasonable to assume that if the total reinforcement degree in a vertical section is less than about $0.4/\sqrt{f_c}$ (f_c in MPa), bending failure will govern the load-carrying capacity. Whether this limit is satisfied is easily checked when doing the bending analysis by requiring that the depth y_0 of the compression zone

$$\frac{y_0}{h} \leq \frac{0.4}{\sqrt{f_c}} \qquad (f_c \text{ in MPa}) \tag{4.10.5}$$

h being the total depth of the section.

For continuous disks the limit in (4.10.5) should be half as big.

If these conditions are satisfied it is only necessary to check the concrete stresses at the supports and load application points. Probably ν-values close to one may be used here. For the bending analysis the ν-value is unimportant when we are dealing with such small reinforcement degrees. It may be put at 1.

Regarding the choice of support moments in continuous disks we refer the remarks in Section 4.8.3.

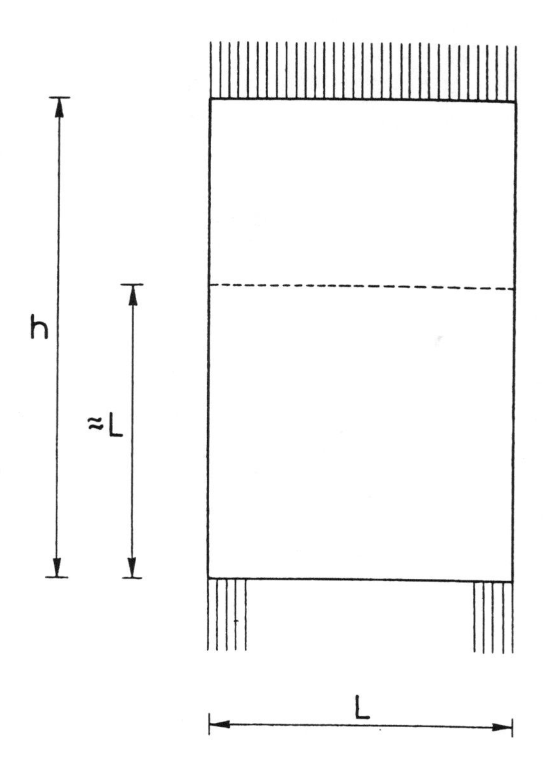

Figure 4.10.3 Very deep beam with uniform load.

For higher reinforcement degrees a complete lower bound analysis could be carried out. However, this will be cumbersome. In practice it will probably be sufficient, besides satisfying the above requirements, to check the shear capacity taking into account crack sliding by the upper bound technique developed in Chapter 5.

4.10.4 Very Deep Disks

For very deep disks, as shown in Figs. 4.10.3 and 4.10.4, the methods derived may be questionable, because the deeper the disk, the bigger the plastic deformations will be at the parts of the disk where yielding will first occur, so the crack widths will also be large at those parts. Bay [31.2], who has treated simply supported disks in the elastic state, recommends that disks with greater depth than span, having a uniformly distributed load at the top face, be designed so that the lowest square part is reinforced as a disk with uniformly distributed load at the top face. The reason for this is that the uppermost part in the elastic state will have almost a uniaxial stress field, and will consequently transfer the load only to the lower square part. Although theoretically the plastic solutions described above have no other limitations than those mentioned previously, it is recommended that this rule be complied with to avoid excessive cracking and plastic deformations before the final collapse load is reached.

A very deep disk loaded at the top with a concentrated load is similarly divided into three parts: I, II, and III (see Fig. 4.10.4). The uppermost part, I, is calculated as a disk loaded with a concentrated force and supported along the entire bottom face by a uniformly distributed reaction. The middle part, II, is assumed to have a uniaxial stress field; and the lowest part, III, is calculated as a disk with a uniformly distributed load at the top. Parts I and III are chosen to be square; that is, this method of calculation applies only for $h \geq 2L$.

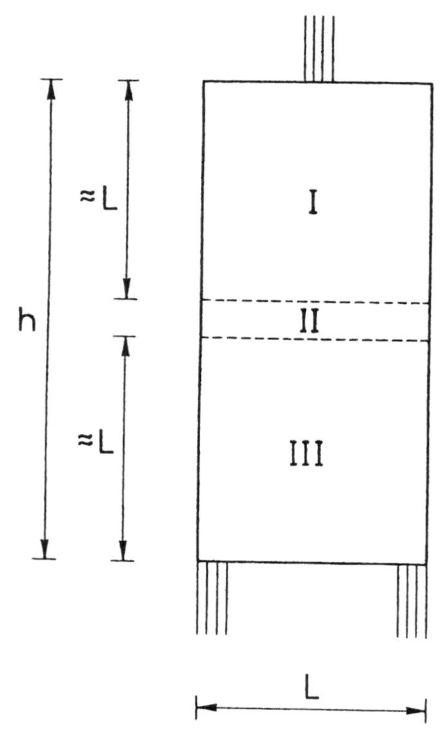

Figure 4.10.4 Very deep beam loaded by a concentrated force.

However, a similar method may be used for $L \le h \le 2L$. The disk is, for example, divided into a lower square part and an upper part, the depth of which will then be between L and $2L$. The upper part is assumed to load the lower part along a distance symmetrically around the line of loading equal to the depth of the upper part.

For very deep disks loaded along the lower face, we similarly assume that only the lowest square part carries the loads. It is, of course, important that the load can be transferred to the top of this part of the disk by supplying a suspension reinforcement.

For very deep disks it is very important to ensure that stability failure does not occur for a smaller load than the desired collapse load. These remarks also apply to disks in general.

4.11 DESIGN ACCORDING TO THE ELASTIC THEORY

We conclude the present chapter on the plastic theory of disks with some remarks on the use of the linear theory of elasticity in general and especially in disk design (cf. Section 2.4.6).

In the early days of reinforced concrete theory the reinforcement design was often based on the stress field in an uncracked and unreinforced structure having the same geometry and loading as the structure to be designed. The reinforcement was designed by investigating a suitable number of sections. In these sections the resultant of the tensile stresses was calculated and a reinforcement having the same resultant in magnitude and position, when utilized to the design value of the yield stress, was provided.

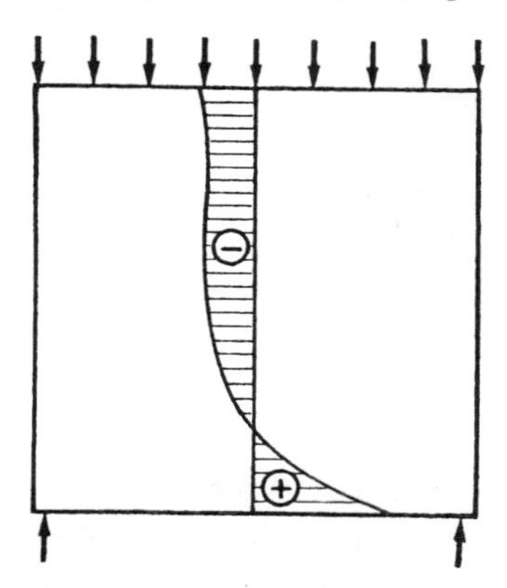

Figure 4.11.1 Normal stress distribution in a deep beam according to the elastic solution.

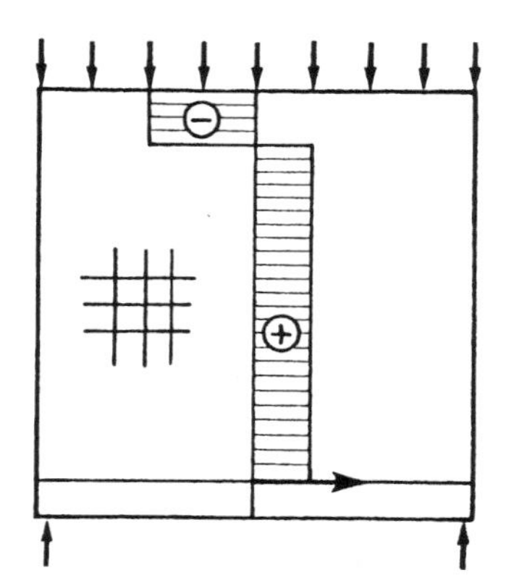

Figure 4.11.2 Normal stress distribution in a deep beam according to the plastic solution.

Instead of analyzing certain sections the reinforcement formulas of Section 2.4 may be used. In principle this method can be considered a lower bound method since the stress field used is an equilibrium solution. This method is often used in practice, i.e., the sectional forces or the stress field to design the reinforcement for are determined by linear elastic, uncracked theory, and the reinforcement is designed by plastic theory or other methods reflecting the behavior at failure. From a lower bound point of view there are, in principle, no objections against this method. Such design methods still have to be used for structures for which a rational failure theory has not yet been developed.

However, the method has some disadvantages. First, it is often very uneconomical, sometimes resulting in the use of several times the amount of reinforcement necessary when compared with that indicated by a rational failure theory. Second, the reinforcement layout often becomes very complicated and therefore very expensive to carry out. However, the method has the advantage of giving the largest amount of reinforcement at points of the structure where cracks will first appear.

The different results obtained by using the two approaches in disk design can be illustrated by considering a deep beam with a depth/span ratio of the order of 1. For uniform loading at the top face, the elastic normal stress distribution in the middle section is sketched in Fig. 4.11.1. The lever arm is approximately half the depth. If the disk

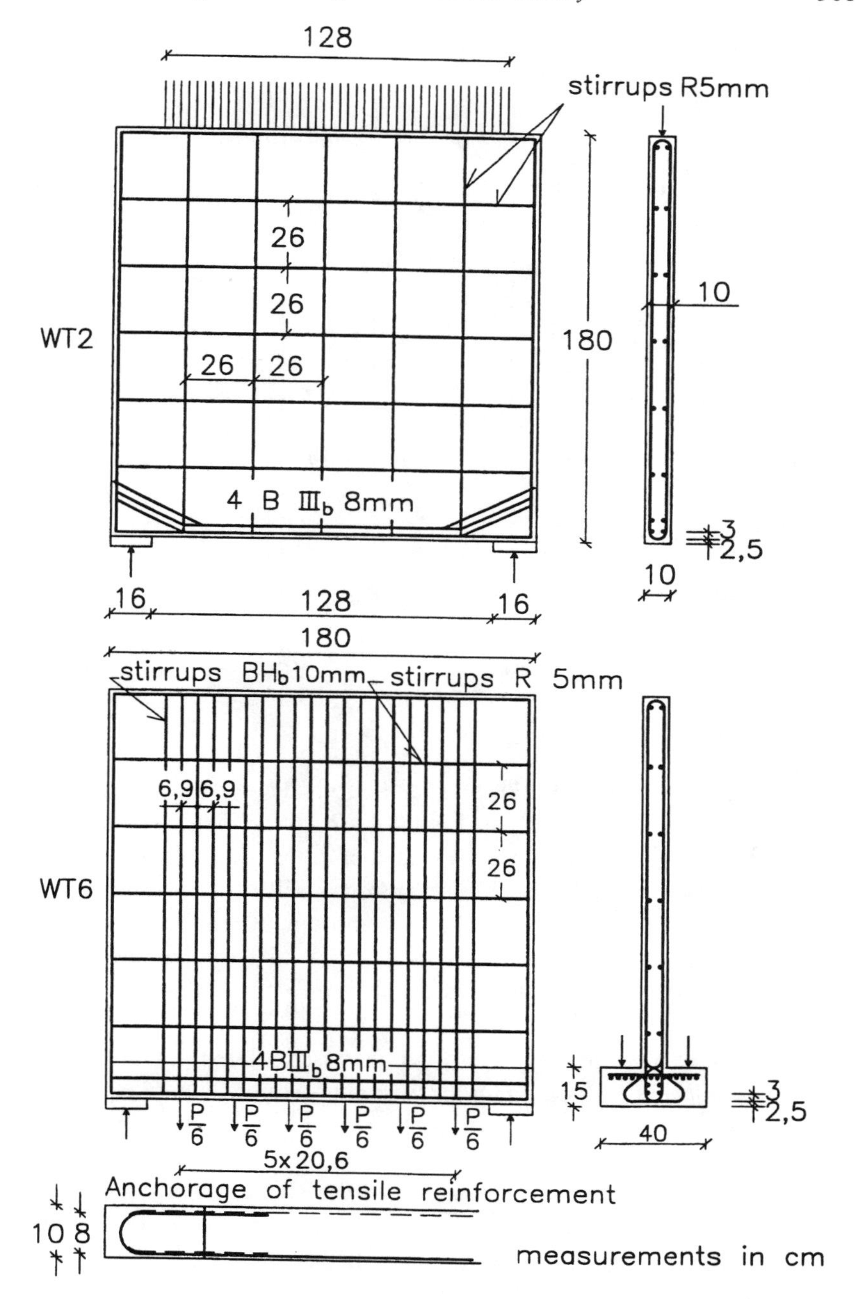

Figure 4.11.3 Deep beam tests by Leonhardt & Walther [66.7]. The reinforcement stress σ_s in the bottom reinforcement as a function of the bending moment M in the middle section (see also next page).

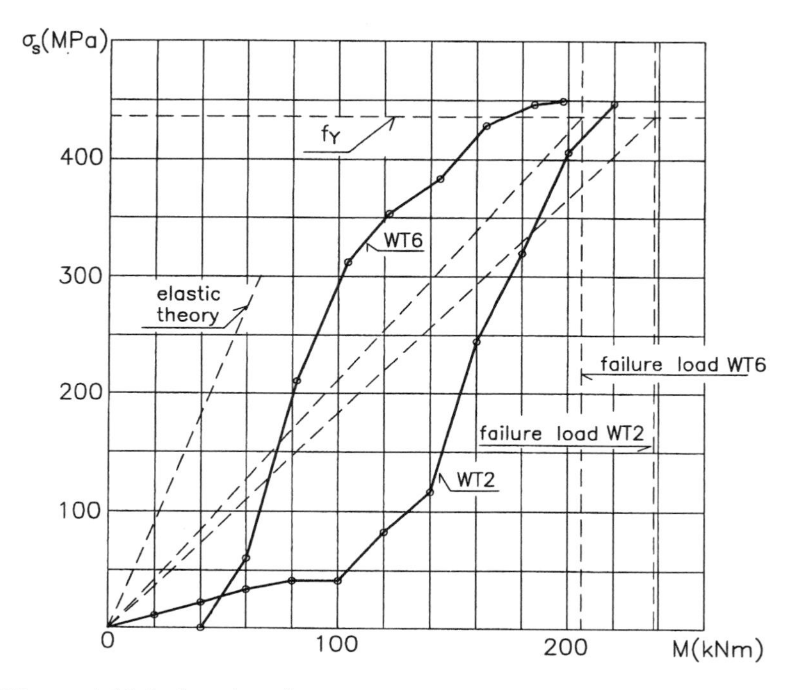

Figure 4.11.3 (continued).

is reinforced according to the plastic theory, the stress distribution at failure will be as shown in Fig. 4.11.2. Even if the uniformly distributed reinforcement corresponds to a rather small reinforcement ratio, its contribution to the failure moment will often be significant.

It appears that the difference in the amount of bending reinforcement between the two methods may be considerable, the elastic method providing up to twice as much reinforcement as the plastic theory.

One might fear that the reinforcement calculated by the plastic theory would be judged to be insufficient if the behavior for the working load is considered. However, it has been clearly demonstrated by experiments (see [69.2]), that even for the working load, the stresses found by using the lever arm of the elastic theory will be far too large; that is, the elastic theory gives a poor description of the behavior for the working load in all respects.

In Fig. 4.11.3 the stress σ_s measured in the bottom reinforcement in two tests carried out by Leonhardt and Walther [66.7] is shown as a function of the bending moment M in the middle section. The deep beams tested are also shown. In the figure the curve named "elastic theory" indicates the stress calculated by using the lever arm of linear elastic theory assuming uncracked sections. It is evident that the stress calculated in this way is completely wrong. A better estimate, but still not very good at least for the first loading, would be simply to scale down linearly from the failure moment to the moment of the working load as indicated in the figure.

This and many other examples show that the elastic theory should be used only if the behavior of the structural element to be designed cannot be analyzed by more rational methods.

Chapter 5

BEAMS

In this chapter the plastic theory of reinforced concrete beams is treated. Emphasis is laid on shear and torsion problems. Pure bending is described only briefly, since the solution to this problem is treated extensively by other methods in the literature [65.12] and in design codes. In the majority of this chapter concrete is identified as a modified Coulomb material with tensile strength equal to zero.

5.1 BEAMS IN BENDING

Load-carrying capacity. Consider the case of a beam with a rectangular cross section having depth h and width b stressed to pure bending. The reinforcement is a tensile reinforcement with area A_s. The effective depth is d (see Fig. 5.1.1). If the tensile reinforcement is stressed to yielding, the stress distribution at failure will be as shown in the figure. By projection,

$$A_s f_Y = y_0 b f_c \tag{5.1.1}$$

Introducing the notation

$$\Phi = \frac{A_s f_Y}{b d f_c} = r \frac{f_Y}{f_c} \tag{5.1.2}$$

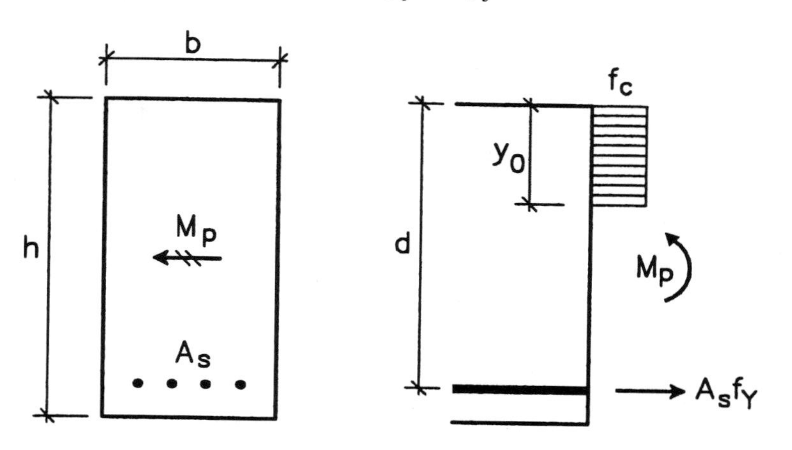

Figure 5.1.1 Normal stress distribution at the yield moment in a rectangular section (normally reinforced section).

where r is the reinforcement ratio and Φ is the degree of reinforcement, we find that

$$\frac{y_0}{d} = \Phi \tag{5.1.3}$$

The yield moment in pure bending M_p can then be determined:

$$M_p = A_s f_Y \left(d - \frac{1}{2} y_0\right) = \left(1 - \frac{1}{2}\Phi\right) A_s f_Y d$$

$$= \left(1 - \frac{1}{2}\Phi\right)\Phi\, b d^2 f_c \tag{5.1.4}$$

The coefficient $(1 - \frac{1}{2}\Phi)$ varies only slightly in cases of a low degree of reinforcement and the yield moment is then mainly proportional to A_s. Proportioning on the basis of a known yield moment should always be made down to smaller areas of reinforcement, since this is on the safe side.

The solution described above is valid only when $y_0 \leq d$ (i.e., $\Phi \leq 1$). If $\Phi > 1$, the stress distribution at failure will be as shown in Fig. 5.1.2. In this case $y_0 = d$ and the reinforcement will not yield. The failure moment will be

$$M_p = \frac{1}{2} b d^2 f_c \tag{5.1.5}$$

Beams with $\Phi \leq 1$ are called normally reinforced beams and beams with $\Phi > 1$ are called overreinforced beams.

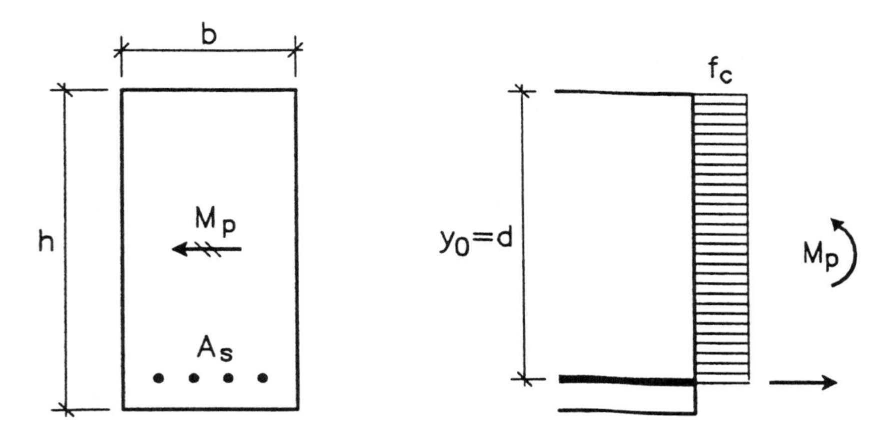

Figure 5.1.2 Normal stress distribution at the failure moment in a rectangular section (overreinforced section).

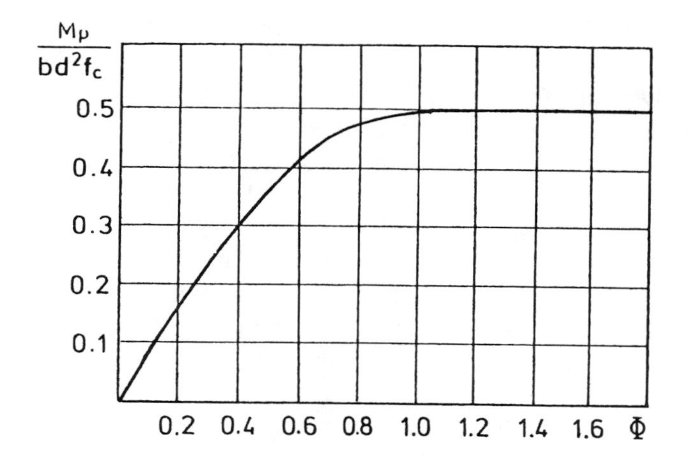

Figure 5.1.3 Yield moment versus reinforcement degree.

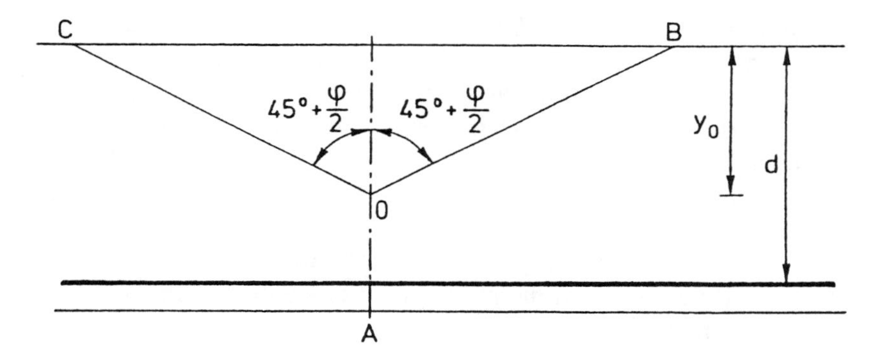

Figure 5.1.4 Failure mechanism at pure bending.

The complete solution can be written

$$M_p = \begin{cases} \left(1 - \frac{1}{2}\Phi\right)\Phi\, bd^2 f_c & \text{for } \Phi \le 1 \\ \frac{1}{2} bd^2 f_c & \text{for } \Phi > 1 \end{cases} \tag{5.1.6}$$

The result is illustrated in Fig. 5.1.3.

To show that the solution is an exact plastic solution, a geometrically admissible strain field corresponding to the stress field has to be found. Many such strain fields exist. One is illustrated in Fig. 5.1.4. It is based on the Rankine field treated in Section 3.5.5 (see

Fig. 3.5.14). The part of the beam outside AOB is rotating as a rigid body at an angle about O, and the part outside AOC is rotating as a rigid body about O at the same angle but in the opposite direction. The strain field in area OBC was described in Section 3.5.5. It is left to the reader to write down the work equation by using the dissipation formula (3.5.44) and to minimize the resulting expression for M_p with respect to y_0.

It is easy, along the same lines, to calculate the yield moment for other sections with tensile reinforcement only and for sections having both tensile and compression reinforcement. It is also easy to calculate the yield moment as a function of a normal force in the section.

In Box 5.1.1 the complete set of formulas for bending in rectangular sections with both tensile and compression reinforcement and with normal force has been given. (In the box M has been used instead of M_p.) The formulas are due to Bent Feddersen [85.2].

Effectiveness factors. The plastic theory of beams in bending is unique in one respect: it is the only known case where it is possible to calculate the effective strength of concrete. If a linear longitudinal strain distribution over the section and a uniaxial stress-strain curve for concrete in compression is assumed, it is possible to calculate the moment-curvature relationship. With a stress-strain curve of the type shown in Fig. 2.1.36, the curve has a maximum which can be defined as the load-carrying capacity of the section. By comparing the results of such calculations with the plastic solutions with f_c replaced by νf_c, ν can be evaluated.

For pure bending of a rectangular section with tensile reinforcement only, ν has been determined in this way by Exner [83.2], using stress-strain curves measured by Wang et al. [78.21]. Some of the results are shown in Fig. 5.1.5. It turns out that ν is a function of the uniaxial compressive strength f_c, the yield stress of the reinforcement f_Y, and the reinforcement ratio r. For small values of r the yield stress is reached in the reinforcement and in this region (normally reinforced beams), ν is independent of r and decreases with increasing concrete strength. The curves have a minimum, and when this point is passed (overreinforced beams), ν increases with increasing r. It appears that ν is rather close to 1 for small reinforcement ratios and low concrete strengths.

For practical purposes ν may be taken as the minimum value. This value may be calculated approximately by the simple formula

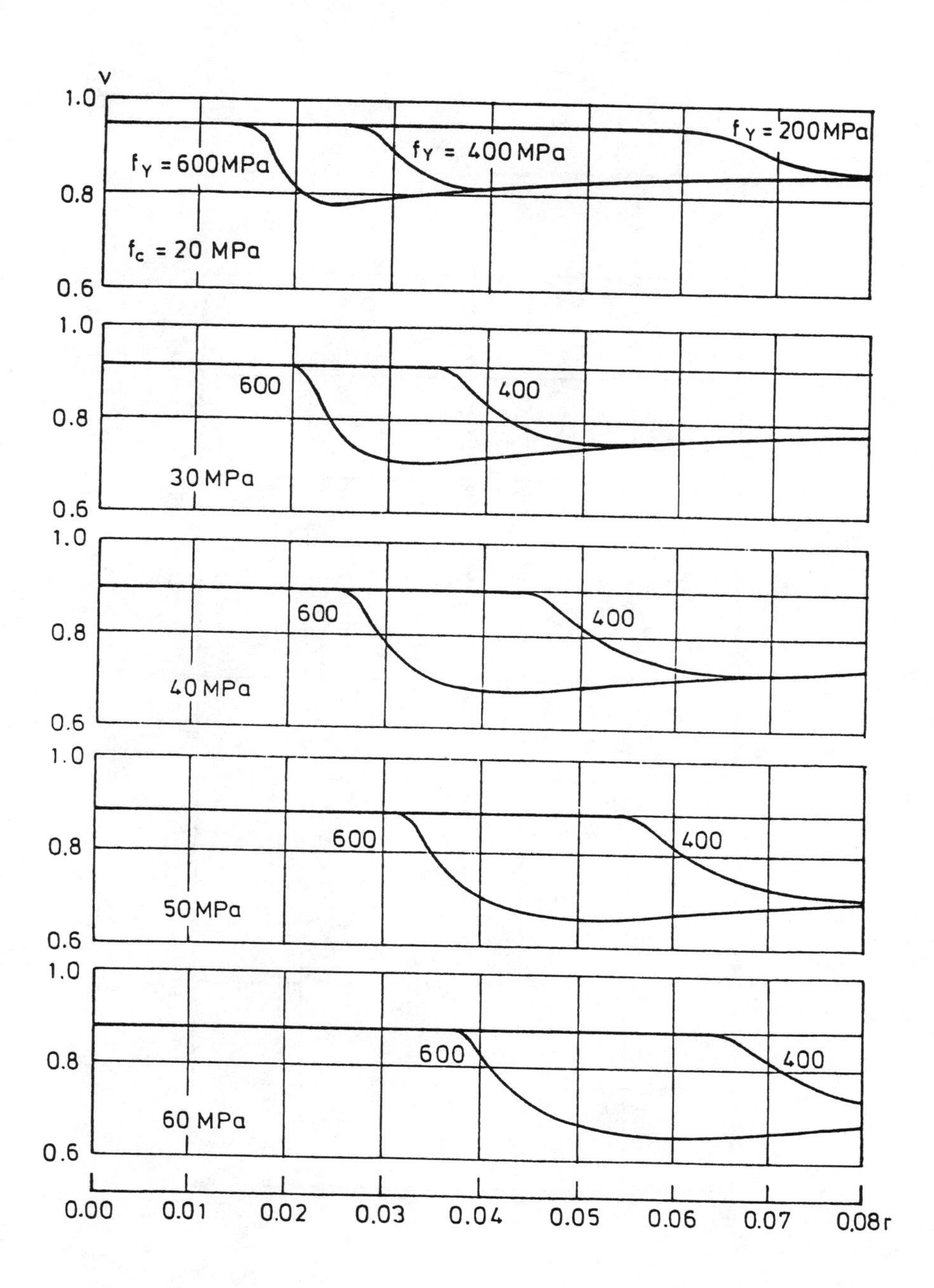

Figure 5.1.5 Effectiveness factor at pure bending of a rectangular section versus yield strength, compressive strength and reinforcement ratio.

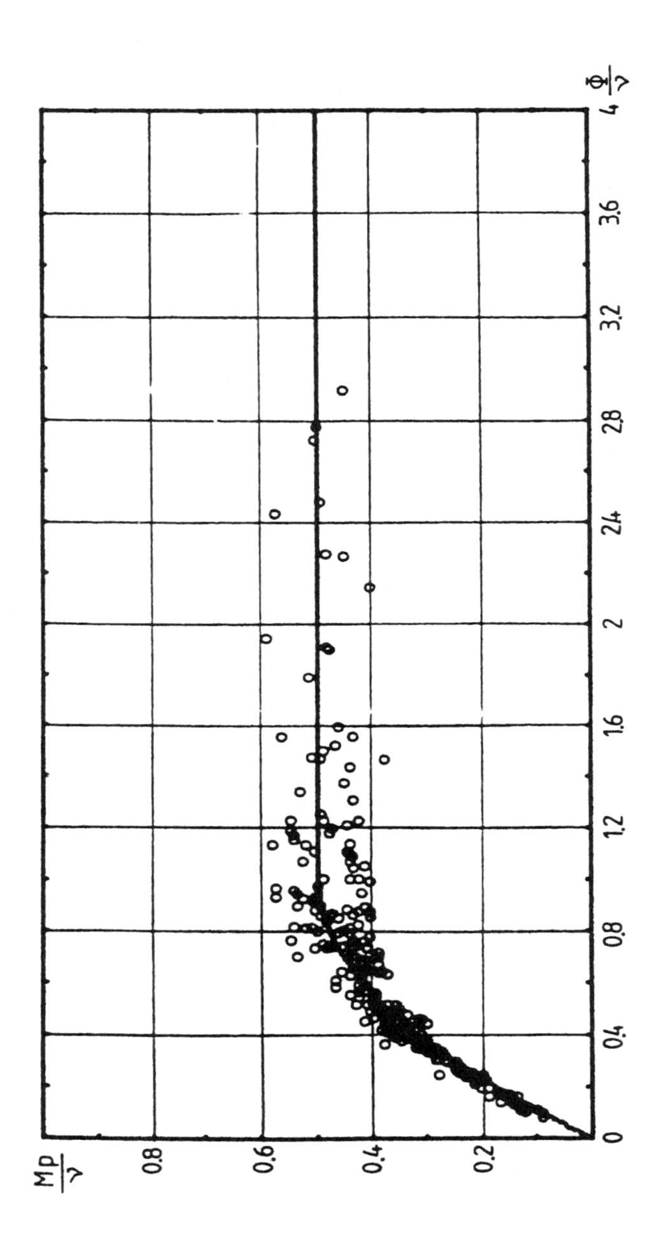

Figure 5.1.6 The plastic solution for pure bending of a rectangular section compared with test results.

Case	1	2	3	4
d_c, $A_{sc}\,(f_{Yc}, f_{Yc}^*)$, d, $A_t\,(f_{Yt}, f_{Yt}^*)$, b	y_o, νf_c, $A_{sc} f_{Yc}$, $A_t f_{Yt}$, $d/2$, $d/2$, N, M	y_o, νf_c, $A_t f_{Yt}$, $d/2$, $d/2$, N, M	y_o, νf_c, $A_{sc} f_{Yc}^*$, $A_t f_{Yt}$, $d/2$, $d/2$, N, M	y_o, νf_c, $A_{sc} f_{Yc}^*$, $d/2$, $d/2$, N, M
$m = \dfrac{M}{bd^2 f_c},\ n = \dfrac{N}{bdf_c},\ \Phi_t = \dfrac{A_t f_{Yt}}{bdf_c},\ \Phi_c = \dfrac{A_{sc} f_{Yc}}{bdf_c},\ \Phi_t^* = \dfrac{A_t f_{Yt}^*}{bdf_c},\ \Phi_c^* = \dfrac{A_{sc} f_{Yc}^*}{bdf_c},\ \ell_c = \dfrac{\Phi_c}{\Phi_t},\ \ell_c^* = \dfrac{\Phi_c^*}{\Phi_t},\ \alpha = \dfrac{d_c}{d}$				
y_o	$\dfrac{d}{\nu}(\Phi + \Phi_c + n)$	d_c	$\dfrac{d}{\nu}(\Phi - \Phi_c^* + n)$	d
m	$(1+\alpha\ell_c)\Phi_t + \frac{1}{2}n - \frac{1}{2\nu}((1+\ell_c)\Phi_t + n)^2$	$\frac{1}{2}\alpha^2\nu + (1-\alpha)\Phi_t + (\frac{1}{2} - \alpha)n$	$(1-\alpha\ell_c^*)\Phi_t + \frac{1}{2}n - \frac{1}{2\nu}((1-\ell_c^*)\Phi_t + n)^2$	$\frac{1}{2}\nu + (1-\alpha)\ell_c^*\Phi_t - \frac{1}{2}n$
Regions of validity	$y_o < d_c$	$y_o = d_c$	$d_c < y_o < d$	$y_o = d$
	$n < \alpha\nu - (1+\ell_c)\Phi_t$	$\alpha\nu - (1+\ell_c)\Phi_t \le n \le \alpha\nu - (1-\ell_c^*)\Phi_t$	$\alpha\nu - (1-\ell_c^*)\Phi_t < n \le \nu - (1-\ell_c^*)\Phi_t$	$\nu - (1-\ell_c^*)\Phi_t < n \le \nu + \Phi_c^* + \Phi_t^*$

* means yield stress in compression

Box 5.1.1 Formulas for bending in rectangular sections with both tensile and compression reinforcement and with normal force.

$$\nu = 0.97 - \frac{f_Y}{5000} - \frac{f_c}{300}, \quad \begin{array}{l} f_Y < 900 \text{ MPa} \\ f_c < 60 \text{ MPa} \end{array} \tag{5.1.7}$$

For most practical cases f_Y will be less than 600 MPa and conservatively we get

$$\nu = 0.85 - \frac{f_c}{300}, \quad \begin{array}{l} f_Y < 600 \text{ MPa} \\ f_c < 60 \text{ MPa} \end{array} \tag{5.1.8}$$

If, for some reason, the ν-value for normally reinforced beams is wanted, it might be taken as

$$\nu = 0.98 - \frac{f_c}{500}, \quad f_c < 60 \text{ MPa} \tag{5.1.9}$$

If formulas (5.1.6) for a rectangular section with tensile reinforcement are modified by replacing f_c with νf_c, we get

$$M_p = \begin{cases} \left(1 - \dfrac{1}{2}\dfrac{\Phi}{\nu}\right)\Phi\, bd^2 f_c & \text{for } \Phi \leq \nu \\[2ex] \dfrac{\nu}{2}\, bd^2 f_c & \text{for } \Phi > \nu \end{cases} \tag{5.1.10}$$

Note that Φ is still defined by (5.1.2).

A comparison with test results using the ν-value determined by (5.1.7) shows good agreement. Good agreement is also found when comparing with generally accepted design code formulas.

In Fig. 5.1.6 almost 300 tests from many different test series have been compared with (5.1.10) using the ν-values of (5.1.7). The scatter is perhaps bigger than one expects in connection with bending tests and bending theories, but it turns out that the coefficient of variation is almost equal to the coefficient of variation when using normal design code formulas, which generally are much more complicated, especially in the overreinforced region. The tests have been treated in [83.4].

The traditional methods lead to a slightly increasing bending capacity for increasing reinforcement degree in the overreinforced region. This is not found neither in plastic theory nor in the tests. It seems that plastic theory is more accurate on this point.

The ν-functions found can in practice be used for all sections, including sections with compression reinforcement and normal forces.

Great simplifications, especially in the overreinforced case, are obtained compared to normal design code practice.

For high strength concrete formula (5.1.7) gives too low values of ν. It was shown in [91.24] that the following formula gives satisfactory results when compared with tests:

$$\nu = 0.97 - \frac{f_Y}{5000} - \frac{f_c}{300} \nless 0.6 \qquad \begin{array}{l} f_c < 110 \text{ MPa} \\ f_Y < 900 \text{ MPa} \end{array} \tag{5.1.11}$$

The formula imposes a lower limit of 0.6 in ν.

5.2 BEAMS IN SHEAR

The plastic theory for the shear strength of beams presented here has been developed primarily by the author and his associates at the Technical University of Denmark. Similar methods have been presented by Thürlimann and his associates in Zürich [76.12; 78.8].

5.2.1 Maximum Shear Capacity, Transverse Shear Reinforcement

Lower bound solutions. Consider a horizontal, simply supported beam loaded by two symmetrical forces P (see Fig. 5.2.1). The beam is assumed to have constant depth. The compression zone is idealized as a stringer carrying a force C (positive as compression) and the tensile zone as a stringer carrying a force T (positive as tension). The shear zone, the web, has a constant thickness b. The length of the zone with constant shear force $V = P$, the *shear span*, is denoted as a, and the distance between the compression and the tensile stringer is denoted as h. In this section we assume that the beam is shear reinforced by vertical bars in the usual form of closed stirrups. Since the plastic theory utilizes the strength of the materials to the utmost degree, open stirrups should not be prescribed when using plastic theory.

To simplify things regarding the boundary conditions for the shear span, we imagine the loads be transferred to the shear zone by some rigid blocks, for instance, in the form of sufficiently strong end beams and sufficiently strong transverse diaphragms.

A lower bound solution requires a statically admissible, safe stress field to be constructed. Consider a homogeneous stress field in the web consisting of a uniaxial compressive stress σ_c in the concrete. The second principal direction forms a constant angle θ with the horizontal x-axis. The stress field is called the *diagonal compression field*. This stress field can be considered as an idealized model of a cracked web, the cracks being parallel to the second principal direction. The stress field carries the following stresses, referred to as the x, y-system shown in Fig. 5.2.1,

$$\sigma_x = -\sigma_c \cos^2\theta \tag{5.2.1}$$

$$\sigma_y = -\sigma_c \sin^2\theta \tag{5.2.2}$$

$$\tau = |\tau_{xy}| = \sigma_c \sin\theta \cos\theta \tag{5.2.3}$$

The relation between the constant shear stress τ and the shear force V is

$$\tau = \frac{V}{bh} \tag{5.2.4}$$

As usual, we assume the stirrups to be closely spaced; that is, the stirrup forces can be replaced by an *equivalent stirrup stress* equal to the forces in stirrups distributed over the concrete area. If the stirrup stress is σ_s, the equivalent stirrup stress is

$$\sigma_y = \frac{A_s \sigma_s}{cb} = r\,\sigma_s \tag{5.2.5}$$

where A_s is the stirrup area crossing the concrete area cb, c being the longitudinal stirrup spacing and r the reinforcement ratio.

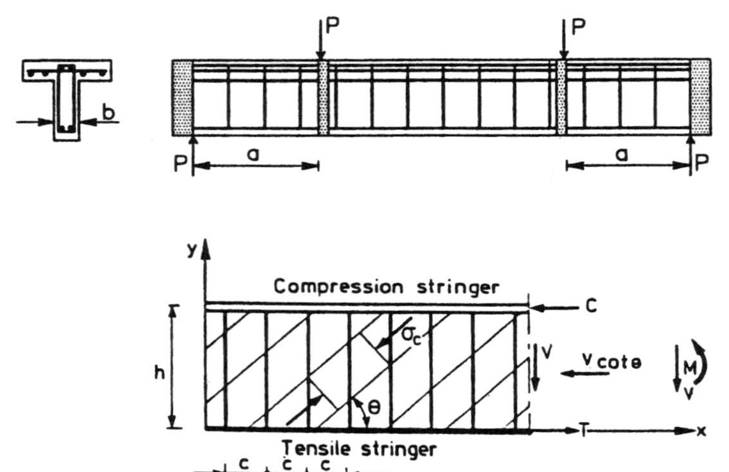

Figure 5.2.1 Beam loaded in shear by concentrated forces. Diagonal compression stress field in the web.

When the stirrups are vertical, the equivalent stirrup stresses σ_x and τ_{xy} are, of course, zero. The total stresses carried by the concrete and the stirrups are thus

$$\sigma_x = -\sigma_c \cos^2\theta \tag{5.2.6}$$

$$\sigma_y = -\sigma_c \sin^2\theta + r\,\sigma_s \tag{5.2.7}$$

$$\tau = \sigma_c \sin\theta \cos\theta \tag{5.2.8}$$

The stress field is statically admissible if the boundary conditions are fulfilled (the weight is disregarded). The boundary conditions along the stringers require the total stress $\sigma_y = 0$. Solving (5.2.8) for σ_c, we get

$$\sigma_c = \frac{\tau}{\sin\theta\cos\theta} = \tau(\tan\theta + \cot\theta) \tag{5.2.9}$$

Inserting this result in (5.2.7) and setting $\sigma_y = 0$, we get

$$r\,\sigma_s = \tau \tan\theta \tag{5.2.10}$$

Furthermore, we get from (5.2.6)

$$\sigma_x = -\tau \cot\theta \tag{5.2.11}$$

The stress field is safe if $\sigma_c \le f_c$, where f_c is the compressive strength of the concrete; and if $\sigma_s \le f_{Yw}$, where f_{Yw} is the yield stress of the stirrups.

If we assume the tensile and the compression stringers to be strong enough, the best lower bound solution is the largest load satisfying the requirements.

$$\sigma_c = \tau(\tan\theta + \cot\theta) \le f_c \tag{5.2.12}$$

$$\sigma_s \le f_{Yw} \tag{5.2.13}$$

If $\sigma_c = f_c$, the *web crushing criterion*, and $\sigma_s = f_{yw}$, we get by solving the two equations for τ and θ the solution

$$\frac{\tau}{f_c} = \sqrt{\psi(1-\psi)} \tag{5.2.14}$$

$$\tan\theta = \sqrt{\frac{\psi}{1-\psi}} \tag{5.2.15}$$

where

$$\psi = \frac{rf_{Yw}}{f_c} \tag{5.2.16}$$

is the *degree of shear reinforcement*. Equation (5.2.14) represents a circle in a $\tau/f_c, \psi$-coordinate system (see Fig. 5.2.2).

The maximum value of τ/f_c is 0.5, which we get for $\psi = 0.5$. For ψ lying in the interval $0 \le \psi \le 0.5$ it is easily verified that the conditions $\sigma_c = f_c$ and $\sigma_s = f_{Yw}$ do give the best lower bound. However, for $\psi > 0.5$ the best lower bound is found for $\sigma_s = 0.5 f_c / r < f_{Yw}$, corresponding to

$$\frac{\tau}{f_c} = \frac{1}{2} \tag{5.2.17}$$

a straight line in the $\tau/f_c, \psi$-diagram. It is found that when ψ runs from 0 to 0.5, θ runs from 0 to 45°. For $\psi > 0.5$ the θ-value is constant at 45°.

The complete lower bound solution for the load-carrying capacity is thus

$$\frac{\tau}{f_c} = \begin{cases} \sqrt{\psi(1-\psi)}\,, & \psi \le \frac{1}{2} \\ \frac{1}{2}\,, & \psi > \frac{1}{2} \end{cases} \tag{5.2.18}$$

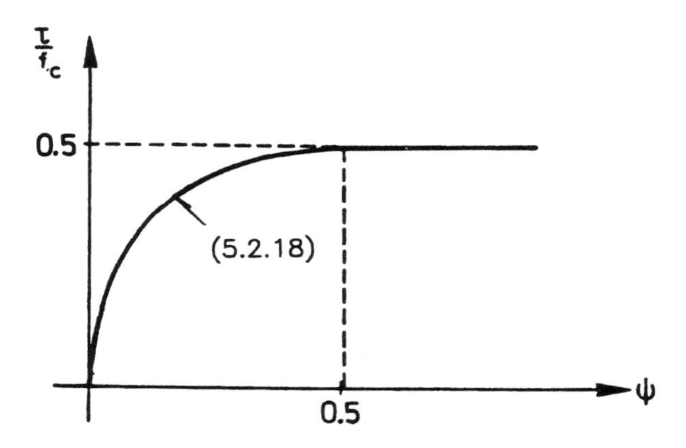

Figure 5.2.2 Lower bound solution for the maximum shear capacity of a beam loaded by concentrated forces.

The solution has to be modified for small ψ-values (see [78.3; 79.17; 81.2]). The modification leads to the same solution for small ψ-values as the upper bound solution presented in the following (see also Fig. 5.2.6). The load-carrying capacity (5.2.18) is, as mentioned, correct only if the tensile and the compression stringer are sufficiently strong.

Because of the presence of compressive stresses $\sigma_x = -\tau\cot\theta$ in normal sections of the web, the stringer forces T and C are not equal only to M/h, where M is the bending moment. By simple moment equations we find that

$$T = \frac{M}{h} + \frac{1}{2}V\cot\theta \tag{5.2.19}$$

$$C = \frac{M}{h} - \frac{1}{2}V\cot\theta \tag{5.2.20}$$

T is thus increased compared with a section with bending only, and C is decreased by the same amount. It is easily verified that $dT/dx = dC/dx = \tau b$, which is the jump condition along the stringers.

The boundary conditions at the vertical end sections are strictly fulfilled only if the shear force is transferred to the shear span as uniformly distributed shear stresses. This may indeed be the case if the load is transferred to the shear span through some rigid blocks. In practice, such conditions are met approximately, if the beam is supported on transverse beams and if the beam is loaded indirectly through transverse beams. If the loads are acting at the top of the beam and the reactions at the bottom, the boundary conditions are not satisfied.

It may be shown (see J. F. Jensen [81.2]) that the solution (5.2.18) is still valid if certain conditions at the support region and the region with the concentrated loads are fulfilled. The T-value at the point with maximum moment is still found to exceed the M/h-value in the regions with concentrated loads. However, experience from many tests seems to indicate that in the case of loading on the top, the T-value does not have to be increased beyond the value $T_{max} = M_{max} / h$ at the point with maximum moment M_{max}. In practice, the T-curve under these circumstances can therefore be taken to be as shown in Fig. 5.2.3. If the load is indirectly transferred to the beam, (5.2.19) and (5.2.20) are valid everywhere.

The lower bound solution for concentrated loading was first given by the author [67.2]. The stress field has also been utilized by Grob and Thürlimann [76.2] to derive solutions for shear capacity in the case where both shear reinforcement and longitudinal reinforcement are yielding (see Section 5.2.4).

The diagonal compression field is analogous to the diagonal tension field introduced by Wagner [29.1] to describe the state of stress in thin-webbed metal I-beams. In [82.7] the analogy was pursued further by deriving a complete set of formulas for stiffened I-beams. For compressed concrete webs, the term was used by Mitchell and Collins [74.2] (cf. also Collins [78.4; 78.5]).

Campbell et al. [80.2] considered the elastic deformations of the stirrups and the concrete struts, thus performing an analysis analogous to that of Wagner [29.1]. The analysis shows that the rigid-plastic approach slightly overestimates the upper limit (5.2.17) for the ultimate shear stress. A statically admissible stress field with trajectories in the form of noncentered fans has been constructed by Müller [78.9].

In the case of uniform loading at the top of the beam, there exists a simple statically admissible stress field consisting of homogeneous, uniaxial stress fields in parallellogram-shaped regions (see Fig. 5.2.4). Considering the section 1-1′ to be the maximum moment point and region 1-1′-2 to be stress free in the web, a homogeneous stress field in region 2-3-2′-1′ must satisfy the condition

$$\tau = |\tau_{xy}| = p\cot\theta \tag{5.2.21}$$

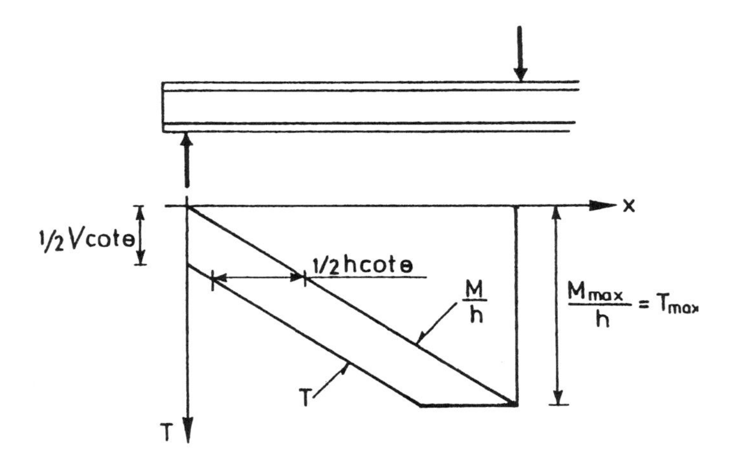

Figure 5.2.3 The tensile stringer force in a beam loaded by concentrated forces.

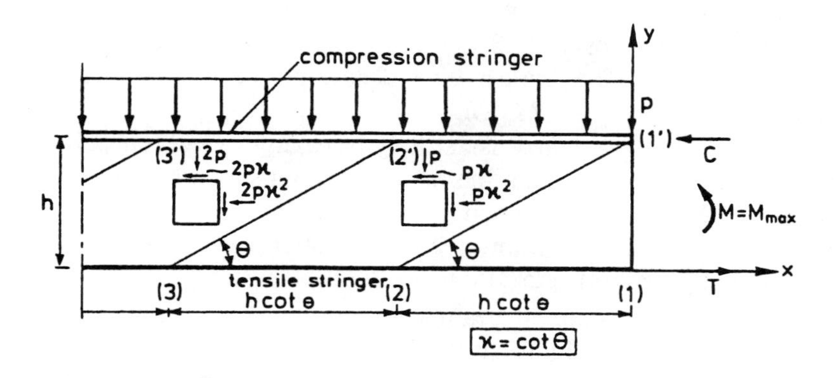

Figure 5.2.4 Diagonal compression stress field in the web of a beam with uniform load.

where p is the compressive normal stress from the load in a horizontal section in the web at the level of the compression stringer. A uniaxial stress field then requires that

$$\sigma_x = -\tau\cot\theta = p\cot^2\theta \tag{5.2.22}$$

$$\sigma_y = -\tau\tan\theta + r\sigma_s = -p + r\sigma_s \tag{5.2.23}$$

The concrete stress is

$$\sigma_c = \tau(\tan\theta + \cot\theta) = p(1+\cot^2\theta) \tag{5.2.24}$$

Using the boundary condition along 1′-2′, $\sigma_y = -p$, in region 1′-2-2′ we get

$$\sigma_s = 0 \tag{5.2.25}$$

$$\sigma_x = -p\cot^2\theta \tag{5.2.26}$$

$$\sigma_y = -p \tag{5.2.27}$$

$$\sigma_c = p(1+\cot^2\theta) \tag{5.2.28}$$

In region 2′-2-3, $\sigma_s = p/r$.

Treating the next parallellogram-shaped region in the same way, we find that all stresses σ_x, σ_y, τ, and σ_c are twice as big (see Fig. 5.2.4). In region 2′-3-3′, $\sigma_s = p/r$, so the boundary condition along 2′-3′ is satisfied. In this way, the whole beam can be treated.

Having constant shear stresses in each region, the stringer forces vary linearly. However, (5.2.19) and (5.2.20) can be used without significant errors. The stringer force T does not have to be increased beyond the value $T_{max} = M_{max}/h$.

The solution presented is, strictly speaking, valid only if the distance from the maximum moment point to a support is an integer times $h \cot\theta$, but a modification of the stress field is possible in other cases [81.2]. Therefore, a lower bound solution can be developed in the following way (cf. [75.3]).

For a simply supported beam with constant shear reinforcement degree ψ, the most highly stressed region of the web is the region near the support. Considering a line such as 2-2′ or 3-3′ to be at the support, we find that

$$\tau = \frac{pa}{h} \tag{5.2.29}$$

$$\sigma_s = \frac{\tau \tan\theta - p}{r} \tag{5.2.30}$$

$$\sigma_x = -\tau \cot\theta \tag{5.2.31}$$

$$\sigma_y = -\tau \tan\theta \tag{5.2.32}$$

$$\sigma_c = \tau(\tan\theta + \cot\theta) \tag{5.2.33}$$

where a is half the free span of the beam.

Setting $\sigma_s = f_{Yw}$ and $\sigma_c = f_c$, we get the load-carrying capacity

$$\frac{\tau}{f_c} = \sqrt{\left(\psi + \frac{p}{f_c}\right)\left[1 - \left(\psi + \frac{p}{f_c}\right)\right]} \tag{5.2.34}$$

and the angle θ determined by

$$\tan\theta = \sqrt{\frac{\psi + p/f_c}{1 - (\psi + p/f_c)}} \tag{5.2.35}$$

The solution corresponds to (5.2.14) and (5.2.15) if ψ is replaced by $\psi + p/f_c$.

Equation (5.2.34) is valid as long as $\psi + p/f_c \leq 0.5$. If $\psi + p/f_c > 0.5$, the load-carrying capacity is constant, $\tau/f_c = 0.5$.

Setting

$$\tau = \frac{pa}{h} \tag{5.2.36}$$

(5.2.34) can be solved with respect to τ. We get

$$\frac{\tau}{f_c} = \frac{1}{2} \frac{\frac{a}{h}}{1 + (a/h)^2} \left[1 - 2\psi + \sqrt{1 + 4\left(\frac{a}{h}\right)^2 \psi (1 - \psi)} \right] \tag{5.2.37}$$

which is valid for $\psi \le \frac{1}{2}(1 - h/a)$. For $\psi > \frac{1}{2}(1 - h/a)$ we have

$$\frac{\tau}{f_c} = \frac{1}{2} \tag{5.2.38}$$

This result is found by inserting $\psi = \frac{1}{2}(1 - h/a)$ into (5.2.37). For $a/h \le 1$, the highest lower bound solution is obtained for $\sigma_s = 0$, that is, by setting $\psi = 0$ in (5.2.37), irrespective of the actual shear reinforcement provided.

Concerning the boundary conditions at the vertical end sections, the same remarks apply as for concentrated loading. For design purposes, however, the stress field can be used with confidence if the beam and support region are designed to carry the stresses according to the solution.

The homogeneous stress field considered is well suited for design purposes, since it gives constant shear reinforcement in each region. The solution presented is valid for all types of beams with uniform loading, for example, clamped beams as well as continuous beams.

The lower bound solution for uniform loading was first given by the author [67.2]. For external loads other than those treated above, lower bound solutions do not yet exist.

Upper bound solutions. Whereas lower bound solutions exist only in certain simple cases and are difficult to deal with for more complicated loadings, it is easy to develop upper bound solutions for even the most complicated cases. Besides the beams treated in the following, it is possible to treat beams with bent-up bars [78.7] [79.5] and prestressed beams with inclined cables even with varying inclinations of the cables [83.1]. Beams with variable depth can also be treated.

Consider again a horizontal stringer beam with two symmetrical forces P and vertical shear reinforcement. The beam has a constant shear reinforcement degree ψ. We shall look for an upper bound solution using a simple displacement field where the central part, I, of the beam gets a vertical displacement u. Part II is not moving. Between parts I and II we have a straight yield line. We assume that the yield lines form an angle β with the horizontal axis (see Fig. 5.2.5).

The work equation assumes the following appearance:

$$Pu = rf_Y bh\cot\beta \cdot u + \tfrac{1}{2} f_c b(1 - \cos\beta)\,\frac{h}{\sin\beta} \cdot u \qquad (5.2.39)$$

The first term on the right-hand side is the dissipation in the stirrups crossing the yield line; the second term is the dissipation in the concrete determined by means of (3.4.91). The stringers do not contribute at all according to our assumptions.

From this equation we find the upper bound solution

$$\frac{\tau}{f_c} = \frac{P}{bhf_c} = \psi\cot\beta + \tfrac{1}{2}(1 - \cos\beta)\,\frac{1}{\sin\beta} \qquad (5.2.40)$$

If τ/f_c is minimized with respect to β, we get the solution

$$\frac{\tau}{f_c} = \sqrt{\psi(1 - \psi)} \qquad (5.2.41)$$

which is identical to the lower bound solution (5.2.14). The angle β is found to be

$$\tan\beta = \frac{2\sqrt{\psi(1 - \psi)}}{1 - 2\psi} \qquad (5.2.42)$$

which may be shown to correspond to

$$\beta = 2\theta \qquad (5.2.43)$$

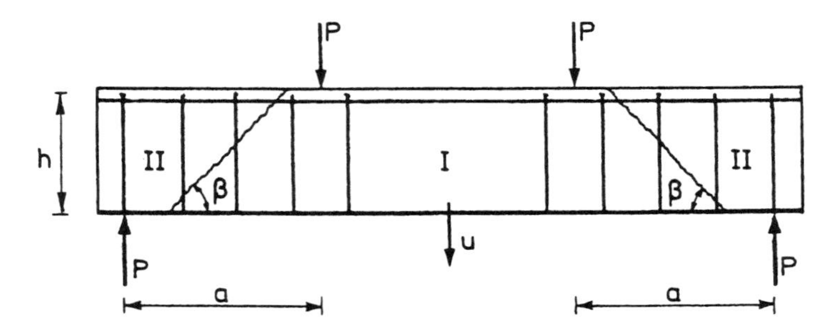

Figure 5.2.5 Shear failure mechanism for a beam loaded by concentrated forces.

where θ is determined by (5.2.15).

For geometrical reasons we must require that

$$\frac{h}{a} \leq \tan\beta \leq \infty \qquad (5.2.44)$$

for the solution (5.2.41) to be valid.

Inserting the value of β determined by $\tan\beta = h/a$ into (5.2.40), we get

$$\frac{\tau}{f_c} = \frac{1}{2}\left[\sqrt{1 + \left(\frac{a}{h}\right)^2} - \frac{a}{h}\right] + \psi\,\frac{a}{h} \qquad (5.2.45)$$

which is the shear capacity when $\tan\beta$ determined by (5.2.42) does not satisfy the first condition (5.2.44). The straight line (5.2.45) is tangent to the circle (5.2.41), and since the inclination is known to be a/h, it is easily constructed.

The value $\beta = \pi/2$ ($\tan\beta = \infty$) corresponds to a vertical yield line and is obtained for $\psi = 0.5$, where $\tau/f_c = 0.5$. An increase in the amount of shear reinforcement beyond this value evidently does not increase the load-carrying capacity, so for $\psi > 0.5$, we have $\tau/f_c = 0.5$.

The solution obtained is shown in Fig. 5.2.6. The solution was derived by Nielsen and Bræstrup [75.3]. The displacement field sketched in Fig. 5.2.7 gives the same upper bound solution, and has often been observed in tests. The displacement field shown in Fig. 5.2.7 can be used to estimate the influence of the stringers, taking into account the "hinges" at the four points H.

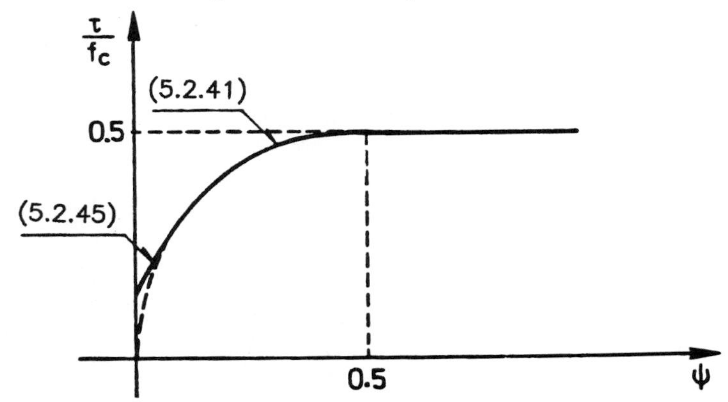

Figure 5.2.6 Upper bound solution for the maximum shear capacity of a beam loaded by concentrated forces.

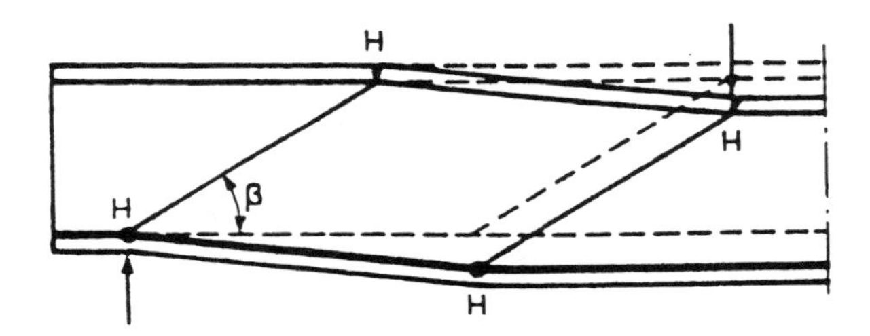

Figure 5.2.7 Alternative shear failure mechanism for a beam loaded by concentrated forces.

Apart from the region where (5.2.45) governs the solution, the upper bound solution coincides with the lower bound solution (5.2.18). A lower bound solution coinciding with (5.2.45) was derived by Jensen et al. [78.3; 79.17; 81.2]. Thus the solution of Fig. 5.2.6 is exact for all values of ψ.

In the foregoing development we have assumed a symmetrical loading arrangement. However, the solution can easily be transformed to a case with only one concentrated force. The relevant displacement field is shown in Fig. 5.2.8. It is seen that the solution found above is valid if τ is calculated on the basis of the shear force in the shear span, shown as a in the figure. It should be noted that a displacement field of the type shown in Fig. 5.2.8 can be used only for statically determined beams. The shear failure for statically indetermined beams will often involve more than one span in the failure mechanism.

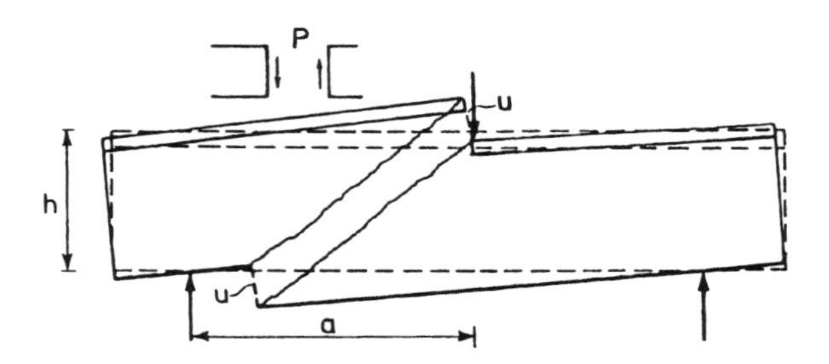

Figure 5.2.8 Failure mechanism for a beam with one concentrated load.

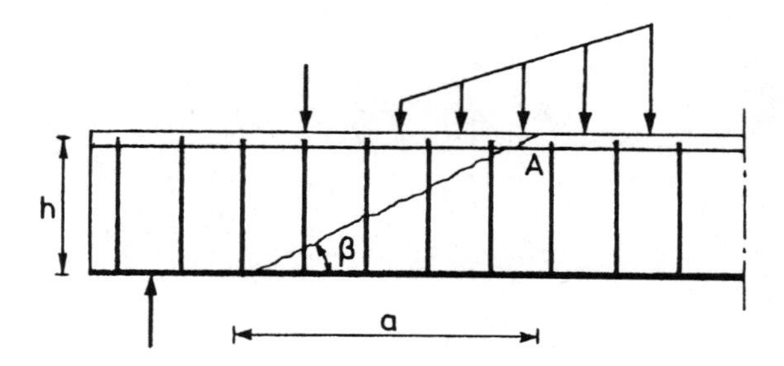

Figure 5.2.9 Shear failure mechanism for a beam with arbitrary loading.

For uniform loading, (5.2.37) and (5.2.38), derived as lower bound solutions for the stringer beam, can also be shown to be upper bound solutions. For a proof, the reader is referred to [75.3].

Displacement fields of the type shown in Fig. 5.2.9 make simple upper bound solutions easily available for arbitrary loading on simply supported beams. For a yield line such as the one shown in Fig. 5.2.9, an upper bound solution is obtained by inserting the value of β determined from $\tan\beta = h/a$ into (5.2.40) and by identifying τ as the shear stress in the section corresponding to point A, where the yield line intersects the compression stringer. The same result is found by inserting the value of a, shown in Fig. 5.2.9, in (5.2.45). An upper bound investigation of a beam with arbitrary loading can therefore take place as follows (cf. Fig. 5.2.10).

A graph of the load-carrying capacity corresponding to yield lines originating from the support is drawn. If the distance from the support to the intersection of the yield line with the compression stringer is called x, the shear stress that can be carried in the section corresponding to x is found by replacing a by x in (5.2.45). In the regions where the solution (5.2.41) gives lower values, this solution governs the load-carrying capacity.

If in the same figure the shear stress corresponding to the load is drawn as a function of x, it is seen that if this diagram lies below the diagram for the load-carrying capacity, the beam is safe (from an upper bound point of view). The load that can be carried can be calculated as the load for which the shear stress diagram touches in one or more points the diagram for the load-carrying capacity. The investigation must, of course, also be made for yield lines originating from the other support.

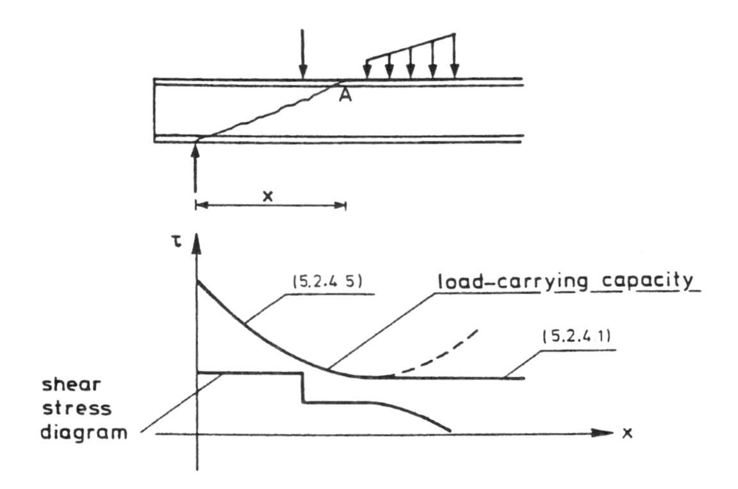

Figure 5.2.10 Upper bound solution for the maximum shear capacity of a beam with arbitrary loading.

Similar investigations can be made for beams with varying degrees of shear reinforcement. In this case yield lines originating from other points than the support must also be investigated.

5.2.2 Maximum Shear Capacity, Inclined Shear Reinforcement

Lower bound solutions. Beams with closely spaced inclined stirrups can be treated in the same way as beams with vertical stirrups. We shall therefore restrict ourselves mainly to giving the results. Defining the reinforcement ratio by

$$r = \frac{A_s}{cb} \tag{5.2.46}$$

c having the meaning shown in Fig. 5.2.11, we find that the total stresses carried by the concrete and steel are

$$\sigma_x = -\sigma_c \cos^2\theta + r\,\sigma_s \cos^2\alpha \tag{5.2.47}$$

$$\sigma_y = -\sigma_c \sin^2\theta + r\,\sigma_s \sin^2\alpha \tag{5.2.48}$$

$$\tau = |\tau_{xy}| = \sigma_c \cos\theta \sin\theta + r\,\sigma_s \cos\alpha \sin\alpha \tag{5.2.49}$$

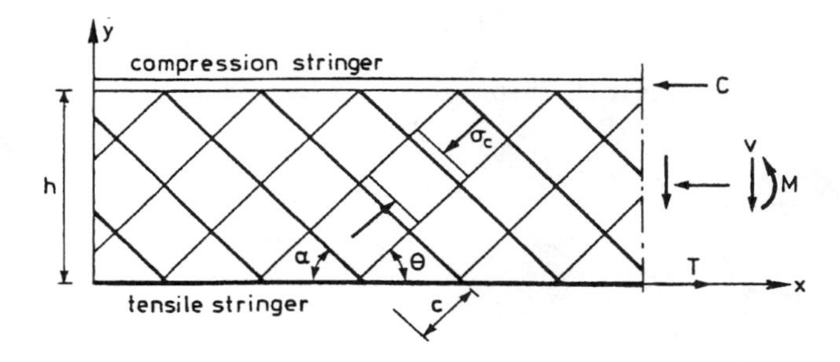

Figure 5.2.11 Diagonal compression stress field in the web of a beam with inclined shear reinforcement.

Setting $\sigma_y = 0$, we get

$$r\sigma_s = \sigma_c \frac{\sin^2\theta}{\sin^2\alpha} \tag{5.2.50}$$

and thus

$$\sigma_c = \frac{\tau}{\sin^2\theta(\cot\theta + \cot\alpha)} \tag{5.2.51}$$

which inserted into (5.2.50) gives

$$r\sigma_s = \frac{\tau}{\sin^2\alpha(\cot\theta + \cot\alpha)} \tag{5.2.52}$$

The formulas for the stringer forces turn out to be

$$T = \frac{M}{h} + \frac{1}{2}V(\cot\theta - \cot\alpha) \tag{5.2.53}$$

$$C = \frac{M}{h} - \frac{1}{2}V(\cot\theta - \cot\alpha) \tag{5.2.54}$$

The load-carrying capacity corresponding to web crushing, $\sigma_c = f_c$, now reads

$$\frac{\tau}{f_c} = \sqrt{\psi\sin^2\alpha(1 - \psi\sin^2\alpha)} + \psi\cos\alpha\sin\alpha \tag{5.2.55}$$

The angle θ is determined by

$$\tan\theta = \sqrt{\frac{\psi\sin^2\alpha}{1 - \psi\sin^2\alpha}} \tag{5.2.56}$$

The solution is valid as long as

$$\psi \sin^2\alpha \le \tfrac{1}{2}(1 + \cos\alpha) \tag{5.2.57}$$

If $\psi \sin^2\alpha > \tfrac{1}{2}(1 + \cos\alpha)$, then

$$\frac{\tau}{f_c} = \frac{1}{2}\cot\frac{\alpha}{2} \tag{5.2.58}$$

and

$$\tan\theta = \cot\frac{\alpha}{2} \tag{5.2.59}$$

The load-carrying capacity is depicted in Fig. 5.2.12.

The solution has to be modified for smaller ψ-values (see below). There exists an optimal stirrup inclination $\alpha = \alpha_M$ giving a maximum value of τ for a given value of ψ. Disregarding the modification for smaller ψ-values, one gets

$$\cot\alpha_M = \sqrt{\psi} \tag{5.2.60}$$

$$\frac{\tau}{f_c} = \sqrt{\psi} \tag{5.2.61}$$

$$\tan\theta = \sqrt{\psi} \tag{5.2.62}$$

In this case the concrete stress is perpendicular to the stirrups. The load-carrying capacity corresponding to $\alpha = \alpha_M$ is also shown in Fig. 5.2.12.

The lower bound solution for inclined shear reinforcement was derived by Nielsen [75.4] and Nielsen and Bræstrup [75.3]. The modification for smaller ψ-values was treated by Jensen [79.17; 81.2].

Upper bound solutions. Considering the same type of displacement field as in the case of transverse stirrups, we get the load-carrying capacity

$$\frac{\tau}{f_c} = \sqrt{\psi \sin^2\alpha\,(1 - \psi \sin^2\alpha)} + \psi \cos\alpha \sin\alpha \tag{5.2.63}$$

The inclination of the yield line is, as before, determined by

$$\beta = 2\theta \tag{5.2.64}$$

where θ is found from (5.2.56).

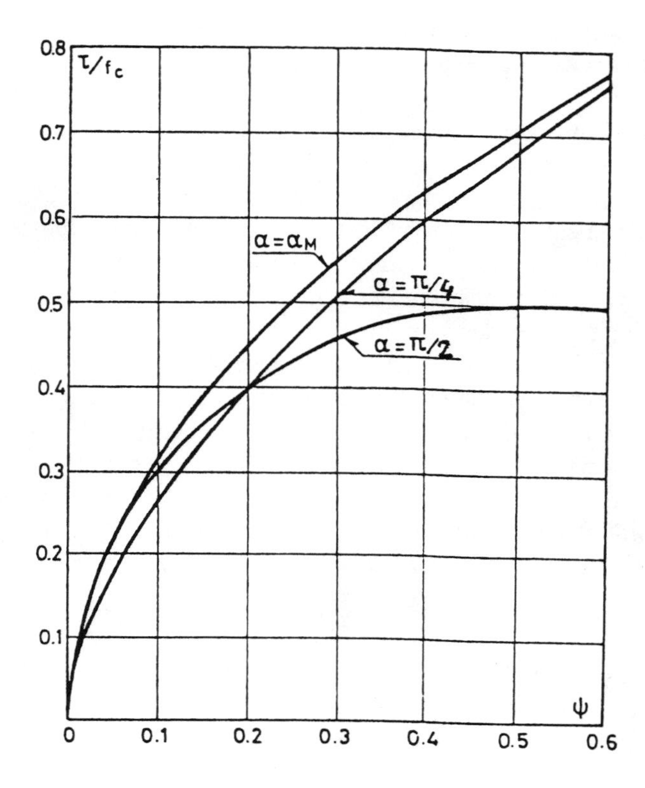

Figure 5.2.12 Maximum shear capacity of beams with inclined shear reinforcement compared to beams with transverse shear reinforcement.

This upper bound solution is valid as long as $\tan\beta \geq h/a$ (see Fig. 5.2.5). If this is not the case, we get the solution (5.2.58) if the yield line is parallel to the stirrups, and for $\tan\beta = h/a$ we find that

$$\frac{\tau}{f_c} = \frac{1}{2}\left[\sqrt{1+\left(\frac{a}{h}\right)^2} - (1 - 2\psi\sin^2\alpha)\frac{a}{h}\right] + \psi\cos\alpha\sin\alpha \tag{5.2.65}$$

The complete solution (5.2.55), (5.2.58), and (5.2.65) has been shown to be exact by Jensen [79.17; 81.2].

The upper bound solution for inclined shear reinforcement was derived by Nielsen and Bræstrup [75.3]. Other types of shear reinforcement (e.g., bent-up bars) can be treated by the same methods (see [78.7; 79.5; 83.1]).

5.2.3 Maximum Shear Capacity, Beams without Shear Reinforcement

Lower bound solutions. Consider first the case of a beam with a rectangular cross section and with concentrated loading. A lower bound solution can be derived for the idealized case shown in Fig. 5.2.13. The beam is assumed to act as an arch, where region *ABDE* is in uniaxial compression. The loads and the tensile force in the reinforcement are transferred to the arch through regions *AEF* and *BCD*, which are under biaxial hydrostatic pressure. The anchoring force is transferred to the concrete by means of an anchor plate. The hydrostatic stress is assumed equal to the uniaxial stress, and both are assumed equal to the concrete strength f_c. Therefore, angle *BDE* is $\pi/2$. It appears that maximum load is obtained when *BC* is as large as possible. Since *D* lies on a circle having *BE* as diameter, the maximum load is obtained when $CD = y_0 = h/2$. Then it is only a matter of geometry to calculate

$$BC = x_0 = \frac{1}{2}\left[\sqrt{a^2 + h^2} - a\right] \tag{5.2.66}$$

The lower bound solution is thus

$$P = b x_0 f_c \tag{5.2.67}$$

which can be written

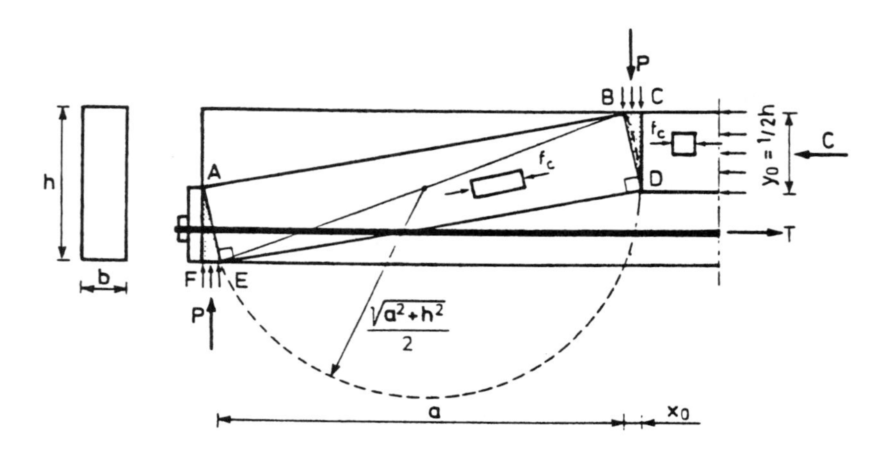

Figure 5.2.13 Lower bound solution for a beam without shear reinforcement.

$$\frac{\tau}{f_c} = \frac{P}{bhf_c} = \frac{x_0}{h} = \frac{1}{2}\left[\sqrt{1+\left(\frac{a}{h}\right)^2} - \frac{a}{h}\right] \tag{5.2.68}$$

This solution was already derived in Section 4.8 in another way. For the solution to be valid for a real beam, we must require, of course, that the loads can be transferred along a length $\geq x_0$, and that the reinforcement can be fully anchored and placed in a position corresponding to the solution above. Solutions corresponding to other positions of the reinforcement have been developed by J. F. Jensen [81.2]. If the beam is a stringer beam with an end beam, of course, the reinforcement may be replaced by a statically equivalent reinforcement in the stringers as demonstrated in Section 4.8.

The tensile reinforcement area $A_{s\ell}$ has to fulfill the condition

$$A_{s\ell} f_{Y\ell} \geq \frac{1}{2} bhf_c \tag{5.2.69}$$

that is,

$$\Phi = \frac{A_{s\ell} f_{Y\ell}}{bhf_c} \geq \frac{1}{2} \tag{5.2.70}$$

where Φ is the *longitudinal reinforcement degree* and $f_{Y\ell}$ is the yield stress of the longitudinal reinforcement. Note that τ is calculated here on the basis of the full depth of the beam.

For not too small values of the shear span τ/f_c will according to (5.2.68) be roughly inversely proportional to a/h. That means if τ_c is the load-carrying capacity for $a/h = K$, then approximately

$$\frac{\tau}{f_c} = \tau_c \frac{K}{a/h} \tag{5.2.71}$$

Such a formula has been used in some codes to take into account the substantial increase in shear capacity for small shear spans. We will use it later when dealing with the practical design of beams without shear reinforcement and beams with only a small amount of shear reinforcement (lightly shear reinforced beams).

Solutions have also been developed for uniform loading. The reader is referred to [81.2].

Upper bound solutions. The upper bound solutions considered in Section 5.2.1 are also valid for beams with a rectangular cross section without shear reinforcement.

Setting $\psi = 0$ in (5.2.45), we get

$$\frac{\tau}{f_c} = \frac{1}{2}\left[\sqrt{1 + \left(\frac{a}{h}\right)^2} - \frac{a}{h}\right] \qquad (5.2.72)$$

which is identical to (5.2.68). The solution is therefore exact.

The yield line runs from the load to the support, as shown in Fig. 5.2.14. The load-carrying capacity τ/f_c is shown as a function of a/h in Fig. 5.2.15.

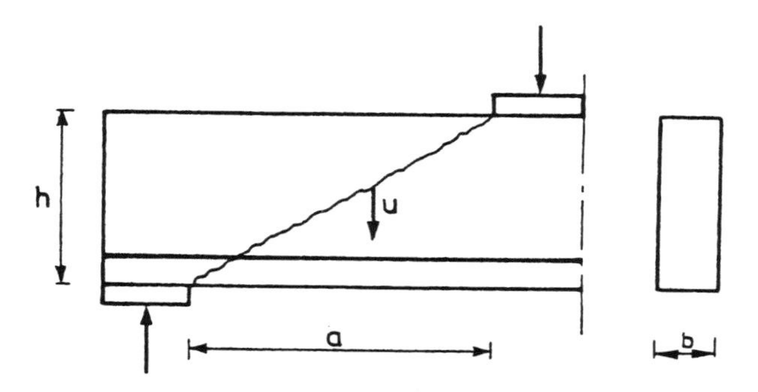

Figure 5.2.14 Failure mechanism for a beam without shear reinforcement.

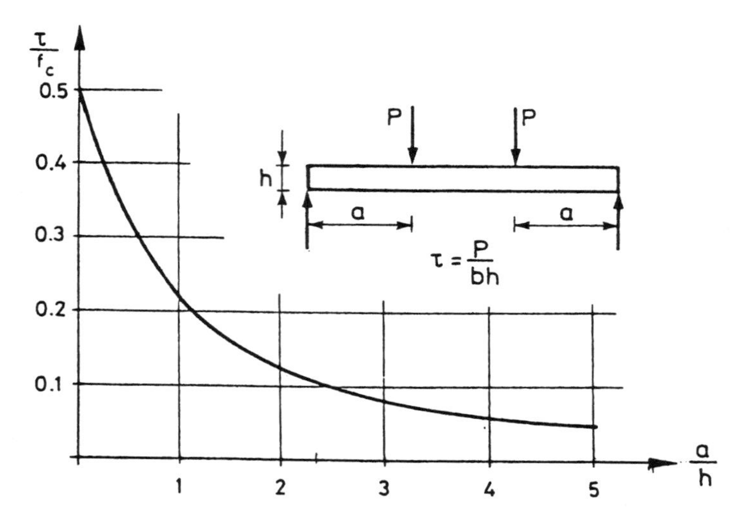

Figure 5.2.15 Maximum shear capacity of a beam without shear reinforcement versus the shear span/depth ratio.

5.2.4 The Influence of Longitudinal Reinforcement on Shear Capacity

Beams with shear reinforcement. In developing solutions for the maximum shear capacity, it has been assumed that the tensile reinforcement and the compression stringer are sufficiently strong. Let us now investigate the case where the tensile stringer governs the load-carrying capacity.

In practical situations it may normally be assumed that the load-carrying capacity of such beams is governed by the bending capacity, although it is not always completely true.

Consider again a stringer beam with concentrated loading (Fig. 5.2.1). If the tensile stringer at the maximum moment point is able to carry the force $T_Y = A_{s\ell} f_{Y\ell}$, where $A_{s\ell}$ is the reinforcement area and $f_{Y\ell}$ the yield stress, the following two equations, (5.2.10) with $\sigma_s = f_{Yw}$ and (5.2.19), determine a statically admissible solution:

$$T_Y = \frac{Pa}{h} + \frac{1}{2} P \cot\theta \tag{5.2.73}$$

$$\tau = \frac{P}{bh} = r f_{Yw} \cot\theta \tag{5.2.74}$$

This means that the tensile reinforcement is yielding at the maximum moment point and the stirrups are yielding along the entire shear span. Introducing

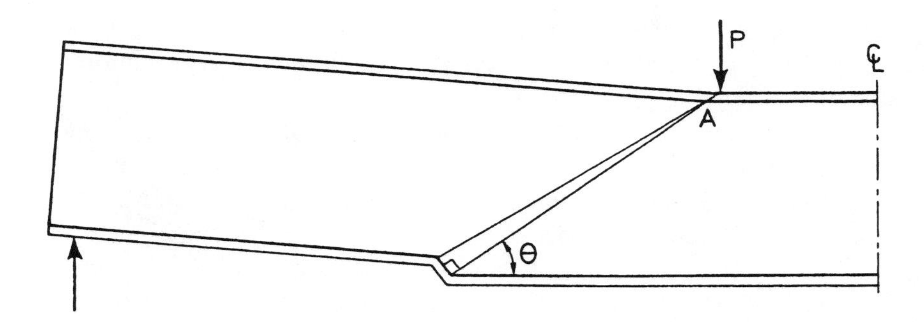

Figure 5.2.16 Failure mechanism for a beam with longitudinal reinforcement at the yield stress.

$$\Phi = \frac{A_{s\ell} f_{Y\ell}}{b h f_c} \tag{5.2.75}$$

and ψ through (5.2.16), the solution of (5.2.73) and (5.2.74) furnishes

$$\frac{\tau}{f_c} = \psi \frac{a}{h}\left[\sqrt{1 + \frac{2\Phi}{\psi (a/h)^2}} - 1\right] = \psi\left[\sqrt{\frac{2\Phi}{\psi} + \left(\frac{a}{h}\right)^2} - \frac{a}{h}\right] \tag{5.2.76}$$

The solution is only a true lower bound solution as long as the concrete stress σ_c in the diagonal compression field is less than f_c.

If the tension reinforcement is curtailed according to (5.2.19) with the value of $\cot\theta$ found from (5.2.73) and (5.2.74), the solution is exact as long as the yield line does not end beyond the support, as pointed out by Grob and Thürlimann [76.2]. A collapse mechanism corresponding to the solution is shown in Fig. 5.2.16. The relative displacement is simply a rotation about point A.

For beams with constant longitudinal reinforcement, the solution is more complicated. It has been derived by J. F. Jensen [81.2], to which the reader is referred.

The reduction in shear capacity according to (5.2.76), compared with the maximum value determined in Section 5.2.1, is illustrated in Fig. 5.2.17 for some values of Φ and for $a/h = 3$. In the figure the solution corresponding to the flexural capacity

$$\frac{\tau}{f_c} = \Phi \frac{h}{a} \tag{5.2.77}$$

is shown by the dashed lines. This solution is, of course, an upper bound solution for constant as well as curtailed longitudinal reinforcement.

The formula (5.2.76) is of limited practical interest for small values of ψ and a/h, since it does not take strut action into account. In Section 4.9 on shear walls a lower bound solution taking strut action into account has been developed.

Beams without shear reinforcement. In Section 5.2.3 we found that in order to obtain the maximum shear capacity of a beam with a rectangular cross section without shear reinforcement and loaded by concentrated forces, the longitudinal reinforcement degree Φ had to satisfy the condition $\Phi \geq \frac{1}{2}$.

What happens if $\Phi < \frac{1}{2}$? Since for $\Phi = \frac{1}{2}$, the moment in the part of the beam with constant bending moment equals the flexural yield moment, it would be natural to expect that the beam reaches the flexural capacity if $\Phi < \frac{1}{2}$. Indeed, this is theoretically the case.

It is easy to calculate a lower bound solution for $\Phi < \frac{1}{2}$. We just have to replace $y_0 = \frac{1}{2}h$ in Fig. 5.2.13 by a value satisfying the condition

$$A_{s\ell} f_{Y\ell} = b\, y_0 f_c \tag{5.2.78}$$

$$y_0 = \frac{A_{s\ell} f_{Y\ell}}{bhf_c} h = \Phi h \tag{5.2.79}$$

Calculating the corresponding value of x_0 and using (5.2.67) to determine the load-carrying capacity, we get

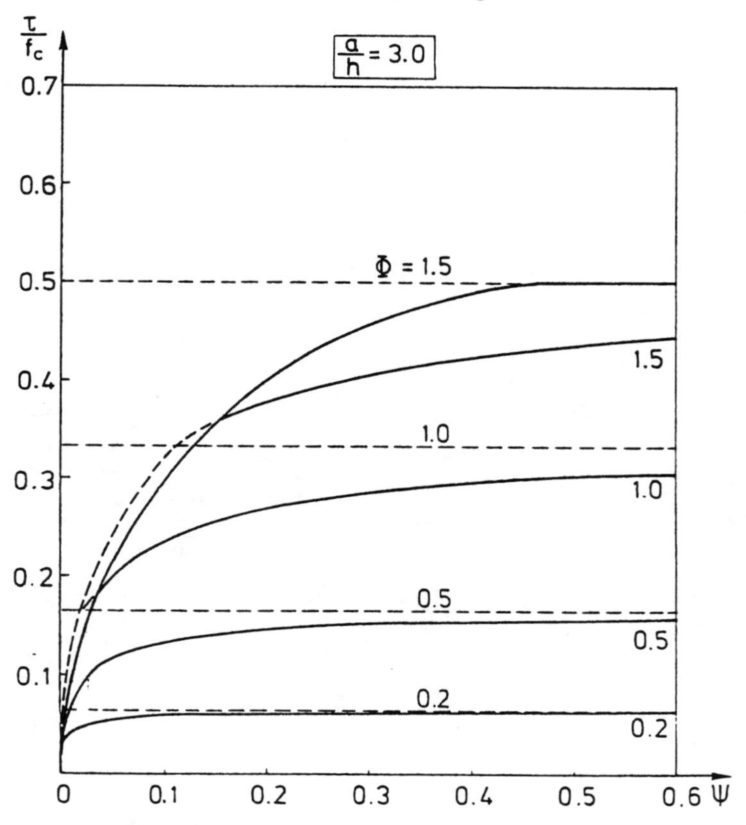

Figure 5.2.17 Shear capacity of a beam loaded by concentrated forces versus longitudinal reinforcement degree and shear reinforcement degree.

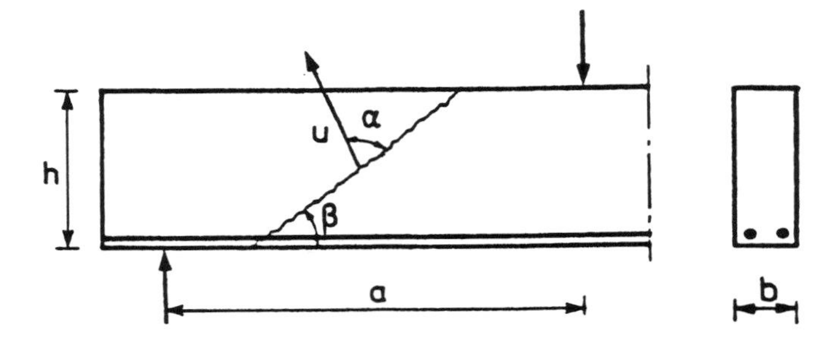

Figure 5.2.18 Failure mechanism for a beam loaded by concentrated forces and with longitudinal reinforcement at the yield stress.

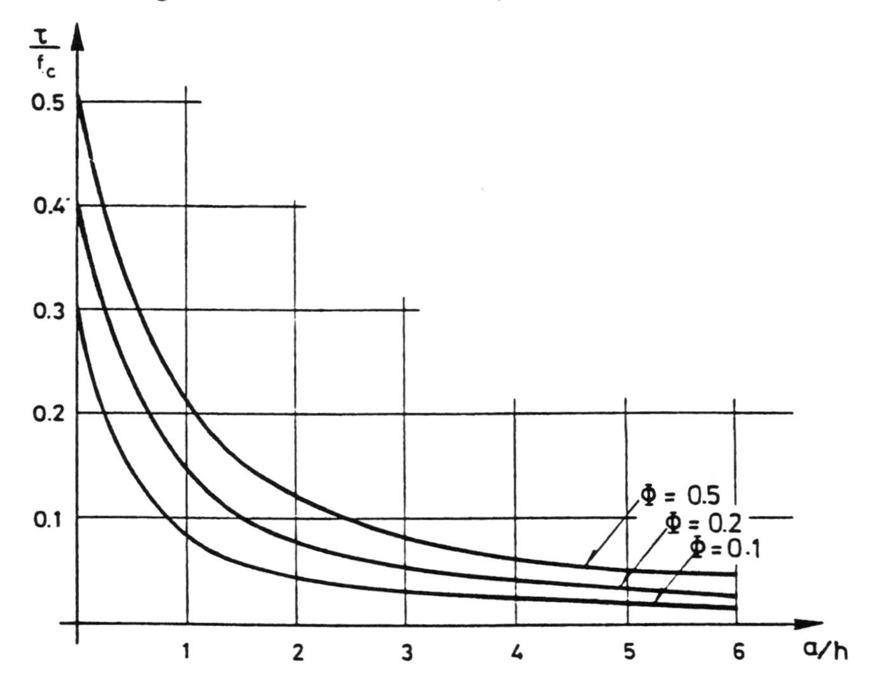

Figure 5.2.19 Shear capacity versus longitudinal reinforcement degree and shear span/depth ratio.

$$\frac{\tau}{f_c} = \frac{1}{2}\left[\sqrt{4\Phi(1-\Phi) + \left(\frac{a}{h}\right)^2} - \frac{a}{h}\right] \qquad \left(\Phi \leq \frac{1}{2}\right) \qquad (5.2.80)$$

For $\Phi = \frac{1}{2}$, this solution is in agreement with (5.2.68). For $\Phi > \frac{1}{2}$, $4\Phi(1-\Phi)$ is replaced by 1.

The same result is found using the upper bound technique on a failure mechanism, where the displacement is not restricted to be transverse but is allowed to form any angle α to the yield line [79.14] (see Fig. 5.2.18). In Fig. 5.2.18 the reaction is considered the active force which sometimes may be preferred. In [79.14] the effect of transverse and inclined stirrups was included in the upper bound solutions. These solutions are of limited practical interest as they often render much too high shear capacity.

Solutions for corbels have been developed by B. C. Jensen [79.16].

In Fig. 5.2.19 τ/f_c according to (5.2.80) is plotted against a/h for different values of Φ.

The solution presented is also valid for the stringer beam. Concerning solutions for uniform loading, see [79.17; 81.2].

5.2.5. Effective Concrete Compressive Strength for Beams in Shear

Beams with shear reinforcement. A fair accordance between theory and test results is obtained only if the theory is modified by the introduction of an *effective strength of the concrete* (cf. Section 2.1.4).

An extensive experimental test program was carried out at the Structural Research Laboratory at the Technical University of Denmark in order to determine the effective strength of concrete in the web (see [76.3; 77.3; 78.1; 80.1]).

The main conclusion of the tests is that for practical purposes, the effectiveness factor ν can be considered a function of f_c only. As an average value, the linear relationship

$$\nu = 0.8 - \frac{f_c}{200} \qquad (f_c \text{ in MPa}) \tag{5.2.81}$$

can be used. A reasonably safe value will be

$$\nu = 0.7 - \frac{f_c}{200} \qquad (f_c \text{ in MPa}) \tag{5.2.82}$$

These two functions and some of the test results are shown in Fig. 5.2.20.

All test beams were T-beams with ordinary or prestressed reinforcement and vertical stirrups. The depth of the stringer beam has been identified with the internal moment lever arm in the section with maximum bending moment. However, it was shown by Chen Ganwei [88.12] that the depth may just as well be taken to be the depth of the stirrups.

There are minor influences from other parameters: the width of the web, the number of longitudinal bars that are supported by stirrups, the type of stirrups, and the concrete cover.

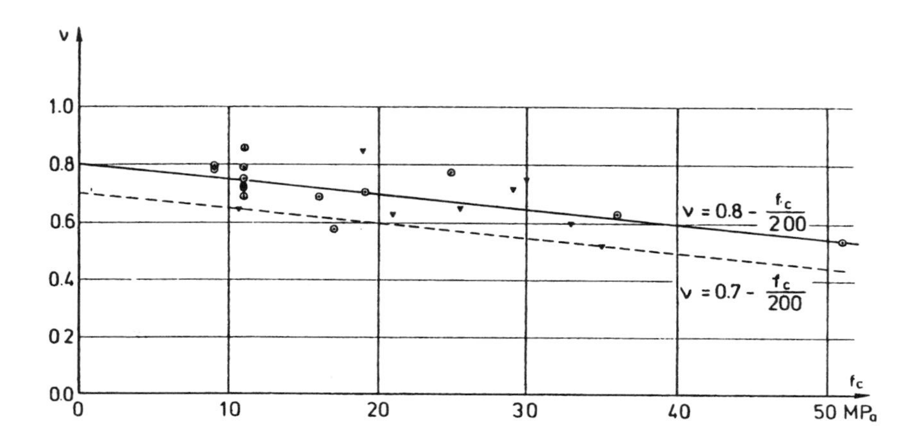

Figure 5.2.20 Effectiveness factor for shear reinforced beams.

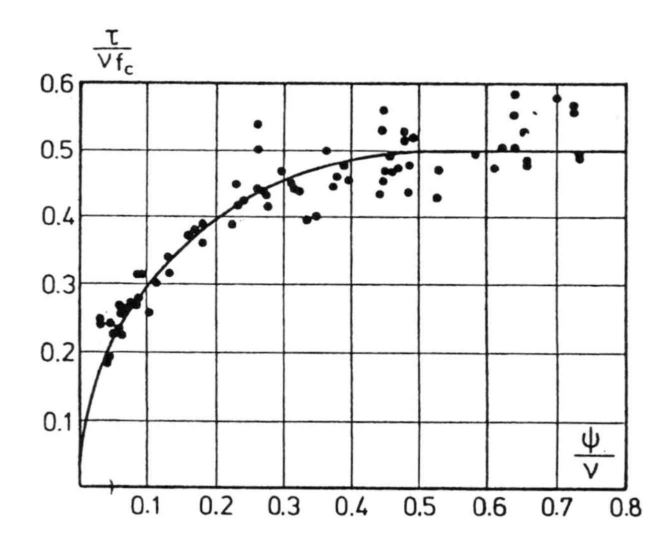

Figure 5.2.21 Maximum shear capacity for shear reinforced beams compared with test results.

To demonstrate the general applicability of the theory, Fig. 5.2.21 shows the results of the shear tests. The ν-value has been calculated from (5.2.81).

As pointed out already in Section 4.6 the formula (5.2.82) is not particularly good for high strength concrete. In Eurocode 2 [91.23] it has been suggested to combine the formula with a lower limit of 0.5, i.e.,

$$\nu = 0.7 - \frac{f_c}{200} \not< 0.5 \qquad (f_c \text{ in MPa}) \tag{5.2.83}$$

Probably for high strength concrete it will be better to use the formula (4.6.6) derived from disk tests. It means that (5.2.82) is only used for normal strength concrete, say $f_c \leq 50$ MPa, while for $50 < f_c < 100$ MPa we may use

$$\nu = \frac{1.9}{f_c^{0.34}} \not> 1 \qquad (50 < f_c < 100 \text{ MPa}) \tag{5.2.84}$$

This formula renders higher ν-values in the whole f_c range. If (5.2.81) and (5.2.84) are used together the transition point is around $f_c = 75$ MPa, i.e., (5.2.81) is used for $f_c < 75$ MPa and (5.2.84) for $75 < f_c < 100$ MPa.

The web thickness used in the Danish tests to arrive at (5.2.81) was rather high compared to the depth of the beam. Rangan [91.25] found that the formula is also valid for I-beams with a rather small web thickness. This is probably a happy coincidence. In the Danish tests the main influence on ν was the internal cracking due to the concentrated forces along the longitudinal bars supported by the stirrups. In Rangan's tests, as in disk tests, the main influence probably stems from internal cracking between the primary cracks in the web (cf. Section 4.6.1).

The application of the theory to lightweight aggregate concrete was demonstrated by Bjørnbak-Hansen and Krenchel [80.3], who found very good correlation with their experiments if a 15 to 20% smaller effectiveness factor than for normal concrete of the same strength is introduced.

The strength prediction of the theory developed is not influenced by any prestress of the beam. This is in accordance with some experimental investigations [76.10], but is disputed by others [74.9]. Some tests carried out by Jensen et al. [78.16] showed that the theory underestimates the shear strength for low degrees of stirrup reinforcement, probably due to the neglected contribution from the relatively large flanges of the test beams.

By treating a large number of tests, Chen Ganwei [88.12] found that ν is higher for prestressed beams. He suggested the formula

$$\nu = \left(0.8 - \frac{f_c}{200}\right)\left(1 + 2.2\,\frac{\sigma_{cp}}{f_c}\right) \ngtr 1 \qquad (f_c \text{ in MPa}) \qquad (5.2.85)$$

Here σ_{cp} is the average concrete stress from the prestressing force, i.e. the prestressing force divided by the whole concrete cross section.

We conclude this section by a remark on minimum reinforcement.

From Fig. 5.2.21 it appears that the test results for beams follow perfectly well the predicted strength if ψ is larger than around 0.05ν.

Beams with small shear reinforcement degrees ψ must be designed by taking into account crack sliding (cf. Section 5.2.7), Lightly Reinforced Beams. In fact, crack sliding may be governing the load-carrying capacity for even larger values of ψ than 0.05ν. However, since in crack sliding solutions the whole depth is active, the effect is less serious than might be envisaged.

A reinforcement degree $\psi = 0.05\nu$ is roughly the same as the minimum reinforcement degree (4.6.18) suggested for disks. Thus we may conclude that plastic theory, based on the web crushing criterion, may be used for small beams, as those used in the tests, if the beam is provided with a transverse minimum shear reinforcement degree determined by

$$\psi = \psi_{min} = \frac{0.16}{\sqrt{f_c}} \qquad (f_c \text{ in MPa}) \qquad (5.2.86)$$

Small beams may be defined as beams with depths up to, say 300-400 mm. In such beams longitudinal concentrated reinforcement plays an important role in governing crack development in the shear zone.

Deeper beams, i.e., beams with a depth more than 300-400 mm, should also be provided with horizontal minimum reinforcement. Experience seems to show that it might be taken at half that determined by (5.2.86).

Beams with a depth more than 1m should be minimum reinforced according to formula (5.2.86) in two perpendicular directions.

Beams without shear reinforcement. Beams without shear reinforcement may be treated using the effective strength concept as shown in Section 4.8.4. However, it is much more satisfactory to develop an alternative theory for these beams. This is done in the following section.

5.2.6 Theory of Beams without Shear Reinforcement

As pointed out in Section 4.8 on strut and tie models and in Section 4.9 on shear walls, the situation regarding the ν-values is much more complicated for beams without shear reinforcement than for beams with shear reinforcement. We refer again to Fig. 4.8.11 comparing formula (5.2.80) with tests on nonshear reinforced beams. These tests show that ν depends on the shear span. Empirical formulas including the influence of the shear span/depth ratio (*a/h* ratio) are already stated in Section 4.8.4.

For prestressed beams the influence of the shear span/depth ratio is much less. In Fig. 5.2.22 the results of a small test series with rectangular beams [77.4] [78.19] are shown. There is a remarkably good agreement with (5.2.80) when a proper ν-value is chosen.

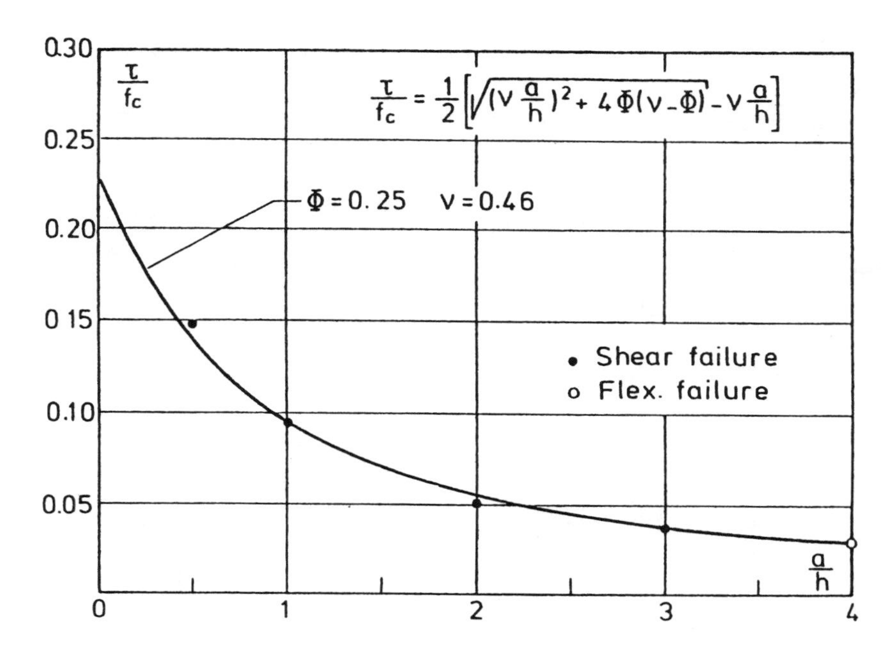

Figure 5.2.22 Maximum shear capacity of prestressed beams without shear reinforcement versus shear span/depth ratio.

For lightweight concrete slab elements the validity of (5.2.80) was investigated in [77.6].

For a long time the state of affairs was rather unsatisfactory since a physical explanation of the a/h-dependence of ν was lacking. Only when the study of sliding in initial cracks was initiated was the mystery unravelled. The work was done by Jin-Ping Zhang [94.10].

In what follows we will explain the main features of the new theory. Consider again the case of two symmetrical point loads on a nonshear reinforced beam with a rectangular cross section (see Fig. 5.2.23). The cracks normally start to develop in the constant moment section. These cracks are vertical. The crack development in the shear span is shown schematically in the figure. The first crack will normally start at the bottom face near the constant moment section at a distance of one crack distance from the load. The next one will be at one more crack distance from the load, etc. The cracks will follow a curved path and are all approximately pointing to the load application point. The load needed to develop these cracks will be higher the less the distance x' to the support (see the curve marked *cracking load* in the figure).

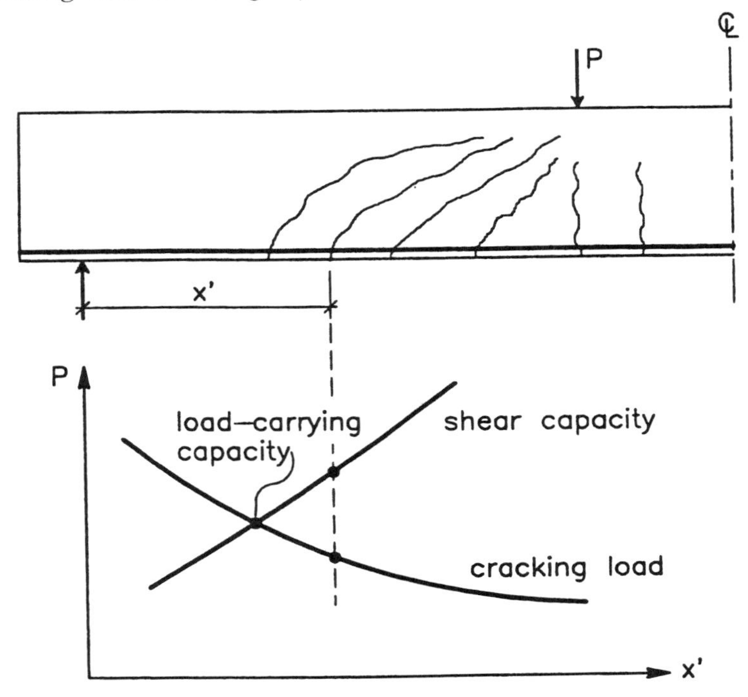

Figure 5.2.23 Crack formation in a beam without shear reinforcement.

The sliding resistance of a crack is, as mentioned before, dramatically reduced compared to the sliding resistance of uncracked concrete. The load needed to develop a sliding failure through a crack will be lower the less the distance is from the support just as in the original plastic solution; the higher the shear span, the flatter the yield line and the lower the load-carrying capacity. The curve marked *shear capacity* in the figure illustrates this. When the two curves in the figure intersect the crack may develop into a yield line and a shear failure takes place. It turns out that the load needed to produce this kind of shear failure sometimes may be much smaller than the load given by the original plastic solution.

The curves in Fig. 5.2.23 do, of course, not always intersect. It happens that the cracking load curve is lower than the shear capacity curve within the actual x'-range. In these cases the load-carrying capacity is determined by the minimum value of the shear capacity curve. Shear failure will take place along a crack originating from the support when the load has reached the value corresponding to the shear capacity. In these cases the failure does not follow immediately after the formation of the crack.

The transformation of a crack to a yield line has been demonstrated clearly by Muttoni [90.20], who measured the relative displacements along a crack. When the crack is formed the relative displacement is mainly perpendicular to the crack. When the crack is transformed into a yield line, there will be a displacement component parallel to the crack.

The cracking moment and the shear capacity of a curved crack can only be calculated in a simple way by introducing some simplifications. The bending capacity of an unreinforced beam may be calculated using an equivalent plastic normal stress distribution as shown in Fig. 5.2.24 (left). Neglecting the depth of the compression zone the cracking moment for an unreinforced beam subjected to pure bending equals

$$M_{cr} = \frac{1}{2} b h^2 f_{tef} \tag{5.2.87}$$

f_{tef} being the effective tensile strength and b the width. To get good agreement with laboratory tests on small specimens we may use

$$f_{tef} = 0.6 f_t \qquad (\nu_t = 0.6) \tag{5.2.88}$$

$$f_t = 0.26 f_c^{2/3} \qquad (f_c \text{ in MPa}) \tag{5.2.89}$$

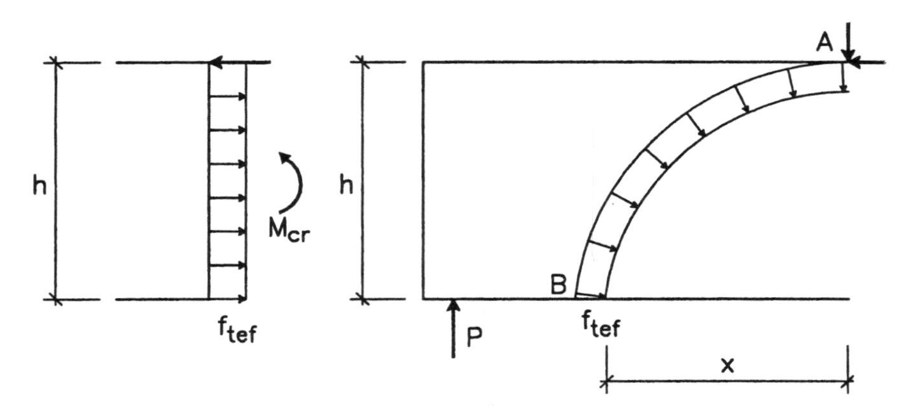

Figure 5.2.24 Stress field at the formation of a crack.

As is well known there is a considerable size effect. It may be taken into account by using a Weibull size effect factor $s(h)$ on f_{tef}. This may be set to

$$s(h) = \left(\frac{h}{0.1}\right)^{-0.3} \quad (h \text{ in m}) \tag{5.2.90}$$

h being the depth in meters. This means that

$$f_{tef} = 0.6 f_t\, s(h) \tag{5.2.91}$$

f_t still being determined by (5.2.89).

For a beam with an inclined crack the cracking load may be determined by a moment equation about point A (see Fig. 5.2.24 right) utilizing that a uniform tensile stress f_{tef} along the curve AB is statically equivalent to a uniform tensile stress f_{tef} along a straight line AB.

Thus

$$M_{cr} = \frac{1}{2} b\left(x^2 + h^2\right) f_{tef} \tag{5.2.92}$$

where the meaning of x is shown in Fig. 5.2.24 and f_{tef} is determined by (5.2.91).

This way of calculating the cracking load is, of course, somewhat primitive. Improvements require a fracture mechanics approach.

To calculate the shear capacity of a curved yield line in a simple way, the curved yield line may be replaced by a yield line composed of straight lines as shown in Fig. 5.2.25. There is a tensile failure along the upper face of the reinforcement and two inclined yield lines with the same inclination. The calculations carried out assume the beams to be overreinforced, i.e., the relative displacement along the yield lines is vertical. The contribution from the tensile failure along the reinforcement may thus be neglected.

It follows that the shear capacity may be determined by formula (5.2.72) replacing a with a-x', x' being the distance from the edge of the support platen to the crack (see Fig. 5.2.25).

The load-carrying capacity and the position of the most dangerous crack may now be determined by equalizing the cracking load and the shear capacity. The accurate calculations are lengthy and must be done numerically. For details the reader is referred to [94.10]. Some approximations suitable for simplifying the calculations will be introduced later.

In Fig. 5.2.26 the results for the point load case have been compared with tests. The ν-value used is

$$\nu = \nu_s \nu_o$$

$$\nu_s = 0.5 \ , \qquad \nu_o = 1.6 f_1(f_c) f_2(h) f_3(r) \qquad (5.2.93)$$

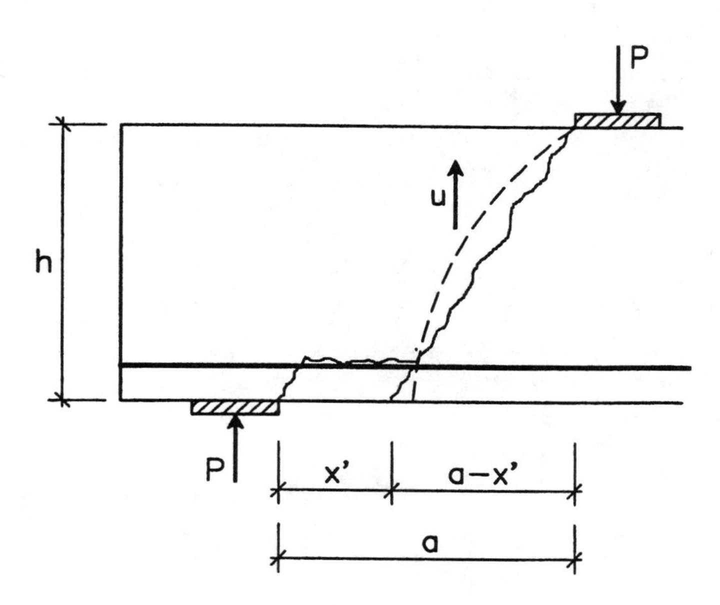

Figure 5.2.25 Shear failure in a beam with point loads.

where ν_s is the sliding reduction factor which has been set at $\nu_s = 0.5$ as before. The functions f_1, f_2 and f_3 have already been stated in Section 4.8.4. These functions represent the rather easily acceptable part of the old ν-formula. Of course, the function $f_4(a/h)$ is not included since the purpose of the new theory is to explain the a/h-dependence in the old ν-formula. The condition for overreinforcement may thus be stated as $\Phi \geq \frac{1}{2} \nu_s \nu_o$, where Φ is the reinforcement degree corresponding to the longitudinal reinforcement [cf. formula (5.2.70)].

The lower curve shown in Fig. 5.2.26 should, strictly speaking, be shown as a number of curves depending on the absolute depth h. However, it turns out that the result is almost independent of h which stems from the fact that the size effect is almost the same for the effective tensile strength used when calculating the cracking moment and for the resulting formula of the shear capacity.

It appears that the theory is rather accurate for $a/h > 2$. The shear capacity is almost constant, i.e., almost independent of the shear span in this interval. The position of the most dangerous crack may be determined by the approximate formula

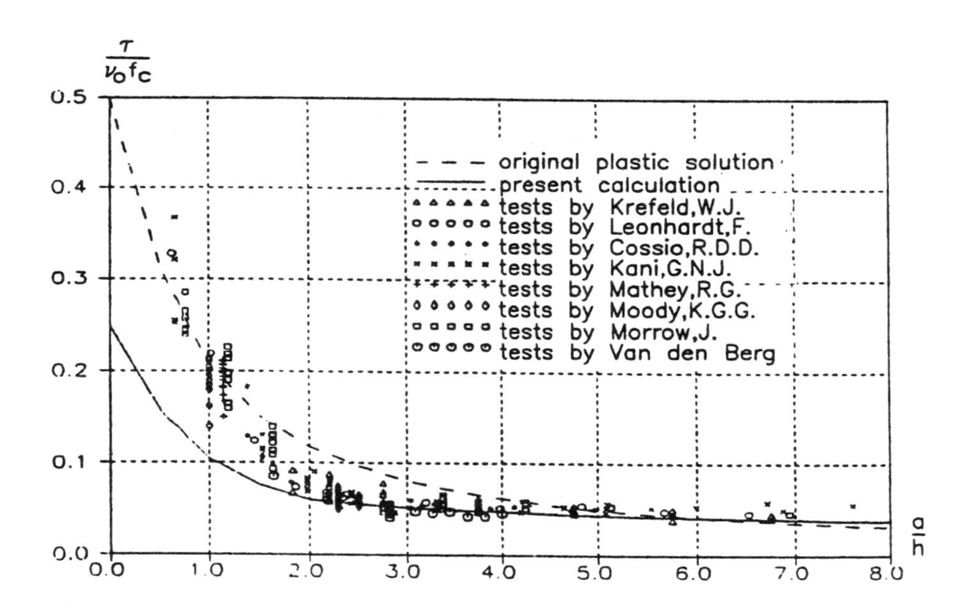

Figure 5.2.26 Comparison of theoretical shear strength with test results for point loads [94.10].

$$\frac{x'}{h} = 0.74\left(\frac{a}{h} - 2\right) \qquad \frac{a}{h} > 2 \qquad (5.2.94)$$

which is in good agreement with the tests taking into account the effect of the finite crack distances. In Fig. 5.2.27 this formula has been compared with some test results. Besides the result of the theory, the figure shows curves where the average crack distance L_{cr} and a maximum crack distance $L_{max} \sim 2\ L_{cr}$ have been added to or subtracted from x'. The scatter is very large. However, for larger shear spans the scatter in x' does not influence the load-carrying capacity to any high degree.

For $0.75 < a/h < 2$ the test results fill up almost the whole interval between the new solution and the original plastic solution using $\nu = \nu_o$ [cf. formula (5.2.93)]. In this region the most dangerous crack is running from the support to the load, i.e., $x' = 0$. The reason why there is such a big scatter in this region is explained in [94.10] as a consequence of the finite crack distances. The most dangerous crack is not formed if a crack has been formed near the support at a distance less than the crack distance. This means that the load-carrying capacity is either determined by the original plastic solution with a yield line running through uncracked concrete from support to load or by a yield line through the crack formed near the support. In both cases the load-carrying capacity is higher than the one we have when the most dangerous crack is formed.

For large beams it is likely that the load-carrying capacity will show less scatter because for such beams the crack distance will be a relatively smaller part of the shear span. For large beams the load-carrying capacity will approach the new solution. This means that in practice the new solution should be applied in this region of a/h-values as well as for $a/h > 2$.

For $a/h < 0.75$ a yield line following a crack is not geometrically possible because the angle between the displacement vector and the yield line will be less than the friction angle. (Remember that a yield line following a crack must be treated as a plane strain problem.) That is why the original solution may be used in this region.

We thus have three regions, one with $a/h < 0.75$ which is not sensitive to cracks developed by normal proportional loading, one transition region $0.75 < a/h < 2$ where a dangerous crack **may** be or **may not** be formed and one region $a/h > 2$ where a dangerous crack is always formed.

Since the shear strength is almost constant for $a/h > 2$ it is easy to find a simple formula covering this region. Conservatively, it may be taken as

$$\frac{\tau_o}{v_o f_c} = 0.045 \tag{5.2.95}$$

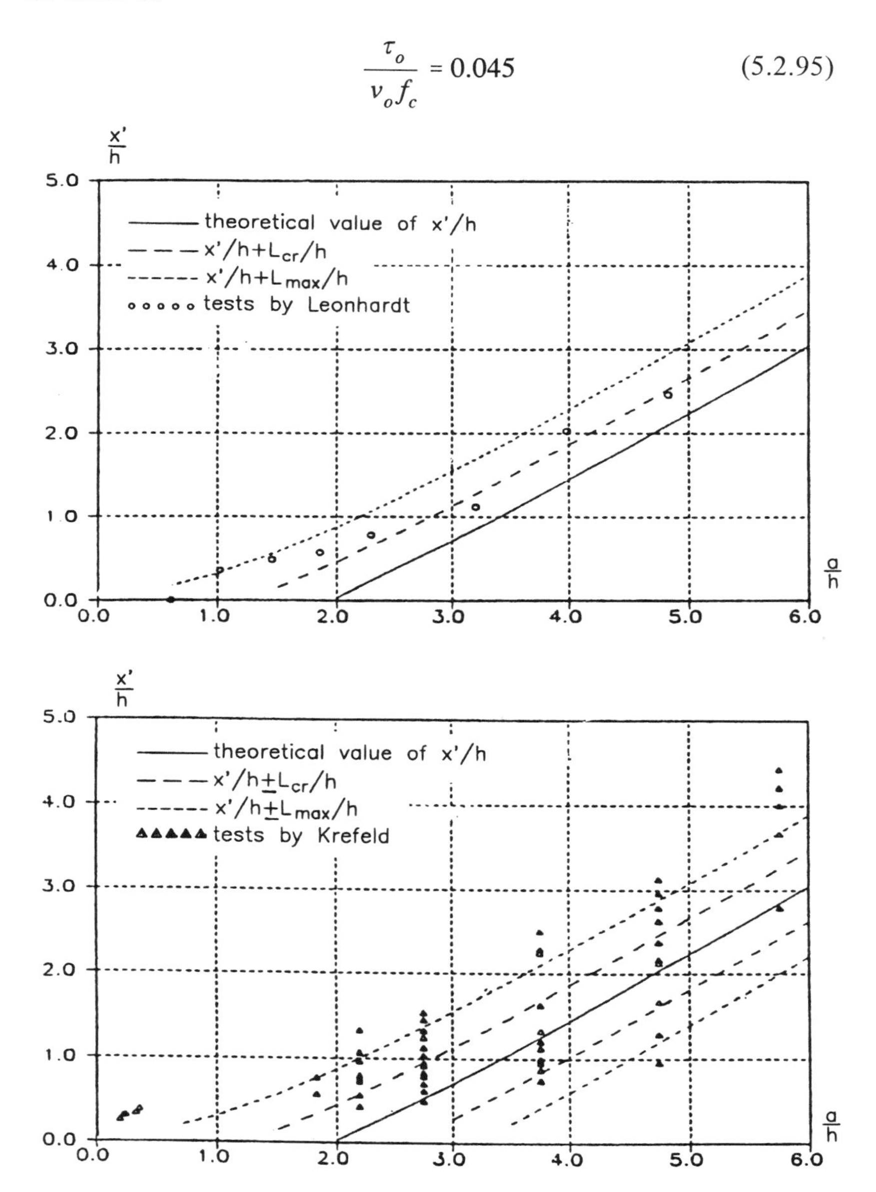

Figure 5.2.27 Theoretical position of shear crack compared with tests [94.10].

denoting the constant value of the shear strength τ_0.

Written out in full length the formula reads

$$\tau_o = 0.068\sqrt{f_c}\left(1 + \frac{1}{\sqrt{h}}\right)(0.15r + 0.58) \qquad \begin{array}{l} f_c < 60 \text{ MPa} \\ 0.08 < h < 0.7 \text{ m} \\ r < 4.5\% \end{array} \qquad (5.2.96)$$

The formula may also be written

$$\tau_o = 0.16\sqrt{f_c}\left(0.354 + \frac{0.354}{\sqrt{h}}\right)(0.72 + 0.19r) \qquad (5.2.97)$$

In this formula $0.16\sqrt{f_c}$ is half the tensile strength determined by $f_t = \sqrt{0.1 f_c} = 0.32\sqrt{f_c}$ (f_c in MPa) and the terms depending on h and r are unity when h = 0.3m and r = 1.5%, respectively.

Thus for h = 0.3m and r = 1.5%, the shear capacity is half the tensile strength, a result which is easy to remember. This is in good agreement with old code formulas. If instead h = 0.15m, we get τ_o = 0.63 f_t which will be a typical value for a slab in a building.

It seems that plastic theory taking into account sliding in cracks has brought us in very close agreement with the old empirical rules, indeed a very satisfactory development.

When the reinforcement is more or less unbonded cracking will be much more beneficial for the load-carrying capacity. It seems that the original plastic solution may be used in this case (see [94.10]).

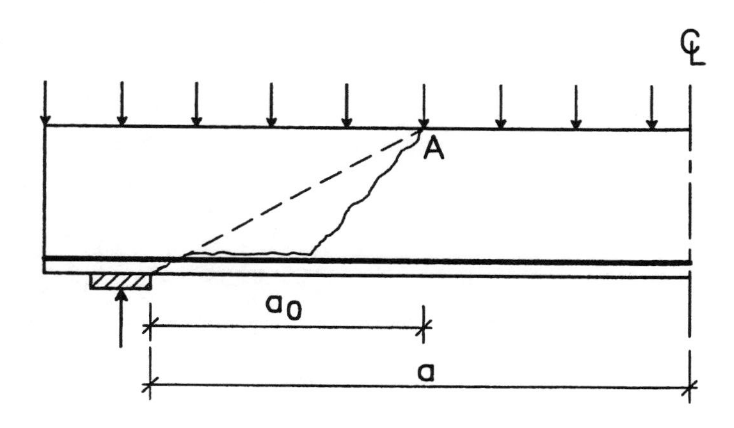

Figure 5.2.28 Cracking in a uniformly loaded beam.

For uniform loading the most dangerous yield line in an uncracked beam in the form of a straight line running from the edge of the support platen is intersecting the top face in the distance (see Fig. 5.2.28)

$$\frac{a_o}{h} = \frac{\left(\frac{a}{h}\right)^2 - 1}{2\,\frac{a}{h}} \tag{5.2.98}$$

and the load-carrying capacity is [cf. formula (5.2.37)]

$$\frac{\tau}{f_c} = \frac{\frac{a}{h}}{1 + \left(\frac{a}{h}\right)^2} \tag{5.2.99}$$

Here τ is the average shear stress at the support.

Calculations for a cracked beam along similar lines as for point loads were also done in [94.10]. The results are shown in Fig. 5.2.29.

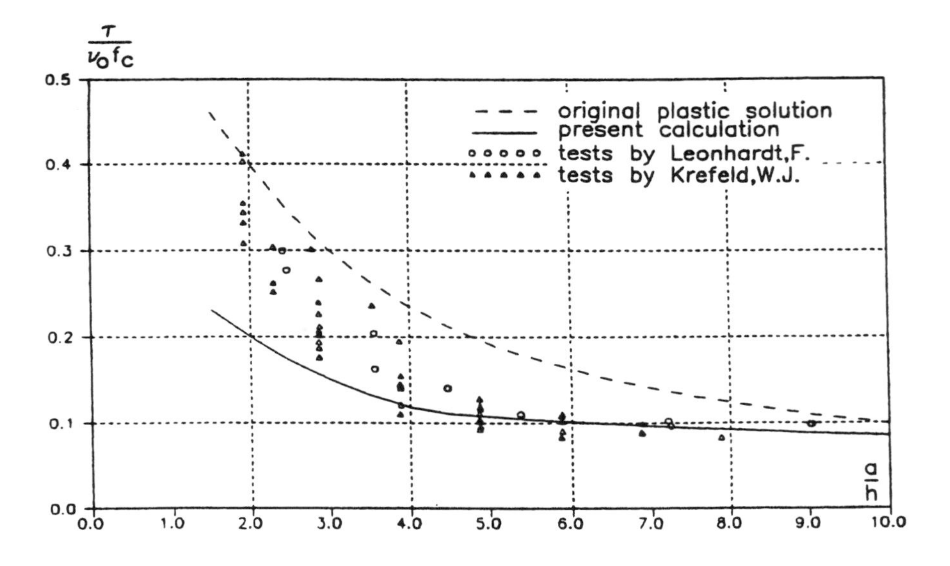

Figure 5.2.29 Comparison of theoretical shear strength with test results for uniform load [94.10].

In these calculations it was assumed that the most dangerous crack ends at the same point as the most dangerous yield line for uncracked concrete, which turned out to be a good approximation when compared with tests. However, as shown below, there exists a more dangerous point A (see Fig. 5.2.28) in which the cracking moment should be calculated. The approximation used leads to roughly a 5-8% overestimation of the load-carrying capacity. Due to the position of the crack it is to be expected that for larger a-values (notice that a is half the span of the beam), where $a_o \approx a/2$, the shear capacity measured by the average stress at the support would be around twice that for point loads since only about half the load is doing work in this mechanism. It is seen from Fig. 5.2.29 that this is also approximately true. The adopted value of ν_o was lower for uniform load, namely

$$\nu_o = 1.2 f_1 f_2 f_3 \tag{5.2.100}$$

i.e., 25% lower than (5.2.93).

When compared with the available tests it was found that for $a/h > 4$ the mean value of the ratio test result/theory was about 1.11. Due to the small overestimation of the load-carrying capacity, we may conclude that the ν-value for uniform load is only around 10% lower than for point loads. This may be neglected for most practical purposes. Thus formula (5.2.93) may be used generally.

For $a/h > 4$ the shear capacity is almost constant at $0.1\ \nu_o f_c$.

For $a/h < 4$ the test results fill up almost the whole interval between the original plastic solution and the new one, taking into account sliding in cracks. The explanation is, of course, the same as for point loads.

The results obtained for point loads and for uniform loads may, of course, be directly used in practical design. For any value of the shear span or half the span, respectively, the shear capacity may be read off the lower curves in Fig. 5.2.26 and Fig. 5.2.29. Remember that, due to the approximation introduced, there are two formulas for the effectiveness factor ν_o, formula (5.2.93) being valid in the point load case and (5.2.100) for uniform load. The curve valid for point loads is, of course, only used for shear spans $a/h > 0.75$. Further, it should be noticed that in Fig. 5.2.29 τ is the shear stress at the support. In most cases it will be reasonable to consider the reactions and the concentrated forces as point loads when calculating the a/h-values.

Jin-Ping Zhang [94.10] also treated prestressed beams with point loads. The calculations were done in the same way as before except that the prestressing was considered as an external compressive normal force at the level of the reinforcement when calculating the cracking load (see Fig. 5.2.30). It is evident that this will enhance the cracking load and consequently the distance from the support to the most dangerous crack will be larger. Hence the load-carrying capacity will be higher.

Also in this case the calculations complied well with the tests.

A most remarkable result is that, if the ν-value is calculated taking into account an effect of the prestressing, the original plastic solution is almost identical to the new solution which considers sliding in cracks.

The formula is

$$\nu = 1.2 f_1(f_c) f_2(h) f_3(r) f_4\left(\frac{\sigma_{cp}}{f_c}\right) \tag{5.2.101}$$

The functions f_1, f_2 and f_3 are as before while the new function f_4 is

$$f_4 = 1 + 2\frac{\sigma_{cp}}{f_c} \tag{5.2.102}$$

where

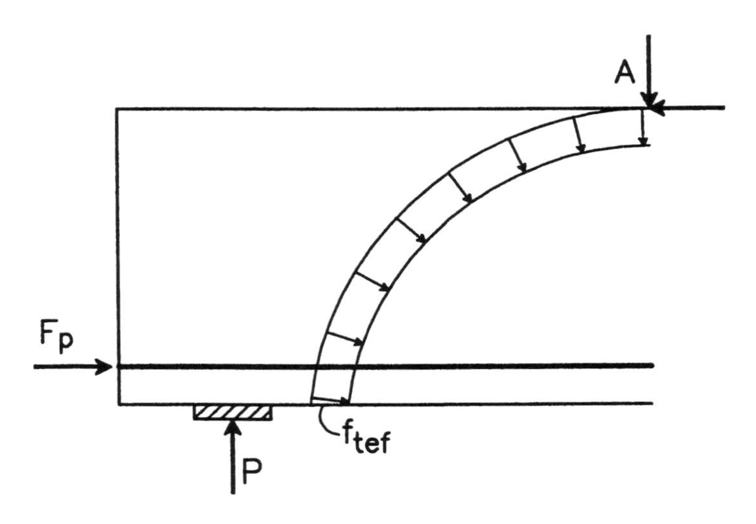

Figure 5.2.30 Cracking in a prestressed beam.

$$\sigma_{cp} = \frac{F_p}{A_c} \tag{5.2.103}$$

F_p is the prestressing force and A_c is the area of the concrete cross section. Thus σ_{cp} is the average normal stress from prestressing, which determines the increase in the effective concrete strength.

In this case, too, it is possible to simplify the results by giving a simple formula for the shear capacity. The formula, which is valid for $a/h > 2$, is

$$\frac{\tau_o}{\nu_o f_c} = 0.038 + 0.25\frac{\sigma_{cp}}{f_c} \qquad \left(\frac{\sigma_{cp}}{f_c} \leq 0.14\right) \tag{5.2.104}$$

When using this formula ν_o is calculated by formula (5.2.93) instead of (5.2.101).

By comparing with formula (5.2.95) it appears that for $\sigma_{cp} \rightarrow 0$ the shear capacity for prestressed beams is lower than for conventional beams. Prestressing reinforcement without prestressing is less effective than conventional reinforcement because of the large strains needed for reaching the yield strain.

According to formula (5.2.104) prestressing will increase the shear capacity with roughly 10% of the average prestress, which is in good agreement with code formulas.

When anchorage takes place by bonding, it is necessary to take into account that the full value of the prestressing force is not obtained along the development length. As shown by Linh Hoang [97.8] when treating hollow-core slabs, the prestressing force may be assumed to grow linearly from zero at the beam end to the normal value at the end of the development length. Linh Hoang also showed that in such a case it is necessary to consider an end failure consisting of a rotation mechanism, the rotation point being at the top face and with a straight tensile yield line emerging from the edge of support. The effective tensile strength may be calculated as before. The small contribution from the bond strength along the support was disregarded in this calculation. Concerning the effect of flanges, see the following.

The theory may, of course, also be used for any external normal force N (see Fig. 5.2.31). The normal force N is transformed to a statically equivalent normal force N' at the level of the tensile reinforcement by means of the equation

$$Ne = N'd \tag{5.2.105}$$

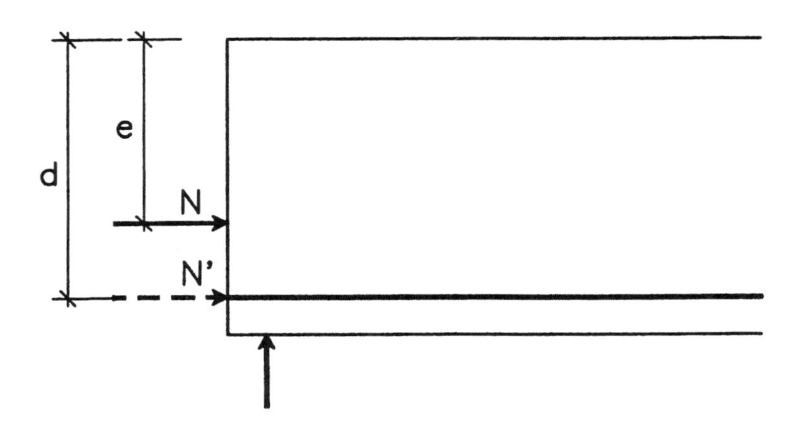

Figure 5.2.31 Beam with an external normal force N transformed to an equivalent normal force N' at the level of the reinforcement.

d being the effective depth which may be set equal to $0.9h$, where h is the total depth of the section. Then in formula (5.2.104) the average stress from N', i.e.,

$$\sigma_{cN'} = \frac{N'}{A_c} \tag{5.2.106}$$

is inserted. When dealing with conventional beams it will be natural to replace the numerical value 0.038 by 0.045 [cf. formula (5.2.95)]. Thus the shear capacity for conventional beams with a normal force may be calculated by means of the formula

$$\frac{\tau_o}{\nu_o f_c} = 0.045 + 0.25 \frac{\sigma_{cN'}}{f_c} \tag{5.2.107}$$

The Jin-Ping Zhang theory may easily be generalized to other loading cases. One crucial point is to determine the point A at the top face at which the most dangerous cracks to be considered will end (see Fig. 5.2.32). The position of A is determined below. When A is known, the cracking moment about this point must be determined as a function of the horizontal projection x of the crack (see Fig. 5.2.32). The cracking moment is calculated in the same way as for rectangular sections taking into account the real sectional area, i.e., by including the contribution from a possible tension flange. The contribution of a compression flange may normally be neglected.

For a rectangular section the cracking moment is as before

$$M_{cr}(x) = \tfrac{1}{2} f_{tef}\, b\left(x^2 + h^2\right) \tag{5.2.108}$$

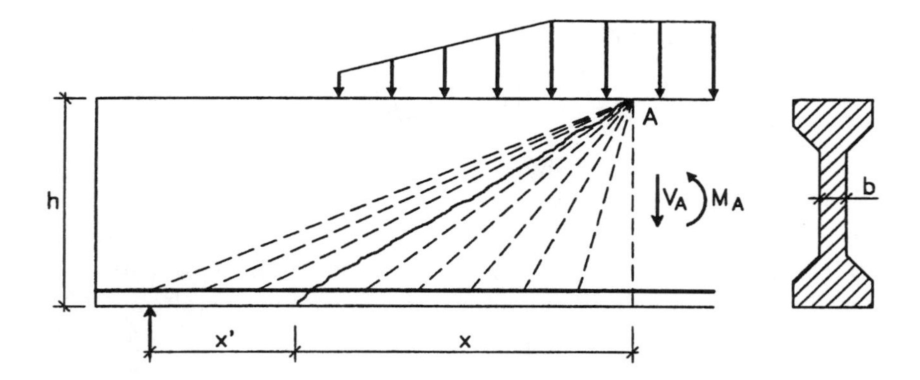

Figure 5.2.32 Arbitrary load on a nonshear reinforced beam.

where f_{tef} is determined by (5.2.91).

The bending moment M_A (P) from the external load in the section containing A is a function of the load which is indicated by introducing a load parameter P. Different values of P scale up and down the moment curve.

The equation for the determination of the load needed for developing a crack with the horizontal projection x therefore is

$$M_A(P) = M_{cr}(x) \tag{5.2.109}$$

This equation is solved with respect to the load parameter P and the solution renders P as a function of x.

The calculation of the shear capacity as a function of x is greatly simplified if the approximation (5.2.71) is used. We may put

$$\tau = \tau_c \frac{2}{x/h} \tag{5.2.110}$$

The constant τ_c is the shear capacity for $x/h = 2$. When compared with Fig. 5.2.26 it appears that we may set

$$\tau_c = 0.059\, \nu_o f_c = 1.3\, \tau_o \tag{5.2.111}$$

In this equation τ_o is the almost constant shear capacity for large shear spans in the point load case. It may be calculated by formula (5.2.96) or (5.2.97). The approximation (5.2.110) loses its accuracy for small values of the shear span and the formula should not be used for $x/h < 0.75$.

The shear force which can be carried at the point A, $V_A(P)$, is, of course, also a function of the load parameter P. This shear force may be calculated by the formula

$$V_A(P) = \tau(x) b h \tag{5.2.112}$$

$\tau = \tau(x)$ being calculated by (5.2.110).

The effect of flanges may safely be neglected. The effect of a compression flange in a T-section has been studied by Linh Hoang [97.7]. He showed that the calculations may be carried out for a rectangular beam, the depth of which is the total depth h of the section and the width of which is the web thickness. The effect of the compression flange could be taken into account in a very simple way by using an enhancement factor $K = 1.08\ t/h + 0.86 \nless 1$, t being the thickness of the flange. This formula was verified up to t/h-values of about 0.35, which means that a T-section may give up to 25% higher load-carrying capacity compared to a rectangular section. In Linh Hoang's treatment of hollow-core slabs referred to above [97.8], the compression flanges were neglected but the area of the tensile flanges was included when calculating the shear force capacity.

When equation (5.2.112) is solved with respect to the load parameter P the solution renders P as a function of x.

Finally, the two expressions obtained for P as a function of x are equalized and the equation solved for x. To find the load-carrying capacity the value of x found is inserted in one of the expressions for P as a function of x and the load-carrying capacity has been calculated.

It might turn out, as before, that the solution corresponds to a point outside the beam region. We return to this case below.

Now we confine our attention to the position of the most dangerous crack, i.e., the determination of point A. Our two equations (5.2.109) and (5.2.112) may be written

$$\begin{aligned} P\, M_A(1) &= M_{cr}(x) \\ P\, V_A(1) &= V_Y(x) = \tau(x) b h \end{aligned} \tag{5.2.113}$$

Here $M_A(1)$ is the bending moment at A for $P = 1$ and $V_A(1)$ the shear force at A for $P = 1$. We have termed the shear force capacity, corresponding to a crack with the horizontal projection x, as $V_Y(x)$. Thus the equation for x may be written

$$\frac{M_{cr}(x)}{M_A(1)} = \frac{V_Y(x)}{V_A(1)} \tag{5.2.114}$$

If this equation has a solution $x = x_o$ corresponding to a crack within the beam region, we derive from (5.2.113)

$$P^2 = \frac{M_{cr}(x_o)\, V_Y(x_o)}{M_A(1)\, V_A(1)} \tag{5.2.115}$$

This equation shows that if x_o was not a function of $M_A(1)$ and $V_A(1)$ the point A should be determined as to make $M_A(1)V_A(1)$ as large as possible. For a uniform load, $M_A(1)V_A(1)$ has its maximum value at a distance $a/\sqrt{3} = 0.58a$ from the middle point of the beam where a, as before, is half the span. Thus the distance from the support is $0.42a$, which is a little less than the distance to the end point of the most dangerous yield line according to plastic theory [cf. formula (5.2.98)].

Now x_o is also depending on the moment and shear force curve, so the point where $M_A(1)V_A(1)$ is maximum is only an approximation to the accurate position. For a uniform load it turns out that A is closer to the support than $0.42a$, around $0.35a$, when reasonable values of f_{tef} and τ_c are assumed. In practice, A must be determined by a trial and error process taking the $M_A(1)V_A(1)$ maximum as a starting point. Alternatively equation (5.2.114) is solved directly. Usually the shear capacity is not very sensitive to the position of A.

When the solution $x = x_o$ to (5.2.114) corresponds to a crack lying partly outside the beam region, the most dangerous crack will originate from the support. A new calculation must be set up where now the end point A of the crack is the unknown. In the equations (5.2.109) and (5.2.112) both M_A and V_A will be functions of x, the horizontal projection of the crack. The conditions are illustrated in Fig. 5.2.33. In this case the shear capacity curve may have a minimum which is the load-carrying capacity for a cracked beam according to plastic theory using an effectiveness factor $\nu = \nu_s\, \nu_o$. It is evident that if the solution corresponding to the intersection point of the shear capacity curve and the cracking load curve is beyond the point corresponding to the minimum value of the shear capacity curve, the load-carrying capacity must be set to the minimum value of the shear capacity curve. This situation is illustrated in Fig. 5.2.33 for a cracking load curve represented by the dotted line.

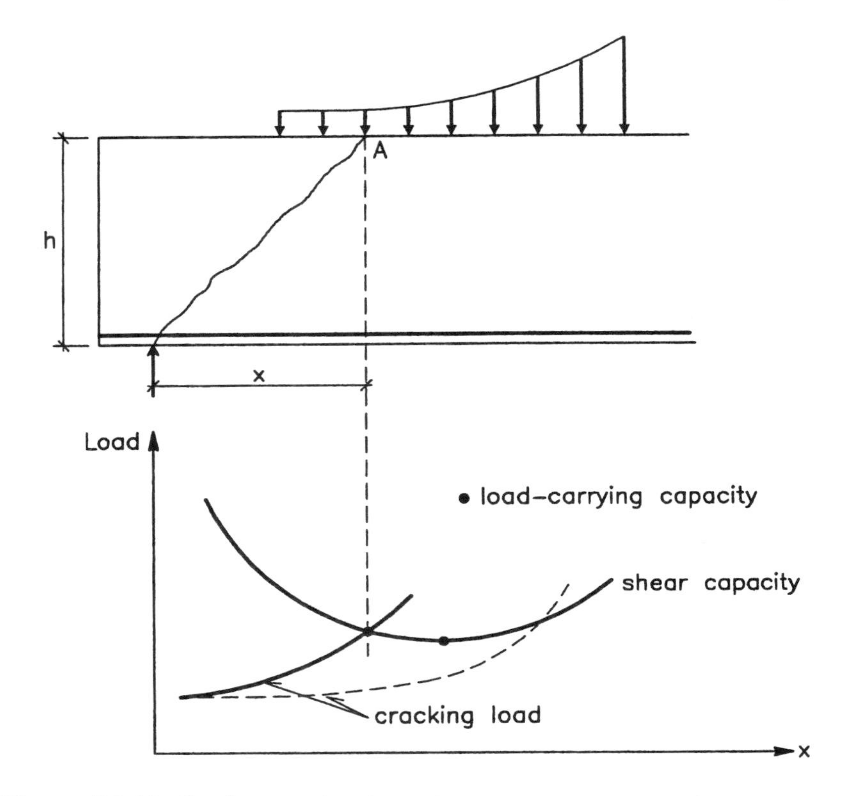

Figure 5.2.33 Crack emerging from the support.

When the effective tensile strength f_{tef} used in the calculation of the cracking load is put to zero, the plastic solution using the effectiveness factor $\nu = \nu_s\ \nu_o$ is obtained as a limiting case.

When large concentrated forces closer to the support than 0.75 h are present and it turns out that the most dangerous crack goes to one of these loads, a normal plastic analysis is carried out using the effectiveness factor $\nu = \nu_o$ (cf. the end of Section 5.2.1).

For a continuous beam the region with positive moments as well as the region with negative moments must be treated along similar lines. In fact the support region with negative moments will often be decisive because the shear forces are large in this region and the bending moments may be large, too.

If the absolute value of the support moment is denoted M_S, the equation for the determination of the load which enables cracking to take place in a skew crack whose end point A is in the distance x from the support (see Fig. 5.2.34) now reads

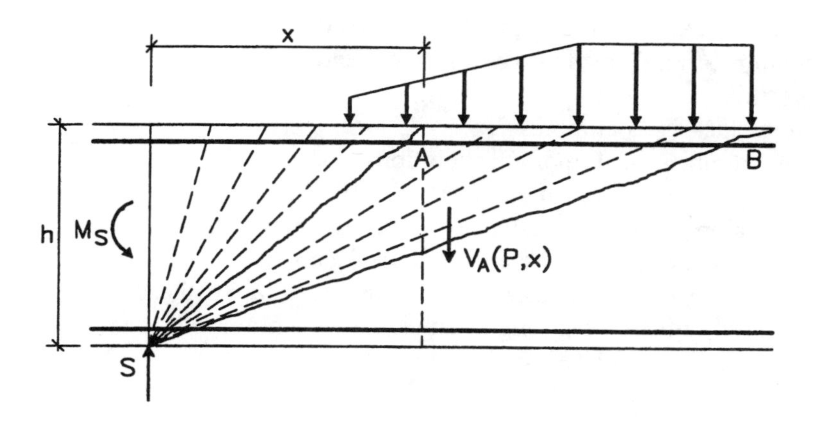

Figure 5.2.34 Arbitrary load on a nonshear reinforced continuous beam.

$$M_S(P) = M_{cr}(x) \tag{5.2.116}$$

When calculating the cracking moment, a moment equation around point S at the support is used (see Fig. 5.2.34).

$M_S(P)$ may possibly be reduced by taking into account a downward directed load on the top face along the length x.

The shear capacity according to formula (5.2.110) now renders the shear stress which may be carried in the section at the distance x from the support. We denote this shear force V_A (P, x) since it will be a function of the load parameter P as well as of x.

The equation determining the load which will cause a shear failure along a crack corresponding to the distance x now reads

$$V_A(P, x) = \tau(x)bh \tag{5.2.117}$$

$\tau = \tau$ (x) still being calculated by formula (5.2.110).

When solving (5.2.116) and (5.2.117) for P two expressions for the load parameter P are obtained as before and x may be found. Inserting the value of x into one of the expressions for P renders the load-carrying capacity.

If x exceeds the distance to point B in Fig. 5.2.34, which indicates the point where the most dangerous yield line ends according to plastic theory, the distance from the support to B must be inserted in the expression for the shear capacity to find the load-carrying capacity.

When treating the region with positive moments, no special attention is paid to the point of zero bending moment. Cracks may pass this point even if according to elementary beam theory the bottom face is in compression.

A number of important solutions for statically indeterminate beams has been given by Linh Hoang [97.4]. He even treated loads acting along the bottom face, which are more dangerous than the loads on the top face as tacitly assumed above.

Normally it may be assumed that the bending stiffness governing the distribution of bending moments in a continuous beam is the bending stiffness corresponding to fully cracked sections. Otherwise shear failure in a crack is not likely to occur. In some cases it may happen that one region, say the region with positive moments, is uncracked while the region with negative moments is cracked. In such cases the bending moment and shear force distribution should be determined using the corresponding stiffnesses. For a beam with a rectangular section and with the same constant amount of reinforcement in the top and bottom the usual moment distribution obtained by assuming uncracked stiffness may be used when calculating the shear capacity. In special cases the designer should consider the most unfavorable combination of stiffnesses when calculating the shear capacity. He may then utilize the elementary fact that a large bending stiffness attracts bending moments and a small bending stiffness repels bending moments.

Replacement of the accurate formula with the simple expression (5.2.110) greatly facilitates the determination of the most dangerous yield line according to plastic theory. We may again refer to Fig. 5.2.10. The load-carrying capacity curve determined by formula (5.2.45) is simply replaced by a curve determined by (5.2.110). (The curve corresponding to (5.2.41) is not relevant for beams without shear reinforcement.)

By drawing the actual shear stress diagram and scaling it up or down to touch the load-carrying capacity curve the end point of the most dangerous yield line is easily found.

The reader may verify that for a uniform load the most dangerous yield line always ends at the distance of one quarter of the total span from the support when using (5.2.110) for the shear capacity [cf. formula (5.2.98)].

We conclude this section by discussing some implications of the Jin-Ping Zhang theory which are of fundamental practical significance. It is evident that the results of the theory depend on the tensile strength of the concrete since the cracking moment is proportional to the tensile strength. By calibrating with experiments good agreement between the theory and tests has been found. The designer may now pose the important question: can we always rely upon this tensile strength? In other words, do all parts of a beam or slab which according to theory are uncracked and thus cannot suffer a dangerous sliding failure along a crack stay uncracked during the whole lifetime of the structure? This is certainly not true in all cases. Cracks may develop because of many other reasons than the loading considered in the design situation. We might get overloading because of dynamic actions or other reasons; we might get cracks from temperature, creep and shrinkage deformations; and we might even have cracks from chemical reactions like alkali-silica reactions which substantially reduce the tensile strength.

Codes seldom give a firm answer to this important question. On the one hand they allow utilization of the shear strength of nonshear reinforced beams and slabs, and on the other hand they recommend that at least important structural elements are always provided with a certain minimum shear reinforcement to improve ductility. The Jin-Ping Zhang theory is of great importance because it allows us to state that even if the tensile strength of concrete disappears completely a sliding strength remains in the cracks provided the crack widths are not too large (cf. Section 2.1.4 and Chapter 8). The implications of this fact are equally important. The most important one is that even if a crack developed in a region which according to calculations should remain uncracked and even if this crack is vulnerable to a shear failure a certain shear strength remains. Another important implication is that a designer who wants to carry out a structure without shear reinforcement but at the same time does not want to rely upon the tensile strength of concrete may do so with confidence. He just has to perform a normal plastic analysis with an effectiveness factor $\nu = \nu_s \nu_o$. For small values of the shear span, the result of such an analysis will either be identical or only a little more conservative than the result which he will find using the Jin-Ping Zhang theory. But for slender beams and slabs the load-carrying capacity may easily be half or less.

The risk of getting an unexpected crack vulnerable to a shear failure cannot be evaluated in a serious way due to lacking knowledge of the large number of effects influencing the short- and long-term crack development. Since nonshear reinforced beams and slabs have very little ductility when subjected to a shear failure these structures should be avoided as far as possible. However, many years' experience shows that it is easier to write such a thing in a book or a code than to carry it out in practice. Often the final decision will be left to the responsible designer, who will make a compromise satisfying his professional experience and integrity, the economy of the structure and the code makers representing the community.

5.2.7 Design of Shear Reinforcement in Beams

Beams with constant depth and arbitrary transverse loading. To illustrate how the plastic theory can be used to formulate practical and simple design recommendations, we will show how design rules for the design of shear reinforcement in slender beams can be formulated [78.20; 80.7]. The design method has been adopted in Eurocode 2, [91.23]. It is natural to base the procedure on the lower bound solutions of the preceeding sections. These solutions can immediately be extended to combinations of uniform loading and concentrated loading by superposition. Furthermore, the stress field for uniform loading is in fact also valid for a loading, the intensity of which varies from region to region but which is constant within each region. Therefore, it does not seem unreasonable, as an approximation, to treat all combinations of continuous loadings and concentrated forces by a lower bound method which is a simple extension of the method presented in the previous sections.

Considering a part of a beam loaded on the top face and with closely spaced transverse shear reinforcement from a maximum moment point to a support, it is suggested that the amount of shear reinforcement be calculated as a function of a nominal shear stress τ', which is found on the basis of the shear stress diagram τ as shown in Fig. 5.2.35. Denoting

$$\kappa = \cot\theta \qquad (5.2.118)$$

the τ'-curve is constant in each of the regions κh into which the span has been subdivided.

Having found the τ'-curve, the amount of transverse shear reinforcement is determined by

$$rf_{Yw} = \frac{\tau'}{\kappa} \qquad (5.2.119)$$

where r is the reinforcement ratio and f_{Yw} is the yield stress of the transverse reinforcement.

Since the average reduction in the shear stress τ to arrive at τ' is $\frac{1}{2}p\kappa$, one might also use

$$\tau' = \tau - \frac{1}{2}p\kappa \qquad (5.2.120)$$

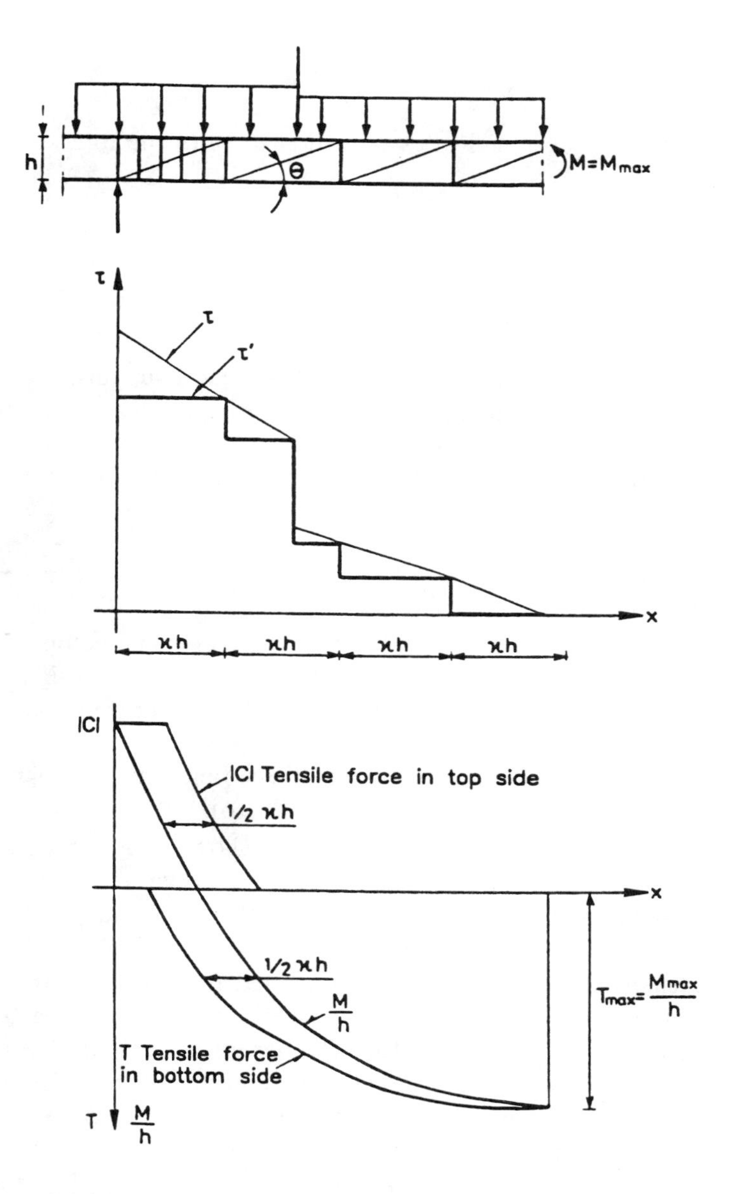

Figure 5.2.35 Design procedure for shear reinforcement.

which means a reduction of $\frac{1}{2}p$ in $r f_{Yw}$.

The procedure is seen to give correct results for concentrated loading and for uniform loading.

The concrete stress σ_c has to satisfy the condition

$$\sigma_c = \tau\left(\kappa + \frac{1}{\kappa}\right) \leq \nu f_c \tag{5.2.121}$$

The stringers are designed in each section to carry the forces

$$T = \frac{M}{h} + \frac{1}{2} V\kappa$$
$$C = \frac{M}{h} - \frac{1}{2} V\kappa \tag{5.2.122}$$

In a continuous beam T will be negative near a support, meaning compression. C will be negative in the same region, meaning tension. V is always inserted as a positive number.

At a simple support there will be tension in both stringers. The reinforcement has to be anchored for a force

$$T_0 = -C_0 = \frac{1}{2} R\kappa \tag{5.2.123}$$

where R is the reaction. This implies that if the diagonal compression stress field is used throughout, the end beam must be designed to carry these tensile forces which are balanced by a uniform compression stress with the resultant $R\kappa$.

It is possible to work with other stress fields at the supports. Some solutions may be found in [81.2] [90.16].

Instead of using (5.2.122), the procedure shown in Fig. 5.2.35 can be used. Here the M/h-curve to give the T-curve is displaced a distance $\frac{1}{2}\kappa h$ in such a direction that the absolute value of M/h is increased.

The tensile reinforcement can be curtailed if desired. Concerning design rules for the proper anchorage of such reinforcement, the reader is referred to [75.4] [90.16]. The design is, of course, carried out using design values of loads and strength parameters. Since the internal moment arm in the section having $M = M_{max}$ has been used to determine experimentally the value of the effective compressive strength, h is set equal to the internal moment arm in the section having $|M| = |M|_{max}$.

For beams with tension and/or compression flanges, shear stresses are of course calculated on the basis of the web thickness. Beams loaded at the bottom or at a level between the top and the bottom can be treated as beams loaded at the top if an extra transverse reinforcement capable of transferring the load to the top is supplied.

Theoretically, κ can be chosen arbitrarily as long as condition (5.2.121) is satisfied and the tensile reinforcement, especially the anchorage at simple supports, is designed according to the κ-value chosen. However, to make sure of proper behavior for the service load the κ-value chosen should not be too large.

The elastic behavior is difficult to calculate. An approximate solution can be found by minimizing the complementary elastic energy determined on the basis of the diagonal compression stress field (cf. Kupfer [64.9]). Such calculations have been carried out in [78.6] and [97.17]. If it is required that the stirrups do not yield for the service load, it is found that κ has to satisfy the conditions[1]

$$1 \le \kappa \le \begin{cases} 2.5 & \text{for beams with constant longitudinal reinforcement} \\ 2.0 & \text{for beams with curtailed longitudinal reinforcement} \end{cases} \tag{5.2.124}$$

The classical shear theory based on the truss analogy developed by Ritter [1899.1], Mörsch [12.1; 22.2], and Johansen [28.2] is equivalent to the value $\kappa = 1$. So even if design rules today normally include a small concrete contribution, a κ-value equal to, for example, 2.5, gives rise to a considerable reduction in the shear reinforcement. One might fear that such a reduction would lead to excessive crack widths in the shear zone for the service load. This is not so. The maximum crack width for all the Danish test beams, where stirrups of smooth bars were used, some even having a high yield stress ($f_Y \sim 500$ MPa), was not on average more than 0.18 mm for a load equal to 0.5 to 0.6 times the ultimate load. However, it is recommended that stirrups of deformed bars be used in the most severe class of environmental influences or that inclined stirrups of smooth or deformed bars be used for which the crack widths are very small. It is also important, of course, that the stirrups be closely spaced.

Effectiveness factors ν are determined as explained above.

[1] Different code committees have assigned different values to these κ-limits through the years, see, for instance, [78.10] [84.11]. The values (5.2.124) were suggested originally and they are now in Eurocode 2 [91.23].

In practice, characteristic values of the concrete strength can be inserted in the formulas for ν. If design values are used, the ν-values will be increased, which is in contradiction with normal safety concepts.

The requirement (5.2.123) regarding the force for which the tensile reinforcement has to be anchored sometimes leads to trouble, and the requirement can, in fact, be decisive for the choice of the κ-value. If a very short anchorage length is wanted, another type of shear reinforcement can be used, consisting of both vertical and horizontal bars in a part of the beam near the support (see the end of this section).

For shear reinforced beams no distinction needs to be made between ordinary beams and prestressed beams except that the ν-value may be taken higher for prestressed beams [cf. formula (5.2.85)]. Since fully prestressed beams are designed to have no cracks for the service load, the limits (5.2.124) can be disregarded in this case. The κ-value can be chosen as the value fully utilizing the concrete strength in the web.

Of course, the minimum shear reinforcement degree (5.2.86) should be kept in mind also for prestressed beams. Effectively this puts a limit on κ of around 5.

The design of shear reinforcement can also be based on the reinforcement formulas developed in Section 2.4.

Consider again the stringer beam (see Fig. 5.2.36). Let us determine the shear reinforcement in a zone with constant shear force V. It is easily seen that a statically admissible stress field can be expressed by

$$\sigma_x = \sigma_y = 0 \ , \qquad \tau_{xy} = \tau = \frac{V}{bh} \tag{5.2.125}$$

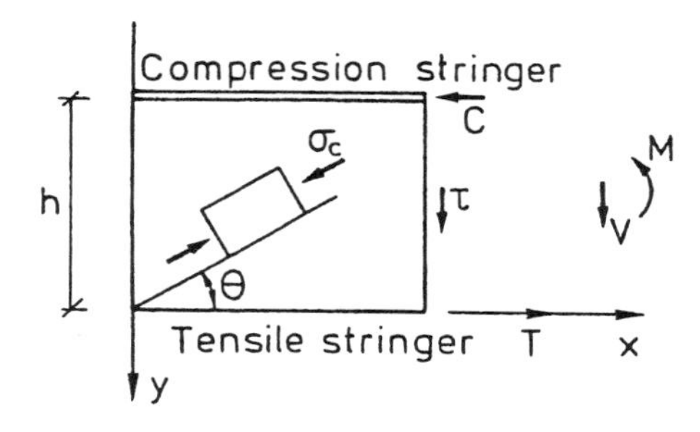

Figure 5.2.36 Beam with constant depth.

The necessary reinforcement in the x- and y-directions and the concrete stress σ_c is determined by (2.4.21)-(2.4.23). We find by replacing γ with κ

$$f_{tx} = \tau\kappa \tag{5.2.126}$$

$$f_{ty} = \frac{\tau}{\kappa} \tag{5.2.127}$$

$$\sigma_c = \tau\left(\kappa + \frac{1}{\kappa}\right) \tag{5.2.128}$$

If the shear zone is reinforced accordingly (i.e., reinforced in both directions x and y), the stringers have to carry the forces

$$T = C = \frac{M}{h} \tag{5.2.129}$$

However, such a shear reinforcement is more expensive than necessary, although it has the advantage of giving zero stringer forces at a simple support, a fact that may facilitate the design of this part of the beam if a small anchorage length of the tensile reinforcement is required.

The shear reinforcement in the x-direction can be avoided since the total force in this direction $V\cot\theta$ can be carried by the stringers. This means that the stringers have to carry the forces

$$\begin{aligned} T &= \frac{M}{h} + \frac{1}{2}V\kappa \\ C &= \frac{M}{h} - \frac{1}{2}V\kappa \end{aligned} \tag{5.2.130}$$

in agreement with (5.2.122). The result implies that we have to reinforce for a tensile stress in the y-direction determined by (5.2.127), to ensure that the stringers can carry the forces (5.2.130) and that the concrete stress σ_c determined by (5.2.128) can be carried by the concrete. Thus we are back to the above procedure.

The total amount of reinforcement can, of course, be minimized with respect to θ [67.2]. The optimum value $\kappa = \cot\theta$ is different from that corresponding to the formulas in Section 2.4 when no x-reinforcement is provided.

The reinforcement volume per unit length of the beam may be written

$$V_s = \frac{|M|}{hf_{Y\ell}} + \frac{1}{2}\frac{V\kappa}{f_{Yw}} + \frac{V}{f_{Yw}\kappa} \tag{5.2.131}$$

Here $f_{Y\ell}$ is the yield strength of the longitudinal reinforcement and f_{Yw} , as before, the yield stress of the shear reinforcement. Extra reinforcement due to bends, etc. is not taken into account. Further the reduction possible when dealing with loadings on the top face is not considered.

The minimum of V_s is found for

$$\kappa = \sqrt{\frac{2f_{Y\ell}}{f_{Yw}}} \tag{5.2.132}$$

It appears that for equal yield stresses the value of κ giving minimum reinforcement is 1.4. The minimum is, however, very flat. Large deviations may be tolerated without influencing the economy in any significant way.

Normally the designer prefers to use the upper κ-limits (5.2.124) when possible, i.e., if this can be done without exceeding the upper value of the concrete stress νf_c.

Beams with normal forces. The problem of finding the shear reinforcement in a beam without normal forces might also be posed in a third way, which is often useful. Suppose that we introduce a constant compressive stress σ_o in the web. If no horizontal reinforcement is provided, the resultant of σ_o

$$N_w = \sigma_o hb \tag{5.2.133}$$

will act upon the end beam and must be equilibrated by tensile forces in the flanges

$$T_o = -C_o = \tfrac{1}{2}N_w \tag{5.2.134}$$

The compressive stress σ_o might be considered as a redundant. Any σ_o will correspond to an equilibrium system.

A reinforcement problem may then be solved by optimizing the reinforcement with respect to σ_o.

Let us use this way of thinking to indicate how the design of reinforcement in a beam with a compression normal force N is done. For the time being we assume for simplicity that the normal force is acting in the middle of the end beam. The equilibrium stress field is illustrated in Fig. 5.2.37. The beam in this case has to carry a constant shear stress τ along a shear span a. Let us, again for simplicity, assume the stringer reinforcement to be constant. Since we only want vertical shear reinforcement we have $\tan\theta = \tau/\sigma_o$.

Using the reinforcement formulas we find the following total reinforcement volume

$$V_s = \begin{cases} \dfrac{\tau a^2 b}{f_{Y\ell}} + \dfrac{\sigma_o hb - N}{f_{Y\ell}} a + \dfrac{\tau^2 ahb}{\sigma_o f_{Yw}} & N \le N_w \\ \dfrac{\tau a^2 b}{f_{Y\ell}} + \dfrac{1}{2} \dfrac{\sigma_o hb - N}{f_{Y\ell}} a + \dfrac{\tau^2 ahb}{\sigma_o f_{Yw}} & N_w < N \le 2\tau ab + N_w \end{cases} \tag{5.2.135}$$

Here $f_{Y\ell}$ is the yield stress of the stringer reinforcement and f_{Yw} the yield stress of the web reinforcement.

When $N < N_w$ both stringers must be reinforced for tension. When $N > N_w$ only the bottom stringer needs tensile reinforcement. When $N = 2\tau ab + N_w$ the bottom stringer is in compression along the whole shear span.

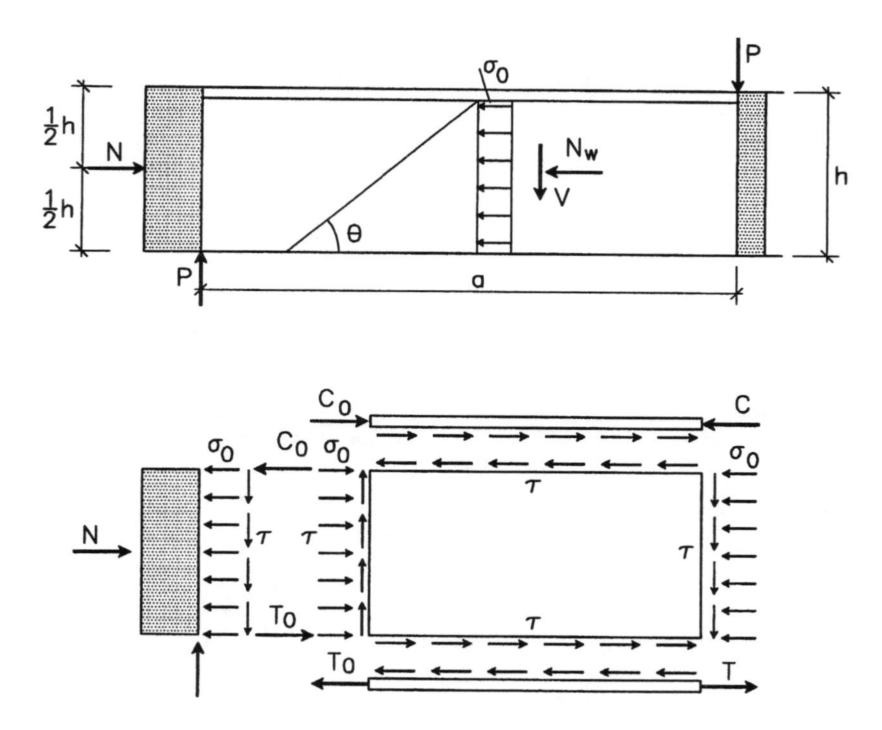

Figure 5.2.37 Beam with normal force subjected to point loads.

If V_s is minimized with respect to σ_o or, what is the same, N_w, we get

$$N \leq N_w \qquad \sigma_o = \tau \sqrt{\frac{f_{Y\ell}}{f_{Yw}}}$$
$$N_w < N \leq 2\tau ab + N_w \qquad \sigma_o = \tau \sqrt{\frac{2f_{Y\ell}}{f_{Yw}}} \tag{5.2.136}$$

Since $\sigma_o / \tau = \cot\theta$ we may introduce the notation

$$\kappa_1 = \sqrt{\frac{f_{Y\ell}}{f_{Yw}}} \tag{5.2.137}$$

$$\kappa_2 = \sqrt{\frac{2f_{Y\ell}}{f_{Yw}}} \tag{5.2.138}$$

Then the result of the minimization may be written

$$N \leq \tau hb\,\kappa_1 \qquad \sigma_o = \tau\kappa_1 \tag{5.2.139}$$

$$\tau hb\,\kappa_1 < N \leq \tau hb\,\kappa_2 \qquad \sigma_o = \frac{N}{hb} \tag{5.1.140}$$

$$\tau hb\,\kappa_2 < N \leq 2\tau ab + \tau hb\,\kappa_2 \qquad \sigma_o = \tau\kappa_2 \tag{5.2.141}$$

In the intermediate case θ is changing from the value corresponding to κ_1 to the value corresponding to κ_2. In the corresponding range of N-values both T_o and C_o are zero. The result implies further that in the upper range of the N-interval considered it pays off to use the normal force to reduce the stringer reinforcement instead of using it to reduce the shear reinforcement.

In fact the minimum is very flat, so the change in V_s by deviating from the minimum value is small.

When $N > 2\tau ab + \tau hb\kappa_2 = N_\ell$ we only have to reinforce the web. This means that minimum reinforcement is found when letting the web carry the normal force exceeding the limit N_ℓ. Thus we have

$$N > N_\ell \qquad \sigma_o = \tau\kappa_2 + \frac{N - N_\ell}{hb} \tag{5.2.142}$$

This can be done as long as the concrete stress in the web does not exceed νf_c, i.e.,

$$\sigma_c = \sigma_o \left(1 + \left(\frac{\tau}{\sigma_o} \right)^2 \right) \le \nu f_c \tag{5.2.143}$$

Using $\sigma_c = \nu f_c$ and solving with respect to σ_o we find the limit (the largest root)

$$\sigma_o = \sigma_{o\max} = \nu f_c \left[\frac{1}{2} + \sqrt{\frac{1}{4} - \left(\frac{\tau}{\nu f_c} \right)^2} \right] \tag{5.2.144}$$

Thus we have

$$N_\ell < N \le \sigma_{o\max} h b + N_\ell - \tau h b \kappa_2$$

$$\sigma_o = \tau \kappa_2 + \frac{N - N_\ell}{hb} \tag{5.2.145}$$

For N exceeding the upper limit in (5.2.145), the part of the normal force exceeding this limit must be carried by the flanges and no additional savings in reinforcement are possible.

Now the whole range of σ_o-values are determined. In any case

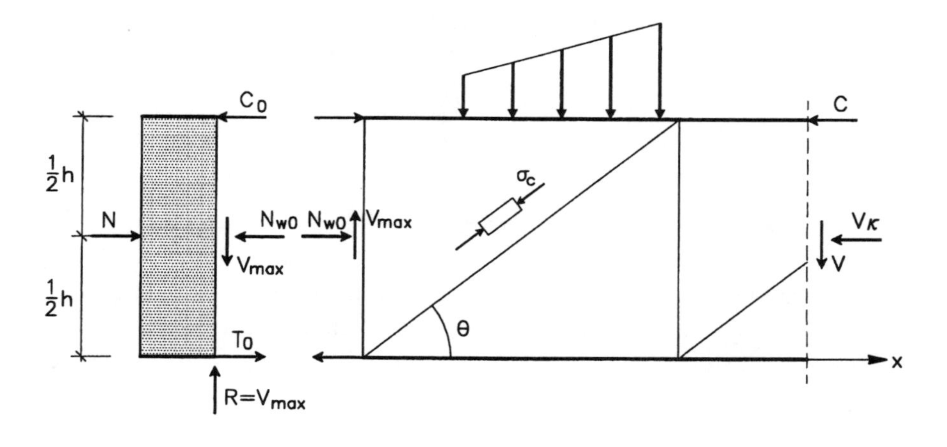

Figure 5.2.38 Beam with normal force and arbitrary transverse load.

$$\kappa = \frac{\sigma_o}{\tau} \tag{5.2.146}$$

and

$$r f_{Yw} = \frac{\tau}{\kappa} \tag{5.2.147}$$

$$\sigma_c = \tau\left(\kappa + \frac{1}{\kappa}\right) = \sigma_o\left(1 + \left(\frac{\tau}{\sigma_o}\right)^2\right) \tag{5.2.148}$$

There might be other choices of κ preferred by the designer. For instance when a substantial part of the tensile stringer is in tension he might prefer to use the limits (5.2.124) valid for beams without normal forces, which should be respected in this case.

In the above calculation it has been tacitly assumed that the concrete stress σ_c does not exceed νf_c neither when $\kappa = \kappa_1$ nor when $\kappa = \kappa_2$.

The above results may easily be extended to a beam with arbitrary, downward directed loading on the top face (see Fig. 5.2.38). We simply use the solution in Fig. 5.2.4, the only difference being that we let the κ-value depend on the normal force.

The free parameter may again be taken to be the normal force in the web at the support. This is $N_{wo} = \sigma_o hb$.

When a proper value of N_{wo} has been selected we have

$$\kappa = \frac{N_{wo}}{V_{max}} \tag{5.2.149}$$

V_{max} being the shear force at the support. Then at any point

$$r f_{Yw} = \frac{\tau'}{\kappa} \tag{5.2.150}$$

τ' having the same meaning as before determining the staggered shear stress diagram (see Fig. 5.2.35). Alternatively, using τ in (5.2.150) instead of τ', rf_{Yw} may be reduced by the average pressure $\frac{1}{2}p$.

Further by moment equations

$$T = \frac{M}{h} + \frac{1}{2}V\kappa - \frac{1}{2}N \tag{5.2.151}$$

$$C = \frac{M}{h} - \frac{1}{2} V\kappa + \frac{1}{2} N \qquad (5.2.152)$$

Finally

$$\sigma_c = \tau\left(\kappa + \frac{1}{\kappa}\right) \leq \nu f_c \qquad (5.2.153)$$

Notice that in formulas (5.2.151)-(5.2.153) it is N and not N_{wo} which has to be used. Further notice that the normal force in the web in an arbitrary section is $V\kappa$ and not N_{wo}. The reason why the normal force in the web is decreasing with decreasing V is that the shear stresses at the top and the bottom face are different.

In fact the only change from the case $N = 0$ to $N \neq 0$ is the extra terms in (5.2.151) and (5.2.152).

The only problem left is the choice of κ.

For small normal forces the usual limits of κ (5.2.124) should, of course, be used. The definition of small normal forces may be

$$N \leq N_\ell \qquad (5.2.154)$$

where N_ℓ is the normal force transforming the tensile flange into a compression flange, i.e., for $N = N_\ell$, $T \leq 0$ everywhere.

This means that for

$$N \leq N_\ell \,, \quad \kappa = \kappa_o \,, \quad N_{wo} = V_{\max}\kappa_o \qquad (5.2.155)$$

κ_o normally being chosen at the limits given by (5.2.124) unless the condition $\sigma_c \leq \nu f_c$ requires a smaller value.

For $N > N_\ell$ the additional normal force may be used to reduce the shear reinforcement, which means that $N - N_\ell$ is added to N_{wo}.

In other words

$$N > N_\ell \qquad N_{wo} = V_{\max}\kappa_o + (N - N_\ell) \qquad \Rightarrow$$

$$\kappa = \frac{N_{wo}}{V_{\max}} = \kappa_o + \frac{N - N_\ell}{V_{\max}} \qquad (5.2.156)$$

This κ-value may be used until the concrete stress in the web exceeds the limit νf_c. This is the case when

$$N_{wo} = N_{wo\max} = \sigma_{o\max} hb \qquad (5.2.157)$$

$\sigma_{o\max}$ being determined by (5.2.144), replacing τ with $\tau_{\max} = V_{\max}/hb$.

According to (5.2.156), $N_{wo} \leq N_{wo\max}$ for

$$N \leq N_{wo\max} - V_{\max}\kappa_o + N_\ell \qquad (5.2.158)$$

which means that the formula (5.2.156) for κ is valid for

$$N_\ell \le N \le N_{wo\,max} - V_{max}\,\kappa_o + N_\ell \tag{5.2.159}$$

Normal forces higher than the right-hand side of this expression must be carried by the flanges and no further reduction in shear reinforcement is possible.

The ν-values may on the safe side be chosen as for bending and shear without normal force. For large normal forces it may be advantageous to use formula (4.6.10) involving the χ-parameter, since in this case the shear reinforcement will be small. This formula has the additional advantage that when there is no shear in the web ν = 1, a quite natural result.

The minimum requirement (5.2.86) for the stirrup reinforcement must, of course, be respected.

In the solution presented we have tacitly assumed that equilibrium is satisfied in the undeformed configuration. When designing a beam column such a simplification normally is not justified.

Exact equilibrium in the deformed configuration requires a number of additional terms in the functions determining the stress field. However, for practical purposes, only one correction is necessary. This is illustrated in Fig. 5.2.39. In the traditional beam column theory the sectional forces in the deformed configuration usually are referred to the directions of a global rectangular coordinate system, say a x, y-system and not to a local, rectangular n, t-system as shown in the figure. This means that the normal force N and the shear force V are not perpendicular and parallel, respectively, to the normal section in the deformed configuration. Expressing the statical equivalence of the true normal force N' and the true shear force V' by N and V we get for small deformation gradients

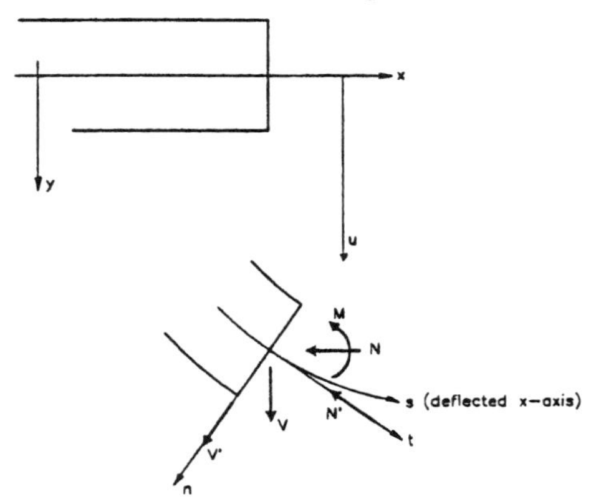

Figure 5.2.39 Sectional forces in a deflected beam.

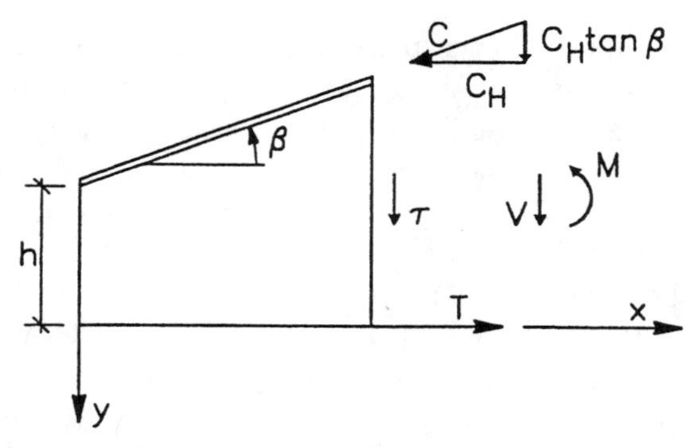

Figure 5.2.40 Beam with variable depth.

$$N' = N$$
$$V' = V + N\frac{du}{dx} \tag{5.2.160}$$

u being the deflection of the beam perpendicular to the undeformed beam axis, the x-axis.

The contribution $N\,du/dx$ to V' is not always small and should be added to the shear force determined in the undeformed configuration when designing the shear reinforcement. Notice that N is positive as compression.

Beams with variable depth. If the beam has variable depth (see Fig. 5.2.40) and constant shear force, we may reinforce for the statically admissible stress field:

$$\sigma_x = 0$$
$$\sigma_y = 2\tau\frac{y}{h}\tan\beta \tag{5.2.161}$$
$$\tau_{xy} = \tau = \frac{1}{bh}\left(V - \frac{M}{h}\tan\beta\right)$$

which is easily seen to satisfy the equilibrium equations and the boundary conditions. The σ_y-stress generally is small and can be neglected. Doing so, (5.2.126)-(5.2.128) are still approximately valid. If in (5.2.130) C is replaced by the horizontal component $C_H \approx M/h$ (see Fig. 5.2.40) and V by $|V - (M/h)\tan\beta|$, formulas (5.2.130) are also approximately valid.

Beams with bent-up bars or inclined prestressing reinforcement. Lower bound solutions for beams with shear reinforcement in the form of bent-up bars only exist in the most primitive form as strut and tie solutions. When using these solutions it is important to check the concrete stresses in the bent-up points. They may be very high if the curvature in the bend is small. Examples may be found in [75.4] and [90.16]. Alternatively, upper bound solutions may be used [78.7] [79.5].

For beams with inclined prestressing reinforcement a simplified method has been described in [83.1]. The idea is well known. The shear stresses to be carried by the concrete are determined as the shear force minus the component along the normal section of the force in the inclined prestressing reinforcement (positive in the direction of the positive direction of the shear force). The stress in the prestressing reinforcement may normally be taken to be the yield stress. When the shear stresses to be carried by the concrete have been determined, the shear reinforcement may be determined as described above. The formulas for the forces in the stringers, of course, must be modified to take into account the forces in the inclined prestressing reinforcement.

Variable θ solutions. In the solutions above we have steadily thought of a stress field in the web corresponding to constant inclination of the uniaxial compression. This is convenient but not necessary. In fact a rather arbitrary set of smooth non-intersecting curves can be chosen as representing the direction of the uniaxial compression. They only have to agree with the sign of the shear stress.

Let us imagine that we have selected a set of such curves and that the inclination of the tangents to these curves have been used when computing the necessary reinforcement. Then for an equilibrium stress field of total stresses σ_x, σ_y and τ_{xy} we have arrived at a set of concrete stresses and reinforcement stresses which, at each place, will be able to carry the stresses σ_x, σ_y and τ_{xy} assuming, of course, $\sigma_c \leq \nu f_c$. Now in practice the reinforcement will be chosen higher or in special cases equal to the required reinforcement. This means that stresses have to be transferred between the concrete and the reinforcement bars, stresses which we have neglected so far. Considering orthogonal reinforcement the stresses in the reinforcement bars σ_{sx} and σ_{sy} may be calculated by the formulas

$$r_x \sigma_{sx} = \sigma_x - \sigma_{cx} \tag{5.2.162}$$

$$r_y \sigma_{sy} = \sigma_y - \sigma_{cy} \tag{5.2.163}$$

Here r_x and r_y are reinforcement ratios and σ_{cx} and σ_{cy} are normal stresses in the concrete, positive as compression, i.e.,

$$\sigma_{cx} = |\tau_{xy}| \cot\theta \tag{5.2.164}$$

$$\sigma_{cy} = |\tau_{xy}| \tan\theta \tag{5.2.165}$$

As before, θ is the angle between the x-axis and the direction of the uniaxial compression.

Only when r_x and r_y are exactly what is required will the reinforcement stresses σ_{sx} and σ_{sy} be equal to the yield stresses. When r_x and r_y are higher than the required values, the reinforcement stresses will be smaller. Furthermore, they will be varying. Using the equilibrium equations we find (no body forces)

$$\frac{\partial r_x \sigma_{sx}}{\partial x} = \frac{\partial \sigma_x}{\partial x} - \frac{\partial \sigma_{cx}}{\partial x} = -\left[\frac{\partial \tau_{xy}}{\partial y} + \frac{\partial \sigma_{cx}}{\partial x}\right] \tag{5.2.166}$$

$$\frac{\partial r_y \sigma_{sy}}{\partial y} = -\left[\frac{\partial \tau_{xy}}{\partial x} + \frac{\partial \sigma_{cy}}{\partial y}\right] \tag{5.2.167}$$

We could also write τ_{cxy} instead of τ_{xy} since only the concrete carries shear stresses.

Since there is an upper limit for the rate by which reinforcement stresses can change, these equations in fact put restrictions both on the equilibrium stress field and the uniaxial compression direction. If the reinforcement ratios are constant we must require

$$\frac{\partial r_x \sigma_{sx}}{\partial x} \leq r_x \frac{f_Y}{\ell} \tag{5.2.168}$$

$$\frac{\partial r_y \sigma_{sy}}{\partial y} \leq r_y \frac{f_Y}{\ell} \tag{5.2.169}$$

f_Y being the yield stress and ℓ the anchorage length. If these conditions are not satisfied, the reinforcement must be increased or bars with larger anchorage length must be used.

In homogeneous stress fields the anchorage takes place only at the boundaries and σ_{sx} and σ_{sy} will be constant in the interior of the structure.

To avoid the transfer of stresses from reinforcement to concrete except along the boundaries, for slender beams it is natural to look for solutions rendering constant r_y for x = const. and $r_x = 0$, i.e., r_y is only a function of x. The last condition may always be fulfilled if σ_x is a compressive stress. Then $\sigma_{cx} = \sigma_x$. Further, it will be natural to look for solutions with σ_c = const. $= \nu f_c$. However, this is only possible when we are dealing with homogeneous stress fields which may be shown in the following way. When $r_x = 0$ and r_y = const. for x = const. the concrete stresses must satisfy the equilibrium equations. For $\sigma_c = \nu f_c$ we have

$$\sigma_x = \sigma_{cx} = -\nu f_c \cos^2\theta$$

$$\sigma_y = \sigma_{cy} = -\nu f_c \sin^2\theta \tag{5.2.170}$$

$$\tau_{xy} = \tau_{cxy} = \nu f_c \sin\theta\cos\theta = \frac{1}{2}\,\nu f_c \sin 2\theta$$

That means

$$\frac{\partial \sigma_x}{\partial x} + \frac{\partial \tau_{xy}}{\partial y} = 2\cos\theta\sin\theta\,\frac{\partial\theta}{\partial x} + \cos^2\theta\,\frac{\partial\theta}{\partial y} - \sin^2\theta\,\frac{\partial\theta}{\partial y}$$

$$= \sin 2\theta\,\frac{\partial\theta}{\partial x} + \cos 2\theta\,\frac{\partial\theta}{\partial y} = 0 \tag{5.2.171}$$

and

$$\frac{\partial \sigma_y}{\partial y} + \frac{\partial \tau_{xy}}{\partial x} = -\sin 2\theta\,\frac{\partial\theta}{\partial y} + \cos 2\theta\,\frac{\partial\theta}{\partial x} = 0 \tag{5.2.172}$$

Thus we have

$$\frac{\partial\theta}{\partial x} = -\cot 2\theta\,\frac{\partial\theta}{\partial y} \tag{5.2.173}$$

$$\frac{\partial\theta}{\partial x} = \tan 2\theta\,\frac{\partial\theta}{\partial y} \tag{5.2.174}$$

These conditions can only be fulfilled if θ = const.

Thus for beams there are no curved strut solutions with constant concrete stress. We must be content to utilize fully the concrete strength only in highly stressed regions.

Curved strut solutions have not been developed so far. They are not appealing to the designer because they lead to a continuously varying reinforcement ratio r_y instead of piecewise constant r_y. Nevertheless, let us consider a simplified curved strut solution for a uniformly loaded beam (see Fig. 5.2.41). Notation is as before.

The simplification consists of choosing a uniform distribution of shear stresses in the sections of the curved strut region. Then r_y will not be exactly constant because of the σ_y-stresses. However, the σ_y-stresses may be taken into account reducing $r_y f_{Yw}$ by $\frac{1}{2}p$ as before. In the solution there is a curved strut domain a-x' and a constant θ domain x' which end at the maximum moment section. In the region x' the stress field is the familiar one already considered (see Fig. 5.2.4). We imagine a constant normal stress σ_o being introduced by the end beam in the region a-x'. The inclination of the strut in the region a-x' is then determined by

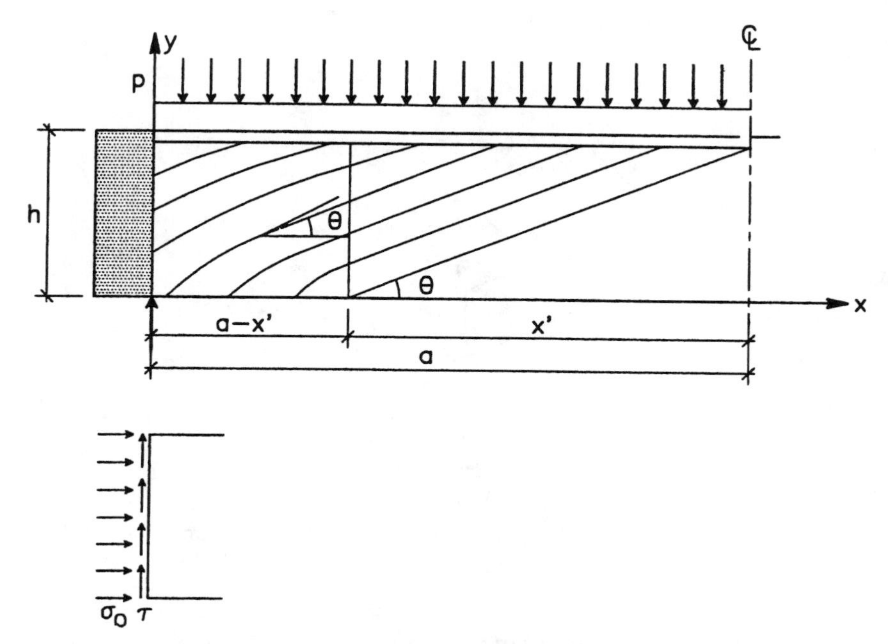

Figure 5.2.41 Curved strut solution for uniformly loaded beam.

$$\cot\theta = \frac{\sigma_o}{\tau} \tag{5.2.175}$$

τ being the shear stress. It is easily shown that the curves describing the strut inclination are parabolas, all being identical. This means that the curves are described by a mother curve which is translated up and down. To make the two regions fit at $x = a - x'$ we must have

$$\frac{px'}{h\,\sigma_o} = \frac{h}{x'} \quad \Rightarrow \tag{5.2.176}$$

$$\frac{x'}{h} = \sqrt{\frac{\sigma_o}{p}} \tag{5.2.177}$$

Notice that, as before, p is the stress at the top face of the web. The condition $x' \leq a$ may impose a limitation on the pressure σ_o.

The necessary shear reinforcement is easily determined.

$$\begin{aligned} &\text{Region } a - x'\text{:} \quad r_y f_{yw} = \frac{\tau^2}{\sigma_o} - \frac{1}{2}p = \frac{\tau}{\cot\theta} - \frac{1}{2}p \\ &\text{Region } x'\text{:} \quad r_y = 0 \end{aligned} \tag{5.2.178}$$

The concrete stresses are

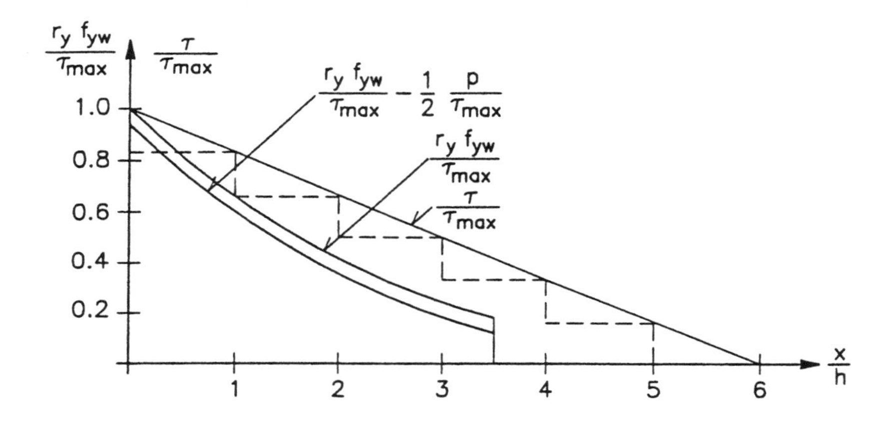

Figure 5.2.42 Curved strut solution compared with constant θ solution.

$$\text{Region } a-x': \quad \sigma_c = \tau(\tan\theta + \cot\theta) = \sigma_o\left(1 + \left(\frac{\tau}{\sigma_o}\right)^2\right)$$

$$\text{Region } x': \quad \sigma_c = \text{const.} = \sigma_o\left(1 + \left(\frac{\tau'}{\sigma_o}\right)^2\right) \tag{5.2.179}$$

where τ' is the shear stress at $x = a\text{-}x'$.

Finally the stringer forces are

$$\text{Region } a-x': \quad T = \frac{M}{h} + \frac{1}{2}\sigma_o hb$$

$$C = \frac{M}{h} - \frac{1}{2}\sigma_o hb \tag{5.2.180}$$

In region x' the tensile stringer force T is constant, equal to the value at the maximum moment. The compression stringer force is decreasing linearly from the value at the maximum moment to the value at $x = a - x'$.

Any value of σ_o may be chosen by the designer. When the limits (5.2.124) must be obeyed, these may restrict the value of σ_o through formula (5.2.176) since x'/h is $\cot\theta$ in region x'. Of course, the requirement $\sigma_c \leq \nu f_c$ also may restrict the value of σ_o . Maximum σ_c generally is at the support $x = 0$.

The solution has been illustrated in Fig. 5.2.42 for the case $a = 6h$ and $p = f_c/12$ ($\nu = 1$). We then have $\tau_{max} = \frac{1}{2}f_c$ and with $\sigma_o = \tau_{max} = \frac{1}{2}f_c$ we get

$\sigma_c = f_c$ at the support. The reinforcement required is shown both with and without reduction by $\frac{1}{2}p$. It is left for the reader to calculate the concrete stresses and the stringer forces. In the figure the curved strut solution has been compared with the constant θ solution. In this case $\cot\theta = 1$ throughout. The curved strut solution offers some savings in the intermediate part of the beam. In the part of the beam where shear reinforcement is not necessary according to the solution, minimum reinforcement according to (5.2.86) must, of course, be supplied. This is not shown in the figure.

Instead of using a constant stress σ_o it would be tempting to use a $\cot\theta = \kappa$ value in each point of the beam utilizing fully the concrete strength. Such a κ-value may be found by solving (5.2.121) for κ with $\sigma_c = \nu f_c$. Since $\tau = \sigma_c \cos\theta \sin\theta = \frac{1}{2}\sigma_c \sin 2\theta$, alternatively θ may be determined by

$$\sin 2\theta = \frac{2\tau}{\nu f_c} \tag{5.2.181}$$

However, when τ is varying along the beam axis the compression stress $\sigma_o = \nu f_c \cos^2\theta = \tau \cot\theta$ will vary which means that the local equilibrium condition expressing horizontal balance will not be fulfilled, at least for a constant τ in the sections. Thus, strictly speaking, horizontal reinforcement in the web must be supplied.

Nevertheless, in [80.7] a method has been suggested where κ is allowed to vary in such a way that $\sigma_c = \nu f_c$ without supplying horizontal reinforcement in the web. The calculation is started in the section with the largest shear stress and θ is calculated by (5.2.181) or by the formula obtained by solving (5.2.121) for κ with $\sigma_c = \nu f_c$.

The value found, say κ_1, is used for calculation of the shear reinforcement in the interval of length $\kappa_1 h$. Then a new value of κ, say κ_2, is determined at the end of the first interval and this κ-value is used in the next interval of length $\kappa_2 h$, etc. To take into account the reduction in shear reinforcement caused by loading on the top face, the smallest shear stress in the interval in question is used when determining the shear reinforcement. When calculating the stringer forces by using piecewise constant κ-values, the stringer forces will have jumps. This is, of course, on the unsafe side. As a practical compromise it is recommended either to use the envelope curve of the stringer force curve determined or to use constant reinforcement in the stringers.

The method is certainly not a lower bound method because of the discontinuities in the σ_x-stresses. It has been compared with upper bound solutions in [84.9] and it was found that the method is reasonably safe from a practical point of view.

Lightly reinforced beams. In Section 5.2.5 we concluded on the basis of the tests that the web crushing criterion could be applied if the shear reinforcement degree satisfies the minimum requirement (5.2.86), i.e.,

$$\psi = \psi_{min} = \frac{0.16}{\sqrt{f_c}} \qquad (f_c \text{ in MPa}) \qquad (5.2.182)$$

Beams with lower shear reinforcement are termed lightly reinforced beams. They cannot be designed according to the plastic methods described above because of the danger of sliding in initial cracks. Such beams should, in fact, only be used for inferior structures or better; they should be avoided as far as possible. However, this is not in agreement with practice for economical reasons. Therefore, it is necessary to treat such beams.

It is rather easy to extend the Jin-Ping Zhang theory to cover the effect of shear reinforcement. The work has been done by Linh Hoang [97.9].

Instead of (5.2.110) for the shear capacity corresponding to a crack with the horizontal projection x, we have now

$$\tau = \tau_c \frac{2}{x/h} + r f_{Yw} \frac{x}{h} \qquad (5.2.183)$$

This is nothing else than a transcription of formula (5.2.45).

For $x/h < 0.75$ formula (5.2.183) is not valid due to the normality condition. For $x/h < 0.75$, the original plastic solution (5.2.45) is used with $\nu = \nu_o$. Thus a discontinuity in shear capacity is present for $x/h = 0.75$.

The cracking moment may be calculated as before, thereby neglecting the influence of the shear reinforcement on the cracking moment.

In Fig. 5.2.43 the cracking load and the shear capacity are schematically shown in the case of a beam with shear reinforcement. To simplify matters we imagine we are in the point load case. As in Fig. 5.2.23 the loads which can be carried are depicted as a function of x', the distance of the crack from the support. The only difference from Fig. 5.2.23 is that now the shear capacity curve may display a minimum value. Thus if the cracking load curve intersects the shear capacity curve for an x'-value larger than that corresponding to the minimum value, the load-carrying capacity will correspond to the intersection point. If, however, the minimum value is reached for an x'-value larger than that corresponding to the intersection point, the load-carrying capacity will be determined by the minimum value of the shear capacity curve. In the last mentioned case the failure will not take place when the crack is formed. Only when the load has been increased to the load corresponding to the minimum value of the shear capacity curve does failure take place. Both cases have been illustrated in the figure.

As in the case of nonshear reinforced beams it may happen that there is no intersection point for x'-values within the beam region or that the minimum value lies outside the beam region. In these cases the load-carrying capacity is determined by inserting into (5.2.183) the largest x-value possible.

The minimum value of shear capacity is easily determined by minimizing τ in (5.2.183) with respect to x/h. The result is

$$\frac{x}{h} = \sqrt{\frac{2\tau_c}{rf_{Yw}}} \tag{5.2.184}$$

and

$$\frac{\tau}{\tau_c} = \sqrt{\frac{8rf_{Yw}}{\tau_c}} \tag{5.2.185}$$

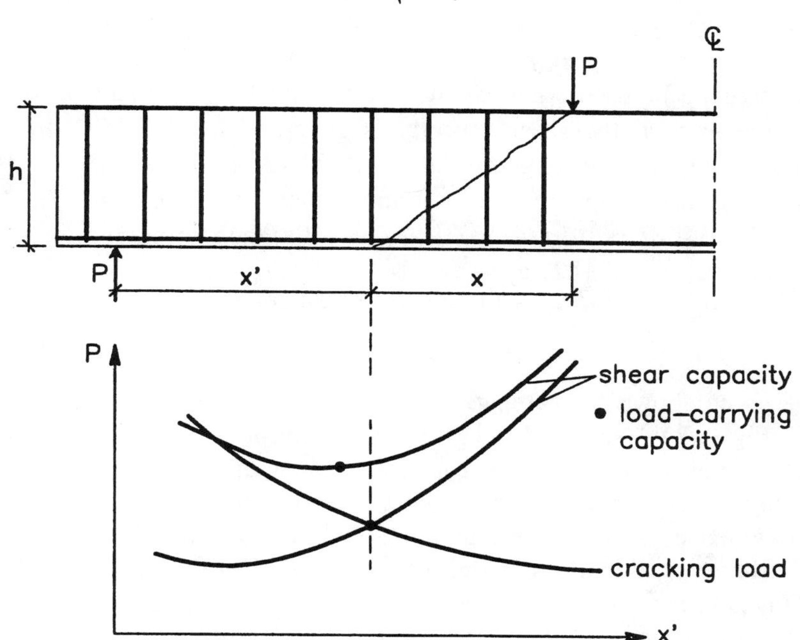

Figure 5.2.43 Crack formation in a beam with shear reinforcement.

Thus the load-carrying capacity of lightly shear-reinforced beams is either determined by (5.2.185), by (5.2.183) inserting the largest possible x-value or by the intersection point between the cracking load curve and the shear capacity curve. The last mentioned value of the load-carrying capacity according to Fig. 5.2.43 will always be higher than the minimum value (5.2.185).

If x/h according to (5.2.184) is found less than 0.75, the load-carrying capacity is determined by (5.2.183), inserting $x/h = 0.75$.

Inserting into formulas (5.2.184) and (5.2.185) $\tau_c = 1.3\ \tau_o$ [cf. formula (5.2.111)], they may be written

$$\frac{x}{h} = \sqrt{\frac{2.6\,\tau_o}{rf_{Yw}}} \tag{5.2.186}$$

$$\frac{\tau}{\tau_o} = \sqrt{\frac{10.4\,rf_{Yw}}{\tau_o}} \tag{5.2.187}$$

In Fig. 5.2.44 the shear capacity has been depicted as a function of the shear reinforcement. By drawing the curve it has been assumed that $a/h = 6$, $f_{tef} = \tau_c = 1.3\ \tau_o$, f_{tef} being the effective tensile strength to be used when calculating the cracking moment. In most of the interval the shear capacity is governed by (5.2.187). In the figure the web crushing criterion (5.2.14) has also been shown. By drawing this curve it has been assumed that $f_c = 16 f_t$, $\tau_o = \frac{1}{2} f_t$ and $\nu = 0.58$. With $f_t = \sqrt{0.1 f_c} = 0.32\sqrt{f_c}$ (f_c in MPa) the minimum reinforcement degree (5.2.182) corresponds to $r f_{Yw} / \tau_o = 1$.

Finally in Fig. 5.2.44 the old empirical formula

$$\tau = \tau_o + rf_{Yw} \tag{5.2.188}$$

has been shown for comparison.

Fig. 5.2.44 shows that lightly shear-reinforced beams may have considerably lower shear capacity than that determined by the diagonal compression stress field because the shear capacity is governed by sliding in initial cracks instead of web crushing.

However, the difference in shear capacity is not as large in reality. This is because we have in both cases used the same reference depth, which is not correct. For crack sliding the effective depth is the total depth while the depth to be used in the web crushing solution is the internal lever arm at the maximum moment. A correct comparison therefore requires the crack sliding solution to be multiplied by the ratio total depth/internal lever arm. Doing so we find the reduction due to crack sliding less dramatic; in fact, in many cases it turns out to be negligible. Thus it might be that the true reason why crack sliding may be neglected, if a certain minimum reinforcement is provided, is not due to the minimum reinforcement but due to the fact that for beams with this and higher shear reinforcement the crack sliding solution agrees reasonably well with the web crushing criterion.

The crack sliding solution intersects the web crushing solution for a ψ-value, which is normally considerably larger than ψ_{min}. Thus it may be reasonable to use the crack sliding solution in the entire ψ-region where it, according to theory, governs the load-carrying capacity.

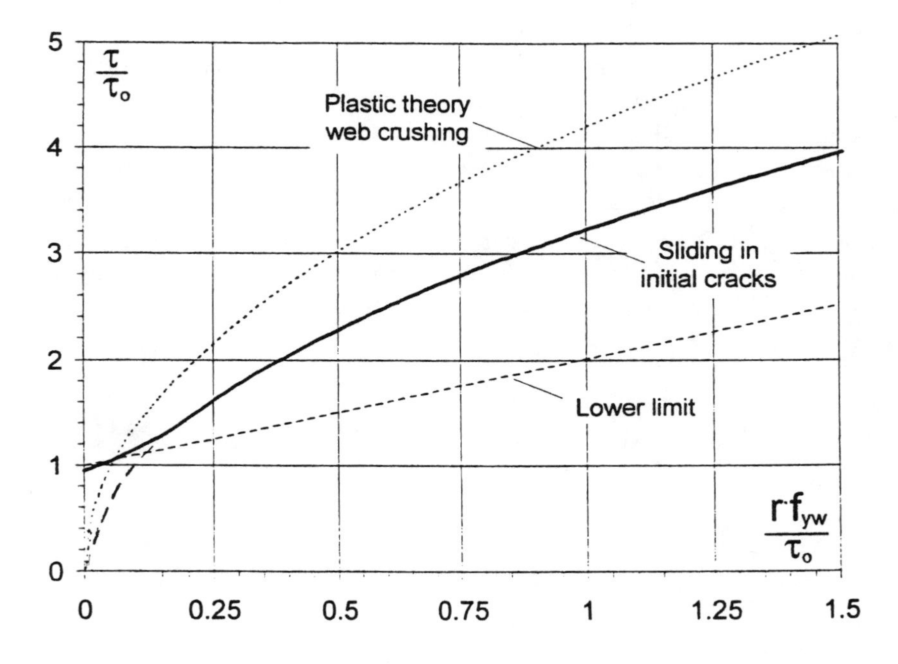

Figure 5.2.44 Shear capacity of lightly shear-reinforced beams.

The results of our efforts are quite simple but in many cases they will be unrealistic. When we are dealing with small shear reinforcement degrees, the stirrup distance may be quite large which means that the assumption tacitly made when deriving (5.2.187), namely that the shear reinforcement is uniformly distributed, often is quite wrong.

The effect of finite stirrup distances has been neglected so far. However, it should be mentioned that Ichinose (see [95.4] with further reference to Japanese research), has suggested that part of the reduced concrete strength in shear problems may be due to the finite stirrup distances.

To accurately take the effect of finite stirrup distances into account, the curve for the shear capacity in Fig. 5.2.43 must be drawn by counting the number of stirrups crossing the cracks. This is rather easy but cumbersome. If the stirrup distance is s and the reinforcement ratio is r, one stirrup crossed adds srf_{Yw}/h to the shear capacity τ. The shear capacity curve will, of course, be discontinuous.

When the minimum value of the shear capacity curve is governing the load-carrying capacity the effect of finite stirrup spacing may be found in the following way. Under that assumption, only cracks running from the bottom of one stirrup to the top of another stirrup have to be considered.

In Fig. 5.2.45 cracks crossing one or two stirrups, respectively, have been shown. Denoting the number of stirrups being crossed by N and the stirrup spacing by $s = \alpha h$ we find

$$
\begin{aligned}
N &= 0 \qquad && \frac{\tau}{\tau_c} = \frac{2}{\alpha} \\
N &= 1 \qquad && \frac{\tau}{\tau_c} = \frac{1}{\alpha} + \alpha \frac{rf_{Yw}}{\tau_c} \\
N &= 2 \qquad && \frac{\tau}{\tau_c} = \frac{2}{3\alpha} + 2\alpha \frac{rf_{Yw}}{\tau_c} \\
&\cdots\cdots && \\
N &= N \qquad && \frac{\tau}{\tau_c} = \frac{2}{(N+1)\alpha} + N\alpha \frac{rf_{Yw}}{\tau_c}
\end{aligned}
\qquad (5.2.189)
$$

The shear capacity has been depicted according to these formulas for $s = 0.75h$, h, and $1.5h$ in Fig. 5.2.46. Notice that this time τ/τ_c has been drawn as a function of rf_{Yw}/τ_c.

The curves demonstrate that the stirrup spacing may have a dramatic effect on the shear capacity. For $s = 1.5h$ there is, according to our assumption, almost no effect of shear reinforcement at all.

Even for $s = 0.75h$ there is a clear reduction.

The conclusion is obvious. For large stirrup spacings the load-carrying capacity must be determined taking into account the stirrup spacing. It should be noticed that even the old empirical formula (5.2.188), which may be written $\tau/\tau_c = 0.77 + rf_{Yw}/\tau_c$, is unsafe when the stirrup distance is larger than about h. To approach the theoretical result stirrup distances of the order $s = 0.4h$ must be used.

As pointed out by Linh Hoang [97.9], a simple, approximate formula for the effect of finite stirrup spacing may be written down. For any yield line it will be on the safe side to reduce the contribution from the average value r of the reinforcement ratio by the contribution from one stirrup. Thus in the work equation (5.2.183) rx is replaced by $r(x - s)$. The minimum value of the shear capacity is then found to be

$$\frac{\tau}{\tau_c} = \sqrt{\frac{8rf_{Yw}}{\tau_c}} - \frac{rf_{YW}}{\tau_c}\frac{s}{h} \qquad (5.2.190)$$

which replaces (5.2.185). Formula (5.2.184) is not changed.

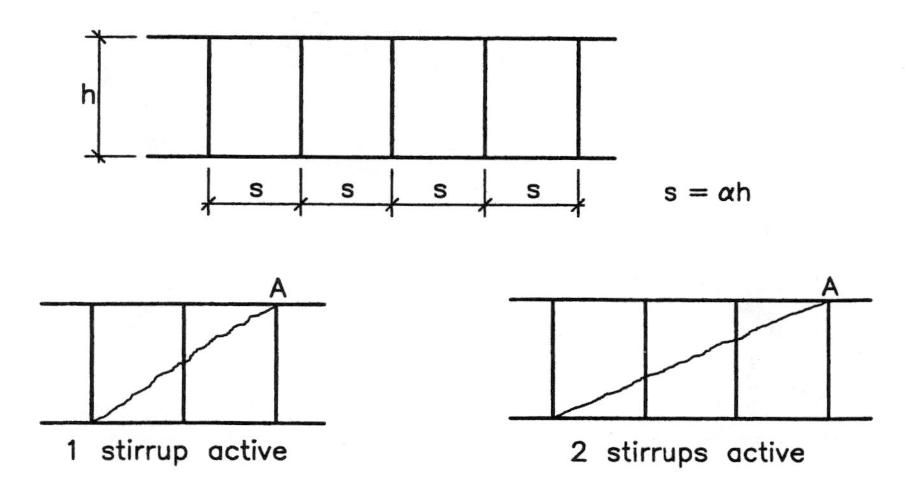

Figure 5.2.45 Yield lines for finite stirrup spacing.

If we disregard whether a dangerous crack has actually been formed or not, which is always a conservative approach, the design of lightly shear-reinforced beams for any loading case is rather easy. We may again refer to Fig. 5.2.10. For constant reinforcement ratio the shear capacity curve in Fig. 5.2.10 is drawn replacing formula (5.2.45) by (5.2.183) and (5.2.41) by the minimum value (5.2.185). Finite stirrup distances may be approximately taken into account as described above. The actual shear stress diagram is then required everywhere to lie below or on the shear capacity curve. If the shear stress diagram just touches the shear capacity curve crack sliding may occur. From the analysis the most dangerous crack also is found and it might be checked whether this crack actually has been formed for a load equal to or smaller than the load determined. If this is not the case a more complicated analysis along the lines described in Section 5.2.6 must be carried out to find a more correct and higher load-carrying capacity.

When carrying out the design we may, as mentioned before, use the total depth h of the section as for nonshear reinforced beams.

In [97.9] the theory has been compared with tests covering even continuous beams. Very good agreement was found.

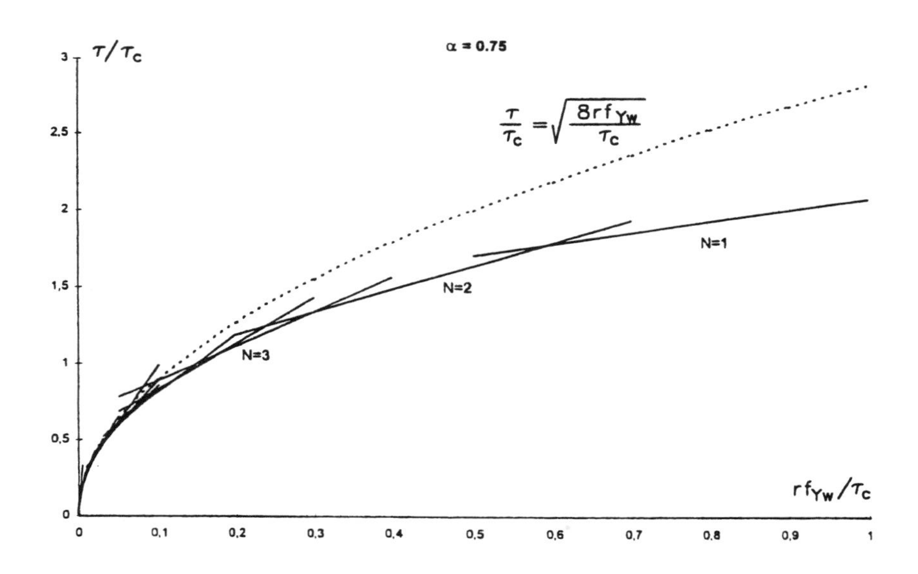

Figure 5.2.46 Shear capacity for finite stirrup spacing.

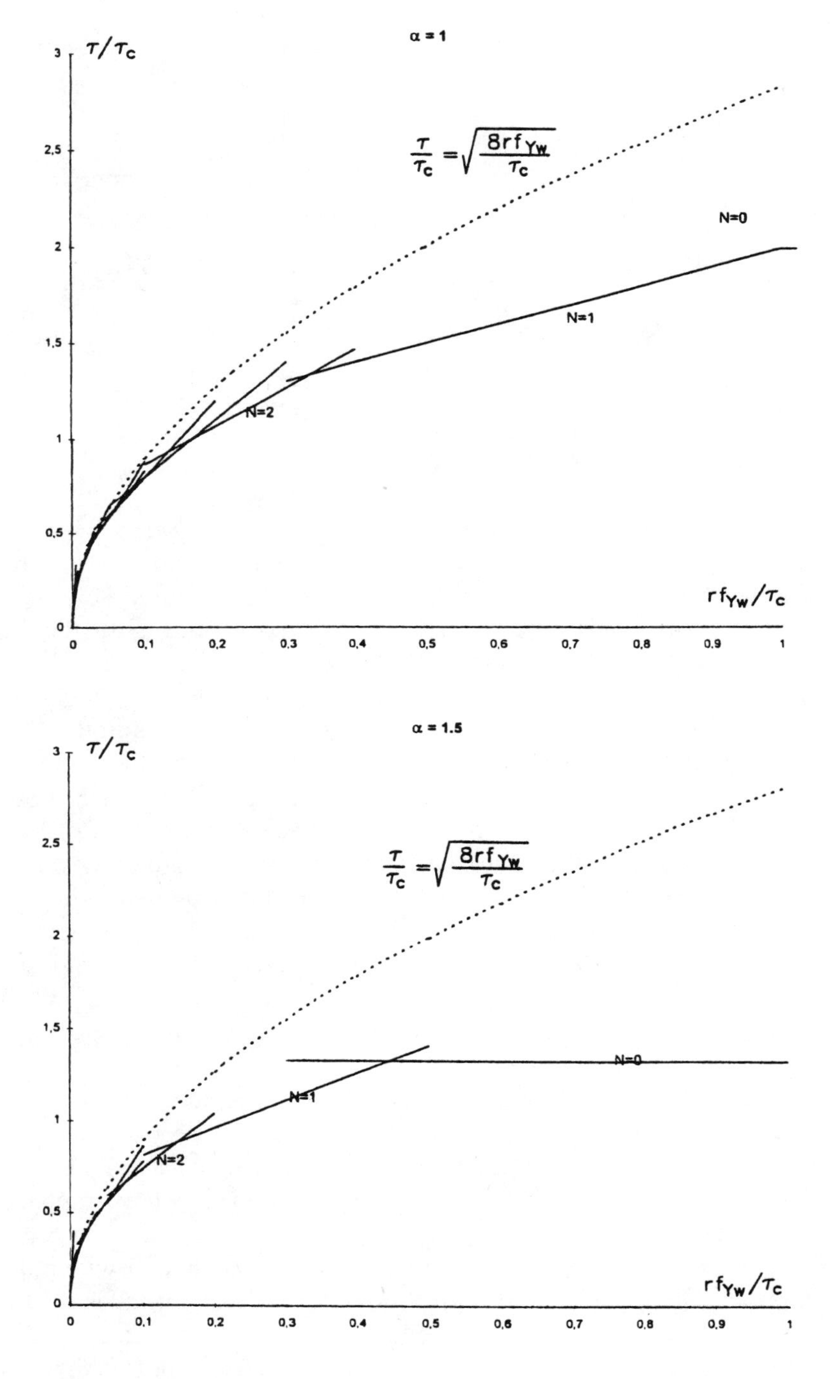

Figure 5.2.46 (continued).

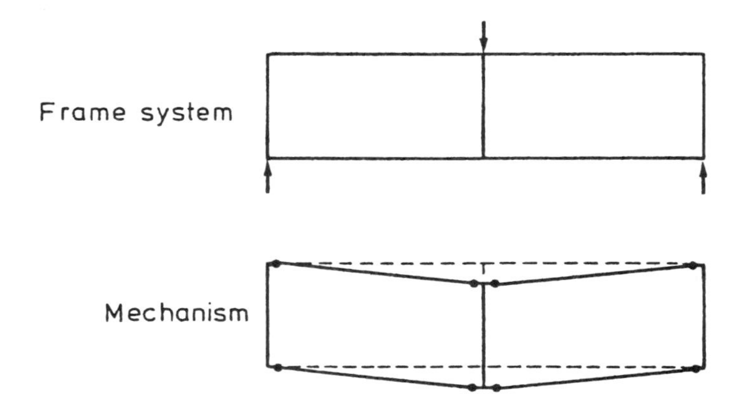

Figure 5.2.47 Frame action in beams with heavy flanges.

Beams with strong flanges. If a beam has stronger flanges than necessary to carry the stringer forces, it would be natural to utilize the bending and shearing strength of the flanges. The simplest way of doing this, when applying a lower bound method, is to superimpose the stress fields given above on a stress field corresponding to ordinary beam action in a frame system composed by the flanges, the end sections, and, if necessary, some compression struts in the shear zone. Figure 5.2.47 illustrates how such a stress field can be determined in a simple example. The bending moments are chosen here in accordance with the mechanism shown in the figure. When the ratio between the bending moments in the hinges has been selected, the values of the bending moments can be determined by the work equation. Having done this, the normal forces and the shear forces can be determined by equilibrium equations. Superimposing the stress fields mentioned, the reinforcement in the flanges and the shear zone can be calculated.

Beams with arbitrary cross section. The rectangular section, sometimes in the form of a rectangular web, has been discussed at some length, which is justified because of its practical importance. There are, however, many other beam sections. They might be triangular, circular or we may have to deal with a complicated bridge section (see. Fig. 5.2.48). The sections may be massive or with holes.

Here the tremendous power of the reinforcement formulas may be utilized. Such sections are calculated by selecting an effective thin-walled section. Some possible choices have been indicated in the figure by the shaded areas. The stresses in these thin-walled sections are calculated by conventional methods or simple equilibrium stress fields are chosen.

When the stress field has been found, the reinforcement formulas immediately provide the necessary reinforcement.

For small beams reinforcement is often only supplied in one layer in the thin-walled section. Then the ν-value used should be chosen conservatively, for instance by using the lower values given in Section 2.1.4. Probably the reinforcement layer is best placed as close to the surface as the required cover allows. It might be reasonable to disregard the cover in this case (see the following section on torsion).

In large structures the thin-walled sections should always be reinforced by at least two layers of reinforcement, one layer along each surface. The stirrups are carried out as closed stirrups wherever possible. In such cases the ν-values valid for rectangular sections reinforced with closed stirrups may be used.

Actually this is the method we are going to develop in the following section when treating beams in torsion and with combined actions.

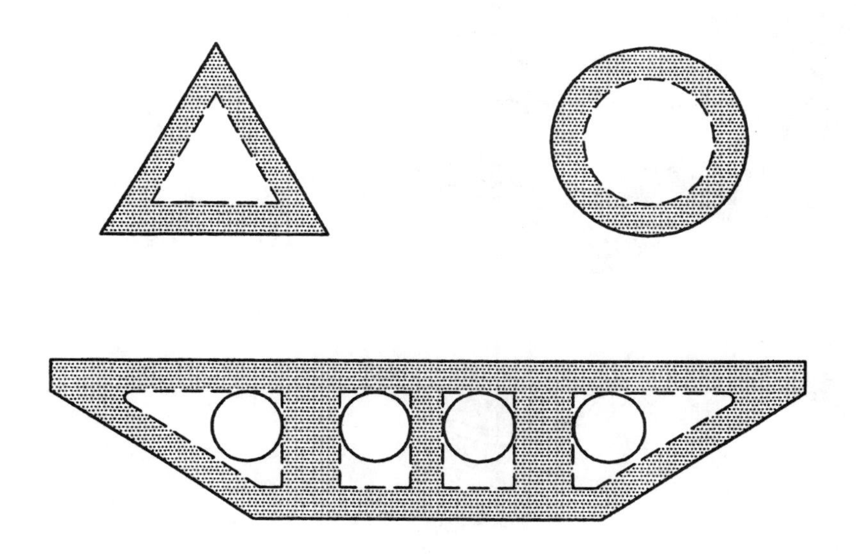

Figure 5.2.48 Beam sections subjected to bending and shear.

5.3 BEAMS IN TORSION

Reinforcement design. An immediate application of the reinforcement formulas of Section 2.4 to torsion problems is possible for a thin-walled, closed section. For such a section a pure shear field

$$\tau_{xs} = \frac{T}{2A_0 t} \tag{5.3.1}$$

T being the torsional moment, A_0 the area within the center line of the section, s the arc length along the center line, and t the thickness, is statically admissible (see Fig. 5.3.1). Formula (5.3.1) is Bredt's formula. Notice that shear force per unit length $H = \tau_{xs}\, t =$ const.

The area of the longitudinal bars and the area of the bars along the center line are determined, for example, by (2.4.14) and (2.4.15) or (2.4.21) and (2.4.22), the former formulas giving a minimum amount of reinforcement. Since we are dealing with pure shear the limits of γ in formulas (2.4.21) and (2.4.22) will normally be $\frac{1}{2} \le \gamma \le 2$. The concrete stress is determined by (2.4.16) or (2.4.23). The concrete stress must, of course, satisfy the condition $\sigma_c \le \nu f_c$.

The same simple stress field is statically admissible in any solid section if (5.3.1) is applied to a thin-walled closed section lying within the concrete area of the section, as pointed out by Rausch [53.3].

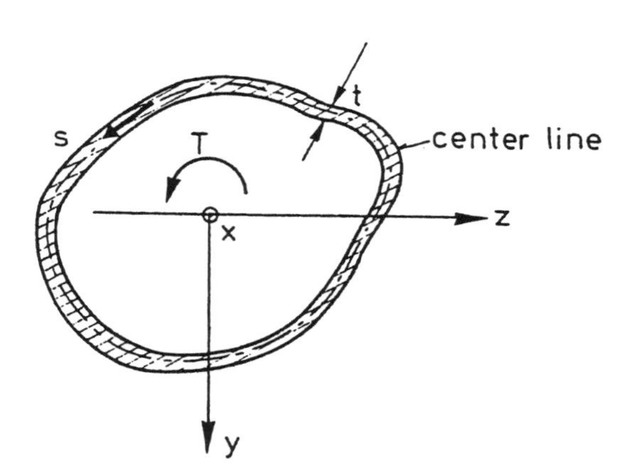

Figure 5.3.1 Thin-walled closed section subjected to torsion.

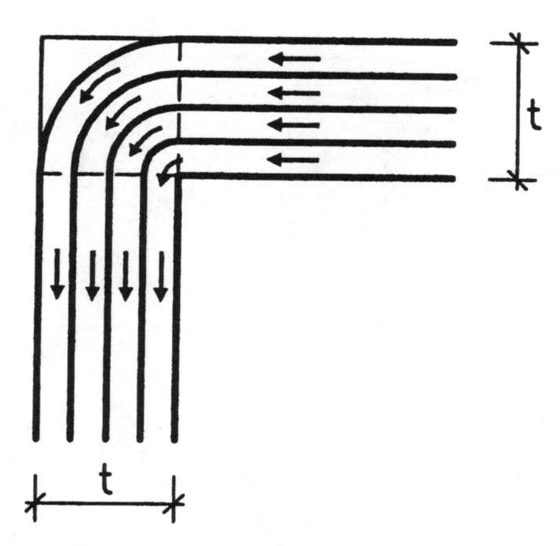

Figure 5.3.2 Transition region in a corner.

The thin-walled section used for design is called the effective section. The section often has corners. The effective section may have corners, too, or it may be chosen as a curved thin-walled section with a periphery having the same tangents as the section given in some distance from the corner. The circular transition region shown in Fig. 5.3.2 provides a simple statically admissible stress field with constant shear stress everywhere. We will soon meet another one with lines of discontinuity.

The thickness t of the thin-walled section must, of course, be so large as to render it possible to satisfy the condition $\sigma_c \leq \nu f_c$. The minimum value of t may be determined by the condition $\sigma_c = \nu f_c$.

In many cases, the reinforcement does not have to be placed so as to give a reinforcement resultant along the centerline. What is required is to preserve statical equivalence. A single reinforcement layer along the centerline should normally not be used (see below). Concerning the longitudinal reinforcement, a statically equivalent reinforcement layout can be used if proper care is taken to design the end sections, as in the case of shear in beams.

Consider, as an example, a rectangular section (see Fig. 5.3.3). If the yield stress in both reinforcement directions is equal to f_Y, we get by means of (2.4.14) and (2.4.15), the following reinforcement areas per unit length:

$$A_{sx} = A_{ss} = \frac{T}{2hbf_Y} \tag{5.3.2}$$

The total amount of longitudinal reinforcement is thus

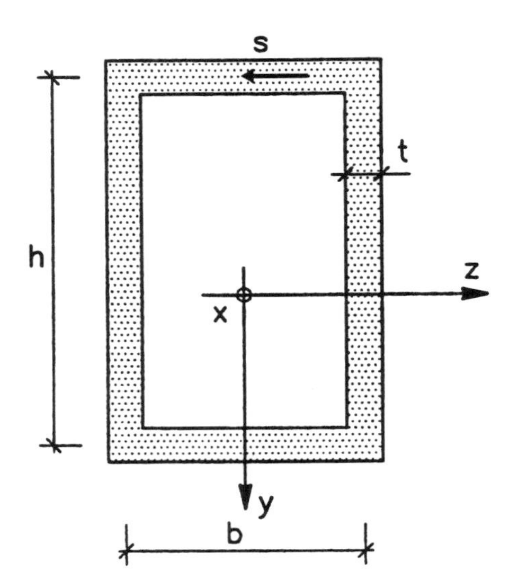

Figure 5.3.3 Thin-walled closed section for the design of a solid rectangular section subjected to torsion.

$$A_\ell = \frac{T(h+b)}{hbf_Y} \tag{5.3.3}$$

which for small sections can be concentrated in the corners, each of the corner bars having an area of $A_\ell/4$. The circumferential reinforcement can for small sections be chosen as one closed stirrup. Regarding its position, see below. For large sections, closed stirrups in the individual wall sections should be used.

For an elongated rectangular section, the bars in both reinforcement directions can be placed outside the thin-walled section as long as both reinforcement layers are placed symmetrically with respect to the middle plane (see Fig. 5.3.4). In this case, we are in fact concerned with pure torsion in a slab, the action of which was studied in Section 2.3.2, to which the reader is referred. Of course, the end sections require special attention. One possible reinforcement solution is shown in Fig. 5.3.4 (right).

In Fig. 5.3.5, thin-walled sections which can be used for reinforcement calculation in some other cases of solid sections have been illustrated.

It appears that as long as the required effective section may be placed in the section in question, the design is independent of possible holes of the section. This has been verified by tests by

Lampert and Thürlimann [68.3] (see also [71.1]). Of course, the stiffness may be influenced by the presence of holes.

Sections composed of rectangular sections may be treated by reinforcing the individual sections as described above. When distributing the torsional moment between the individual sections one may be guided by the torsional stiffnesses of the individual sections, i.e., the torsional moment is distributed approximately according to its stiffness. Irrespective of the design assumptions, the circumferential reinforcement in a section is, if relevant, prolonged into the neighboring section as to make possible a stiff reinforcement arrangement.

Alternative methods dealing with torsion problems according to the plastic theory have been described by Thürlimann and his associates [76.13; 78.8].

The yield conditions for disks can, of course, also be utilized to determine lower bound solutions for the load-carrying capacity in pure torsion for a given section. Solutions of this kind have been presented by the author [69.2; 71.1], Lampert [70.2], and Lampert and Thürlimann [71.3].

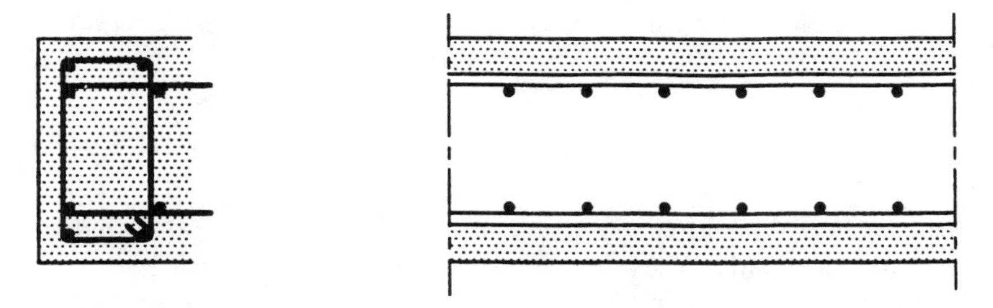

Figure 5.3.4 Thin-walled section in a slab subjected to torsion.

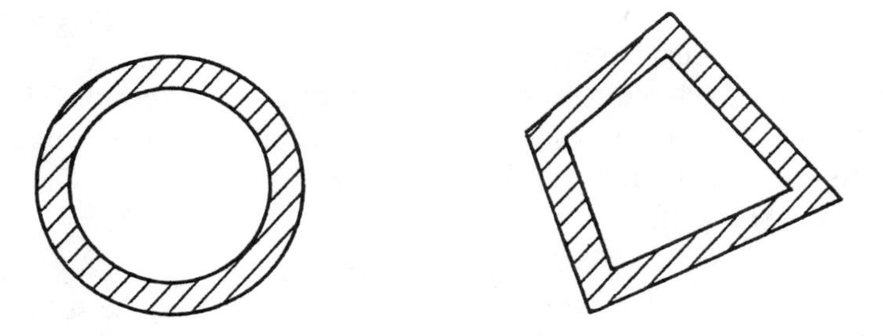

Figure 5.3.5 Examples of thin-walled closed sections for the design of corresponding solid sections subjected to torsion.

Consider again as an example the thin-walled, closed section in Fig. 5.3.1. If the reinforcement is homogeneous, corresponding to the tensile strength of the walls $f_{t\ell}$ in the longitudinal direction and the tensile strength f_{ts} in the circumferential direction, then, according to (2.2.20), the shear strength is

$$f_v = \sqrt{f_{t\ell} f_{ts}} \tag{5.3.4}$$

The yield moment in torsion T_p is thus [cf. (5.3.1)]

$$T_p = 2A_0 t \sqrt{f_{t\ell} f_{ts}} \tag{5.3.5}$$

Formula (2.2.20) is valid only as long as the concrete compression strength is not decisive [i.e., the condition (2.2.22) is considered fulfilled]. For the rectangular section in Fig. 5.3.3, (5.3.5) reads

$$T_p = 2hbt \sqrt{f_{t\ell} f_{ts}} \tag{5.3.6}$$

If, as before, A_ℓ is the total area of longitudinal reinforcement, $f_{Y\ell}$ the yield strength of the longitudinal reinforcement, A_{ss} the reinforcement area per unit length in the circumferential direction, and f_{Ys} the yield strength of the circumferential reinforcement, then

$$f_{t\ell} = \frac{A_\ell f_{Y\ell}}{2t(h+b)} \tag{5.3.7}$$

$$f_{ts} = \frac{A_{ss} f_{Ys}}{t} \tag{5.3.8}$$

As mentioned before, A_ℓ need not being uniformly distributed. Different statically equivalent distributions can be used.

For rectangular, thin-walled sections it has been shown by Müller [76.4] that an upper bound solution exists which under certain assumptions gives the same result as the lower bound solution presented here. The mechanism in the individual walls is a simple extension of the beam mechanism shown in Fig. 5.2.16.

An attempt to create a rational method for the overreinforced case ($\sigma_c = \nu f_c$) based on a description of the strain field as done in bending problems has been made by Mitchell and Collins [74.2] (see also [79.12] and [79.13]).

It should be pointed out that, as usual in shear problems, the stiffness is dramatically reduced in the fully cracked state compared to the stiffness in the uncracked state. The stiffness in the fully cracked state may be found by minimizing the elastic energy with respect to the angles of the uniaxial compressions in the concrete. For rectangular sections the result may be found in [78.6]. An approximate solution is obtained by calculating the elastic energy per unit length using the stress field of the plastic solution and equalizing it to $\frac{1}{2} T d\Theta/dx$, Θ being the rotation of the section. Then the torsional stiffness $GJ_v = T/(d\Theta/dx)$ may be determined. For a thin-walled square box section with side lengths a, thickness t everywhere and reinforced in a longitudinal direction and in the direction perpendicular to that corresponding to the reinforcement ratio r, the torsional stiffness in the fully cracked state is found to be

$$GJ_v = \frac{G_c a^3 t}{2 + \frac{1}{rn}} \tag{5.3.9}$$

Here the shear modulus of concrete $G_c \cong \frac{1}{2} E_c$, E_c being Young's modulus and $n = E_s / E_c$, E_s being Youngs's modulus for the reinforcement. In the uncracked state $GJ_v = G_c a^3 t$. For instance for $r = 1\%$, $n = 10$ it appears that GJ_v in the cracked state is 12 times smaller than in the uncracked state.

Corner problems. Although the pure shear field considered is an exact equilibrium solution, then, as soon as the uniaxial compression fields develop and as soon as the reinforcement in the circumferential direction is not uniformly distributed over the thickness, things get more complicated. Probably it is not worthwhile to go into details about the stress field belonging to different reinforcement solutions and it will also be very difficult. However, a few important points must be dealt with.

When the concrete compression exists along a curved surface the turning of the concrete stresses outside the reinforcement cannot take place without the help of tensile stresses in the concrete in sections parallel to the surface. This is illustrated in Fig. 5.3.6. In a corner these tensile stresses, possibly combined with the bursting stresses resulting from shear transfer between reinforcement and concrete along a crack distance, often spall off the concrete cover around the corner. In some cases the bursting stresses spall off the whole cover.

It is not possible to predict accurately whether the cover spalls off or not. Therefore, whenever the direction of the concrete stress is changed the stirrups should be placed as close to the surface as the required cover allows. This means that there must be a reinforcement layer near the surface and that the periphery of the effective section is governed by the position of this layer. In practice the outer surface of the stirrups may be taken as the periphery of the effective section.

Thus, along a curved surface curved stirrups provide a compression stress field perpendicular to the surface which changes the direction of the compressive stresses parallel to the surface.

If only one layer of reinforcement is used, it is placed eccentrically in the effective thin-walled section (see Fig. 5.3.7).

If there are two reinforcement layers the second layer may be placed along the internal periphery of the effective section.

The outer reinforcement layer in a right angled corner may, according to these rules, be carried out as shown in the figure to the right. However, in practice the reinforcement shown in Fig. 5.3.8 will be preferred in this case. For small beams (see the figure to the left) a longitudinal bar is placed in the corner and the stirrups are bent around this bar. The stress field in the corner is now much more complicated compared to the stress field present when the concrete compression direction is changed in a smooth way. Obviously there will be severe stress concentrations around the corner bar. These complications must be covered by the effectiveness factors. In the figure to the right there are two layers of reinforcement in the effective section and the stirrups are closed. This layout is, of course, rather obvious if the effective section is the real section of a thin-walled box girder. In this case, too, the stress field in the corner is more complicated since two diagonal compression fields are combined. Because of stress concentrations this leads to higher compression stresses than we are facing outside the corner. However, since the stress field is triaxial in the corner, it will be natural to include the cover in the effective section in this case and the effectiveness factors may be calculated as for disks.

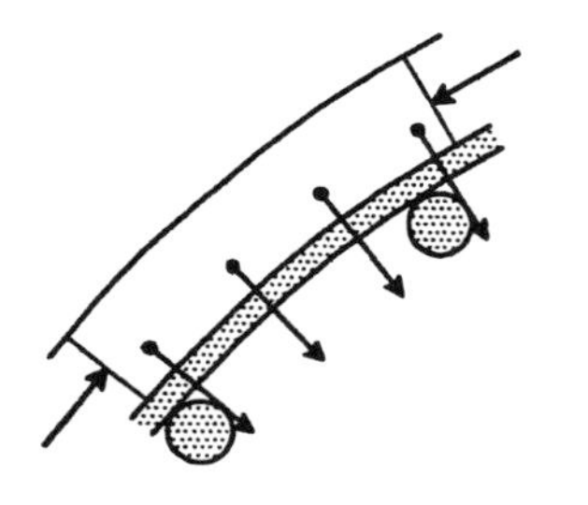

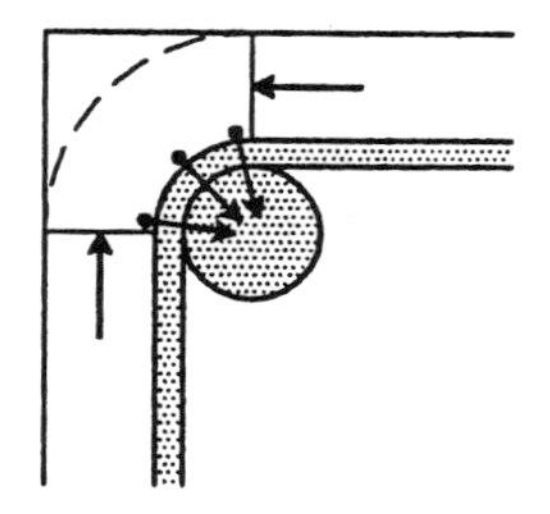

Figure 5.3.6 Change of direction of uniaxial compression.

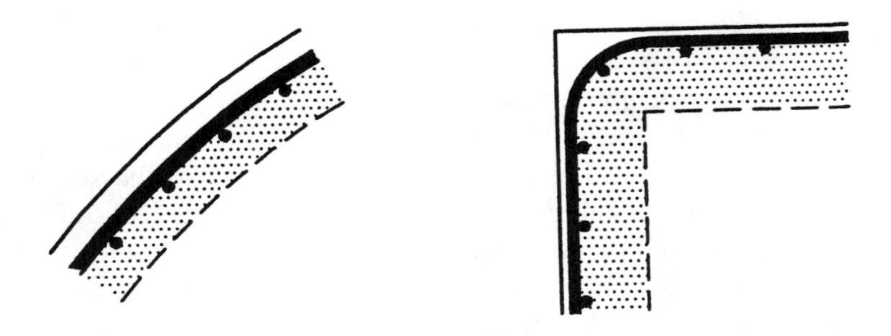

Figure 5.3.7 Effective sections limited by the outer stirrups.

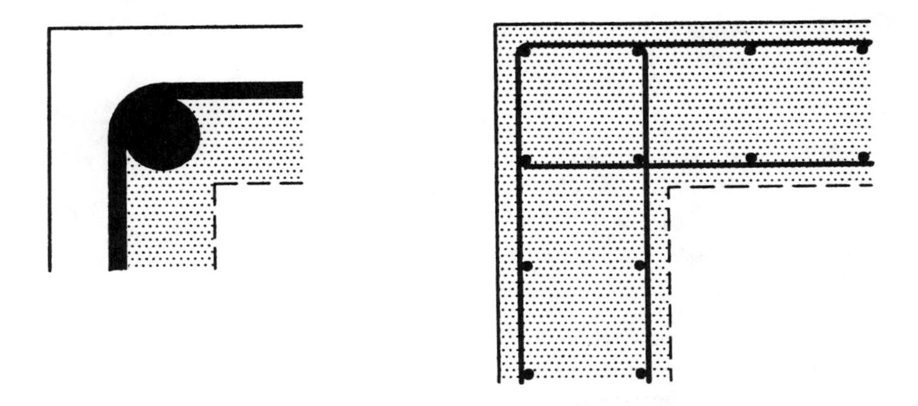

Figure 5.3.8 Reinforcement solutions in right angled corners.

Torsion capacity of rectangular sections. The majority of torsion studies have dealt with the rectangular section. In relation to plastic theory important work was done by Bent Feddersen [83.6] [84.2] [90.1]. What is known about effectiveness factors solely stems from this case and from torsion in slabs (cf. Chapter 7). When the necessary thickness of the effective section is large we may work with the stress field shown in Fig. 5.3.9 to get the correct limiting value when the effective section is the whole section within the outer periphery of the stirrup. The shear stress τ is constant within each trapezoidal part of the effective section. Along the edges of the trapezoidal areas there are lines of discontinuity under 45° with the sides of the section. There is only one closed stirrup with outer lengths a and b ($b \geq a$). Longitudinal bars with equal area are only shown in the corners but any longitudinal bar arrangement rendering the same center of gravity may be assumed. As pointed out later, bars must always be present in the corners.

The shear stresses originate from diagonal compression fields with the uniaxial stress σ_c in the trapezoidal areas. To make the solution an exact equilibrium solution the stirrup reinforcement should be uniformly distributed in the trapezoidal areas and anchored at the ends of the individual trapezoidal areas. This reinforcement solution has been illustrated in the figure to the right, the anchorage taking place by internal anchor plates. Such a reinforcement layout is, of course, highly theoretical and will not be carried out in practice. Indeed using a large number of stirrups in the form of closed stirrups the ideal situation may be approached. If the angle of the compression direction with the beam axis is θ, the overall equilibrium equations are

$$2t\,\sigma_c \sin^2\theta = \frac{2T_{Yw}}{s} \tag{5.3.10}$$

$$4t\left(\frac{1}{2}(a+b) - t\right)\sigma_c \cos^2\theta = T_{Y\ell} \tag{5.3.11}$$

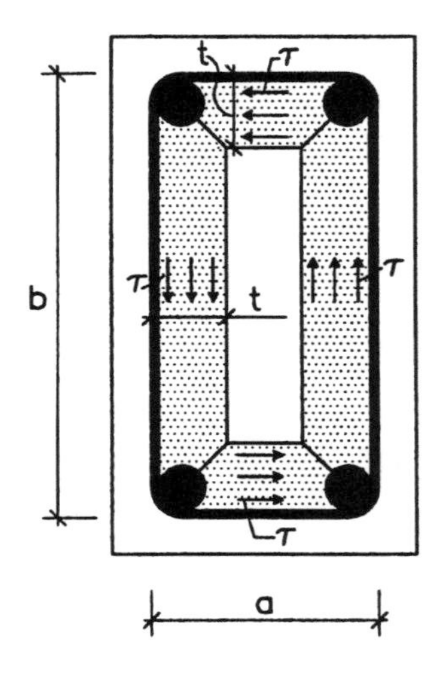

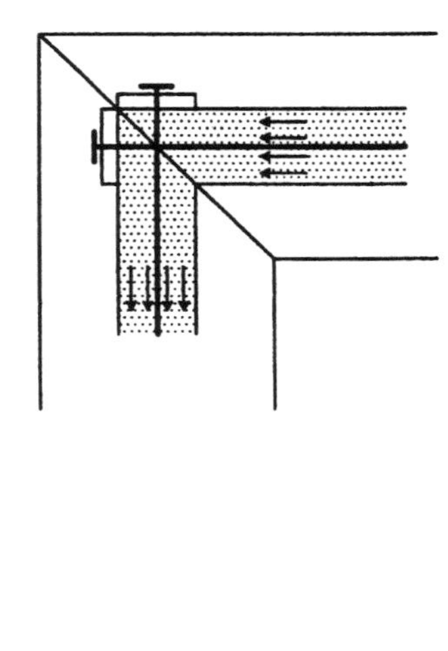

Figure 5.3.9 Rectangular section subjected to torsion.

Here T_{Yw} is the yield force corresponding to one stirrup cross section, s the stirrup distance and $T_{Y\ell}$ is the total yield force of the longitudinal reinforcement. The shear stress is

$$\tau = \sigma_c \cos\theta \sin\theta \tag{5.3.12}$$

and the torsional moment carried by the shear stress τ is

$$T = \tau a^2 t\left[\left(1-2\frac{t}{a}\right)\left(\frac{b}{a}-\frac{t}{a}\right)+\left(1-\frac{t}{a}\right)\left(\frac{b}{a}-2\frac{t}{a}\right)\right]$$

$$+\ \tau a t^2\left[1+\frac{b}{a}-\frac{4}{3}\frac{t}{a}\right] \tag{5.3.13}$$

In the case $b = a$ and $t=\frac{1}{2}a$ we have simply

$$T = \frac{1}{3}\tau a^3 \tag{5.3.14}$$

Since $\tau_{max}=\frac{1}{2}f_c$, for $b = a$ we arrive at the simple formula

$$T_{max} = \frac{1}{6}f_c a^3 \tag{5.3.15}$$

When $t=\frac{1}{2}a$, $\tau=\tau_{max}=\frac{1}{2}f_c$ and b is large compared with a we find

$$T_{max} = \frac{1}{4}f_c a^2 b \tag{5.3.16}$$

This means a torsional moment per unit length in the b-direction equal to $\frac{1}{4}f_c a^2$, which is twice that determined in Section 2.3.2 under the same assumptions (formula (2.3.16) with $a=\frac{1}{2}h$ and $\tau_{cxy}=\frac{1}{2}f_c$). The reason is that even if the resultants of the shear stresses at the ends of a long section are small, they have a large moment arm, so in fact they contribute half the torsional moment of the section. In the slab theory the resultants of these stresses are approximated by concentrated forces along the edges. These forces are extremely important and bear the name, the Kirchhoff boundary forces. We will become familiar with them in Chapter 6.

Defining the reinforcement degrees

$$\Phi_w = \frac{T_{Yw}}{stf_c} \qquad \Phi_\ell = \frac{T_{Y\ell}}{4t\left(\frac{1}{2}(a+b)-t\right)f_c} \tag{5.3.17}$$

the equilibrium equations may be written

$$\sigma_c \sin^2\theta = \Phi_w f_c \tag{5.3.18}$$

$$\sigma_c \cos^2\theta = \Phi_\ell f_c \tag{5.3.19}$$

These equations have the same formal structure as (2.2.16) and (2.2.17) valid for disks; the only difference being that in the disk equations σ_c and θ are the only unknowns, while here the unknowns are t, σ_c and θ.

Probably one would expect maximum load-carrying capacity to be reached when t is selected such that the reinforcement is yielding, at least for small reinforcement degrees, and that $\sigma_c = f_c$. Let us see what comes out of this assumption.

By adding the equilibrium equations we get for $\sigma_c = f_c$

$$\Phi_w + \Phi_\ell = 1 \tag{5.3.20}$$

or

$$\frac{T_{Yw}}{stf_c} + \frac{T_{Y\ell}}{4t\left(\frac{1}{2}(a+b) - t\right)f_c} = 1 \tag{5.3.21}$$

This equation may be solved for t.

If $t \le \frac{1}{2}a$, Φ_w and Φ_ℓ are calculated by (5.3.17) and by means of (5.3.18) and (5.3.19) we get

$$\tan^2\theta = \frac{\Phi_w}{\Phi_\ell} \tag{5.3.22}$$

Since $\tau = f_c \cos\theta\sin\theta$ we find as in the disk case

$$\frac{\tau}{f_c} = \sqrt{\Phi_w \Phi_\ell} \tag{5.3.23}$$

The torsional moment capacity may be found by formula (5.3.13).

If there is no solution rendering $t \le \frac{1}{2}a$, the beam is overreinforced. Denoting the values of Φ_w and Φ_ℓ for $t = \frac{1}{2}a$ by Φ_w' and Φ_ℓ', respectively, i.e.,

$$\Phi_w' = \frac{2T_{Yw}}{saf_c} \qquad \Phi_\ell' = \frac{T_{Y\ell}}{abf_c} \tag{5.3.24}$$

we have in the overreinforced case

$$\Phi_w' + \Phi_\ell' > 1 \tag{5.3.25}$$

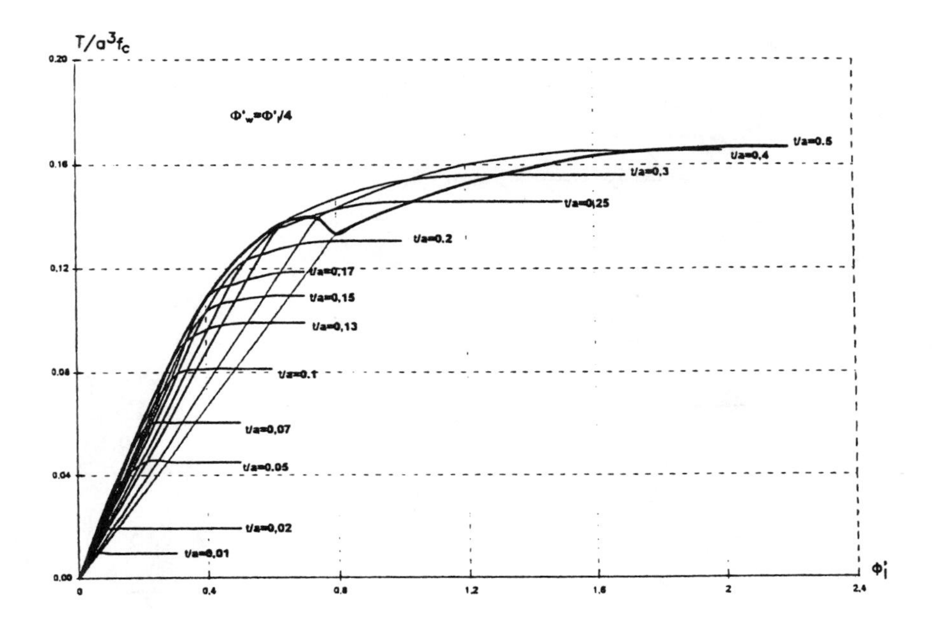

Figure 5.3.10 Torsional capacity of a square section, $\Phi_w' = \Phi_\ell'/4$.

The solution in the overreinforced case may be found using the disk solution (cf. Section 2.2.2). If Φ_i' indicates the smaller of the reinforcement degrees Φ_w' and Φ_ℓ' we get

$$\frac{\tau}{f_c} = \sqrt{\Phi_i'(1-\Phi_i')} \qquad \Phi_i' \le 0.5 \qquad (5.3.26)$$

$$\frac{\tau}{f_c} = \frac{1}{2} \qquad \Phi_i' > 0.5 \qquad (5.3.27)$$

The torsional moment capacity is calculated by (5.3.13) inserting $t = \frac{1}{2}a$.

However, it turns out that determining t in this way does not always lead to maximum load-carrying capacity. Sometimes a lower t-value renders higher load-carrrying capacity.

For fixed t the load-carrying capacity may be determined by once again using the disk solution. The reinforcement degrees Φ_w and Φ_ℓ are calculated by means of (5.3.17). Then if Φ_i indicates the smaller of Φ_w and Φ_ℓ we have

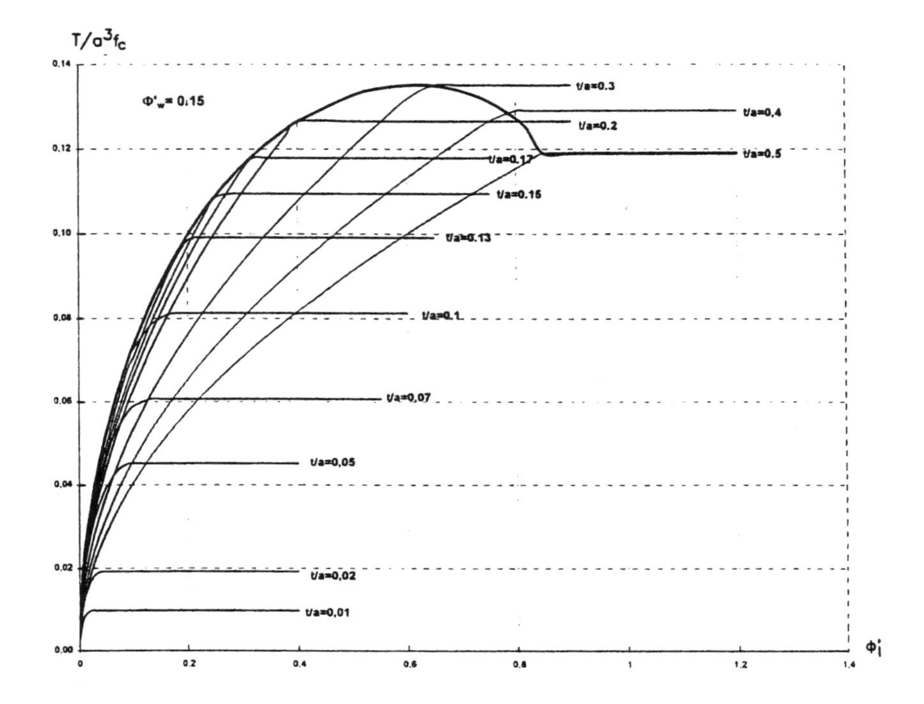

Figure 5.3.11 Torsional capacity of a square section, $\Phi_w' = 0.15$.

$$\frac{\tau}{f_c} = \begin{cases} \sqrt{\Phi_w \Phi_\ell} & \Phi_w + \Phi_\ell \le 1 \\ \sqrt{\Phi_i(1-\Phi_i)} & \Phi_w + \Phi_\ell > 1 \quad \Phi_i \le 0.5 \\ \dfrac{1}{2} & \Phi_w + \Phi_\ell > 1 \quad \Phi_i > 0.5 \end{cases} \tag{5.3.28}$$

When τ has been determined the torsional capacity is calculated by (5.3.13).

The relationship between the reinforcement degrees are easily seen to be

$$\Phi_w = \frac{\Phi_w'}{2\dfrac{t}{a}} \qquad \Phi_\ell = \frac{\Phi_\ell'}{4\dfrac{a}{b}\dfrac{t}{a}\left[\dfrac{1}{2}\left(1+\dfrac{b}{a}\right)-\dfrac{t}{a}\right]} \tag{5.3.29}$$

In Fig. 5.3.10 the solution found by determining t by the condition $\sigma_c = f_c$ has been shown with a thick line in the case $a = b$ and $\Phi_w' = \Phi_\ell'/4$. It turns out that the curve has a minimum for $\Phi_\ell' = 0.8$ where $\Phi_w' + \Phi_\ell' = 1$ and $t = 0.5a$. The shear stress τ is ever decreasing when Φ_ℓ' is growing from zero to 0.8 being $\tau \cong 0.4 f_c$ when $\Phi_\ell' = 0.8$, which explains the minimum. Also shown in the figure are the load-carrying capacities for fixed t-values. These have been drawn with thin lines. It appears that around the minimum value higher load-carrying capacity may be obtained for $t \cong 0.3a$ instead of $t = 0.5a$.

The mathematics needed for determining the optimum value of t is much too involved for practical use. It is sufficient to draw curves for different t-values and in this way determine an approximate envelope curve.

In Fig. 5.3.11 the torsional capacity of a square section has been depicted for constant $\Phi_w' = 0.15$. Again the maximum value of the load-carrying capacity is not obtained for $t = 0.5a$ but for $t \cong 0.3a$.

Whenever the torsional capacity curve displays a descending branch an envelope curve must be determined using fixed t-values.

Another approach suggested in [90.1] (see also [92.22]), is to limit the thickness to a value less than $0.5a$. This approach works well but it gives a little less limiting value of the load-carrying capacity for high reinforcement degrees. The difference is, however, without practical significance.

Effectiveness factors. Because of the corner problems effectiveness factors for pure torsion are normally lower than for shear in beams.

For rectangular sections with one reinforcement layer Bent Feddersen [90.1] found that a rather complicated ν-formula is necessary to render good agreement with tests. His formula reads

$$\nu = \begin{cases} 0.50 + \dfrac{k}{20} + \dfrac{d_c}{200k} - \dfrac{f_c}{300} - \dfrac{f_Y}{8000} & k \le 4 \\[2ex] 0.74 - \dfrac{f_c}{300} - \dfrac{f_Y}{8000} & k > 4 \end{cases} \tag{5.3.30}$$

Here $k = b/a \ge 1$ (see Fig. 5.3.9) is the ratio between the side lengths. They are the real side lengths reduced by 2 times the cover. The compressive strength of concrete f_c and the yield strength of the reinforcement f_Y are in MPa. Finally d_c is the diameter of the corner bars in mm.

The formula was verified for $f_c \le 60$ MPa, $f_Y \le 1000$ MPa and $d_c \le 30$ mm. The coefficient of variation on the ratio theoretical value/test value for 111 tests was 13% and the mean value 0.94.

The formula is of the same type as those developed for bending (cf. Section 5.1), except for the terms depending on the side ratio and the diameter of the corner bars, respectively.

It appears that increasing k-values leads to increasing ν-value, which means that the corner problems are decreasing for increasing side ratios. Further it appears that an increasing diameter of the corner bars leads to increasing ν-value. Thus it is important to supply bars in the corners.

Rasmussen and Ibsø [92.22] in a master thesis found that rather good agreement with tests may be obtained, even for high strength concrete, using the simpler formula

$$\nu = \begin{cases} 0.67 - \dfrac{f_c}{250} & f_c \le 25 \text{ MPa} \\ \dfrac{2.85}{\sqrt{f_c}} & 25 \text{ MPa} < f_c \le 110 \text{ MPa} \end{cases} \tag{5.3.31}$$

Cover is disregarded.

By comparing with 119 tests including their own they found the coefficient of variation on the ratio theoretical value/test value to be 16% and the mean value 1.

Sections composed of rectangular sections reinforced with closed stirrups in the individual sections may be treated using the ν-formulas for disks. The cover may be included.

Design codes normally restrict the thickness of the effective section. It appears from the above that this is not necessary.

5.4 COMBINED BENDING, SHEAR, AND TORSION

In the case of combined bending, shear, and torsion in a thin-walled closed section, the necessary amount of reinforcement can also be determined by means of the reinforcement formulas of Section 2.4. Considering as a simple example a box section subjected to a bending moment M_z, a shear force V_y, and a torsional moment T (see Fig. 5.4.1), one statically admissible stress field can be found using the Navier distribution of the normal stresses σ_x from the bending moment, the corresponding Grashof distribution of the shear stresses τ_{xs} from the shear force V_y, and the Bredt distribution (5.3.1) of the shear stresses τ_{xs} from the torsional moment T. However, a more suitable statically admissible stress distribution is found by distributing the normal stresses from M_z uniformly, for instance, along the top and bottom flanges. The corresponding shear stress diagram is then linear in the individual walls bearing in mind that when the normal stress distribution is selected the shear stresses may be determined; for instance, by the local equilibrium equation expressing equilibrium along the beam axis. If N is the normal force per unit length in the x-direction and H the shear force per unit length we have

$$\frac{\partial N}{\partial x} + \frac{\partial H}{\partial s} = 0 \tag{5.4.1}$$

N is positive as tension and H is positive in the direction of s, the arc length along the center line.

The loading may, of course, be such that transverse stresses, i.e., stresses in sections perpendicular to the s-direction, may be important. For instance, in a bridge girder the top flange may act as a slab. An equilibrium solution is determined by considering a part of the beam limited by two normal sections in the distance dx and loaded by the external load and the difference in shear forces. Any redundant forces or moments are chosen so as to minimize reinforcement. For cylindrical shells this method was suggested by Lundgren as early as in 1949 (see [49.2]). Having determined the stress distribution, the reinforcement formulas immediately give the necessary amount of reinforcement.

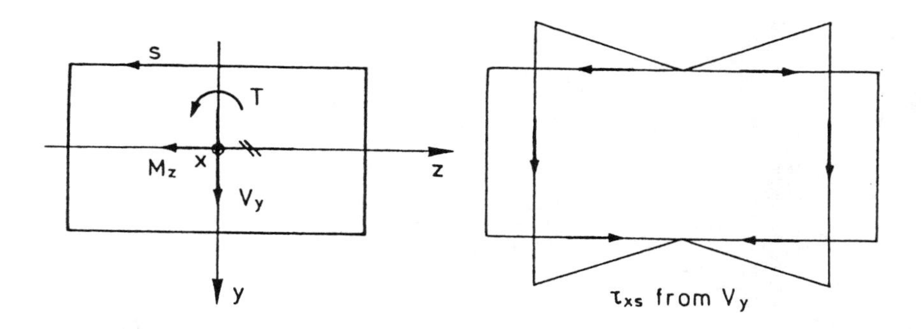

Figure 5.4.1 Thin-walled closed rectangular section subjected to combined bending, shear, and torsion.

In the ultimate limit state warping stresses may be taken into account if they are beneficial, otherwise they may be neglected. For closed sections warping stresses are normally unimportant contrary to open sections, where they should be taken into account. In plastic theory this may be done by subdividing the load into one part carried by pure torsion and one part carried by warping stresses. Then the minimum reinforcement solution should be looked for.

Other thin-walled closed sections can be treated in a similar way. For a solid section, the same method as described for pure torsion can be used; that is, a thin-walled section lying within the concrete area is selected for carrying the stresses.

Consider as an example a solid rectangular section (see Fig. 5.4.2). If reinforcement is provided in the longitudinal direction and in the circumferential direction and if the yield stress of the steel is the same in both directions, then for a section subjected to a torsional moment T, the reinforcement formulas (2.4.14) and (2.4.15) require the total longitudinal reinforcement to carry a force $P_\ell = A_\ell f_Y$, which can be calculated by means of (5.3.3). For small sections the corresponding reinforcement area can be placed as one-fourth of the total area in each corner.

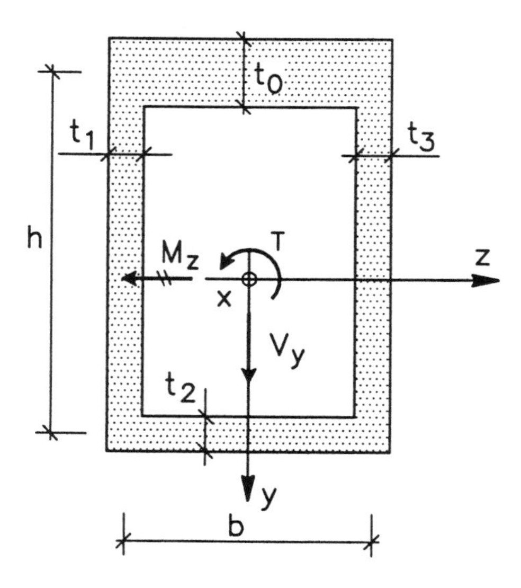

Figure 5.4.2 Thin-walled closed section for the design of a solid rectangular section subjected to combined bending, shear, and torsion.

Small bending moments M_z [i.e., $M_z \leq \frac{1}{2}T(h+b)/b$] can be carried by moving a part of the reinforcement in the compression zone to the tension zone; that is, the force in the longitudinal reinforcement at the top flange can be reduced by M_z/h, and the force in the longitudinal reinforcement at the bottom flange must be increased by M_z/h. If $M_z > \frac{1}{2}T(h+b)/b$, the longitudinal reinforcement at the top flange can be chosen as zero, while the force in the longitudinal reinforcement at the bottom flange still has to be increased by M_z/h.

If the section is subjected to an additional shear force V_y, the reinforcement can be determined by adding the shear stress in the thin-walled section from V_y to the stresses from T and M_z. In one of the vertical walls, the shear stresses from V_y add to the shear stresses from T, and in the other one, they subtract from the shear stresses from T. The concrete stress in the individual walls can be determined by means of the formulas of Section 2.4. The quantities t_0, t_1, t_2, and t_3, the meaning of which is shown in Fig. 5.4.2 must, of course, be fixed at values making it possible to satisfy the condition $\sigma_c \leq \nu f_c$ in each wall. Normally the cover may be taken into account in the compression zone, because in this zone the cover is not likely to spall off. In the walls subjected to shear stresses, the cover should be disregarded if only one layer of reinforcement is used. The reinforcement is placed as close to the surface as allowed by the required cover. If two layers of reinforcement are used in each wall the cover may be taken into account.

If only one layer of reinforcement is used in the walls, the ν-value should be taken as the lower of ν for torsion and ν for shear in disks. If two layers of reinforcement are used in each wall in the form of closed stirrups, the ν-values may be calculated as for shear in disks. In this case it may be advantageous to use formula (4.6.10) involving the χ-parameter, particularly for a compression flange, since due to the compressive stresses, the χ-value will be low and correspondingly the ν-value high.

Other solid sections may be treated in a similar way. More complicated stress fields may be used to minimize the reinforcement consumption. Some solutions have been outlined in [90.1]. The design method presented has been developed by the author [71.1; 79.10] and Clyde [79.11].

Of course, the yield conditions for disks can also be used to determine lower bound solutions for the load-carrrying capacity for a section (see [71.1]). Yield conditions of this type were first presented by Lampert [70.2] and Lampert and Thürlimann [71.3].

Consider as an example a thin-walled, rectangular section. The four individual walls are equally and homogeneously reinforced. Let us for simplicity assume that when the section is acted upon by a bending moment with a moment vector parallel to the sides b, the wall area bt is greater than or equal to the necessary compression zone. Then the yield moment in bending will be

$$M_p = (hb + h^2)tf_{t\ell} \tag{5.4.2}$$

If the section is acted upon by a bending moment M and a torsional moment T at the same time, a statically admissible stress field can be found by superimposing the stress fields shown in Fig. 5.4.3. We therefore get

$$\sigma = \frac{M}{(hb + h^2)t} \tag{5.4.3.}$$

$$\tau = \frac{T}{2hbt} \tag{5.4.4}$$

Since the compression zone will not be decisive, the combinations of σ and τ that can be carried will be determined by the yield condition (2.2.41), which gives

$$-\left[f_{t\ell} - \frac{M}{(hb + h^2)t}\right]f_{ts} + \left(\frac{T}{2hbt}\right)^2 = 0 \tag{5.4.5}$$

Introducing the yield moment in bending M_p and the yield moment in torsion T_p (5.3.6), we find the yield condition

$$\frac{M}{M_p} + \left(\frac{T}{T_p}\right)^2 = 1 \tag{5.4.6}$$

The yield condition has been drawn in Fig. 5.4.4. The curve is a parabola.

The longitudinal reinforcement can be concentrated in the corners without changing the values of M_p and T_p and without changing the yield condition. If the reinforcement of the individual walls is different, the yield condition will be more complicated, but it can be found in the same way. The yield condition (5.4.6) will also be valid for other thin-walled closed sections having one cell, and for the corresponding massive sections.

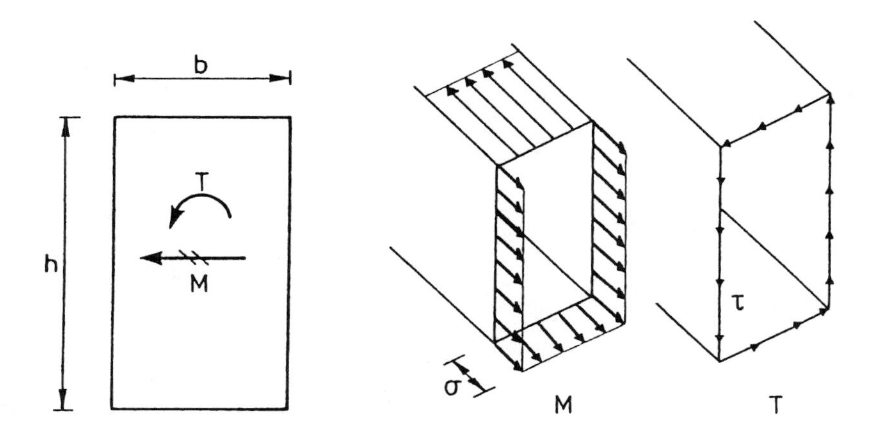

Figure 5.4.3 Stress fields for combined bending and torsion of a thin-walled, closed rectangular section.

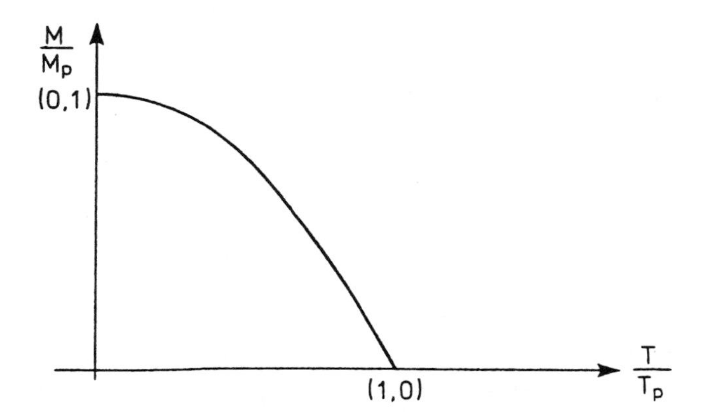

Figure 5.4.4 Yield condition for combined bending and torsion.

Tests confirming the design method outlined have been carried out by Lampert and Thürlimann [69.4]. They have been treated in [71.1] according to the methods described here.

Chapter 6

SLABS

As stated in Chapter 4, slabs are structural members bounded by two parallel planes having a distance, the thickness, which is small compared with the other dimensions. The external loads are perpendicular to the planes.

The plastic theory of reinforced concrete slabs is now rather old, the pioneering works being by Ingerslev in 1921 [21,1] and Johansen, beginning his writings on that subject in 1931 [31,1].

6.1 STATICAL CONDITIONS

6.1.1 Internal Forces in Slabs

In the main part of this chapter we shall assume that the internal forces per unit length in a normal section are as shown in Fig. 6.1.1, that is, a bending moment m_b, a torsional moment m_v, and a shear force v, all measured per unit length. The bending moment per unit length (abbreviated in the following to "the bending moment") is positive when it results in tension at the "bottom", which is assumed to be defined beforehand. The torsional moment per unit length (abbreviated in the following to "the torsional moment") is positive when the corresponding vector is an outward-directed normal to the section under consideration. The sign for the shear force is defined in relation to the coordinate system used (see below). These assumptions regarding the internal forces are discussed in more detail in Section 6.11.

6.1.2 Equilibrium Conditions

Although the transformation formulas for moments are identical to

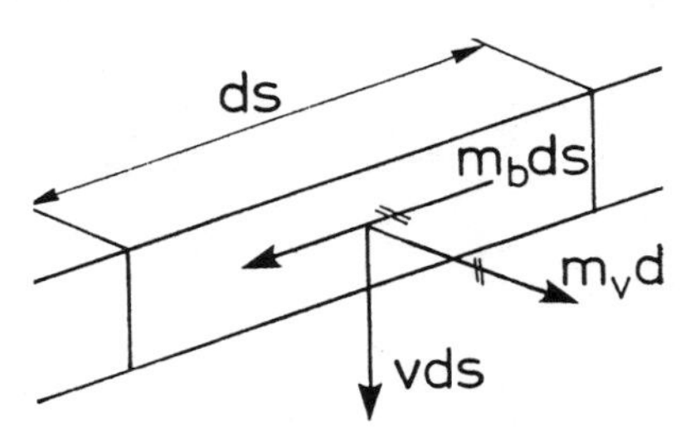

Figure 6.1.1 Internal forces in slabs.

those valid in plane stress situations, we state the formulas for the special sign conventions used here. The *principal sections* are the two sections in which the torsional moments are zero and the bending moments extreme. The two sections are at right angles to each other. The bending moments in these sections, the *principal moments*, are denoted m_1 and m_2.

In an arbitrary section forming an angle α with the section with the bending moment m_1, the bending moment m_b, and the torsional moment m_v are determined as follows:

$$m_b = m_1 \cos^2 \alpha + m_2 \sin^2 \alpha = \tfrac{1}{2}(m_1 + m_2) + \tfrac{1}{2}(m_1 - m_2)\cos 2\alpha \quad (6.1.1)$$

$$m_v = \tfrac{1}{2}(m_1 - m_2)\sin 2\alpha \quad (6.1.2)$$

Graphically, these expressions are described by means of a Mohr's circle (see Fig 6.1.2). The angle α is considered positive in a clockwise direction.

The greatest principal moment is the *first principal moment*, and the corresponding section is the *first principal section*. The smallest principal moment is the *second principal moment*, and the corresponding section is the *second principal section*. The torsional moments in two sections at right angles to each other are equal but have opposite signs.

If the shear forces v_1 and v_2 in two sections at right angles to each other are known, then the shear force v in a section that forms an angle α with the section with v_1 is equal to

$$v = v_1 \cos \alpha + v_2 \sin \alpha \quad (6.1.3)$$

The equilibrium equations in Cartesian coordinates may be found from equilibrium equations for an element $dx \cdot dy$, as shown in Fig. 6.1.3. The sign convention for the shear forces is apparent from this figure.

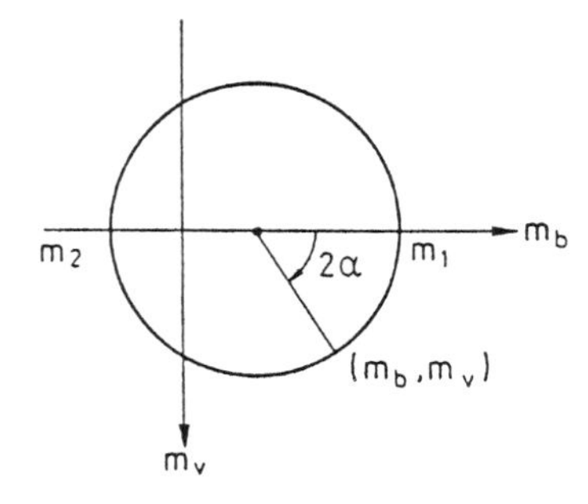

Figure 6.1.2 Mohr's circle for bending and torsional moments in slabs.

The conditions of equilibrium are found to be

$$v_y = \frac{\partial m_y}{\partial y} - \frac{\partial m_{xy}}{\partial x} \tag{6.1.4}$$

$$v_x = \frac{\partial m_x}{\partial x} - \frac{\partial m_{xy}}{\partial y} \tag{6.1.5}$$

$$\frac{\partial v_x}{\partial x} + \frac{\partial v_y}{\partial y} = -p \tag{6.1.6}$$

where use has been made of the requirement that $m_{xy} = -m_{yx}$, and where p is the load intensity. By elimination of v_x and v_y,

$$\frac{\partial^2 m_x}{\partial x^2} - 2\frac{\partial^2 m_{xy}}{\partial x \partial y} + \frac{\partial^2 m_y}{\partial y^2} = -p \tag{6.1.7}$$

It appears that the internal forces are fully determined by the values of m_x, m_y and m_{xy}, which are the *generalized stresses* of the slab theory presented here.

In polar coordinates r and θ, the corresponding equations are as follows:

$$rv_r = \frac{\partial(rm_r)}{\partial r} - \frac{\partial m_{r\theta}}{\partial \theta} - m_\theta \tag{6.1.8}$$

$$v_\theta = \frac{1}{r}\frac{\partial m_\theta}{\partial \theta} - \frac{\partial m_{r\theta}}{\partial r} - 2\frac{m_{r\theta}}{r} \tag{6.1.9}$$

$$\frac{\partial(rv_r)}{\partial r} + \frac{\partial v_\theta}{\partial \theta} = -pr \tag{6.1.10}$$

$$\frac{1}{r}\frac{\partial^2 (rm_r)}{\partial r^2} - \frac{2}{r^2}\frac{\partial^2 (rm_{r\theta})}{\partial r \partial \theta} + \frac{1}{r^2}\frac{\partial^2 m_\theta}{\partial \theta^2} - \frac{1}{r}\frac{\partial m_\theta}{\partial r} = -p \tag{6.1.11}$$

The symbols used for common types of loads in the following illustrations are shown in Fig. 6.1.4.

6.1.3 Lines of Discontinuity

Figure 6.1.5 shows two slabs parts, I and II, meeting along a line of discontinuity l. The statical conditions can be expressed as follows

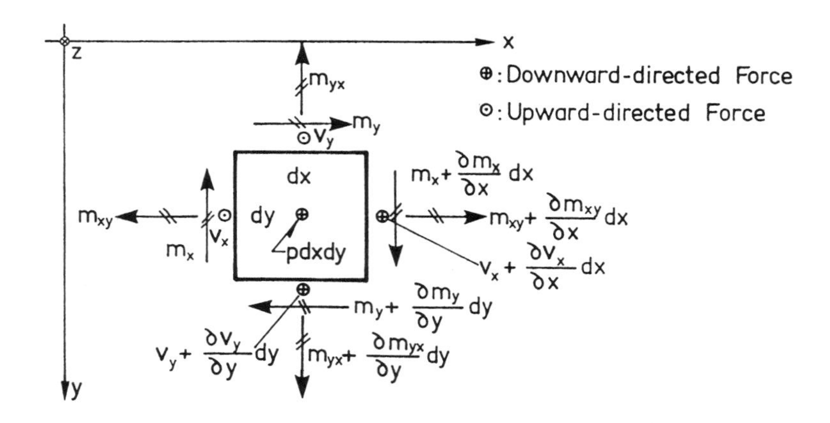

Figure 6.1.3 Slab element with loading and internal forces.

in a local *n*, *t* - system, where the *t*-axis is tangential to *l*:

$$
\begin{aligned}
m_n^{\mathrm{I}} &= m_n^{\mathrm{II}} \\
m_{nt}^{\mathrm{I}} &= m_{nt}^{\mathrm{II}} \\
v_n^{\mathrm{I}} &= v_n^{\mathrm{II}}
\end{aligned}
\tag{6.1.12}
$$

The last condition can be rewritten

$$
\frac{\partial m_n^{\mathrm{I}}}{\partial n} - \frac{\partial m_{nt}^{\mathrm{I}}}{\partial t} = \frac{\partial m_n^{\mathrm{II}}}{\partial n} - \frac{\partial m_{nt}^{\mathrm{II}}}{\partial t} \tag{6.1.13}
$$

where the partial derivatives are calculated at the point *O* of the line of discontinuity.

6.2 GEOMETRICAL CONDITIONS

6.2.1 Strain Tensor in a Slab

As in the theory of elasticity for thin slabs, it is assumed that the strain field can be described by one function $w(x, y)$, the displacement perpendicular to the plane of the slab (x, y). Thus the displacement w is assumed to be independent of the coordinate z. It is further assumed

⊗ Downward-directed concentrated force
⊙ Upward-directed concentrated force
— Line load

Figure 6.1.4 Symbols for common types of loads.

that a normal to the plane of the slab is inextensible, remains straight, and is normal to the deflected surface (the Kirchhoff assumptions). This yields the following displacements u_x and u_y in the directions of the x-axis and the y-axis, respectively:

$$u_x = -z\frac{\partial w}{\partial x} = z\omega_y, \qquad u_y = -z\frac{\partial w}{\partial y} = -z\omega_x \tag{6.2.1}$$

where ω_x and ω_y are the Cartesian components of the rotation of the normal. Consequently, the strain tensor is determined by

$$\begin{aligned}
&\varepsilon_x = \frac{\partial u_x}{\partial x} = -z\frac{\partial^2 w}{\partial x^2} = z\kappa_x, \qquad \varepsilon_y = \frac{\partial u_y}{\partial y} = -z\frac{\partial^2 w}{\partial y^2} = z\kappa_y \\
&2\varepsilon_{xy} = \gamma_{xy} = \frac{\partial u_x}{\partial y} + \frac{\partial u_y}{\partial x} = -2z\frac{\partial^2 w}{\partial x \partial y} = -2z\kappa_{xy} \\
&\varepsilon_z = \varepsilon_{xz} = \varepsilon_{yz} = 0
\end{aligned} \tag{6.2.2}$$

In these expressions κ_x, κ_y, and κ_{xy} denote the curvature in the x-direction, the curvature in the y-direction, and the torsion with regard to the x- and y- directions, respectively. The strain field is fully determined if κ_x, κ_y, and κ_{xy} are known. These quantities are the *generalized strains* of the slab theory presented here. The Kirchhoff assumptions imply that the effects of transverse shear and strains along the slab normal are neglected.

In the theory of elasticity these assumptions lead to excellent results when the deflections are small compared with the depth of the slab. When this is not the case, it is necessary to include strains in the x, y-plane, whereby the normal forces n_x, n_y, and shear forces n_{xy} in the direction of the middle surface of the slab are introduced. These forces are for the time being neglected (see also Section 6.11).

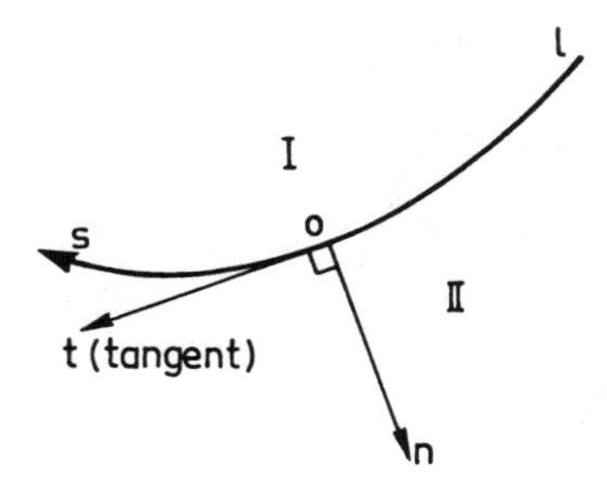

Figure 6.1.5 Local coordinate system at a line of discontinuity.

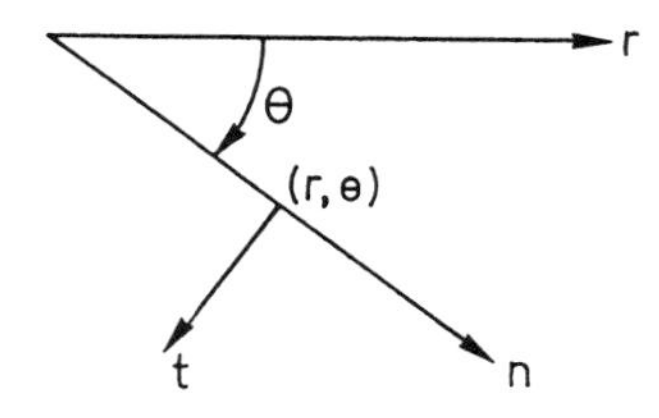

Figure 6.2.1 Polar coordinate system and local rectangular coordinate system.

There exist two sections at right angles to each other, in which the torsion is equal to zero and the curvatures are extreme. These values are called the *principal curvatures* and are denoted κ_1 and κ_2. The corresponding directions are the *principal directions*. If these are given, the curvature and the torsion in an arbitrary direction n, forming an angle α with the direction having the curvature κ_1, are given by

$$\kappa_n = \kappa_1 \cos^2 \alpha + \kappa_2 \sin^2 \alpha \tag{6.2.3}$$

$$\kappa_{nt} = \tfrac{1}{2}(\kappa_1 - \kappa_2)\sin 2\alpha \tag{6.2.4}$$

These equations can also be described graphically by means of Mohr's circle.

For later use, the expressions for the curvature and the torsion are given in polar coordinates (see Fig 6.2.1):

$$\kappa_r = -\frac{\partial^2 w}{\partial n^2} = -\frac{\partial^2 w}{\partial r^2} \tag{6.2.5}$$

$$\kappa_\theta = -\frac{\partial^2 w}{\partial t^2} = -\frac{1}{r^2}\frac{\partial^2 w}{\partial \theta^2} - \frac{1}{r}\frac{\partial w}{\partial r} \tag{6.2.6}$$

$$\kappa_{r\theta} = \frac{\partial^2 w}{\partial n \partial t} = \frac{1}{r}\frac{\partial^2 w}{\partial r \partial \theta} - \frac{1}{r^2}\frac{\partial w}{\partial \theta} \tag{6.2.7}$$

6.2.2 Conditions of Compatibility

The curvatures κ_x and κ_y and the torsion κ_{xy} cannot be selected arbitrarily. The conditions of compatibility for these quantities can be found by expressing the conditions necessary for the determination of $\partial w / \partial x$ and $\partial w / \partial y$ by integration of

$$\frac{\partial}{\partial x}\left(\frac{\partial w}{\partial x}\right)dx + \frac{\partial}{\partial y}\left(\frac{\partial w}{\partial x}\right)dy = -\kappa_x dx + \kappa_{xy} dy \tag{6.2.8}$$

and

$$\frac{\partial}{\partial x}\left(\frac{\partial w}{\partial y}\right)dx + \frac{\partial}{\partial y}\left(\frac{\partial w}{\partial y}\right)dy = \kappa_{xy}dx - \kappa_y dy \tag{6.2.9}$$

In other words, the quantities on the right-hand sides must be total differentials. This yields the conditions of compatibility:

$$\frac{\partial \kappa_x}{\partial y} = -\frac{\partial \kappa_{xy}}{\partial x} \tag{6.2.10}$$

$$\frac{\partial \kappa_y}{\partial x} = -\frac{\partial \kappa_{xy}}{\partial y} \tag{6.2.11}$$

In polar coordinates, the corresponding equations are

$$\frac{1}{r}\frac{\partial \kappa_r}{\partial \theta} = -\frac{\partial \kappa_{r\theta}}{\partial r} - \frac{2}{r}\kappa_{r\theta} \tag{6.2.12}$$

$$\frac{\partial \kappa_\theta}{\partial r} - \frac{1}{r}\left(\kappa_r - \kappa_\theta\right) = -\frac{1}{r}\frac{\partial \kappa_{r\theta}}{\partial \theta} \tag{6.2.13}$$

6.2.3 Lines of Discontinuity, Yield Lines

It has been tacitly assumed above that $w(x, y)$ is differentiable. However, in the theory of plasticity it is necessary to operate with displacements $w(x, y)$ that do not have continuous partial first derivatives along certain lines.

It is assumed that w is continuous, but along continuous curves in the x, y-plane, discontinuities are permitted in $\partial w / \partial n$, and possibly also in $\partial^2 w / \partial n^2$, where n is a coordinate axis in the direction of the normal to the curve (see Fig. 6.2.2). This entails the assumption that $\partial w / \partial n$ has a limiting value for $n \rightarrow 0$ from both sides of the line of discontinuity, although not necessarily the same limiting value. Outside the lines of discontinuity w is assumed to be at least three times differentiable, with continuous, partial third-order derivatives.

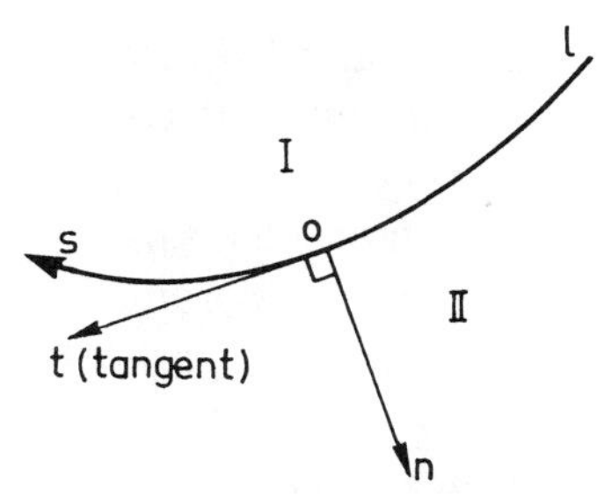

Figure 6.2.2 Local coordinate system along a line of discontinuity in slope.

The jump $(\partial w/\partial n)^{\text{I}} - (\partial w/\partial n)^{\text{II}}$ in the slope of the surface of deflection is denoted θ_n, which can, for example, be described as a function of the length of arc s, measured along the line of discontinuity. A geometrical line of discontinuity is as usual called a *yield line*. The signature for a yield line is shown in Fig. 6.5.3.

6.3 VIRTUAL WORK, BOUNDARY CONDITIONS

6.3.1 Virtual Work

The statical and geometrical conditions introduced above show that the theory established is a two-dimensional theory, since all quantities are described as a function of the coordinates of the middle surface. In Cartesian coordinates the stress field is determined by means of three generalized stresses m_x, m_y, and m_{xy}, which must satisfy the equilibrium condition (6.1.7). The strain field is determined by means of three generalized strains κ_x, κ_y, and κ_{xy}, which have to satisfy the conditions of compatibility (6.2.10) and (6.2.11). These strain measures satisfy the obvious requirements that they vanish when the slab is displaced as a rigid body and that they determine the deformed shape of the slab uniquely. These strain measures further permit the definition of a principle of virtual work, which, in Cartesian coordinates, can be expressed as follows:

$$\iint \left(m_x \kappa_x + m_y \kappa_y + 2 m_{xy} \kappa_{xy} \right) dx\,dy = \iint p w\,dx\,dy - \int m_n \frac{\partial w}{\partial n} ds + \int \left(v_n - \frac{\partial m_{nt}}{\partial s} \right) w\,ds \tag{6.3.1}$$

Here the integral on the left-hand side and the first term on the right-hand side are extended over the entire slab, and the last two terms along the edge of the slab. It can be shown that this equation is valid when the conditions of equilibrium and compatibility are satisfied. It will be seen that all terms can be interpreted as work. The component along the t-axis of the rotation of the normal is $-\partial w/\partial n$, where the n, t-coordinate system is a Cartesian coordinate system of moving axes (see Fig. 6.3.1). The last term is of particular interest. A torsional moment $m_{nt}(s)$ along an edge PQ (see Fig 6.3.2) will be seen to be statically equivalent to a shear force

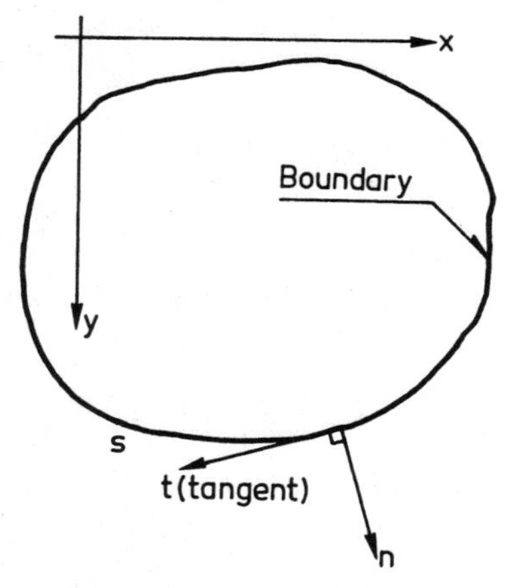

Figure 6.3.1 Arbitrary slab with global coordinate system and local coordinate system at a boundary point.

$$v_n^* = -\frac{\partial m_{nt}}{\partial s} \tag{6.3.2}$$

together with a concentrated force m_{nt}^P, acting at P, and a concentrated force m_{nt}^Q, acting at Q. That this is so can be seen simply from the fact that the work along the edge can be written in the form given in (6.3.1) and from statical considerations alone.

In the formulation (6.3.1) it is assumed that m_{nt} is continuous as a function of s. If this is not the case, the concentrated forces mentioned must be taken into account in the virtual work equation.

A line of discontinuity in the surface of deflection provides a contribution $m_n\theta_n$ per unit length in the internal work, where θ_n, as stated

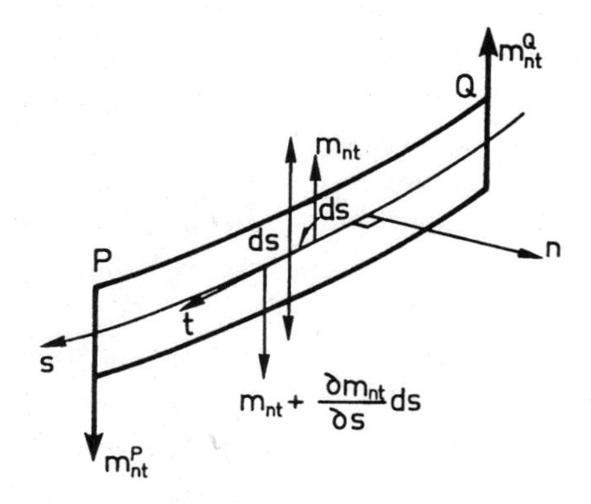

Figure 6.3.2 Statical equivalence of torsional moments at a boundary.

earlier, is defined by

$$\theta_n = \left(\frac{\partial w}{\partial n}\right)^{\text{I}} - \left(\frac{\partial w}{\partial n}\right)^{\text{II}} \tag{6.3.3}$$

and m_n is the bending moment in a section with the direction of the tangent to the line of discontinuity, with the normal n (see Fig. 6.2.2). The complete virtual work equation thus becomes

$$\iint \left(m_x \kappa_x + m_y \kappa_y + 2 m_{xy} \kappa_{xy}\right) dx\, dy + \int m_n \theta_n ds$$
$$= \iint pw\, dx\, dy - \int m_n \frac{\partial w}{\partial n} ds + \int \left(v_n - \frac{\partial m_{nt}}{\partial s}\right) w\, ds \tag{6.3.4}$$

where the last integral on the left-hand side is extended over all lines of discontinuity. Later we shall have use for the virtual work equation in polar coordinates, which can be written

$$\iint \left(m_r \kappa_r + m_\theta \kappa_\theta + 2 m_{r\theta} \kappa_{r\theta}\right) r\, dr\, d\theta + \int m_n \theta_n ds$$
$$= \iint pwr\, dr\, d\theta - \int m_n \frac{\partial w}{\partial n} ds + \int \left(v_n - \frac{\partial m_{nt}}{\partial s}\right) w\, ds \tag{6.3.5}$$

6.3.2 Boundary Conditions

The boundary conditions in the theory of slabs can be derived from the principle of virtual work. The statical and geometrical conditions are formulated in such a way that, with regard to the work along the edge, a slab cannot differentiate between the shear forces v_n and the shear forces v_n^* originating from the statical equivalence of torsional moments along the edge; it is therefore natural to expect that it can only be required that the statical equivalence of v_n and m_{nt}, found by means of internal forces, corresponds to the statical equivalence of the values of the shear forces $\widetilde{v}_n$ and the torsional moments $\widetilde{m}_{nt}$, applied at the edge.

It will immediately be seen that this is correct since, as can be shown in the usual way, this formulation of the boundary conditions gives a unique solution. As there is thus only one solution (if any) with this formulation of the boundary condition, agreement between $\widetilde{v}_n$ and

v_n and between $\tilde{m}_{nt}$ and m_{nt} (i.e., the applied forces and moments and the internal forces and moments) can be obtained only in special cases.

The boundary conditions can thus be expressed as follows if the bending moment $\tilde{m}_n$, the torsional moment $\tilde{m}_{nt}$, and the shear force $\tilde{v}_n$ are applied at the edge:

$$\tilde{m}_n = m_n \tag{6.3.6}$$

$$\tilde{v}_n - \frac{\partial \tilde{m}_{nt}}{\partial s} = v_n - \frac{\partial m_{nt}}{\partial s} \tag{6.3.7}$$

In the special case of a free edge we obtain

$$v_n - \frac{\partial m_{nt}}{\partial s} = 0 \tag{6.3.8}$$

and for an edge loaded with a line load $\bar{p}$,

$$\bar{p} = v_n - \frac{\partial m_{nt}}{\partial s} \tag{6.3.9}$$

The geometrical part of the boundary conditions consists of prescribing w or $\partial w / \partial n$ in such a way as to obtain a total of two boundary conditions corresponding to the number of boundary terms in the virtual work equation.

In the theory of elasticity, use is often made of the fact that the alteration of a system of forces to a statically equivalent system entails only a change in the stress field in a small zone near the point of alteration of the system of forces (Saint-Venant's principle). As far as slabs are concerned, the thinner the slab, the smaller will be the change in the stress field. It is reasonable to conclude that the same will apply in the plastic case. The formulation of the boundary conditions mentioned above, which is due to Kirchhoff, means that torsional moments can exist on a free edge provided only that these moments, together with the shear forces, have a statical equivalence of zero.

In order to get a deeper understanding of the boundary conditions, we shall make the following further considerations. The reason for the applicability of the Kirchhoff boundary conditions is that in a slab, one may have large torsional moments close to a boundary even if the boundary itself is never subjected to torsional moments. Consider as an example a beam with a cross section in the form of a long, narrow rectangle, subjected to pure torsion. In such a section the distribution of the shear stresses will be as sketched in Fig.

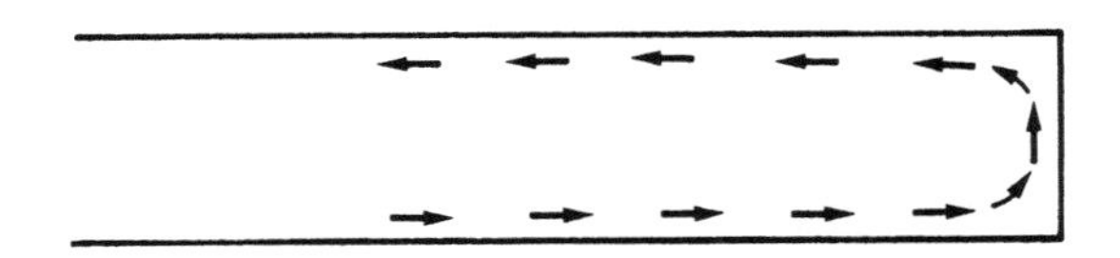

Figure 6.3.3 Shear stresses in a slab element subjected to torsional moments.

6.3.3. Thus, in this particular slab, one finds large values of torsional moments per unit length in the entire slab except in a small boundary strip.

Any slab which according to the Kirchhoff boundary conditions has torsional moments along the boundary will have similar stress distributions in the boundary zone. For an elastic slab, the width of the boundary strip will be of the order of magnitude of half the slab thickness. For a reinforced concrete slab the width will be strongly dependent on the reinforcement layout.

The boundary zone might be said to have the special task to transform the boundary load to the internal forces found by means of the slab theory. In fact, the Kirchhoff boundary conditions can be developed by considering the special behavior of a narrow boundary strip.

Imagine such a narrow boundary strip to be a beam having the usual internal forces, a bending moment M and a shear force V. Assume for the moment that the boundary is straight. The loading on the beam comes from the external loads on the slab boundary and the loads from the internal forces in the slab (see Fig. 6.3.4). The equilibrium equations for the beam are

$$\frac{dV}{ds} = -\widetilde{v}_n + v_n \qquad (6.3.10)$$

$$\frac{dM}{ds} = V + \widetilde{m}_{nt} - m_{nt} \qquad (6.3.11)$$

If we put $M = 0$, we get from (6.3.11)

$$V = -\widetilde{m}_{nt} + m_{nt} \qquad (6.3.12)$$

and from (6.3.10) it is found that

$$v_n - \frac{\partial m_{nt}}{\partial s} = \widetilde{v}_n - \frac{\partial \widetilde{m}_{nt}}{\partial s} \qquad (6.3.13)$$

This means that if the bending moment M in the boundary beam is put equal to zero, the Kirchhoff boundary condition appears as an equilibrium equation for the boundary beam. From this we get the additional information that the shear force in the boundary beam has the value given by (6.3.12). The result will be the same even for a curved boundary beam.

The assumption $M = 0$ is quite natural since the bending moment in a narrow boundary strip resulting from bending moments in a slab must necessarily be small. Statically, however, there is no reason why M should not be different from zero. In a reinforced concrete slab this is possible if a concentrated beam reinforcement is provided in the boundary zone.

If the boundary is supported in such a way that $\tilde{m}_{nt} = 0$, then the reaction per unit length r equals $\tilde{v}_n$, and the value can be calculated be means of (6.3.13):

$$r = v_n - \frac{\partial m_{nt}}{\partial s} \qquad (6.3.14)$$

According to (6.3.12), the boundary zone in this case has to carry a shear force $V = m_{nt}$. If the support is a beam, then bending moments and shear forces in the beam are determined as for a beam with a load per unit length equal to r.

If the boundary is loaded by a line load $\bar{p}$, then $\tilde{v}_n = \bar{p}$ and $\tilde{m}_{nt} = 0$. Condition (6.3.13) gives

$$\bar{p} = v_n - \frac{\partial m_{nt}}{\partial s} \qquad (6.3.15)$$

The boundary zone again has to carry a shear force $V = m_{nt}$.

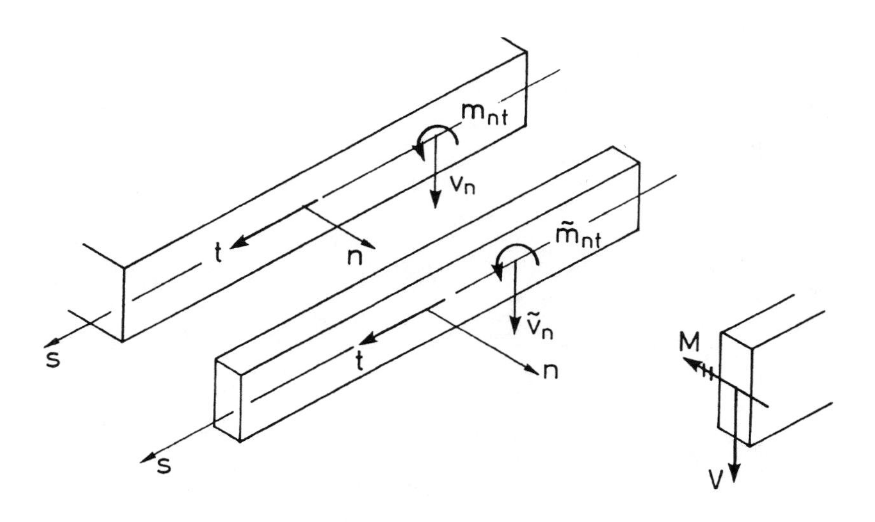

Figure 6.3.4 Boundary beam used in alternative derivation of Kirchhoff boundary conditions.

If the boundary zone itself is considered to be a supporting beam, and if $\widetilde{v}_n = \widetilde{m}_{nt} = 0$, then, the bending moment M and the shear force V in the boundary beam are determined by (6.3.10) and (6.3.11) with $\widetilde{v}_n = \widetilde{m}_{nt} = 0$. It appears that there is a difference in shear force in the case where the boundary zone itself is a supporting beam, and the case where the boundary is supported on another beam. The sum of the shear forces in the boundary zone and in the boundary beam is of course the same in the two cases.

Finally, we shall write down the boundary condition at a corner loaded by a concentrated force P (see Fig. 6.3.5). In this case, the Kirchhoff boundary condition reads

$$P = m_v^{(2)} - m_v^{(1)} + \int v\, ds \tag{6.3.16}$$

The first two terms on the right-hand side are the end-point values found by transforming the torsional moments into shear forces. The last term is the sum of the shear forces along a curve l approaching the corner in the limit. The latter is not always zero, as is often assumed. An example is given in Section 6.5.4.

To facilitate understanding of the illustrations in the following sections, the notation shown in Fig. 6.3.6 is used throughout to indicate various boundary conditions.

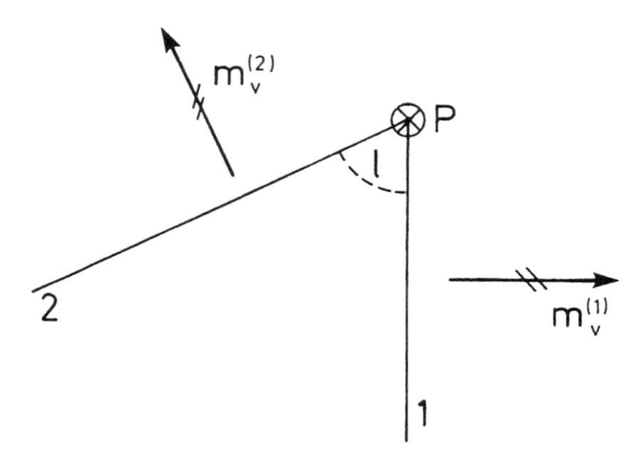

Figure 6.3.5 Corner with a concentrated load and torsional moments.

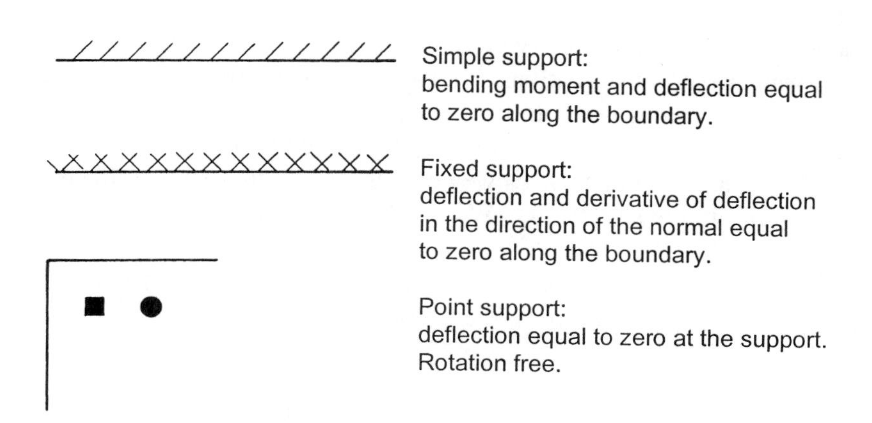

Figure 6.3.6 Notation for frequently occurring boundary conditions.

6.4 CONSTITUTIVE EQUATIONS

6.4.1 Plastic Strains in Slabs

The flow rule (1.1.9) cannot be applied to reinforced concrete slabs without a few comments. The reason for this is that extension of the middle surface will generally arise when there are only bending and torsional moments in normal sections because the neutral axis does not lie in the middle surface. In fact, the flow rule should, strictly speaking, be derived on the basis of a more general yield condition $f(m_x, m_y, m_{xy}, n_x, n_y, n_{xy}) = 0$, which also contains normal and shear forces n_x, n_y and n_{xy} lying in the middle surface, since then, in accordance with the assumptions regarding small deformations, it should be required that the extensions ε_x, ε_y and ε_{xy} are equal to zero. This would mean that in, for example, pure bending, great normal forces would usually arise, which would result in a much greater load-carrying capacity than is actually measured in tests, because the corresponding bending moments for low degrees of reinforcement are much greater than the yield moment corresponding to vanishing normal forces. The fact that this membrane effect is not too pronounced in practice is naturally due to the deflections of the slab, which allow, to some degree, the extensions necessary for statical pure bending and torsion. For certain types of slabs (e.g., laterally restrained slabs), however, such a greatly increased load-carrying capacity is found as a result of this compressive membrane effect, but the load-carrying capacity is exhausted by stability failure caused by extensions of the middle surface, after which the load-carrying capacity falls to approximately the capacity calculated in the usual way (see Section 6.11)

From these considerations it appears more realistic to calculate with normal forces equal to zero than with values corresponding to zero extensions of the middle surface; this has been confirmed by tests in a large number of cases. However, it must be admitted that this is a weak point in the slab theory. We will try to improve on this point in Section 6.11.

If the flow rule (1.1.9) is now applied together with the yield conditions (2.3.62) and (2.3.63), the following expressions are obtained:

$$\begin{aligned} &m_y \geq -\eta m_x + \eta m_{px} - m'_{py} \\ &\kappa_x = \lambda(m_{py} - m_y) \\ &\kappa_y = \lambda(m_{px} - m_x) \qquad (\lambda \geq 0) \\ &\kappa_{xy} = \lambda m_{xy} \end{aligned} \tag{6.4.1}$$

$$\begin{aligned} &m_y \leq -\eta m_x + \eta m_{px} - m'_{py} \\ &\kappa_x = -\lambda(m'_{py} + m_y) \\ &\kappa_y = -\lambda(m'_{px} + m_x) \qquad (\lambda \geq 0) \\ &\kappa_{xy} = \lambda m_{xy} \end{aligned} \tag{6.4.2}$$

By means of the yield condition it can easily be verified that

$$\kappa_x \kappa_y = \kappa_{xy}^2 \tag{6.4.3}$$

which means that one principal curvature is zero.

The formulas are not valid in the apexes *A* and *C* of the yield surface (see Fig. 6.4.1). Neither are they valid along the edge, the projection of which on the m_x, m_y-plane is *BD*. Along the edge, use can be made of each positive linear combination of the strain vectors corresponding to the limiting positions of the normals to the two conical surfaces that intersect along the edge. The following therefore applies here:

$$\begin{aligned} \kappa_x &= \lambda_1 (m_{py} - m_y) + \lambda_2 (-m'_{py} - m_y) \\ \kappa_y &= \lambda_1 (m_{px} - m_x) + \lambda_2 (-m'_{px} - m_x) \\ \kappa_{xy} &= \lambda_1 m_{xy} + \lambda_2 m_{xy} \end{aligned} \tag{6.4.4}$$

where λ_1 and λ_2 are arbitrary nonnegative factors. Here $\kappa_x \kappa_y \leq \kappa_{xy}^2$.

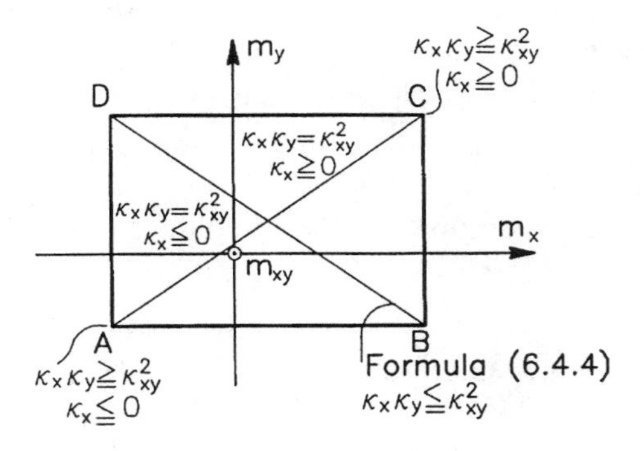

Figure 6.4.1 Plastic strains in an orthotropic slab.

The admissible strain vectors in the isotropic case can be seen from Fig. 6.4.2. In this case, at corner *C*, all vectors can be used that correspond to positive principal curvatures; this can be expressed by $\kappa_x \kappa_y \geq \kappa_{xy}^2$, and, for example, $\kappa_x \geq 0$. Correspondingly, at corner *A* we have $\kappa_x \kappa_y \geq \kappa_{xy}^2$ and $\kappa_x \leq 0$. The same conditions can be seen to apply in the general case at the apexes *A* and *C* (see Fig. 6.4.1). A complete description of the flow rule in the general case is thus obtained.

The above-mentioned conditions for the strains at the singular points of the yield surface, together with the relationship (6.4.3), facilitate a survey of the strains belonging to the various zones of the m_x, m_y-plane. This is shown in Fig. 6.4.1.

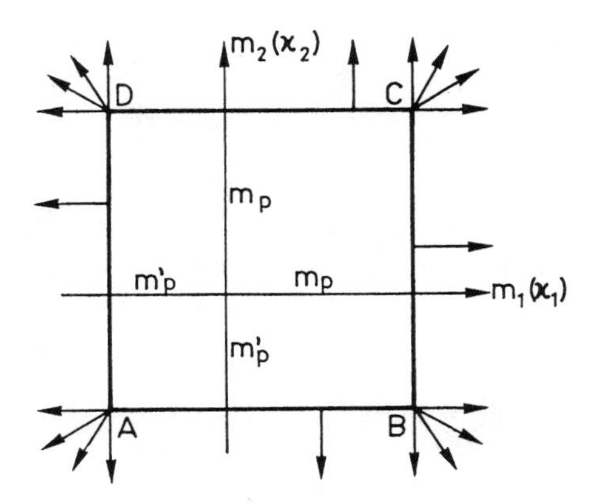

Figure 6.4.2 Plastic strains in an isotropic slab.

A simple connection exists between the plastic strains in a special case of an orthotropic slab and an isotropic slab. Let $\overline{m} = (a,b,c)$ represent a moment field that satisfies the yield condition for an isotropic slab $m_{px} = m'_{px} = m_{py} = m'_{py} = m_p$. Then $\overline{m}^* = (a, \mu b, \sqrt{\mu} c)$ will satisfy the yield condition in the case $m_{px} = m'_{px} = m_p$, $m_{py} = m'_{py} = \mu m_p$ (see Section 2.3.2). A comparison between the corresponding expressions for the plastic strains shows that if $\overline{\kappa} = (\alpha, \beta, \gamma)$ is the strain vector corresponding to $\overline{m}$, then $\overline{\kappa}^* = (\alpha, (1/\mu)\beta, (1/\sqrt{\mu})\gamma)$, except for some factor, is the strain vector corresponding to $\overline{m}^*$.

The same relationship exists between the strain vectors in the isotropic case $m_{px} = m_{py} = m_p$, $m'_{px} = m'_{py} = m'_p$, and in the case $m_{px} = m_p$, $m_{py} = \mu m_p$, $m'_{px} = m'_p$, $m'_{py} = \mu m'_p$. In the following, the latter case is called "the simple affine case."

6.4.2 Dissipation Formulas

In regions where one principal curvature is zero, there exists a very simple formula for the dissipation W per unit area of the slab. Consider, for example, the region where (6.4.1) is valid. Here we find the following dissipation per unit area

$$\begin{aligned} W &= m_x \kappa_x + m_y \kappa_y + 2 m_{xy} \kappa_{xy} \\ &= \lambda m_x (m_{py} - m_y) + \lambda m_y (m_{px} - m_x) + 2\lambda m_{xy}^2 \end{aligned} \quad (6.4.5)$$

Utilizing the yield condition (2.3.62), we get

$$W = m_{px}\kappa_x + m_{py}\kappa_y = m_{px}|\kappa_x| + m_{py}|\kappa_y| \quad (6.4.6)$$

In the region where (6.4.2) is valid one finds in a similar way that

$$W = m'_{px}|\kappa_x| + m'_{py}|\kappa_y| \quad (6.4.7)$$

Expressions (6.4.6) and (6.4.7) are also seen to be valid in apexes C and A, respectively. Along the edge a more complicated expression is found, which will not be written down here.

In the special case of an isotropic slab, a general formula similar to (4.4.8) valid for disks can be established. The formula reads

$$W = \tfrac{1}{2}(m_p + m'_p)(|\kappa_1| + |\kappa_2|) + \tfrac{1}{2}(m_p - m'_p)(\kappa_1 + \kappa_2) \quad (6.4.8)$$

The dissipation formulas can also be developed in the following way, which in addition gives some useful information about the internal forces. If the principal curvature that is not zero is equal to κ, and if the corresponding line of principal curvature forms the angle α with the x-axis, the following is obtained from (6.2.3) and (6.2.4):

$$\kappa_x = \kappa \cos^2\alpha, \quad \kappa_y = \kappa \sin^2\alpha, \quad \kappa_{xy} = -\kappa \sin\alpha \cos\alpha \tag{6.4.9}$$

Considering first the case where $\kappa_x > 0$ (i.e., $\kappa > 0$) formulas (6.4.1) apply, and solution of these with respect to the moments yields (assuming $\lambda > 0$)

$$\begin{aligned} m_x &= m_{px} - \frac{\kappa \sin^2\alpha}{\lambda} \\ m_y &= m_{py} - \frac{\kappa \cos^2\alpha}{\lambda} \\ m_{xy} &= -\frac{\kappa \sin\alpha \cos\alpha}{\lambda} \end{aligned} \tag{6.4.10}$$

These expressions give the bending moment m_b in the section at right angles to the line with curvature κ

$$\begin{aligned} m_b &= m_x \cos^2\alpha + m_y \sin^2\alpha - 2m_{xy}\sin\alpha \cos\alpha \\ &= m_{px} \cos^2\alpha + m_{py} \sin^2\alpha \end{aligned} \tag{6.4.11}$$

which means that

$$W = (m_{px} \cos^2\alpha + m_{py} \sin^2\alpha)\kappa = m_{px}\kappa_x + m_{py}\kappa_y \tag{6.4.12}$$

Correspondingly, for $\kappa < 0$,

$$m_b = -(m'_{px} \cos^2\alpha + m'_{py} \sin^2\alpha) \tag{6.4.13}$$

$$W = m'_{px}|\kappa_x| + m'_{py}|\kappa_y| \tag{6.4.14}$$

We have reached the significant result that the bending moment in the section perpendicular to the line with the principal curvature $|\kappa|$ can be calculated as if the principal directions were coinciding with the directions of the reinforcement, which is obviously not the case except for points of the yield surface corresponding to $m_{xy} = 0$. Moreover, in this section there will usually be torsional moments, which can be found to be

$$\kappa > 0: \quad m_v = (m_{px} - m_{py}) \sin\alpha \cos\alpha \tag{6.4.15}$$

$$\kappa < 0: \quad m_v = (m'_{py} - m'_{px}) \sin\alpha \cos\alpha \tag{6.4.16}$$

This means that the torsional moment can be calculated on the basis of the same assumptions as those valid for the bending moments.

The formulas for the bending and torsional moments were proposed by Johansen [43.1; 62.1] on an intuitive basis. The agreement between Johansen's proposal and the yield conditions was demonstrated by the author [64.3].

The formulas for bending and torsional moments can be calculated in a more direct manner in the following way. Let us consider a concrete disk reinforced in two directions, the x- and y- directions, at right angles to each other. The reinforcement area in the x-direction is A_{sx} per unit length measured in the y-direction. The reinforcement area in the y-direction is A_{sy} per unit length measured in the x-direction. The yield stress of the reinforcement is f_Y. The forces in the reinforcement are determined by the longitudinal strain in the longitudinal direction of the reinforcement. The tensile strength of the concrete is assumed to be zero.

The disk is assumed to be loaded to uniaxial strain in the n-direction, forming angle α with the x-direction (see Fig. 6.4.3). Therefore, the concrete will crack in sections perpendicular to the n-direction, and the reinforcement will be stressed to the yield stress f_Y . The stresses in a section perpendicular to the n-direction therefore only get contributions from the stresses in the reinforcement. The axial force N and the shear force V, both measured per unit length in the direction of the t-axis, can be calculated by the transformation formulas for plane stresses. We have

$$N = A_{sx} f_Y \cos^2\alpha + A_{sy} f_Y \sin^2\alpha \tag{6.4.17}$$

$$V = -(A_{sx} - A_{sy}) f_Y \sin\alpha\cos\alpha \tag{6.4.18}$$

that is, when $A_{sx} \neq A_{sy}$ and the n-direction does not coincide with the directions of reinforcement, an axial force N as well as a shear force V must be applied in order to create uniaxial strain in the n-direction.

If the figure represents the reinforcement at the bottom side of a slab, with the deformation pure bending in the n-direction, the bending moment m_b and the torsional moment m_v , in a section perpendicular to the n-direction, can with good approximation be calculated as

$$m_b = m_{px} \cos^2\alpha + m_{py} \sin^2\alpha \tag{6.4.19}$$

$$m_v = (m_{px} - m_{py}) \sin\alpha\cos\alpha \tag{6.4.20}$$

where m_{px} and m_{py} are the yield moment for pure bending in the x-direction and the yield moment for pure bending in the y-direction, respectively, both corresponding to positive moments.

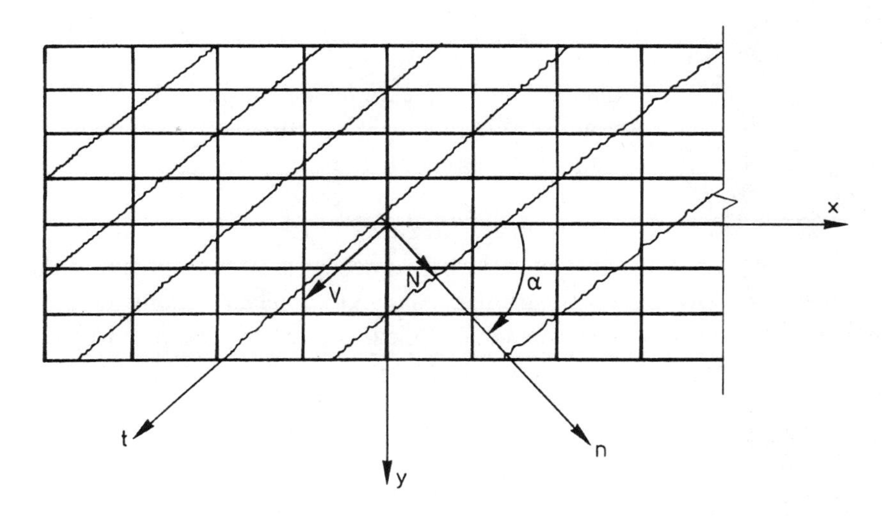

Figure 6.4.3 Disk in uniaxial strain.

6.5 EXACT SOLUTIONS FOR ISOTROPIC SLABS

6.5.1 Various Types of Yield Zones

Three types of yield zones are considered. These can be characterized as follows, denoting the positive yield moment m_p and the negative yield moment m'_p:

1. Both principal moments are equal to m_p or $-m'_p$ all over the slab.
2. The principal moments are given by $m_1 = m_p$, $m_2 = -m'_p$.
3. The principal moments satisfying the conditions $m_1 = m_p$, $-m'_p < m_2 < m_p$.

Yield zone of type 1. In a yield zone of type 1, there are equal principal moments, either m_p or $-m'_p$. Suppose the direction of the principal sections to be described by the angle $\psi(x, y)$ between the first principal section and the x-axis (ψ positive in a clockwise direction).

In a yield zone of type 1, we find from (6.1.1) and (6.1.2) that

$$\begin{aligned} m_y &= m_p \cos^2\psi + m_p \sin^2\psi = m_p \\ m_x &= m_p \\ m_{xy} &= 0 \end{aligned} \tag{6.5.1}$$

It will be seen that the equilibrium equation (6.1.7) is satisfied only for $p = 0$. The shear forces are similarly zero.

Each section is thus a principal section, and the load intensity is zero. Thus a yield zone of this type arises only when a slab part is subjected solely to a bending moment along the edge.

Yield zone of type 2. In a yield zone of type 2, we have $m_1 = m_p$, $m_2 = -m'_p$ (see Fig. 6.5.1). In order to accept such a yield zone in a given problem, it must naturally be shown that the conditions of compatibility and the geometrical and statical conditions can be fulfilled.

Let us assume that the lines of principal sections, which are identical with the coordinate lines, are produced by

$$x = f(\alpha, \beta) \qquad y = g(\alpha, \beta) \tag{6.5.2}$$

where x and y are Cartesian coordinates and α and β are the curvilinear coordinates. The functions f and g denote two functions of α and β. The arc differential ds corresponding to the differentials $d\alpha$ and $d\beta$ is determined by

$$ds^2 = A^2 d\alpha^2 + B^2 d\beta^2 \tag{6.5.3}$$

where

$$A^2 = \left(\frac{\partial f}{\partial \alpha}\right)^2 + \left(\frac{\partial g}{\partial \alpha}\right)^2, \qquad B^2 = \left(\frac{\partial f}{\partial \beta}\right)^2 + \left(\frac{\partial g}{\partial \beta}\right)^2 \tag{6.5.4}$$

The condition of orthogonality is expressed by

$$\frac{\partial f}{\partial \alpha}\frac{\partial f}{\partial \beta} + \frac{\partial g}{\partial \alpha}\frac{\partial g}{\partial \beta} = 0 \tag{6.5.5}$$

The radii of curvature of the coordinate curves β = constant and α = constant are $1/R_\alpha$ and $1/R_\beta$, respectively, and are determined as

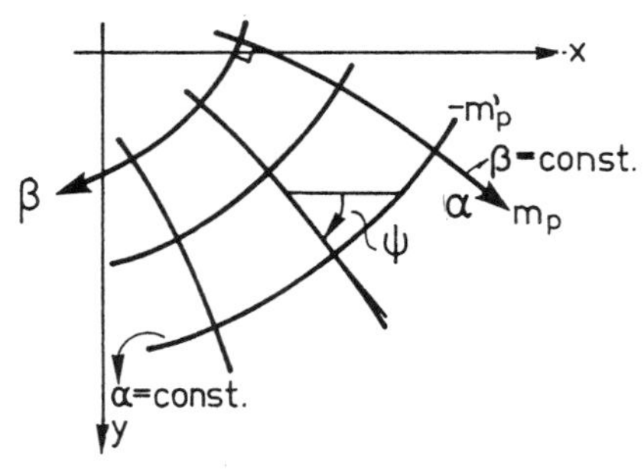

Figure 6.5.1 Yield zone of type 2 in Cartesian coordinates.

follows:

$$\frac{1}{R_\alpha} = -\frac{1}{AB}\frac{\partial A}{\partial \beta}, \qquad \frac{1}{R_\beta} = -\frac{1}{AB}\frac{\partial B}{\partial \alpha} \tag{6.5.6}$$

The signs are selected in such a way that a positive curvature corresponds to the coordinate curve in question, being concave in the positive direction of the other coordinate curve.

The equilibrium conditions have the following appearance (see e.g., [64.4]):

$$ABv_\alpha = \frac{\partial Bm_\alpha}{\partial \alpha} - \frac{\partial A}{\partial \beta}m_{\alpha\beta} - \frac{\partial Am_{\alpha\beta}}{\partial \beta} - \frac{\partial B}{\partial \alpha}m_\beta \tag{6.5.7}$$

$$ABv_\beta = \frac{\partial Am_\beta}{\partial \beta} - \frac{\partial B}{\partial \alpha}m_{\alpha\beta} - \frac{\partial Bm_{\alpha\beta}}{\partial \alpha} - \frac{\partial A}{\partial \beta}m_\alpha \tag{6.5.8}$$

$$\frac{\partial Bv_\alpha}{\partial \alpha} + \frac{\partial Av_\beta}{\partial \beta} = -ABp \tag{6.5.9}$$

Insertion of $m_\beta = m_p$, $m_\alpha = -m'_p$, and $m_{\alpha\beta} = 0$ yields

$$v_\alpha = -(m_p + m'_p)\frac{1}{AB}\frac{\partial B}{\partial \alpha} = (m_p + m'_p)\frac{1}{R_\beta} \tag{6.5.10}$$

$$v_\beta = (m_p + m'_p)\frac{1}{AB}\frac{\partial A}{\partial \beta} = -(m_p + m'_p)\frac{1}{R_\alpha} \tag{6.5.11}$$

Then, by insertion in the equilibrium equation (6.5.9), we obtain

$$(m_p + m'_p)\left[\frac{\partial}{\partial \alpha}\left(\frac{1}{A}\frac{\partial B}{\partial \alpha}\right) - \frac{\partial}{\partial \beta}\left(\frac{1}{B}\frac{\partial A}{\partial \beta}\right)\right] = ABp \tag{6.5.12}$$

Since

$$\frac{\partial}{\partial \alpha}\left(\frac{1}{A}\frac{\partial B}{\partial \alpha}\right) + \frac{\partial}{\partial \beta}\left(\frac{1}{B}\frac{\partial A}{\partial \beta}\right) = 0 \tag{6.5.13}$$

which is Gauss's equation, we get

$$\frac{1}{AB}\frac{\partial}{\partial \alpha}\left(\frac{1}{A}\frac{\partial B}{\partial \alpha}\right) = \frac{1}{2}\frac{p}{m_p + m'_p}, \qquad \frac{1}{AB}\frac{\partial}{\partial \beta}\left(\frac{1}{B}\frac{\partial A}{\partial \beta}\right) = -\frac{1}{2}\frac{p}{m_p + m'_p} \tag{6.5.14}$$

In the case $p = 0$, these equations may, by means of (6.5.6), be written as follows (assuming that $R_\alpha \neq 0$ and $R_\beta \neq 0$):

$$\frac{1}{A}\frac{\partial R_\beta}{\partial \alpha} = -1, \qquad \frac{1}{B}\frac{\partial R_\alpha}{\partial \beta} = -1 \tag{6.5.15}$$

which shows that the orthogonal net of principal sections is a Hencky net, since (6.5.15) is exactly Hencky's second theorem [23.4; 50.2] for slip line fields. An equivalent to Hencky's first theorem may also be established [62.2], and the well-known methods for the determination of slip line fields may be used [23.3; 50.2].

If the lines with the principal moment m_p form the angle ψ with a fixed x-axis (see Fig. 6.5.1), then in Cartesian coordinates we get the following:

$$\begin{aligned} m_x &= \tfrac{1}{2}(m_p - m_p') - \tfrac{1}{2}(m_p + m_p')\cos 2\psi \\ m_y &= \tfrac{1}{2}(m_p - m_p') + \tfrac{1}{2}(m_p + m_p')\cos 2\psi \\ m_{xy} &= \tfrac{1}{2}(m_p + m_p')\sin 2\psi \end{aligned} \tag{6.5.16}$$

In polar coordinates, we get

$$\begin{aligned} m_r &= \tfrac{1}{2}(m_p - m_p') - \tfrac{1}{2}(m_p + m_p')\cos 2\psi \\ m_\theta &= \tfrac{1}{2}(m_p - m_p') + \tfrac{1}{2}(m_p + m_p')\cos 2\psi \\ m_{r\theta} &= \tfrac{1}{2}(m_p + m_p')\sin 2\psi \end{aligned} \tag{6.5.17}$$

where the meaning of $\psi(r,\theta)$ in this case is shown in Fig. 6.5.2.

Inserting these expressions in the equilibrium equations, we get a partial differential equation for the determination of ψ as shown by Johansen [43.1; 62.1]. However, only trivial solutions to these equations are known.

Yield zone of type 3. In yield zones of type 3 there is yielding only in the first or second principal section. For example, there may be yielding in the first principal section with the moment m_p, and in the second principal section, the bending moments satisfy everywhere the condition that $-m_p' \leq m_2 \leq m_p$, where the equality holds at certain points only. One of the principal curvatures κ_2 is thus zero. Therefore,

$$\kappa_1 \kappa_2 = \kappa_x \kappa_y - \kappa_{xy}^2 = 0 \tag{6.5.18}$$

$$\kappa_1 + \kappa_2 > 0 \text{ or } \kappa_1 + \kappa_2 < 0 \quad \text{everywhere} \tag{6.5.19}$$

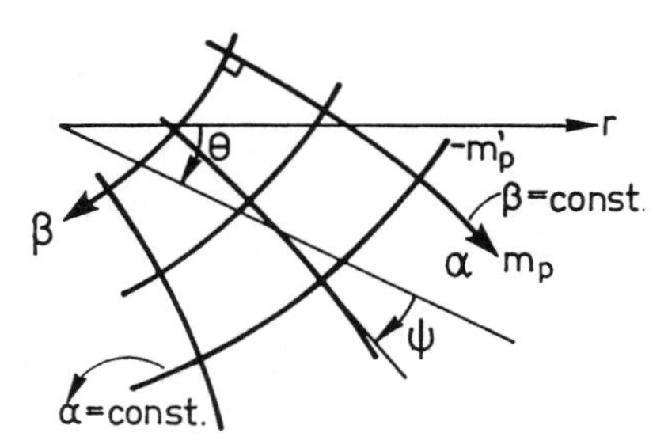

Figure 6.5.2 Yield zone of type 2 in polar coordinates.

The general solution of (6.5.18) is developable surfaces. The curves along which $\kappa_2 = 0$ are straight lines (generatrices). The solutions to (6.5.18) are conical surfaces, cylindrical surfaces, and tangential surfaces. A number of solutions involving yield zones of type 3 have been developed by Mansfield [57.2; 60.4] and the author [62.3; 64.3].

Yield lines. There is naturally nothing to prevent the yield condition from being satisfied purely along a straight line. As the number of statical conditions in the slab would thereby be increased, the number of geometrical constraints must be reduced, which is done by permitting discontinuities in the slope of the surface of deflection. Such a line discontinuity, which, as mentioned in Section 6.2.3, is called a yield line, can be regarded as a limiting case of uniform curvature κ over a certain length δ in a single direction, where the curvature $\kappa \to \infty$ and, simultaneously the length $\delta \to 0$, while the total change of slope along δ is constant.

Along the yield line there is the bending moment m_n equal to m_p or $-m_p'$ and the torsional moment $m_{nt} = 0$. When m_n is differentiable in the neighborhood of the yield line, and the yield condition is satisfied, v_n will also be equal to zero, as can be seen from the equilibrium equation (6.1.4).

If there is a jump θ_n in the derivatives $\partial w / \partial n$, then

$$\theta_n = \left(\frac{\partial w}{\partial n}\right)^{\mathrm{I}} - \left(\frac{\partial w}{\partial n}\right)^{\mathrm{II}} \tag{6.5.20}$$

and the dissipation per unit length in the yield line is

$$W_1 = | m_n || \theta_n | \tag{6.5.21}$$

If θ_n is to be expressed by the jump in the derivatives in the direction of the x- and y- axes (Fig. 6.5.3), we get

$$|\theta_n| = \sqrt{\theta_x^2 + \theta_y^2} \tag{6.5.22}$$

where

$$\theta_x = \left(\frac{\partial w}{\partial x}\right)^{\mathrm{I}} - \left(\frac{\partial w}{\partial x}\right)^{\mathrm{II}} = \theta_n \cos\alpha \tag{6.5.23}$$

$$\theta_y = \left(\frac{\partial w}{\partial y}\right)^{\mathrm{I}} - \left(\frac{\partial w}{\partial y}\right)^{\mathrm{II}} = \theta_n \sin\alpha \tag{6.5.24}$$

θ_x and θ_y being the jump in the derivatives in the x- and y- directions, respectively. If the discontinuity is found along a curve C, the contribution to the dissipation can be expressed as follows:

$$\Delta D = \int_C |m_n| \sqrt{\theta_x^2 + \theta_y^2}\, ds \tag{6.5.25}$$

The circular fan. One of the most common yield zones is the circular fan, which is a combination of a yield zone of type 3 and a circular yield line (see Fig. 6.5.4).

We therefore have the moment field

$$m_\theta = m_p, \quad m_r = -m_p' \quad at \quad r = a \tag{6.5.26}$$

If point P is given a unit displacement downward, the following deflections are obtained:

$$w = 1 - \frac{r}{a} \tag{6.5.27}$$

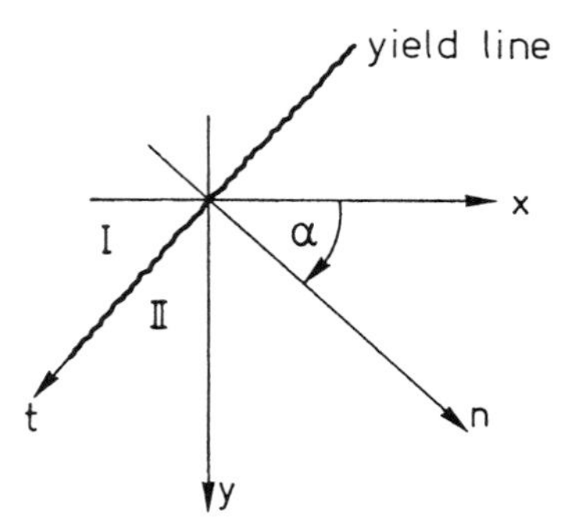

Figure 6.5.3 Local coordinate system in a yield line.

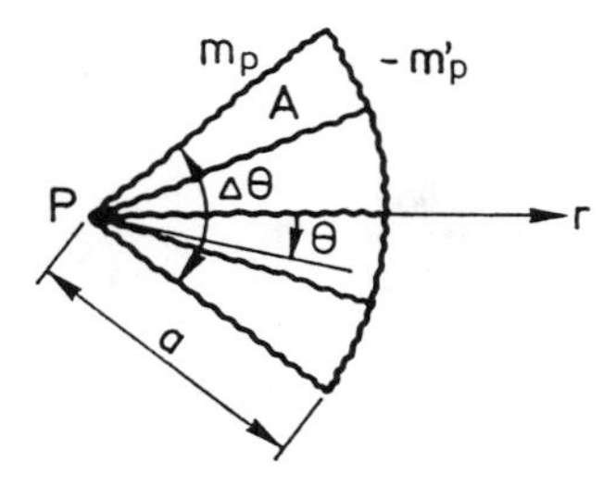

Figure 6.5.4 The circular fan.

According to (6.2.5) to (6.2.7), the curvatures and torsion are

$$\kappa_r = \kappa_{r\theta} = 0$$
$$\kappa_\theta = \frac{1}{ar} \tag{6.5.28}$$

The dissipation ΔD can now be calculated:

$$\begin{aligned} \Delta D &= \iint_A m_\theta \kappa_\theta \, r dr d\theta + \int_0^{\Delta\theta} (m_r)_{r=a} \left(\frac{\partial w}{\partial r}\right)_{r=a} a d\theta \\ &= \int_0^a r\,dr \int_0^{\Delta\theta} m_p \frac{1}{ra} d\theta + \int_0^{\Delta\theta} m'_p \frac{1}{a} a d\theta = \Delta\theta (m_p + m'_p) \end{aligned} \tag{6.5.29}$$

This expression does not include possible contributions from yield lines along the limiting radii.

6.5.2 Boundary Conditions

Boundary conditions for yield lines. A number of cases will first be considered in which the yielding is concentrated in lines in which the moment is equal to one of the yield moments. Referring to Fig. 6.5.5, we have $m_n = m_p$ or $m_n = -m'_p$.

If it is assumed that m_n is differentiable in the neighborhood of the yield line, it will, as mentioned before, be seen from (6.1.4) that $v_n = 0$. If the yield line ends at a free edge that forms an angle α with the yield line, the bending moment at the edge is required to be zero. Mohr's circle at point A is therefore as shown in Fig. 6.5.5, from which it will be seen that the torsional moment m_v at the edge is

$$m_v = -m_p \tan(\tfrac{\pi}{2} - \alpha) = -m_p \cot \alpha, \qquad 0 \le \alpha \le \pi \tag{6.5.30}$$

and that the smallest principal moment is

$$m_2 = -\frac{m_p}{\sin^2 \alpha} + m_p = -m_p \cot^2 \alpha \tag{6.5.31}$$

If it is required that $|m_2| \le m_p'$, the following condition is obtained:

$$\tan^2 \alpha \ge \frac{m_p}{m_p'} \tag{6.5.32}$$

For $m_p = m_p'$, the special result is obtained that

$$|\tan \alpha| \ge 1 \tag{6.5.33}$$

that is,

$$\frac{\pi}{4} \le \alpha \le 3\frac{\pi}{4} \tag{6.5.34}$$

If $m_p' = 0$, then $\alpha = \pi/2$.

Provided that condition (6.5.32) is satisfied, the yield condition will not be violated at the point where the yield line intersects the edge. When $\alpha \ne \pi/2$, there will be a torsional moment on the edge, which is admissible according to the Kirchhoff boundary conditions (6.3.6) and (6.3.7). Condition (6.5.32) is naturally also valid when the yield line ends at a simply supported edge.

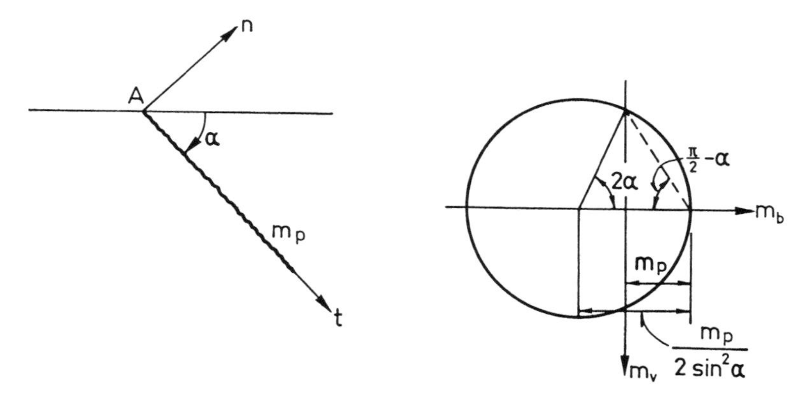

Figure 6.5.5 Yield line crossing a free edge.

Next we consider a corner to which a positive yield line runs (see Fig. 6.5.6). If the yield line is required to be a principal section and the bending moment must, at the same time, be zero at the two edges that intersect each other at the corner, it will be seen that it is necessary to operate with two different Mohr's circles at the point, as shown in Fig. 6.5.6.

Condition (6.5.32) is then valid for both angles α and β; that is, we have the following requirements at the corner:

$$\tan^2\alpha \geq \frac{m_p}{m'_p}, \qquad \tan^2\beta \geq \frac{m_p}{m'_p} \tag{6.5.35}$$

If $m_p = m'_p$, both angles must therefore be greater than or equal to 45°. If the corner is right-angled, the yield line must therefore lie in the bisector of the angle.

It should be noted that if the requirement that the yield line be a principal section is abandoned, then naturally it is possible to construct only one Mohr's circle at the corner, but this will normally entail torsional moments in the yield line.

Boundary conditions for yield zones. A yield zone of type 1 that is limited by a straight line will now be considered. As Mohr's circle at every point in the yield zone is reduced to a point, the boundary must be loaded with a bending moment m_p or $-m'_p$.

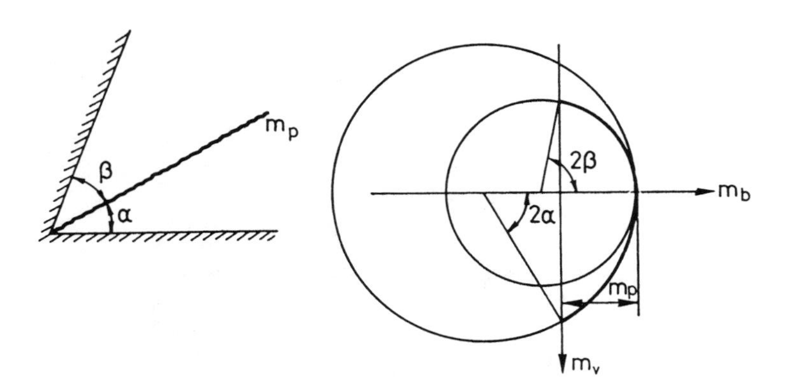

Figure 6.5.6 Yield line entering a corner.

For a yield zone of type 2 that is limited by a free edge, Mohr's circle will be as shown in Fig. 6.5.7. As the bending moment m_b at the edge has to be zero, we get

$$m_b = m_p \cos^2\alpha - m_p' \sin^2\alpha = 0 \tag{6.5.36}$$

or

$$\tan^2\alpha = \frac{m_p}{m_p'} \tag{6.5.37}$$

For $m_p' = m_p$, it is found that $\alpha = \pi/4$, and for $m_p' = 0$, it is found that $\alpha = \pi/2$.

At the edge there is a torsional moment

$$m_v = -m_p \cot\alpha \tag{6.5.38}$$

For $m_p = m_p'$ (i.e., $\alpha = \pi/4$), this becomes

$$m_v = -m_p \tag{6.5.39}$$

The same conditions hold when the boundary is simply supported.

Next, a yield zone of type 3 will be considered (see Fig. 6.5.8). Here there is a yield zone with a bending moment m_p along straight lines. The bending moment in the second principal section is assumed to be equal to $-m_p'$ along a curve which forms right angles with the lines with the moment m_p (which is necessary if the yield condition is not to be violated). Mohr's circle for point O is as shown in the figure.

As previously, the bending moment at the edge is found to be

$$m_b = m_p \sin^2\alpha - m_p' \cos^2\alpha = 0 \tag{6.5.40}$$

that is,

$$\tan^2\alpha = \frac{m_p'}{m_p} \tag{6.5.41}$$

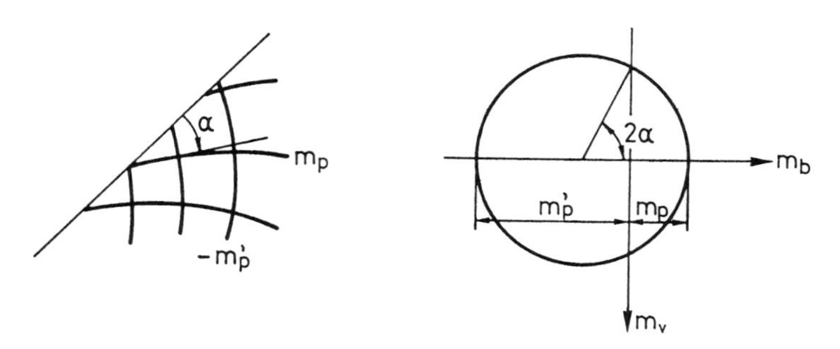

Figure 6.5.7 Yield zone of type 2 crossed by a free edge.

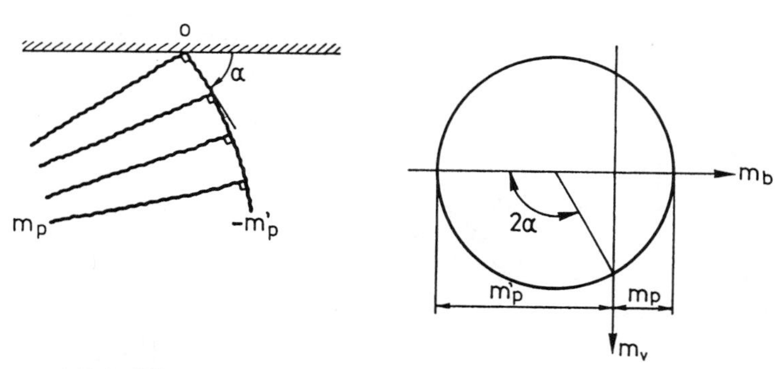

Figure 6.5.8 Yield zone of type 3 crossed by a free edge.

For $m_p = m'_p$ it is found that $\alpha = \pi/4$. For $m'_p = 0$, we get $\alpha = 0$. At the edge, the following torsional moment occurs at point O:

$$m_v = m'_p \cot\alpha \tag{6.5.42}$$

that is, for $m_p = m'_p$, $\alpha = \pi/4$, we get $m_v = m_p$.

6.5.3 A Survey of Exact Solutions

There exists a substantial number of exact solutions, a survey of which is given in Fig. 6.5.9. Solutions *a, b,* and *c* were given by Johansen [43.1; 62.1]. Solution *b* contains the solution $P = 2\pi(m_p + m'_p)$, which is valid for a concentrated force acting on a circular slab with fixed or simply supported edges as special cases. This solution was already given by Ingerslev [21.1; 23.1]. As shown by Haythornthwaite and Shield [58.4], this solution is valid for an arbitrary fixed slab *g*. Solution *d* was given by Prager [52.2], and solutions *e* and *f* by Wood [61.5]. Johansen gave solution *f* as an upper bound solution [43.1; 62.1]. Solutions *h* through *r* are the author's [62.2; 63.10; 64.3]. Solutions *p, q,* and *r* have equivalents in the slip line theory [20.2; 28.4; 31.3; 50.3; 51.2] (see Section 3.6).

The clamped square slab defied solution for a long time. In fact, it was claimed that the problem had no solution according to the present plastic theory [68.4]. However, in 1974 it was shown by Fox [74.3] that the exact load-carrying capacity in the case $m_p = m'_p$ is $pa^2/m_p = 42.851$. The solution turned out to be rather complicated, containing one region *CAE* with a yield zone of type 3 and one region *AED* with a yield zone of type 2. Finally, there is a rigid part *EDB*. The regions are shown in Fig. 6.5.10. Fox [72.2] also solved the rectangular simply supported slab with a concentrated force. Finally, a whole class of exact solutions was developed by Massonnet [67.5].

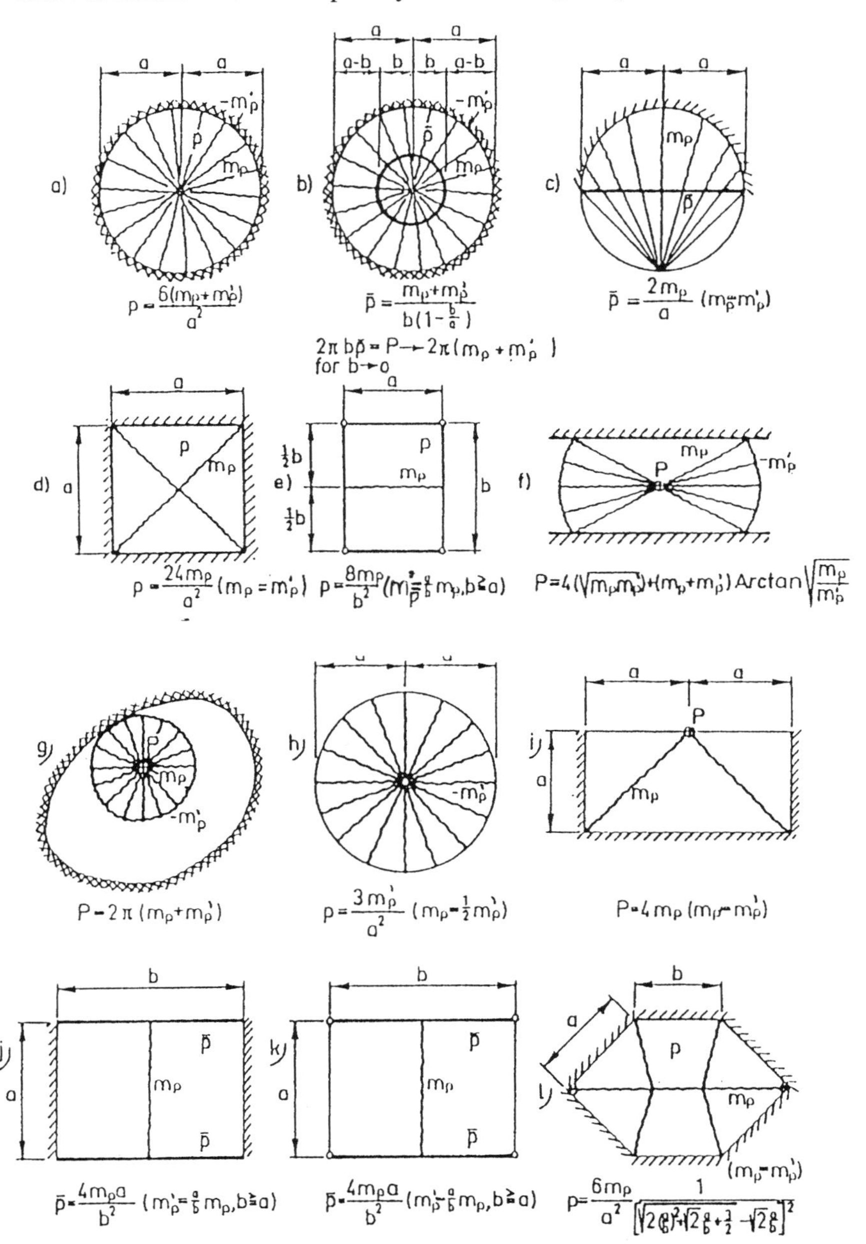

Figure 6.5.9 Survey of exact solutions for isotropic slabs.

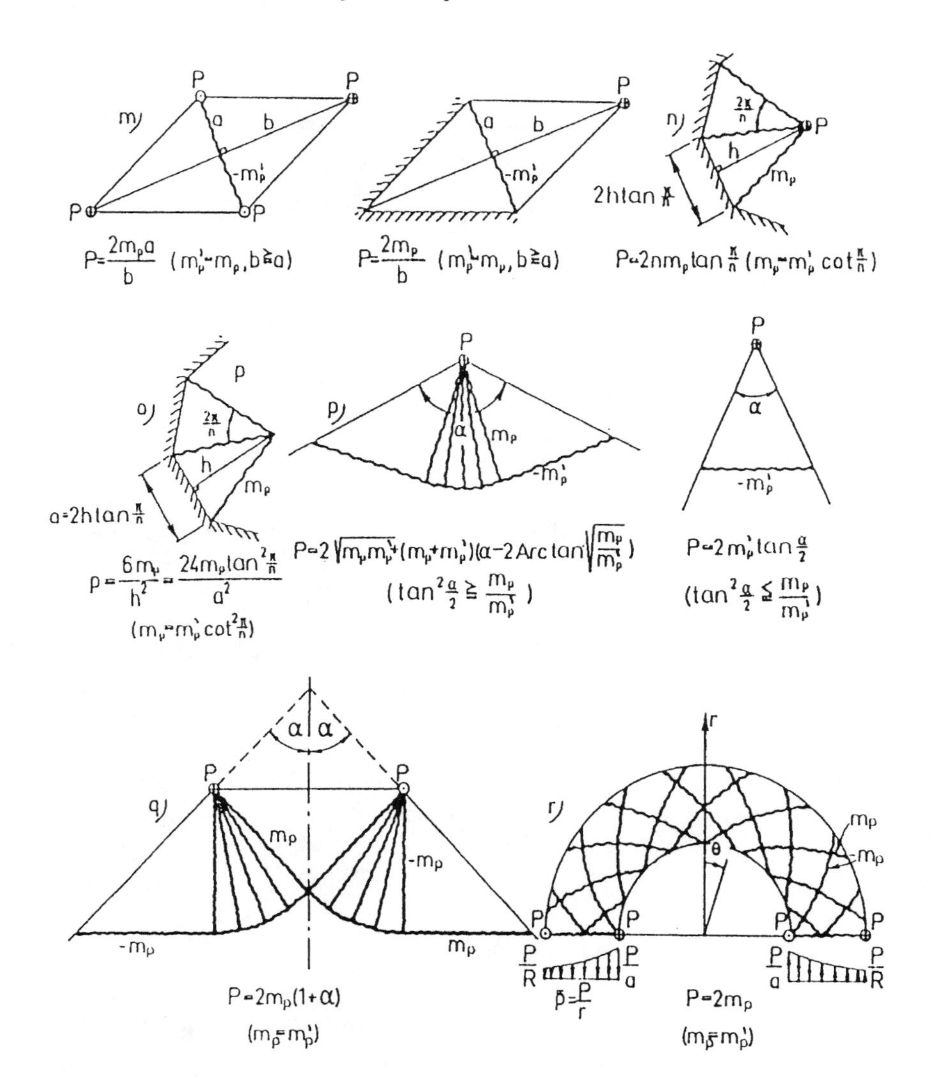

Figure 6.5.9 (continued)

It is not possible to deal with all these solutions here, and it should suffice to consider a number of illustrative examples. Most of the moment fields used are covered by only a few simple solutions to the equilibrium equations, and these solutions will therefore be dealt with first.

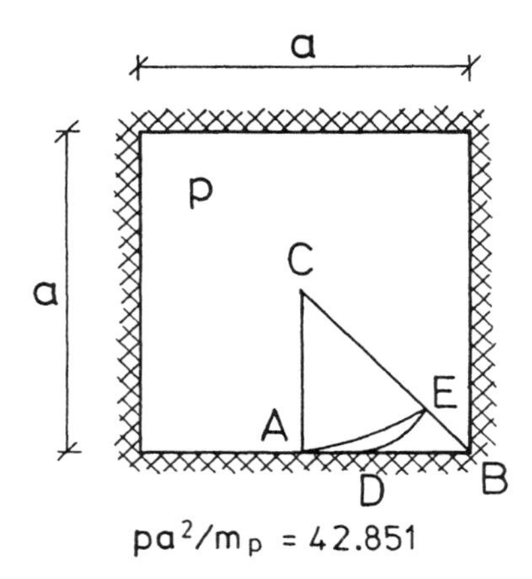

Figure 6.5.10 Exact solution for a clamped square slab with uniform load.

6.5.4 Illustrative Examples

Simple statically admissible moment fields. Consider the moment field

$$m_\theta = m_p, \qquad m_r = -m'_p, \qquad m_{r\theta} = 0 \tag{6.5.43}$$

It can easily be verified that the equilibrium condition (6.1.11) is satisfied with $p = 0$, and from (6.1.8) and (6.1.9) it is found that

$$v_\theta = 0, \qquad v_r = -\frac{m_p + m'_p}{r} \tag{6.5.44}$$

It will be seen that the absolute value of the total shear force in circles with their centers at the pole $(r, \theta) = (0, 0)$ is constant and equal to $2\pi(m_p + m'_p)$, corresponding to a concentrated force $P = 2\pi(m_p + m'_p)$ at the pole. If $-\alpha \le \theta \le \beta$, the force $P = (\alpha + \beta)(m_p + m'_p)$ will result.

The moment field may be summarized as follows:

$$\begin{aligned} &m_r = -m'_p, \qquad m_\theta = m_p, \qquad m_{r\theta} = 0 \\ &v_r = -\frac{m_p + m'_p}{r}, \qquad v_\theta = 0 \\ &p = 0 \\ &P = (\alpha + \beta)(m_p + m'_p) \qquad \text{at} \quad (r, \theta) = (0,0) \end{aligned} \tag{6.5.45}$$

Now consider the moment field determined by

$$m_\theta = m_p, \qquad m_{r\theta} = 0 \tag{6.5.46}$$

and the requirement that the bending moment m_x (see Fig. 6.5.11) must equal zero. This yields

$$m_x = m_p \sin^2\theta + m_r \cos^2\theta = 0 \tag{6.5.47}$$

or

$$m_r = -m_p \tan^2\theta, \qquad -\frac{\pi}{2} < \theta < \frac{\pi}{2} \tag{6.5.48}$$

This moment field also satisfies equilibrium condition (6.1.11) with $p = 0$, and we get the shear forces

$$v_\theta = 0, \qquad v_r = -\frac{m_p}{r}\frac{1}{\cos^2\theta} \tag{6.5.49}$$

These values lead to

$$v_x = v_r \cos\theta = -\frac{m_p}{r}\frac{1}{\cos\theta} = -\frac{m_p}{x} \tag{6.5.50}$$

If $-\alpha \leq \theta \leq \beta$, the moment field will correspond to a concentrated force

$$\mathrm{P} = \lim_{x\to 0}\frac{m_p}{x}x(\tan\alpha + \tan\beta) = m_p(\tan\alpha + \tan\beta) \quad \text{at } (r,\theta) = (0,0) \tag{6.5.51}$$

Further, it is calculated that

$$m_{xy} = \left(m_p + m_p \tan^2\theta\right)\sin\theta\cos\theta = m_p \tan\theta = m_p\frac{y}{x} \tag{6.5.52}$$

If an edge at $x = a$ is loaded by a line load $\overline{p}$, the boundary condition (6.3.9) furnishes

$$\overline{p} = \left[\frac{\partial m_x}{\partial x} - 2\frac{\partial m_{xy}}{\partial y}\right]_{x=a} = -2\frac{m_p}{a} \tag{6.5.53}$$

The moment field may be summarized as follows:

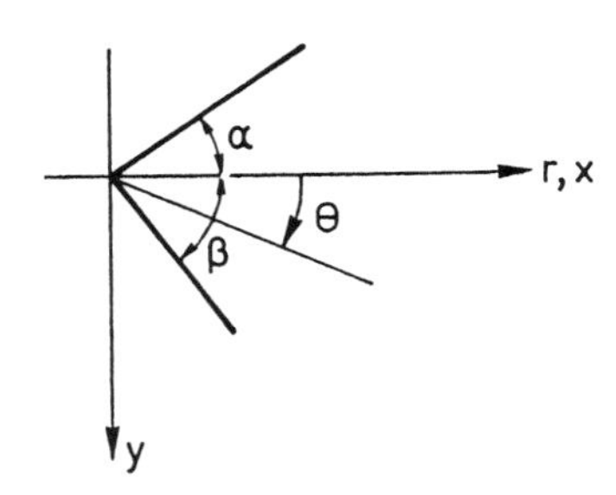

Figure 6.5.11 Notation used in connection with the moment field (6.5.54).

$$
\begin{aligned}
&m_r = -m_p \tan^2\theta, \qquad m_\theta = m_p, \qquad m_{r\theta} = 0 \\
&v_r = -\frac{m_p}{r}\frac{1}{\cos^2\theta}, \qquad v_\theta = 0 \\
&p = 0 \\
&\bar{p}_{x=a} = -\frac{2m_p}{a} \\
&P = m_p(\tan\alpha + \tan\beta) \qquad \text{at} \quad (r,\theta) = (0,0)
\end{aligned}
\tag{6.5.54}
$$

Consider next a triangle ABC subjected to a uniform load p. We try to find a statically admissible moment field satisfying the following three conditions (see Fig. 6.5.12):

1. The bending moment equal to m_p in sections AC and CB
2. The bending moment equal to zero in section AB
3. The shear force and the torsional moment equal to zero in sections AC and CB

By making

$$m_\theta = m_p, \qquad m_{r\theta} = 0 \tag{6.5.55}$$

the following is obtained from (6.1.9):

$$v_\theta = 0 \tag{6.5.56}$$

and the equilibrium equation (6.1.11) becomes

$$\frac{\partial^2 (rm_r)}{\partial r^2} = -pr \tag{6.5.57}$$

This equation has the solution

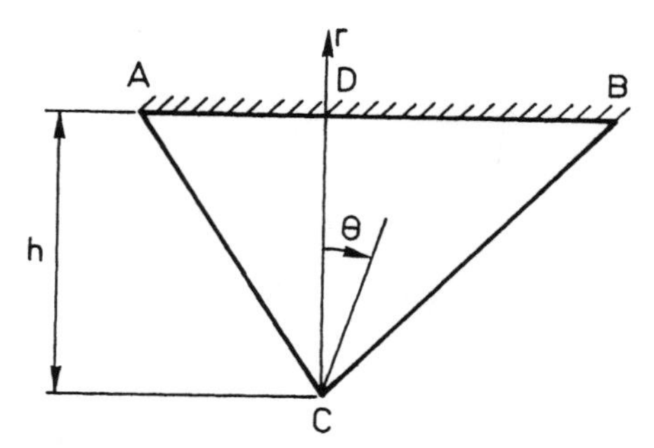

Figure 6.5.12 Triangular slab element, cf. the moment field (6.5.64).

$$m_r = -\tfrac{1}{6} pr^2 + c_1(\theta) + \frac{c_2(\theta)}{r} \tag{6.5.58}$$

where $c_1(\theta)$ and $c_2(\theta)$ are arbitrary functions of θ. In order to get finite moments when $r \to 0$, $c_2(\theta)$ is made equal to zero.

The bending moment along *AB* is

$$m_b = m_p \sin^2\theta + \left[-\frac{1}{6} p \frac{h^2}{\cos^2\theta} + c_1(\theta) \right] \cos^2\theta \tag{6.5.59}$$

Solution of the equation $m_b = 0$ for $c_1(\theta)$ results in

$$c_1(\theta) = m_p \tag{6.5.60}$$

where we have utilized a moment equation about the support for the entire slab giving

$$m_p = \tfrac{1}{6} ph^2 \tag{6.5.61}$$

where h is height *CD* of the triangle. Equation (6.5.58) then becomes

$$m_r = -\tfrac{1}{6} pr^2 + m_p = m_p \left[1 - \left(\frac{r}{h} \right)^2 \right] \tag{6.5.62}$$

If it is required that $-m_p' \le m_r \le m_p$, the following condition is obtained:

$$\tan^2\theta \le \frac{m_p'}{m_p} \tag{6.5.63}$$

The moment field may be summarized as follows:

$$
\begin{aligned}
m_r &= m_p\left[1-\left(\frac{r}{h}\right)^2\right], \qquad m_\theta = m_p, \qquad m_{r\theta} = 0 \\
v_r &= \frac{-3m_p}{h^2}r, \qquad v_\theta = 0 \\
p &= \frac{6m_p}{h^2}
\end{aligned}
\tag{6.5.64}
$$

It can be verified that $rv_r \to 0$ for $r \to 0$, as it must, since there is no point load at C.

Consider finally a trapezoid $ABCD$ (see Fig. 6.5.13) subjected to a uniform load p. As previously, the following three conditions have to be satisfied:

1. The bending moment equal to m_p in sections AB, BC and CD
2. The bending moment equal to zero in section AD
3. The shear force and the torsional moment equal to zero in sections AB, BC, and CD

This problem is solved in the same way as the preceding one. However, the calculations are more tedious, so only the results will be given:

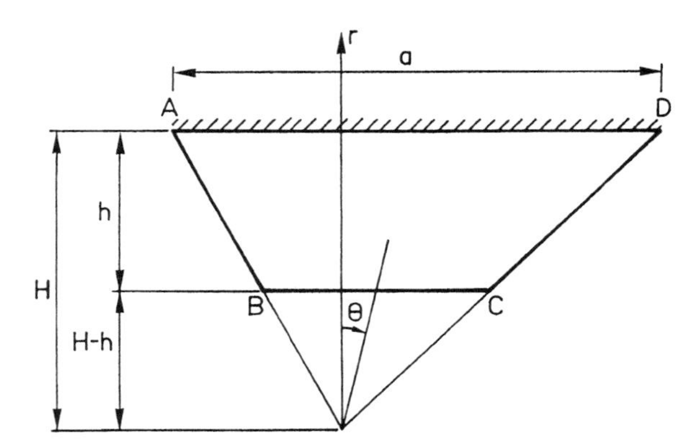

Figure 6.5.13 Slab element in the form of a trapezoid, cf. the moment field (6.5.65).

$$\begin{aligned}
m_r &= -\frac{1}{6}pr^2 + \frac{1}{2}ph^2\left[1 - \frac{2}{3}\left(\frac{h}{H}\right)\right] + \frac{1}{2}p\frac{(H-h)^2}{\cos^2\theta} - \frac{1}{3}p\frac{(H-h)^3}{r\cos^3\theta} \\
m_\theta &= m_p, \qquad m_{r\theta} = 0 \\
v_r &= -\frac{1}{2}pr + \frac{1}{2}p(H-h)^2\frac{1}{r\cos^2\theta}, \qquad v_\theta = 0 \\
m_p &= \frac{1}{2}ph^2\left[1 - \frac{2}{3}\left(\frac{h}{H}\right)\right]
\end{aligned} \tag{6.5.65}$$

The requirement that $-m_p' \le m_r \le m_p$, leads to

$$\tan^2\theta \le \frac{m_p'}{m_p} \tag{6.5.66}$$

Simply supported circular slab subjected to uniform load. We try to make $m_{r\theta} = 0$ and $m_\theta = m_p$ throughout, corresponding to yielding in all radial sections (yield zone of type 3), whereby the conditions of symmetry are fulfilled. The equilibrium equation (6.1.11) then gives (polar coordinate system with pole at center of circle) (see Fig. 6.5.14):

$$\frac{1}{r}\frac{d^2(rm_r)}{dr^2} = -p \tag{6.5.67}$$

Hence

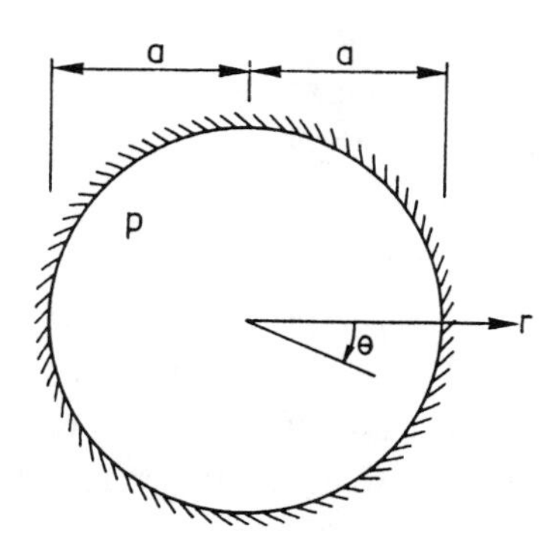

Figure 6.5.14 Simply supported circular slab with uniform load.

$$m_r = -\tfrac{1}{6}pr^2 + c_1 + \frac{c_2}{r} \tag{6.5.68}$$

For $r = a$, where a is the radius of the circle, it is necessary that $m_r = 0$, and c_2 must be zero in order to give finite moments for $r \to 0$. This gives

$$c_1 = \tfrac{1}{6}pa^2 \tag{6.5.69}$$

that is,

$$m_r = -\tfrac{1}{6}pr^2 + \tfrac{1}{6}pa^2 \tag{6.5.70}$$

m_r is greatest for $r = 0$. If the yield condition must not be violated, the following is obtained for determination of the load-carrying capacity:

$$m_p = \tfrac{1}{6}pa^2 \tag{6.5.71}$$

or

$$p = \frac{6m_p}{a^2} \tag{6.5.72}$$

It can be seen directly that $rv_r \to 0$ for $r \to 0$, as should be the case with uniform load.

By this means, a safe moment field is obtained. By making

$$w = 1 - \frac{r}{a} \tag{6.5.73}$$

which gives curvatures in radial sections only, it will be seen that the moment field corresponds to a geometrically admissible strain field. Similarly, if the slab is fixed, then

$$p = \frac{6(m_p + m_p')}{a^2} \tag{6.5.74}$$

Simply supported circular slab with circular line load. For $r<b$ (see Fig. 6.5.15), $v_r = 0$, whereby (6.1.8) gives

$$\frac{d(rm_r)}{dr} - m_\theta = 0 \tag{6.5.75}$$

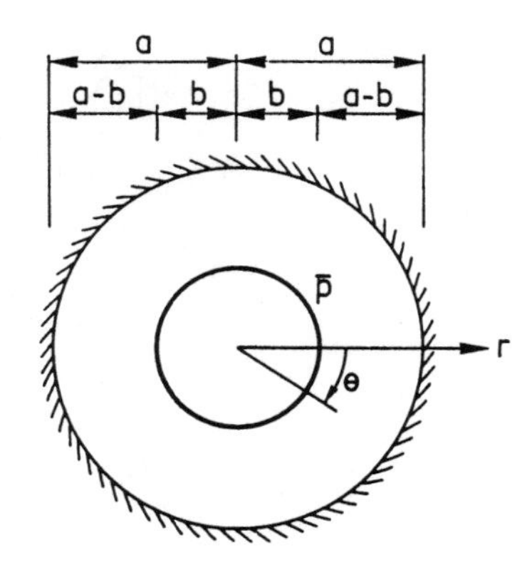

Figure 6.5.15 Simply supported circular slab with circular line load.

By insertion of $m_\theta = m_p$,

$$m_r = m_p \tag{6.5.76}$$

This moment field thus corresponds to a yield zone of type 1. For reasons of symmetry, when $r > b$,

$$v_r = -\bar{p}\frac{b}{r} \tag{6.5.77}$$

If $m_\theta = m_p$ is assumed here as well, we obtain from (6.1.8)

$$\frac{d(rm_r)}{dr} - m_p = -\bar{p}b \tag{6.5.78}$$

$$m_r = m_p - \bar{p}b + \frac{c_1}{r} \tag{6.5.79}$$

If $m_r = m_p$ for $r = b$, then $c_1 = \bar{p}b^2$, that is,

$$m_r = m_p - \bar{p}b\left(1 - \frac{b}{r}\right) \tag{6.5.80}$$

If it is then required that $m_r = 0$ for $r = a$, the follwing results:

$$\bar{p} = \frac{m_p}{b(1 - b/a)} \tag{6.5.81}$$

Correspondingly, in the case of a fixed edge,

$$\bar{p} = \frac{m_p + m'_p}{b(1 - b/a)} \tag{6.5.82}$$

The moment field found obviously corresponds to a geometrically admissible strain field. Such a strain field can be derived from, for example, (6.5.73):

$$w = 1 - \frac{r}{a} \tag{6.5.83}$$

If $b \to 0$ and at the same time the total load $\bar{p} \cdot 2\pi b = P$ remains constant, it will be seen that

$$P \to 2\pi(m_p + m'_p) \tag{6.5.84}$$

which is thus the load-carrying capacity of a circular fixed slab loaded at the center with a concentrated force.

Semicircular slab subjected to a line load. Consider now a semicircular slab, simply supported along the perimeter and subjected to a line load $\bar{p}$ along the diameter (see Fig. 6.5.16). It is assumed that $m_p = m'_p$. In this case the moment field (6.5.54) is applicable. Then we have

$$m_r = -m_p \tan^2\theta, \qquad m_\theta = m_p, \qquad m_{r\theta} = 0 \tag{6.5.85}$$

$$v_r = -\frac{m_p}{r}\frac{1}{\cos^2\theta}, \qquad v_\theta = 0 \tag{6.5.86}$$

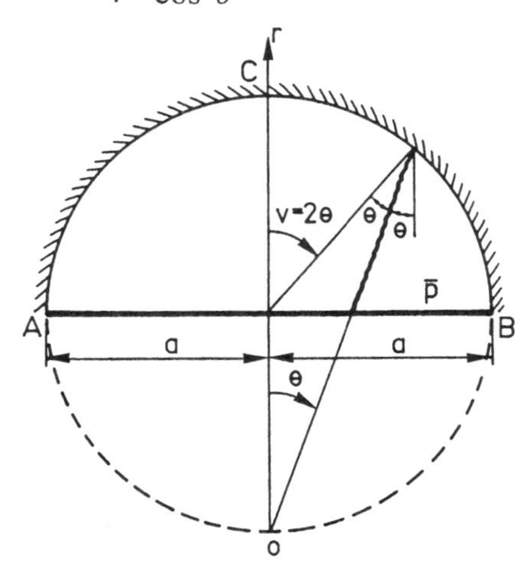

Figure 6.5.16 Semicircular slab subjected to a line load.

$$\bar{p} = -\bar{p}_{x=a} = 2\frac{m_p}{a} \tag{6.5.87}$$

where (6.5.87) determines the load-carrying capacity. We also have to verify that the bending moment m_b along the circular boundary is zero. We find that

$$m_b = m_p \sin^2\theta - m_p \tan^2\theta \cos^2\theta = 0 \tag{6.5.88}$$

The torsional moment m_v along the circular boundary is

$$m_v = -\left(m_p + m_p \tan^2\theta\right)\sin\theta\cos\theta = -m_p \tan\theta = -m_p \tan\tfrac{1}{2} v \tag{6.5.89}$$

(see Fig. 6.5.16).

The shear force along the circular boundary v is found by means of (6.1.3) to be

$$v = v_r \cos\theta = -\frac{1}{r}\left(m_p \tan^2\theta + m_p\right)\cos\theta \tag{6.5.90}$$

As $r = 2a\cos\theta$ along this edge, we get

$$v = -\frac{m_p}{2a}(1 + \tan^2\theta) = -\frac{m_p}{2a}\left(1 + \tan^2\frac{1}{2}v\right) \tag{6.5.91}$$

From (6.3.9) we find the reaction r per unit length:

$$r = v - \frac{\partial m_v}{\partial s} = v - \frac{dm_v}{a\,dv} = -\frac{m_p}{2a}\left(1 + \tan^2\frac{1}{2}v\right) + \frac{m_p}{2a}\left(1 + \tan^2\frac{1}{2}v\right) = 0 \tag{6.5.92}$$

The reaction is thus zero along the simply supported edge.

At points A and B there is the torsional moment m_p on the loaded edge AB, and $-m_p$ on the circular edge. These give a total upward-directed force $R = 2m_p$ at each of the corners A and B. The slab is thus carried at these points only.

The projection equation yields

$$\bar{p} \cdot 2a - 4m_p = 0 \tag{6.5.93}$$

which is seen to be fulfilled.

It will immediately be seen that the safe moment field derived corresponds to a geometrically admissible strain field arising from a downward displacement of the point of intersection O of the yield lines. The surface of deflection is then produced by a cone with the yield lines as generatrices and the circular support as directrix.

Rectangular slab subjected to two line loads. Consider a rectangular slab $ABCD$ simply supported along the edges AB and CD and subjected to two line loads $\bar{p}$ along BC and AD (Fig. 6.5.17). It is assumed that $AB = a$ and $BC = b$, and that $m_p' = am_p/b$ and $b \geq a$.

An upper bound solution is readily obtainable by taking a failure mechanism that corresponds to a single yield line along the y-axis. This gives the upper bound solution

$$\bar{p} = \frac{4m_p a}{b^2} \tag{6.5.94}$$

Then we try to construct a statically admissible moment field. It is natural to make

$$m_x = m_p - \frac{4m_p}{b^2}x^2 \tag{6.5.95}$$

which corresponds to a parabolic cylinder. Taking

$$m_{xy} = -\frac{4\lambda}{ab}xy, \qquad m_y = 0 \tag{6.5.96}$$

the equilibrium condition (6.1.7) with $p = 0$ is satisfied provided that

$$-\frac{8m_p}{b^2} + \frac{8\lambda}{ab} = 0, \qquad \text{that is,} \quad \lambda = m_p\frac{a}{b} \tag{6.5.97}$$

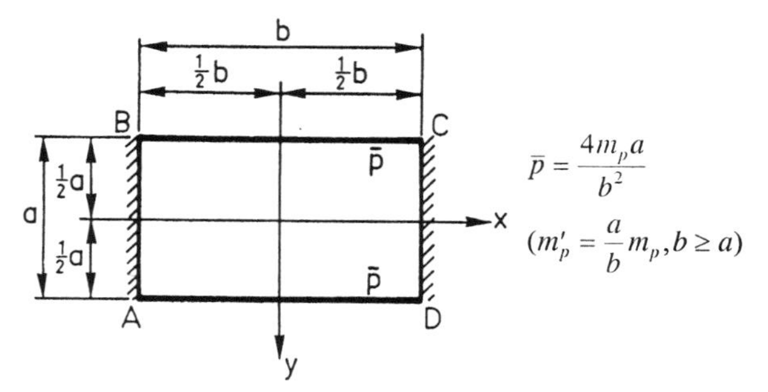

Figure 6.5.17 Rectangular slab subjected to two line loads.

which gives

$$m_{xy} = -\frac{4m_p}{b^2} x y \tag{6.5.98}$$

Now the boundary condition (6.3.9) is fulfilled along AD and BC if

$$\bar{p} = \left[\frac{\partial m_y}{\partial y} - 2 \frac{\partial m_{xy}}{\partial x} \right]_{y=1/2a} = \frac{4 m_p a}{b^2} \tag{6.5.99}$$

which agrees with the upper bound solution (6.5.94).

The reaction r per unit length along AB and CD is

$$r_{x=-1/2b} = \left[\frac{\partial m_x}{\partial x} - 2 \frac{\partial m_{xy}}{\partial y} \right]_{x=-1/2b} = 0 \tag{6.5.100}$$

which means that the load corresponding to this moment field is carried at the corners A, B, C, and D only, by four concentrated forces

$$R = 2 \left| m_{xy} \right|_{x=1/2b,y=1/2a} = 2m_p \frac{a}{b} \tag{6.5.101}$$

It now remains only to verify that the yield condition is not violated anywhere. It is only necessary to investigate the principal moments along AD and BC. Here we have

$$\left. \begin{matrix} m_1 \\ m_2 \end{matrix} \right\} = \frac{1}{2}\left(m_x + m_y\right) \pm \sqrt{\frac{1}{4}\left(m_x - m_y\right)^2 + m_{xy}^2}$$

$$= \frac{1}{2}\left(m_p - \frac{4m_p}{b^2} x^2 \right) \pm \sqrt{\frac{1}{4}\left(m_p - \frac{4m_p}{b^2} x^2 \right)^2 + \left(\frac{2m_p a}{b^2} x \right)^2}$$

$$= \frac{1}{2}\left(m_p - \frac{4m_p}{b^2} x^2 \right) \pm \frac{1}{2} \sqrt{m_p^2 + \frac{8m_p^2}{b^2} x^2 \left[2\left(\frac{a}{b} \right)^2 - 1 \right] + \frac{16m_p^2}{b^4} x^4}$$

From this expression it can be seen that

$$-\frac{a}{b} m_p \le \begin{matrix} m_1 \\ m_2 \end{matrix} \le m_p \tag{6.5.103}$$

The solution (6.5.94) is therefore exact. As the reactions along *AB* and

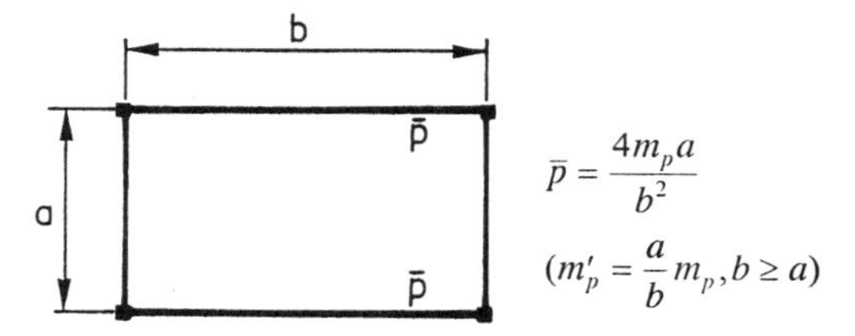

Figure 6.5.18 Rectangular slab supported on columns subjected to two line loads.

CD are zero, the solution is also exact in the case shown in Fig. 6.5.18, where the slab is supported on four columns.

When $m'_p = 0$, it may be shown that the load-carrying capacity is zero. Zero load-carrying capacity is always obtained when a slab having $m'_p = 0$ is loaded on a free boundary. This may be shown by using the upper bound theorem on a simple yield line pattern having negative yield lines infinitely close to the boundary [64. 3] (see Section 6.6.7).

Hexagonal slab subjected to uniform load. The hexagonal slab shown in Fig. 6.5.19 is subjected to a uniform load p. It is assumed that $m_p = m'_p$. First, an upper bound for the load-carrying capacity is sought. By taking moments about the supports for slab parts I and II (see Fig. 6.5.19) formed by the yield lines shown, we obtain

$$\text{Part I:} \quad m_p a = \tfrac{1}{6} pax^2$$

$$\text{Part II:} \quad m_p b = \tfrac{1}{2} pb(\tfrac{1}{2}a^2) - 2 \cdot \tfrac{1}{3} p\left(x\sqrt{2} - \frac{a}{\sqrt{2}}\right)\left(\tfrac{1}{2}a^2\right) \tag{6.5.104}$$

The solution of these equations with respect to x and p yields

$$\frac{x}{a} = \sqrt{2\left(\frac{a}{b}\right)^2 + \sqrt{2}\frac{a}{b} + \frac{3}{2}} - \sqrt{2}\frac{a}{b} \tag{6.5.105}$$

$$p = \frac{6m_p}{a^2} \frac{1}{\left[\sqrt{2(a/b)^2 + \sqrt{2}(a/b) + \frac{3}{2}} - \sqrt{2}(a/b)\right]^2} \tag{6.5.106}$$

The same value would be found by using the upper bound technique (see Section 6.6), so solution (6.5.106) is really an upper bound solu-

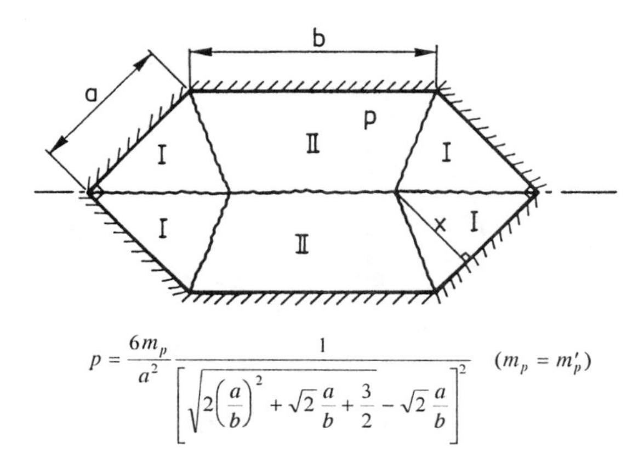

Figure 6.5.19 Simply supported hexagonal slab with uniform load.

tion. A possible deflection surface is found by giving the yield line between slab parts II a unit downward displacement. The dissipation can be calculated from (6.5.21), but a simpler method is developed in Section 6.6.

In this case the moment fields (6.5.64) and (6.5.65) apply. From (6.5.105) it appears that if $0 \le b < \infty$, then $a/2 \le x \le \sqrt{3}a/\sqrt{2}$, which means that conditions (6.5.63) and (6.5.66) are fulfilled; so solution (6.5.106) is an exact solution. As in many of the cases treated earlier, only a small part of the slab has to be reinforced at the top.

Concentrated force at a corner. Turning now to the moment field (6.5.45), we combine such a field with two pure torsion fields. The slab under consideration is shown in Fig. 6.5.20, and is loaded by a concentrated force at point *B*. It is first assumed $m_p = m'_p$ and $\alpha \ge \pi/2$. In slab part *ABE*, $m_x = m_p$, $m_y = -m_p$, and $m_{xy} = 0$ are selected, and in slab part *BFC*, $m_x = m_p$, $m_y = -m_p$, and $m_{xy} = 0$. In slab part *BEF*, the moment field (6.5.45), is selected (i.e., $m_r = -m_p$, $m_\theta = m_p$, and $m_{r\theta} = 0$).

It will be seen that these moment fields can be combined in this way without violating the boundary conditions. The moment fields correspond to a downward force $2m_p + 2m_p\beta$ at point *B*, where the load $2m_p$ comes from triangles *ABE* and *BFC*, and the load $2m_p\beta$ [see (6.5.45)] comes from slab part *BEF*. Then

$$P = 2m_p(1+\beta) = 2m_p\left(1+\alpha-\frac{\pi}{2}\right) \tag{6.5.107}$$

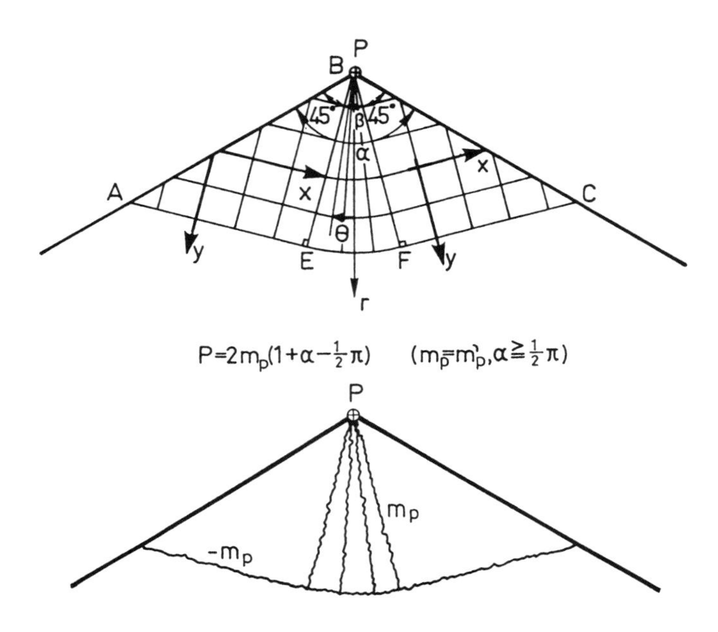

Figure 6.5.20 Concentrated force at a corner, $\alpha \geq \pi/2$.

The moment field also corresponds to a geometrically admissible failure mechanism, as demonstrated in Fig. 6.5.20. This solution requires that $\alpha \geq \pi/2$. If $\alpha \leq \pi/2$, the exact solution is as shown in Fig. 6.5.21. This corresponds to

$$P = 2\,m'_p \tan\frac{\alpha}{2} \tag{6.5.108}$$

A statically admissible moment field is

$$\begin{aligned} m_y &= -m'_p \\ m_x &= m'_p \tan^2\frac{\alpha}{2} \leq m'_p = m_p \\ m_{xy} &= 0 \end{aligned} \tag{6.5.109}$$

This moment field yields the bending moment m_b at the edges:

$$m_b = -m'_p \cos^2\left(\frac{\pi}{2} - \frac{\alpha}{2}\right) + m'_p \tan^2 \frac{\alpha}{2} \sin^2\left(\frac{\pi}{2} - \frac{\alpha}{2}\right) = 0 \qquad (6.5.110)$$

and the torsional moment m_v:

$$|m_v| = \frac{1}{2}\left(m'_p \tan^2 \frac{\alpha}{2} + m'_p\right) \sin\alpha = m'_p \tan\frac{\alpha}{2} \qquad (6.5.111)$$

which is statically equivalent to the force

$$P = 2m'_p \tan\frac{\alpha}{2} \qquad (6.5.112)$$

In this case, full reinforcement corresponding to m_p is not required, only $m_p = m'_p \tan^2 \frac{1}{2}\alpha \leq m'_p$.

If $m_p \neq m'_p$, the following load-carrying capacity is obtained:

$$P = 2\sqrt{m_p m'_p} + \left(m_p + m'_p\right)\left(\alpha - 2 \operatorname{Arctan}\sqrt{\frac{m_p}{m'_p}}\right) \qquad (6.5.113)$$

This formula is valid if $m_p \leq m'_p \tan^2(\alpha/2)$. If $m_p \geq m'_p \tan^2(\alpha/2)$, the load-carrying capacity is

$$P = 2m'_p \tan\frac{\alpha}{2} \qquad (6.5.114)$$

corresponding to (6.5.112).

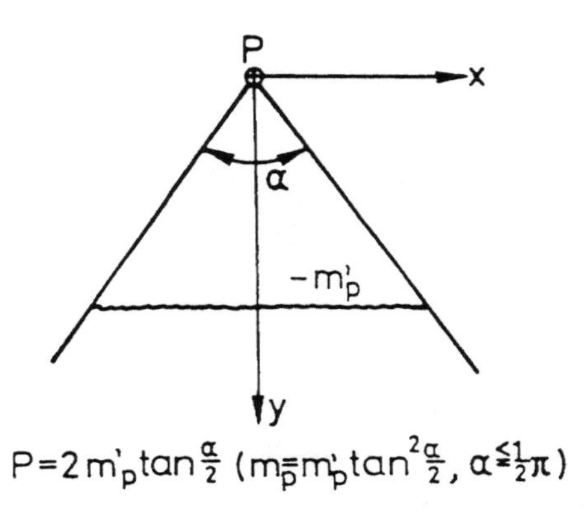

Figure 6.5.21 Concentrated force at a corner, $\alpha < \pi/2$.

The solutions in this section were given by Johansen [43.1; 62.1] as upper bound solutions, and Wood [61.5] developed the exact solution (6.5.107) in the special case $\alpha = \pi$.

Ring-shaped slab under torsion. It is assumed that $m'_p = m_p$. By introducing in (6.5.17), $\sin 2\psi = 1$, $\cos 2\psi = 0$, corresponding to $\psi = \pi/4$, equilibrium equation (6.1.11) with $p = 0$ is satisfied. Then, if $m'_p = m_p$, the following is obtained from (6.5.17):

$$\begin{aligned} m_r &= 0 \\ m_\theta &= 0 \\ m_{r\theta} &= m_p \end{aligned} \qquad (6.5.115)$$

It will be seen directly from equilibrium equations (6.1.8) and (6.1.9) that $v_r = 0$ and $v_\theta = 2m_p / r$ throughout. This boundary condition and the boundary condition $m_{r\theta} = m_p$ thus correspond to, for example, the case shown in Fig. 6.5.22 where a ring-shaped slab is subjected to torsion by four concentrated forces $P = 2m_p$, and by two diametrically opposite line loads $\overline{p} = 2m_p / r = P / r$. The yield lines are logarithmic spirals. The solution is equivalent to Nadai's solution of a thick-walled tube under internal pressure [28.4; 31.3; 50.3].

It is obvious that the yield condition is not violated anywhere, but it still remains to be proved that a geometrically admissible strain field can be found. If

$$\kappa_\theta = \kappa_r = 0, \quad \kappa_{r\theta} = \frac{a^2}{r^2} \quad \text{(a constant)} \qquad (6.5.116)$$

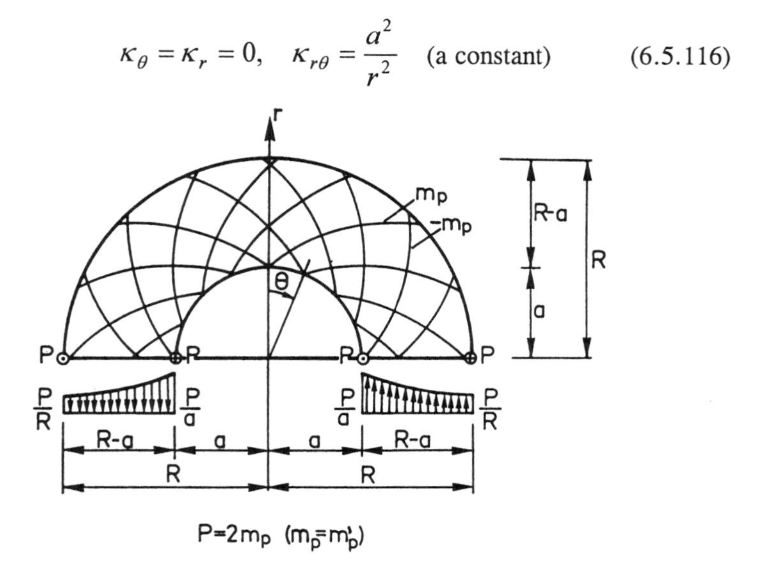

Figure 6.5.22 Ring-shaped slab under torsion.

then we have the curvatures $\pm a^2 / r^2$ in the lines of principal sections, and the conditions of compatibility (6.2.12) and (6.2.13) are fulfilled, which means that the strain field is geometrically admissible.

The corresponding surface of deflection w is found by solving the differential equations [see (6.2.5) to (6.2.7)]:

$$\frac{\partial^2 w}{\partial r^2} = 0, \quad \frac{1}{r^2}\frac{\partial^2 w}{\partial \theta^2} + \frac{1}{r}\frac{\partial w}{\partial r} = 0, \quad \frac{1}{r}\frac{\partial^2 w}{\partial r \partial \theta} - \frac{1}{r^2}\frac{\partial w}{\partial \theta} = \frac{a^2}{r^2} \qquad (6.5.117)$$

A particular integral is easily proved to be

$$w = -a^2\theta \qquad (6.5.118)$$

corresponding to a right helicoid.

Rectangular slab subjected to uniform load. The boundary conditions and statically admissible moment fields derived in the foregoing now permit us to prove that the solution for the rectangular slab given by Ingerslev in 1921 [21.1; 23.1] is only an exact solution when $m'_p = 3m_p$ and to give a moment field throughout the slab.

With a uniform load and isotropic reinforcement at the top and bottom of the slab, the yield line pattern shown in Fig. 6.5.23 is determined by a single parameter, x. Calculations corresponding to those made for the hexagonal slab yield (see also Section 6.6.7).

$$\frac{x}{b} = \frac{1}{2}\frac{a}{b}\left(\sqrt{\left(\frac{a}{b}\right)^2 + 3} - \frac{a}{b}\right) \qquad (a \le b) \qquad (6.5.119)$$

$$p = \frac{24m_p}{a^2}\frac{1}{\left[\sqrt{(a/b)^2 + 3} - a/b\right]^2} \qquad (6.5.120)$$

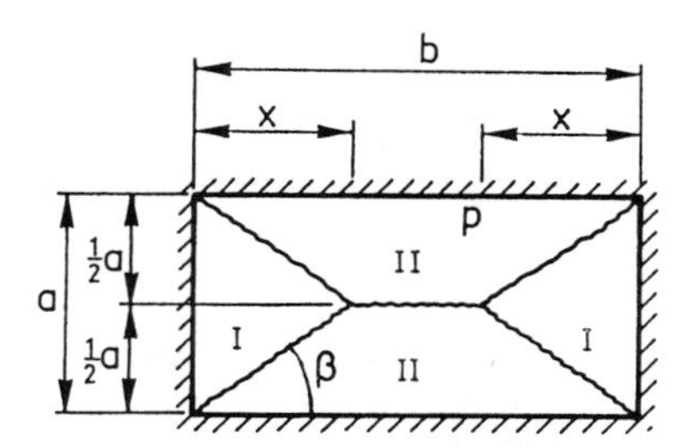

Figure 6.5.23 Rectangular slab with uniform load.

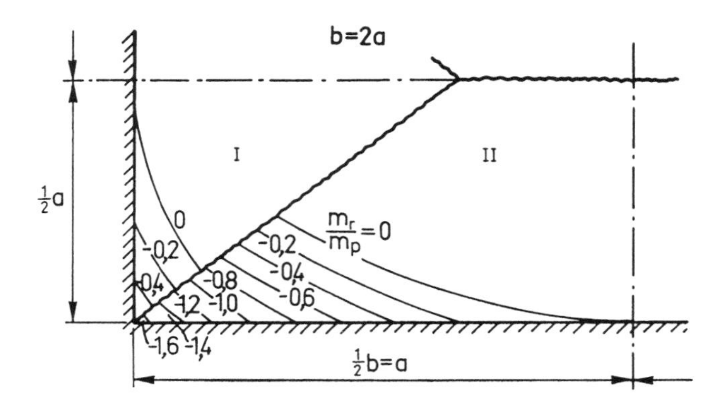

Figure 6.5.24 Lines of constant negative principal moment in a rectangular slab with uniform load.

When b/a traverses the interval 1 to ∞, the angle β (see Fig. 6.5.23) traverses the interval 45 to 30°. Hence, when $m_p = m_p'$, the boundary conditions (6.5.35) are fulfilled only in the case $a = b$ (i.e., a square slab). In all other cases, the yield condition will be violated at the corner.

A moment field that satisfies the equilibrium equations and the boundary conditions and that has $m_b = m_p$ in the yield lines is (6.5.64) and (6.5.65). By means of these expressions it is possible to calculate the necessary magnitude of m_p' throughout the slab corresponding to the load-carrying capacity (6.5.120).

Figure 6.5.24 shows lines of constant negative principal moments for $b/a = 2$. It also shows which zones have negative moments according to the above-mentioned moment fields, and gives an impression of the extent to which the yield condition is violated for $b/a = 2$ if $m_p' = m_p$. If reinforcement corresponding to these negative moments is introduced at the top of the slab, Ingerslev's solution will, of course, be exact.

In order to give an impression of the minimum value of the negative moment, Fig. 6.5.25 shows the lowest value of m_r/m_p (which is found at the corner of slab part II at the support, independent of whether another moment field is selected outside the yield lines) as a function of the aspect ratio b/a.

It is found that

$$\left(\frac{m_r}{m_p}\right)_{\min} = -\cot^2\beta = -\left[\sqrt{\left(\frac{a}{b}\right)^2 + 3} - \frac{a}{b}\right]^2 \qquad (6.5.121)$$

When b/a traverses the interval $1 \leq b/a < \infty$, β traverses as mentioned the interval $\pi/6 \leq \beta \leq \pi/4$. Hence

$$-3 \leq \left(\frac{m_r}{m_p}\right)_{\min} \leq -1 \tag{6.5.122}$$

From this it will be seen that Ingerslev´s solution is always exact for $m_p' = 3m_p$.

For another statically admissible moment field in this case, see Section 6.7. Here it is also shown that the difference between the solution (6.5.120) and the best known lower bound solution in the case $m_p = m_p'$ is very small.

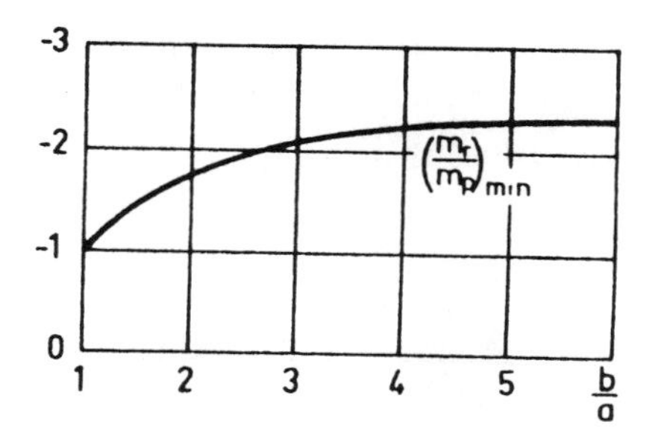

Figure 6.5.25 Minimum value of bending moment in a rectangular slab with uniform load.

6.6 UPPER BOUND SOLUTIONS FOR ISOTROPIC SLABS

6.6.1 The Work Equation Method and the Equilibrium Method

There are two methods that can be used to determine upper bounds for the load-carrying capacity: the work equation method and the equilibrium method. When the work equation method, which was developed by Johansen [43.1; 62.1] and Gvozdev [49.1], is applied, a geometrically admissible yield line pattern, usually a simple one consisting of straight yield lines, is selected. The work dissipated in the yield lines is equated with the work done by the external load, and the solution of this equation, the work equation, furnishes an upper bound for the load-carrying capacity. The reinforcement can, of course, also be determined in this way.

It is natural to look for the yield line pattern that gives the lowest value of the load. If the load is minimized with respect to the geometrical parameters determining the location of the yield lines, the exact solution (i.e., the true load-carrying capacity) cannot always be obtained, but one does get the best answer possible from the yield line pattern selected. In order to show that a solution obtained in this way is exact, it is necessary to find a safe moment field, that is, a moment field that corresponds to points within or on the yield surface.

The primary idea in the equilibrium method, which was introduced by Ingerslev [21.1; 23.1] and further developed by Johansen [43.1; 62.1], is to avoid the minimization process in the work equation, which may lead to tedious calculations. Although the name of the method might give the impression that it concerns a lower bound method (i.e., a method that gives a safe value of the load-carrying capacity), this is not actually the case. The equilibrium considerations concern only the slab parts between the yield lines, the supports, and so on. It is therefore natural to require that the equilibrium method gives the same answer as the work equation method and the minimizing process. The circumstances under which this is possible are investigated in the following. It turns out that in order to get the same answer by the two methods, it is necessary to introduce a set of concentrated forces in the yield lines called *nodal forces*. The first nodal force theory is due to Johansen [43.1; 62.1].

The following sections contain a nodal force theory developed by the author [62.2; 64.3; 65.3]. The theory serves the purpose of giving an understanding of the relationship between the two upper bound methods. Alternative theories explaining the limitations of the Johansen theory have been given by Nylander [60.2; 63.4], Kemp [65.4], Wood [65.5], Jones [65.6], Morley [65.7], Møllmann [65.8], and Harder [88.11].

6.6.2 The Relationship between the Work Equation Method and the Equilibrium Method

Bending and torsional moments in the neighborhood of yield lines. Consider a normal section II-II through a point P (see Fig. 6.6.1) situated at the distance δx and δy from a point Q through which the normal section I-I passes. The two normal sections mentioned form the angle $\delta\theta$ with each other. In section I-I, there is the bending moment $m_b = m_n$ and the torsional moment m_{nt}. By means of the laws of transformation of moments, the bending moment $m_b' = m_b + \delta m_b$ in section II-II is found to be

$$m'_b = m_b + \delta m_b = m_n + \frac{\partial m_n}{\partial x}\delta x + \frac{\partial m_n}{\partial y}\delta y - 2m_{nt}\delta\theta \tag{6.6.1}$$

This result is now applied to the case shown in Fig. 6.6.2. Here there is a straight line *AB* from which another straight line *CD* is generated by certain variations δx, δy, and $\delta\theta$. The bending moment

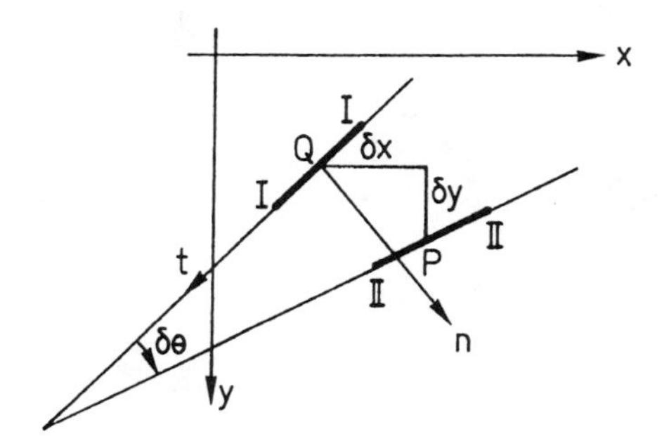

Figure 6.6.1 Normal sections having a small distance and at a small angle to each other.

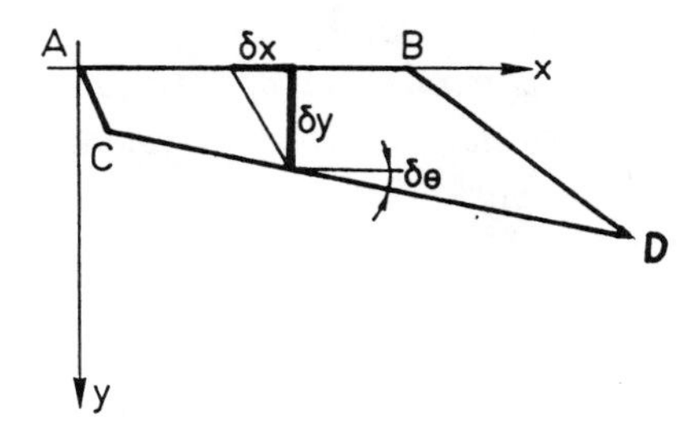

Figure 6.6.2 Two slab sections having a small distance at a small angle to each other.

in *AB* is constant and equal to $m_b = m_y$. The following variation is obtained by means of (6.6.1):

$$\delta m_b = \frac{\partial m_y}{\partial y}\delta y - 2m_{yx}\delta\theta \tag{6.6.2}$$

It follows from (6.6.2) that if

$$\partial m_y / \partial y = m_{xy} = 0 \tag{6.6.3}$$

then

$$\delta m_b = 0 \tag{6.6.4}$$

Furthermore, it will be seen that if these conditions are not fulfilled, then $\delta m_b = 0$ will not be obtained for arbitrary values of δy and $\delta\theta$. Thus, if AB is a yield line, then according to the yield condition $m_{xy} = 0$. If, in addition $\partial m_y / \partial y = 0$, then $\delta m_b = 0$ for arbitrary values of δy and $\delta\theta$.

In the following the term "internal yield line" is used to indicate a yield line, the position of which is unknown. External yield lines are yield lines whose position is fixed, for example, yield lines along fixed edges.

A stationary moment field in a slab part limited by yield lines, supports, and so on, is defined as a moment field that satisfies the following requirements along the limiting internal yield lines (see Fig. 6.6.3):

$$\frac{\partial m_n}{\partial n} = 0, \quad m_{nt} = 0 \tag{6.6.5}$$

It is now possible to investigate in detail the relationship between the two upper bound methods, proceeding by proving the following important theorem:

If the equilibrium equations for each slab part that has been formed by the yield lines are satisfied in such a way that a stationary moment field may be found in each slab part, the corresponding load is identical to the load obtained by performing the minimization process for the yield line pattern considered.

The load intensity is determined by a positive parameter λ and a function $p(x, y)$ (x and y Cartesian coordinates), so that the load intensity at an arbitrary point is $\lambda p(x, y)$. As the equilibrium equations are presumed to be satisfied for each slab part, the virtual work equation is valid. Let w be the deflections, m_b the bending moment in the yield lines (m_p or $-m'_p$), and θ the discontinuities in slope along the yield lines. Then from (6.3.4):

$$\lambda \iint pw\,dx\,dy = \int m_b \theta\,ds \tag{6.6.6}$$

where the integration on the left-hand side is performed over the entire slab, and the integration on the right-hand side along all yield lines.

A small variation of the yield line pattern is now performed. This changes the deflections to w', and the bending moments to m_b + δm_b, where the variation δm_b is calculated in accordance with the stat-

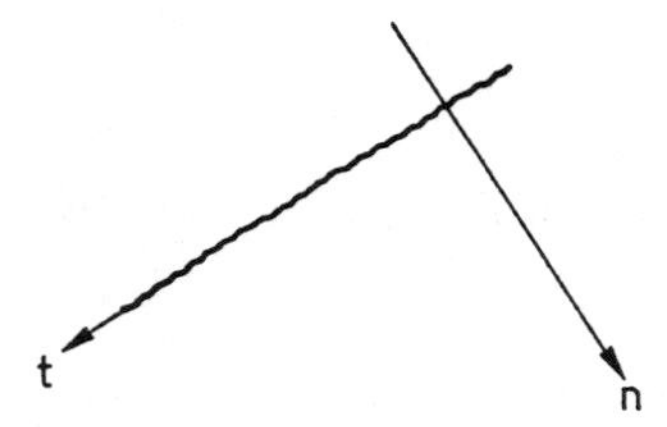

Figure 6.6.3 Local coordinate system in a yield line.

ionary moment field that can be associated with each slab part. Finally, the discontinuities θ are changed to θ'.

Maintaining the external load, which is permissible because the equilibrium equations are still satisfied, the following is obtained:

$$\lambda \iint p w' dx dy = \int \left(m_b + \delta m_b\right) \theta' ds' \tag{6.6.7}$$

where the integration is now performed along the new yield lines. According to (6.6.4), $\delta m_b = 0$, so (6.6.7) now gives

$$\lambda \iint p w' dx\, dy = \int m_b \theta' ds' \tag{6.6.8}$$

Equations (6.6.6) and (6.6.8), which represent virtual work equations, may now be imagined as the work equations corresponding to two infinitely close yield line patterns, since the bending moments in each equation are equal to the yield moments. However, as the equations correspond to the same load (which means that $\delta\lambda = 0$), the theorem has been proved.

This theorem is the key to an understanding of the equilibrium method. If the equilibrium equations are satisfied in such a way that the conditions (6.6.5) are fulfilled along the yield lines, and it is known that there exists a stationary moment field (although such a moment field does not have to be found for each slab part), the load obtained by solution of the equilibrium equations is equal to the load obtained by the work equation method and the minimization process.

Using the equilibrium condition (6.1.5) it will be seen that the requirements (6.6.5) are equivalent to

$$v_n = m_{nt} = 0 \tag{6.6.9}$$

It is thus a sufficient requirement for the application of the equilibrium method that the conditions of equilibrium for each slab part are

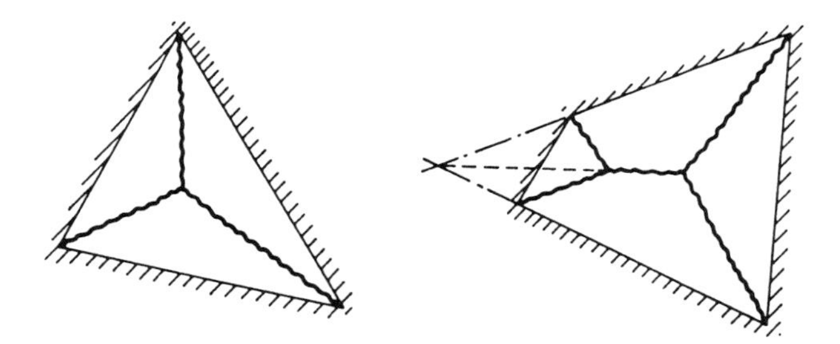

Figure 6.6.4 Slabs with internal yield lines enabling condition (6.6.9) to be satisfied.

satisfied with $v_n = m_{nt} = 0$ along each internal yield line.

Three independent equilibrium equations generally exist for each slab part. However, supports, symmetry, and so on, may reduce the number of equations that have to be satisfied to ensure the existence of a stationary moment field. For instance, if a slab part is limited by internal yield lines and a support (see Fig. 6.6.4), only one equilibrium condition, a moment equation about the support, has to be satisfied.

It is not always possible to satisfy the condition (6.6.9), for instance, at intersections between yield lines of different signs. The fact necessitates the introduction of certain forces, *nodal forces*, at such points. This problem is treated in the following section. The existence of a stationary moment field has already been demonstrated in several cases in Section 6.5.4.

6.6.3 Nodal Forces

The derivation of the nodal forces is based on one simple equation which expresses the relationship between bending and torsional moments in two slab sections forming an angle α (see Fig. 6.6.5). A moment equation immediately furnishes the following:

$$m_v^{(1)} + m_v^{(2)} = \left(m_b^{(1)} - m_b^{(2)}\right)\cot\alpha \tag{6.6.10}$$

where m_v represents torsional moments and m_b denotes bending moments.

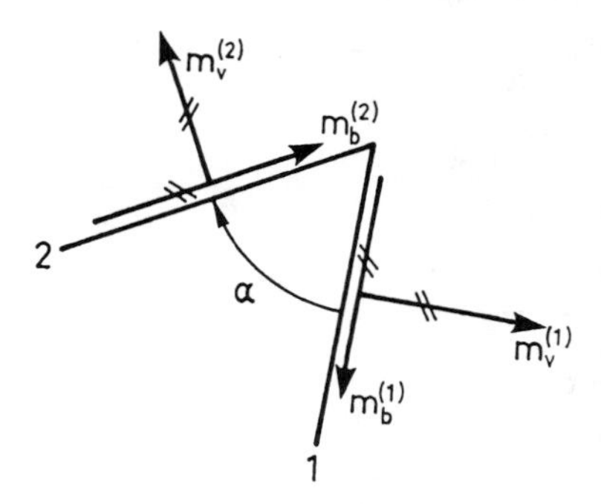

Figure 6.6.5 Slab sections with bending and torsional moments forming an angle with each other.

Nodal forces of type 1. It is convenient to distinguish between two types of nodal forces. *Nodal forces of type 1* are defined as the concentrated forces that appear in the thin plate theory when torsional moments are transformed into statically equivalent shear forces (Kirchhoff's boundary conditions, see Section 6.3.2). For instance, at a free boundary, the boundary condition (6.3.8).

$$v_n - \frac{\partial m_{nt}}{\partial s} = 0 \tag{6.6.11}$$

must be satisfied (see Fig. 6.6.6)

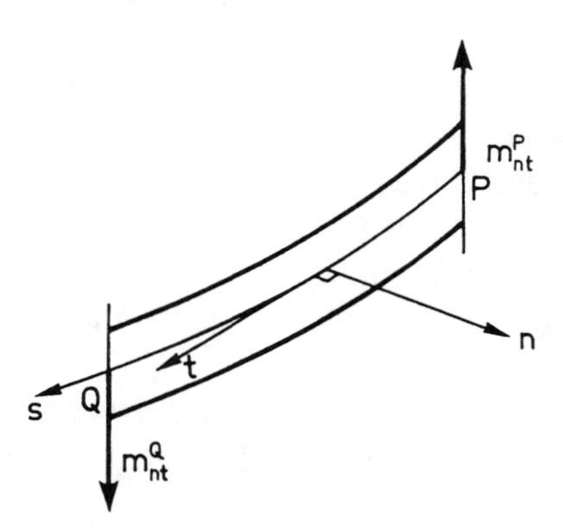

Figure 6.6.6 Free edge with shear forces and torsional moments statically equivalent to zero giving rise to concentrated forces at the end points of the edge.

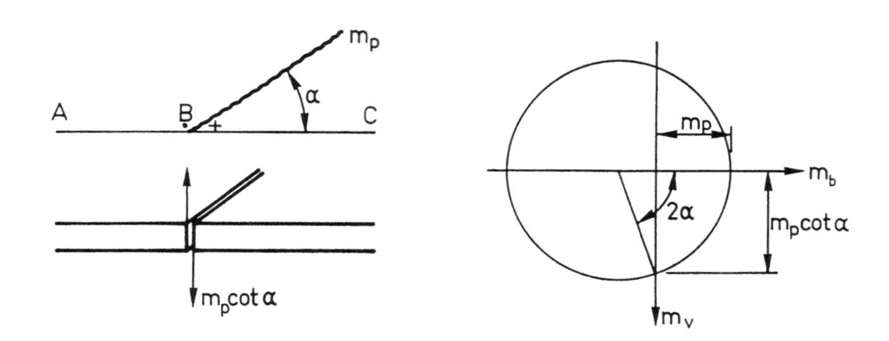

Figure 6.6.7 Nodal force at an edge point crossed by a yield line.

The left-hand side is the statical equivalence of v_n and m_{nt}. In addition, two concentrated forces equal to the torsional moment at the ends of the boundary are obtained. Such concentrated forces must be taken into account when equilibrium of slab parts with free boundaries is considered [see, for example, the boundary AC (Fig. 6.6.7)]. At point B, a yield line with the bending moment m_p ends.

If equilibrium of the slab parts containing the free boundaries AB or BC is now considered, condition (6.6.11) must be considered as satisfied, but at point B there may be concentrated forces if there is a torsional moment. The torsional moment at the boundary is, apart from the sign, determined from (6.5.30):

$$m_v = m_p \cot \alpha \tag{6.6.12}$$

This result can also be obtained by (6.6.10) having no torsional moments in the yield line and no bending moments at the boundary. Then the concentrated forces $m_p \cot \alpha$, as shown in Fig. 6.6.7, must be taken into account in the equilibrium equations. Mohr's circle is also shown in the figure. It appears that the nodal force in the acute angle will always act downward. This is the most important example of nodal forces of type 1.

Another example is shown in Fig 6.6.8. Here a yield line ends at a corner. The nodal forces shown are obtained directly from (6.5.30) or (6.6.10). The case is interesting because it has two different Mohr's circles, one in each angle α and β, a case that has already been considered in Section 6.5.2.

As many examples showing the application of the Kirchhoff boundary condition have been given, no more cases of nodal forces of type

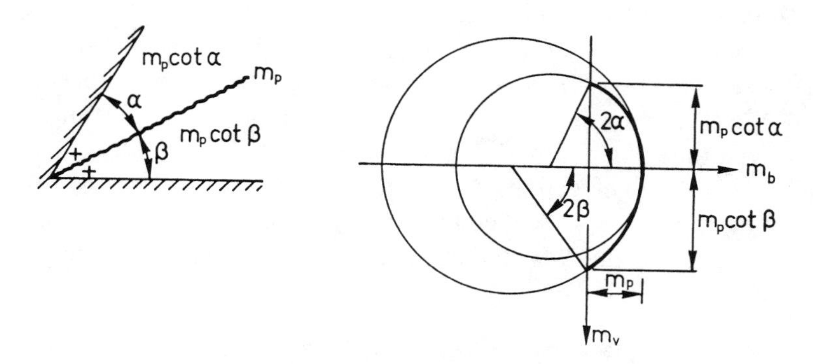

Figure 6.6.8 Nodal forces at a corner with a yield line.

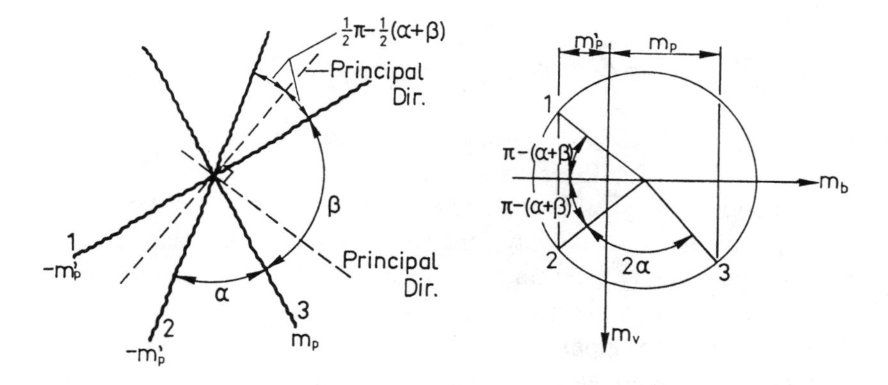

Figure 6.6.9 Mohr's circle at a point of intersection between yield lines of different signs.

1 will be treated. A few more examples will be shown in Fig. 6.6.14. That nodal forces of type 1 really have a physical significance has also been noticed by Clyde [79.15].

Nodal forces of type 2. Nodal forces of type 2 appear at intersections between yield lines of different signs. If yield lines of the same sign intersect, the Mohr's circle reduces to a point, and there is no difficulty in satisfying the condition (6.6.9).

Figure 6.6.9 shows a case of general character, where two negative yield lines and a positive yield line intersect. The Mohr's circle is then determined as shown in the figure. It appears that generally there are torsional moments in the yield lines, preventing fulfillment of condition (6.6.9), that secure a stationary value of the load found by the equilibrium method. It is easily seen that if the "disturbed zone" (i.e., the zone in which the yield lines contain torsional moments and shear forces) can be made infinitely small, it will still be possible to obtain stationary values of the load. Considering one of the yield lines (see Fig. 6.6.10), it is assumed that the torsional moment varies from zero at a distance Δs from the point of intersection to the value $(m_{yx})_{x=0}$ at the intersection; this value being determined by the mode of intersection. The statical equivalence Q of the shear forces v_y in the interval $0 \le x \le \Delta s$, where the x-axis is in the direction of the yield line, is

$$Q = \int_0^{\Delta s} v_y \, dx = \int_0^{\Delta s} \left(\frac{\partial m_y}{\partial y} - \frac{\partial m_{xy}}{\partial x} \right) dx = \left(\frac{\partial m_y}{\partial y} \right)_{mean} \Delta s + \left(m_{xy} \right)_{x=0} \quad (6.6.13)$$

and

$$\lim_{\Delta s \to 0} Q = \left(m_{xy} \right)_{x=0} = -\left(m_{yx} \right)_{x=0} \quad (6.6.14)$$

assuming $\partial m_y / \partial y$ to be finite. In other words, if the "disturbed zone" is made infinitely small, the statical equivalence of the shear forces in a yield line will be equal to the torsional moment at the point of intersection, while the statical equivalence of the torsional moment reduces to zero. The force Q determined by means of (6.6.14) must be taken into account when equilibrium equations are established. The sign is easily seen to be the opposite of the sign of the Kirchhoff force coming from the statical equivalence of the torsional moment if the yield line is considered as a boundary.

In the angle between two yield lines 1 and 2 (see Fig. 6.6.11) the total force is therefore

$$Q_{12} = m_v^{(1)} - m_v^{(2)} \quad (6.6.15)$$

that is, the difference between the torsional moments in the yield lines. It is evident that the sum of all forces at a point is zero. In the case shown in Fig. 6.6.12 the nodal forces are therefore as shown. By means of (6.6.10) we immediately get

$$m_v^{(1)} + m_v^{(2)} = \left(m_p + m_p' \right) \cot \alpha \quad (6.6.16)$$

$$m_v^{(2)} + m_v^{(3)} = -\left(m_p + m_p' \right) \cot \beta \quad (6.6.17)$$

$$m_v^{(1)} + m_v^{(3)} = 0 \quad (6.6.18)$$

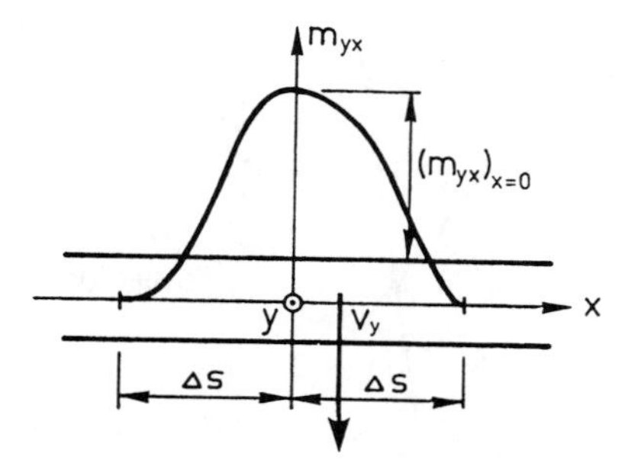

Figure 6.6.10 Torsional moments at the point of intersection between yield lines of different signs.

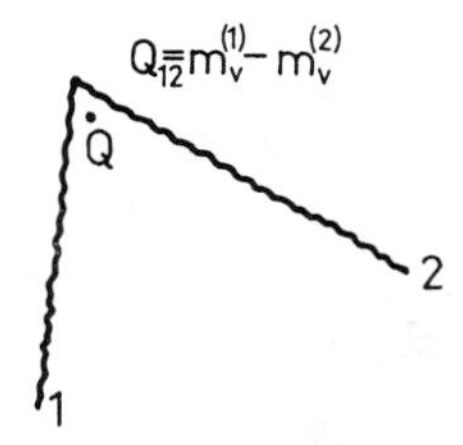

Figure 6.6.11 Nodal force expressed by torsional moments.

which lead to

$$m_v^{(1)} - m_v^{(2)} = \left(m_p + m'_p\right)\cot\beta \tag{6.6.19}$$

$$m_v^{(2)} - m_v^{(3)} = \left(m_p + m'_p\right)\cot\alpha \tag{6.6.20}$$

which exactly determine the nodal forces in all angles (see Fig. 6.6.12). If one or more of the yield lines in Fig. 6.6.12 are missing, the nodal forces in each new and greater angle are found by summation of the nodal forces in the angles that make the new, greater angle. The forces defined in this section are called *nodal forces of type 2*. The results are in agreement with those of Johansen[1], [43.1; 62.1], but the derivation presented here is simpler.

The two types of nodal forces may act together. Consider as an example a case in which a negative and a positive yield line intersect at a simple support. The Mohr's circle is determined when the angles α and ψ are known (see Fig. 6.6.13). Generally, there will be torsional

[1] This is not true for orthotropic slabs (see Section 6.8).

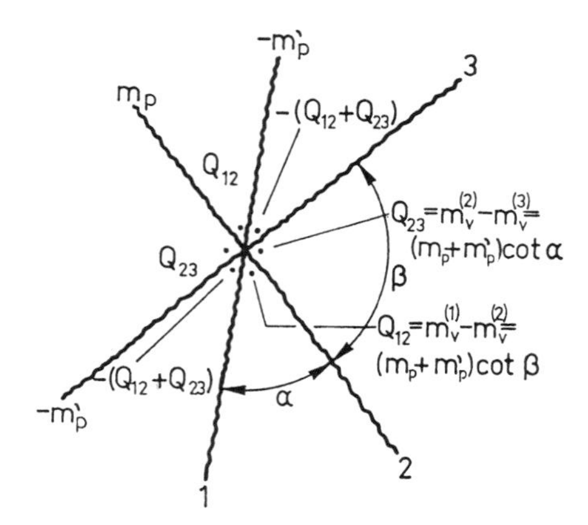

Figure 6.6.12 Nodal forces at a point of intersection between yield lines of different signs.

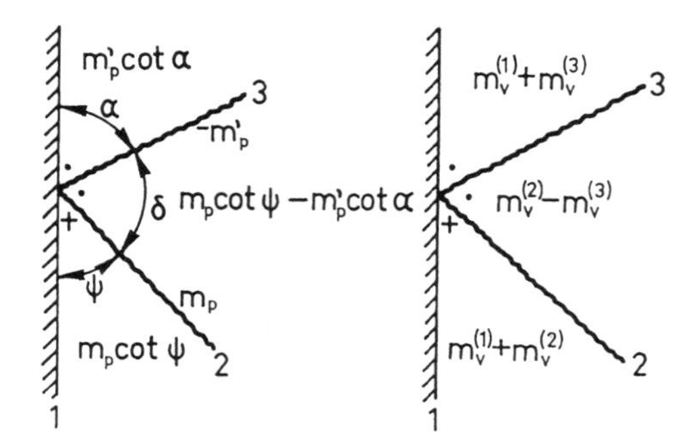

Figure 6.6.13 Nodal forces at boundary with yield lines of different signs.

moments at the boundary and in the yield lines. Consequently there are nodal forces of type 1 at the boundary and nodal forces of type 2 in the yield lines.

The forces expressed by the torsional moments are as shown in the figure. By means of (6.6.10) the nodal forces in angles α and ψ can immediately be calculated. Utilizing the fact that the sum is zero, the nodal forces shown in Fig. 6.6.13 are then obtained.

A fixed edge may be treated in the same way, in which case the result shown in Fig. 6.6.14 is obtained. Figure 6.6.14 summarizes the results obtained.

6.6.4 Calculations by the Equilibrium Method

For the establishment of the equilibrium equations, it is necessary to recall the conditions $v_n = m_{nt} = 0$, which apply along internal yield lines. It will now be demonstrated how the nodal force theory described permits an understanding of the relationship between the equilibrium method and the work equation method, and also how it indicates whether a yield line pattern can be calculated by means of both methods, or only one of them.

Consider first a square slab subjected to a uniform load p. It is assumed that $m'_p = 0$. The yield line pattern has one degree of freedom; that is, the location of the yield lines is determined by one geometrical parameter, for example, angle α (see Fig. 6.6.15). By the work equation method it is found that there is a minimum for $\tan \frac{1}{2}\alpha = 0.19$, which corresponds to $p = 22m_p / a^2$. To calculate this yield line pattern by the equilibrium method, it is first necessary to find out how many equilibrium equations have to be satisfied to ensure that conditions (6.6.9) can be fulfilled. As all yield lines are internal yield lines, there are, in principle, three equations for each slab part in Fig. 6.6.15.

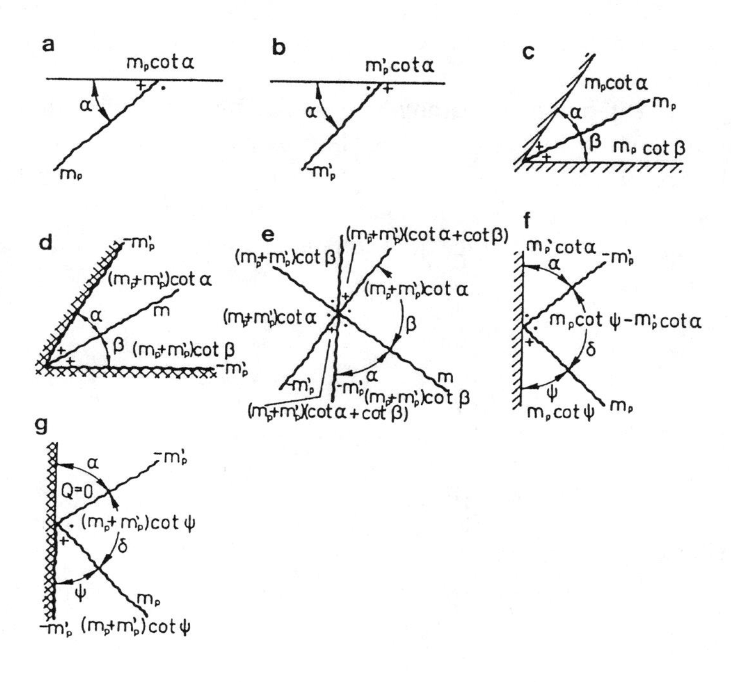

Figure 6.6.14 Survey of nodal forces.

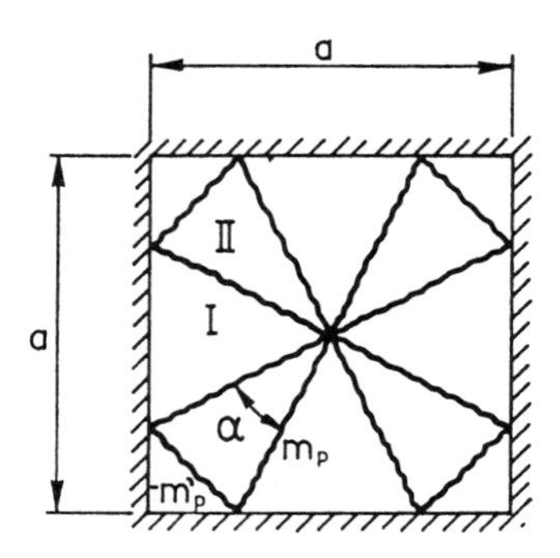

Figure 6.6.15 Yield line pattern for a square slab with one free parameter.

However, because of symmetry, this number reduces to two for the slab parts of type II, and because of the support, only one equation has to be fulfilled for the slab part of type I. There are therefore three equations, but there are only two unknown parameters, one geometrical parameter, α, and the load, p. It appears impossible to satisfy three equations, and this is indeed the case. The load calculated by satisfying only two equilibrium equations (e.g., moment equations about the axes of rotation) does not correspond to the load obtained by the work equation method.

One more geometrical parameter is gained by considering the yield line pattern shown in Fig. 6.6.16. Here there is agreement between the number of equilibrium equations to be satisfied and the number of unknown parameters. The solution of the equilibrium equations is the same as that obtained by the work equation method and the minimization process.

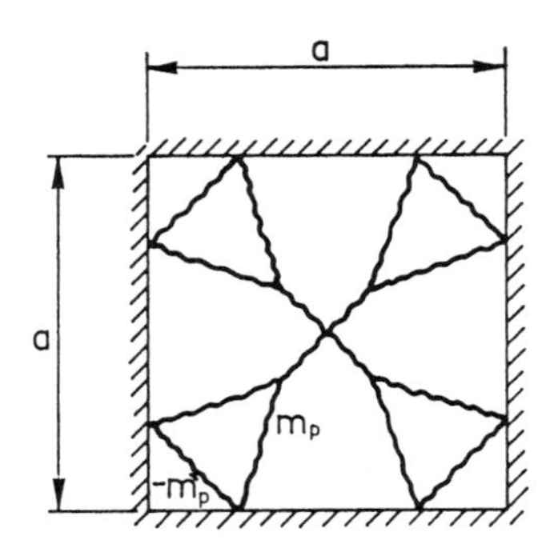

Figure 6.6.16 Yield line pattern for a square slab with two free parameters.

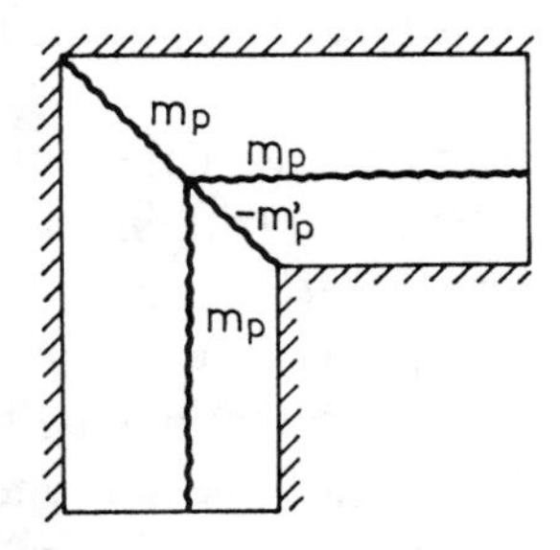

Figure 6.6.17 Yield line pattern in L-shaped slab.

This example has been used by Wood [61.5] and Nylander [63.4] to demonstrate a contradiction in Johansen's theory. The theorem stating that the nodal forces at points where yield lines of the same sign intersect are zero is not valid in this case. If the moment equations about the axes of rotation are checked in the case shown in Fig. 6.6.15, using the values of the parameter α and the load p obtained by the work equation method and the minimization process, it will be found that equilibrium can be maintained only if there are shear forces along the yield lines (i.e., nodal forces must be added at the intersection). The explanation is, as stated above, the lack of agreement between the number of unknown parameters and the number of equilibrium equations to be satisfied to fulfill conditions (6.6.9).

Another case will now be considered where calculations according to the nodal force theory are impossible (see Fig. 6.6.17). In this case, three positive and one negative yield lines intersect at a point. It is obvious that a mode of intersection such as this, in which Mohr's circle is not defined, is statically inadmissible. Therefore, the nodal forces that ensure a stationary value of the load in relation to the work equation method cannot be defined, and the nodal force theory cannot be applied.

Of course, it is possible by means of the work equation method and the minimization process to find out what forces have to be added at the point of intersection in order to satisfy the equilibrium conditions with a load found by the work equation method, but it has not been proved that formulas for these forces can be developed with knowledge of only the mode of intersection. It is naturally also possible to introduce unknown nodal forces and to calculate the values giving the lowest load, but this procedure does not differ in principle from the work equation method. In such cases, therefore, the work equation method may just as well be applied. These examples illustrate the nature of yield line patterns that must be calculated by means of the work equation method.

The nodal force theory presented here is basically quite different from that of Johansen. Whereas Johansen considers the nodal forces to be the statical equivalence of torsional moments and shear forces, in the theory as described here they are either the statical equivalence of torsional moments only (nodal forces of type 1) or the statical equivalence of shear forces only (nodal forces of type 2), and the length of the yield line that contributes to nodal forces of type 2 is infinitely small. It has been explained why the theorem of zero nodal forces in intersections between yield lines of the same sign is not always valid. The same restriction applies to all formulas for nodal forces. Apart from this, the resulting formulas are the same for isotropic slabs as those of Johansen.

6.6.5 Geometrical Conditions

By considering only yield line patterns with straight yield lines, it is possible to give the geometrical conditions a lucid form, which at the same time facilitates the sketching of geometrically possible failure mechanisms. As the plastic deformations in this case are zero outside the yield lines, there are plane slab parts here. Considering now two slab parts I and II (see Fig. 6.6.18) which are separated by a yield line l, it can be seen that yield line l must pass through the point of intersection O of the axes of rotation. This follows from the fact that the axes of rotation and the yield line are secants between three planes – the original plane of the slab and the two planes that slab parts I and II form after the deformation – and from the fact that the secants between three planes pass through the same point if the planes intersect at all.

6.6.6 The Work Equation

In the case of straight yield lines, the work equation is most conveniently established by calculating the work done on each slab part and thereafter summing over all slab parts. As the work done by an arbitrary system of forces when it is rotated is equal to the moment about the axis of rotation times the angle of rotation, the work equation may be written

$$\sum_j M_{Ej}\omega_j = \sum_j M_{Ij}\omega_j \tag{6.6.21}$$

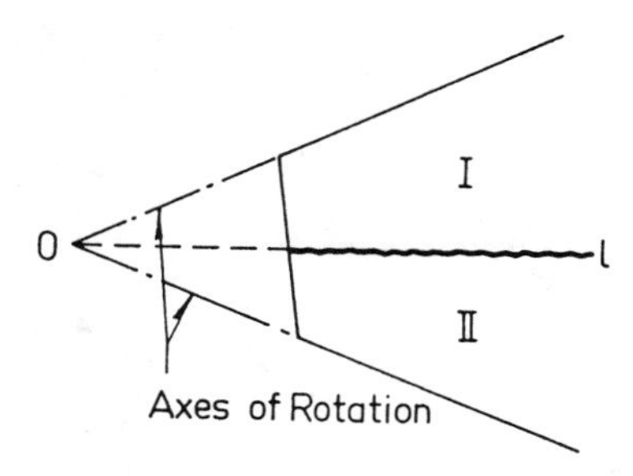

Figure 6.6.18 Point of intersection between axes of rotation and a yield line.

where M_{Ej} = moment about the axis of rotation of the external load acting on the j th slab part

M_{Ij} = moment about the axis of rotation of the bending moments acting on the j th slab part

ω_j = rotation of the j th slab part

Since the work performed by the bending moments in the yield lines has been written at the right-hand side of (6.6.21), the work has to be counted as positive when being negative. When summed over all slab parts, the work calculated in this way will become equal to the dissipation in the yield lines.

When establishing the work equation it is sometimes convenient to draw the moment vectors in the yield lines as shown in Fig. 6.6.19 for a positive and a negative yield line, respectively. This may facilitate getting the correct sign of all terms in (6.6.21).

To illustrate the calculation of the individual terms in (6.6.21), consider as an example a positive yield line *ABCD* in a slab part *ABCD* rotating about *AD* (see Fig. 6.6.20). The moment about *AD* is seen to be $AD \cdot m_p$ and to have the direction *AD*. If the rotation vector is assumed to have the direction *DA* (the slab part moves downward), the contribution to the right-hand side of (6.6.21) from the slab part is positive and equal to $AD \cdot m_p\ \omega$, where ω is the angle of rotation.

To facilitate the setup of the work equation, the moment of a uniformly distributed load p for frequent cases is given in Fig. 6.6.21. An application of this method of calculation is shown in the following examples.

6.6.7 Examples

Square slab supported on two adjacent edges. As an example of the application of both the equilibrium method and the work equa-

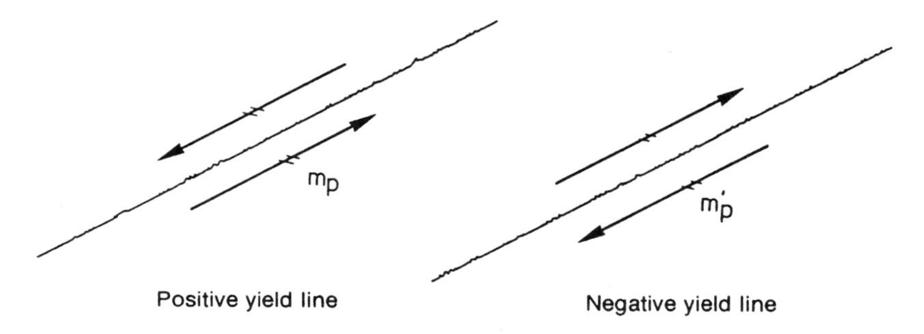

Figure 6.6.19 Notation for positive and negative yield lines, respectively.

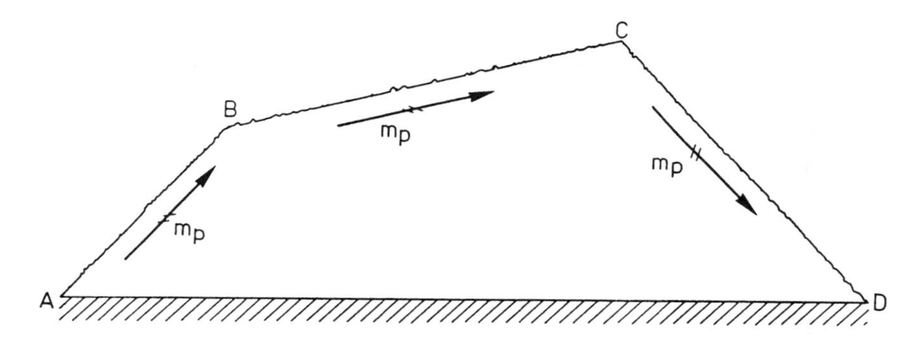

Figure 6.6.20 Slab part limited by a simple supported edge and yield lines.

tion, the load-carrying capacity of the isotropic slab shown in Fig.6.6.22 will be determined. The slab is subjected to a uniformly distributed load p.

For reasons of symmetry it may be expected that the "best" solution of all yield line patterns arising from a positive yield line originating from B is a positive yield line through point D. This is, however, not the case, the reason being that it results in torsional moments in the corner, the statical equivalence of which is a downward-directed force $2m_p$ at point D. This yield line is therefore possible only if such a force belongs to the loading. If this is not the case, the torsional moment at point D must be zero. A yield line like that shown in the figure must then be investigated. The yield line pattern is determined by a single parameter x.

The parameter x is first determined by means of the work equation.

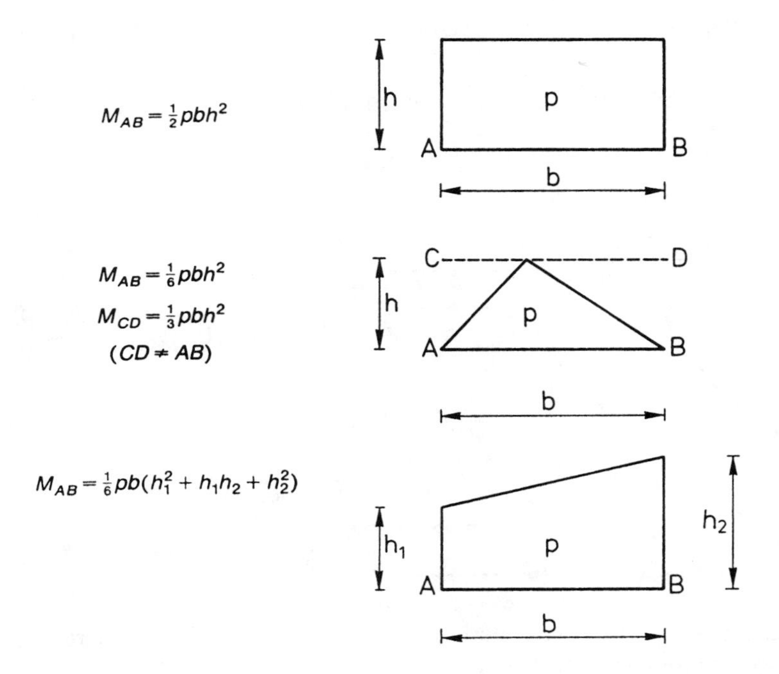

Figure 6.6.21 Moments of uniform load on simple slab parts.

A displacement field is used corresponding to the point E being given a unit downward displacement. Slab part I rotates at an angle $\omega_{\mathrm{I}} = 1/x$ about AB, and slab part II at an angle $\omega_{\mathrm{II}} = 1/a$ about BC. Application of (6.6.21) now yields

$$m_p a \cdot \frac{1}{x} + m_p x \cdot \frac{1}{a} = \frac{1}{6} pax^2 \cdot \frac{1}{x} + \frac{1}{6} pxa^2 \cdot \frac{1}{a} + \frac{1}{2} p(a-x)a^2 \cdot \frac{1}{a} \tag{6.6.22}$$

p is thus

$$p = \frac{6m_p}{a^2} \frac{a^2 + x^2}{3ax - x^2} \tag{6.6.23}$$

If $dp/dx = 0$, the following quadratic equation is obtained for determination of x:

$$3\left(\frac{x}{a}\right)^2 + 2\left(\frac{x}{a}\right) - 3 = 0 \tag{6.6.24}$$

which has the following applicable solution:

$$\frac{x}{a} = 0.72 \tag{6.6.25}$$

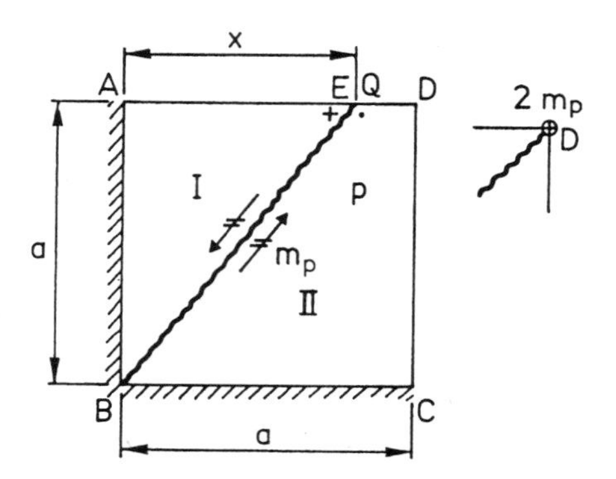

Figure 6.6.22 Yield line pattern with positive yield lines in a square slab supported on two adjacent edges.

Introduction of this solution in the expression for p yields

$$p = \frac{5.55}{a^2} m_p \tag{6.6.26}$$

This solution appears to violate the conditions of symmetry. However, it can be shown that the yield line symmetrical to BE results in the same values of p when this arises together with BE, whereby symmetry is preserved.

The quantity x is now determined by the equilibrium method. In the yield line BE, v and m_v are set equal to zero. According to Fig. 6.6.14, the following nodal forces have to be considered at point E:

$$Q = \pm m_p \cot \alpha = \pm m_p \frac{x}{a} \tag{6.6.27}$$

By taking the moment about the supports AB and BC, it is found that

$$\text{Part I:} \quad m_p a = \tfrac{1}{6} pax^2 + m_p \frac{x}{a} x$$

$$\text{Part II: } m_p x = \tfrac{1}{6} pxa^2 + \tfrac{1}{2} p(a-x)a^2 - m_p \frac{x}{a} a \qquad (6.6.28)$$

Elimination of m_p from the equations gives exactly the same quadratic equation for determination of x, and the same load-carrying capacity p as found above.

In Section 6.7 a lower bound solution is presented.

It is seen that a negative yield line as shown in Fig. 6.6.23 is also possible. If the free corner is displaced δ downward, we have with the notation of Fig. 6.6.23:

$$\frac{1}{6} plh^2 \cdot \frac{\delta}{h} = m'_p l \cdot \frac{\delta}{h} \qquad (6.6.29)$$

$$p = \frac{6m'_p}{h^2} \qquad (6.6.30)$$

A minimum for p is obtained when h is as large as possible, that is,

$$h = \frac{\sqrt{2}}{2} a \qquad (6.6.31)$$

corresponding to the negative yield line being a diagonal in the square.

We thus obtain

$$p = \frac{12m'_p}{a^2} \qquad (6.6.32)$$

The two upper bound solutions give the same load-carrying capacity if $12m'_p = 5.55m_p$ (i.e., $m'_p / m_p = 5.55/12 = 0.46$). We have then found the upper bound solution

$$p = \begin{cases} \dfrac{5.55m_p}{a^2} & \text{for } m'_p \geq 0.46m_p \\[2ex] \dfrac{12m'_p}{a^2} & \text{for } m'_p < 0.46m_p \end{cases} \qquad (6.6.33)$$

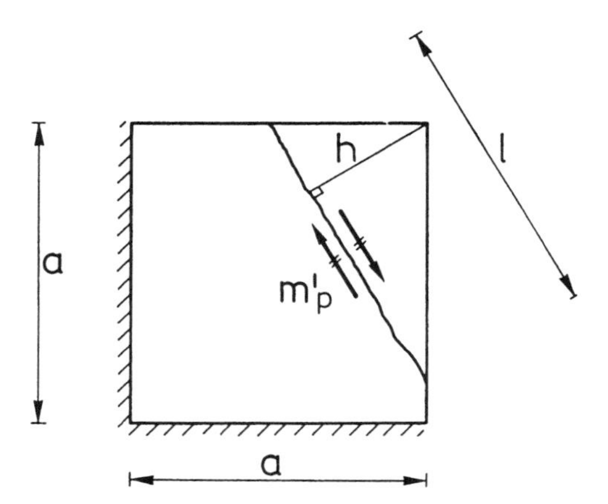

Figure 6.6.23 Yield line pattern with one negative yield line in a square slab supported on two adjacent edges.

Rectangular slab supported along all edges. As a further example of the application of the equilibrium method, an isotropic, rectangular slab subjected to a uniformly distributed load p, as shown in Fig. 6.6.24, is now considered. The case of simple supports on all four edges is calculated first. It is assumed that $a \leq b$.

A possible yield line pattern is obtained by assuming that the slab parts rotate about the supports, so that yield lines originate at the corners. The yield line pattern is determined by one parameter x.

The moment about AB gives

$$m_p a = \tfrac{1}{6} p a x^2 \qquad (6.6.34)$$

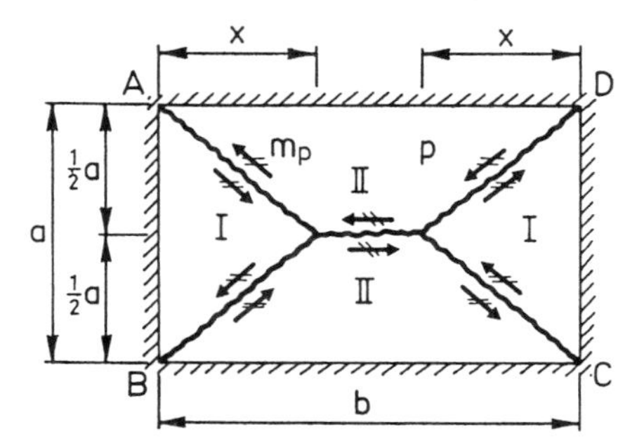

Figure 6.6.24 Yield line pattern in simply supported rectangular slab.

The moment about BC yields

$$m_p b = \tfrac{1}{2} pb\left(\tfrac{1}{2}a\right)^2 - 2 \cdot \tfrac{1}{3} px\left(\tfrac{1}{2}a\right)^2 \tag{6.6.35}$$

Elimination of m_p results in the quadratic equation

$$\left(\frac{x}{b}\right)^2 + \left(\frac{a}{b}\right)^2 \frac{x}{b} - \frac{3}{4}\left(\frac{a}{b}\right)^2 = 0 \tag{6.6.36}$$

which has the following applicable solution:

$$\frac{x}{b} = \frac{1}{2}\frac{a}{b}\left(\sqrt{3 + \left(\frac{a}{b}\right)^2} - \frac{a}{b}\right) \tag{6.6.37}$$

When b is lying in the interval $a \leq b < \infty$, x is lying in the interval $\frac{1}{2}a \leq x < (\sqrt{3}/2)a$. By inserting x/b in the first equation, the load-carrying capacity becomes

$$p = \frac{24m_p}{a^2} \frac{1}{\left[\sqrt{3 + (a/b)^2} - a/b\right]^2} \qquad (a \leq b) \tag{6.6.38}$$

This solution was given by Ingerslev in 1921 [21.1; 23.1].

If the slab is fixed, the following equilibrium equations are obtained instead, the notation being as shown in Fig. 6.6.25:

$$\begin{aligned} am_p(1+i_1) &= \tfrac{1}{6} pax_1^2 \\ am_p(1+i_3) &= \tfrac{1}{6} pax_3^2 \\ bm_p(1+i_2) &= \tfrac{1}{2} pbx_2^2 - \tfrac{1}{3} p(x_1+x_3)x_2^2 \\ bm_p(1+i_4) &= \tfrac{1}{2} pbx_4^2 - \tfrac{1}{3} p(x_1+x_3)x_4^2 \end{aligned} \tag{6.6.39}$$

Further, the following condition is obtained:

$$x_2 + x_4 = a \tag{6.6.40}$$

From the last two equilibrium equations it follows that

$$\frac{1+i_2}{1+i_4} = \frac{x_2^2}{x_4^2}, \quad \frac{x_4}{\sqrt{1+i_4}} = \frac{x_2}{\sqrt{1+i_2}} = \frac{x_2+x_4}{\sqrt{1+i_4}+\sqrt{1+i_2}} = \frac{a}{\sqrt{1+i_4}+\sqrt{1+i_2}} \tag{6.6.41}$$

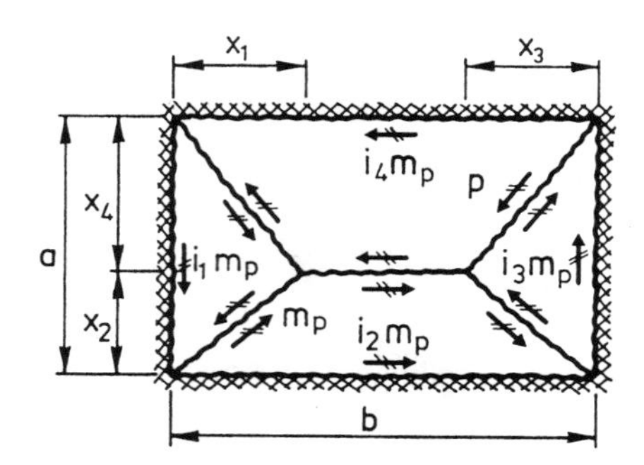

Figure 6.6.25 Yield line pattern in a clamped rectangular slab.

By introduction of

$$a_r = \frac{2a}{\sqrt{1+i_2} + \sqrt{1+i_4}} \tag{6.6.42}$$

the following is obtained:

$$x_2 = \tfrac{1}{2} a_r \sqrt{1+i_2}\,, \quad x_4 = \tfrac{1}{2} a_r \sqrt{1+i_4} \tag{6.6.43}$$

From the first two equilibrium equations, we get

$$\frac{x_1}{\sqrt{1+i_1}} = \frac{x_3}{\sqrt{1+i_3}} = x_r \tag{6.6.44}$$

The quantity b_r, analogous to a_r, is then introduced:

$$b_r = \frac{2b}{\sqrt{1+i_1} + \sqrt{1+i_3}} \tag{6.6.45}$$

which changes the equilibrium equations to the following two equations:

$$\begin{aligned} m_p &= \tfrac{1}{6} p x_r^2 \\ m_p b_r &= \tfrac{1}{2} p b_r \left(\tfrac{1}{2} a_r\right)^2 - 2 \cdot \tfrac{1}{3} p x_r \left(\tfrac{1}{2} a_r\right)^2 \end{aligned} \tag{6.6.46}$$

which are identical to those applying to the simply supported slab. The load-carrying capacity is thus determined by

$$p = \frac{24 m_p}{a_r^2} \frac{1}{\left[\sqrt{3 + (a_r / b_r)^2} - a_r / b_r\right]^2} \qquad (a_r \le b_r) \tag{6.6.47}$$

Instead of the expressions derived for the load-carrying capacity, it is possible to apply an approximation formula given by Johansen [49.3; 72.3], which is, moreover, on the safe side in relation to the upper bound solution. The approximation formula is as follows:

$$p = \frac{8m_p}{a_r b_r}\left(1 + \frac{a_r}{b_r} + \frac{b_r}{a_r}\right) \tag{6.6.48}$$

This expression has the added advantage of being symmetrical in a_r and b_r. In fact, this is indeed a lower bound solution (see Section 6.7.2).

Notice the special result from (6.6.38) for $b = a$:

$$p = \frac{24m_p}{a^2} \tag{6.6.49}$$

which is the load-carrying capacity of a square slab. Notice also that $b \to \infty$ leads to

$$p \to \frac{8m_p}{a^2} \tag{6.6.50}$$

meaning that in this case the effect of the supports along the short sides disappears. The slab acts as a simply supported one-way slab.

We shall also develop the result (6.6.38) by means of the upper bound technique. If the yield line separating slab parts II is given a downward displacement δ, the rotation of slab parts I is

$$\omega_{\mathrm{I}} = \frac{\delta}{x} \tag{6.6.51}$$

and the rotation of slab parts II is

$$\omega_{\mathrm{II}} = \frac{2\delta}{a} \tag{6.6.52}$$

The work equation therefore reads

$$2 \cdot \frac{1}{6} pax^2 \cdot \frac{\delta}{x} + 2\left[\frac{1}{6} p \cdot 2x \cdot \left(\frac{1}{2}a\right)^2 + \frac{1}{2} p(b - 2x)\left(\frac{1}{2}a\right)^2\right] \cdot \frac{2\delta}{a}$$
$$= 2m_p a \cdot \frac{\delta}{x} + 2m_p b \cdot \frac{2\delta}{a} \tag{6.6.53}$$

TABLE 6.1

$\dfrac{a}{b}$	$\dfrac{pa^2}{m_p}$ (6.6.38)	$\dfrac{pa^2}{m_p}$ (6.6.58)
1	24	24
0.75	18.5	18.7
0.5	14.2	14.4
0.25	10.6	10.9
0	8	8

giving

$$p = \frac{12m_p}{a^2}\frac{a^2 + 2bx}{3bx - 2x^2} \tag{6.6.54}$$

The best upper bound solution is found by minimizing with regard to x. Determination of the extremum for an expression of the form $T(x)/N(x)$ is seen to be identical with putting $T(x)/N(x) = T'(x)/N'(x)$, where the prime denotes differentiation with regard to x. We then have

$$\frac{a^2 + 2bx}{3bx - 2x^2} = \frac{2b}{3b - 4x} \tag{6.6.55}$$

giving

$$\left(\frac{x}{b}\right)^2 + \left(\frac{a}{b}\right)^2 \frac{x}{b} - \frac{3}{4}\left(\frac{a}{b}\right)^2 = 0 \tag{6.6.56}$$

which is identical to (6.6.36).

Writing (6.6.54) in the form

$$p = \frac{12m_p}{a^2}\frac{a^2 + 2bx}{3bx - 2x^2} = \frac{12m_p}{a^2}\frac{2b}{3b - 4x} = \frac{6m_p}{x^2} \tag{6.6.57}$$

we find the same solution (6.6.38) as before. The minimum determined for p is very flat. For example, setting $x = \frac{1}{2}a$, corresponding to yield lines at an angle of 45° from the corners, we get for the simply supported slab.

$$p = \frac{24m_p}{a^2}\frac{1 + a/b}{3 - a/b} \quad (a \le b) \tag{6.6.58}$$

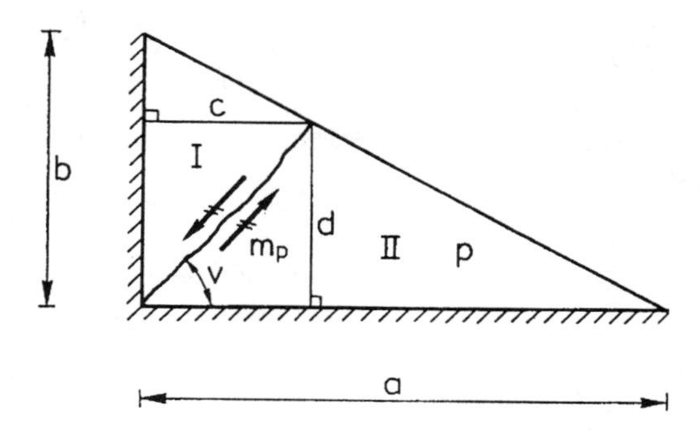

Figure 6.6.26 Yield line pattern in triangular slab simply supported on two adjacent edges.

In Table 6.1 this solution has been compared with the minimized solution (6.6.38) for different values of a/b. It is seen that the difference between the solutions is negligible.

Triangular slab with uniform load. Consider a triangular slab simply supported on two adjacent edges and with a free edge. The load is a uniformly distributed load p. A geometrically possible yield line pattern is shown in Fig. 6.6.26. A positive yield line radiates from the intersection of the supports. The yield line pattern has one free geometrical parameter, for example, the angle v, as shown. However, we set up the work equation with the distances c and d as shown in the figure as parameters. If the intersection of the yield line with the free edges is displaced δ downward, we have

$$\omega_{\mathrm{I}} = \frac{\delta}{c}$$
$$\omega_{\mathrm{II}} = \frac{\delta}{d} \tag{6.6.59}$$

The work equation then reads

$$\frac{1}{6} pbc^2 \cdot \frac{\delta}{c} + \frac{1}{6} pad^2 \cdot \frac{\delta}{d} = m_p d \cdot \frac{\delta}{c} + m_p c \cdot \frac{\delta}{d} \tag{6.6.60}$$

that is,

$$p = 6m_p \frac{c/d + d/c}{bc + ad} \tag{6.6.61}$$

Expressing that twice the area of the slab is $ab = bc + ad$ and introducing

$$\tan v = \frac{d}{c} \tag{6.6.62}$$

we have

$$p = \frac{6m_p}{ab}(\tan v + \cot v) \tag{6.6.63}$$

The extremum for p is found by expressing

$$\frac{dp}{dv} = \frac{6m_p}{ab}\left[1 + \tan^2 v - \left(1 + \cot^2 v\right)\right] = 0 \tag{6.6.64}$$

giving

$$\tan^4 v = 1 \tag{6.6.65}$$

or, having $0 \leq v \leq \pi/2$,

$$v = \frac{\pi}{4} \tag{6.6.66}$$

It is easily verified that we are dealing with a minimum. If we put $v = \pi/4$ in expression for p, we get

$$p = \frac{12m_p}{ab} \tag{6.6.67}$$

It is seen that the yield line bisects the angle from which it radiates. It may be shown to be valid even if the angle is not $\pi/2$.

Line load on a free edge. A slab is loaded with a line load $\bar{p}$ along a free, rectilinear edge. The slab has the positive yield moment m_p while the negative yield moment $m_p' = 0$ (the slab has no topside reinforcement).

We consider the yield line pattern shown in Fig. 6.6.27 having one free geometrical parameter x. Giving the center of the edge a displacement δ, we have the work equation

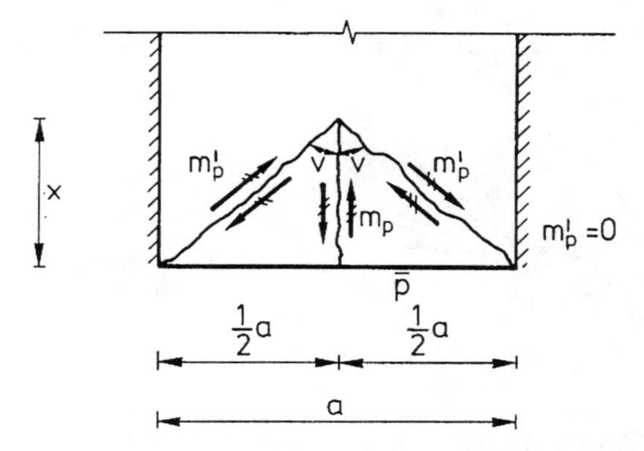

Figure 6.6.27 Yield line pattern for a slab with a line load on a free edge.

$$\tfrac{1}{2}\,\bar{p}a\delta = 2m_p x\cos\nu\cdot\frac{\delta}{x\sin\nu} \tag{6.6.68}$$

from which we have

$$\bar{p} = \frac{4m_p}{a}\cot\nu = \frac{8m_p}{a^2}x \tag{6.6.69}$$

The lowest value of $\bar{p}$ is obtained for $x \to 0$, leading to $\bar{p} \to 0$. This being an upper bound solution, the load-carrying capacity of the slab is zero. This means that in case of a free edge with line load there should always be a top-side reinforcement.

Concentrated load. We consider a slab with the positive yield moment m_p and the negative yield moment m'_p. The slab is loaded with a concentrated force P. A geometrically possible yield line pattern is shown in Fig. 6.6.28. Positive yield lines radiate from the concentrated force P. The yield line pattern is limited by a closed, polygonal, negative yield line.

If the concentrated force is displaced δ downward, we have the work equation

$$P\cdot\delta = \sum\left(m_p + m'_p\right)\Delta s\cdot\frac{\delta}{r\sin\beta} \tag{6.6.70}$$

summarizing over all triangular slab parts. Setting all sides $\Delta s \to 0$ for all triangles, we have

$$\sum\left(m_p + m'_p\right)\Delta s\frac{\delta}{r\sin\beta} \to \left(m_p + m'_p\right)\int_0^{2\pi}\frac{\delta}{\sin^2\beta}\,d\varphi \tag{6.6.71}$$

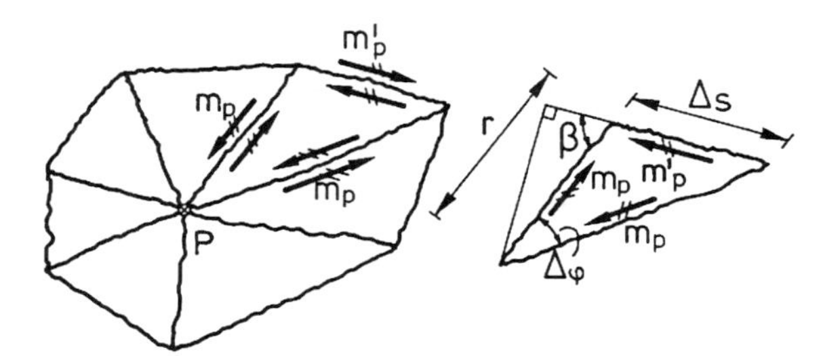

Figure 6.6.28 General yield line pattern for a slab with a concentrated force.

from which

$$P = \left(m_p + m'_p\right) \int_0^{2\pi} \frac{1}{\sin^2 \beta} d\varphi \tag{6.6.72}$$

The lowest value of P is obtained for $\beta = \pi/2$, that is, the negative yield line is a circle. For $\beta = \pi/2$ we have

$$P = 2\pi \left(m_p + m'_p\right) \tag{6.6.73}$$

Note that the radius of the circle is indefinite. The yield line pattern is shown in Fig. 6.6.29.

We have dealt with this solution in Section 6.5.4, having shown that the solution is an exact solution for a circular slab with fixed edges loaded with a concentrated force in the center. For $m'_p = 0$ we get $P = 2\pi m_p$ as the exact solution for a simply supported circular slab without top reinforcement and loaded with a concentrated force at the center. These solutions are illustrated in Fig. 6.6.30.

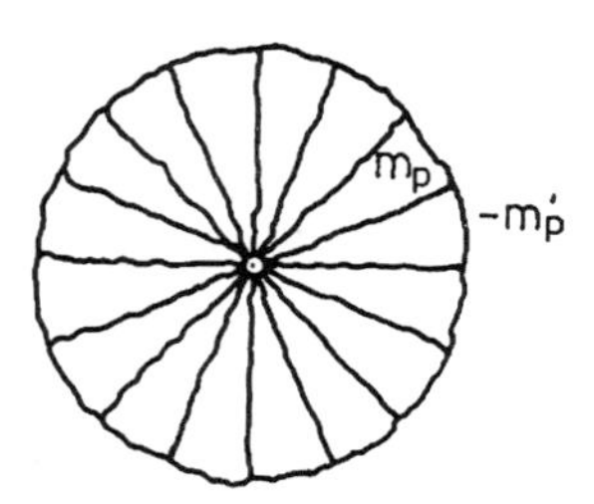

Figure 6.6.29 Optimized yield line pattern for a slab with a concentrated force.

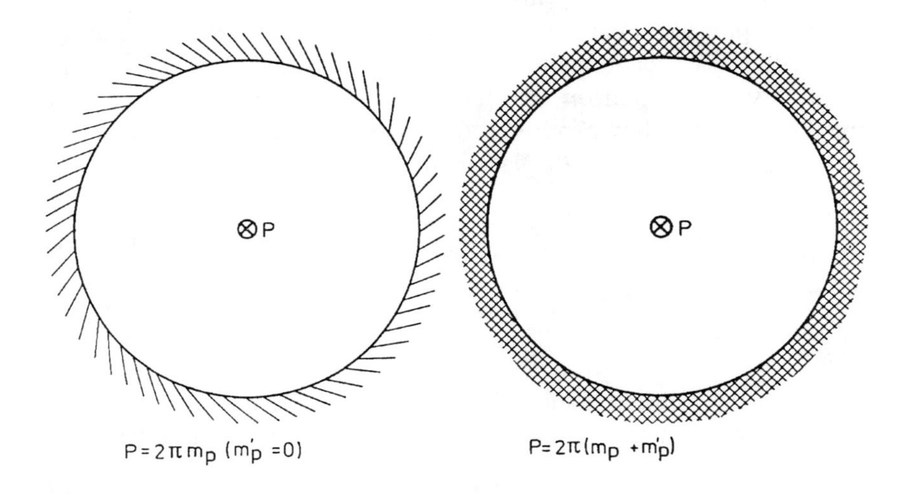

Figure 6.6.30 Concentrated force solutions for circular slabs.

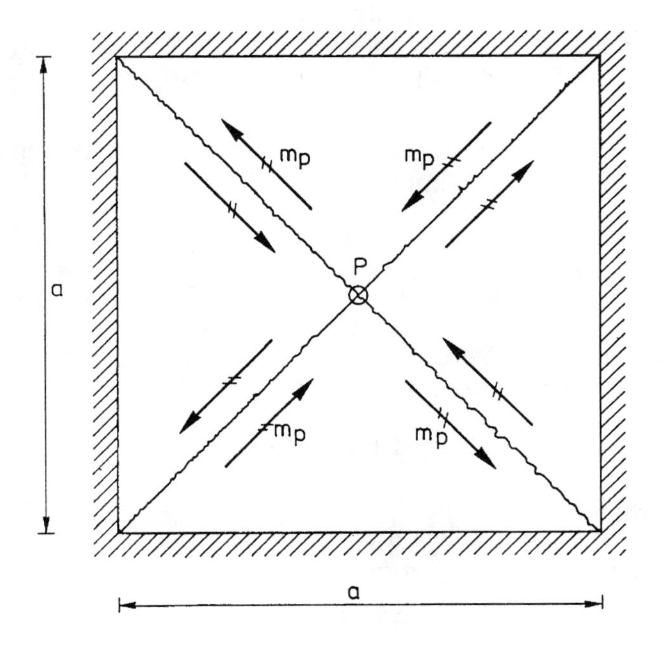

Figure 6.6.31 Yield line pattern for a simply supported square slab with a concentrated force.

Simply supported square slab with a concentrated load. A simply supported square slab with side lengths a is in the intersection of the diagonals loaded with a concentrated force P. The slab has the positive yield moment m_p and the negative yield moment m'_p. The yield line pattern shown in Fig. 6.6.31 yields the following work equation, corresponding to a displacement δ of the force P:

$$P \cdot \delta = 4m_p a \cdot \frac{2\delta}{a} \tag{6.6.74}$$

From this we obtain

$$P = 8m_p \tag{6.6.75}$$

Note that the solution is independent of the side length a.

The upper bound solution from the preceding example gives

$$P = 2\pi\left(m_p + m'_p\right) \tag{6.6.76}$$

The two yield line patterns give the same load-carrying capacity for $m'_p = 0.27m_p$. It is noted that the last-mentioned solution is correct only when there is top-side reinforcement in the entire slab. If only the areas near the edge have top-side reinforcement, a yield circle with $m'_p = 0$ can be formed and then the load-carrying capacity is only $P = 2\pi m_p$.

From the yield line patterns investigated in this connection we can state the upper bound solution

$$P = \begin{cases} 8m_p & \text{for } m'_p \geq 0.27m_p \\ 2\pi\left(m_p + m'_p\right) & \text{for } m'_p < 0.27m_p \end{cases} \tag{6.6.77}$$

From the exact solutions of Section 6.5.3 it appears that $P = 8m_p$ is an exact solution for $m'_p = m_p$.

6.6.8 Practical Use of Upper Bound Solutions

In practical calculations of reinforced concrete slabs it is often difficult to find the best solution corresponding to a selected yield line pattern, nor it is necessary to do so, since the work equation generally provides a good approximation as long as the yield line pattern is, in principle, correct. A check may be made on the applicability of the yield line pattern by deriving equilibrium equations for the individual slab parts, and these equations also provide information as to how the yield line pattern should be altered in order to give a better solution.

The general experience is that rather big differences (100%) in the results from the individual equilibrium equations can be tolerated without affecting the result of the work equation to any significant level. To illustrate that we do have a “flat extreme”, Fig. 6.6.32 shows m_p drawn as a function of x from one of the examples above.

When estimating yield line patterns it is pertinent to take advantage of the fact that

1) Restrained edges repel yield lines
2) Concentrated forces and column supports often attract yield lines
3) Corners in holes in a slab attract yield lines

When designing slabs by means of upper bound solutions a difficulty often arises concerning the amount and extension of the top reinforcement. Even a simply supported slab needs some top reinforcement near the supports because of torsional moments. This need also appears from upper bound calculations when the conditions at the supports are examined in more detail. Consider as examples the yield line patterns in corners shown in Fig. 6.6.33. In Section 6.6.3 we found the downward-directed corner reactions shown in the figure. We also found negative principal moments. In the angle α we have, according to (6.5.31), $m_2 = -m_p \cot^2\alpha$; in the angle β we have $m_2 = -m_p \cot^2\beta$. When $\alpha = \beta = \pi/4$, $m_2 = -m_p$ in both angles.

If the slab is restrained one finds that the negative principal moment will be $m_2 = -(m_p + m_p')\cot^2\alpha - m_p'$ in the angle α and $m_2 = -(m_p + m_p')\cot^2\beta - m_p'$ in the angle β. When $\alpha = \beta = \pi/4$ and $m_p = m_p'$, then $m_2 = -3m_p$ in both angles. This moment cannot be

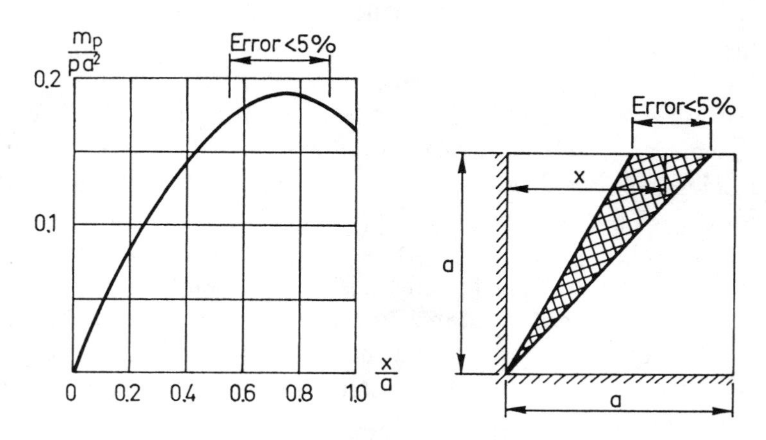

Figure 6.6.32 Example used to demonstrate the flat extreme for yield line patterns.

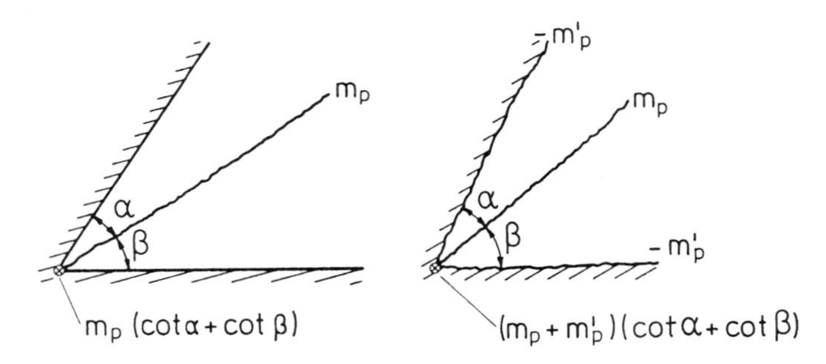

Figure 6.6.33 Nodal forces at corners.

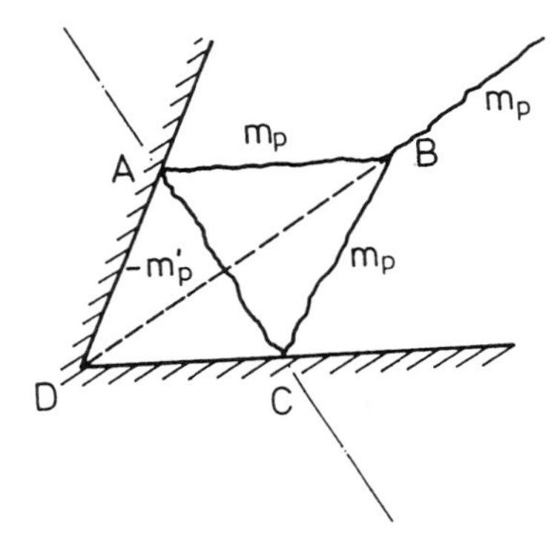

Figure 6.6.34 Yield line pattern with corner lever.

absorbed since the top-side reinforcement corresponds to the yield moment m_p, so the yield line pattern in question is statically impossible in this case.

Even for a simply supported reinforced concrete slab, reinforcement in the top side is thus required in order that the corner reaction can be absorbed. If the top-side reinforcement near the corner is not sufficient, the yield line pattern in the neighborhood of the corner, shown in Fig. 6.6.33, will not be possible.

A yield line pattern giving a better value of the load-carrying capacity if the top-side reinforcement is not sufficiently large is shown in Fig. 6.6.34. Here the slab behaves as if there is a fixed edge along *AC*, *AC* being a negative yield line. Slab part *ABC* was denoted a corner lever by Johansen [43.1; 62.1]. If upward displacements in the corner are not prevented, the corner will rise and slab part *ABCD* will rotate about *AC* as a rigid body. Then yield line *AC* will not be found.

The signifiance of the corner levers increases the more acute the corner angle and the less the top-side reinforcement at a simple support. Normally, the significance of the corner levers is greater for restrained than for simply supported slabs.

Yield line patterns with corner levers are more difficult to calculate than those where the yield lines run entirely into the corners. For rectangular slabs with simple supports the reduction in load-carrying capacity proves insignificant, even when there is no top-side reinforcement in the neighborhood of the supports. To prevent severe cracking in the serviceability state, however, a certain top-side reinforcement is always supplied, for example, by bending up a certain amount of reinforcement in the neighborhood of the supports.[2] For a square slab with uniformly distributed load and simply supported along all sides, the simple yield line pattern with the yield lines running entirely into the corners gives $p = 24m_p/a^2$ according to a previous example. If $m'_p = 0$, one finds that the most dangerous yield line pattern with corner levers gives $p = 22m_p/a^2$, that is, only a little above 8% reduction of the load-carrying capacity (see [43.1; 62.1]).

In Fig. 6.6.35, a case where the edge of the corner lever falls outside the slab is shown. For the slab shown we have without taking notice of the corner lever, according to a previous example, $p = 12m_p/a^2$, while the yield line pattern shown in Fig. 6.6.35 with $m'_p = 0$ gives the minimized value $p = 10m_p/a^2$, that is, a reduction of about

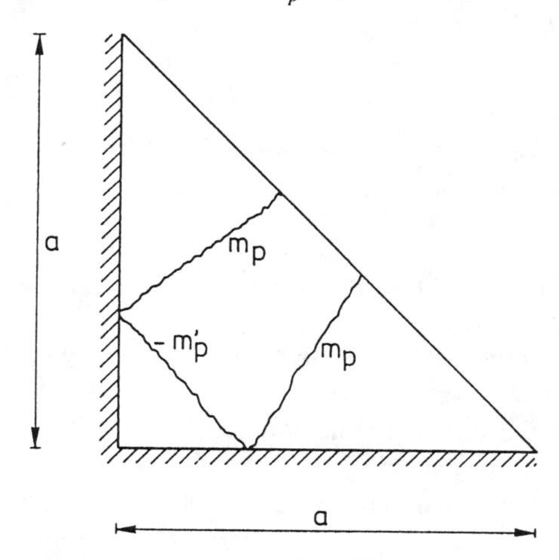

Figure 6.6.35 Yield line pattern with significant influence of corner lever.

[2] The positive yield moment will decrease in areas with bent-up reinforcement, and this must be taken into account by thorough calculations.

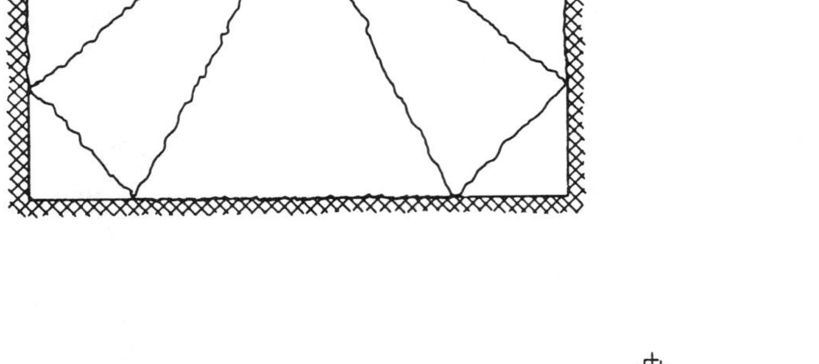

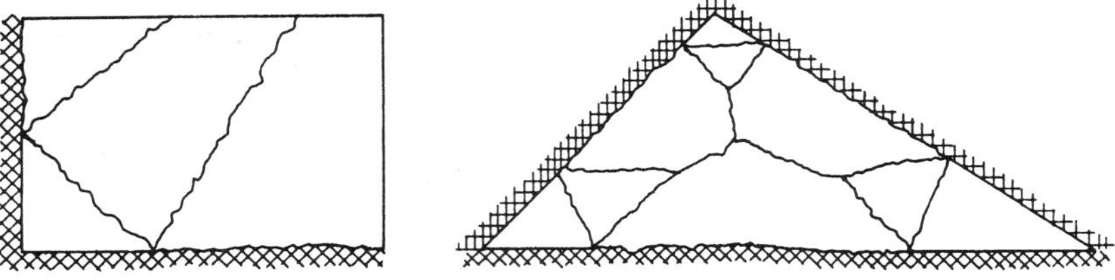

Figure 6.6.36 Yield line patterns with significant influence of corner levers.

17% (see [49.3; 72.3]).

In Fig. 6.6.36 some other examples of yield line patterns with corner levers are shown. For a restrained triangular slab with uniformly distributed load and corner angles 120°, 30°, and 30°, we have as an example about 30% reduction of the load-carrying capacity compared with the solution where the yield lines run entirely into the corner.

If only part of the top face is reinforced the question arises as to the extension of this reinforcement. Normally it is a laborious task to determine the extent by upper bound methods. It is much easier using lower bound methods (see Section 6.7).

In principle the extent of the top reinforcement can be determined by considering yield line patterns involving the top-side reinforcement as well as yield line patterns not or only partly involving the top reinforcement. For instance, a rectangular slab $a \cdot b$ with fixed edges, reinforced in the top side only in strips of width c near the edges, may be calculated using two yield line patterns: one which would be valid if the whole top side was reinforced and one valid for a simply supported slab $(a-2c) \cdot (b-2c)$, i.e., a yield line pattern in the interior of the slab with yield lines along the edges $m_p' = 0$. By requiring these yield line patterns to render the same load-carrying capacity, c may be determined. This is seldomly done for well-known slab types like rectangular slabs. Instead one relies upon rules by experience, c being chosen to, for instance, 1/5 of the short span when dealing with restrained edges and 1/7 of the short span for simple supports.

If one deals with a slab system consisting of many slabs, yield line patterns involving more than one slab must in principle be considered. Full load on the neighbor slabs and minimum load on the slab in question might lift up the slab, activating only the top reinforcement. Such loading combinations are often decisive for the extension of the top reinforcement. A practical solution to this problem is to consider for each slab two loading combinations:

1) Full load with bending moments along the edges equal to the yield moments.
2) Minimum load with bending moments along the edges equal to the yield moments.

Case 1) is treated by a normal yield line calculation. Using upper bound methods, case 2) may be treated considering the yield moments along the edges as the active load.

A typical yield line pattern for case 2) in a rectangular slab is shown in Fig. 6.6.37. The negative yield lines are emerging from the corners and meet a negative yield line with $m_p' = 0$ along the zone without top reinforcement. The displacement is upward directed, i.e., it will be resisted by the load on the slab.

The extension of top reinforcement near column supports may be determined in a similar way. Consider Fig. 6.6.38 showing a column support in a slab having a uniform load p. The slab has only top reinforcement in a circular area with radius c around the column. In practice, of course, often a square

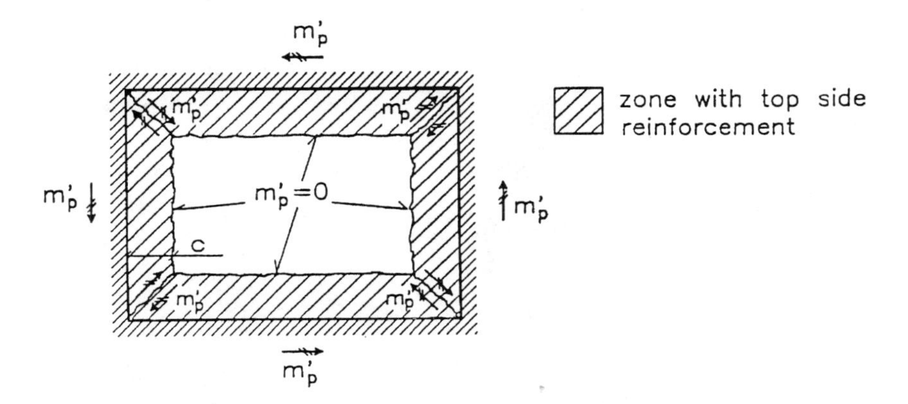

Figure 6.6.37 Typical yield line pattern for determining the extension of top-side reinforcement.

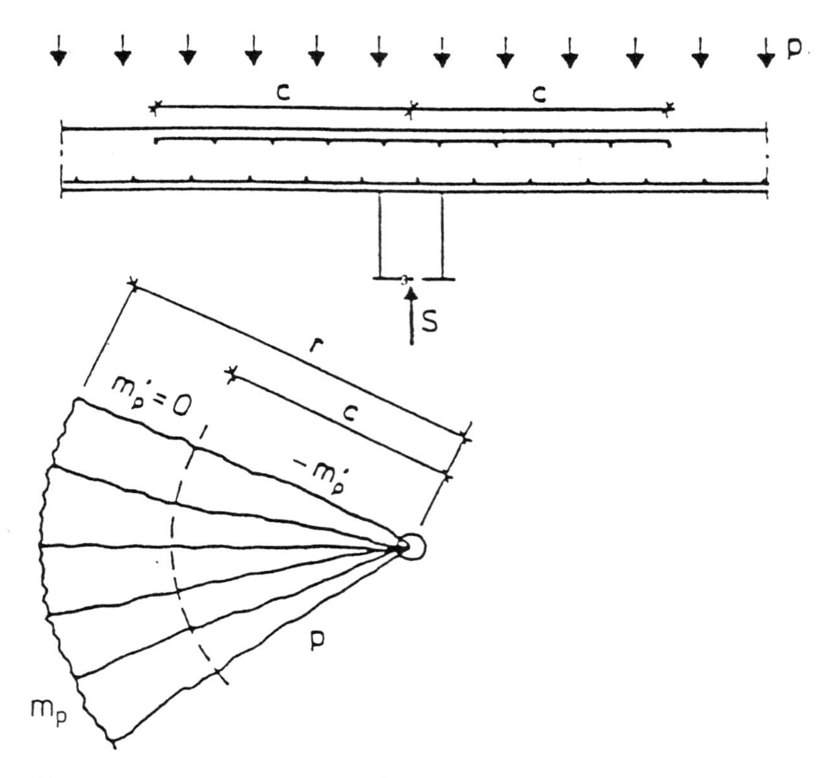

Figure 6.6.38 Column support in a slab with uniform load.

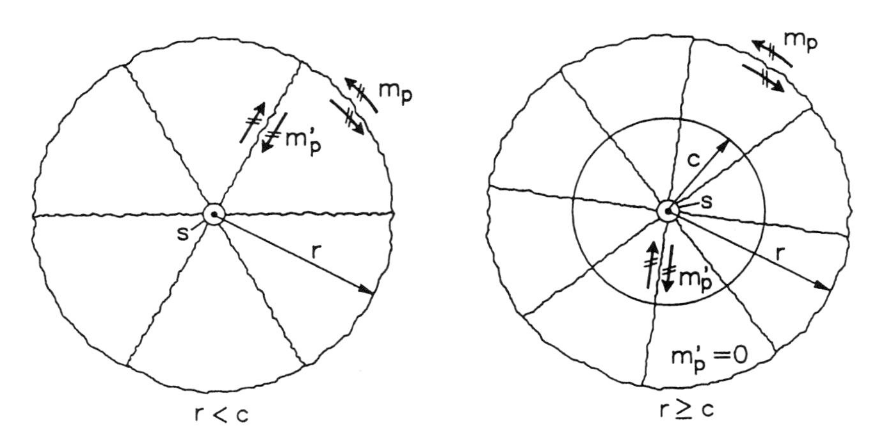

Figure 6.6.39 Yield line patterns near a column support.

with side lengths $2c$ is used. Simple yield line patterns may in this case be found considering the column force S to be the active force. We simplify the calculations by considering S to be a point force.

In Fig. 6.6.39 two yield line patterns with straight negative yield lines through the column and a positive circular yield line with radius r are shown. In the figure to the left $r < c$ and to the right $r \geq c$.

When $r < c$ one easily finds that the load-carrying capacity is

$$S = 2\pi\left(m_p + m_p'\right) \tag{6.6.78}$$

which is a limiting value for $r \to 0$.

When $r \geq c$ one gets

$$S = 2\pi m_p + 2\pi m_p' \frac{c}{r} + \frac{1}{3}\pi r^2 p \tag{6.6.79}$$

The minimum is found for

$$r = \sqrt[3]{\frac{3m_p' c}{p}} \tag{6.6.80}$$

For this solution to be valid, $c^2 < 3m_p'/p$ $(r \geq c)$ is required. Inserting the r - value (6.6.80) into (6.6.79) and equalizing the two load-carrying capacities (6.6.78) and (6.6.79), c may be determined. The result is

$$c = \sqrt{\frac{8}{9}\frac{m_p'}{p}} \tag{6.6.81}$$

If c is less than this value the load-carrying capacity is governed by (6.6.79), otherwise by (6.6.78).

Notice that the condition $c^2 \leq 3m_p'/p$ is always fulfilled when c is determined by (6.6.81).

If the slab thickness is increased in a circular area near the column, things are getting more complicated (see Fig. 6.6.40). Here we have three different sets of yield moments. We have the normal ones m_p and m_p' in the slab outside the column region. We have increased values m_{pc} and m_{pc}' in a region with increased top-side reinforcement but corresponding to the normal slab thickness outside the column region. Finally we have increased values m_{pb} and m_{pb}' corresponding to the increased slab thickness. We may also take into account the extension of the column. In the figure the length $2a$ is the column diameter and p_c is the uniform pressure delivered by the column reaction S.

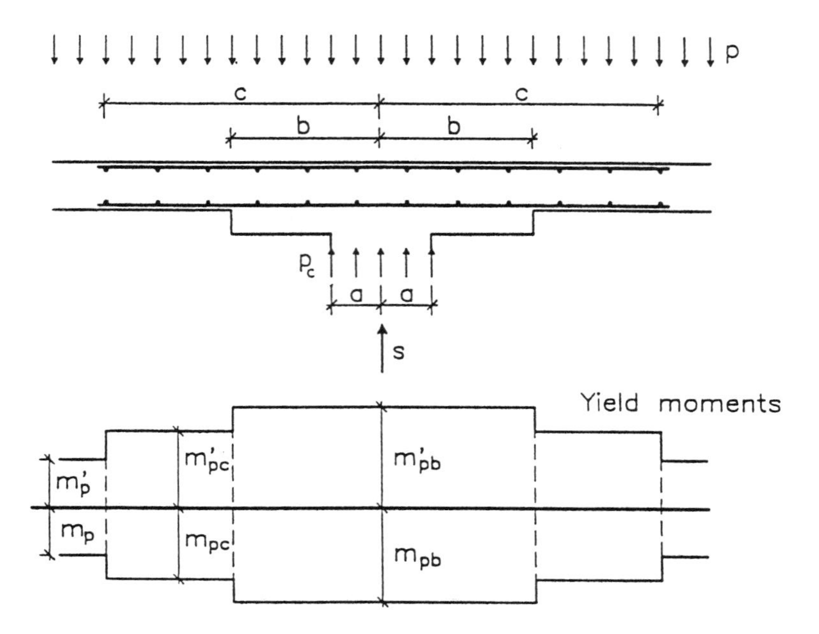

Figure 6.6.40 Column support with increased thickness of the slab around the column. The normal reinforcement outside the column region is not shown.

By considering three yield line patterns of the same type as before we find

$$m_{pb} + m'_{pb} = \frac{S}{2\pi}\left(1 - \sqrt[3]{\frac{p}{p_c}}\right) \tag{6.6.82}$$

$$c = \frac{p}{3}\frac{r_2^3 - r_1^3}{m'_{pc} - m'_p} \tag{6.6.83}$$

$$b = \frac{p}{3}\frac{r_1^3 - r_0^3}{m'_{pb} - m'_{pc}} \tag{6.6.84}$$

Here

$$r_0 = \sqrt[3]{\frac{p_c}{p}}\,a \tag{6.6.85}$$

$$r_1^2 = \frac{2}{p}\left(\frac{S}{2\pi} - m_{pc} - m'_{pc}\right) \tag{6.6.86}$$

$$r_2^2 = \frac{2}{p}\left(\frac{S}{2\pi} - m_p - m'_p\right) \tag{6.6.87}$$

Strictly speaking the yield line patterns used are not geometrically possible in the region of the slab just above the column. This is however of minor importance.

The parameters r_0, r_1 and r_2 are the radii of the positive yield lines in the different regions.

Note that taking into account the finite extension of the column a reduction of the required yield moments is obtained.

Columns with other sections, not too far from a circle, may be treated by transforming to a circle with the same area.

This solution was derived by Johansen [43.1; 62.1] using the equilibrium method.

When slabs are supported on beams and columns and in other cases, too, one needs to find the reactions. The shear forces also might need to be checked. In principle it is not possible to find these quantities by the upper bound methods unless a statically admissible moment field corresponding to the solution is constructed. This is normally impossible by using hand calculations.

In some common cases reasonable values of the reactions may be determined by distributing the load carried by the adjacent slab parts, into which the slab has been subdivided by the yield lines, uniformly along the supports. Account must be taken of the nodal forces and it must be required that the yield line pattern is rather accurate, Johansen [43.1; 49.3; 62.1; 72.3].

Consider as an example a simply supported triangular slab part with uniform load p (see Fig. 6.6.41 left). The total load carried

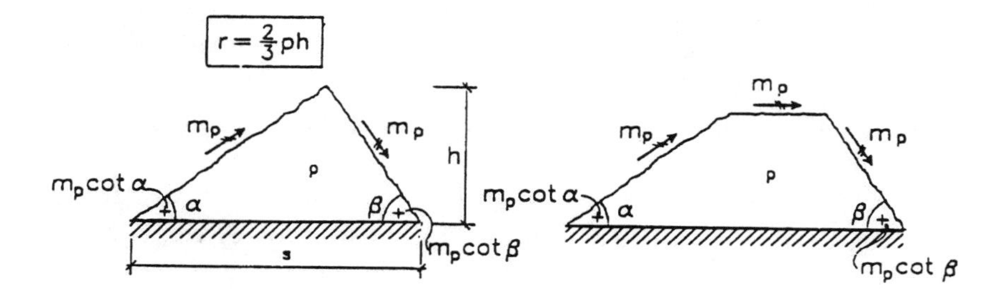

Figure 6.6.41 Slab parts used to determine reactions.

by the slab is $\frac{1}{2}psh$. The nodal forces are downward directed and they have the sum $m_p(\cot\alpha + \cot\beta) = m_p\, s/h$. Then the reaction per unit length along the support is

$$r = \tfrac{1}{2}\,ph + 2\frac{m_p}{h} \tag{6.6.88}$$

Utilizing the equilibrium equation

$$m_p s = \tfrac{1}{6}\,psh^2 \tag{6.6.89}$$

we find

$$r = \tfrac{2}{3}\,ph \tag{6.6.90}$$

If the edge is restrained we have instead of (6.6.88):

$$r = \tfrac{1}{2}\,ph + 2\frac{m_p + m_p'}{h} \tag{6.6.91}$$

In the same way the trapezoidal slab may be treated. Instead of deriving a formula it is easier to write down the projection equation in the individual cases.

When determining column reactions extreme care must be taken to make sure that the yield line pattern is accurate. Yield line calculations for slabs with column supports often give weird results. If no extra reinforcement is placed around the columns, one may experience that the yield lines are infinitely close to the columns for the most dangerous yield line patterns (cf. Example 6.8.1). Since the column forces may often be estimated by guessing the area from which they take the load it is recommended to always start the yield line calculations for column supported slabs by checking the columns by means of the formulas for local failure around a concentrated force. If a global yield line pattern is found to require less yield moments it might be a local extremum and we should look for other more dangerous yield line patterns. Alternatively extra reinforcement is supplied near the columns, as described above.

Another problem often met in practice is that the actual loading case for a slab is a combination of loading cases each of which has a known upper bound solution. In this case, the following combination theorems may be useful.

Let us consider a loading case, characterized by a load parameter p_1 and another loading case, characterized by a load parameter p_2. Let p_{1p} be the upper bound solution found by the work equation for an arbitrarily chosen yield line pattern when the load corresponds to p_1. We then have

$$p_{1p} W_{E1} = D \tag{6.6.92}$$

where W_{E1} is the external work for $p_1 = 1$ and D is the dissipation.

Let p_{2p} be the upper bound solution found by the work equation for the same yield line pattern. With analogous notation we have

$$p_{2p} W_{E2} = D \tag{6.6.93}$$

Of course, we assume that p_1 and p_2 perform positive work at the displacements corresponding to the yield line pattern considered. Similar considerations apply in the following.

If both loads act simultaneously the same yield line pattern gives

$$p_1 W_{E1} + p_2 W_{E2} = D \tag{6.6.94}$$

or

$$\frac{p_1}{p_{1p}} + \frac{p_2}{p_{2p}} = 1 \tag{6.6.95}$$

For an arbitrary number of loads we get

$$\frac{p_1}{p_{1p}} + \frac{p_2}{p_{2p}} + \cdots = 1 \tag{6.6.96}$$

Equations (6.6.95) and (6.6.96) are exact if we choose the correct yield line pattern corresponding to the combined loading case. Inserting instead of p_{1p} the correct load-carrying capacity in the loading case corresponding to p_1, the first term of (6.6.95) or (6.6.96) will be greater (or perhaps unchanged), according to the upper bound theorem. Analogous conditions apply to the other terms. It is then seen that if p_{1p}, p_{2p}, and so on, denote the load-carrying capacity in the individual loading cases, (6.6.95) or (6.6.96) becomes an approximation formula on the safe side for determination of upper bound solutions for combined loads.

If, on the other hand, the task is to determine the dimensions of the slab, for example, characterized by a yield moment m_p, we have for determination of the necessary yield moment m_p for an arbitrarily chosen yield line pattern and for the load p_1 an equation of the form

$$p_1 W_{E1} = m_{p1} D_1 \tag{6.6.97}$$

where D_1 is the internal work corresponding to $m_p = 1$.

For the load p_2 we have for the same yield line pattern the necessary yield moment m_{p2}:

$$p_2 W_{E2} = m_{p2} D_1 \tag{6.6.98}$$

When p_1 and p_2 act simultaneously we have for the same yield line pattern the necessary yield moment m_p:

$$p_1 W_{E1} + p_2 W_{E2} = m_p D_1 \tag{6.6.99}$$

It is thus seen that

$$m_p = m_{p1} + m_{p2} \tag{6.6.100}$$

or

$$\frac{m_{p1}}{m_p} + \frac{m_{p2}}{m_p} = 1 \tag{6.6.101}$$

For several loading cases we have

$$m_p = m_{p1} + m_{p2} + \cdots \tag{6.6.102}$$

or

$$\frac{m_{p1}}{m_p} + \frac{m_{p2}}{m_p} + \cdots = 1 \tag{6.6.103}$$

Equations (6.6.101) and (6.6.103) are exact if we choose the yield line pattern corresponding to the correct yield line pattern for the combined loading case considered.

Inserting, instead of m_{1p}, the correct yield moment in the loading case corresponding to p_1, the first term in (6.6.101) or (6.6.103) becomes greater. Analogous conditions apply to the other terms. It is seen that if m_{1p}, m_{2p}, and so on, signify the necessary yield moments for the individual loading cases, (6.6.101) or (6.6.103) is a conservative approximation formula for determination of the necessary yield moment for combined loadings. The superposition law used in this way is consequently conservative.

A great number of upper bound solutions have been developed by Johansen [32.2; 49.3; 72.3]. His collection of formulas considerably facilitates the practical use of the upper bound method.

We conclude this section by an important remark about the serviceability limit state.

When designing slabs by plastic methods it is extremely important to check the deflections. When only load-carrying capacity is considered the slabs may often be built very thin, but then, frequently, the stiffness will be unsatisfactory.

In the fully cracked state, a rough estimate of the deflections and the stresses in the serviceability limit state may be found using the results of linear elastic slab theory assuming uncracked concrete. If the reinforcement is isotropic but different in top and bottom the average value in case of fixed edges may be used when calculating the bending stiffness to be inserted in the linear elastic solutions and when calculating the stresses. For orthotropic reinforcement the reinforcement to use must be estimated in the individual cases.

Example 6.6.1 In this example we want to show a practical design of an isotropic, rectangular slab. The slab is shown in Fig. 6.6.42. It is restrained along three edges and free along the fourth. The slab must carry a uniform load $p = 9.7$ kN/m², which includes the dead weight, and it must carry the line loads shown in the figure. They all have the magnitude $\bar{p} = 3.0\,\text{kN/m}$. The yield moment m_p for positive moments is sought when the yield moment for negative moments $m_p' = \frac{2}{3} m_p$.

The yield line pattern is estimated as shown. It consists of yield lines under 45° from the corners, i.e., corner levers are neglected. The rotations of the slab parts I, II and III are

$$\omega_{\text{I}} = \omega_{\text{II}} = \omega_{\text{III}} = \frac{\delta}{3.50}$$

δ being the deflection of the midpoint of the free edge.

The work equation reads:

$$\left(\frac{1}{2}\cdot 9.7\cdot 1.00\cdot 3.50^2 + \frac{1}{6}\cdot 9.7\cdot 3.50\cdot 3.50^2 + \frac{1}{2}\cdot 3.0\cdot 3.00^2\right)\frac{\delta}{3.50} +$$

$$\left(\frac{1}{2}\cdot 9.7\cdot 1.00\cdot 3.50^2 + \frac{1}{6}\cdot 9.7\cdot 3.50\cdot 3.50^2 + \frac{1}{2}\cdot 3.0\cdot 2.50^2 +\right.$$

$$\left. 3.0\cdot 2.00\cdot 2.50\right)\frac{\delta}{3.50} + \left(\frac{1}{6}\cdot 9.7\cdot 7.00\cdot 3.50^2 + \frac{1}{2}\cdot 3.0\cdot 3.00^2\right.$$

$$\left. + \frac{1}{2}\cdot 3.0\cdot 2.50^2\right)\frac{\delta}{3.50}$$

$$= 2\cdot\left(m_p + \frac{2}{3}m_p\right)\cdot 4.50\cdot\frac{\delta}{3.50} + \left(m_p + \frac{2}{3}m_p\right)\cdot 7.00\cdot\frac{\delta}{3.50}$$

By solving for m_p we find

$$m_p = 17.20\,\text{kNm/m}$$

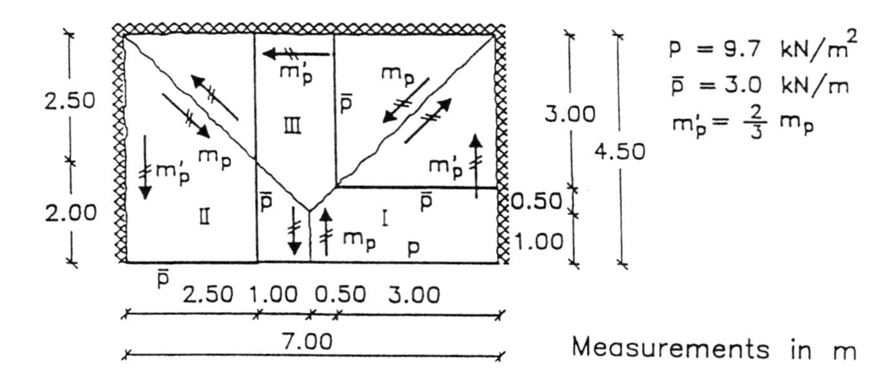

Figure 6.6.42 Slab to be designed using the upper bound method.

Thus the slab must be reinforced in the bottom face in two perpendicular directions corresponding to the yield moment $m_p = 17.20$ kNm/m and similarly in the top face for a yield moment $m_p' = \frac{2}{3} \cdot 17.20\,\text{kNm/m} = 11.46$ kNm/m.

To control the estimate of the yield line pattern, equilibrium of the individual slab parts is controlled. The nodal forces in the intersection point of the positive yield lines are zero and the nodal forces in the point where the positive yield line intersects the free edge are also zero because the yield line is perpendicular to the edge. The nodal forces in the corners from which the positive yield lines emerge do not play any role in the equilibrium equations to follow.

Moment equation about the restrained edge for slab part I:

$$\frac{1}{2} \cdot 9.7 \cdot 1.00 \cdot 3.50^2 + \frac{1}{6} \cdot 9.7 \cdot 3.50 \cdot 3.50^2 + \frac{1}{2} \cdot 3.0 \cdot 3.00^2$$
$$= \left(m_p + \frac{2}{3} m_p \right) \cdot 4.50$$

or

$$7.50 m_p = 142.7\,\text{kNm} \Rightarrow m_p = 19.0\,\text{kNm/m}$$

Moment equation for slab part II:

$$\frac{1}{2} \cdot 9.7 \cdot 1.00 \cdot 3.50^2 + \frac{1}{6} \cdot 9.7 \cdot 3.50 \cdot 3.50^2 + \frac{1}{2} \cdot 3.0 \cdot 2.50^2$$
$$+ 3.00 \cdot 2.00 \cdot 2.50 = \left(m_p + \frac{2}{3} m_p \right) \cdot 4.50$$

or

$$7.50 m_p = 154.6\,\text{kNm} \Rightarrow m_p = 20.6\,\text{kNm/m}$$

Moment equation for slab part III:

$$\frac{1}{6}\cdot 9.7\cdot 7.00\cdot 3.50^2+\frac{1}{2}\cdot 3.0\cdot 3.00^2+\frac{1}{2}\cdot 3.0\cdot 2.50^2$$
$$=\left(m_p+\frac{2}{3}m_p\right)\cdot 7.00$$

or

$$11.67m_p = 161.5\,\text{kNm} \Rightarrow m_p = 13.85\,\text{kNm/m}$$

As expected the equilibrium equations do not give the same value as the work equation, but the differences are so small that the work equation may be assumed to yield a rather accurate value of m_p.

It is left for the reader to change the yield line pattern to obtain values of m_p for the different slab parts which approach the work equation value.

Notice that the values of m_p obtained by the equilibrium equations may be used to write down the work equation. This is also valid when the equilibrium equations contain contributions from nodal forces since these contributions will cancel when summing over the slab parts. Using the results of the equilibrium equations we get the work equation.

$$2\cdot 7.50\cdot m_p\cdot\frac{\delta}{3.50}+11.67m_p\cdot\frac{\delta}{3.50}$$
$$=142.7\cdot\frac{\delta}{3.50}+154.6\cdot\frac{\delta}{3.50}+161.5\cdot\frac{\delta}{3.50}$$
$$m_p = 17.20\ \text{kNm/m}$$

Thus it is demonstrated that the work equation gives a weighted average of the values obtained by the equilibrium equations.

If top reinforcement is only supplied near the supports the extension may be determined as explained above.

For practical purpose the yield line pattern is sufficiently accurate to be used for determining the reactions. Account must here be taken of the nodal forces as shown above.

To the practising engineer: remember to check the stiffness in the serviceability limit state. How to obtain a rough estimate is described above.

6.7 LOWER BOUND SOLUTIONS

6.7.1 Introduction

In the following the lower bound theorem is used to determine lower bounds for the load-carrying capacity. First, a class of solutions for rectangular slabs is developed (see [79.20; 81.1]); then the strip method, introduced by Hillerborg [56.2; 59.1; 74.1], is described briefly. As stated earlier, a lower bound solution consists in finding a statically admissible moment field that corresponds to moments within or on the yield surface (i.e., a safe moment field).

6.7.2 Rectangular Slabs with Various Support Conditions

As the equilibrium equations in Cartesian coordinates,

$$\frac{\partial^2 m_x}{\partial x^2} - 2\frac{\partial^2 m_{xy}}{\partial x\,\partial y} + \frac{\partial^2 m_y}{\partial y^2} = -p \tag{6.7.1}$$

contains partial derivatives of the second order, it is natural for constant p to consider such functions that have constant second derivatives, that is, the parabolic cylindrical surfaces for m_x and m_y, and hyperbolic paraboloids for m_{xy}.

Consider therefore a slab with a moment distribution of the type

$$\begin{aligned} m_x &= a + bx + cx^2 \\ m_y &= d + ey + fy^2 \\ m_{xy} &= g + hx + my + ixy \end{aligned} \tag{6.7.2}$$

For a rectangular slab $k \cdot l$ with a uniform load intensity p, line loads K_1, K_2, K_3 and K_4 and bending moments m_1, m_2, m_3 and m_4 along the boundaries and concentrated reactions S_A, S_B, S_C, and S_D at the corners (see Fig. 6.7.1), the following conditions have to be satisfied. The equilibrium equation (6.7.1) furnishes the condition

$$c - i + f = -\tfrac{1}{2}p \tag{6.7.3}$$

The boundary conditions which are of the type $(m_x)_{x=\frac{1}{2}k} = -m_1$, $\left(\partial m_x/\partial x - 2\,\partial m_{xy}/\partial y\right)_{x=\frac{1}{2}k} = K_1$, and $S_A = -\left(2m_{xy}\right)_{(x,y)=(\frac{1}{2}k,\frac{1}{2}l)}$, give the equations

$$a + \tfrac{1}{2}bk + \tfrac{1}{4}ck^2 = -m_1 \tag{6.7.4}$$

$$a - \tfrac{1}{2}bk + \tfrac{1}{4}ck^2 = -m_3 \tag{6.7.5}$$

$$d + \tfrac{1}{2}el + \tfrac{1}{4}fl^2 = -m_2 \tag{6.7.6}$$

$$d - \tfrac{1}{2}el + \tfrac{1}{4}fl^2 = -m_4 \tag{6.7.7}$$

$$b - 2m + (c - i)k = K_1 \tag{6.7.8}$$

$$b - 2m - (c - i)k = -K_3 \tag{6.7.9}$$

$$e - 2h + (f - i)l = K_2 \tag{6.7.10}$$

$$e - 2h - (f - i)l = -K_4 \tag{6.7.11}$$

$$2g + hk + ml + \tfrac{1}{2}ikl = -S_A \tag{6.7.12}$$

$$2g - hk + ml - \tfrac{1}{2}ikl = S_B \tag{6.7.13}$$

$$2g - hk - ml + \tfrac{1}{2}ikl = -S_C \tag{6.7.14}$$

$$2g + hk - ml - \tfrac{1}{2}ikl = S_D \tag{6.7.15}$$

A slab supported on four edges. Consider as a first example a uniformly loaded, rectangular slab supported on all four edges (see Fig. 6.7.2). One or more of the edges may be fixed; that is, the moments m_1, m_2,.... can be different from zero.

In this case the conditions (6.7.3)-(6.7.7) have to be satisfied. Solving the last four equations for b, c, e, and f and choosing a symmetrical distribution for the torsional moments (i.e., $g = h = m = 0$), we get the moment distribution:

$$\begin{aligned}
m_x &= a - \frac{1}{2}(m_1 - m_3)\left(\frac{x}{\frac{1}{2}k}\right) - \left[a + \frac{1}{2}(m_1 + m_3)\right]\left(\frac{x}{\frac{1}{2}k}\right)^2 \\
m_y &= d - \frac{1}{2}(m_2 - m_4)\left(\frac{y}{\frac{1}{2}l}\right) - \left[d + \frac{1}{2}(m_2 + m_4)\right]\left(\frac{y}{\frac{1}{2}l}\right)^2 \\
m_{xy} &= \frac{1}{4}ikl\left(\frac{x}{\frac{1}{2}k}\right)\left(\frac{y}{\frac{1}{2}l}\right)
\end{aligned} \tag{6.7.16}$$

The equilibrium condition (6.7.3) now reads

$$\frac{4\left[a + \frac{1}{2}(m_1 + m_3)\right]}{k^2} + \frac{4\left[d + \frac{1}{2}(m_2 + m_4)\right]}{l^2} + i = \frac{1}{2}p \tag{6.7.17}$$

We thus get six of seven parameters a, d, m_1, m_2, m_3, m_4 and i as

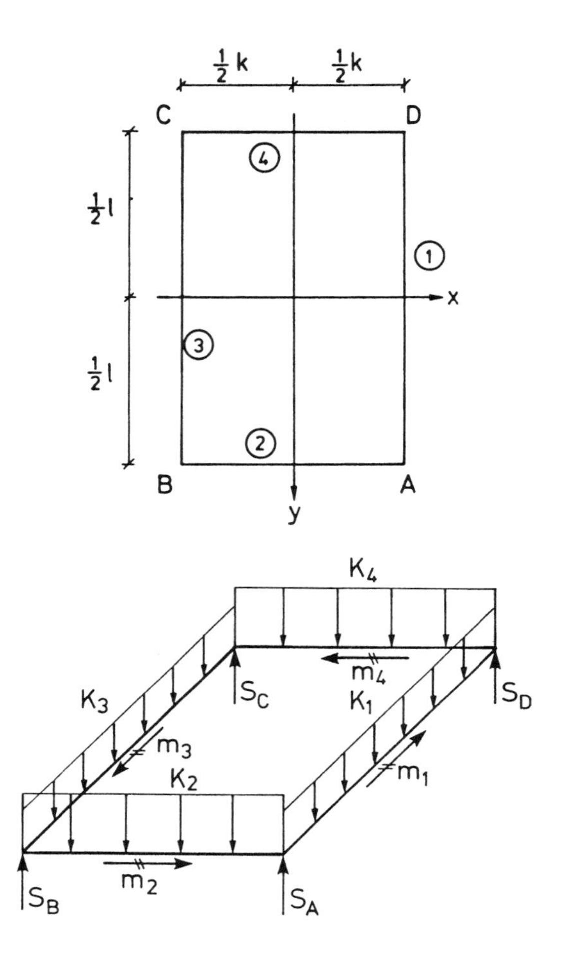

Figure 6.7.1 Rectangular slab element, coordinate system and notation for edge loadings.

free parameters. Having fixed six parameters, the seventh can be determined by (6.7.17).

The reactions along the boundaries and the corner reactions are determined by (6.7.8)-(6.7.15). Setting $r_1 = -K_1$, and so on [i.e., r_1 is the upward-directed reaction per unit length along the boundary 1], we get by utilizing the equilibrium equation (6.7.17):

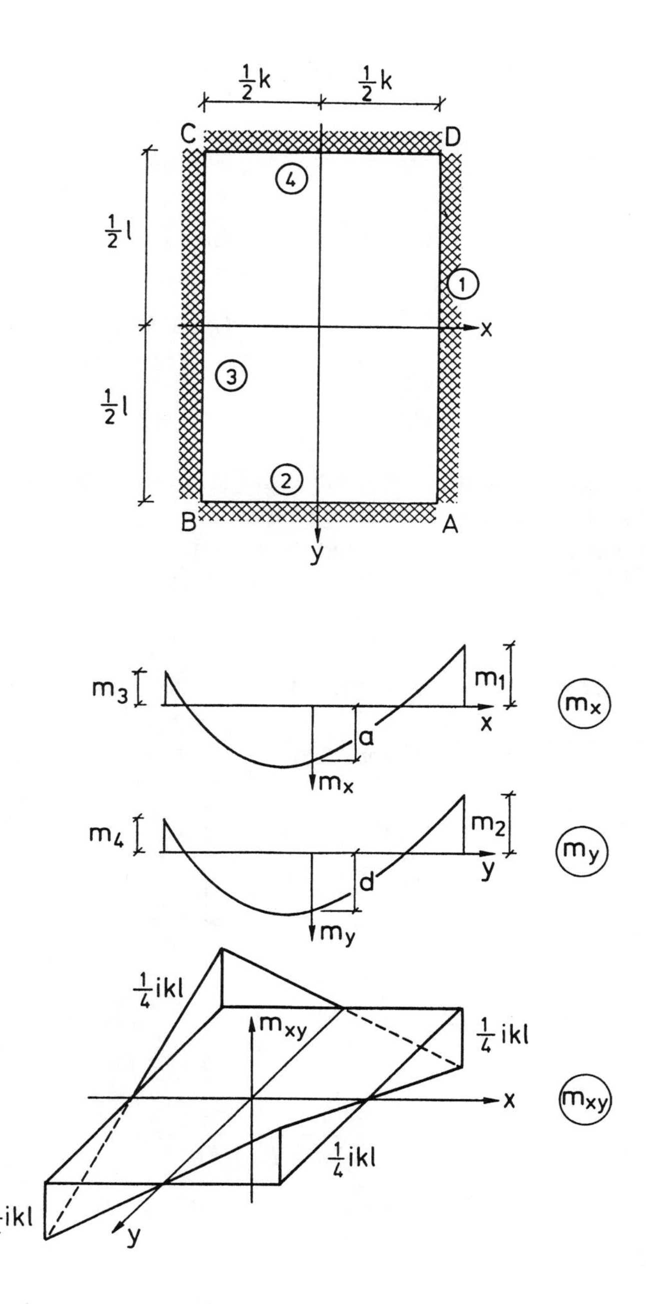

Figure 6.7.2 Slab supported on four edges.

$$
\begin{aligned}
r_1 &= \frac{1}{2}pk + \frac{m_1 - m_3}{k} - 4\left[d + \frac{1}{2}(m_2 + m_4)\right]\frac{k}{l^2} \\
r_2 &= \frac{1}{2}pl + \frac{m_2 - m_4}{l} - 4\left[a + \frac{1}{2}(m_1 + m_3)\right]\frac{l}{k^2} \\
r_3 &= \frac{1}{2}pk - \frac{m_1 - m_3}{k} - 4\left[d + \frac{1}{2}(m_2 + m_4)\right]\frac{k}{l^2} \\
r_4 &= \frac{1}{2}pl - \frac{m_2 - m_4}{l} - 4\left[a + \frac{1}{2}(m_1 + m_3)\right]\frac{l}{k^2}
\end{aligned}
\qquad (6.7.18)
$$

$$
S_A = S_B = S_C = S_D = -\frac{1}{2}ikl \qquad (6.7.19)
$$

The moment distribution is sketched in Fig. 6.7.2, where the meaning of the various parameters also appears.

Consider as an example a simply supported, isotropic slab having $m_p = m'_p$. We then have $m_1 = m_2 = m_3 = m_4 = 0$. In particular, set

$$
a = d = \tfrac{1}{4}ikl = m_p \qquad (6.7.20)
$$

which corresponds to the moment field

$$
\begin{aligned}
m_x &= m_p\left[1 - \left(\frac{x}{\frac{1}{2}k}\right)^2\right] \\
m_y &= m_p\left[1 - \left(\frac{y}{\frac{1}{2}l}\right)^2\right] \\
m_{xy} &= m_p\left(\frac{x}{\frac{1}{2}k}\right)\left(\frac{y}{\frac{1}{2}l}\right)
\end{aligned}
\qquad (6.7.21)
$$

Then (6.7.17) gives

$$
m_p = \frac{1}{8}\frac{pkl}{1 + k/l + l/k} \qquad (6.7.22)
$$

The principal moments become

$$\left.\begin{array}{l} m_1 \\ m_2 \end{array}\right\} = \frac{1}{2}\left(m_x + m_y\right) \pm \sqrt{\frac{1}{4}\left(m_x - m_y\right)^2 + m_{xy}^2}$$

$$= \frac{1}{2}\left[2m_p - 4m_p\left(\left(\frac{x}{k}\right)^2 + \left(\frac{y}{l}\right)^2\right)\right]$$

$$\pm \sqrt{\frac{1}{4}\left[4m_p\left(\left(\frac{x}{k}\right)^2 - \left(\frac{y}{l}\right)^2\right)\right]^2 + \left(\frac{4m_p}{kl}xy\right)^2} \qquad (6.7.23)$$

$$= \begin{cases} m_p \\ m_p - 4m_p\left[\left(\frac{x}{k}\right)^2 + \left(\frac{y}{l}\right)^2\right] \end{cases}$$

It will be seen that both m_1 and m_2 lie within the yield surface at all points when $m_p = m'_p$. The solution is thus a lower bound solution.

Throughout the slab, one of the principal moments is equal to the positive yield moment. The reason the solution does not correspond to that of Ingerslev [see (6.6.38)] is that the principal sections do not follow the yield lines in Ingerslev's solution.

The solution (6.7.22) which was given by Wood [55.4] is thus a lower bound solution. It is interesting to note that this corresponds completely to Johansen's approximation formula (6.6.48). A comparison of Ingerslev's solution with (6.7.22) shows that the difference is a maximum of 1.5% [55.4].

The Danish Code DS 411; [73.2] contains a slab formula [64.10] which is a special case of the solution presented here. The code formula corresponds to the choice $i = [a + d + \frac{1}{2}(m_1 + m_2 + m_3 + m_4)]/kl$, which may be shown to give close results to the upper bound solutions.

As is well known, there are negative moments at the corners of a simply supported slab in the elastic region corresponding to i being positive and the corner reactions being downward directed. Therefore, i should always be chosen positive. Some practical guidelines concerning the choice of the free parameters are given in [81.1].

A slab supported on three edges. The next example is a uniformly loaded rectangular slab having one edge free (see Fig. 6.7.3). One or more of the other edges may be fixed; that is, the moment m_1,

m_3, and m_4 can be different from zero. In this case the conditions (6.7.3)-(6.7.7) with $m_2 = 0$ and (6.7.10) with $K_2 = 0$ have to be satisfied. Choosing the distribution of the torsional moments to be symmetrical about the y-axis ($g = m = 0$), we find the moment distribution:

$$m_x = a - \frac{1}{2}(m_1 - m_3)\left(\frac{x}{\frac{1}{2}k}\right) - \left[a + \frac{1}{2}(m_1 + m_3)\right]\left(\frac{x}{\frac{1}{2}k}\right)^2$$

$$m_y = d + \frac{1}{2}m_4\left(\frac{y}{\frac{1}{2}l}\right) - \left[d + \frac{1}{2}m_4\right]\left(\frac{y}{\frac{1}{2}l}\right)^2 \qquad (6.7.24)$$

$$m_{xy} = -\left[\frac{1}{4}m_4\frac{k}{l} + d\frac{k}{l} + \frac{1}{4}ikl\right]\left(\frac{x}{\frac{1}{2}k}\right) + \frac{1}{4}ikl\left(\frac{x}{\frac{1}{2}k}\right)\left(\frac{y}{\frac{1}{2}l}\right)$$

The equilibrium equation (6.7.3) reads

$$\frac{4\left[a + \frac{1}{2}(m_1 + m_3)\right]}{k^2} + \frac{4\left[d + \frac{1}{2}m_4\right]}{l^2} + i = \frac{1}{2}p \qquad (6.7.25)$$

We get in this case five of six parameters a, d, m_1, m_3, m_4 and i as free parameters. Having fixed five parameters, the sixth can be determined by (6.7.25). The reactions per unit length and the corner reactions can be determined as before, remembering that $r_2 = 0$.

For a simply supported slab, S_A and S_B will be upward directed while S_C and S_D will be downward directed in the elastic state. Therefore, negative moments will be found at corners C and D and the free

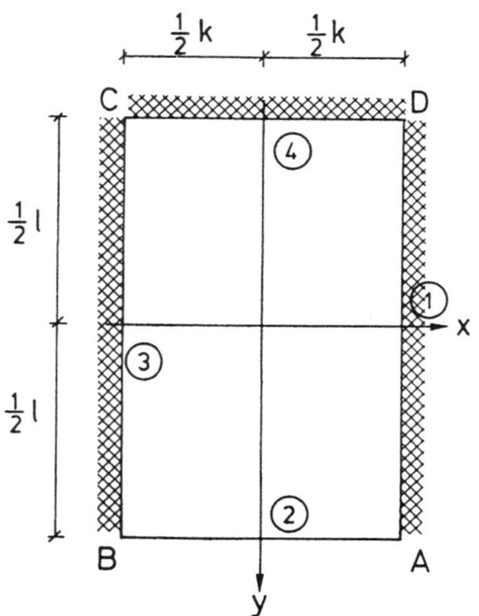

Figure 6.7.3 Slab supported on three edges.

parameters should be chosen accordingly. Some more detailed guidelines are given in [81.1].

Consider as an example the special case of a simply supported isotropic slab with $m_p = m'_p$. We then have $m_1 = m_3 = m_4 = 0$. In particular, set

$$d = 0, \qquad a = \tfrac{1}{2} ikl = m_p \tag{6.7.26}$$

which corresponds to the moment field

$$\begin{aligned} m_x &= m_p\left(1 - \left(\frac{x}{\frac{1}{2}k}\right)^2\right) \\ m_y &= 0 \\ m_{xy} &= -\frac{1}{2} m_p\left(\frac{x}{\frac{1}{2}k}\right) + \frac{1}{2} m_p\left(\frac{x}{\frac{1}{2}k}\right)\left(\frac{y}{\frac{1}{2}l}\right) \end{aligned} \tag{6.7.27}$$

and

$$m_p = \frac{pkl}{4 + 8(l/k)} \tag{6.7.28}$$

The distribution of the torsional moments is sketched in Fig. 6.7.4. As m_x is independent of y, it is only necessary to investigate the principal moments along the line $y = -\frac{1}{2}l$. Thus

$$\begin{aligned} \left.\begin{matrix} m_1 \\ m_2 \end{matrix}\right\} &= \frac{1}{2}\left(m_p - \frac{4m_p}{k^2}x^2\right) \pm \sqrt{\frac{1}{4}\left(m_p - \frac{4m_p}{k^2}x^2\right)^2 + \frac{4m_p^2}{k^2}x^2} \\ &= \begin{cases} m_p \\ -4m_p\left(\dfrac{x}{k}\right)^2 \end{cases} \end{aligned} \tag{6.7.29}$$

It will be seen that the yield condition has not been violated anywhere. Along the lines $x = 0$ and $y = -\frac{1}{2}l$, the greatest principal moment is equal to the yield moment m_p, while at corners C and D, the smallest principal moment is equal to the negative yield moment $m'_p = m_p$. Solution (6.7.28) is therefore a lower bound solution.

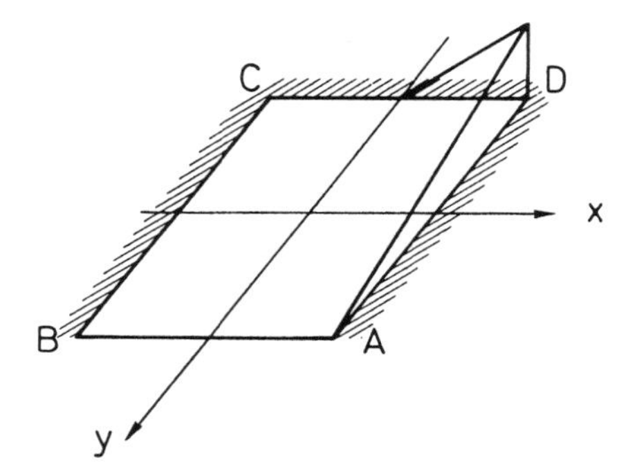

Figure 6.7.4 Torsional moments in a slab supported on three edges.

It is not as good as the previous solution for a rectangular slab. In this case, an approximate upper bound solution by Johansen [49.3; 72.3] has the following appearance:

$$m_p = \frac{pkl}{8 + 6(l/k)} \qquad \text{for } l \le k$$
$$m_p = \frac{pkl}{4 + 2(k/l) + 8(l/k)} \qquad \text{for } l \ge k \tag{6.7.30}$$

For example, for $k = l$ the following is then obtained from (6.7.28) and (6.7.30):

$$m_p = \tfrac{1}{12} pk^2, \qquad m_p = \tfrac{1}{14} pk^2 \tag{6.7.31}$$

The former formula thus gives an approximately 15% higher value of m_p. It turns out that a better solution can be obtained by assuming a torsional moment field of the form $m_{xy} = 2z\,x/k$, where z is a constant. The reader is referred to [64.3].

A slab supported on two adjacent edges. The uniformly loaded slab is now imagined to have two edges free (see Fig. 6.7.5). The supports may be fixed; that is, the moments m_3 and m_4 can be different from zero. In this case, the conditions (6.7.3) and (6.7.4) with $m_1 = 0$, (6.7.5) and (6.7.6) with $m_2 = 0$, (6.7.7) and (6.7.8) with $K_1 = 0$, (6.7.10) with $K_2 = 0$, and (6.7.12) with $S_A = 0$ have to be satisfied. We find that

$$
\begin{aligned}
m_x &= a + \frac{1}{2}m_3\left(\frac{x}{\frac{1}{2}k}\right) - \left[a + \frac{1}{2}m_3\right]\left(\frac{x}{\frac{1}{2}k}\right)^2 \\
m_y &= d + \frac{1}{2}m_4\left(\frac{y}{\frac{1}{2}l}\right) - \left[d + \frac{1}{2}m_4\right]\left(\frac{y}{\frac{1}{2}l}\right)^2 \\
m_{xy} &= \frac{1}{4}ikl + \left(a + \frac{1}{4}m_3\right)\frac{l}{k} + \left(d + \frac{1}{4}m_4\right)\frac{k}{l} \\
&\quad - \left[\frac{1}{4}m_3\frac{l}{k} + a\frac{l}{k} + \frac{1}{4}ikl\right]\left(\frac{y}{\frac{1}{2}l}\right) \\
&\quad - \left[\frac{1}{4}m_4\frac{k}{l} + d\frac{k}{l} + \frac{1}{4}ikl\right]\left(\frac{x}{\frac{1}{2}k}\right) + \frac{1}{4}ikl\left(\frac{x}{\frac{1}{2}k}\right)\left(\frac{y}{\frac{1}{2}l}\right)
\end{aligned}
\tag{6.7.32}
$$

The equilibrium equation (6.7.3) reads

$$
\frac{4\left[a + \frac{1}{2}m_3\right]}{k^2} + \frac{4\left[d + \frac{1}{2}m_4\right]}{l^2} + i = \frac{1}{2}p \tag{6.7.33}
$$

Thus four of the five parameters a, d, m_3, m_4, and i are free parameters. Having fixed four parameters, the fifth can be determined by (6.7.33).

The reactions per unit length and the corner reactions can be determined as before, remembering that $r_1 = r_2 = S_A = 0$. For a simply

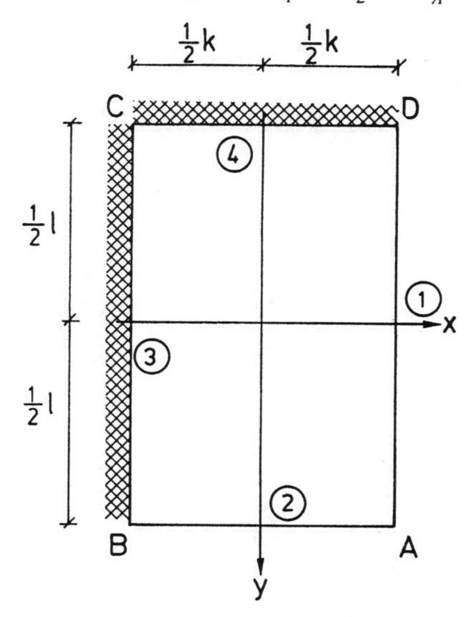

Figure 6.7.5 Slab supported on two adjacent edges.

supported slab, the corner reactions S_B and S_D are upward directed and S_C downward directed in the elastic state, and the free parameters should be chosen accordingly. Some more detailed guidelines are given in [81.1].

In the following we specialize to the case of an isotropic square slab with $m_p = m'_p$ (i.e., $k = l$). As remarked by Hillerborg [56.2], the load can be carried by torsional moments only. Setting $a = d = m_3 = m_4 = 0$, we get by (6.7.33), $i = \frac{1}{2}p$ (i.e., the moment field):

$$m_{xy} = \frac{1}{8}pk^2 - \frac{1}{8}pk^2\left(\frac{y}{\frac{1}{2}k}\right) - \frac{1}{8}pk^2\left(\frac{x}{\frac{1}{2}k}\right) + \frac{1}{8}pk^2\left(\frac{x}{\frac{1}{2}k}\right)\left(\frac{y}{\frac{1}{2}k}\right) \tag{6.7.34}$$

The distribution is sketched in Fig. 6.7.6.

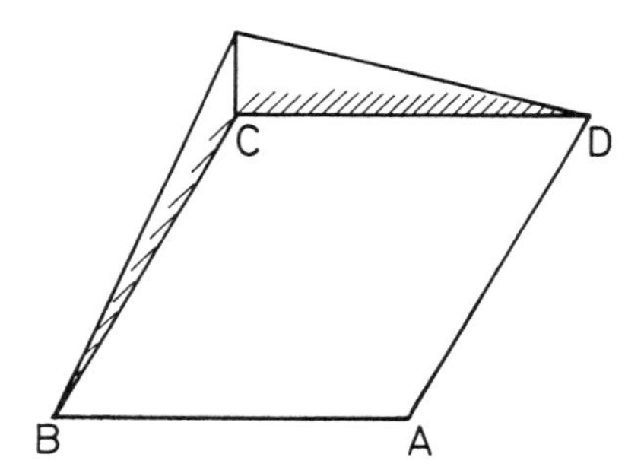

Figure 6.7.6 Torsional moments in a slab supported on two adjacent edges.

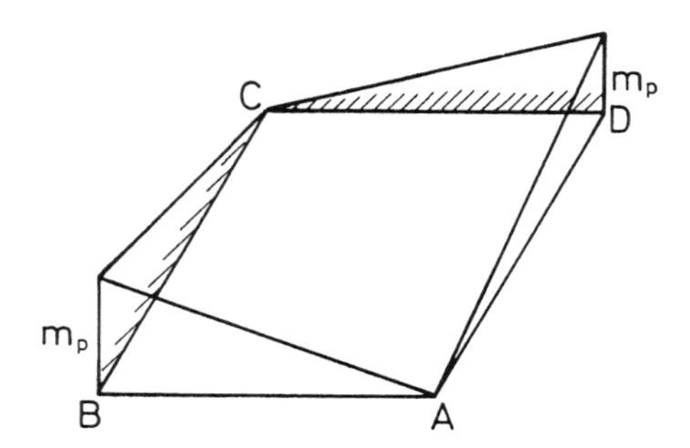

Figure 6.7.7 Torsional moments in a slab supported on two adjacent edges.

The maximum value is found in the corner C. Equating this value with m_p, we get the load-carrying capacity

$$p = \frac{2m_p}{k^2} \tag{6.7.35}$$

It is easy to improve the solution considerably. Assuming a distribution of the torsional moments as shown in Fig. 6.7.7, one finds [64.3] the load-carrying capacity:

$$p = \frac{4m_p}{k^2} \tag{6.7.36}$$

This moment field corresponds to the slab being carried at points B and D only.

It is possible to obtain an even better solution by considering a torsional moment distribution as shown in Fig. 6.7.8. It may be shown [64.3] that by optimizing the value of the constant z, one finds the load-carrying capacity

$$p = \frac{4.50m_p}{k^2} \tag{6.7.37}$$

The best known upper bound solution corresponds to the yield line pattern shown in Fig. 6.6.22, which gave

$$p = \frac{5.55}{a^2} m_p \tag{6.7.38}$$

corresponding to an approximately 20% higher load-carrying capacity than the best known lower bound solution.

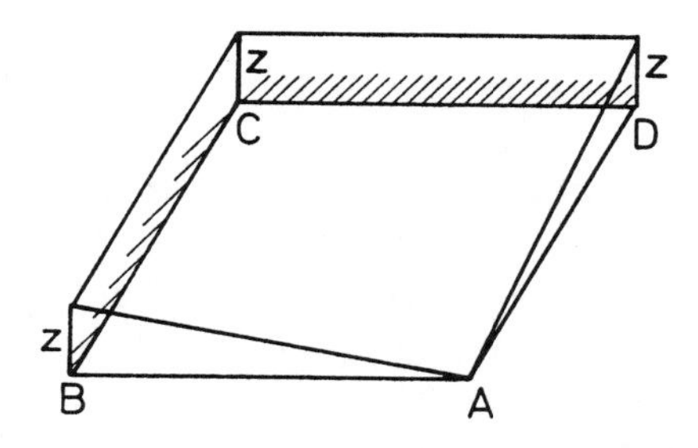

Figure 6.7.8 Torsional moments in a slab supported on two adjacent edges.

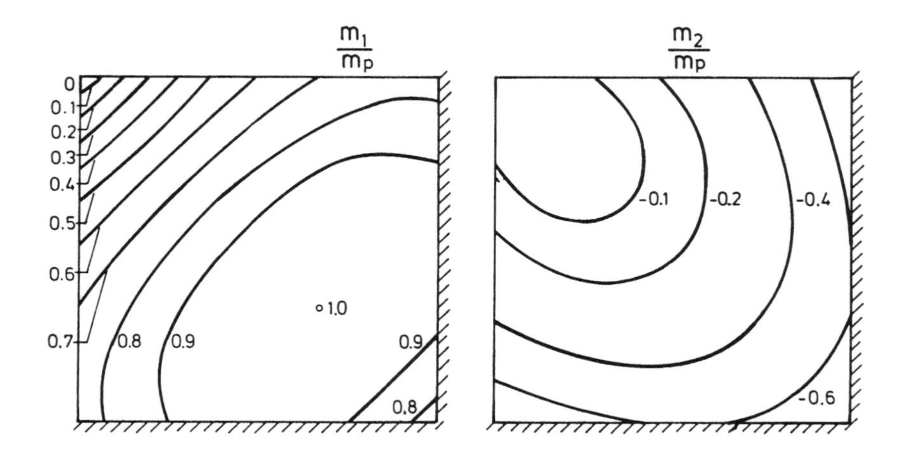

Figure 6.7.9 Lines of constant principal moments in a slab supported on two adjacent edges.

The extent of possible negative reinforcement may, as mentioned before, be difficult, if not impossible, to determine by upper bound methods. On the other hand, it is obtained automatically in lower bound calculations. Fig. 6.7.9 shows the distribution of the positive and the negative principal moments according to the solution (6.7.37) [64.3].

Hillerborg's solution (6.7.35) results in less than half the carrying capacity obtained from the best lower bound solution. If the slab is reinforced isotropically and homogeneously, the solution will be an expensive one. However, if the reinforcement is varied in accordance with the moment distribution, the conditions will naturally not be so bad.

A slab supported along one edge and on two columns. The uniformly loaded slab is now imagined to have one edge supported and two corners supported on columns (see Fig. 6.7.10). The conditions that have to be satisfied are (6.7.3) and (6.7.4) with $m_1 = 0$, (6.7.5) and (6.7.6) with $m_2 = 0$, (6.7.7) with $m_4 = 0$, (6.7.8) with $K_1 = 0$, (6.7.10) with $K_2 = 0$, and (6.7.11) with $K_4 = 0$. The torsional moment distribution is chosen to be symmetrical about the x-axis ($g =$

$h = 0$).

The moment field is found to be

$$
\begin{aligned}
m_x &= a + \frac{1}{2}m_3\left(\frac{x}{\frac{1}{2}k}\right) - \left[a + \frac{1}{2}m_3\right]\left(\frac{x}{\frac{1}{2}k}\right)^2 \\
m_y &= d\left[1 - \left(\frac{y}{\frac{1}{2}l}\right)^2\right] \\
m_{xy} &= -\left[\frac{1}{4}m_3\frac{l}{k} + a\frac{l}{k} - d\frac{k}{l}\right]\left(\frac{y}{\frac{1}{2}l}\right) - d\frac{k}{l}\left(\frac{x}{\frac{1}{2}k}\right)\left(\frac{y}{\frac{1}{2}l}\right)
\end{aligned}
\tag{6.7.39}
$$

The equilibrium equation (6.7.3) now gives

$$
\frac{4\left[a + \frac{1}{2}m_3\right]}{k^2} = \frac{1}{2}p \tag{6.7.40}
$$

This equation contains two parameters only, which means that in this equation only one variable is free. Note, however, that the parameter d can be chosen arbitrarily. It affects only the ratio between the moments m_y and m_{xy}, without affecting the load on the slab.

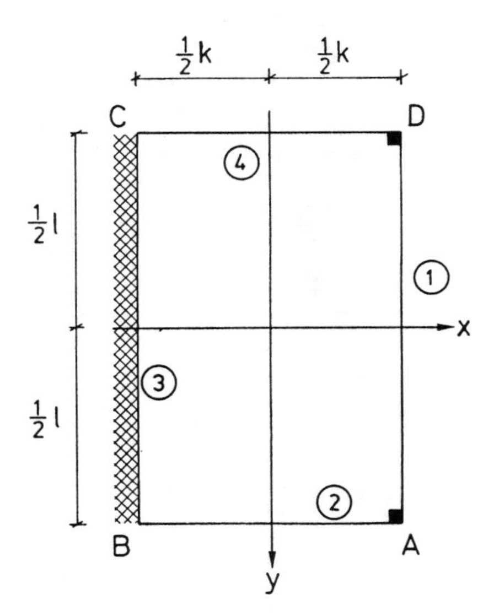

Figure 6.7.10 Slab supported along one edge and on two columns.

The reactions can be determined as before. In particular, the column reactions are found by means of (6.7.12) and (6.7.15). One gets

$$S_A = S_D = \frac{1}{4}pkl - \frac{1}{2}m_3\frac{l}{k} \tag{6.7.41}$$

having utilized (6.7.40). This result is also found immediately by means of a moment equation for the whole slab.

For a simply supported slab, the corner reactions S_B and S_C can have both signs in the elastic state, depending on the ratio l/k. The free parameters should be chosen accordingly. Some more detailed guidelines are given in [81.1].

A slab supported on two edges and on a column. A uniformly loaded rectangular slab is supported as shown in Fig. 6.7.11, on two adjacent edges and on a column. The conditions that have to be satisfied are in this case (6.7.3) and (6.7.4) with $m_1 = 0$, (6.7.5) and (6.7.6) with $m_2 = 0$, (6.7.7) and (6.7.8) with $K_1 = 0$, and (6.7.10) with $K_2 = 0$. The moment distribution becomes

$$\begin{aligned}
m_x &= a + \frac{1}{2}m_3\left(\frac{x}{\frac{1}{2}k}\right) - \left[a + \frac{1}{2}m_3\right]\left(\frac{x}{\frac{1}{2}k}\right)^2 \\
m_y &= d + \frac{1}{2}m_4\left(\frac{y}{\frac{1}{2}l}\right) - \left[d + \frac{1}{2}m_4\right]\left(\frac{y}{\frac{1}{2}l}\right)^2 \\
m_{xy} &= g - \left[\frac{1}{4}m_4\frac{k}{l} + d\frac{k}{l} + \frac{1}{4}ikl\right]\left(\frac{x}{\frac{1}{2}k}\right) \\
&\quad - \left[\frac{1}{4}m_3\frac{l}{k} + a\frac{l}{k} + \frac{1}{4}ikl\right]\left(\frac{y}{\frac{1}{2}l}\right) + \frac{1}{4}ikl\left(\frac{x}{\frac{1}{2}k}\right)\left(\frac{y}{\frac{1}{2}l}\right)
\end{aligned} \tag{6.7.42}$$

The equilibrium equation (6.7.3) assumes the form

$$\frac{4\left[a + \frac{1}{2}m_3\right]}{k^2} + \frac{4\left[d + \frac{1}{2}m_4\right]}{l^2} + i = \frac{1}{2}p \tag{6.7.43}$$

This equation contains five parameters, four of which can be chosen arbitrarily. In addition, the parameter g in (6.7.42) is free.

The reactions can be determined as in the previous examples. In particular, the column reaction S_A is found to be

$$S_A = \frac{1}{4}pkl - \frac{1}{2}m_4\frac{k}{l} - \frac{1}{2}m_3\frac{l}{k} - 2g \tag{6.7.44}$$

having utilized (6.7.43).

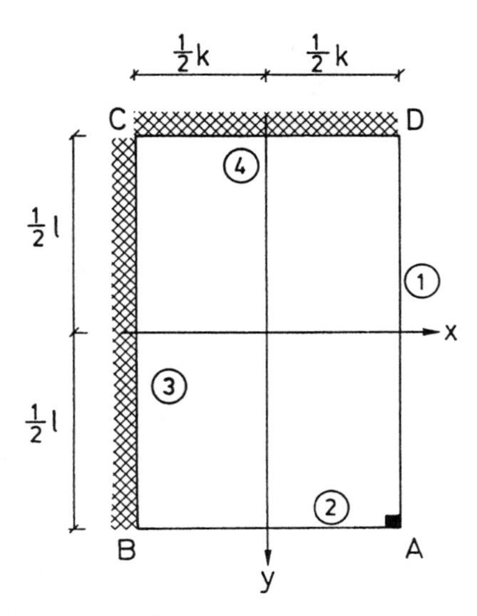

Figure 6.7.11 Slab supported on two edges and on a column.

For a simply supported slab, the corner reactions S_B and S_D are upward directed, S_C is downward directed in the elastic state. The free parameters should be chosen accordingly. Some guidelines are given in [81.1].

Other solutions. The examples above are not the only practical slab problems that can be solved by the moment distribution considered. First, line loads on the boundaries can be incorporated in the solutions. In Fig. 6.7.12, two more examples, which can be easily solved, are shown. The first case is a rectangular slab supported on four corner columns, and the second case is a rectangular slab supported on one edge and on one column. In [81.1] it is explained how even more complicated slabs with more complicated loadings can be treated along similar lines.

6.7.3 The Strip Method

Hillerborg has suggested a lower bound method, the strip method. The idea behind this method is that the slab is imagined to carry the loads as two sets of beams at right angles to each other. Indeed, if m_{xy} is put equal to zero in (6.7.1), the following result is obtained:

$$\frac{\partial^2 m_x}{\partial x^2} + \frac{\partial^2 m_y}{\partial y^2} = -p \tag{6.7.45}$$

which can be broken down into the following equations:

$$\frac{\partial^2 m_x}{\partial x^2} = -p_x, \quad \frac{\partial^2 m_y}{\partial y^2} = -p_y, \quad p_x + p_y = p \tag{6.7.46}$$

These equations correspond to the equilibrium condition in the beam theory. If (6.7.46) can be satisfied together with the statical boundary conditions, then (6.7.45) and the statical boundary conditions will also be satisfied. The division of the load into p_x and p_y need not be the same throughout the slab.

The idea of dividing the slab into two beam systems at right angles to each other is, of course, old. It has, in fact, often been utilized to determine approximate solutions for slabs. It is almost self-evident that such simplified assumptions regarding the behavior of a slab lead to a reasonable consumption of reinforcement only if the reinforcement is varied over the slab to correspond to the moments. On the other hand, if this is done consistently throughout the slab, the strip method appears to furnish an exact solution (cf. Fernando and Kemp [75.7]).

Hillerborg's method is not as general in its application as the yield line theory; in fact, it has to be altered and adjusted according to the various type of slabs. This is indeed what Hillerborg [59.1] himself did for, among other arrangements, slabs supported on piles. It is suggested that individual strips, which here are statically indeterminate, are calculated according to the theory of elasticity in the hope of achieving a proper behavior for the working load.

A generalized strip-deflection method has been proposed by Fernando and Kemp [78.22]. The idea is that the load is divided into parts p_x and p_y in such a way that the elastic deflections of the x and y strips are identical at any point.

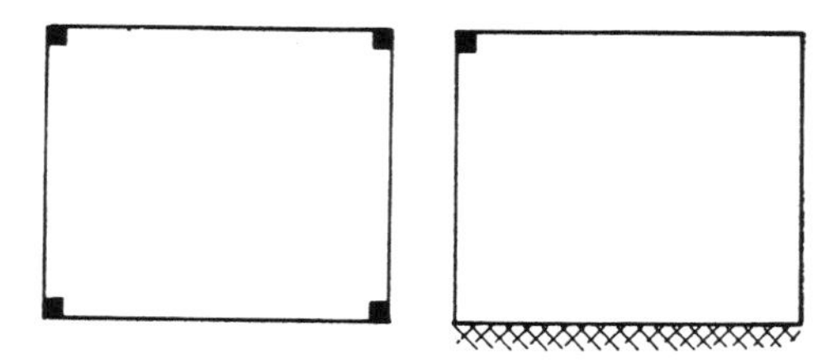

Figure 6.7.12 Examples of slab problems easily solved by the moment field (6.7.2).

Neglect of the twisting moment in design can also be exploited to reduce the statical indeterminacy of the slab by the introduction of a fourth equilibrium equation, involving moments of moments, corresponding to the local twisting of a slab element (cf. Gurley [79.6]).

The application of the theory will be illustrated by some examples. For further studies the reader is referred to the work of Hillerborg [56.2; 59.1; 74.1] (see also [68.1; 68.2]).

Square slab with uniform load. As an example, let us consider a simply supported square slab with uniformly distributed load p. If $p_x = p_y = \frac{1}{2}p$, m_x and m_y correspond to parabolic cylindrical surfaces, as illustrated in Fig. 6.7.13. The maximum values of m_x and m_y are $\frac{1}{8} \cdot \frac{1}{2} p \cdot a^2 = \frac{1}{16} pa^2$.

For a slab which is homogeneously reinforced, this lower bound solution will, however, be an expensive one, since the volume of reinforcement will be 50% higher than according to the upper bound solution of the yield line theory $m_p = \frac{1}{24} pa^2$. In the actual case the average moment in both directions is $\frac{2}{3} \cdot \frac{1}{16} pa^2 = \frac{1}{24} pa^2$, meaning that if it were possible to adjust the reinforcement fully to the moment curve the volume of reinforcement would be reduced to what corresponds to homogeneous reinforcement according to the upper bound solution.

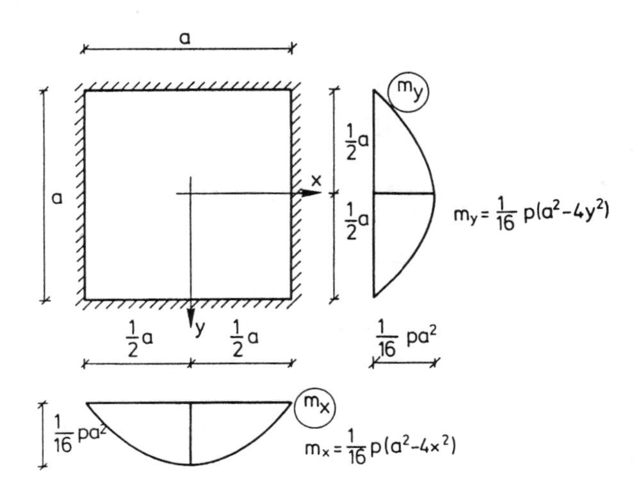

Figure 6.7.13 Strip solution for square slab with uniform load.

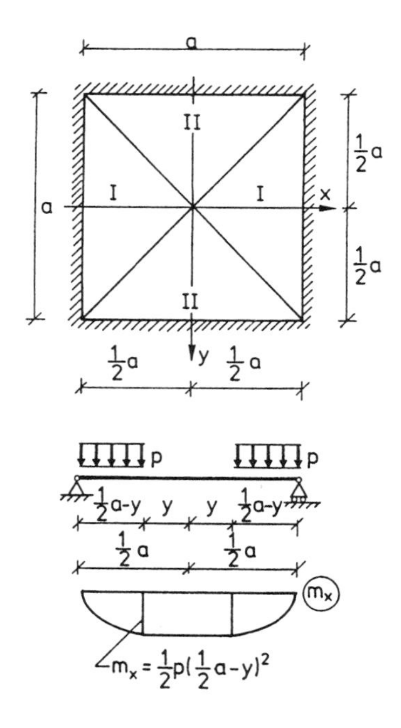

Figure 6.7.14 Strip solution for square slab with uniform load.

If, instead, we arrange the load so that with the notation of Fig. 6.7.14 we set

$$p_x = p, \quad p_y = 0 \quad \text{in the slab parts I}$$
$$p_y = p, \quad p_x = 0 \quad \text{in the slab parts II}$$

the strips in the x-direction have the moment distribution shown in the figure. The moment distribution in the y-direction will be analogous. In this case the maximum values of m_x and m_y would be $\frac{1}{8}pa^2$, twice as large as before.

In Fig. 6.7.15 a corresponding subdivision of a uniformly distributed load on a rectangular slab is shown. Arrows are used to indicate the direction in which the slab is considered load bearing. As seen, the middle part of the slab in this case is calculated as a simply supported beam with a span equal to the short side.

One-way slab with a hole. A slab is simply supported along two parallel edges with the distance 4.00 m. The slab is loaded with a uniform load $p = 4$ kN/m². In the slab a hole is left open with the position and dimensions as shown in Fig. 6.7.16. The necessary reinforcement around the hole is to be calculated by the strip method. The slab thickness has been chosen so that the slab can carry the bending moment 12.50 kNm/m at a maximum.

A statically admissible moment distribution is found by superimposing two loading cases. In the first case the slab parts I, II, I are assumed to carry as a beam in the x-direction. The slab part II is loaded with a downward-directed load $p = 4$ kN/m², and the slab parts I with a uniformly distributed upward-directed load p_r per unit area, that is in equilibrium with the load on slab part II. To determine p_r we have the equation (see Fig. 6.7.16)

$$p \cdot 2.00 = 2p_r a$$

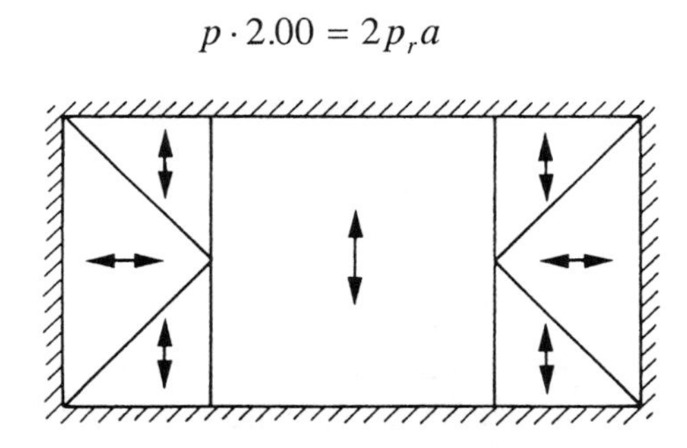

Figure 6.7.15 Subdivision of load in a rectangular slab.

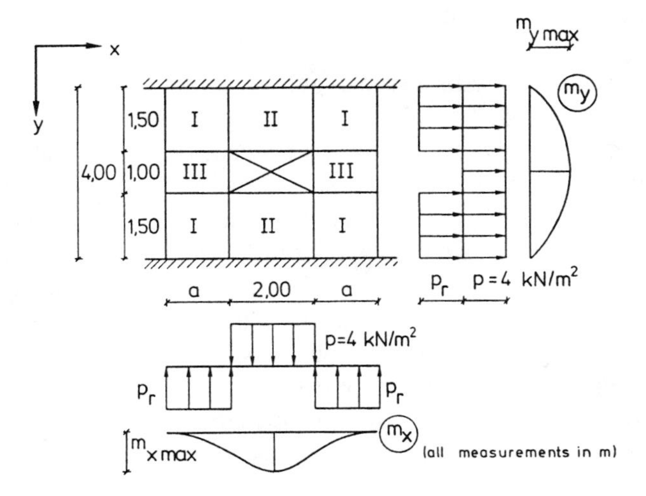

Figure 6.7.16 Strip solution for a slab with a hole.

In the second case slab parts I, III, I are assumed to carry as a beam in the y-direction. The load on slab parts I is a downward-directed load $p+p_r$ per unit area, the load on slab part III is a downward-directed load p per unit area.

The maximum moment in the first loading case is, inserting a in meters,

$$m_{x\max} = 4.00 \cdot 1.00\left(\tfrac{1}{2}a + 0.50\right) \qquad \text{(kNm / m)}$$

The moments m_x are positive all over. The maximum moment in the other loading case is

$$m_{y\max} = \frac{1}{8} \cdot 4.00 \cdot 4.00^2 + p_r \cdot 1.50 \cdot 0.75 = 8.00 + \frac{4.50}{a} \qquad \text{(kNm / m)}$$

The moments m_y are positive all over.

Choosing a so that $m_{y\max}$ is equal to the maximum moment 12.50 kNm/m, we get

$$a = 1.00 \text{ m}$$

and thus

$$m_{x\max} = 4.00 \cdot 1.00\left(\tfrac{1}{2} \cdot 1.00 + 0.50\right) = 4.00 \text{ kNm / m}$$

Consequently, the slab is to be reinforced around the hole so that a 1.00 m-wide strip in the y-direction is reinforced to carry 12.50 kNm/m.

Outside the reinforced zone the slab is to be reinforced for the moment $\frac{1}{8} \cdot 4.00 \cdot 4.00^2 = 8.00$ kNm / m in the y-direction. In practice, of course, a certain minimum reinforcement should always be placed perpendicular to the load-carrying direction.

Triangular slab with a free edge. A triangular slab is restrained along two edges and free along the third edge (see Fig. 6.7.17). The slab is loaded with a uniformly distributed load. The slab is subdivided by lines l_1 and l_2 into three slab parts I, II, and III. Slab parts I and III have load-carrying direction perpendicular to the restraint, that is, the individual strips carry as restrained beams. The load-carrying direction of slab part II is parallel to the free edge, that is, the individual strips carry as simply supported beams. The coordinate systems shown in Fig. 6.7.17 are inserted into slab parts II and III.

The shear forces v_x^{II} in slab part II along line l_2 are determined as the reaction for these strips. By transformation formulas for the shear forces, see (6.1.3), it is seen that the shear force v_x^{III} in the slab part III

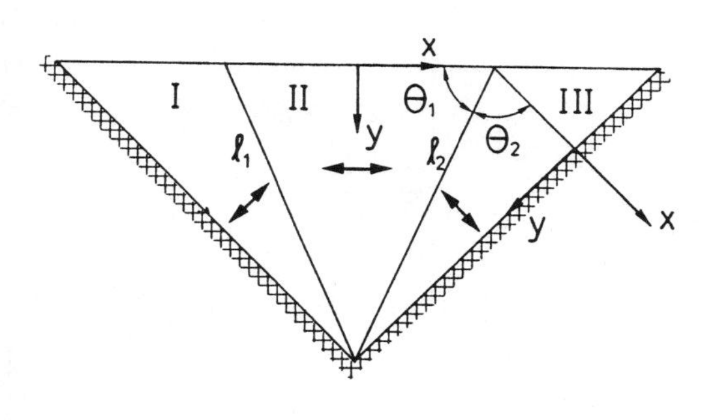

Figure 6.7.17 Triangular slab with a free edge.

along the line l_2 is determined by

$$v_x^{\text{III}} = v_x^{\text{II}} \frac{\sin\theta_1}{\sin\theta_2} \tag{6.7.47}$$

where the significance of the angles θ_1 and θ_2 is seen from Fig. 6.7.17. Then slab part III can be calculated as restrained beams loaded with the given load as well as with the reactions from slab part II.

The boundary conditions along the free edge in slab parts I and III are automatically satisfied when the strips in the slab parts that are limited by the free edge are calculated as restrained beams which are free at one end and subjected to a uniformly distributed load.

Angular slab. A slab is simply supported along edges *GA*, *BC*, and *ED* and is free along *AB, CD, EF,* and *FG*. The load is a uniformly distributed load 4.00 kN/m² (see Fig. 6.7.18).

Slab part *HBDE*, which is assumed to carry in the direction *BD*, is simply supported along *ED* and "supported" by a uniformly distributed reaction p_R per unit area in slab part *HBCF*. Besides, there is a uniform load per unit area over the entire slab part. Slab part *ABCG* is assumed to carry as a simply supported beam supported along *AG* and *BC*. The load is p per unit area over one half, and p_R per unit area over the other half.

Load-carrying direction *BD*: The "reaction" p_R is calculated by taking moments about *ED*

$$\tfrac{1}{2}\cdot 4.00\cdot 6.00^2 = p_R\cdot 3.00\cdot 4.50$$

$$p_R = \frac{\tfrac{1}{2}\cdot 4.00\cdot 6.00^2}{3.00\cdot 4.50} = 5.33 \text{ kN / m}^2$$

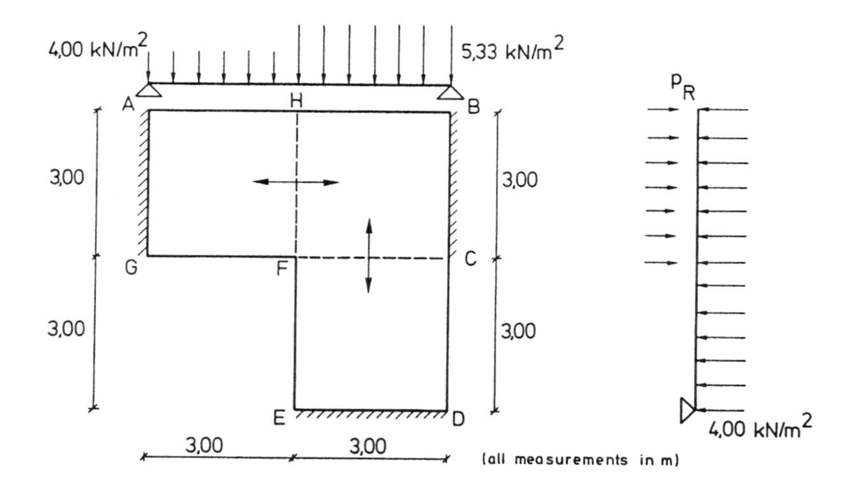

Figure 6.7.18 Strip solution for L-shaped slab with uniform load.

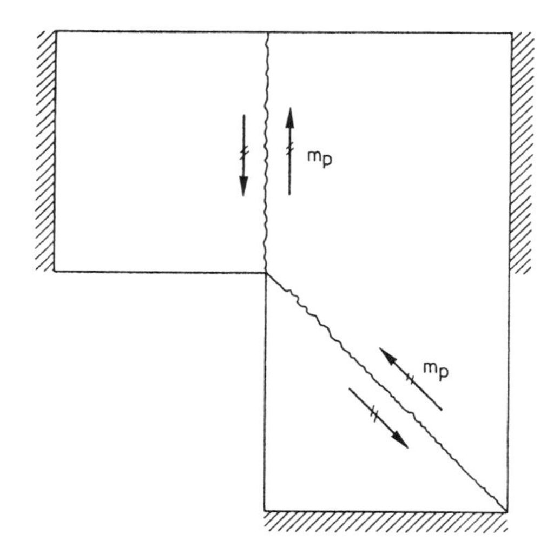

Figure 6.7.19 Simple yield line pattern for L-shaped slab.

The distance x from AB to the point with maximum moment is found to be $x = 4.00$ m, from which the maximum moment m_{max} is

$$m_{max} = 5.33 \cdot 3.00 \cdot 2.50 - \tfrac{1}{2} \cdot 4.00 \cdot 4.00^2 = 8.00 \text{ kNm / m}$$

Load-carrying direction *AB*: The reaction *r* per unit length along *BC* is

$$r = \frac{1}{2} \cdot 4.00 \cdot 6.00 + 1.33 \cdot 3.00 \cdot \frac{4.50}{6.00} = 15.00 \text{ kN / m}$$

The maximum moment is found in the distance x = 2.81 m from *BC*. The maximum moment is

$$m_{\text{max}} = 15.00 \cdot 2.81 - \tfrac{1}{2} \cdot 5.33 \cdot 2.81^2 = 21.11 \text{ kNm / m}$$

A rough comparison with the yield line theory is obtained by calculating the yield line pattern shown in Fig. 6.7.19, which gives m_p = 12.00 kNm/m. Comparing the volume of reinforcement under the assumption that the slab is isotropically and homogeneously reinforced according to the yield line theory and with contant reinforcement in the two load-carrying directions according to the solution of the strip method, we find in this case that the strip method solution is the cheapest. The saving amounts to about 20%. However, it has not been taken into consideration that, in practice, a minimum reinforcement should be placed perpendicular to the load-carrying directions in the strip solution.

Line load on a free edge. The slab shown in Fig. 6.7.20 is loaded on the edge with a line load $\overline{p} = 4.00$ kN / m. The dead load of the slab is disregarded. We want to limit the bending moments in the slab to ± 12.00 kNm/m.

The load on the strips in the direction of the *y*-axis is, besides the given load, assumed to be an upward-directed reaction *p* per unit area

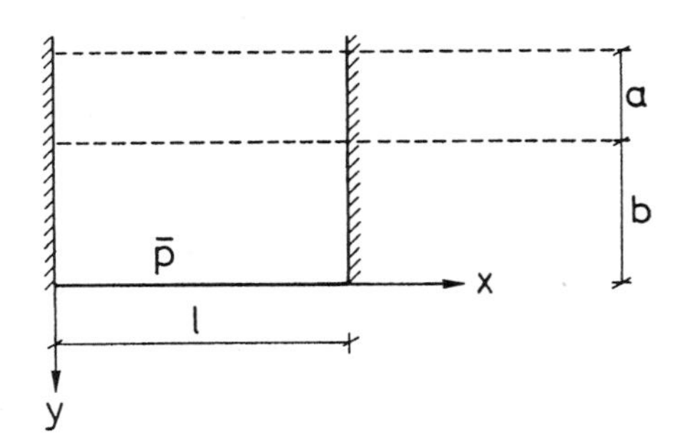

Figure 6.7.20 Line load on a free edge.

over a length b and a downward-directed reaction p per unit area over a length a (see Fig. 6.7.21).

The moment about A (see Fig. 6.7.21) requires $\frac{1}{2}pb^2 = pa\left(b + \frac{1}{2}a\right)$, giving

$$b = \left(1 + \sqrt{2}\right)a \tag{6.7.48}$$

Projection on the vertical

$$\bar{p} + pa = pb \Rightarrow p = \frac{\bar{p}}{\sqrt{2}a} \tag{6.7.49}$$

From this we obtain

$$\left|m_y\right|_{\max} = pa^2 = \frac{\sqrt{2}}{2}\bar{p}a \tag{6.7.50}$$

The strips in the direction of the x-axis have the moment curves shown in Fig. 6.7.22. It is seen that

$$\left|m_x\right|_{\max} = \frac{1}{8}pl^2 = \frac{\sqrt{2}}{16}\frac{\bar{p}}{a}l^2 \tag{6.7.51}$$

If it is required that

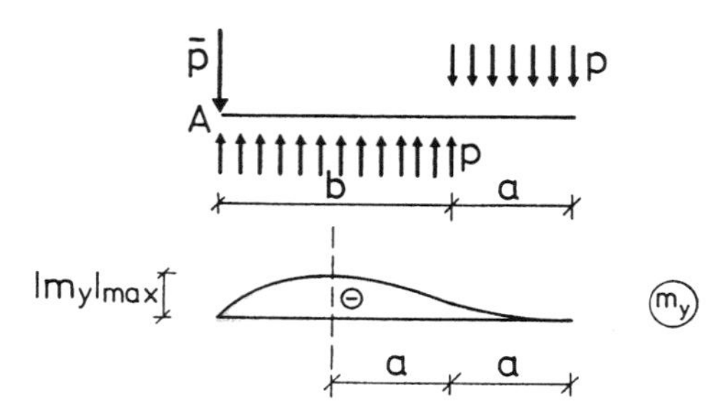

Figure 6.7.21 Strip solution for a line load on a free edge.

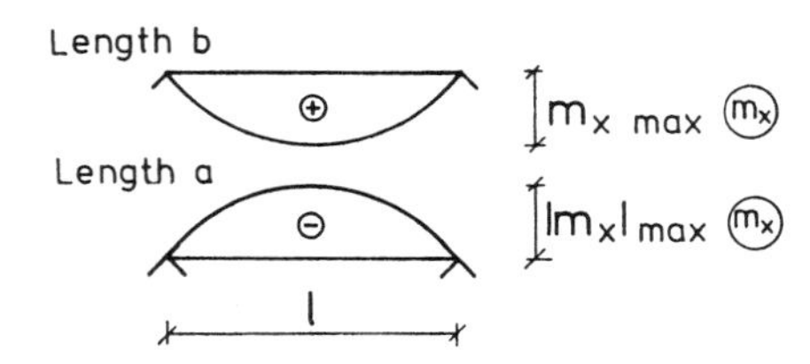

Figure 6.7.22 Moments m_x for a line load on a free edge.

$$\left|m_y\right|_{\max} = \left|m_x\right|_{\max} \tag{6.7.52}$$

we get

$$\frac{\sqrt{2}}{2}\bar{p}a = \frac{\sqrt{2}}{16}\frac{\bar{p}}{a}l^2 \tag{6.7.53}$$

that is,

$$a = \frac{\sqrt{2}}{4}l, \qquad b = \frac{\sqrt{2}}{4}\left(1+\sqrt{2}\right)l \tag{6.7.54}$$

from which we get

$$\left|m_x\right|_{\max} = \left|m_y\right|_{\max} = \tfrac{1}{4}\bar{p}l \tag{6.7.55}$$

If, instead, we want to limit $\left|m_x\right|_{\max}$ to m_{ux}, we get

$$a = \frac{\sqrt{2}\bar{p}l^2}{16m_{ux}} \tag{6.7.56}$$

giving

$$\left|m_y\right|_{\max} = \frac{(\bar{p}l)^2}{16m_{ux}} \tag{6.7.57}$$

If, instead, we finally want to limit $\left|m_y\right|_{\max}$ to m_{uy}, we get

$$a = \frac{\sqrt{2}m_{uy}}{\bar{p}} \tag{6.7.58}$$

giving

$$\left|m_x\right|_{\max} = \frac{(\bar{p}l)^2}{16m_{uy}} \tag{6.7.59}$$

In the example $\bar{p} = 4.00$ kN / m and assuming $l = 4.00$ m, we get

$$a = \frac{\sqrt{2}}{4}l = \frac{\sqrt{2}}{4}\cdot 4.00 = 1.41 \text{ m}$$

$$b = \left(1+\sqrt{2}\right)a = \left(1+\sqrt{2}\right)1.41 = 3.40 \text{ m}$$

$$\left|\mathrm{m_x}\right|_{\max} = \left|m_y\right|_{\max} = \tfrac{1}{4}\bar{p}l = \tfrac{1}{4}\cdot 4.00\cdot 4 = 4.00 \text{ kN} < 12.00 \text{ kN}$$

Setting $\left|m_x\right|_{\max} = m_{ux} = 12.00$ kN, we get

$$a = \frac{\sqrt{2}pl^2}{16m_{ux}} = \frac{\sqrt{2}\cdot 4.00\cdot 4^2}{16\cdot 12.00} = 0.47 \text{ m}$$

$$b = \left(1+\sqrt{2}\right)a = \left(1+\sqrt{2}\right)\cdot 0.47 = 1.13 \text{ m}$$

$$\left|\text{m}_{\text{y}}\right|_{max} = \frac{(\bar{p}l)^2}{16m_{ux}} = \frac{(4.00\cdot 4)^2}{16\cdot 12.00} = 1.33 \text{ kN} < 12.00 \text{ kN}$$

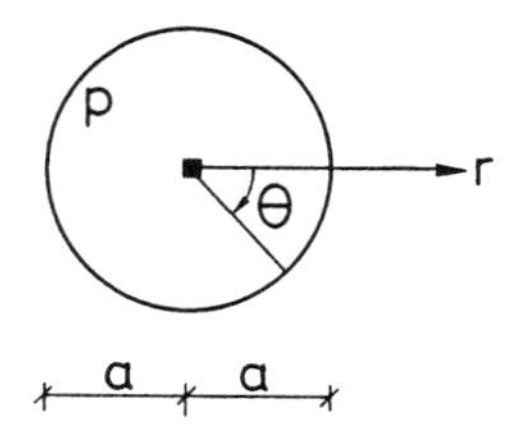

Figure 6.7.23 Circular slab supported on a column in the center.

Slabs supported on a column. To be able to treat concentrated forces in strip solutions, one has to find a moment field in a slab with uniformly distributed load and supported on a column.

Consider first a circular slab with radius a supported at the center by a column. For simplicity, the column is treated as a point support (see Fig. 6.7.23). Because of the rotational symmetry, $v_\theta = m_{r\theta} = 0$. v_r is statically determined and by a projection equation it is directly found as

$$rv_r = \tfrac{1}{2}pa^2 - \tfrac{1}{2}pr^2 \tag{6.7.60}$$

Setting $m_\theta = \text{const} = -m_p'$ in conformity with the solution of the yield line theory for an isotropically and homogeneously reinforced slab, we get for determination of m_r:

$$\frac{\partial rm_r}{\partial r} = \tfrac{1}{2}pa^2 - \tfrac{1}{2}pr^2 - m_p' \tag{6.7.61}$$

giving

$$m_r = \tfrac{1}{2}pa^2 - \tfrac{1}{6}pr^2 - m_p' + \frac{c}{r} \tag{6.7.62}$$

where c is a constant. If we want finite moments for $r = 0$, $c = 0$ is required. Setting $m_r = 0$ for $r = a$, we get

$$m'_p = \tfrac{1}{3} pa^2 \tag{6.7.63}$$

which is also found directly by a moment equation or by the work equation for the yield line pattern shown in Fig. 6.7.24. We then find that

$$m_r = \tfrac{1}{6} pa^2 \left[1 - \left(\frac{r}{a} \right)^2 \right] \tag{6.7.64}$$

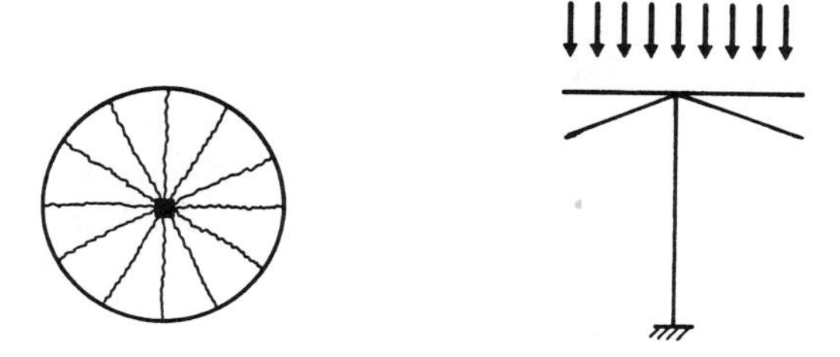

Figure 6.7.24 Yield line pattern for a circular slab supported on a column in the center.

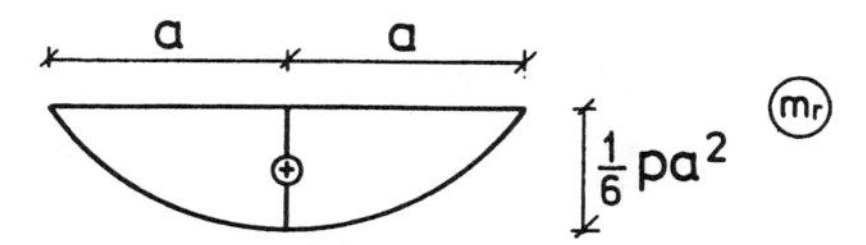

Figure 6.7.25 Bending moments m_r in a circular slab supported on a column in the center.

which has been drawn in Fig. 6.7.25. The column reaction is

$$S = \pi pa^2 \tag{6.7.65}$$

Note that

$$m_r(0) + m'_p = \frac{S}{2\pi} \tag{6.7.66}$$

corresponding to an upper bound solution with an infinitely small failure circle (see Section 6.6.7)

The solution requires reinforcement at the bottom side of the slab ($m_r > 0$). If the reinforcement is not supplied, a yield line pattern of the type shown in Fig. 6.7.26 would be more dangerous than that shown in Fig. 6.7.24.

By increasing the reinforcement in part of the top side, the bottom-side reinforcement may be avoided. It is left for the reader to verify that the following moment field satisfies the equilibrium equations:

$$\left.\begin{aligned} m_r &= -\tfrac{1}{6}pr^2 \\ m_\theta &= -\tfrac{1}{2}pa^2 \end{aligned}\right\} \quad r \le \tfrac{2}{3}a$$

$$\left.\begin{aligned} m_r &= \tfrac{1}{2}pa^2 - \tfrac{1}{6}pr^2 - \tfrac{1}{3}\frac{pa^3}{r} \\ m_\theta &= 0 \end{aligned}\right\} \quad r \ge \tfrac{2}{3}a \qquad (6.7.67)$$

In this solution m_θ is the negative or zero throughout and $|m_\theta|$ is the constant equal to $S/2\pi$ for $r \le \frac{2}{3}a$. The radial moment is negative throughout. The maximum value of $|m_r|$ is obtained for $r = \frac{2}{3}a$ and equals $\frac{2}{27}pa^2$, which is much less than $|m_\theta|$. According to this solution the slab may be reinforced isotropically in the top side for the moment $m_p' = S/2\pi$. No reinforcement is required in the bottom side.

Alternatively, radial and circumferential reinforcement may be used. When radial reinforcement is used, the necessary number of bars for any r-value may be determined by

$$\frac{N}{N_O} = \frac{rm_r}{(rm_r)_O} \qquad (6.7.68)$$

In this equation N_O is the number of bars required to carry the maximum value of $rm_r = (rm_r)_O$ and N the number of bars required for an arbitrary r-value. Of course absolute values are used in this equation. This implies that, when radial reinforcement is used, rm_r is governing the reinforcement and not m_r.

Considering instead the square slab (see Fig. 6.7.27) with uniformly distributed load p and supported in the middle on a column, it is natural to assume that if the slab is reinforced in the entire top side corresponding to the average moment in sections parallel to the sides through the column, the load can be carried. That is,

$$m_p' = \tfrac{1}{2}pb^2 = \frac{S}{8} \qquad (6.7.69)$$

where S is the column reaction $4pb^2$. Further, it is natural to assume that a bottom-side reinforcement is required at the column corresponding to

$$m_p + m_p' = \frac{S}{2\pi} \qquad (6.7.70)$$

that is,

$$m_p = \frac{S}{2\pi} - \frac{S}{8} = 0.136pb^2 = 0.034S \tag{6.7.71}$$

The moments can with approximation be assumed to be distributed according to a parabolic surface as for the circular slab. These rules are confirmed by a close investigation by Hillerborg (see [74.1]).

Correspondingly, for a rectangular slab the solution shown in Fig. 6.7.28 is found.[3] In the figure m'_{px} and m'_{py} are the negative yield moments in the coordinate directions, and m_{px} and m_{py} are the positive yield moments.

These moment distributions make it possible to calculate slabs with concentrated forces and column-supported slabs by the strip method. As an example, consider the slab in Fig. 6.7.29, which is assumed loaded with a uniformly distributed load p per unit area. First the slab is calculated as a one-way slab supported along AF, and by a "reaction" p_R uniformly distributed over slab part $BCDE$ in the center of which the column is placed. Then the moments are determined in conformity with the results above in slab part $BCDE$ loaded with the downward-directed uniformly distributed load p_R and supported by the column.

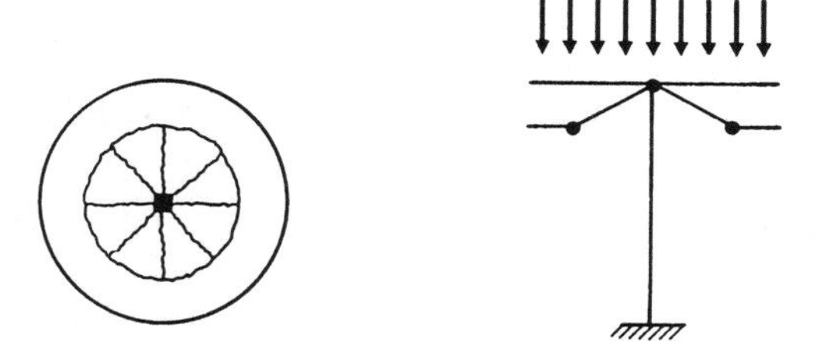

Figure 6.7.26 Yield line pattern for a slab with insufficient bottom reinforcement.

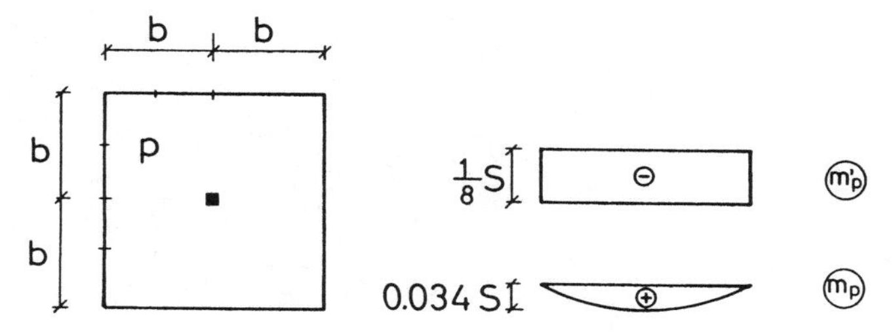

Figure 6.7.27 Square slab supported on a column in the center.

[3] The results are in agreement with the affinity theorem (cf. Section 6.8.1).

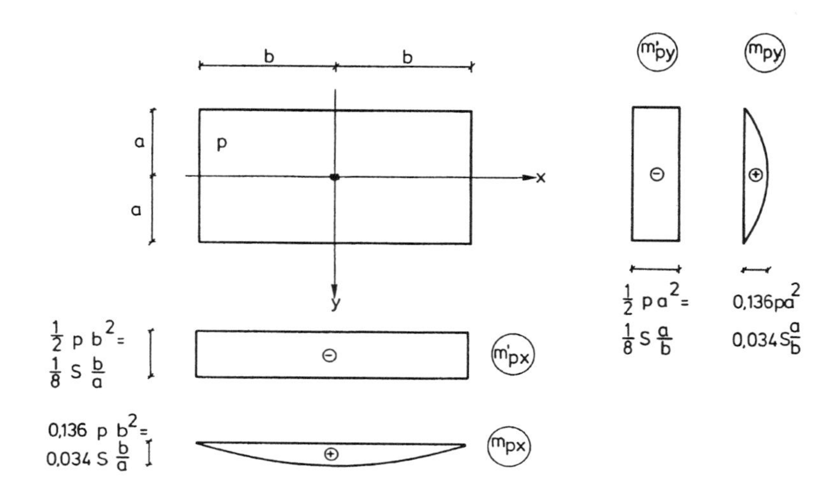

Figure 6.7.28 Rectangular slab supported on a column in the center.

Superimposing the moment distributions found gives the desired statically admissible moment distribution. Fig. 6.7.30 shows a couple of other examples where a similar calculation procedure can be used.

Concentrated force on simply supported slab. Let us apply the results from the preceding example to the slab in Fig. 6.7.31, which is simply supported on two opposite parallel edges and loaded with a concentrated force P (halfway) between the two sides. First the concentrated force P is supposed to be carried by a slab $a \cdot b$ loaded with

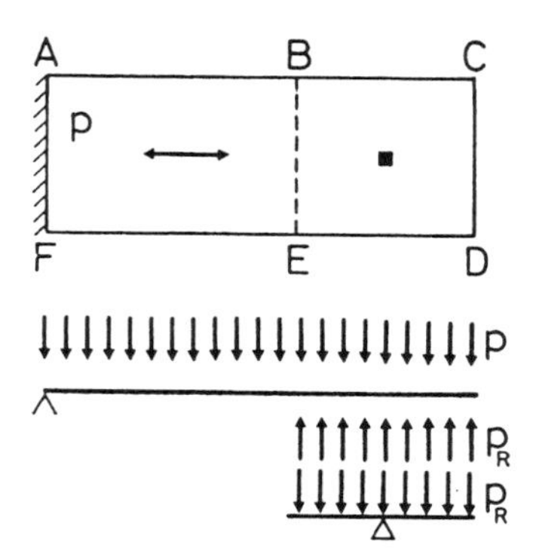

Figure 6.7.29 Example of strip solution for a slab supported on a column.

P in the center and "supported" by a uniformly distributed reaction P/ab. Then slab $a \cdot b$ is calculated as a simply supported slab with a uniform load P/ab with the span a.

The results are shown in Fig. 6.7.31. The bottom-side reinforcement in the y-direction will be designed for a maximum moment $Pa/4b$, and the bottom-side reinforcement in the x-direction will be designed for a moment $Pb/8a$. Note that reinforcement in the top side is also required corresponding to the moment $0.034Pa/b$ in the y-direction and $0.034Pb/a$ in the x-direction.

If, for example, we want $m_{x\,max} = \mu m_{y\,max}$, we get

$$\frac{1}{8}P\frac{b}{a} = \mu \cdot \frac{1}{4}P\frac{a}{b} \tag{6.7.72}$$

or

$$b = \sqrt{2\mu}a \tag{6.7.73}$$

Flat slab. Flat slab structures with columns in a rectangular network can be calculated in a simple way by the strip method when using the foregoing moment fields. Consider the example shown in Fig. 6.7.32, which is a square slab simply supported along the edges and simply supported in the center by a column. The load is a uniformly distributed load of 10 kN/m^2.

In solving the problem we need the moment distribution in a simply supported beam loaded as shown in Fig. 6.7.33. If, for example, we put $M = 0$ in the center, we have

$$q_R = \frac{\frac{1}{2}ql^2}{c\left(l - \frac{1}{2}c\right)} \tag{6.7.74}$$

$$M_{max} = \frac{1}{8}ql^2\left(\frac{l-c}{l-\frac{1}{2}c}\right)^2 \tag{6.7.75}$$

Half the load is assumed carried in the direction LM by a slab having simple supports along LI and MP and a "support" in the form of a uniformly distributed reaction in the part $ADHE$, which is then assumed to transfer the load to supports AE and DH and to slab part $BCGF$.

The other half of the load is assumed carried correspondingly in the other direction. The sizes of the different slab parts can be chosen freely.

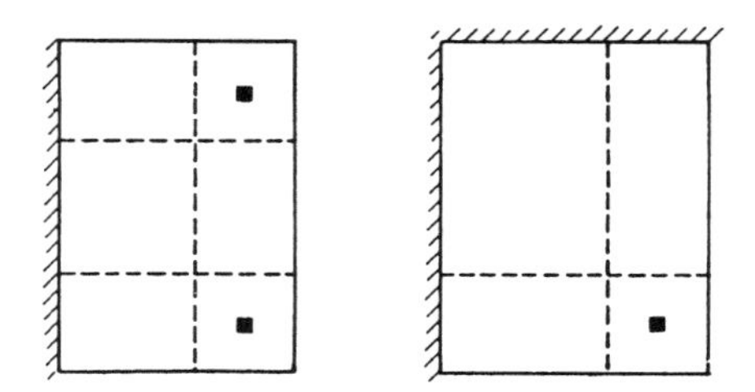

Figure 6.7.30 Example of strip solutions for slabs supported on columns.

For a strip in the direction *LM* we get from (6.7.74) and (6.7.75):

$$m_{\max} = \tfrac{1}{8} \cdot 5.00 \cdot 4^2 \left(\frac{2.5}{3.25} \right)^2 = 5.92 \text{kNm / m}$$

$$q_{R1} = \frac{\tfrac{1}{2} \cdot 5.00 \cdot 4^2}{1.5 \cdot 3.25} = 8.20 \text{ kN / m}^2$$

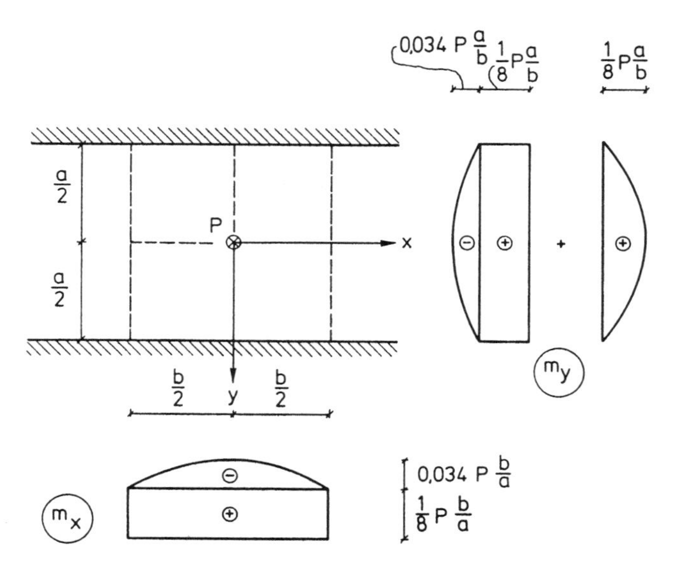

Figure 6.7.31 Strip solution for a concentrated force on a simply supported slab.

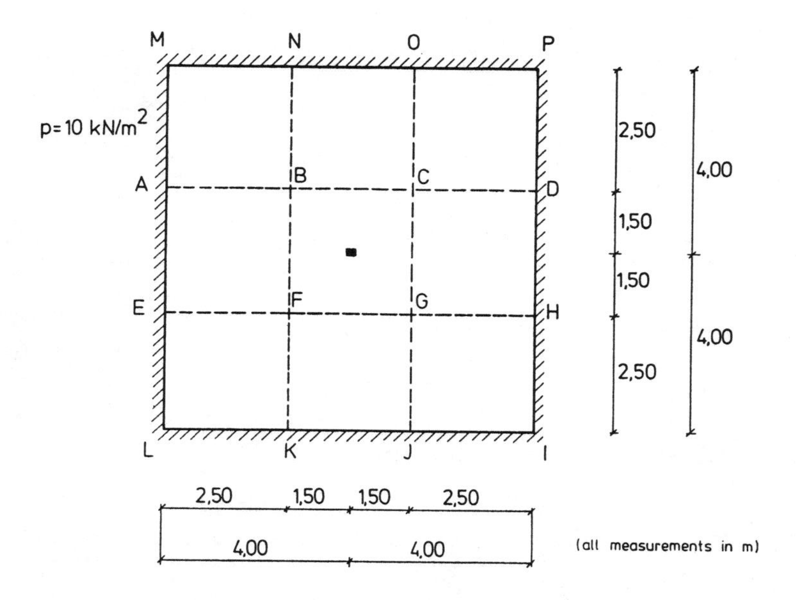

Figure 6.7.32 Square slab with uniform load simply supported along the edge and on a column in the center.

For strip *ADHE:*

$$m_{\max} = \tfrac{1}{8} \cdot 8.20 \cdot 4^2 \cdot \left(\frac{2.5}{3.25}\right)^2 = 9.70 \text{ kNm / m}$$

$$q_{R2} = \frac{\tfrac{1}{2} \cdot 8.20 \cdot 4^2}{1.5 \cdot 3.25} = 13.46 \text{ kN / m}^2$$

The column reaction

$$S = 2 \cdot 13.46 \cdot 3^2 = 242.30 \text{ kN}$$

Slab part *BCGF* will be designed for the moments

$$m'_p = \tfrac{1}{8} S = \tfrac{1}{8} \cdot 242.30 = 30.29 \text{ kNm / m}$$

$$m_p = 0.034 S = 0.034 \cdot 242.30 = 8.24 \text{ kNm / m}$$

The moments for which the whole slab will be designed are shown in Fig. 6.7.34 for one direction. In the direction perpendicular to this the same reinforcement applies.

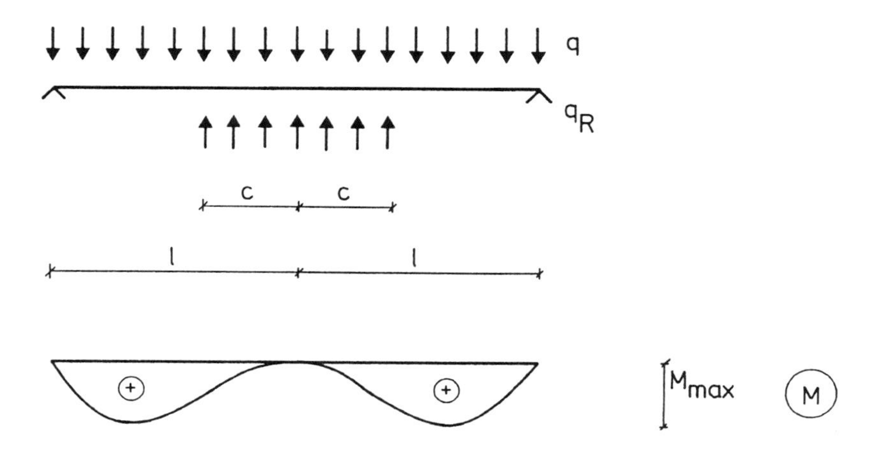

Figure 6.7.33 Bending moments in strips.

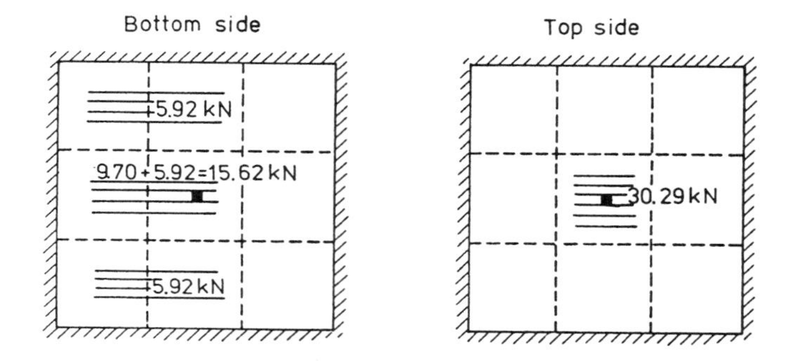

Figure 6.7.34 Reinforcement layout for a square slab with uniform load simply supported along the edge and on a column in the center.

The Johansen solution for this case (see [49. 3; 72. 3]) gives in this case 18% smaller reinforcement volume than according to the strip method, provided that the reinforcement according to the strip method is not cut off in correspondence with the moment distribution. It is noted especially that the Johansen solution only gives about half as much reinforcement in the top side above the column, and that the extension of the area with reinforcement in the top side is a little less.

The area around the column, of course, has to be checked for punching shear.

6.7.4 Some Remarks Concerning the Reinforcement Design

Having found a statically admissible moment distribution it is only a matter of simple calculations to determine the necessary amount of reinforcement. Formulas giving a minimum amount of reinforcement at each point can be found in Section 2.4.4. Since the reinforcement is often kept constant throughout parts of the slab, it may be convenient to work with (2.4.77).

Slab solutions, including those given here, generally have torsional moments along the boundaries in agreement with the Kirchhoff boundary conditions. It is important to realize that in such cases it may be necessary to reinforce not only the top and bottom faces of the slab, but also the boundary itself.

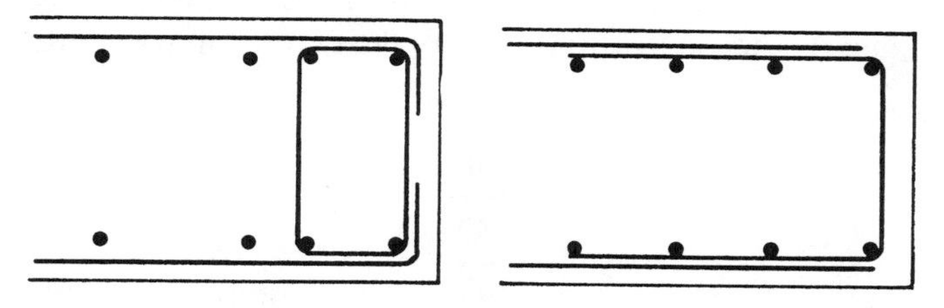

Figure 6.7.35 Reinforcement along an edge with torsional moments.

The boundary can, as shown in Section 6.3.2, be treated as a beam that has to carry a shear force at each point of the boundary equal to the torsional moment at the point (the Kirchhoff force) which corresponds to a reinforcement as shown to the left in Fig. 6.7.35. However, it is to be expected that well-anchored stirrups as shown to the right in Fig. 6.7.35 would also serve the purpose, since this arrangement corresponds to the ordinary torsional reinforcement layout in beams. This problem is treated in more detail in Chapter 7.

Along simply supported and fixed edges the above-mentioned boundary reinforcement is generally not necessary, because the torsional moments can be imagined to decrease to zero along some narrow boundary strips without significantly influencing the load (see [81. 1]). This transformation of the moment field also greatly affects the amount of reinforcement for fixed slabs since, after the transformation, maximum values of the bending moments and of the torsional moments no longer coincide along the boundaries.

6.8 ORTHOTROPIC SLABS

The yield conditions and the flow rule in the general case of an orthotropic slab have been established in Section 2.3. In the following it is shown how to obtain upper and lower bound solutions on the basis of these yield conditions. In particular, it is shown how to transform solutions for isotropic slabs to orthotropic slabs.

6.8.1 The Affinity Theorem

It has been proved that if the moment field $\overline{m} = (m_x, m_y, m_{xy}) = (f, g, h)$ is safe in the isotropic case, the moment field $\overline{m}^* = (f, \mu g, \sqrt{\mu} h)$ is safe in what was termed the simple affine case (see Section 6.4.1). It is assumed that f, g and h are functions of the Cartesian coordinates x and y, and that these functions satisfy the equilibrium condition, that is,

$$\frac{\partial^2 f}{\partial x^2} - 2\frac{\partial^2 h}{\partial x\,\partial y} + \frac{\partial^2 g}{\partial y^2} = -p \tag{6.8.1}$$

It is further assumed that the actual statical boundary conditions are fullfilled.

Now, on the basis of the given slab, let us form a slab that is affine to this, with the affinity

$$x^* = x, \qquad y^* = \sqrt{\mu} y \tag{6.8.2}$$

where x^* and y^* are also Cartesian coordinates. Consider the following moment field in the affine slab:

$$\begin{aligned} m_{x^*} &= f(x,y) = f\left(x^*, \frac{y^*}{\sqrt{\mu}}\right) \\ m_{y^*} &= \mu g(x,y) = \mu g\left(x^*, \frac{y^*}{\sqrt{\mu}}\right) \\ m_{x^* y^*} &= \sqrt{\mu} h(x,y) = \sqrt{\mu} h\left(x^*, \frac{y^*}{\sqrt{\mu}}\right) \end{aligned} \tag{6.8.3}$$

Then

$$
\begin{aligned}
&\frac{\partial^2 m_{x^*}}{\partial x^{*2}} = \frac{\partial^2 f}{\partial x^2} \\
&\frac{\partial m_{y^*}}{\partial y^*} = \frac{\partial m_{y^*}}{\partial y}\frac{\partial y}{\partial y^*} = \mu\frac{\partial g}{\partial y}\frac{1}{\sqrt{\mu}} = \sqrt{\mu}\frac{\partial g}{\partial y} \\
&\frac{\partial^2 m_{y^*}}{\partial y^{*2}} = \frac{\partial}{\partial y}\left(\sqrt{\mu}\frac{\partial g}{\partial y}\right)\frac{\partial y}{\partial y^*} = \frac{\partial^2 g}{\partial y^2} \\
&\frac{\partial^2 m_{x^*y^*}}{\partial x^*\partial y^*} = \frac{\partial^2 h}{\partial x\,\partial y}
\end{aligned}
\tag{6.8.4}
$$

which means that the equilibrium condition is also fulfilled for the affine slab if we make $p^*=p$. The shear forces in the affine slab become

$$
\begin{aligned}
v_{x^*} &= \frac{\partial m_{x^*}}{\partial x^*} - \frac{\partial m_{x^*y^*}}{\partial y^*} = \frac{\partial f}{\partial x} - \frac{\partial h}{\partial y} = v_x \\
v_{y^*} &= \frac{\partial m_{y^*}}{\partial y^*} - \frac{\partial m_{x^*y^*}}{\partial x^*} = \sqrt{\mu} v_y
\end{aligned}
\tag{6.8.5}
$$

A line element that forms the angle α with the x-axis in the original slab will, in the affine slab, form the angle α^* with the x^*-axis, determined by $\tan\alpha^* = \sqrt{\mu}\tan\alpha$. The shear force v_{n^*} in a section that forms the angle α^* with the x^*-axis is thereby found to be $(-\pi/2 \le \alpha \le \pi/2)$:

$$
\begin{aligned}
v_{n^*} &= v_{y^*}\cos\alpha^* - v_{x^*}\sin\alpha^* \\
&= \sqrt{\mu} v_y \frac{1}{\sqrt{1+\mu\tan^2\alpha}} - v_x \frac{\sqrt{\mu}\tan\alpha}{\sqrt{1+\mu\tan^2\alpha}} \\
&= (v_y\cos\alpha - v_x\sin\alpha)\chi = \chi v_n
\end{aligned}
\tag{6.8.6}
$$

where

$$
\chi = \frac{\sqrt{\mu}}{\sqrt{\cos^2\alpha + \mu\sin^2\alpha}} = \sqrt{\mu\cos^2\alpha^* + \sin^2\alpha^*}
\tag{6.8.7}
$$

and where v_n is the shear force in the corresponding section of the original slab. A line load must therefore be multiplied by the same quantity χ in order to satisfy the equilibrium conditions. Finally, if we consider a concentrated force as a load uniformly distributed over a small area, it will immediately be seen that the force must be multiplied by $\sqrt{\mu}$ at the transition to the affine slab.

It can also be verified that a straight line with a bending moment equal to one of the yield moments m_p or m'_p and the torsional moment equal to zero will, by the affinity, be transformed into a straight line with a bending moment equal to that determined by (6.4.11) or (6.4.13), and a torsional moment determined by (6.4.15) or (6.4.16), depending on whether the yield line is positive or negative. In brief, a yield line is transformed into a yield line.

Correspondingly, it can be shown that if the conditions of compatibility (6.2.10) and (6.2.11) are satisfied for an isotropic slab with $\overline{\kappa} = (\kappa_x, \kappa_y, 2\kappa_{xy}) = (\alpha, \beta, \gamma)$, then the conditions of compatibility are also satisfied for the affine slab with $\overline{\kappa}^* = (\kappa_{x^*}, \kappa_{y^*}, 2\kappa_{x^*y^*}) = \left(\alpha, (1/\mu)\beta, (1/\sqrt{\mu})\gamma\right)$, which, as shown in Section 6.4.1, also represents a normal in the point of the affine yield surface that is, by the affinity, transferred from the point with the normal (α, β, γ). It will be seen that these curvatures and this torsion arise when $w^* = w$, that is, that the deflection at corresponding points is the same. The same result naturally also applies if the surface of deflection contains lines of discontinuity. The foregoing considerations lead to the conclusion that lower bound solutions and exact solutions for isotropic slabs can be transferred to corresponding solutions for orthotropic slabs provided that the laws of transformation mentioned are obeyed.

It will now be shown that this affinity theorem is also valid for upper bound solutions. Formulation of the work equation for isotropic slabs gives

$$\iint_A (m_x\kappa_x + m_y\kappa_y + 2m_{xy}\kappa_{xy})\,dx\,dy + \int_S |m_n|\,|\theta_n|\,ds$$
$$= \iint_A pw\,dx\,dy \qquad (6.8.8)$$

where m_n is equal to m_p along the positive yield lines, and equal to $-m'_p$ along negative yield lines. θ_n is obtained from (6.5.23), that is,

$$|\theta_n| = \frac{|\theta_x|}{|\sin\beta|} \qquad (\beta \neq 0) \tag{6.8.9}$$

where β is the angle between the yield line and the x-axis. Correspondingly for the affine slab

$$\begin{aligned} &\iint_{A^*} (m_{x^*}\kappa_{x^*} + m_{y^*}\kappa_{y^*} + 2m_{x^*y^*}\kappa_{x^*y^*})\, dx^*\, dy^* \\ &+ \int_{S^*} |m_{n^*}||\theta_{n^*}|\, ds^* \\ &= \iint_{A^*} p^* w^* dx^* dy^* \end{aligned} \tag{6.8.10}$$

If the foregoing relationship between the moment and curvature fields in a yield zone in the isotropic slab and those in the affine slab is now introduced, it will be seen that the integrants in the first term of (6.8.8) and (6.8.10) are equal at corresponding points. Furthermore, taking, for example, a positive yield line that forms the angle β^* with the x^*-axis in the affine slab we get

$$m_{n^*} = m_p\left(\sin^2\beta^* + \mu\cos^2\beta^*\right) = m_p \frac{\mu}{\cos^2\beta + \mu\sin^2\beta} \tag{6.8.11}$$

and further,

$$\begin{aligned} |\theta_{n^*}| &= \frac{|\theta_{x^*}|}{|\sin\beta^*|} = \frac{|\theta_x|\sqrt{1+\mu\tan^2\beta}}{\sqrt{\mu}\,|\tan\beta|} \\ &= |\theta_n|\frac{\sqrt{\cos^2\beta + \mu\sin^2\beta}}{\sqrt{\mu}} \end{aligned} \tag{6.8.12}$$

$$\begin{aligned} ds^* &= \frac{|dx^*|}{|\cos\beta^*|} = |dx|\sqrt{1+\mu\tan^2\beta} \\ &= ds\sqrt{\cos^2\beta + \mu\sin^2\beta} \end{aligned} \tag{6.8.13}$$

It is assumed in these calculations that β^* is different from $\pi/2$ and 0, but it can easily be proved that the results are also valid for these values of β^*. From this it will be seen that all terms in the work equation for the affine slab are $\sqrt{\mu}$ times the corresponding terms in the work equation for the isotropic slab when $p^* = p$.

It has been assumed here that the boundary terms in the work equation are zero, and that the slab is not subjected to concentrated forces or line loads. However, it can be shown that the laws of transformation remain as found above.

Thus, the following important affinity theorem is obtained:

An upper bound solution, a lower bound solution, or an exact solution for an isotropic slab is valid for an affine slab provided that the following laws of transformation are observed:

$$
\begin{aligned}
x^* &= x, \qquad y^* = \sqrt{\mu}\, y \\
p^* &= p \\
P^* &= \sqrt{\mu}\, P \\
\bar{p}^* &= \chi \bar{p} \\
w^* &= w \\
m_{x^*} &= m_x, \qquad m_{y^*} = \mu m_y, \qquad m_{x^* y^*} = \sqrt{\mu}\, m_{xy} \\
v_{n^*} &= \chi v_n
\end{aligned}
\tag{6.8.14}
$$

where χ is determined from (6.8.7). All quantities belonging to the affine slab are indicated with an asterisk.

The affinity theorem is illustrated in Fig 6.8.1. This theorem can also be applied the opposite way, for tranformation of solutions applying to orthotropic slabs to solutions for isotropic slabs. The affinity theorem was established by Johansen [43.1;62.1] for upper bound solutions. The extension to exact solutions and lower bound solutions was given by the author [64.3].

As an example of the application of the affinity theorem, consider an orthotropic rectangular slab (see Fig. 6.8.2) which is simply supported and subjected to a uniformly distributed load. It is assumed that we are dealing with the simple affine case, so that the affinity theorem can be applied. A lower bound solution for this slab also applies to an isotropic slab with $b = b^*$ and $a = a^*/\sqrt{\mu}$. If m_p is the positive yield moment in the isotropic slab, then according to (6.7.22), the following lower bound solution is obtained for this slab:

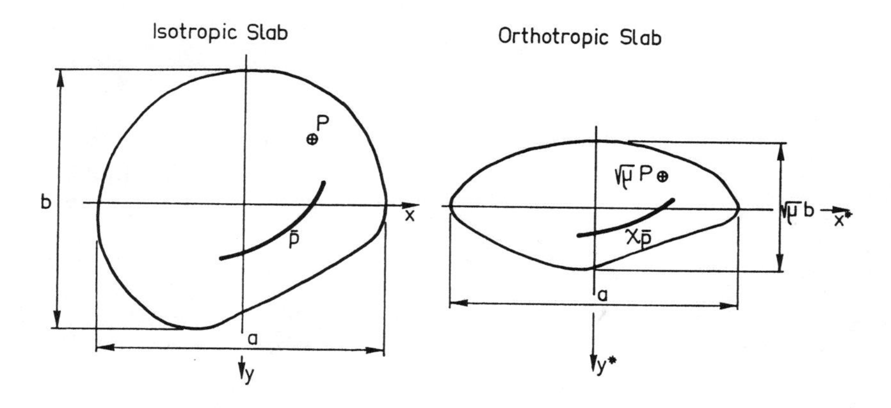

Figure 6.8.1 Affine slabs.

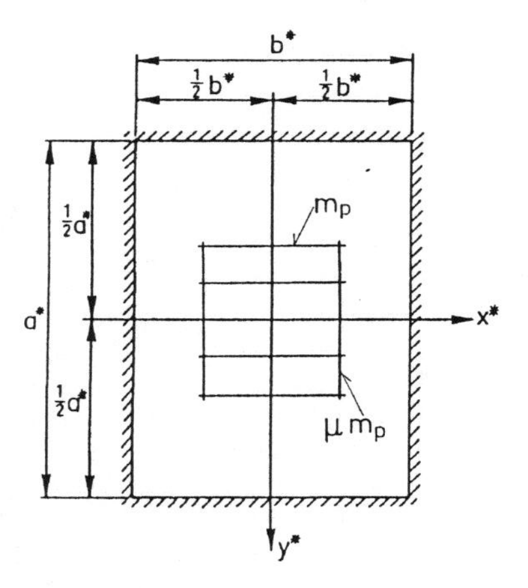

Figure 6.8.2 Rectangular orthotropic slab with uniform load.

$$p = \frac{8m_p\sqrt{\mu}}{a^*b^*}\left(1 + \frac{a^*}{\sqrt{\mu}b^*} + \frac{b^*\sqrt{\mu}}{a^*}\right) \tag{6.8.15}$$

which is therefore a lower bound solution for the given slab. The corresponding moment field in the orthotropic slab is found by means of (6.7.21), together with the affinity theorem, to be

$$m_{x^*} = m_p - \frac{4m_p}{b^{*2}} x^{*2}$$

$$m_{y^*} = \mu\left(m_p - \frac{4m_p}{a^{*2}} y^{*2} \right) \tag{6.8.16}$$

$$m_{x^*y^*} = \frac{4\sqrt{\mu}\, m_p}{a^* b^*} x^* y^*$$

An upper bound solution may immediately be found using the solution in Section 6.6.7. In this case two formulas must be written, one for $a^* \leq b^*/\sqrt{\mu}$ and one for $a^* > b^*/\sqrt{\mu}$. It is left for the reader to write down the expressions.

The minimum value of the reinforcement for a slab with constant thickness was determined by Johansen [49. 3]. The total reinforcement is proportional to $m_p(1+\mu)$. The minimum is found for

$$\mu = \frac{1}{3\left(\dfrac{a^*}{b^*}\right)^2 - 2} \qquad a^* \geq b^* \tag{6.8.17}$$

and for this value of μ

$$m_p = \frac{1}{8} p b^{*2} \left(1 - \frac{2}{3}\left(\frac{b^*}{a^*}\right)^2 \right) \qquad a^* \geq b^* \tag{6.8.18}$$

As usual the minimum is very flat, so this solution is seldom used. In practice μ is chosen by other criteria, for instance, minimum reinforcement requirements regarding cracking behavior for the working load.

6.8.2 Upper Bound Solutions

Upper bound solutions can be found in the same manner as for isotropic slabs by means of simple yield line patterns consisting of straight lines. The bending moment in a positive yield line is calculated from (6.4.11) and the torsional moment from (6.4.15) (see Fig 6.8.3). For a negative yield line, (6.4.13) and (6.4.16) apply. The work equation can be formulated in exactly the same way as in the case of isotropic slabs.

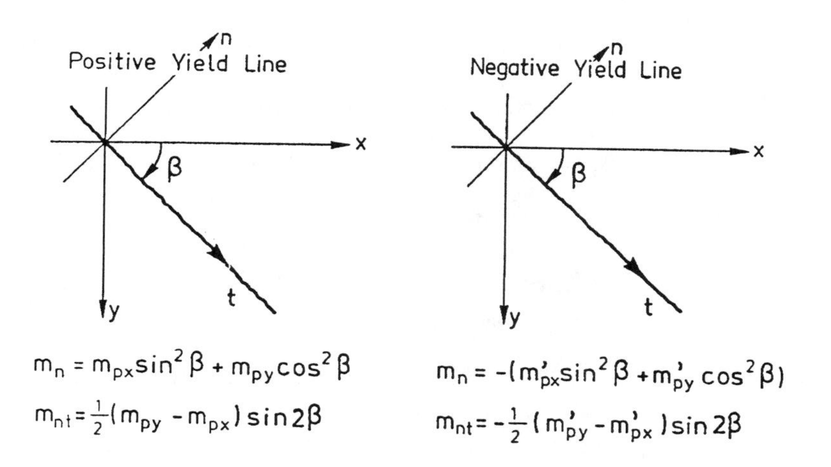

Figure 6.8.3 Moments in yield lines in orthotropic slabs.

A stationary moment field is defined here as a moment field that satisfies the equilibrium conditions and, along the yield lines, fulfills the conditions that the bending and torsional moments have the value determined by the angle formed by the yield lines and the directions of the reinforcement and that the shear forces are equal to zero. Then the equilibrium method can be transferred without difficulty to orthotropic slabs, the theorem in Section 6.6.2 being valid as it stands. The fact that it is possible to obtain stationary values for the loading in the same way as for isotropic slabs can be seen by formulating the equations corresponding to (6.6.6) and (6.6.7). Equation (6.6.6) can be transferred directly:

$$\lambda \iint pw\,dx\,dy = \int m_b \theta\,ds \qquad (6.8.19)$$

where $m_b = m_n$ is calculated as described in Fig 6.8.3. If the yield line pattern is now altered slightly, the following is obtained, corresponding to (6.6.7):

$$\lambda \iint pw'\,dx\,dy = \int (m_b + \delta m_b)\theta'\,ds' \qquad (6.8.20)$$

Considering as an example a positive yield line, we have

$$m_b = m_{px}\sin^2\beta + m_{py}\cos^2\beta \qquad (6.8.21)$$

$$m_v = -\frac{1}{2}\left(m_{px} - m_{py}\right)\sin 2\beta \tag{6.8.22}$$

If β is now altered by a small amount $\delta\beta$, (6.6.2) yields

$$\delta m_b = -2m_v\delta\beta = \left(m_{px} - m_{py}\right)\sin 2\beta\ \delta\beta \tag{6.8.23}$$

It is found from (6.8.21) that

$$\begin{aligned}\delta m_b &= 2m_{px}\sin\beta\cos\beta\ \delta\beta - 2m_{py}\cos\beta\sin\beta\ \delta\beta \\ &= (m_{px} - m_{py})\sin 2\beta\ \delta\beta\end{aligned} \tag{6.8.24}$$

that is, the same result as (6.8.23). This means that a slight alteration of the yield line pattern will result in an alteration of the bending moments in the yield lines that makes the bending moment in the new yield lines correspond to the new directions of the yield lines, provided that the moment field is stationary. Equation (6.8.20) can therefore be rewritten

$$\lambda \iint pw'\,dx\,dy = \int m_b'\,\theta'\,ds' \tag{6.8.25}$$

where m_b' are the yield moments corresponding to the new directions.

From this it can also be concluded that, for orthotropic slabs, (6.8.19) and (6.8.25) can be interpreted as the work equations corresponding to the two adjacent yield line patterns. As the equations result in the same load, the yield line pattern corresponds to a stationary solution.

The nodal force concept can also be transferred without difficulty to orthotropic slabs, and formulas for these slabs can be derived in the same way as for isotropic slabs. However, it must be remembered that there are torsional moments[4] in the yield lines of orthotropic slabs, which means that in (6.6.15) $m_v^{(1)}$, for example, means the difference between the torsional moment in section (1), determined by means of Mohr's circle for the point of intersection of the yield lines, and the torsional moment in yield line (1), determined by means of (6.4.15) or (6.4.16), depending on whether the yield line is positive or negative. However, these formulas for nodal forces are difficult to work with, and as it is usually possible to transfer the calculation of an orthotropic slab to an isotropic slab by means of the affinity theorem, the nodal

[4] These have been neglected by Johansen [43.1; 62.1] in developing formulas for nodal forces. His formulas are therefore incorrect in the case of orthotropic slabs.

force concept need not to be dealt with more thoroughly here. The most important point to note is that, as in the case of an isotropic slab, an intersection between yield lines having the same sign does not give rise to nodal forces (unless there exists no stationary moment field); this will be realized by noting that the moments in the yield line in this case are produced by means of the Mohr's circle that, for a positive yield line, corresponds to the principal moments m_{px} and m_{py} and for a negative yield line, to $-m'_{px}$ and $-m'_{py}$. This means that the above-mentioned difference between the torsional moments is zero.

In the practical formulation of the work equation, the following formula for the work done by the moments in a yield line will be of use. Even though only the bending moments do work in an angular rotation in the yield line, in calculating the work for an arbitrary slab part, it is naturally possible to include the work done by the torsional moments, since their contribution vanishes in summation over all the slab parts. Fig. 6.8.4 shows a slab, the axis of rotation of which is the line l. The vector of rotation ω can be imagined to be determined by a downward displacement δ at point P. By this means ω's components along the x- and y- axes (see Fig. 6.8.4) are found to be

$$\omega_x = \frac{\delta}{h_y}, \qquad \omega_y = \frac{\delta}{h_x} \tag{6.8.26}$$

If we now, for example, consider a positive yield line of length L (see Fig. 6.8.4), it is known that the moment vector in the yield line, with opposite sign, is statically equivalent to the two moment vectors shown in the figure. As the x-axis and the y-axis coincide with the directions of the reinforcement, the contribution of this yield line to the work will become [see the work equation (6.6.21)]

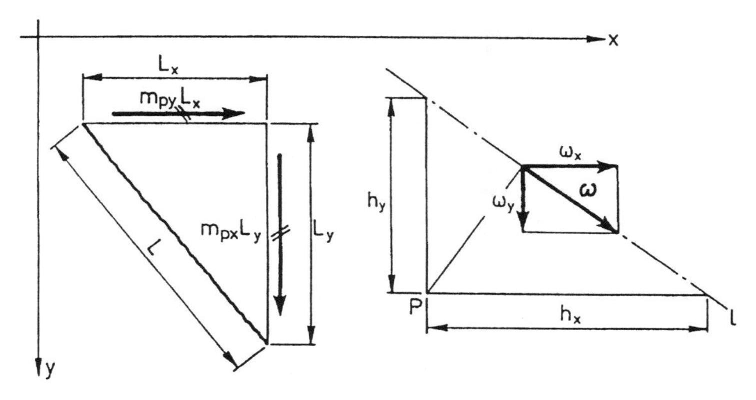

Figure 6.8.4 Notation used in work equation for orthotropic slabs.

$$\Delta W_I = m_{px} L_y \frac{\delta}{h_x} + m_{py} L_x \frac{\delta}{h_y} \tag{6.8.27}$$

If the work is then calculated for all yield lines in the slab part in question in this way, and summation over all slab parts is carried out in the same way as for isotropic slabs, it will not be much more difficult to calculate an upper bound solution for an orthotropic slab than for an isotropic slab.

It appears that the theory presented here, which is based on the yield conditions derived for orthotropic slabs, contains the method originally proposed intuitively by Johansen [43.1; 62.1] as a special case. It is interesting to note that this proposal, which has been followed by many authors, really can be given a sound theoretical foundation by means of a consistent theory of plasticity.

Example 6.8.1 In this example we want to write down the work equation for an orthotropic slab. The slab shown in Fig. 6.8.5 is simply supported along AB and supported on columns in C and D. The load is uniform, equal to $p = 10 kN/m^2$.

The reinforcement along the short span corresponds to a yield moment m_p. Along the long span it is $1/2\, m_p$.

First we estimate the yield line pattern to be that shown in Fig. 6.8.6.

The axis of rotation for the slab part II is under 45° with the edges of the slab. The distances necessary for the calculation are shown in the figure.

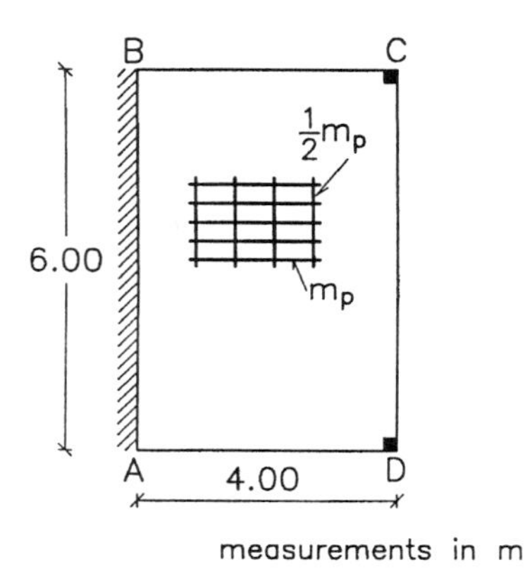

Figure 6.8.5 Orthotropic slab simply supported along one edge and on two columns.

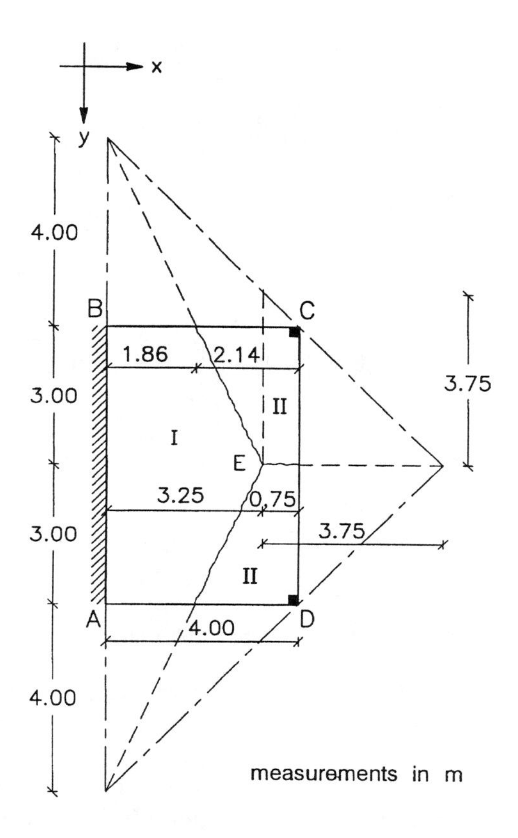

Figure 6.8.6 Yield line pattern in orthotropic slab.

If point E, the intersection point of the yield lines, is displaced δ downward, the individual slab parts have the rotations

$$\omega_{\mathrm{I}} = \frac{\delta}{3.25}$$

$$\omega_{x\mathrm{II}} = \omega_{y\mathrm{II}} = \frac{\delta}{3.75}$$

The work done by the load on slab parts II is calculated using the components along lines parallel to the coordinate axes through the columns.

The work equation reads:

$$m_p \cdot 6.00 \cdot \frac{\delta}{3.25} + 2\left(\frac{1}{2} m_p \cdot 2.14 \cdot \frac{\delta}{3.75} + m_p \cdot 3.00 \cdot \frac{\delta}{3.75}\right) =$$

$$\left(\frac{1}{6} \cdot 10.0 \cdot 14.00 \cdot 3.25^2 - 2 \cdot \frac{1}{6} \cdot 10.0 \cdot 4.00 \cdot 1.86^2\right) \cdot \frac{\delta}{3.25} +$$

$$2\left(\frac{1}{2} \cdot 10.0 \cdot 0.75 \cdot 3.00^2 + \frac{1}{6} \cdot 10.0 \cdot (2.14 - 0.75) \cdot 3.00^2\right) \cdot \frac{\delta}{3.75} +$$

$$2\left[\frac{1}{2} \cdot 10.0 \cdot 3.00 \cdot 0.75^2 + \right.$$

$$\left. 10.0 \cdot \frac{1}{2} \cdot 3.00 \cdot (2.14 - 0.75)\left[0.75 + \frac{1}{3}(2.14 - 0.75)\right]\right] \cdot \frac{\delta}{3.75}$$

Solving for m_p we get

$$m_p = 27.0 \text{ kNm/m}$$

A positive yield line parallel to the long edges in the middle of the slab gives

$$m_p = \frac{1}{8} \cdot 10.0 \cdot 4.00^2 = 20.0 \text{ kNm/m}$$

However, none of these yield line patterns is the most dangerous one. The conditions around the columns turn out to be decisive if the reinforcement is constant.

The column forces are statically determinate because of the symmetry. A moment equation about the simple support yields the column force

$$S = \frac{1}{4} \cdot 10.0 \cdot 4.00 \cdot 6.00 = 60 \text{ kN}$$

This means that in the corners with column support we have a torsional moment $\frac{1}{2} S = 30$ kN and therefore principal moments ± 30 kN, which are larger than the m_p-value found above.

Obviously reinforcement is necessary in the top side. If the top side is reinforced in the same way as the bottom side, the torsional yield moment is by means of the yield conditions for slabs in Section 2.3 found to be $\sqrt{0.5}\, m_p = 0.707 m_p$. If this is put equal to the torsional moment in the corner, 30 kN, we have

$$0.707 m_p = 30 \text{ kN} \quad \Rightarrow \quad m_p = 42.4 \text{ kNm/m}$$

The slab should be reinforced along the short span for $m_p = 42.4$ kNm/m and for $\frac{1}{2} m_p = 21.2$ kNm/m along the long span.

This result demonstrates that yield line patterns cannot always be chosen rather freely when one is dealing with column supports.

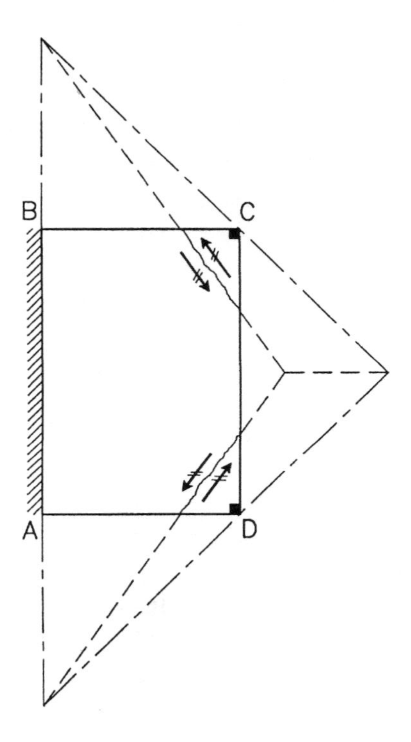

Figure 6.8.7 Yield lines in an orthotropic slab approaching the columns.

Of course it is possible to put extra reinforcement near the columns to prevent the conditions around them being decisive.

The extension of stronger zones near the columns may be determined by using upper bound methods as explained in Section 6.6.8. It may of course also be determined using lower bound solutions (see Section 6.7).

Similar results will be found when yield line patterns of the type shown in Fig. 6.8.7 are calculated. Here the yield lines are approaching the columns.

If the rotation axes for the slab parts II are kept under 45° we get m_p = 40.0 kNm/m. This is a limiting value when the yield lines are infinitely close to the columns. The most dangerous yield line pattern of this type renders m_p = 42.2 kNm/m.

The reinforcement design is not finished in this case by calculating the top and bottom reinforcement of the slab. Evidently it is extremely important to supply a special edge reinforcement along the free edges, particularly near the columns. The solution offered by plastic theory to this problem is described in Chapter 7.

6.9 ANALYTICAL OPTIMUM REINFORCEMENT SOLUTIONS

We shall conclude this chapter on the plastic slab theory by describing briefly some topics related to reinforced concrete slab theory still under development. In this section we deal with analytical optimum reinforcement solutions and in the next and final sections we deal with numerical methods and membrane action.

It is a natural task for a designer to look for one or another kind of optimal solution. A fundamental question in the plastic theory for reinforced concrete slabs is to find the absolute minimum of the reinforcement volume for a given slab with a prescribed load. Considerable progress in answering this question has been gained by the work of Morley, Lowe and Melshers, Rozvany, and others. A review paper containing most of the available information has been written by Rozvany and Hill [76.5].

If the slab thickness is prescribed and if variation of the compressive zones in the concrete is neglected, it may be shown by using the Drucker-Shield criterion for the minimum volume of a plastic structure [56.5] that one has to look for a constant principal curvature field throughout the slab, to which it is possible to assign a principal moment field corresponding in direction and sign to the curvature field. For many important cases the curvature field is the same for a wide class of load configurations on the same slab.

Morley [66.1] gave a solution for the simply supported square slab, which is illustrated in Fig. 6.9.1. In region *BDFH*, the two principal curvatures are positive and equal. In the triangular regions, the principal curvatures are equal and have opposite signs. A load acting in

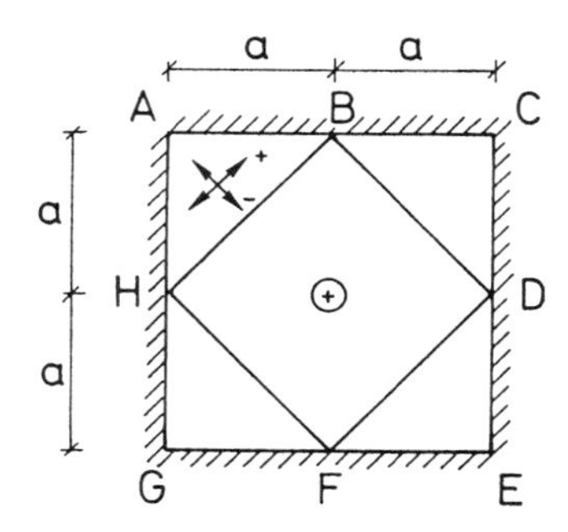

Figure 6.9.1 Optimum reinforcement solution for a simply supported square slab.

region *BDFH* is transferred to the infinitely narrow beams *BD*, *DF*, *FH* and *HB* by strip action. The strips can be arbitrarily selected. A load acting in the triangular regions can, for instance, be carried by strips lying under 45° to the edges and spanning from support to support. A great number of solutions of this kind have been given by Rozvany and Hill [76.5].

It will be seen that the reinforcement has to be rather artificially arranged. A special problem arises if concentrated reinforcement bands are required in theoretically infinitely narrow beams, since this might cause problems concerning the concrete stresses. In any case, the optimal solutions are extremely useful as a basis for comparison with the kind of solutions which for one reason or another are preferred by the designer.

6.10 NUMERICAL METHODS

The development of electronic computers has opened up new possibilities for finding approximate solutions to structural problems. To find lower bound solutions in the plastic theory, one needs to create a sufficiently wide class of statically admissible stress fields and to find the one corresponding to the greatest load factor. For example, statically admissible stress fields can be created by means of the finite element method, where the stress field within each element is expressed by a number of parameters. Equilibrium requirements within the element, possible continuity requirements along the element boundaries, and the statical boundary conditions lead to a set of linear equations.

If the yield conditions are linearized, one gets a set of linear constraints, which together with the equilibrium equations constitute a linear programming problem for the determination of the greatest load that can be carried by the slab. A similar method can be used to determine optimal reinforcement arrangements both in cases where the reinforcement is allowed to vary from point to point and in cases where the reinforcement arrangement is subject to certain geometrical constraints.

In Fig. 6.10.1 a solution obtained by Pedersen [74.4; 79.3] for the clamped square slab uniformly loaded ($m_p = m'_p$) is illustrated. The finite element used was a rectangular element with bending moments varying as a parabolic cylindrical surface and twisting moments varying as a hyperbolic paraboloid (i.e., the load intensity within each element was assumed to be constant). This element enables exact fulfillment of the equilibrium equations.

The linearized yield conditions used were the following simple ones:

$$
\begin{aligned}
m_x \pm m_{xy} &\le m_p \\
-m_x \pm m_{xy} &\le m'_p \\
m_y \pm m_{xy} &\le m_p \\
-m_y \pm m_{xy} &\le m'_p
\end{aligned}
\tag{6.10.1}
$$

These equations were checked at the corners and in the middle of the element. However, for the solutions obtained, the correct yield conditions (2.3.62) and (2.3.63) were checked in a finer mesh, and the solution was scaled down, if needed, to fulfill the correct yield condition in all check points. The figure shows the load-carrying capacity obtained as a function of the mesh size. Also, the total computer time is shown for some of the calculations.

As mentioned in Section 6.5.3, the exact solution is $pa^2/m_p = 42.851$, which means that the best numerical solution deviates only a few percent from the exact one. The first calculations of this kind reported in the literature were those of Wolfensberger [64.1], whose procedure was very similar to that described above. Anderheggen and Knöpfell [72.4], Ceradini and Gavarini [65.9], Gavarini [66.3], and Sacchi [66.4] have adopted a similar approach.

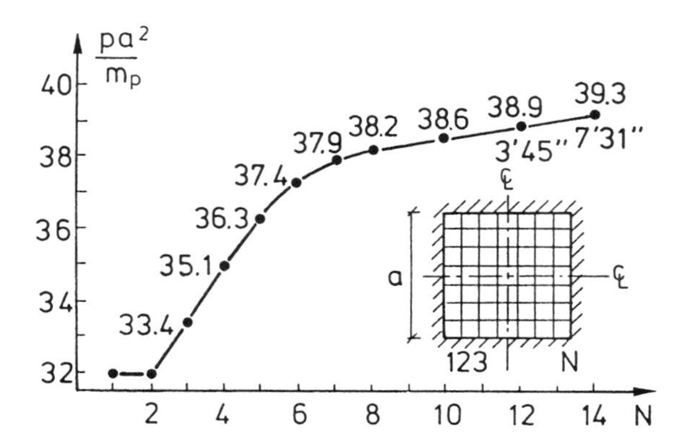

Figure 6.10.1 Finite element solution for clamped square slab with uniform load.

It is not necessary to fulfill the equilibrium equations exactly to obtain good solutions. In fact, practical experience shows that it may be advantageous to deal with approximate equilibrium equations in the same manner as in the common finite element solutions with compatible elements. Similar solutions may be obtained for many other plastic problems. Results may be found for disks, rotationally symmetric plain concrete problems, and structures subjected to earth pressure [79.18; 79.19].

Instead of using linear programming for the determination of the load-carrying capacity of reinforced concrete slabs, Chan [72.5] has applied quadratic programming, which, however, led to considerably longer computer time. Another approach has been used by Bäcklund [73.1], who determined upper and lower bounds by following the complete behavior of the slab when the load grows from zero to the ultimate value.

Linear programming methods require a large computer and computer time far exceeding that required for linear elastic calculations. Nevertheless, it is to be expected that in the future commercial programs based on the plastic theory of reinforced concrete will be in operation.

A beginning was made in Denmark in the late 70s with the program RUPTUS [81.8]. However, the program was seldom used, and it declined. Programs are now available for optimization using the stringer method [94.15], cf. Section 4.6.3.

An ingenious optimization method has been developed by Anderheggen and associates, see [94.16; 94.17; 95.10; 95.11; 95.13; 96.4; 96.5; 96.10]. In this method one starts with the linear elastic solution and approaches the optimal solution by superimposing self-equilibrating stress fields.

6.11 MEMBRANE ACTION

6.11.1 Membrane Effects in Slabs

The classical slab theory presented neglects the fact that the strain field, corresponding to bending and twisting moments only, results in strains in the slab middle surface, and these strains do not generally satisfy the compatibility equations and the geometrical boundary conditions. This leads to in-plane forces in the slab, already in the early stages of cracking.

The rigid plastic theory in its standard formulation (first-order theory) neglects effects of changes in geometry. Since slabs often are

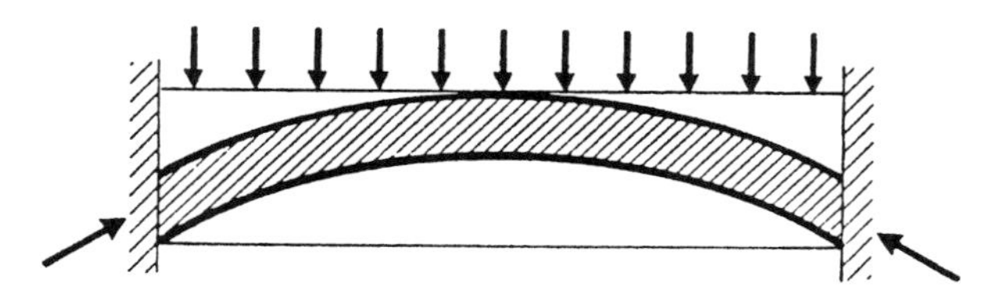

Figure 6.11.1 Illustration of the compressive membrane effect (dome effect).

rather flexible structures, the changes in geometry sometimes have a considerable effect on the load-carrying capacity. These effects are called membrane effects, and one speaks about a compressive membrane effect, which often predominates at small deflections, and of a tensile membrane effect, which dominates at larger deflections.

When the compressive membrane effect is dominating, the slab carries the load like a dome with very little depth. This is illustrated in Fig. 6.11.1.

A uniformly loaded, simply supported square slab may have a load-deflection relationship of a type shown in Fig 6.11.2. Instead of yielding under constant load, one hardly observes anything peculiar at the load corresponding to the rigid plastic first-order theory, the real collapse load generally being somewhat higher. Small degrees of reinforcement lead to relatively higher collapse loads compared to the rigid plastic first-order load than do higher degrees of reinforcement.

Quite different behavior is observed for a clamped slab if horizontal displacements are prevented along the edges. A typical load-deflection curve is shown in Fig. 6.11.3. Failure here is by a snap-through action, after which the load approximately reaches the rigid

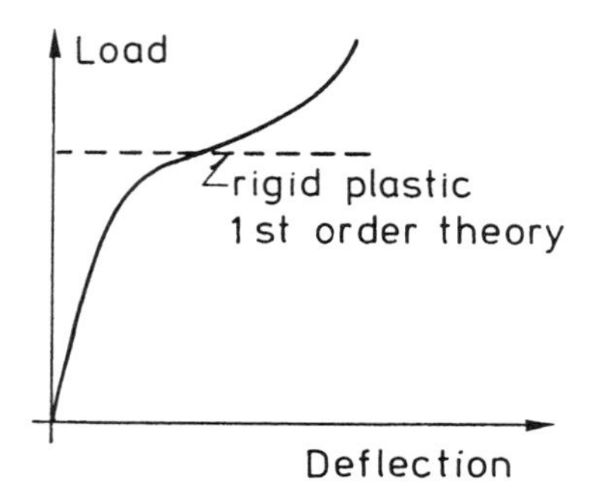

Figure 6.11.2 Membrane action in simply supported slabs.

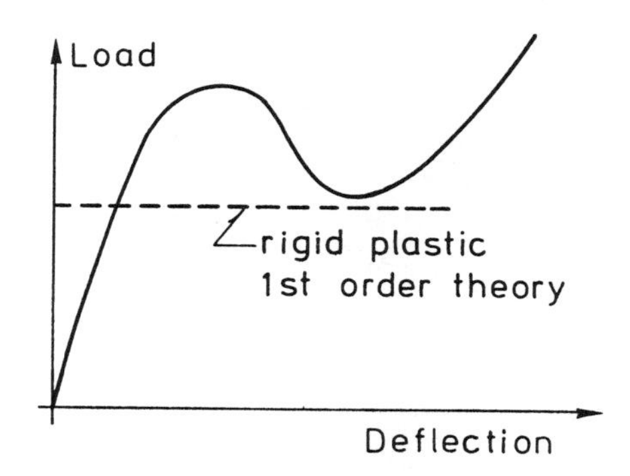

Figure 6.11.3 Membrane action in clamped slabs.

plastic first-order load. Finally, the load is again increased through tensile membrane action. The maximum load may far exceed the rigid plastic first-order load. Factors of 2 or 3 are common.

A theoretically correct determination of the full load-deflection curve taking into account the elastic deformations, cracking of the concrete, realistic constitutive equations of the concrete until failure, and the effect of changes in geometry is extremely complicated and may only be obtained using advanced computer programs.

Estimates of the effect of changes in geometry can, however, be obtained relatively simply by means of a series of upper bound calculations assuming the form of the deflected slab to be known. For instance, a circular slab loaded at the center by a point load can be assumed to deflect as a cone, similar to the deflection cone found by first-order rigid plastic theory. Similarly, a square slab can be assumed to deflect into a pyramidal form.

Having fixed the deflected form, it is a relatively simple task by means of the usual upper bound technique to calculate the load corresponding to the deflected form assumed. The load-carrying capacity, of course, turns out to be a function of the deflection.

As shown by Calladine [68.5]. the calculations in several cases turn out to be very much simpler using the three-dimensional theory instead of the two-dimensional theory usually adopted in slab theory.

For a simply supported slab and a clamped slab, the load deflection curve obtained in this way may be of the type shown in Figs. 6.11.4 and 6.11.5, respectively. The maximum load found for the clamped slab will not be reached in practice because of the neglected elastic deformations. If the elastic properties are taken into account, it is

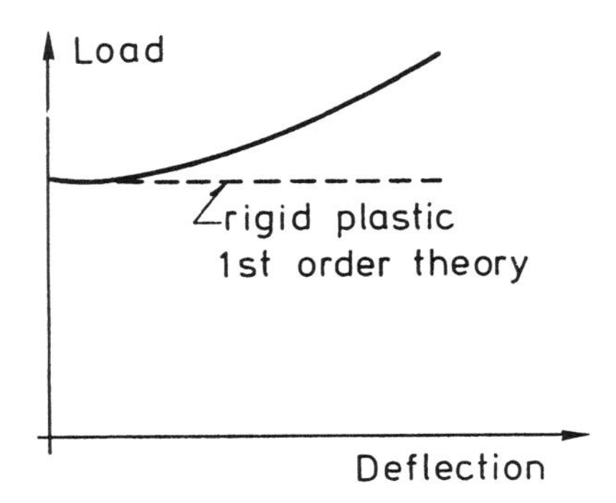

Figure 6.11.4 Load-deflection curve for simply supported slabs obtained through a series of upper bound solutions.

possible to calculate load-deflection curves that correspond very well with experimental observations (cf. Bræstrup and Morley [80.5]).

It is important to bear in mind that the generalized strain measures of plasticity are not the total strains, but the increments or rates. However, at the incipient collapse of a rigid plastic structure, which is the situation considered at yield point analysis, the plastic deformations are the first and only ones to occur, and the distinction becomes immaterial. This is why the term "strain rate" has been avoided throughout, and the analyses are carried out in terms of conventional "small strains". On the other hand, when deriving the load-deflection relationship discussed above, it is necessary to distinguish between the actual values of the strains and their increments at a given time. Fail-

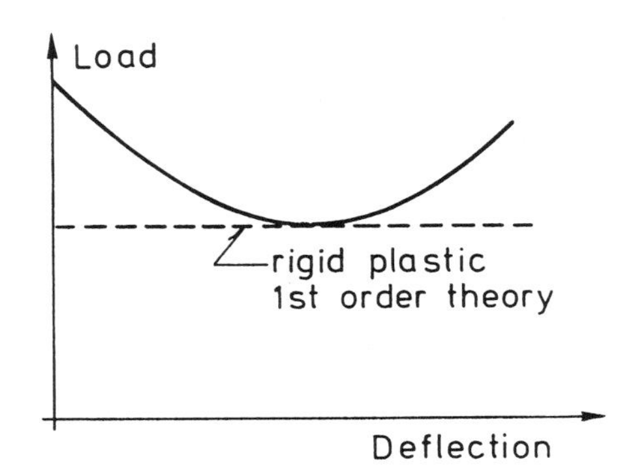

Figure 6.11.5 Load-deflection curve for clamped slabs obtained through a series of upper bound solutions.

ure to do so amounts to application of the deformation theory of plasticity, which neglects the unloading of concrete under numerically decreasing compressive strains. As a result, the deflection corresponding to a given level of compressive membrane forces is overestimated by a factor of 2 (cf. Bræstrup[80.4]).

Because of the great effect of the elastic deformations on the load-carrying capacity of clamped slabs, the rigid plastic theory cannot be used directly with confidence in practice. Since large reserves in load-carrying capacity are inherent in the effect of changes in geometry, one of the most urgent needs of slab research has for a long time been to create a reliable design method capable of utilizing these reserves. Such a method is presented in the following.

Although Johansen [43.1; 62.1] was aware of the tensile membrane action, one of the first to demonstrate the great effect of restrained edges was Ockleston [55.1], who in a test series on a condemned building became aware of a breakdown of the rigid plastic first-order theory for internal slab parts. Several research workers since that time have studied the problem theoretically and experimentally, among them Wood [61. 5], Christiansen [63.11; 82.6; 83.14], Park [64.6; 64.7], Birke [75.10], Eyre and Kemp [83.15] and Christiansen and Frederiksen [83.16]. An upper bound analysis of the type described above was performed by, among others, Sawczuk [64. 8; 65.10], Janas and Sawczuk [66.5], Morley [67.6], Janas [68.6], and as mentioned earlier, by Calladine [68.5]. Literature surveys have been performed by Bäcklund [72.6] and Bræstrup [80.4]. The work by the reasearch group headed by the author has been described in [85.8; 86.3; 88.3; 89.3]. A review, mainly taken from [88.3], is given in the following.

6.11.2 Unreinforced One-way Slabs

Consider a simple example, an unreinforced one-way slab horizontally restrained and loaded by two symmetrical, transversal line loads $\overline{p}$ (see Fig 6.11.6). First we consider the material to be a rigid plastic material, i.e., the displacements before plastic failure is considered to be zero. A simple lower bound solution is found by considering an arch action as illustrated in the figure. The region *ABDE* is in uniaxial compression. The loads and the horizontal reaction *C* are transferred to the arch through the regions *AEF* and *BCD*, which are in biaxial hydrostatic pressure.

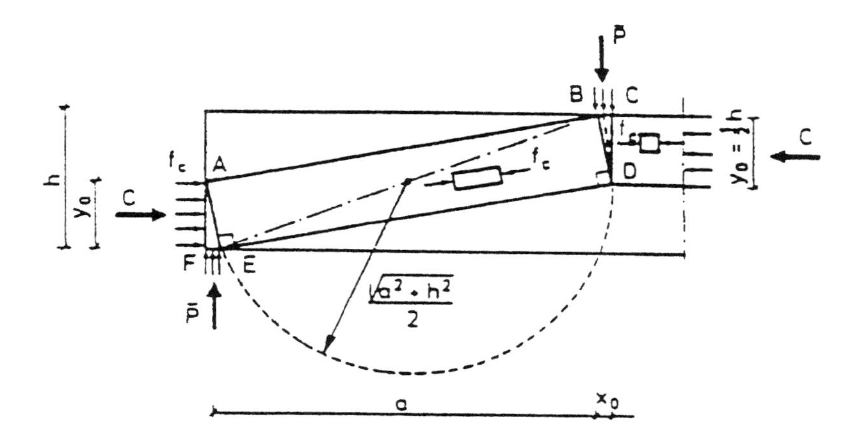

Figure 6.11.6 One-way slab loaded by two line loads.

The hydrostatic stress is assumed equal to the uniaxial stress, and both are assumed equal to the concrete strength f_c. Therefore, angle *BDE* is $\pi/2$. It appears that maximum load is obtained when *BC* is as large as possible. Since *D* lies on a circle having *BE* as the diameter, the maximum load is obtained when $CD = y_0 = h/2$. Then it is only a matter of geometry to calculate

$$\mathrm{BC} = \mathrm{x}_0 = \frac{1}{2}\left(\sqrt{\mathrm{a}^2 + \mathrm{h}^2} - \mathrm{a}\right) \tag{6.11.1}$$

The lower bound solution is thus

$$\overline{p} = x_0 f_c \tag{6.11.2}$$

which can be written

$$\frac{\overline{p}}{hf_c} = \frac{x_0}{h} = \frac{1}{2}\left(\sqrt{1 + \left(\frac{a}{h}\right)^2} - \frac{a}{h}\right) \tag{6.11.3}$$

This solution has already been considered previously (cf. Section 5.2.3.)

If a >> h the solution can be written

$$\overline{p}\,a = \frac{1}{4} f_c h^2 \tag{6.11.4}$$

which simply expresses that the external moment $\overline{p}\,a$ is equal to the moment produced by the two compressive zones of depth $y_0 = \frac{1}{2}h$.

The moment

$$m_m = \frac{1}{4} f_c h^2 \qquad (6.11.5)$$

is called the *membrane moment* and it plays an important role in what follows.

The lower bound solution presented is an exact plastic solution, since a failure mechanism giving an upper bound solution equal to the lower bound solution is easily found as shown in Fig. 6.11.7. The shaded areas represent lines of discontinuity in the longitudinal displacement giving compression along half the depth while the remaining part of the section has longitudinal displacement discontinuities giving tension, i.e., the stress is zero according to our assumptions.

It is easily seen that for any transversal load the load-carrying capacity may be determined by putting the maximum value of the bending moment, calculated for a simply supported slab having the same length as the given slab, equal to the membrane moment.

Now in reality the slab is, of course, not rigid. If the deflection before plastic collapse is reached is δ as shown in Fig. 6.11.8, it is easily seen that in the solution above we only have to replace h by $h-\delta$, which means that the load-carrying capacity is determined by

$$\bar{p}\, a = \frac{1}{4} f_c (h - \delta)^2 \qquad (6.11.6)$$

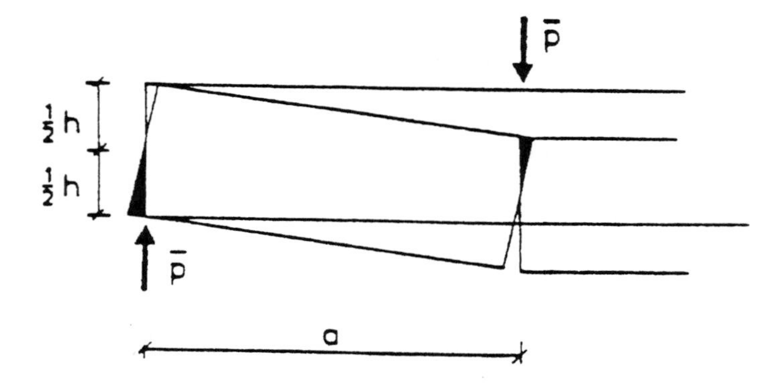

Figure 6.11.7 Failure mechanism for one-way slab.

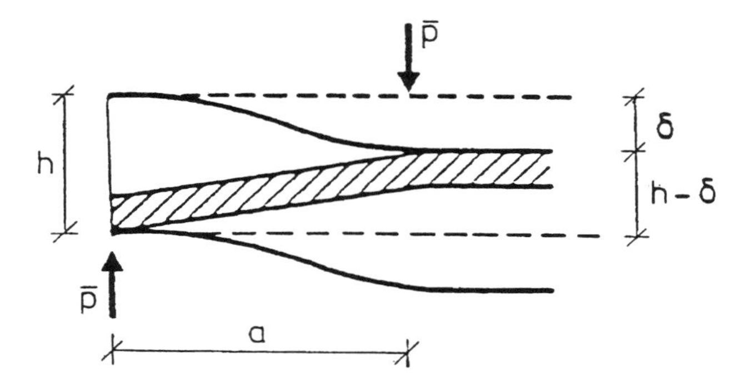

Figure 6.11.8 Deflected one-way slab.

If the load-carrying capacity for $\delta = 0$ is called $\bar{p}_0$, we get the relationship shown in Fig. 6.11.9 between $\bar{p}/\bar{p}_0 = f(\delta/h)$ and δ/h.

The reduction of load-carrying capacity because of the deflection is the same in other loading cases.

The solution for the deflected slab is also, for a class of deflected forms, an exact plastic solution.

6.11.3 Work Equation

To be able to establish the work equation in an upper bound solution, consider first a yield line between two slab parts I and II, i.e., a line with displacement discontinuities as illustrated in Fig 6.11.10.

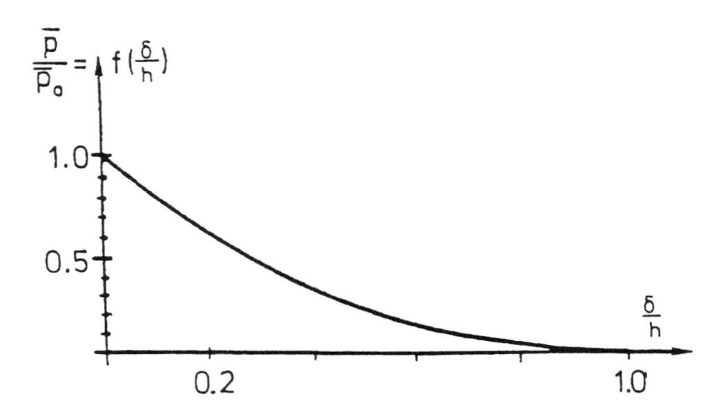

Figure 6.11.9 Load-carrying capacity as a function of deflection.

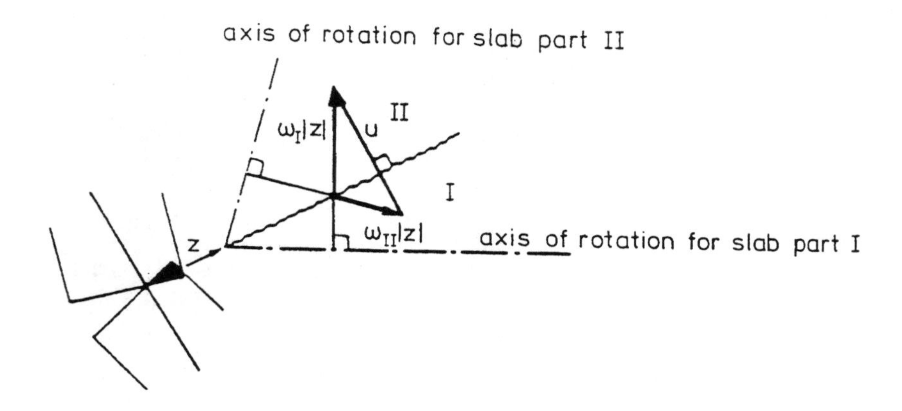

Figure 6.11.10 Yield line between two slab parts.

Consider an element of area dA in compression lying at a distance $|z|$ from the plane formed by the axes of rotation for the two slab parts. The contribution from dA to the dissipation in the yield line is $dD = f_c\, u\, dA$, u being the relative displacement or displacement discontinuity at dA.

The displacement discontinuity u can be found as shown in Fig 6.11.10 as a function of the rotation ω_I and ω_{II} of the slab parts I and II, respectively. We introduce the quantities

$$dD_{Ic} = \pm f_c\, \omega_I\, |z|\, dA_I \tag{6.11.7}$$

$$dD_{IIc} = \pm f_c\, \omega_{II}\, |z|\, dA_{II} \tag{6.11.8}$$

dA_I being the projection of dA on a plane perpendicular to the slab plane along the axis of rotation for slab part I and dA_{II} the projection on a plane perpendicular to the slab plane along the axis of rotation for slab part II. The + sign in (6.11.7) shall be used in the formula if the work done by the compression force $f_c\, dA$ on the slab part I is negative when being displaced by $\omega_I |z|$. If the work is positive the − sign must be used. Similar rules apply in (6.11.8). The dissipation f_c udA is, of course, always positive and normally, as in Fig. 6.11.10, the + sign shall be used in both formulas.

By integrating over the compression zone, we get the dissipations along the yield line

$$D_{Ic} = \pm\omega_I \, S_{Ic} \, f_c \tag{6.11.9}$$

$$D_{IIc} = \pm\omega_{II} \, S_{IIc} \, f_c \tag{6.11.10}$$

where

$$S_{Ic} = \int |z| dA_I \tag{6.11.11}$$

$$S_{IIc} = \int |z| dA_{II} \tag{6.11.12}$$

S_{Ic} is the statical moment about the axis of rotation for slab part I of the projection of the compression zone on a plane perpendicular to the slab plane along the axis of rotation for slab part I. S_{IIc} has similar meaning. By summation over all yield lines and all slab parts the dissipation in the concrete can be calculated.

If the slab is reinforced, the dissipation of the reinforcement in the yield line must be included. Consider a horizontal reinforcement band forming an angle α with the normal to the yield line. If the reinforcement per unit length, measured perpendicular to the reinforcement bars, is A_{s1} and the displacement discontinuity in the direction of the normal to the yield line is u, the dissipation dD_s along an infinitesimal length ds of the yield line is (see Fig. 6.11.11),

$$dD_s = A_{s1}\cos^2\alpha \cdot f_Y \, u \, ds \tag{6.11.13}$$

f_Y being the yield stress of the reinforcement.

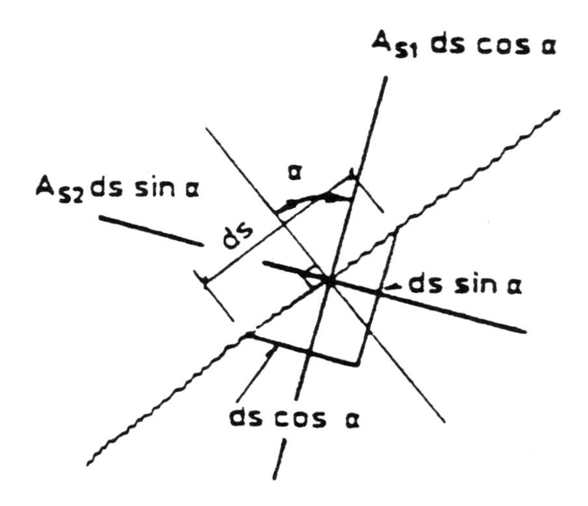

Figure 6.11.11 Reinforcement crossing a yield line.

As for the concrete the dissipation can be expressed as a function of the rotations of the slab parts meeting along the yield line.

Referring again to Fig. 6.11.10 we find

$$dD_{Is} = \pm A_{s1}\cos^2\alpha\ f_Y\omega_I|z|\,ds_I \tag{6.11.14}$$

$|z|$ now being the distance from the plane of the axes of rotation to the reinforcement and ds_I the projection of ds on the axis of rotation for slab part I. Similarly

$$dD_{IIs} = \pm A_{s1}\cos^2\alpha\ f_Y\omega_{II}|z|\,ds_{II} \tag{6.11.15}$$

If there are several reinforcement bands the dissipation is found by summation over all bands. If there is only one more reinforcement band perpendicular to A_{s1} and having the area A_{s2} per unit length we get

$$dD_{Is} = \pm\left(A_{s1}\cos^2\alpha + A_{s2}\sin^2\alpha\right)f_Y\omega_I|z|\,ds_I \tag{6.11.16}$$

$$dD_{IIs} = \pm\left(A_{s1}\cos^2\alpha + A_{s2}\sin^2\alpha\right)f_Y\omega_{II}|z|\,ds_{II} \tag{6.11.17}$$

In the special case $A_{s1} = A_{s2} = A_s$, i.e., the slab is isotropically reinforced, we get

$$dD_{Is} = \pm A_s f_Y \omega_I|z|\,ds_I \tag{6.11.18}$$

$$dD_{IIs} = \pm A_s f_Y \omega_{II}|z|\,ds_{II} \tag{6.11.19}$$

By introducing the statical moments we get similar formulas as for the concrete dissipation. For instance, in the isotropic case we have

$$D_{Is} = \pm\omega_I S_{Is} f_Y \tag{6.11.20}$$

$$D_{IIs} = \pm\omega_{II} S_{IIs} f_Y \tag{6.11.21}$$

where

$$S_{Is} = \int A_s|z|\,ds_I \tag{6.11.22}$$

$$S_{IIs} = \int A_s|z|\,ds_{II} \tag{6.11.23}$$

6.11.4 Unreinforced Square Slabs

Let us consider a simple example, namely an unreinforced square

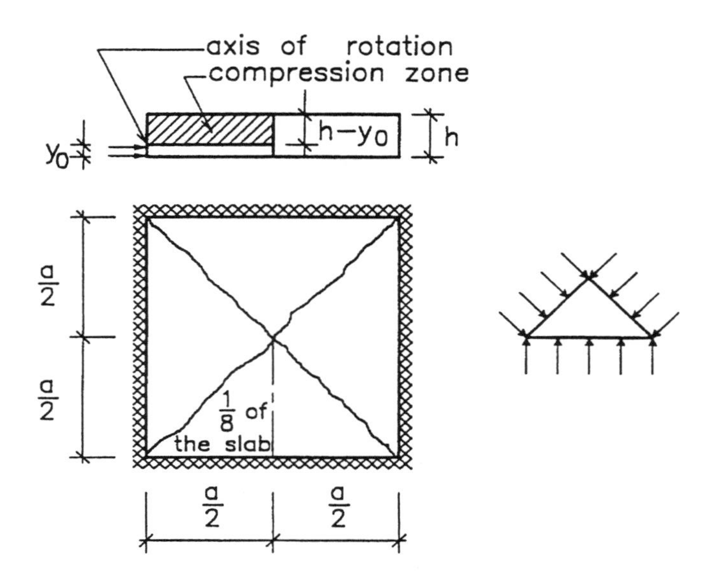

Figure 6.11.12 Square slab restrained along all edges.

slab loaded by a uniform load p per unit area. The slab is horizontally restrained along all edges, (see Fig. 6.11.12).

Let, for the rigid plastic square slab, the axes of rotation for the slab parts formed by the classical yield line pattern consisting of yield lines along the diagonals be situated along the supports in a distance y_0 from the bottom face. It appears that along the yield lines we get a compression zone extending a depth $h-y_0$ from the top face. Along the supports we get compression zones extending a depth y_0 from the bottom face.

For a unit downward displacement of the center of the slab, all slab parts get the rotation

$$\omega = \frac{1}{\frac{1}{2}a} = \frac{2}{a} \tag{6.11.24}$$

The work equation for 1/8 of the slab gets the form

$$\frac{2}{a}\cdot f_c\cdot\frac{1}{2}\cdot\frac{1}{2}a(h-y_0)^2+\frac{2}{a}\cdot f_c\cdot\frac{1}{2}\cdot\frac{1}{2}a\,y_0^2$$
$$=\frac{1}{6}p\cdot\frac{1}{2}a\left(\frac{1}{2}a\right)^2\cdot\frac{2}{a} \quad (6.11.25)$$

The first term on the left-hand side is the contribution from the yield line using (6.11.9), the second term is the contribution from the support. The work done by the load has been calculated as the moment of the load about the axis of rotation times the rotation.

Solving for p and minimizing with respect to y_0 we find, as for the one-way slab, $y_0=\frac{1}{2}h$ and

$$p=\frac{6f_c h^2}{a^2} \quad (6.11.26)$$

It appears that in the yield line solution (6.6.49) for a simply supported slab corresponding to this yield line pattern

$$p=\frac{24m_p}{a^2} \quad (6.11.27)$$

m_p being the positive yield moment, we just have to replace m_p by the membrane moment $\frac{1}{4}f_c h^2$.

Further, it turns out that for $y_0=\frac{1}{2}$ the area of the compression zone along the support for 1/8 of the slab, $\frac{1}{2}a y_0$, equals the projected area of the compression zone in the yield line $\frac{1}{2}a(h-y_0)$. This is also seen directly by minimizing the statical moments of the type $\int|z|\,dA_l$, etc. appearing in the work equation.

This "area rule" is equivalent to a simple horizontal projection equation for a triangular slab part as illustrated in Fig. 6.11.12.

Now we turn to find the effect of a deflection. Let us assume that the deflected slab has the same form as the displacement field used above in the classical yield line solution. If the deflection in the center is δ, the compression zone in the yield lines will get the form shown in Fig. 6.11.13, if δ is not too big, and if we neglect the rotation of the vertical limiting lines of the compression area in the yield lines.

The "area rule" gives

$$(h-\delta-y_0)\cdot\frac{1}{2}a+\frac{1}{2}\cdot\frac{1}{2}a\cdot\delta=\frac{1}{2}a\,y_0 \quad (6.11.28)$$

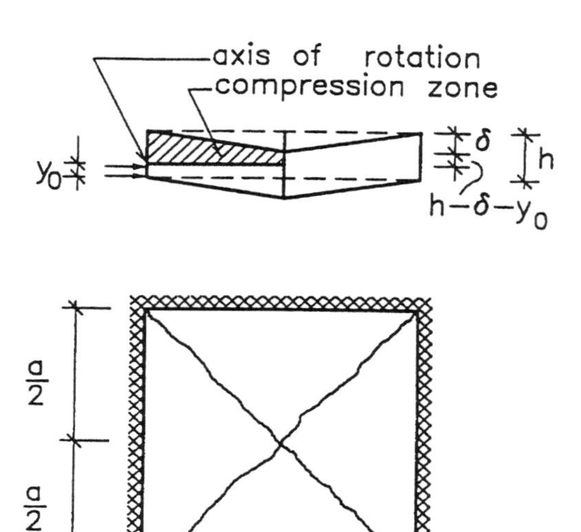

Figure 6.11.13 Compression zone in deflected slab.

leading to

$$y_0 = \frac{1}{2}h - \frac{1}{4}\delta \tag{6.11.29}$$

The work equation obtains the form

$$\begin{aligned}
&\frac{2}{a}\cdot f_c\left[\frac{1}{2}\cdot\frac{1}{2}a(h-\delta-y_0)^2\right.\\
&\left.+\frac{1}{2}\cdot\frac{1}{2}a\cdot\delta\left(h-\delta-y_0+\frac{1}{3}\delta\right)\right]\\
&+\frac{2}{a}\cdot f_c\cdot\frac{1}{2}\cdot\frac{1}{2}a\,y_0^2 = \frac{1}{6}p\cdot\frac{1}{2}a\cdot\left(\frac{1}{2}a\right)^2\cdot\frac{2}{a}
\end{aligned} \tag{6.11.30}$$

Introducing y_0 by (6.11.29) we find

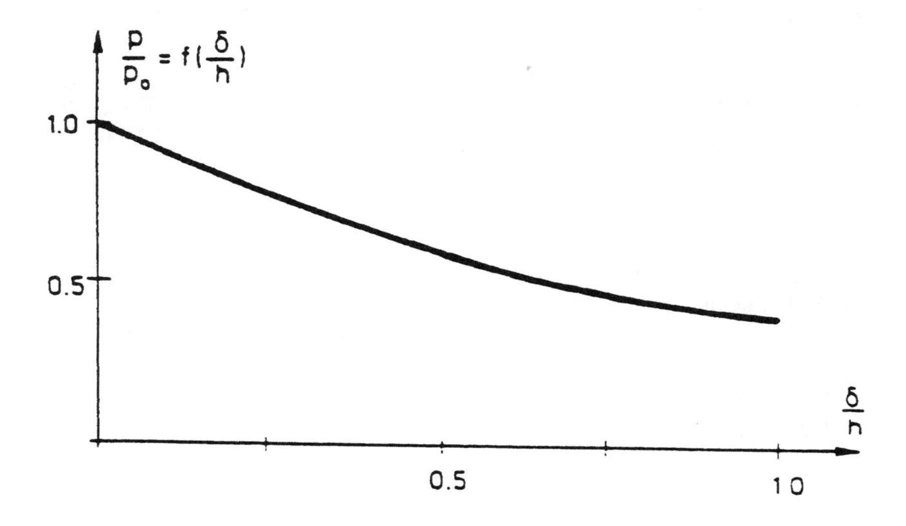

Figure 6.11.14 Load-carrying capacity as a function of deflection.

$$p = \frac{6f_c h^2}{a^2}\left(1 - \frac{\delta}{h} + \frac{5}{12}\left(\frac{\delta}{h}\right)^2\right) \tag{6.11.31}$$

This result is only valid for $h - \delta - y_0 = \frac{1}{2}h - \frac{3}{4}\delta \geq 0$, i.e., $\delta \leq \frac{2}{3}h$. For $\delta > \frac{2}{3}h$ a similar calculation gives

$$p = \frac{6f_c h^2}{a^2}\frac{2}{3}\left(2\left(\frac{\delta}{h}\right)^2 + 6\left(\frac{\delta}{h}\right) + 3 - 2\sqrt{\frac{\delta}{h}\left(\frac{\delta}{h} + 2\right)^3}\right) \tag{6.11.32}$$

If p_0 is the load-carrying capacity for $\delta = 0$ we get a relationship $p/p_0 = f\left(\frac{\delta}{h}\right)$ versus δ/h as shown in Fig. 6.11.14.

It is rather important to notice that this curve is rather flat in the region $\delta = 0.5h$ to $\delta = h$, which is the range of the deflections found in tests. The consequence is that exact knowledge of the deflection at maximum load is much less important for slabs horizontally restrained along all edges than for one-way slabs, assuming the result for a square slab to be typical for that kind of slabs, which turns out to be the case.

For the slabs shown in Fig. 6.11.15 it is found that the reduction in load-carrying capacity because of the deflection is the same as for the square slab. This is the case also for other slabs for which the yield lines are intersecting in a point.

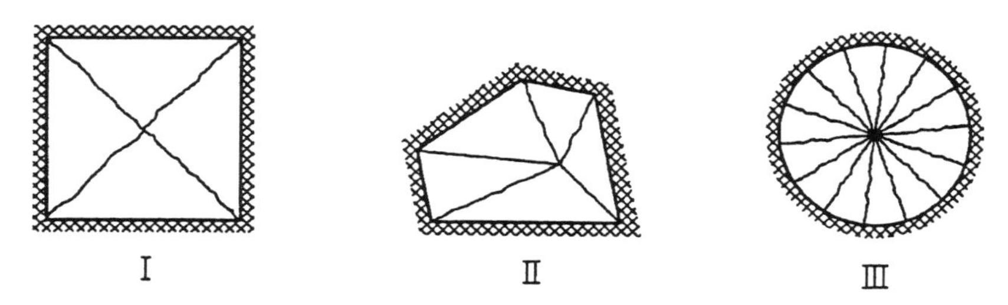

Figure 6.11.15 Slabs with the same load-carrying capacity as a function of deflection as the square slab.

As before, it is found that the load-carrying capacity for the rigid plastic slab can be found by replacing in the usual yield line solution m_p by the membrane moment $\frac{1}{4} f_c h^2$.

6.11.5 Unreinforced Rectangular Slabs

Consider now a rectangular slab $a \cdot b$ horizontally restrained along all edges. The yield line pattern shown in Fig. 6.11.16 has one unknown geometrical parameter x. The slab is uniformly loaded by the load p per unit area. For the rigid plastic slab it turns out, as in previous cases, that the load-carrying capacity may be found by replacing in the usual yield line solution for a simply supported slab the yield moment m_p by the membrane moment $\frac{1}{4} f_c h^2$. The parameter x has the same value as in the usual yield line solution.

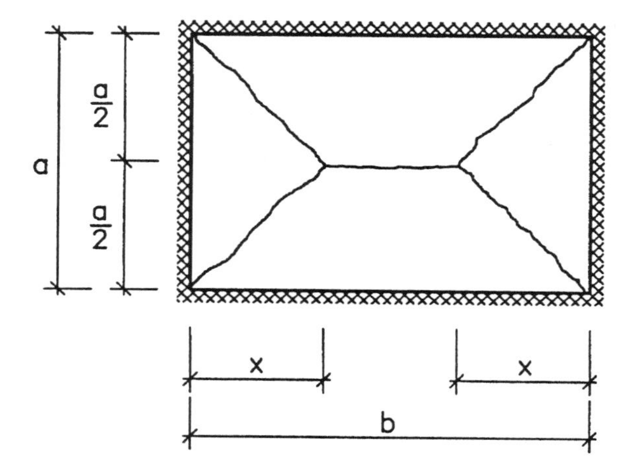

Figure 6.11.16 Yield line pattern for rectangular slab.

Since the yield line solutions is, see (6.6.38),

$$p = \frac{24m_p}{a^2\left[\sqrt{3+\left(\frac{a}{b}\right)^2} - \frac{a}{b}\right]^2} \tag{6.11.33}$$

the load-carrying capacity for the rigid plastic slab is

$$p = \frac{24 \cdot \frac{1}{4} f_c h^2}{a^2\left[\sqrt{3+\left(\frac{a}{b}\right)^2} - \frac{a}{b}\right]^2} = \frac{6 f_c h^2}{a^2\left[\sqrt{3+\left(\frac{a}{b}\right)^2} - \frac{a}{b}\right]^2} \tag{6.11.34}$$

The load-carrying capacity of the deflected slab may be found by similar calculations as carried out for the square slab. The calculations are a little more tedious and will not be given here. The reader is referred to [86.3]. It is assumed that the deflected shape has a similar form as the displacement field used in the yield line solution. The minimization procedure is more complicated than for the rigid plastic case and the optimized value of x is not exactly equal to the value obtained from the yield line solution. However, it may be shown that the error by applying the x-value from the yield line solution is small. The results shown below are based on this approximation.

In Fig. 6.11.17 the reduction of the load because of the deflection is shown. The notation is the same as before, p meaning the load-carrying capacity and p_0 the load-carrying capacity for $\delta = 0$. The deflection δ is the deflection of the slab along the yield line parallel to the sides b. The special cases a/b = 1 and a/b = 0 have been treated above.

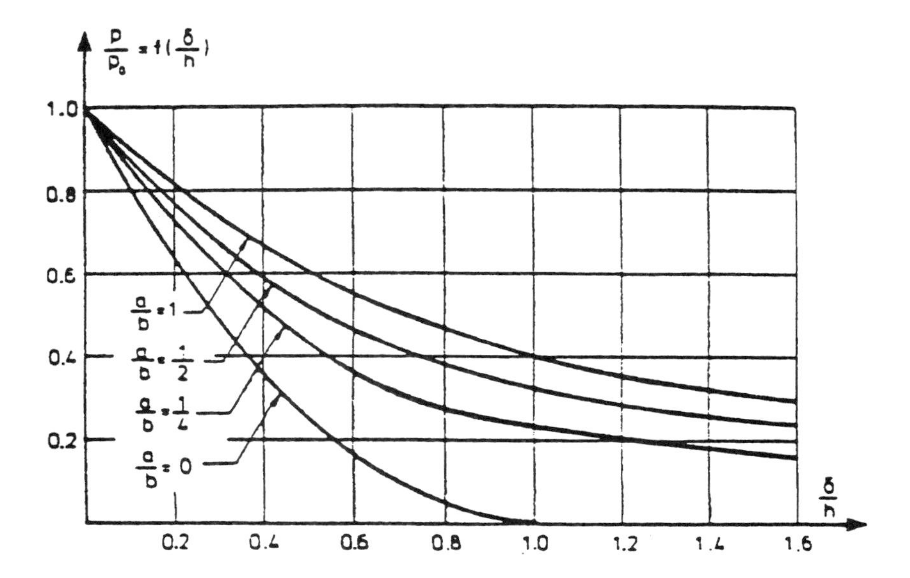

Figure 6.11.17 Load-carrying capacity as a function of deflection.

6.11.6 The Effect of Reinforcement

The effect of reinforcement may be taken into account in the work equation as described above. Results for rectangular slabs may be found in [85.8], [86.3] and [89.3].

However, for small degrees of reinforcement a much simpler approach may be justified. For small degrees of reinforcement, the depth of the compressive zones for bending and torsional moments only is small. Therefore, by a lower bound consideration, it is natural to expect that the load-carrying capacity for a reinforced slab simply is the sum of the load-carrying capacity for an unreinforced slab by dome action and the load-carrying capacity found by a normal yield line calculation. This idea was introduced by Christiansen and Frederiksen [83.16].

If P_M is the load-carrying capacity by dome action for the unreinforced slab and P_J is the load-carrying capacity found by a normal yield line calculation, we have the load-carrying capacity

$$P = P_M + P_J \tag{6.11.35}$$

The comparison between tests and theory referred to below is carried out using this relation.

6.11.7 Comparison with Tests

Since the curves describing the reduction of the load- carrying capacity because of the deflection are rather flat at the deflection at the ultimate load measured in tests, it turns out that for slabs horizontally restrained along all edges it is only necessary to deal with two groups of slabs:

1) Slabs with rigid horizontal restraints, i.e., slabs for which no horizontal displacement takes place along the edges.

2) Slabs with normal horizontal restraints, i.e., slabs surrounded by slabs of similar geometry, etc. as the slab considered or slabs having equivalent horizontal restraints.

Slabs of type 1 reach the ultimate load at a deflection $\delta \approx 0.5h$ while slabs of type 2 reach the ultimate load at a deflection $\delta \approx 0.8h$.

Only rectangular slabs of type 2 will be treated here. For other slabs and slabs of type 1, the reader is referred to [86.3].

A detailed description of the tests may be found in Christiansen and Frederiksen [83.16], Andreasen and Nielsen [86.3], Andreasen [85.8] or in the original papers.

Using the effectiveness factor

$$\nu = \frac{2.0}{\sqrt{f_c}} \qquad (f_c \text{ in MPa}) \tag{6.11.36}$$

in fact very good agreement is found. Both when the deflection at ultimate load is set to the average value in all tests, $\delta/h = 0.851$ and when $f\left(\frac{\delta}{h}\right)$ is set to 0.430, which is the reduction factor for a/b = 1 and $\delta = 0.85h$, the coefficient of variation on the ratio of the ultimate total loads $P_{test} \,/\, P_{theory}$ is found to be about 12%.

The range of the parameters covered by the tests is

$$
\begin{aligned}
& b \leq 2a \\
& \Phi + \Phi_c < 0.18 \\
& 13 < f_c < 45 \text{ MPa} \\
& \lambda = \frac{a+b}{2h} \qquad \text{between } 20 \text{ and } 40
\end{aligned}
\tag{6.11.37}
$$

Here a is the short side length and b the long one. Φ is the bottom reinforcement degree and Φ_c the top reinforcement degree. Finally, as before, h is the slab thickness.

It is a remarkable fact that even up to λ about 40, the compressive membrane effect is governing the load-carrying capacity.

Tests with orthotropic reinforcement are rare. Ockleston [55.1], [58.6] has carried out two in-situ tests of this kind. The same procedure for calculating the theoretical loads as above gives good agreement.

Comparisons between theory and test results for slabs with rigid horizontal restraints also give good agreement, see [85.8] or [86.3]. Tests by Wood [61.5], Park [64.6], Brotchie & Holley [71.6] and Powell [56.7] have been used in the analysis.

A number of tests for rectangular slabs having horizontal restraints along three sides only have been reported in [83.17], [71.4], [71.5] and [56.7] . It turns out that these slabs may exhibit a considerable dome effect. Such slabs may be treated in a similar way as slabs with horizontal restraints along all edges (see [85.8]).

Since the load-carrying capacity of one-way slabs are much more sensitive to deflection, these slabs cannot be treated by the method described here. The reader is referred to [91.20].

Most of the tests reported in the literature have been done with uniform or almost uniform loads.

Tests with concentrated loads have been performed at the Technical University of Denmark [91.3]. They showed that concentrated loads may be treated in the same way as uniform loads.

6.11.8 Conclusion

It has been shown that dome effects in slabs having horizontal restraints along all edges may be calculated in a very simple way. For slabs with normal horizontal restraints, i.e., slabs surrounded by slabs of similar geometry, etc. as the slab considered or slabs having equivalent horizontal restraints, the load-carrying capacity including dome effect may be calculated in the following way:

1) The load-carrying capacity P_J is calculated by a normal yield line calculation.

2) The load-carrying capacity P_M by dome action is calculated by a normal yield line calculation for an isotropic slab identical to the given slab but with simple supports along all edges. In the yield line solution the positive yield moment per unit length m_p is replaced by the reduced membrane moment $m_m = \frac{1}{4} \nu f_c h^2 \cdot f\left(\frac{\delta}{h}\right)$, f_c being the cylinder compression strength of the concrete, h the slab thickness and ν the effectiveness factor. For slabs with normal horizontal restraints

$$\nu = \frac{2}{\sqrt{f_c}} \qquad (f_c \text{ in MPa}) \qquad (6.11.38)$$

The reduction factor $f\left(\frac{\delta}{h}\right)$ takes into account the reduction because of the deflection, δ being the deflection. The reduction factor may be taken from Fig. 6.11.17, remembering that the curves for a/b = 1 are valid for slabs of the types shown in Fig. 6.11.15. For rectangular slabs having the side length ratio between $\frac{1}{2}$ and 2, $f\left(\frac{\delta}{h}\right)$ might be taken to be

$$f\left(\frac{\delta}{h}\right) \cong 0.43 \qquad (6.11.39)$$

If optimization is carried out on a yield line pattern with negative yield lines, the negative yield moment m_p' should be put equal to zero.

3) The total load-carrying capacity P is

$$P = P_J + P_M \qquad (6.11.40)$$

The reader may now ask why the large reserves in load-carrying capacity inherent in the compressive membrane effects are seldomly utilized in practice.

Two answers may be given to this question. First, a simple theory, which can be used for design purposes, has only recently become available.

Second, one often experiences that for slab systems, in which the individual slabs may have a full load at the same time, the reinforcement which is saved in one slab has to be used in the other slabs to carry the horizontal reactions from the slab in question. So, in fact, in many cases there will be only small differences in the total reinforcement consumption compared with traditional solutions. Only when the maximum live load is confined to one or a few slabs at a time are considerable savings possible. Bridge slabs may belong to this group. Hopefully, in the future, designers will learn how to take into account the beneficial effects of dome action in such cases.

Example 6.11.1 Consider as an example a square slab with side lengths a and with normal horizontal restraints. The slab is loaded by a concentrated load P in the middle and a constant uniform load p. The positive and negative yield moments are m_p and m'_p, respectively.

The slab and the yield line pattern used are shown in Fig. 6.11.18.

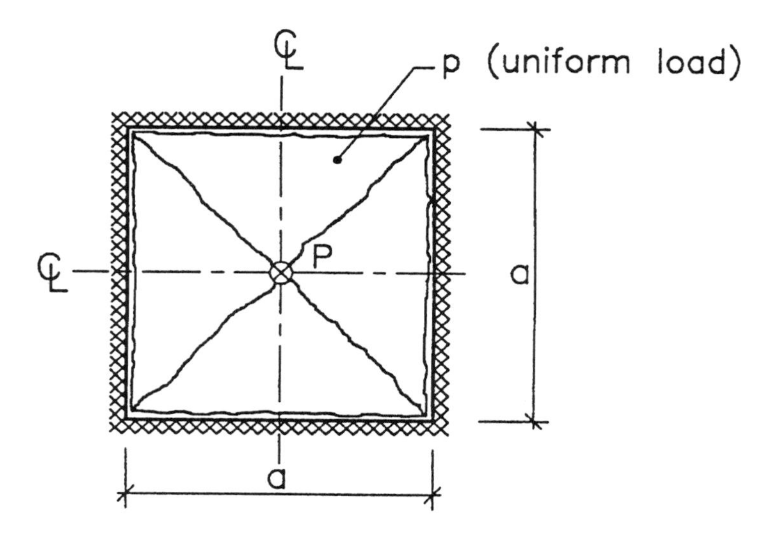

Figure 6.11.18 Loading and yield lines in a square slab.

To find the load-carrying capacity P_J a normal yield line calculation is carried out. This leads to

$$\frac{1}{8}P_J + \frac{1}{24}p\,a^2 = m_p + m'_p$$

assuming that both the concentrated load and the constant uniform load p can be carried.

The load-carrying capacity by dome action P_M is calculated by replacing m_p in the above formula with $\frac{1}{4}\nu f_c h^2 f\left(\frac{\delta}{h}\right)$ and putting $p = m'_p = 0$. This leads to

$$\frac{1}{8}P_M = \frac{1}{4}\nu f_c h^2 f\left(\frac{\delta}{h}\right)$$

where $f\left(\frac{\delta}{h}\right)$ may be taken as 0.43.

The total load-carrying capacity is $P = P_J + P_M$.

An estimate of the uniform load which may be carried by dome action in a square slab may be found setting $f_c = 25$ MPa when calculating ν. The load is found to be

$$P_M \cong h^2 f_c$$

(no ν in this equation)

Chapter 7

PUNCHING SHEAR OF SLABS

7.1 INTRODUCTION

Punching shear in slabs is a two-dimensional analog to shear in beams. However, punching shear is much less critical than beam shear. Failure of slabs by punching occurs mainly in the presence of very highly concentrated loads (e.g., wheel loads on bridge slabs, or at columns).

According to the building codes, slabs are normally designed against punching shear as follows. Consider a cylindrical control surface around the punch (loaded area or column) in a distance proportional to the slab depth (see Fig. 7.1.1). Then the average shear stress τ on the control surface must not exceed a particular design strength which is often taken to be proportional to the tensile strength. The CEB-FIP Model Code [76.11; 78.10] places the control surface at a distance of half the depth h. We will follow this choice here.

The introduction of the tensile strength reflects the common belief that this parameter governs shear failure. This is hardly correct. Indeed, it is not possible to associate a meaningful failure mechanism with the existence of tensile stresses on the control surface. The failure observed in practice occurs in a failure surface (see Fig. 7.1.1) involving sliding as well as separation failure. We would therefore not expect the punching shear failure to be very well described by the analysis offered by the building codes.

On the other hand, experience shows that the average shear stress τ is a good design variable, its ultimate value being fairly independent of the geometry of the problem. It is therefore interesting to note, as will be shown in the following, that this feature is predicted by application of the theory of plasticity. An upper bound analysis is presented, using the failure mechanism sketched in Fig. 7.1.1 and suggested by tests.

Kinnunen and Nylander [60.3] introduced a different approach to punching, based on a collapse mode which is basically a flexural failure. This and other methods were reviewed by Bræstrup [78.17].

In this chapter concrete is identified as a modified Coulomb material.

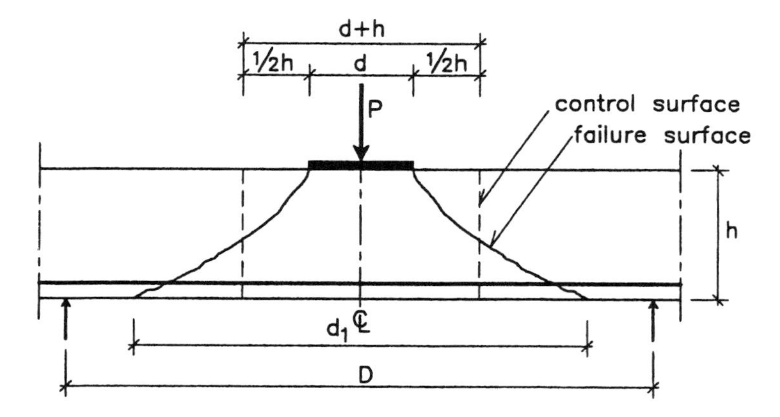

Figure 7.1.1 Notation and definition of control surface.

7.2 INTERNAL LOADS

7.2.1 Concentric Loading, Upper Bound Solution

The failure mechanism. Consider a concrete slab annularly supported and loaded by a circular punch. The slab is reinforced in such a way that flexural failure is prevented. We assume a failure mechanism consisting in the punching out of a solid of revolution, the rest of the slab remaining rigid (see Fig. 7.1.1).

The relative displacement in the failure surface is perpendicular to the main reinforcement. Hence the main reinforcement does not contribute to the load-carrying capacity, dowel action being neglected (cf. Section 2.2.1). The justification for assuming a punching failure mechanism without yielding of the main reinforcement lies in the fact that if the main reinforcement was yielding, the slab would fail in flexure at a lower load, since flexural failure does not involve the dissipation of energy in a failure surface in the concrete passing through the whole depth of the slab.

The generatrix of the axisymmetric failure surface is described by the function $r = r(x)$ and sketched in Fig. 7.2.1. Punch diameter and slab depth are termed d and h, respectively. The relative displacement is u, directed at an angle α to the generatrix.

By the assumed failure mechanism, the strain in the circumferential direction is zero; hence the generatrix may be regarded as a yield line in plane strain. The normality condition then requires that $\alpha \geq \varphi$. Thus the analysis is valid only for $D \geq d + 2h \tan\varphi$, D being the diameter of the annular support.

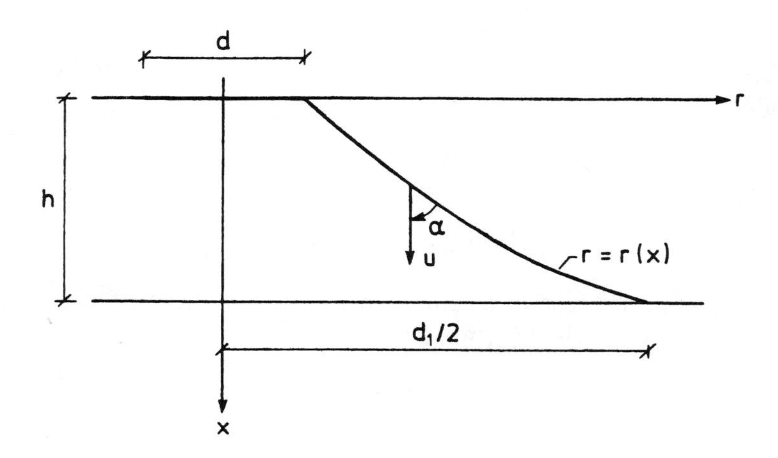

Figure 7.2.1 Failure surface generatrix.

Upper bound solution. An upper bound P for the ultimate punching load is found by equating the work done by the load to the dissipation in the failure surface. The external work is

$$W_E = Pu \tag{7.2.1}$$

The dissipation is found by integration over the failure surface:

$$D = \int W_A dA \tag{7.2.2}$$

where W_A is given by (3.4.86) disregarding the factor b; that is,

$$W_A = \frac{1}{2} f_c u [\ell - m \sin\alpha] \tag{7.2.3}$$

where

$$\ell = 1 - 2 \frac{f_t}{f_c} \frac{\sin\varphi}{1 - \sin\varphi} \tag{7.2.4}$$

$$m = 1 - 2 \frac{f_t}{f_c} \frac{1}{1 - \sin\varphi} \tag{7.2.5}$$

The area element can then be taken as

$$dA = 2\pi r \frac{dx}{\cos\alpha} \tag{7.2.6}$$

The work equation yields

$$Pu = \int_0^h \frac{1}{2} f_c u(\ell - m\sin\alpha) 2\pi r \frac{dx}{\cos\alpha} \tag{7.2.7}$$

Introduction of the relation $\tan\alpha = dr/dx = r'$ leads to the upper bound:

$$P = \pi f_c \int_0^h F(r, r') dx \tag{7.2.8}$$

where

$$F(r, r') = r\left[\ell\sqrt{1 + (r')^2} - m r'\right] \tag{7.2.9}$$

In the special case of a failure surface in the shape of a truncated cone with half-angle α_0, we get

$$r = \frac{d}{2} + x\tan\alpha_0 \tag{7.2.10}$$

$$r' = \tan\alpha_0 \tag{7.2.11}$$

$$F(r, r') = \left(\frac{d}{2} + x\tan\alpha_0\right) \frac{\ell - m\sin\alpha_0}{\cos\alpha_0} \tag{7.2.12}$$

and the upper bound becomes

$$P = \pi f_c \int_0^h F(r, r') dx \tag{7.2.13}$$

$$P = \pi f_c \frac{h}{2} \frac{(d\cos\alpha_0 + h\sin\alpha_0)(\ell - m\sin\alpha_0)}{\cos^2\alpha_0} \tag{7.2.14}$$

The shape of the failure surface that corresponds to the lowest upper bound is determined by calculus of variations. The problem amounts to finding the function $r(x)$ that minimizes the integral $\int_0^h F(r, r') dx$. The Euler equation is (see, e.g., [66.6, p. 206])

$$F - r' F_{r'} = c_1 \tag{7.2.15}$$

c_1 being a constant. The function F is given by (7.2.9); hence

$$r\left(\ell\sqrt{1 + (r')^2} - m r'\right) - r' r\left[\ell \frac{r'}{\sqrt{1 + (r')^2}} - m\right] = c_1 \tag{7.2.16}$$

This may be reduced to

$$1 + (r')^2 = \frac{r^2}{c^2} \tag{7.2.17}$$

or

$$r'' = \frac{r}{c^2} \tag{7.2.18}$$

where

$$c = \frac{c_1}{\ell} \tag{7.2.19}$$

The complete solution to (7.2.18) is

$$r = a\cosh\frac{x}{c} + b\sinh\frac{x}{c} \tag{7.2.20}$$

which describes a catenary curve.

This function satisfies (7.2.17) provided that $c^2 = a^2 - b^2$, and without loss of generality we may take

$$c = \sqrt{a^2 - b^2} \tag{7.2.21}$$

The constants a and b are determined by the boundary conditions; $r = d/2$ for $x = 0$ yields $a = d/2$. If $2r = d_1$ for $x = h$, then b is found from the equation

$$\frac{d_1}{2} = a\cosh\frac{h}{c} + b\sinh\frac{h}{c} \tag{7.2.22}$$

The angle α between the displacement u and the failure surface is given by

$$\tan\alpha = r' = \frac{a}{c}\sinh\frac{x}{c} + \frac{b}{c}\cosh\frac{x}{c} \tag{7.2.23}$$

The minimum value $\tan\alpha = b/c$ is obtained at $x = 0$. Since $\alpha \geq \varphi$ we find that $b/c \geq \tan\varphi$. For some values of d, h, and d_1, this condition cannot be satisfied by the function given by (7.2.20). In such cases, the failure surface generatrix will consist of a straight line in combination with the catenary curve, as sketched in Fig. 7.2.2. The generatrix $r = r(x)$ is then

$$r = \begin{cases} \dfrac{d}{2} + x\tan\varphi & \text{for } 0 \leq x \leq h_0 \quad (7.2.24) \\[2ex] a\cosh\dfrac{x - h_0}{c} + b\sinh\dfrac{x - h_0}{c} & \text{for } h_0 \leq x \leq h \quad (7.2.25) \end{cases}$$

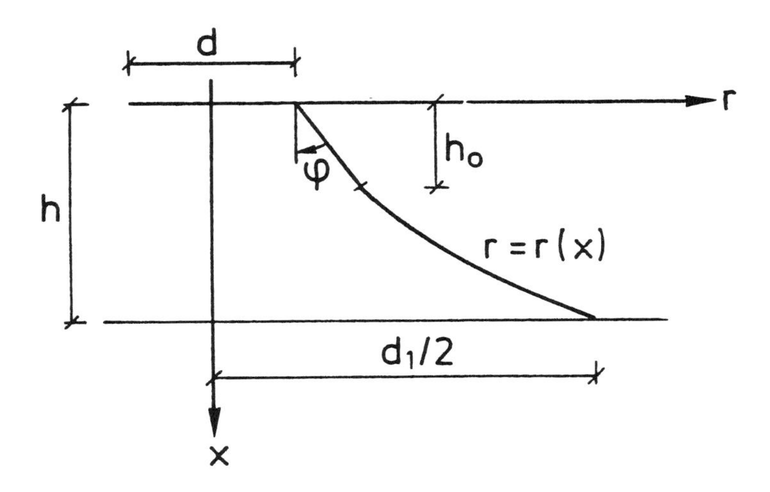

Figure 7.2.2 Failure surface generatrix with a straight line in combination with a catenary curve.

The four constants h_0, a, b, and c are determined by the equations

$$c = \sqrt{a^2 - b^2} \tag{7.2.26}$$

$$a = \frac{d}{2} + h_0 \tan\varphi \qquad (x = h_0) \tag{7.2.27}$$

$$\tan\varphi = \frac{b}{c} \qquad (x = h_0) \tag{7.2.28}$$

$$\frac{d_1}{2} = a\cosh\frac{h - h_0}{c} + b\sinh\frac{h - h_0}{c} \qquad (x = h) \tag{7.2.29}$$

The lowest upper bound for the ultimate load is found from (7.2.8) and (7.2.9), the function $r(x)$ being given by (7.2.24) and (7.2.25).

Thus P is the sum of two parts:

$$\boldsymbol{P} = \boldsymbol{P}_1 + \boldsymbol{P}_2 \tag{7.2.30}$$

where P_1 and P_2 are the contributions from the conical surface and the catenary of revolution, respectively.

P_1 is found from (7.2.14) with $\alpha_0 = \varphi$ and $h = h_0$:

$$P_1 = \pi f_c \frac{h_0}{2} \frac{(d\cos\varphi + h_0 \sin\varphi)(1 - \sin\varphi)}{\cos^2\varphi} \tag{7.2.31}$$

Equations (7.2.8) and (7.2.9) yield

$$P_2 = \pi f_c \int_{h_0}^{h} r\left(\ell\sqrt{1+(r')^2} - m r'\right) dx \tag{7.2.32}$$

Here $r\ dx = c^2 d(r')$ by (7.2.18); hence

$$\begin{aligned}
&\int F(r,r')dx \\
&= \ell \int \sqrt{1+(r')^2}\, c^2 d(r') - m\int r\,dr \\
&= \ell c^2\left[\frac{r'}{2}\sqrt{1+(r')^2} + \frac{1}{2}\ln\left(r' + \sqrt{1+(r')^2}\right)\right] - \frac{1}{2} m r^2 + C_1 \\
&= \ell c^2 \frac{r'}{2}\frac{r}{c} + \frac{1}{2}\ell c^2 \ln\left(r' + \frac{r}{c}\right) - \frac{1}{2} m r^2 + C_1 \\
&= \frac{1}{2}\ell c^2 \ln\left[\frac{a+b}{c}\left(\cosh\frac{x-h_0}{c} + \sinh\frac{x-h_0}{c}\right)\right] + \frac{1}{2} r(\ell c r' - m r) + C_1 \\
&= \frac{1}{2}\ell c(x-h_0) + \frac{1}{2} r\left[\ell\left(b\cosh\frac{x-h_0}{c} + a\sinh\frac{x-h_0}{c}\right)\right. \\
&\quad \left. - m\left(a\cosh\frac{x-h_0}{c} + b\sinh\frac{x-h_0}{c}\right)\right] + C
\end{aligned} \tag{7.2.33}$$

where (7.2.17) and (7.2.25) have been used, and

$$C = C_1 + \frac{1}{2}\ell c^2 \ln\frac{a+b}{c} \tag{7.2.34}$$

Thus

$$\begin{aligned}
P_2 = \pi f_c &\left[\frac{1}{2}\ell c(h-h_0) + \frac{1}{2}\frac{d_1}{2}\left(\ell\left(b\cosh\frac{h-h_0}{c} + a\sinh\frac{h-h_0}{c}\right)\right.\right. \\
&\left.\left. - m\left(a\cosh\frac{h-h_0}{c} + b\sinh\frac{h-h_0}{c}\right)\right) - \frac{1}{2}a(\ell b - m a)\right]
\end{aligned} \tag{7.2.35}$$

Using (7.2.26) and (7.2.29), this may, as shown by Kærn and Jensen [76.8], be reduced to

$$P_2 = \frac{1}{2}\pi f_c \left[\ell c(h - h_0) + \ell \left(\frac{d_1}{2}\sqrt{\left(\frac{d_1}{2}\right)^2 - c^2} - ab \right) - m\left(\left(\frac{d_1}{2}\right)^2 - a^2 \right) \right] \tag{7.2.36}$$

The solutions presented above were developed by Bræstrup et al. [76.6], Uwe Hess et al. [78.11], and Bræstrup [78.17]. The theory has been extended to punching of hollow spheres by Morley [79.9].

Analytical results. The values of $\tan\varphi$, f_c, and f_t, as well as the punch diameter d, support diameter D, and slab thickness h, are assumed to be given. The shape of the optimal failure surface and the corresponding least upper bound may be found by iteration on an electronical computer. The strategy of the iteration process is as follows:

1. Assume a value of the opening diameter d_1, that is, the diameter of the intersection of the failure surface with the bottom of the slab.
2. Assume that $h_0 = 0$.
3. Determine the constant a from (7.2.27) and assume a value of the constant b.
4. Calculate c by (7.2.26) and the radius r_1 given by the right-hand side of (7.2.29). If r_1 is not sufficiently close to $d_1/2$, change the value of b and repeat the step.
5. If $h_0 = 0$, check if $b/c \geq \tan\varphi$. If this is not the case, assume a value $h_0 \neq 0$ and repeat the process from Step 3. If $h_0 \neq 0$, check if (7.2.28) is satisfied. If not, change the value of h_0 and repeat the process from Step 3.
6. Determine P by (7.2.30).

By this procedure a failure surface and an upper bound P are obtained corresponding to the choice of d_1.

The value of d_1 that gives the least upper bound is strongly dependent on the assumed tensile concrete strength. For $f_t = 0$, P decreases with increasing d_1, which means that the failure surface will always extend all the way to the support (i.e., $d_1 = D$).

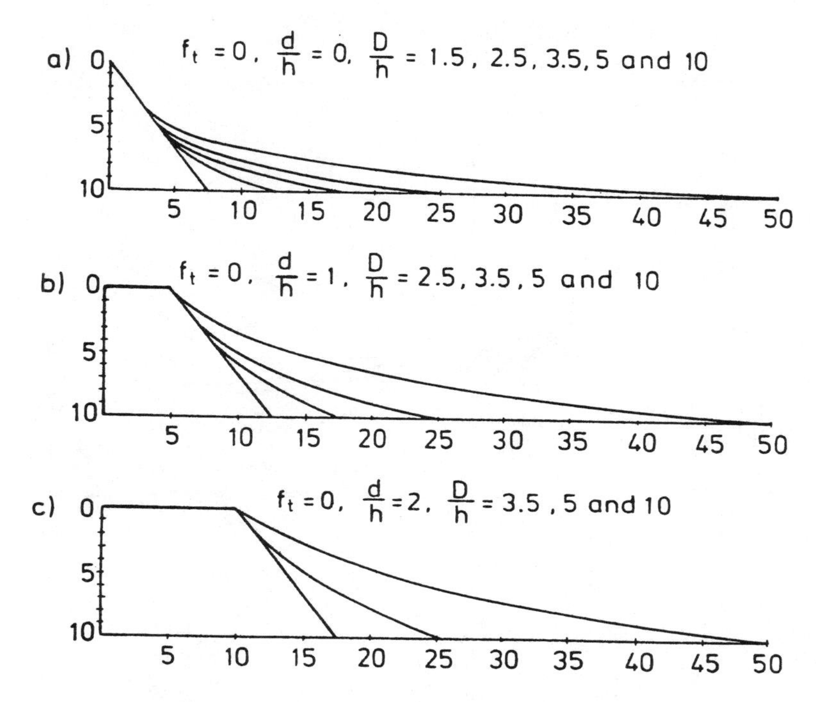

Figure 7.2.3 Failure surfaces for different values of punch diameter and slab diameter with the tensile strength equal to zero.

Examples of failure surface generatrices are shown in Fig. 7.2.3. The angle of friction is assumed to be φ with $\tan\varphi = 0.75$ (see Section 2.1.3). This angle is used in all the calculations below. The shape of the surface is determined by the relative punch diameter d/h and the relative support diameter D/h. As a non-dimensional load parameter it turns out to be very convenient to use the quantity τ/f_c, where $\tau = P/\pi(d + h)h$ is the nominal shear stress on a cylindrical surface with diameter $d + h$ (the control surface, Fig. 7.1.1).

The load parameter is plotted as a function of the relative support diameter in Fig. 7.2.4a. The result is rather insensitive to the value of φ assumed, except for the smallest support diameters. The load approaches zero asymptotically as the support diameter increases toward infinity.

Fig. 7.2.4a may, of course, also be used to find the load-carrying capacity as a function of any opening diameter $d_1 = D$ for $f_t = 0$. Because of that Fig. 7.2.4a has numerous practical applications (see below).

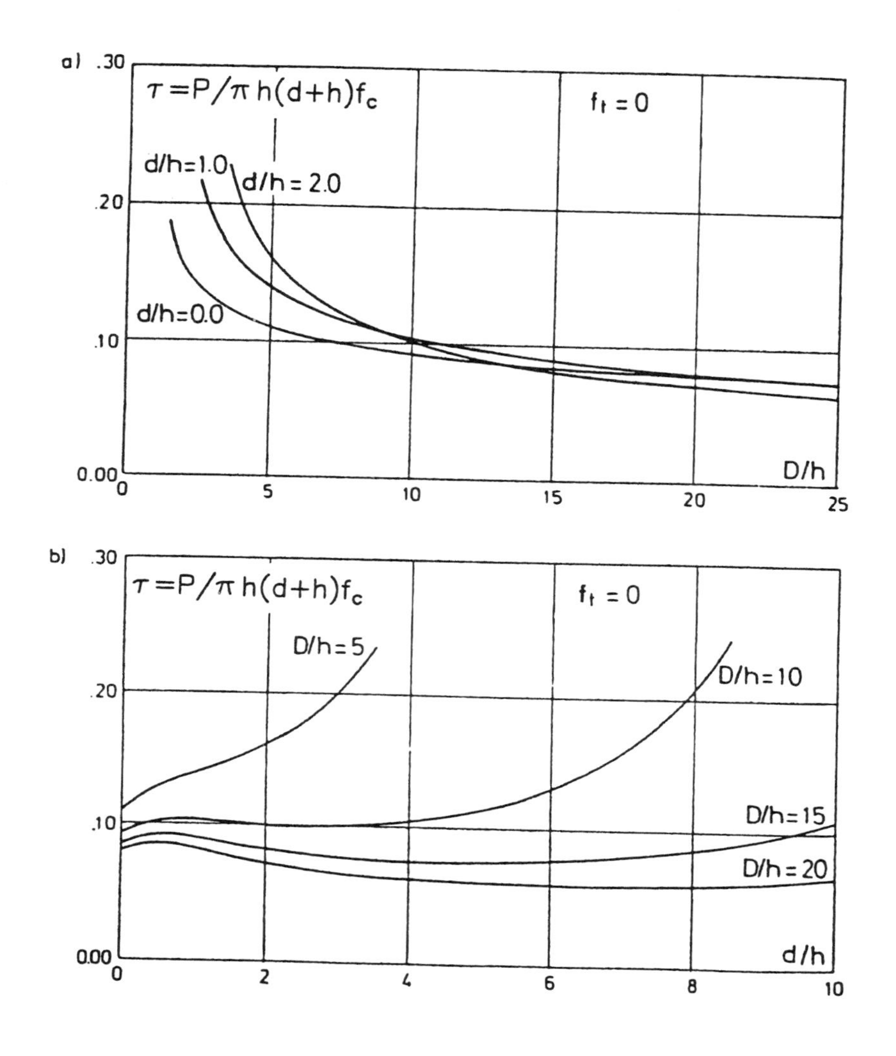

Figure 7.2.4 Punching shear capacity versus slab diameter and punch diameter, the tensile strength being zero.

Figure 7.2.4b shows the load parameter as a function of the relative punch diameter. Note (in both figures) that when the punch diameter is not too large compared with the support diameter, the load parameter is fairly independent of the relative punch diameter. This fact corroborates a design method based on the nominal shear stress τ defined above.

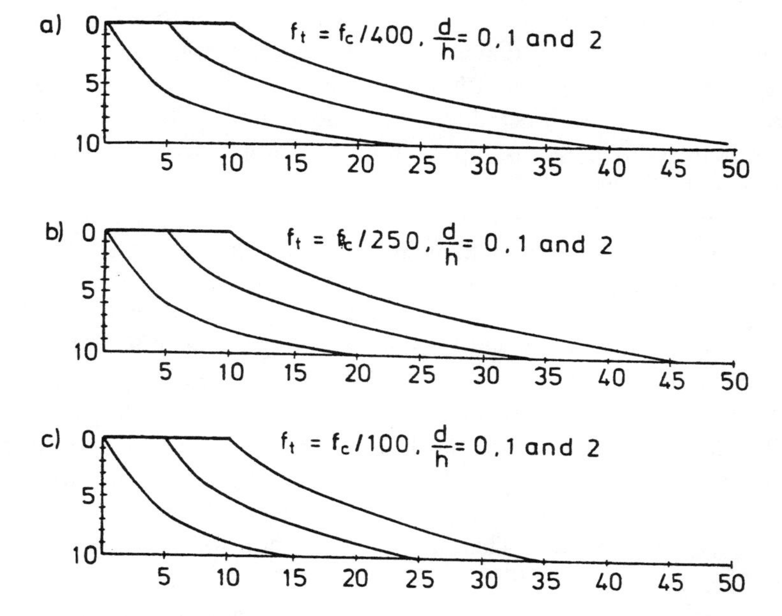

Figure 7.2.5 Failure surfaces for different values of tensile strength and punch diameter.

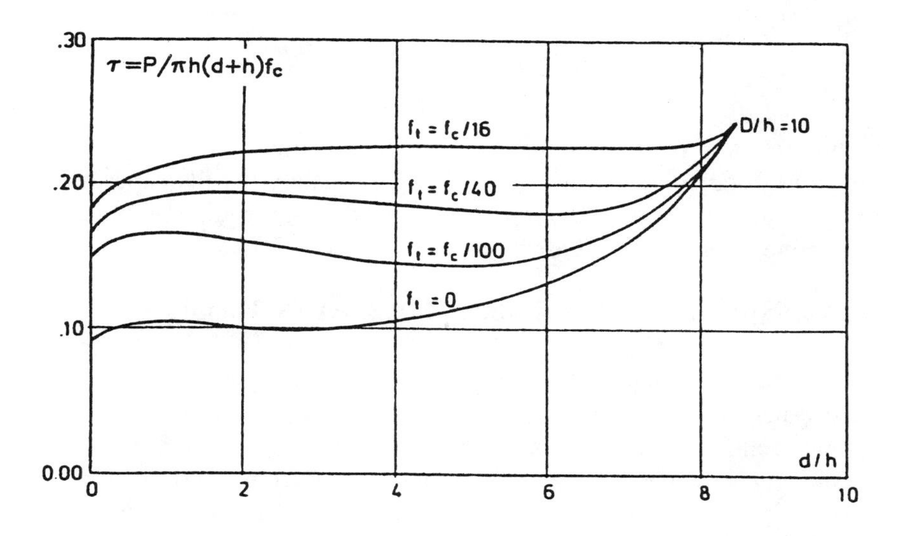

Figure 7.2.6 Punching shear load versus punch diameter for different values of tensile strength.

The introduction of a finite tensile strength leads to a finite value of the optimum opening diameter d_1. This value is found by iteration. As a starting point $d_1 = d + 2h\tan\varphi$ is chosen. The support diameter must not be less than this value in order to comply with the condition $\alpha \geq \varphi$. The corresponding upper bound is found by the procedure described by steps 2 through 6 in the previous list. The diameter d_1 is then increased until a minimum of P is found or $d_1 = D$. When the support diameter is greater than the optimum opening diameter, the ultimate load becomes independent of the support diameter.

Figure 7.2.5 shows examples of optimum failure generatrices. The support diameter is supposed to be great enough so as not to interfere with the solution. The optimal diameter turns out to be approximately a linear function of the punch diameter.

The load parameter is still fairly independent of the relative punch diameter, as seen in Fig. 7.2.6. From this figure it appears that theoretically the tensile strength has a considerable influence on the load-carrying capacity. This influence is not found in tests (see below).

It is worthy of note that the ultimate load is independent of the tensile strength, provided that the support diameter has the value $D = d + 2h\tan\varphi$. In this case the failure surface degenerates into a truncated cone and the ultimate load is given by (7.2.14) with $\alpha_0 = \varphi$, that is

$$P = \pi f_c h(d + h\tan\varphi)\,\frac{1 - \sin\varphi}{\cos\varphi} \qquad (7.2.37)$$

Thus P is proportional to f_c. This means that if the condition $D = d + 2h\tan\varphi$ is satisfied, at least approximately, by the experimental setup, punching tests may be used to determine the compressive concrete strength (cf. Jensen and Bræstrup [76.7]).

7.2.2 Experimental Verification, Effectiveness Factors

Failure surface. Punching tests on slabs generally do not allow close inspection of the failure surface because it is disturbed by the main reinforcement. This inconvenience is not present if the punching is achieved by pulling out a disk embedded in a plain concrete block. Figure 7.2.7 shows the result of a pull-out test done by Hess [75.5]. Opposite faces of the block were tested simultaneously, which means that there was no annular counterpressure (support). The shape of the failure surface agrees reasonably well with the theoretical predictions shown in Fig. 7.2.5. In [76.6], additional examples are cited.

In applying the theory of plasticity to concrete, we would prefer to neglect the tensile strength, in order to take into account the very limited ductility of concrete in tension. This would mean that the theoretical failure surface always extends to the support, and in the case of no support the specimen splits along a plane at the embedment depth. These results are obviously at variance with experience. Thus realistic predictions are obtained only at the cost of introducing some tensile strength. We get an idea of the appropriate amount by comparing the diameters of observed failures with the theoretical opening diameter. In [76.6] it was found that a tensile strength $f_t = f_c/400$ would give realistic results. This extremely low value indicates that the effectiveness factor of the tensile strength is very small.

Slabs, on the other hand, are always supported at a reasonable distance from the loaded area. Therefore, the tensile concrete strength can prudently be neglected, and the optimum failure surface assumed to extend all the way to the nearest support.

Ultimate load. As pointed out by Dragosavic and van den Beukel [74.5], it is extremely important when analyzing punching shear test results to exclude all bending failures. Many failures of slabs with concentrated loads are terminated by a punching failure when the crack widths have become sufficiently large. However, when calculating the bending capacity, using, for instance, yield line theory, or when studying the strains in the reinforcement, it may become clear that one is dealing primarily with failure by yielding of the reinforcement and that punching shear is a secondary phenomenon.

Figure 7.2.7 Punching shear failure surface in a pull-out test.

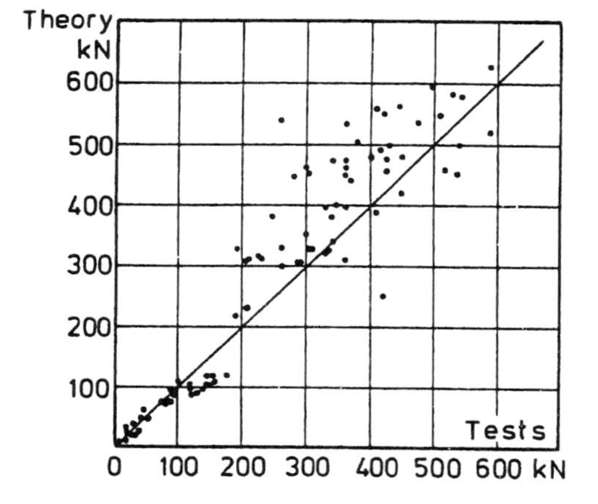

Elstner + Hognestad [56.6]
Base [59.2]
Kinnunen + Nylander [60.3]
Taylor + Hayes [65.11]
Dragosavic + van den Beukel [74.5]
Kærn + Jensen [76.8]

Figure 7.2.8 Calculated punching shear loads compared with test results.

A number of tests have been compared with the theoretical results in Fig. 7.2.8 [78.11]. The tensile strength in this diagram has been put equal to zero and the effectiveness factor on the compressive strength equal to

$$\nu = \frac{4.22}{\sqrt{f_c}} \ngtr 1 \quad (f_c \text{ in MPa}) \tag{7.2.38}$$

The scatter is seen to be considerable. In the ratio test result/-theory the coefficient of variation is about 21%. However, in the individual series, the scatter is much smaller, which means that a considerable part of the scatter stems from the different ways used to measure the compressive strength.

The ν-value (7.2.38) is rather high, in fact one of the highest ones found in shear problems in plastic theory. This might indicate that the load-carrying capacity is not seriously influenced by cracks.

This is easy to understand if the tests are made on simply supported circular slabs. In these slabs the cracks will mainly be radial, i.e., they are not interfering with the punching failure surface. For restrained slabs, circumferential cracks will develop from the top face and these cracks are more likely to interfere with the failure surface. This problem will be discussed below when treating edge and corner loads.

As already mentioned, it appears from Fig. 7.2.4a that, except for small D-values, the load-carrying capacity is rather independent of

D for realistic values of this parameter. Thus in most practical cases one might take the average shear stress τ in the control surface to be a constant independent of D as well as of d. This constant value may, according to Fig. 7.2.4a, be taken to be approximately $0.08\, f_c$, which means that the ultimate value of the average shear stress in the control surface $\tau = \tau_u$ is

$$\tau = \tau_u = 0.08\,\nu f_c \tag{7.2.39}$$

Inserting ν according to formula (7.2.38) we get

$$\tau = \tau_u = 0.08\,\frac{4.22}{\sqrt{f_c}}\,f_c = 0.34\,\sqrt{f_c} \tag{7.2.40}$$

According to the expression for f_t often used in this book, $f_t = \sqrt{0.1 f_c} = 0.32\,\sqrt{f_c}$ (f_c in MPa), we get approximately

$$\tau = \tau_u \cong f_t \tag{7.2.41}$$

This result is quite interesting. It means that when the control surface is a cylindrical surface at a distance of $\frac{1}{2}h$ from the loaded area, punching failure will take place when the average shear stress in the control surface equals f_t. This simple result may be said to justify our choice of control surface. Of course, many other choices could have been made (and are made) leading to other values of τ_u but the simple result (7.2.41), which is easy to remember, gives some preference to the particular control surface used here.

The result (7.2.41) is in a way rather confusing, because it may lead to the perception that punching failure loads are proportional to the tensile strength of concrete. It appears from our analysis that this is not so. In fact, the effective tensile strength in punching shear is very small as stated above. It is evident that we have arrived at this paradoxical result because the effectiveness factor ν is inversely proportional to $\sqrt{f_c}$ which takes us to a load-carrying capacity proportional to $\sqrt{f_c}$, i.e., proportional to the tensile strength.

The large scatter related to the ν-formula (7.2.38) naturally encourages us to look for improvements. It was shown by T. G. Sigurdsson [91.26] that the formula

$$\nu = \frac{1.47}{\sqrt{f_c}}\left(1 + \frac{0.48}{\sqrt{h}}\right)(1 + 0.125\,r) \ngtr 1 \tag{7.2.42}$$

reduces the coefficient of variation to 16%. In this formula f_c is in MPa, the depth of the slab h is in m and the reinforcement ratio r in the face of the slab opposite to the punch is in %. As for nonshear reinforced beams there is a scale effect, i.e., increasing depth leads to decreasing ν. If the reinforcement ratio is different in the usual two perpendicular reinforcement directions x and y, we may take $r = \sqrt{r_x r_y}$.

The formula (7.2.42) was verified for f_c up to about 50 MPa, slab depths h up to 0.75 m and reinforcement ratios r up to 3.7%.

As before, when ν has been calculated the value at failure of the average shear stress in the control surface is calculated by means of (7.2.39).

It appears, once again, that plastic theory conforms very well with well-established empirical formulas, when a proper value of the effectiveness factor is selected.

We conclude this section with a remark about lower bound analysis for punching shear. Such an analysis has not yet been carried out. However, it is rather evident what the solution qualitatively will be when $f_t = 0$. An axisymmetric shell containing the failure surface and being under compression will transfer the load to the supports. Although we do not yet know the solution in detail, it is evident that it is important to anchor the reinforcement well at the supports.

7.2.3 Practical Applications

The above results may be used for slabs with variable thicknesses. In the case shown in Fig. 7.2.9 the thickness is locally reduced. It is evident that the thickness to apply in the punching analysis is the reduced thickness h because the failure surface will only pass through the region with reduced thickness.

In Fig. 7.2.10 a slab with increased thickness around a column support is shown. A failure surface originating from a circle with the diameter d and extending to the opening diameter D may be analyzed using the above results. The load-carrying capacity is determined using the values D, d and h stated in the figure.

In fact, in the case shown only two failure surfaces have to be checked, namely those originating from the points A and B, respectively, as indicated in the figure. This is because AB is assumed to be a straight line. If the slab thickness between A and B is less than shown in Fig. 7.2.10, more failure surfaces have to be checked. However, the procedure is the same.

The opening diameter D may normally be chosen as the longest distance to the supports.

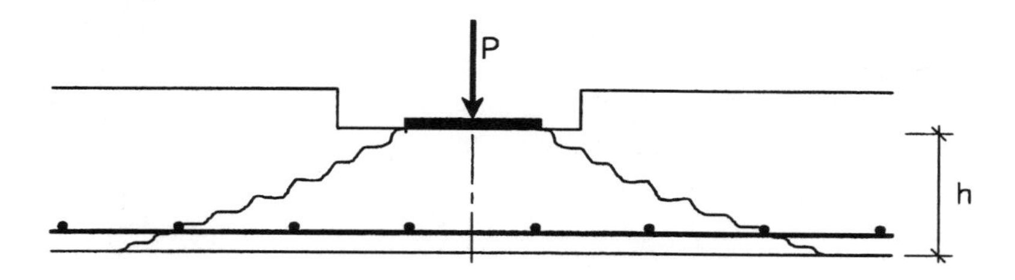

Figure 7.2.9 Slab with locally reduced thickness.

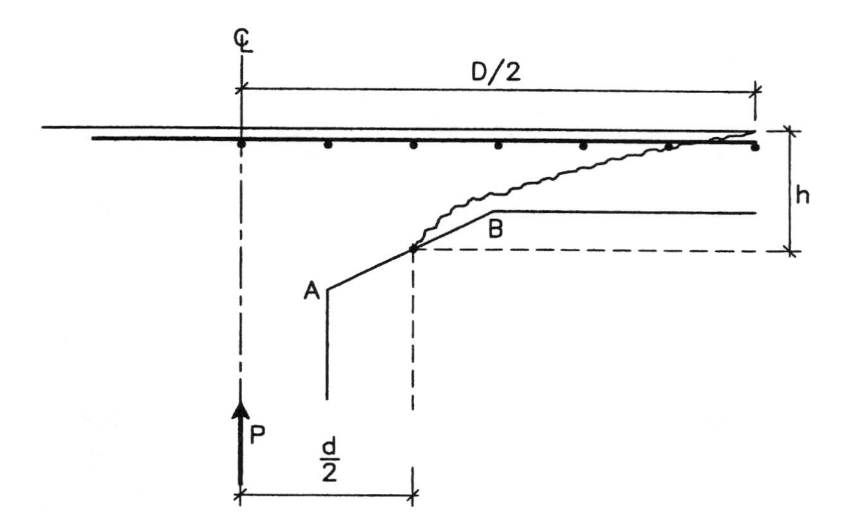

Figure 7.2.10 Slab with variable thickness.

When D, d and h have been determined, Fig. 7.2.4a may be used to find the average shear stress in the control surface at failure.

In practice many forms of loaded areas other than the circular area, and many types of slabs other than circular slabs are met.

The upper bound solution is, however, easy to extend to other cases. This is normally done by determining the shortest control perimeter lying at a distance not less than $\frac{1}{2}h$ from the loaded area.

Fig. 7.2.11 shows some examples of control perimeters. Notice that the control perimeter for a rectangular area is the same as the control perimeter for a circular area with the same perimeter as the rectangle.

If the loaded area is not convex the envelope curve should be used instead, because it leads to a smaller length of the control perimeter. Such a case has been illustrated in the figure.

When the length u of the control perimeter has been determined, the load-carrying capacity is simply

$$P = \tau_u u h \tag{7.2.43}$$

The value τ_u of the average shear stress at failure along the control surface uh is determined in the same way as before.

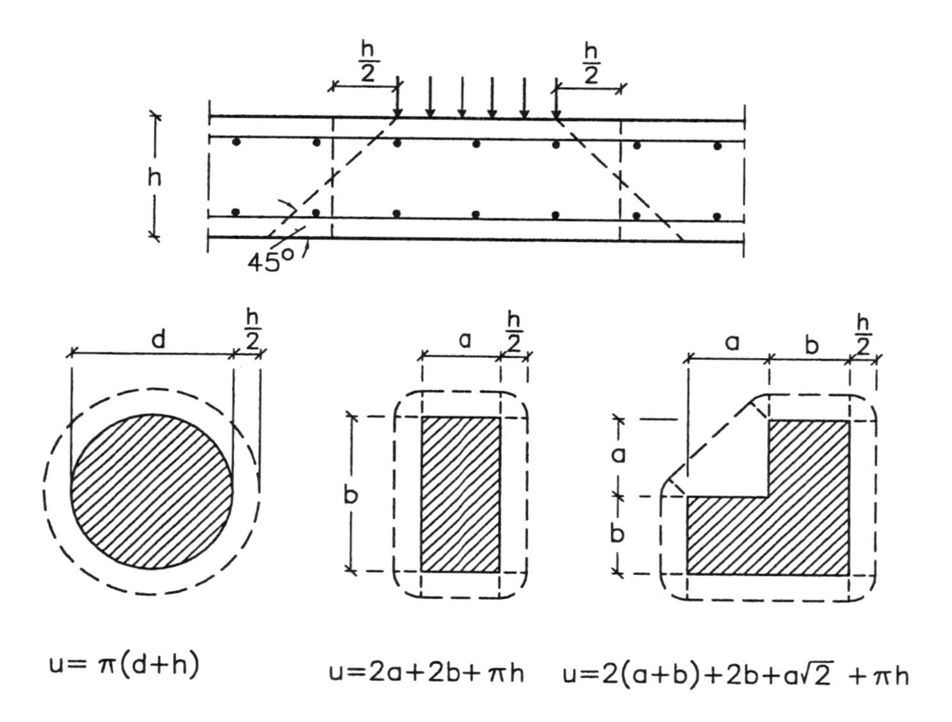

Figure 7.2.11 Control perimeters.

The failure surface may be envisaged by taking the generatrix of the axisymmetric failure surface, putting it into a vertical plane perpendicular to the periphery of the loaded area at an arbitrary point and carrying it around the loaded area. If the periphery of the loaded area is drawn on the loaded face the failure surface will, of course, originate from this curve. In a corner the plane of the generatrix is rotated about a vertical line through the corner. The surface generated will be a geometrically possible failure surface for the loaded area in question. If the parameter D is needed it may normally be taken to be the longest distance to the support as before. If d is needed it may be calculated as the diameter of a circle with the same perimeter as the area in question.

The use of formula (7.2.43) does not exactly lead to the upper bound solution for the failure surface described above, but the deviation is considered small.

This simple extension of the upper bound analysis probably will work well for loaded areas not too far from a circle. Problems may be encountered when the area is far from a circle. Consider as an example an elongated rectangular area. We may still get an upper bound in the way described but it may not be a good approximation of the real failure load. The reason is that the failure surface along the longer sides of the rectangle might be closer to the beam solution than to the punching solution, i.e., the generatrix of the failure surface may approach a straight line in a part of the longer edge instead of being a catenary curve. This is only serious if cracks are developing parallel to the longer sides. Such cracks may reduce the contribution from the longer sides to roughly half the contribution without such cracks (cf. Section 2.1.4). The problem has not yet been thoroughly studied in relation to plastic theory, so the designer is left with empirical code formulas taking this aspect into consideration (see, for instance, [78.10] and [93.10]).

Similar comments may be made regarding the effect of holes near the loaded area. One may take advantage of the rule of thumb that holes at a distance of more than, say, 6 times the slab thickness may be disregarded. Often it is practical to reinforce around the hole in such a way that the concentrated load and other loads may be transferred around the hole by supplying extra reinforcement (cf. Section 4.7.3). Then the hole may be completely neglected. Upper bound solutions taking into account the presence of a hole by disregarding a part of the failure surface may sometimes be used. General rules are not appropriate at this stage of development.

We conclude this section with a remark about the case where the load is transferred to the slab by a wall. Here it might be necessary to consider the stiffness of the wall compared with the stiffness of the slab. Often the wall will be so stiff that the load is transferred exclusively at the ends of the wall, which makes the punching problem more serious than if the load is uniformly distributed along the whole wall/slab section.

7.2.4 Eccentric Loading

Above we have described what is known in relation to plastic theory regarding concentric loading. In practice eccentric loading is, however, often met.

This problem arises, for instance, when a slab is supported on a column which may, and normally will, transfer bending moments to the slab.

The problem is, of course, easily solved if local beams are designed to carry the loads to the supports. This solution should always be considered. In the following we envisage the slab to carry the moments without extra beams.

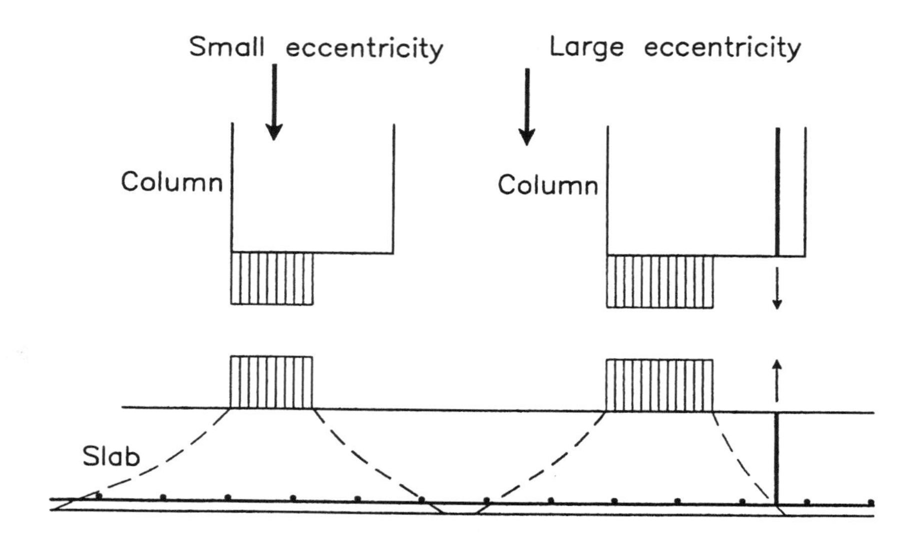

Figure 7.2.12 Column/slab connection with normal force and bending moment.

Eccentric loads may reduce the load-carrying capacity considerably. One might think that only the normal force is important for the punching shear analysis because the bending moment in the column may be carried by bending and torsional moments in the slab. This is actually quite wrong, which may be illustrated by a simple example.

Consider Fig. 7.2.12, which shows two cases where a column delivers an eccentric normal force to a slab.

In the case to the left the eccentricity is small. The effect of moving the force from a concentrical position to an eccentrical one is only to move the loaded area and reduce it. The load-carrying capacity will be reduced because the control perimeter is shorter but the reduction is not significant. In the case to the right the eccentricity is so large that tensile reinforcement in the column is necessary. The compression force is now larger than the normal force in the column and the load-carrying capacity may be significantly reduced. An estimate may be made by calculating the compression zone in the column. This zone is then used as the loaded area in a punching analysis which is carried out as explained above. The resultant of the compression stresses in the compression zone is the punching load. This calculation may not be very accurate because the failure mechanism is different from the usual one. In the actual case the failure surface along one edge of the column section, shown schematically in the figure, would be rather steep and the contribution from this part of the failure surface would be larger than normal.

The anchorage of the tensile reinforcement must, of course, be checked by a similar calculation.

A solution to the problem of eccentric loading adopted frequently rests on the assumption that the failure takes place around the column without involving an internal failure surface between the compression zone and the tensile reinforcement. This may be correct.

It is customary to consider the same control surface as the one used for concentric loading. The vertical shear stresses along the control surface are assumed to vary linearly in the same way as the normal stresses in a linear elastic beam subjected to a bending moment and a normal force. Then, in fact, the usual type of formula may be used to calculate the shear stresses. Failure is supposed to be reached when the maximum shear stress in the control surface attains the shear stress $\tau = \tau_u$, i.e., the same value as for concentric loading.

There are several objections to this method. First of all, to make the calculation in agreement with tests it is necessary to reduce the moment contributions in an arbitrary way by an empirical factor, typically of the order 0.5. This is not only due to the assumed linear elastic stress distribution but also due to the fact that part of the moments may be carried by bending and torsional moments in the slab, i.e., the moments are not carried fully by the vertical shear stresses in the control surface. What is even worse is that concentric loading, by using this method, must be defined as a load, the position of which corresponds to a uniform shear stress distribution in the control surface. This may be completely wrong, which is clearly demonstrated by an example in Section 7.3 dealing with edge and corner loads.

Therefore this method cannot be recommended even if it has rendered rather good results when modified as described (see for instance [74.5]).

The stress distribution from the combined action may be rather complicated. So it is worthwhile to look for a simplified method.

One such method has already been described and used in Chapter 6 (see Section 6.6.8). There we dealt with upper bound solutions and proved that if P_{1p} is the load-carrying capacity for one load action and P_{2p} the load-carrying capacity for another load action, etc., then if P_1 is the actual value of the load parameter in the first loading case, P_2 the load parameter in the second loading case, etc., the structure will be safe for the combined loading if

$$\frac{P_1}{P_{1p}} + \frac{P_2}{P_{2p}} + \ldots \leq 1 \qquad (7.2.44)$$

When dealing with lower bound solutions the validity of (7.2.44) is evidently due to the convexity of the yield surface.

When applying (7.2.44) to eccentric loading we only have to calculate the load-carrying capacity for a concentric load and for pure moment loading. These load-carrying capacities may be determined by either lower bound or upper bound methods.

The applicability of the linear interaction formula (7.2.44) was studied by Larsen [98.1] in relation to plastic theory. The formula rendered rather good results in the case of internal loads. Regarding edge and corner loads the reader is referred to Section 7.3.5.

The calculations which have to be performed are best illustrated by a simple example. Consider a simply supported rectangular slab $2\ell_1 \cdot 2\ell_2$ (Fig. 7.2.13). The slab is isotropically reinforced parallel to the edges in both the top face and the bottom face, and the corresponding yield moments m_p and m_p', respectively, are equal. The loading is a concentrated force in the middle point. The force is distributed over the rectangular area $2a \cdot 2b$ with sides parallel to the supports of the slab. The loaded area is considered a rigid block which might be materialized as a column monolithically cast together with the slab. This means that we will not be concerned with the stress field within the rigid block.

The exact solution for a square slab with a point load in the middle and with $m_p = m_p'$ was determined in Section 6.6.7. The load-carrying capacity was found to be $P = 8m_p$.

The corresponding yield line pattern is shown in Fig. 6.6.31. A statically admissible moment field in each of the four triangles into which the slab is subdivided by the yield lines may be taken to be the one summarized in formula (6.5.54).

The load may also be envisaged to be carried by four torsion fields, one in each of the squares appearing when subdividing the slab into four equal squares with sides parallel to the supports. Apart from sign the torsional moments are equal in the four squares. When the torsional moments are set equal to the torsional yield moment t_p, each of the four squares will carry the load $2t_p$. Thus this lower bound solution furnishes the load-carrying capacity $P = 8t_p$. Since for small reinforcement degrees $m_p = m_p' \cong t_p$ [cf. formula (2.3.18)] we find the same load-carrying capacity as before.

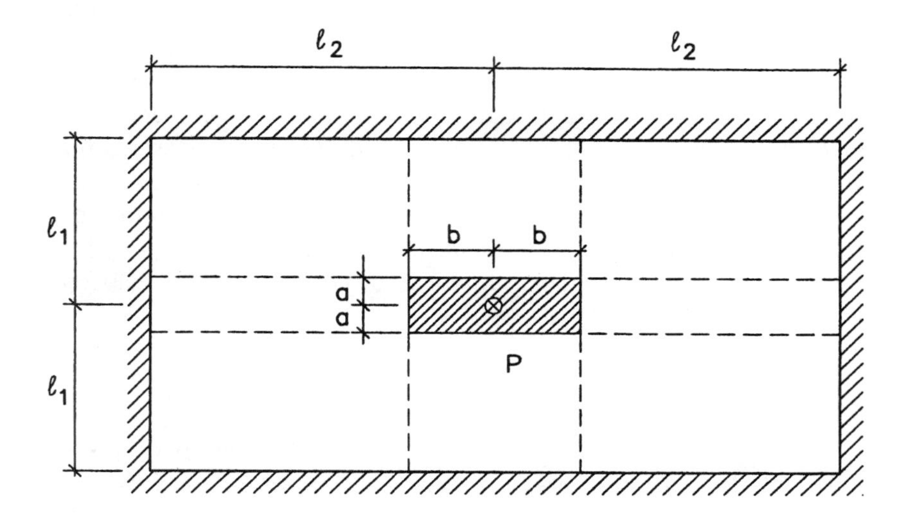

Figure 7.2.13 Concentrated load on a rectangular area $2a \cdot 2b$ in a rectangular, simply supported slab $2\ell_1 \cdot 2\ell_2$.

In fact, the torsion field solution may be used as a lower bound solution for any position of the load on the slab.

We elaborate more on the calculation of the torsional yield moment in Section 7.3. It will appear there, as already mentioned in Section 2.3 dealing with the yield conditions for slabs, that for higher reinforcement degrees $t_p < m_p$ even more so because the ν-value for torsion is less than that for bending.

The torsion fields have been schematically illustrated in Fig. 7.2.14. The load is carried by the vertical shear stresses (equivalent to the Kirchhoff forces) which are part of the shear flow produced by the torsional moments (cf. Section 5.3 on torsion in beams). Since the vertical shear stresses may require a length up to half the slab thickness, the length a (see Fig. 7.2.14) should, when the limit of t_p is reached, not be less than half the slab thickness. Strictly speaking the vertical shear stresses require vertical reinforcement. However, since punching shear is always checked it will probably not be necessary to provide vertical reinforcement to carry the vertical shear stresses.

It turns out that when a is about $h/2$, the maximum load which can be carried by the torsion fields will be approximately equal to the load which can be carried by punching shear. When a is less than $h/2$, punching shear will be decisive; when a is larger, the torsional capacity will be decisive.

The torsion field solution may easily be extended to our example in Fig. 7.2.13. This means that in the corners of the loaded area $2a \cdot 2b$ we may carry concentrated loads $2t_p$. To these loads we may add the loads which may be carried by strip action in two perpendicular strips of widths $2a$ and $2b$, respectively. It is easily found that the load which can be carried by strip action is $4m_p\, a/\ell_2 + 4m_p\, b/\ell_1$. Putting, on the safe side, $m_p \cong t_p$ we find the load-carrying capacity by concentric loading

$$P = P_p = \left(8 + 4\,\frac{a}{\ell_2} + 4\,\frac{b}{\ell_1}\right) t_p \tag{7.2.45}$$

Of course, the load-carrying capacity may also be governed by punching shear. In this case

$$P = P_p = \tau_u (4a + 4b + \pi h)\, h \tag{7.2.46}$$

The smaller of (7.2.45) and (7.2.46) determines the load-carrying capacity $P = P_p$ by concentric loading.

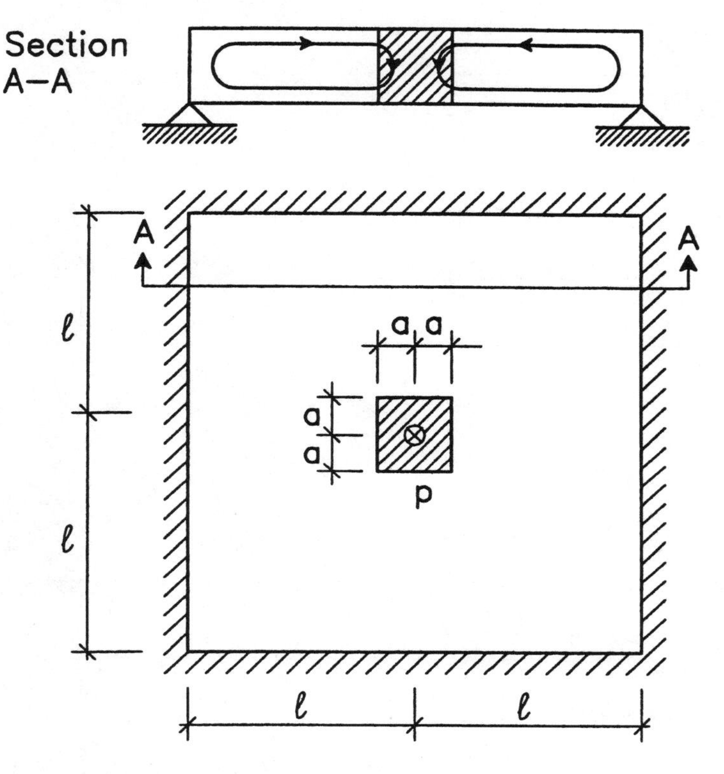

Figure 7.2.14 Concentrated load on a square slab carried by four torsion fields.

Now we consider the area $2a{\cdot}2b$ being acted upon by a moment M parallel to the side $2\ell_1$ (see Fig. 7.2.15). The moment M may partly be carried by four concentrated forces X in the corners, giving rise to torsion fields as in the case of concentric loading except now the forces have opposite directions to the left- and to the right-hand sides of the loaded area. Further a part of M may be carried by pure torsion in the strip of width $2b$ and by bending in a strip of width $2a$. Taking the vertical shear stresses into account in the area $2b{\cdot}h$, the total moment which may be carried by the strip with pure torsion will be $4t_p{\cdot}2b$. The shear flow is schematically illustrated in the figure. The loads on the strip with bending are also shown in the figure. The reaction R along the supports will be

$$R = \frac{M - 4t_p{\cdot}2b - 2X{\cdot}2b}{2\ell_2} \tag{7.2.47}$$

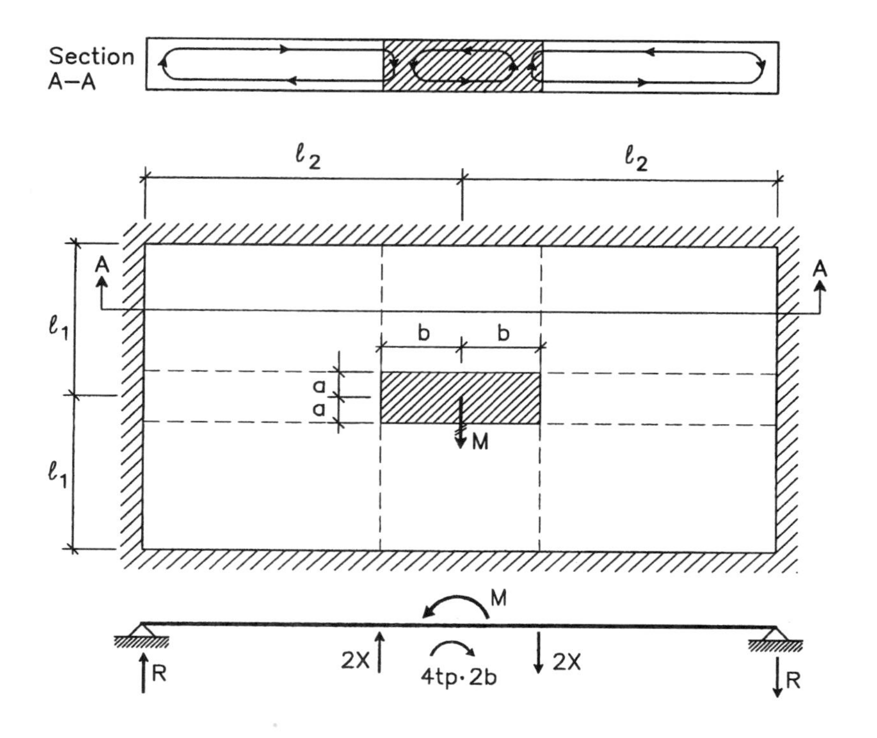

Figure 7.2.15 Concentrated moment on a rectangular slab.

This means that the maximum bending moment m_b in the strip of width $2a$ is

$$m_b = \frac{R(\ell_2 - b)}{2a} \cong \frac{R\ell_2}{2a} = \frac{M}{4a} - 2t_p\frac{b}{a} - X\frac{b}{a} \qquad (7.2.48)$$

having assumed $b << \ell_2$. Setting as before $m_p = m_p' \cong t_p$ and $m_b = t_p = X/2$, we find the load-carrying capacity

$$M = M_p = 4a\left(1 + 4\frac{b}{a}\right)t_p \qquad (7.2.49)$$

The length b in Fig. 7.2.15 should not be less than h because we have vertical shear stresses both from the part of the strip with pure torsion and from the torsional fields. The length a should not be less than $h/2$. If b is near the limit, the section with pure torsion should be reduced to $2b - h$ and the torsional capacity should be calculated by the methods developed for pure torsion of beams (cf. Section 5.3). We leave it for the reader to re-establish the equilibrium solution when the torsional capacity of the strip with pure torsion is changed from $2t_p \cdot 2b$ to the smaller value found by the beam theory. The reduction is not only due to the recommended reduction of the section but also because $2t_p$ times the length of the section, where t_p is determined by the slab theory, for small lengths is larger than the result from beam theory.

If the loaded area is now acted upon by a normal force $P = N$ and a bending moment $M = Ne$ corresponding to a normal force with the eccentricity e (see Fig. 7.2.16) formula (7.2.44) renders

$$\frac{N}{P_p} + \frac{Ne}{M_p} = 1 \qquad (7.2.50)$$

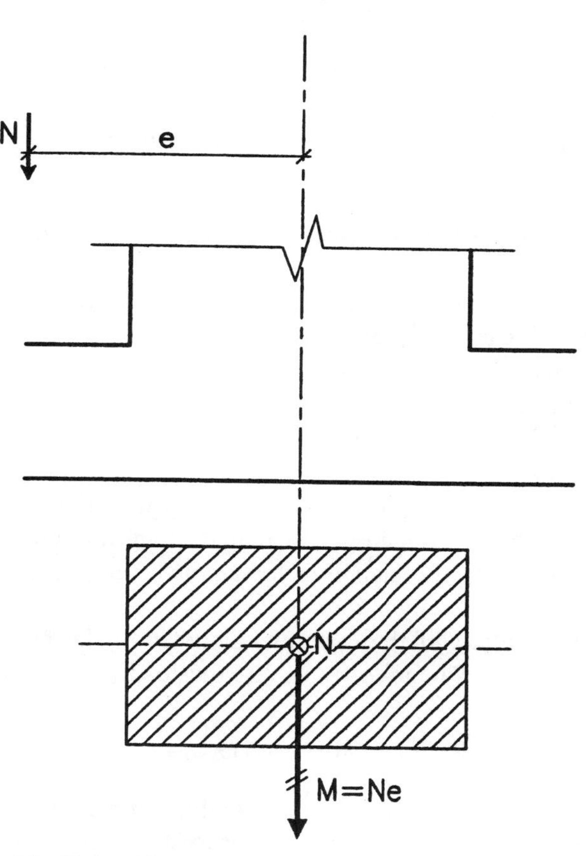

Figure 7.2.16 Slab with eccentrical load.

or

$$N = \frac{P_p}{1 + \frac{P_p}{M_p} e} \tag{7.2.51}$$

This formula determines the load-carrying capacity by an eccentric load. We repeat that P_p is the load-carrying capacity by a concentric loading and M_p is the load-carrying capacity by a pure moment loading.

The procedure is easily generalized to eccentricities in two directions.

Further, the procedure may be generalized to eccentric loads on any area and any slab, provided that the load-carrying capacity in the basic loading cases may be found.

The reduction in load-carrying capacity caused by eccentricities may, as mentioned, be rather dramatic. The designer must pay attention to this fact and check the load-carrying capacity in the basic loading cases, considering failure by yielding as well as by punching.

Upper bound solutions for pure moment loading have been developed in [74.5] and [98.1].

7.2.5 The Effect of Counterpressure and Shear Reinforcement

The presence of a uniform counterpressure, as when a column is punching a slab carrying a distributed load, has an effect very similar to that of tensile strength. If the load per unit area is $p = \chi f_c$, the solution depends on the factor χ in much the same way as on the ratio f_t / f_c. This goes for the shape of the failure surface as well as the ultimate load.

Consider as an example a slab with constant thickness h acted upon by a concentrated load P distributed along a circular area with diameter d. Further, the slab is acted upon by a counterpressure p (see Fig. 7.2.17).

The driving force, if the opening diameter of the failure surface is D, is $P - \pi D^2 p/4$ which means that the load-carrying capacity P is determined by

$$P - \pi D^2 p/4 = \tau_u(D)\, \pi (d + h) h \tag{7.2.52}$$

where $\tau_u(D)$ is the ultimate shear stress along the control surface.

As indicated, $\tau_u(D)$ will be a function of the opening diameter D. By choosing different values of D and using Fig. 7.2.4a for reading the corresponding value of $\tau_u(D)$, the opening diameter D may be determined as the value of D rendering the smallest load-carrying capacity. One should not forget when using the results of Fig. 7.2.4a to replace f_c by νf_c, where ν is determined by (7.2.38) or (7.2.42).

When the punch is not circular one may transform the actual punch area to an equivalent circular one with the same perimeter.

Now we turn our attention to the effect of shear reinforcement. This is normally carried out as a number of closed stirrups bent around the longitudinal slab reinforcement or by bent-up bars.

The effect of a continuously distributed shear reinforcement is exactly the same as that of a distributed counterpressure. Let the shear reinforcement degree be $\psi = s_Y/f_c$, where s_Y is the yield force per unit area perpendicular to the reinforcing bars, and let the bars be perpendicular to the slab face. Then the quantity ψ will correspond to the factor χ, introduced above.

For increasing values of χ or ψ the failure surface will approach a truncated cone, that is, the contribution to the load-carrying capacity from the concrete will approach (7.2.37) and the contribution from the reinforcement will approach

$$\frac{\pi}{4}\left[(d+2h\tan\varphi)^2-d^2\right]s_Y \tag{7.2.53}$$

The load-carrying capacity as a function of s_Y qualitatively will therefore have the appearance shown in Fig. 7.2.18.

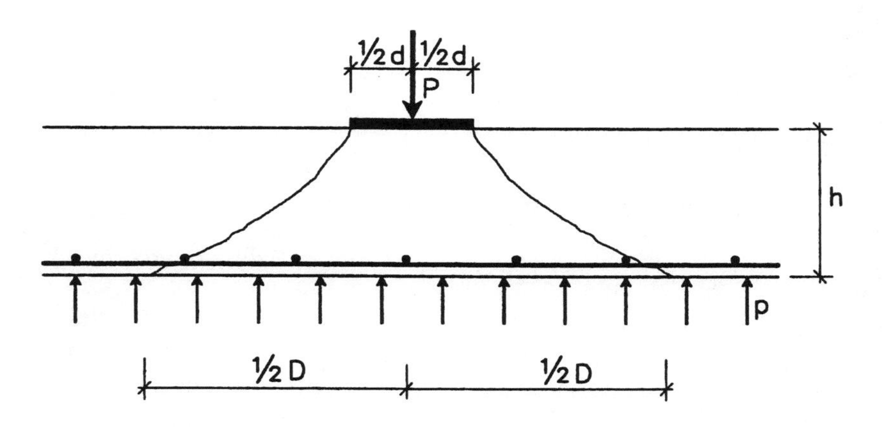

Figure 7.2.17 Slab with counterpressure.

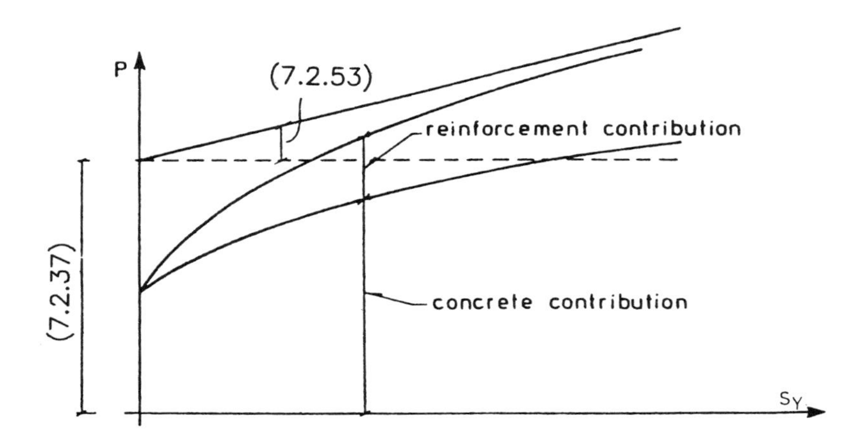

Figure 7.2.18 Punching shear load versus shear reinforcement intensity.

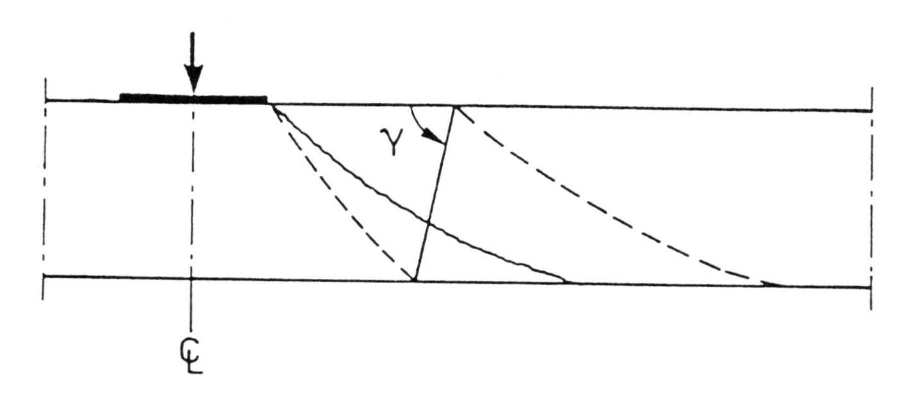

Figure 7.2.19 Reinforcement bar crossing a failure surface.

In practice, the shear reinforcement - if any - will normally be confined to a region around the punch (or column) and not distributed over the entire slab. In that case, the reinforcement will not affect the shape of the failure surface, but enhance the ultimate load by the contribution $S_Y \sin\gamma$, S_Y being the total yield force of the reinforcement and γ the angle of the bar with the slab face. This is only correct, however, provided that a lower upper bound cannot be obtained with a failure surface completely inside or completely outside the shear reinforcement (see Fig. 7.2.19).

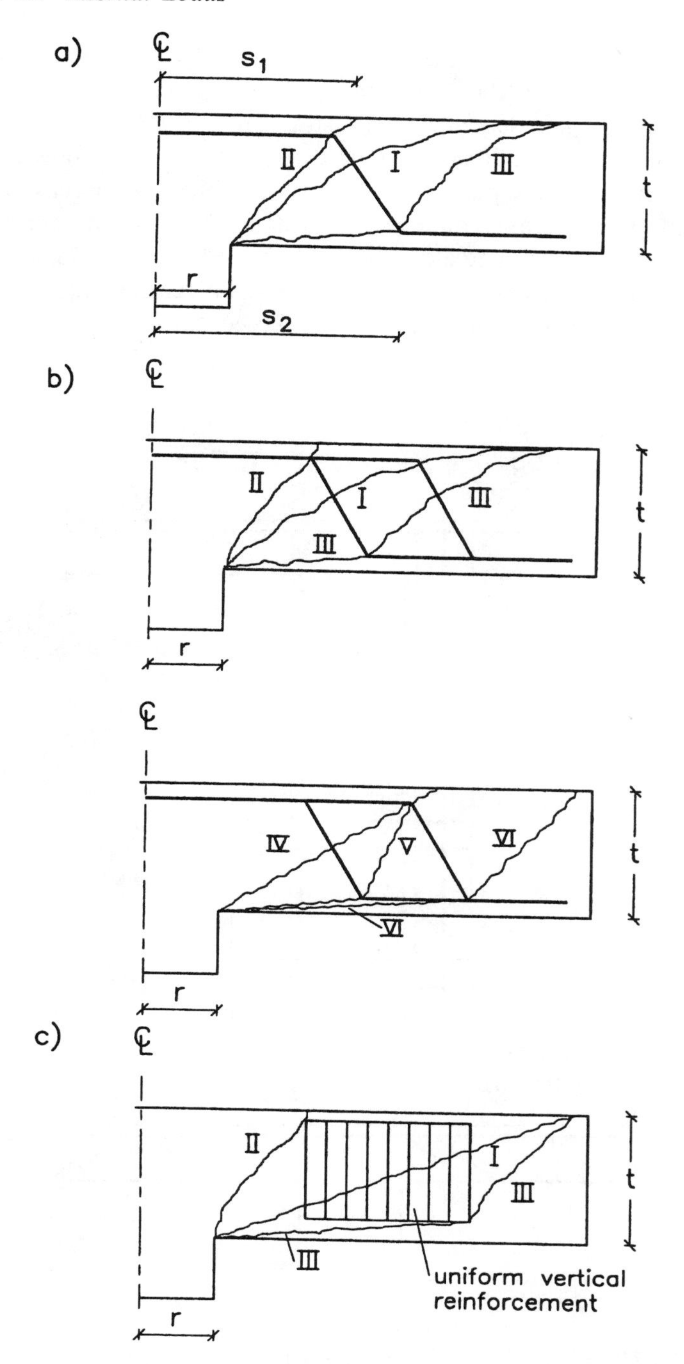

Figure 7.2.20 Failure mechanisms in slabs with shear reinforcement.

When searching for the load-carrying capacity of a shear reinforced slab, one therefore has to investigate several failure mechanisms. It is adequate to illustrate the procedure by some examples. In Fig. 7.2.20a a case with bent-up bars is shown. We envisage the bars be placed in a pattern which is in reasonable agreement with the axisymmetric conditions. Mechanism number I is the usual mechanism treated above. The load-carrying capacity is found by adding the contribution from the shear reinforcement to the load carrying capacity for nonshear reinforced slabs.

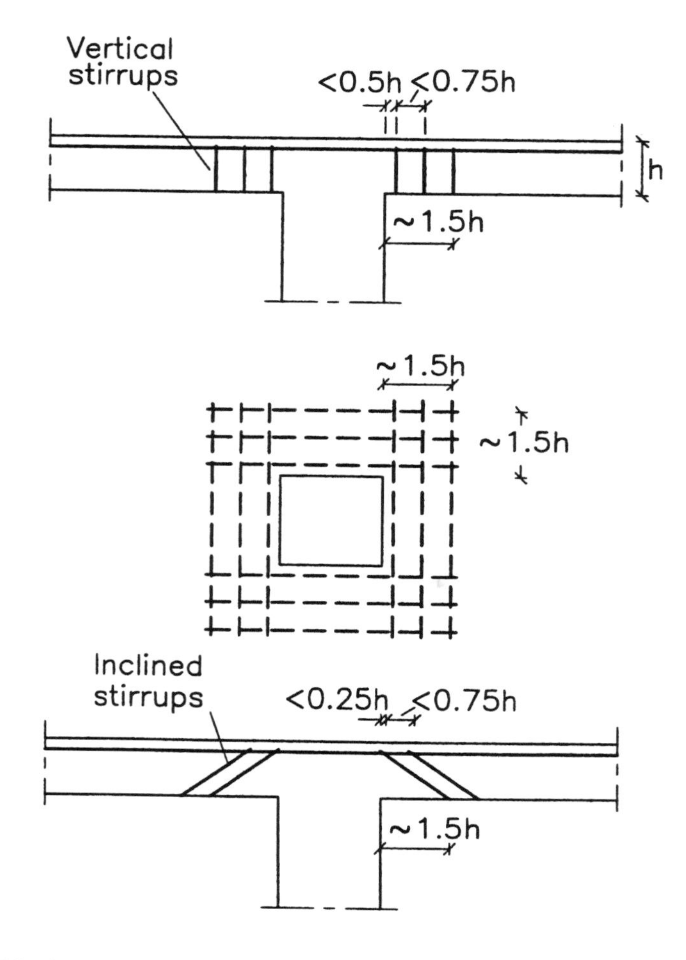

Figure 7.2.21 Guidelines for the position of shear reinforcement in slabs. Vertical or inclined stirrups.

Mechanism number II has no contribution from the shear reinforcement, since the failure surface only cuts horizontal reinforcement. The load-carrying capacity is determined by using Fig. 7.2.4a with $D = 2s_1$, $d = 2r$ and a slab thickness $h = t$. The meaning of the parameters s_1, r and t appears from the figure.

Mechanism III is composed of a flat part contained in the cover and a part through the slab originating from the bending point along the bottom face. There is no reinforcement contribution. To find the concrete contribution from the failure surface in the cover Fig. 7.2.4a is used with $D = 2s_2$, $d = 2r$ and a slab thickness $h = c$, c being the cover. The concrete contribution from the remaining part can be found by using Fig. 7.2.4a with D corresponding to the actual support conditions, $d = 2s_2$ and a slab thickness $h = t\text{-}c$.

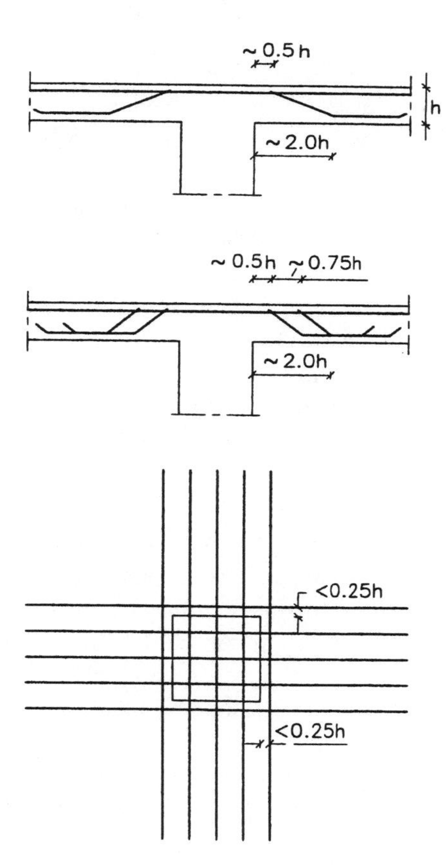

Figure 7.2.22 Guidelines for the position of shear reinforcement in slabs. Bent-up bars.

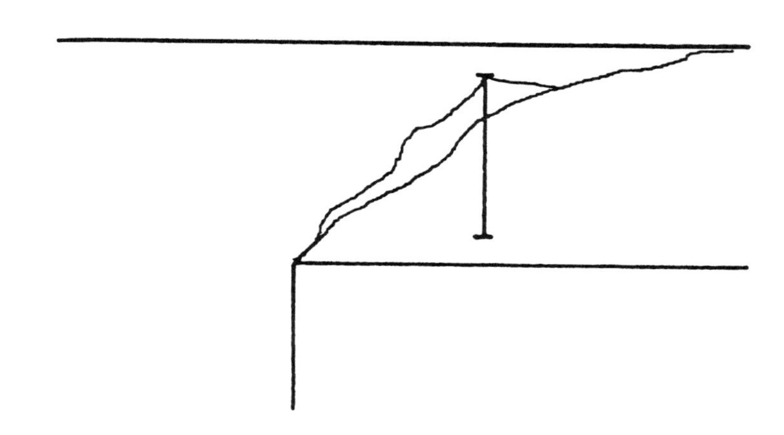

Figure 7.2.23 Failure surface passing a stirrup without crossing it.

In Fig. 7.2.20b and Fig. 7.2.20c a number of mechanisms are shown which must be investigated when there are two sets of bent-up bars and a region with uniform vertical shear reinforcement, respectively.

Before starting the verification of the load-carrying capacity a lay-out of shear reinforcement must be chosen. Some guidance may be found in Fig. 7.2.21 and Fig. 7.2.22.

These drawings were suggested by Kinnunen and Nylander when preparing the CEB-FIP Model Code 1978 [78.10].

Some codes limit the enhancement of shear capacity due to shear reinforcement. This is probably unnecessary when a careful calculation in the above spirit is done and when the design of the bending and torsion reinforcement in the slab as a whole is carefully carried out.

In tests it is sometimes found that the shear reinforcement does not reach the yield point, which has led some codes to also limit the yield stress which may be taken into account. That some shear reinforcement does not yield is probably due to the fact that the failure surface may pass by the reinforcement without activating the reinforcement and without influencing the concrete contribution significantly (see Fig. 7.2.23).

When designing the shear reinforcement it is thus important to make sure that there is sufficient distance from the failure surface to the anchorage points of the shear reinforcement or to the bending points of the longitudinal bars. In fact this condition is easier to fulfill when the shear reinforcement consists of bent-up bars instead of stirrups. It seems to give bent-up bars some preference as shear reinforcement when punching shear is considered. If care is taken in this respect the usual yield strengths may be utilized. When the above phenomena are taken into account when comparing with tests, quite good agreement with the calculation method is obtained. The tests were treated by A. Ø. Brynjarsson [85.6].

7.3 EDGE AND CORNER LOADS

7.3.1 Introduction

For internal concentrated loads far from supports, free edges and other concentrated loads, it is natural to assume that the shear stress distribution around the loaded area is not very sensitive to how the bending and torsion reinforcement of the slab are carried out. The redistribution of shear stresses necessary to give rise to a punching shear failure therefore is small. Accordingly, for practical purposes the same calculation method for checking punching shear may be used for internal loads independent of the bending and torsion reinforcement of the slab.

Quite the contrary is true for loads at or near an edge or a corner. Consider as an illustrative example a load P acting in a right angled corner (see Fig. 7.3.1).

If the slab is isotropically reinforced parallel to the edges in top and bottom and if $m_p = m_p'$ it appears from Section 6.5.3 that the exact solution is $P = 2m_p = 2t_p$, i.e., P equals two times the torsional yield moment, which for small reinforcement degrees equals the bending yield moment. The slab will carry the load by means of the Kirchhoff forces along the boundaries exclusively. This is illustrated in Fig. 7.3.1 (left).

Now imagine that the slab is only reinforced in the top side, i.e., $m_p = 0$ (although we did not say it, the slab is horizontal and the load is downward). Now according to the solutions in Section 6.5.3 the load-carrying capacity is $P = \pi m_p'/2$. It is easily shown that in this case the load is carried by uniformly distributed shear stresses in circumferential sections. In the r,θ-system shown in the figure to the right we have everywhere $m_r = m_p' = \text{const}$ and $m_\theta = m_{r\theta} = 0$. The shear stresses are indicated in the figure.

This is the kind of shear stress distribution we would expect prior to a punching failure in the corner. No part of the load is carried by Kirchhoff forces.

It appears that by changing the reinforcement of the slab the shear stress distribution is fundamentally changed. Therefore we may suspect that a substantial redistribution of shear stresses may be required for edge and corner loads before a punching shear failure may take place. Thus for edge and corner loads the shear capacity cannot be analyzed without taking into consideration the bending and torsion reinforcement.

The example in Fig. 7.3.1 (right) also demonstrates that the resultant of the shear forces along any circle with the centre in the corner is by no means a force in the corner. Only when the bending moments and the shear forces are taken together do we get a resultant force in the corner. Thus it would have no meaning to require in a punching shear analysis the vertical shear stresses along a circular control perimeter to be statically equivalent to the force P (we referred to such an example in Section 7.2.4).

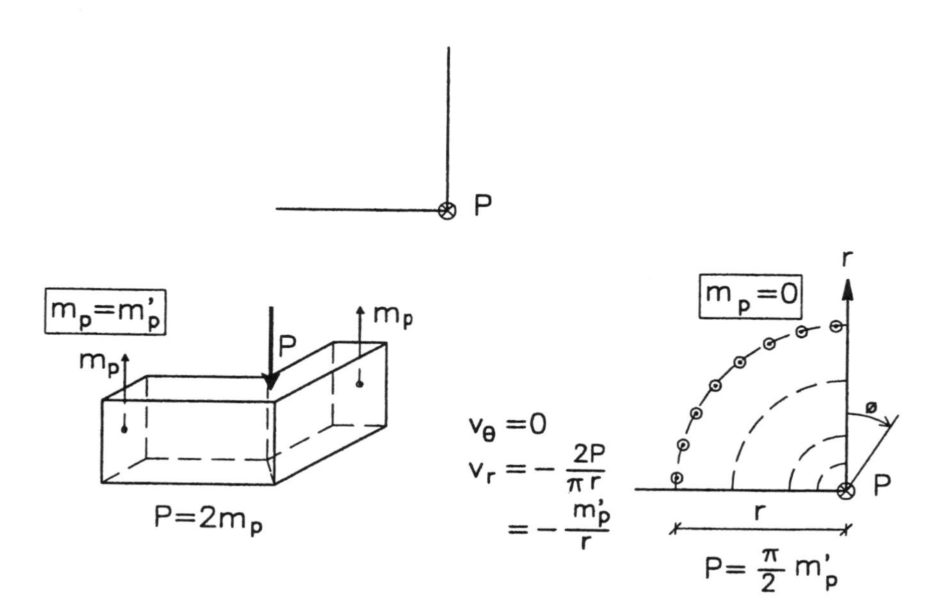

Figure 7.3.1 Distribution of shear forces in a slab subjected to a concentrated force in a right angled corner.

For loads near or at free edges the type of failure mechanism used for internal concentrated loads is valid with small modifications. As before, the length of the control perimeter should be minimized. For a loaded area near a free edge the smallest control perimeter often may be found by drawing lines perpendicular to the free edge touching the control perimeter for internal loads. Fig. 7.3.2 shows two examples of control perimeters for edge loads.

The plastic theory has not yet been developed to such a degree of perfection that makes it possible to treat edge and corner problems using failure mechanisms involving shear reinforcement, for instance, along the free edges. Fortunately these shortcomings turn out not to be serious regarding the practical applications. The reason is that in many important cases edge and corner loads are carried mainly by torsional moments (cf. the example above) and the load-carrying capacity is determined by the upper limit of the torsional capacity and not by punching shear. One finds that when the lengths b and a shown in Fig. 7.3.2 (right) are larger than about half the slab thickness, the load-carrying capacity by punching will exceed the torsional capacity even without any shear reinforcement. However, when there are torsional moments along an edge, the edge should always be reinforced just as a beam in torsion should always be reinforced along the whole periphery. Then also for smaller values of b and a the punching load normally will be larger than the torsional capacity. The conclusion is remarkable: the dimensions of the loaded area are rather insignificant for edge and corner loads when edge reinforcement is supplied. The proof, both by theory and experiment, is given below.

When performing the calculations the designer should not take into account the concrete cover on the edge reinforcement. Also he should make sure that the concentrated force is applied on an area not involving the cover.

Even if a punching shear analysis is not topical for edge and corner loads when edge reinforcement is supplied, the designer will normally check the punching shear resistance in the traditional way. There are no objections to that if he also performs the analysis recommended in the following. Edge reinforcement and other kinds of shear reinforcement may, until the plastic theory has been further developed, be taken into account in a primitive way by including reinforcement bars which have the right direction relative to the concentrated force and are crossing a truncated cone with, say, a 45° angle with the slab middle surface drawn with the control perimeter in the slab middle surface as generatrix.

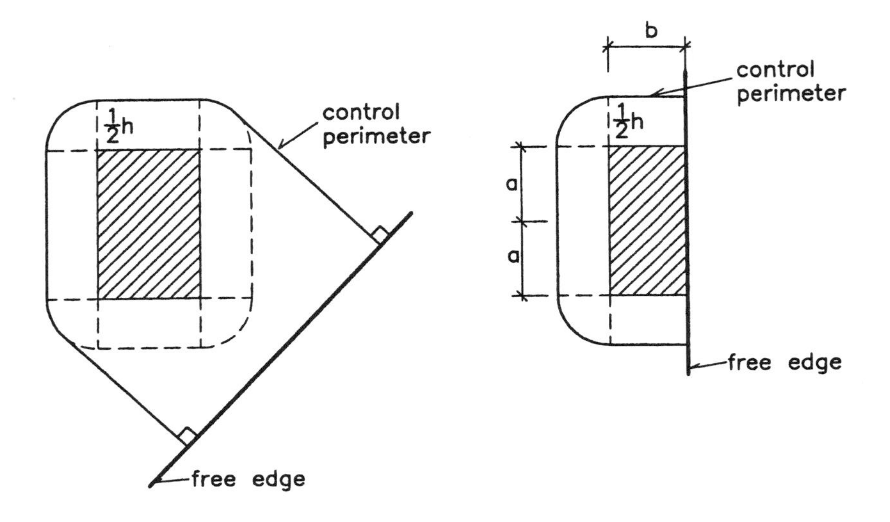

Figure 7.3.2 Control perimeter for loads near or at a free edge.

7.3.2 Corner Load

The corner load case is an important one from a practical point of view. It is also very instructive from a theoretical point of view. Therefore a detailed treatment is given.

We consider first a right angled corner. The slab is isotropically reinforced parallel to the edges in top and bottom and the yield moments are equal, i.e., $m_p = m_p'$. According to Section 6.3.2, where the Kirchhoff boundary conditions are developed, the edge will be loaded as a beam which has a shear force equal to the torsional moment at the point in question.

The way the slab carries a concentrated load in a corner is illustrated in Fig. 7.3.3. Here a square slab is loaded in all four corners by loads P. In two opposite corners the loads are downward, in the other two corners they are upward. The loading on one of the Kirchhoff edge beams is also shown. The torsional moment t_p per unit length balances the loads $P/2$ with opposite directions at the ends of the beam when

$$P = 2t_p \tag{7.3.1}$$

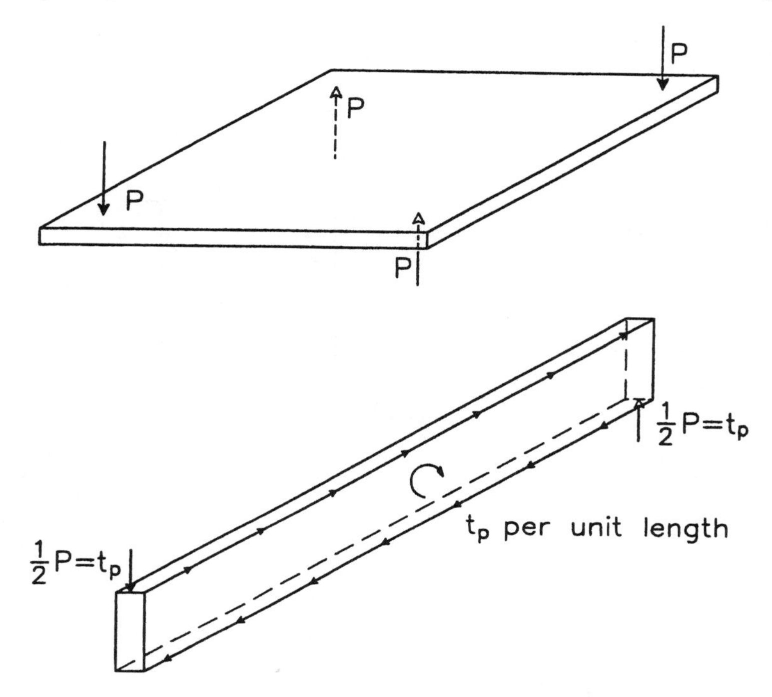

Figure 7.3.3 Square slab loaded by concentrated forces in the corners. Forces on the Kirchhoff edge beams.

which is the solution we have already stated above. Thus this is a way to subject a slab to pure torsion.

It is evident that the shear force in the edge beam is $t_p = \frac{1}{2}P$ in agreement with the general theorem.

As mentioned before for small reinforcement degrees, $m_p = m_p' \cong t_p$, but for higher reinforcement degrees, $t_p < m_p$. Thus when determining the load-carrying capacity for a corner load t_p should be used instead of m_p.

The torsional yield moment t_p was determined in Section 2.3.2. Because of its importance for edge and corner loads the calculation is summarized in Box 7.3.1. The notation is as before. The stress field is denoted σ_x, σ_y and τ_{xy} and index c means concrete stresses. The compressive strength of concrete is f_c. When applying the results f_c must as usual be replaced by νf_c, where ν is the effectiveness factor. This is determined below. The slab thickness is h and a is the depth of the compression zones. Finally, A_s is the reinforcement area per unit length (in each of the four reinforcement directions) and Φ_o is the reinforcement degree.

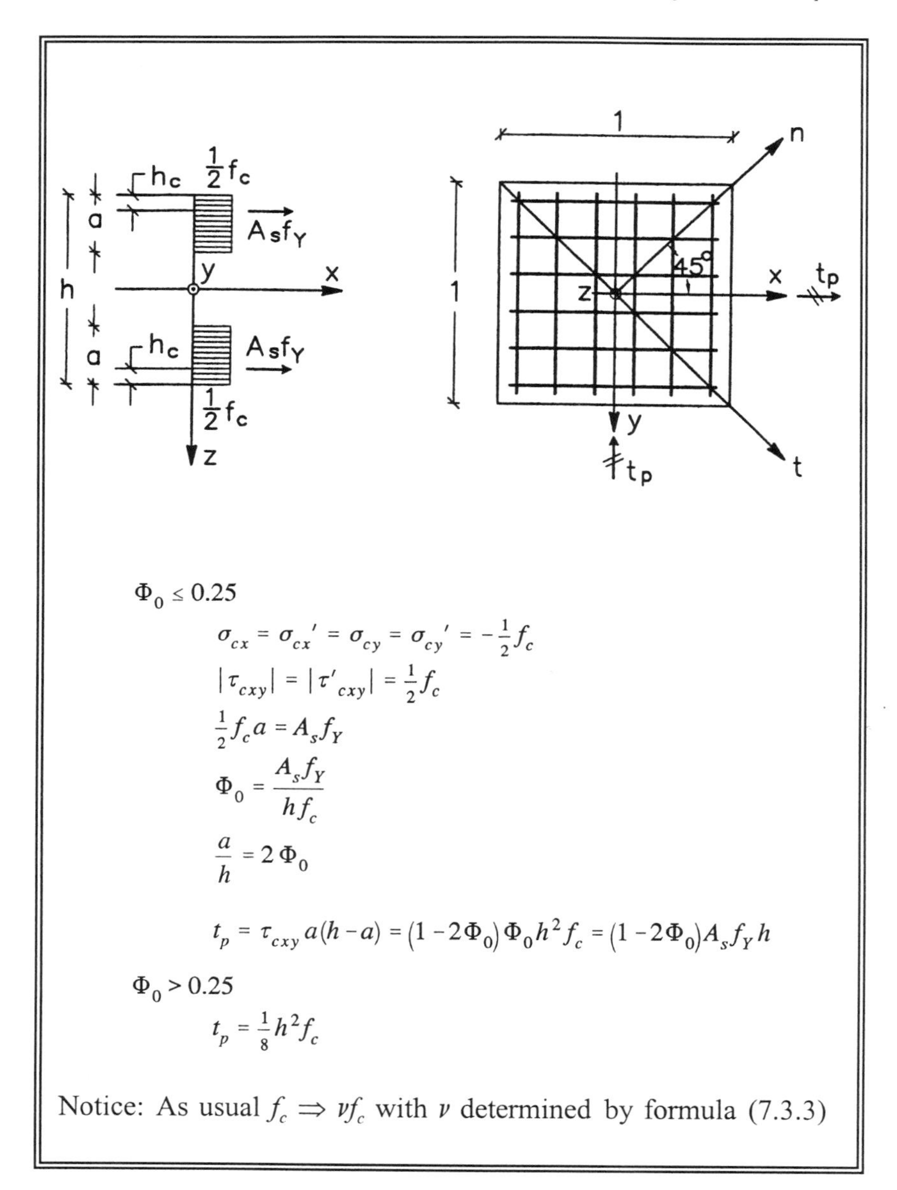

$\Phi_0 \leq 0.25$

$$\sigma_{cx} = \sigma_{cx}' = \sigma_{cy} = \sigma_{cy}' = -\frac{1}{2} f_c$$

$$|\tau_{cxy}| = |\tau'_{cxy}| = \frac{1}{2} f_c$$

$$\frac{1}{2} f_c a = A_s f_Y$$

$$\Phi_0 = \frac{A_s f_Y}{h f_c}$$

$$\frac{a}{h} = 2\,\Phi_0$$

$$t_p = \tau_{cxy}\, a(h-a) = (1 - 2\Phi_0)\,\Phi_0 h^2 f_c = (1 - 2\Phi_0) A_s f_Y h$$

$\Phi_0 > 0.25$

$$t_p = \frac{1}{8} h^2 f_c$$

Notice: As usual $f_c \Rightarrow \nu f_c$ with ν determined by formula (7.3.3)

Box 7.3.1 Calculation of the torsional yield moment t_p in sections perpendicular to the reinforcement directions. Isotropic reinforcement in top and bottom. Same reinforcement in top and bottom.

The maximum value of t_p is

$$t_{p\max} = \frac{1}{8} h^2 f_c \tag{7.3.2}$$

which is reached when $a = h/2$. Thus there is an important difference between the maximum value of a bending yield moment (cf. Section 2.3.2), and the maximum value of a torsional yield moment. The slab will be overreinforced for torsion for roughly half the value of A_s leading to overreinforcement in bending. The implications are important. Whenever torsion is an important part of the moment field carrying a load, the concrete strength will limit the load-carrying capacity for a much lower reinforcement degree than if pure bending is the governing factor.

We may now pose the interesting question: "Will it always be possible by reinforcing the edges of the slab for shear to reach the torsional capacity of the slab?" The answer is yes. In other words, if the edges are reinforced for shear so that they can carry a shear force t_p, the load-carrying capacity is governed by the torsional capacity of the slab and not by one or another kind of punching failure.

A test series demonstrating this fact was carried out at the Technical University of Denmark by J. F. Jensen et al. in 1981 [81.6]. The slabs were subjected to pure torsion as in Fig. 7.3.3. They were square with side length 1430 mm and the thickness was 140 mm. The loads were transferred to the corners through a square steel plate with side length 100 mm. The edges of the steel plate were parallel to the slab edges and the distance from the slab edge to the edge of the steel plate was 25 mm. The slabs were reinforced with 6 mm or 8 mm bars at top and bottom. The bars were parallel to the edges of the slab and had a nominal yield strength of 420 MPa. The cover on the top and bottom reinforcement was about 10 mm. The edge reinforcement, if any, was placed 40 mm from the vertical slab face.

The bar distance varied between 71.4 mm and 125 mm.

The different series may be shortly characterized as follows:

Series A: Only one test. The edge beams were shear reinforced as a real beam with closed stirrups. $f_c \sim 20$ MPa.

Series B: 6 tests. The edge beams were reinforced by carrying the slab reinforcement around the edge (i.e., the whole slab reinforcement consisted of closed stirrups). $f_c \sim 20$ MPa.

Series E: 4 tests as series B. $f_c \sim 12$ MPa.

Series G: 2 tests. The slab reinforcement bars were placed close together two by two. Only one of the two bars were carried around the edge. $f_c \sim 20$ MPa.

Series H: 2 tests. Slab reinforcement in the middle of the slab. The edge beams were shear reinforced as in series A. $f_c \sim 20$ MPa.

Series I: 3 tests as in series B. $f_c \sim 40$ MPa.

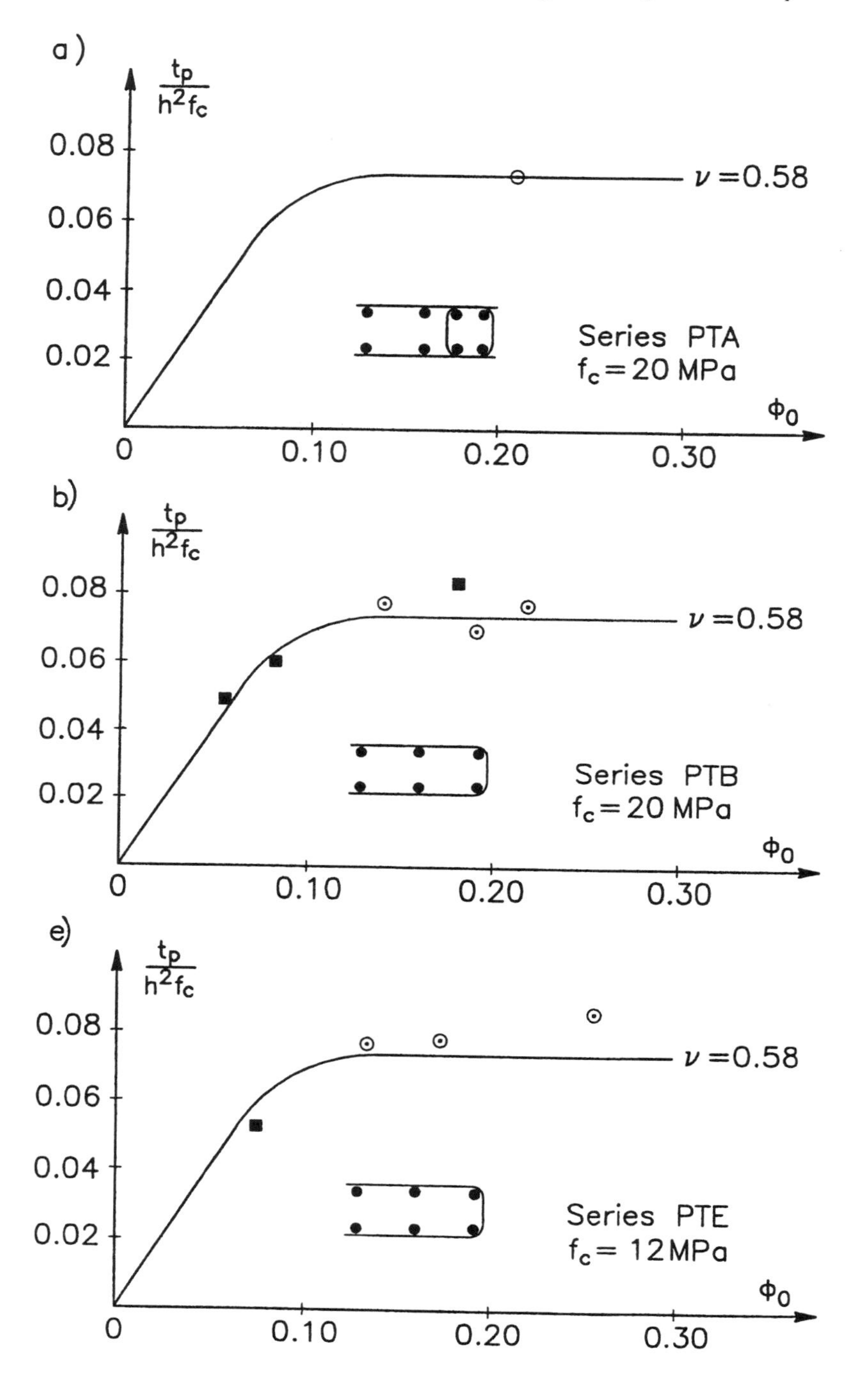

Figure 7.3.4 Results of torsion tests on slabs with reinforced edges.

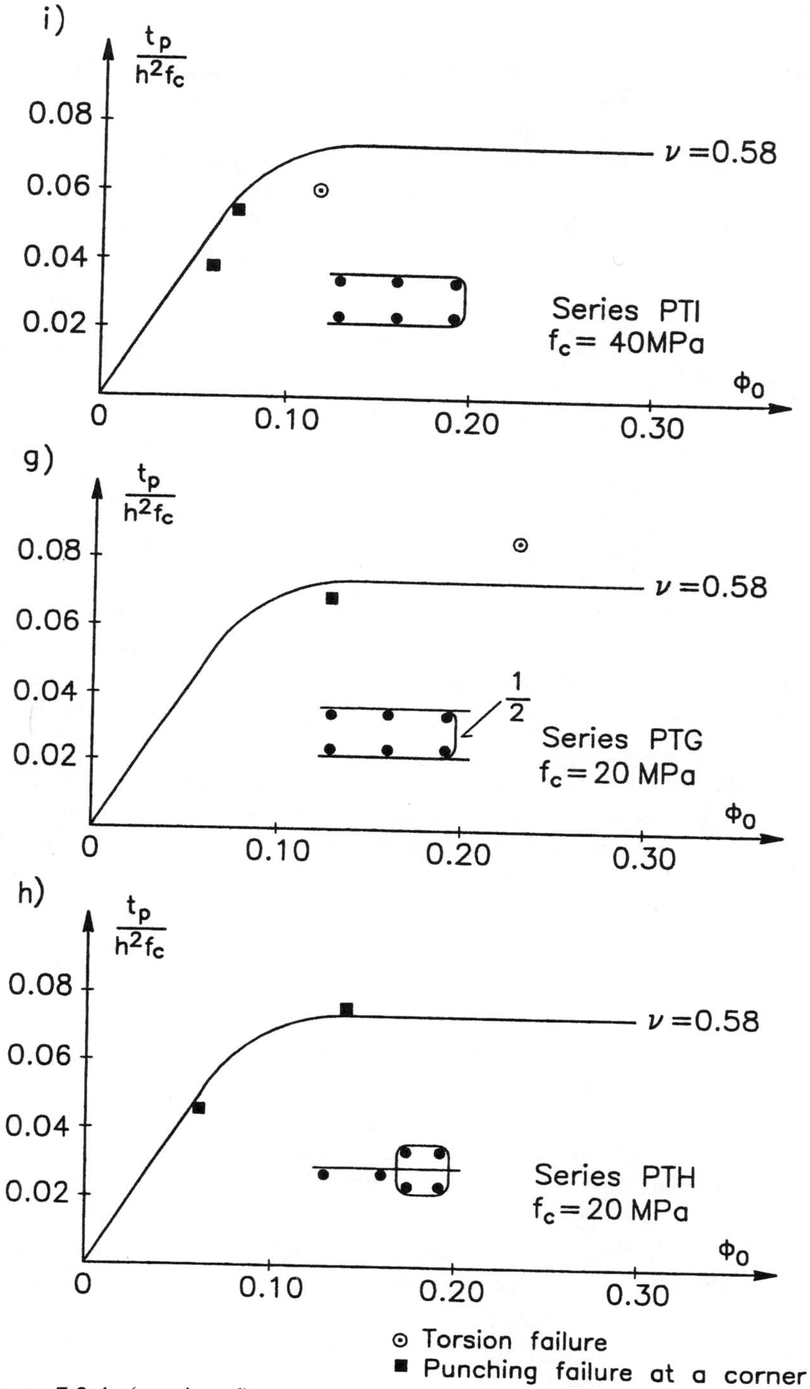

Figure 7.3.4 (continued).

The test results are depicted in Fig. 7.3.4. The figure shows the test value of $t_p\ /h^2 f_c$ as a function of the reinforcement degree Φ_o. The curves have been drawn using $\nu = 0.58$ although the concrete strength varied from around 12 MPa to 40 MPa.

It appears that in all cases the torsional capacity was reached. In some cases the failure was a local shear failure near the concentrated force but in many cases the slab failed in torsion. For high reinforcement degrees concrete crushing was sometimes observed in the whole slab at the same time.

From the tests we may conclude that if the edges are shear reinforced, no punching shear analysis is necessary. The load-carrying capacity will be governed by the torsional capacity. It seems that it is not required to let all the slab reinforcement continue around the edge but the tests are not conclusive on this point. Until further evidence about the possibilities of reducing the shear reinforcement in the Kirchhoff edge beams is obtained, it is recommended to reinforce the edge beams by carrying the slab reinforcement around the edge as was the concern about reinforcing a section for pure torsion (cf. Section 5.3). Of course, instead U-stirrups may be supplied. These stirrups must have the same area and distance as the slab reinforcement and must be carried at least an anchorage length into the slab.

The ν-value found in the tests is considerably lower than the ν-value for bending. On the basis of the tests we tentatively may conclude that for torsion, ν may be taken as 70% of the value for bending, i.e.,

$$\nu = 0.7\left(0.97 - \frac{f_c}{300} - \frac{f_Y}{5000}\right) \quad (f_c \text{ and } f_Y \text{ in MPa}) \tag{7.3.3}$$

The reason why ν for torsion is lower is, of course, that we have stressed tensile reinforcement crossing the cracks parallel to the uniaxial compression stresses. The formula (7.3.3) is in good agreement with the second formula in (5.3.30), which is valid for torsion in beams with a depth/length ratio larger than 4. In beams the cover is disregarded so the ν-value found by (5.3.30) is a little higher than that determined by (7.3.3).

When the load-carrying capacity is determined by the torsional capacity, the size of the loaded area (column dimensions, size of loading plates, etc.) is likely to be rather insignificant. Since the width of the Kirchhoff beam in the limit is of the order of half the slab thickness, the loaded area should in practical design not have dimensions in the slab plane less than half the slab thickness. In Fig. 7.3.5 showing the shear flow in a section parallel to one of the slab edges, the side lengths a and b of a rectangular area with sides parallel to the slab edges should consequently not be less than about $h/2$.

If a punching shear analysis is carried out using the control perimeter shown in the figure for a square with side length a, it will be found that when a is about $h/2$ the punching shear load will be about the same as the load determined by the torsional capacity, when no account is taken of the edge reinforcement. Although an upper bound analysis taking advantage of the edge reinforcement cannot yet be carried out rigorously, it is clear that a punching shear analysis for edge-reinforced slabs with corner loads is not relevant.

In the test series referred to above tests with slabs without edge reinforcement were also performed.

These tests may be briefly described as follows:

Series C: 5 tests. No edge reinforcement. $f_c \sim 20$ MPa.
Series D: 5 tests. No edge reinforcement, but the bars were welded together in the cross points near the edge. $f_c \sim 20$ MPa.
Series F: 4 tests as in series C. $f_c \sim 12$ MPa.

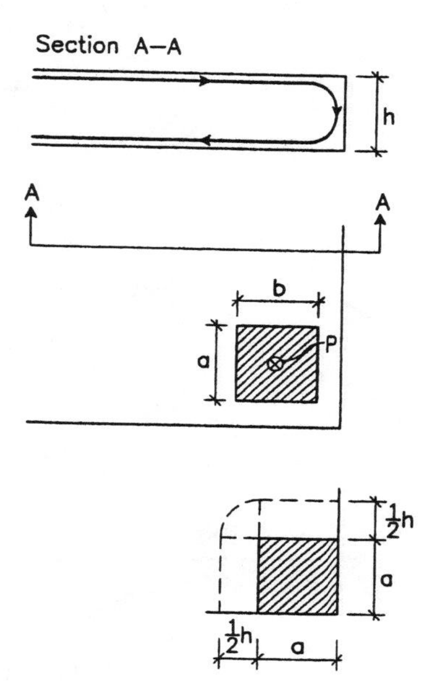

Figure 7.3.5 Concentrated load at a corner.

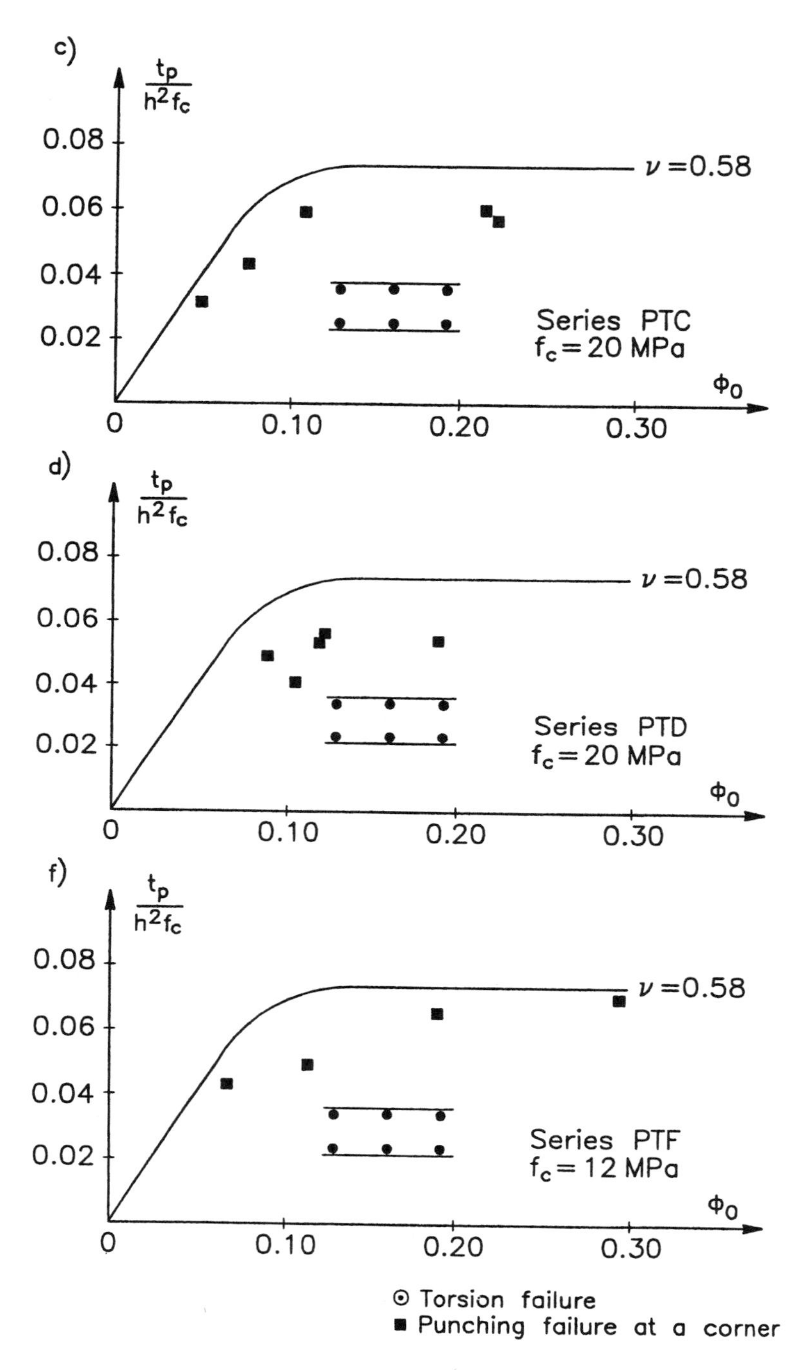

Figure 7.3.6 Results of torsion tests on slabs without edge reinforcement.

The results are shown in Fig. 7.3.6. All these slabs suffered a local shear failure at the concentrated force. Only for a very low concrete strength and a very high reinforcement degree did the load-carrying capacity approach the torsional capacity.

Such slabs should never be carried out in practice. Nevertheless, if they are, they should be analyzed using the beam shear strength for nonshear reinforced beams (cf. Section 7.3.4).

7.3.3 Edge Load

The corner load solution of the previous section may immediately be extended to the edge load case.

In Fig. 7.3.7 we have a concentrated load in the middle of a free edge. The slab is rectangular and simply supported along the other three edges.

The exact solution for $\ell_1 = \ell_2$ was given in Section 6.5.3 when the slab is isotropically reinforced parallel to the edges in top and bottom, and when the yield moments are equal, i.e., $m_p = m_p'$. The load-carrying capacity is $P = 4m_p$. Similar to what we found in Section 7.2.4 the load may be carried by two torsion fields in the rectangular areas $\ell_1 \cdot \ell_2$. These torsion fields carry the load

$$P = 4t_p \tag{7.3.4}$$

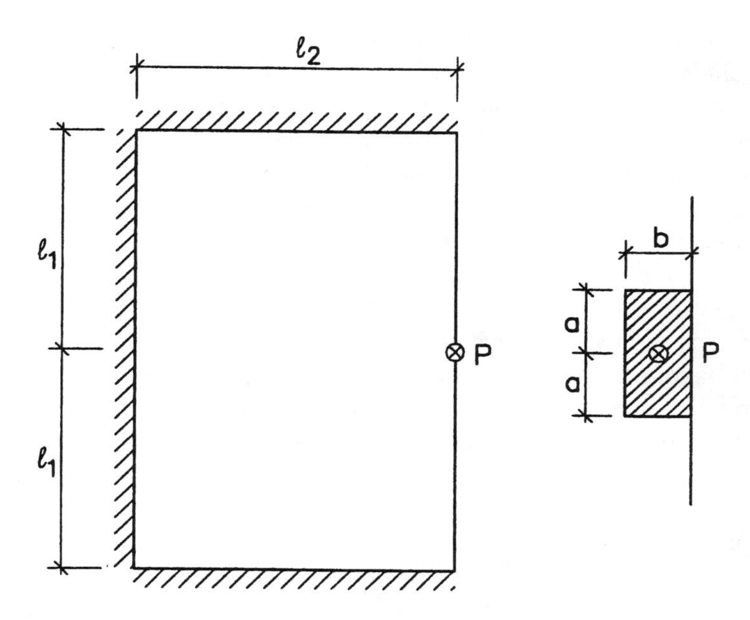

Figure 7.3.7 Edge load on a simply supported rectangular slab.

As before, when the reinforcement degrees are low we have $m_p \cong t_p$, so the solution (7.3.4) is equivalent to the solution $P = 4m_p$. For higher reinforcement degrees $t_p < m_p$ and t_p should be used instead of m_p. Regarding the determination of t_p the reader is referred to Box 7.3.1.

The torsion field solution may be used for any position of the load on the free edge.

If the load is transferred to the slab by a rectangular area $b \cdot 2a$ (see Fig. 7.3.7), the side lengths a and b should not be less than $h/2$. If so, the dimensions of the loaded area are probably quite insignificant. A punching shear analysis is not relevant.

As in Section 7.3.2 we may conclude that if the reinforcement is carried around the edge no punching shear analysis is necessary. The load-carrying capacity will be governed by the torsional capacity.

7.3.4 General Case of Edge and Corner Loads

The problems solved above are important from a practical point of view but special because the concentrated forces could be carried by torsional fields exclusively. In this way it was possible to reduce the shear problems to an analysis of the shear forces in the Kirchhoff edge beams.

In more general cases, when failure is by yielding of the reinforcement, one part of the load is carried by the Kirchhoff forces and another part by the usual internal shear forces. This means that a substantial redistribution of the shear stresses may be required before a true punching shear failure takes place. It is not known whether the ductility of the punching shear failure is sufficient to make topical a traditional punching shear analysis.

Fortunately this is not serious from a practical point of view. In most cases one is dealing with small reinforcement degrees and the designer wishes to have failure by yielding.

It is rather easy, in a conservative way, to make sure that a premature punching shear failure does not take place before failure by yielding. What one has to do is to determine the shear force distribution when failure takes place by yielding and make sure that these shear forces may be resisted. This is done by using the plastic theory of bending and torsion of slabs, e.g., yield line theory (cf. Chapter 6). The shear force distribution found should, of course, not be confused with the shear force distribution at a true punching shear failure.

In many cases the exact solution is known and it is easy to find the shear forces. When a lower bound solution is known this may be taken as a basis. More difficulties are encountered when only upper bound solutions are used. However, even in that case a rough idea of the shear force distribution around a concentrated force may be found by means of equilibrium equations for the individual slab parts into which the yield line pattern has subdivided the slab.

In most edge and corner problems the control of the shear capacity in the above meaning may be reduced to the calculation of the punching shear resistance in an angle α (see Fig. 7.3.8) and to find the necessary edge reinforcement.

An approximate analysis may be carried out as follows. First we transform the real loaded area to an area limited by part of a circle and the edges of the slab. We require the real loaded area to be equal to the area of a circle with the center in the corner *O*.

In Fig. 7.3.8 the loaded area is *OABC*. The area of *OABC* is denoted *A*.

Then we have the following equation for the determination of the radius *r* of the circle

$$\frac{1}{2}\alpha r^2 = A \quad \Rightarrow \quad r = \sqrt{\frac{2A}{\alpha}} \tag{7.3.5}$$

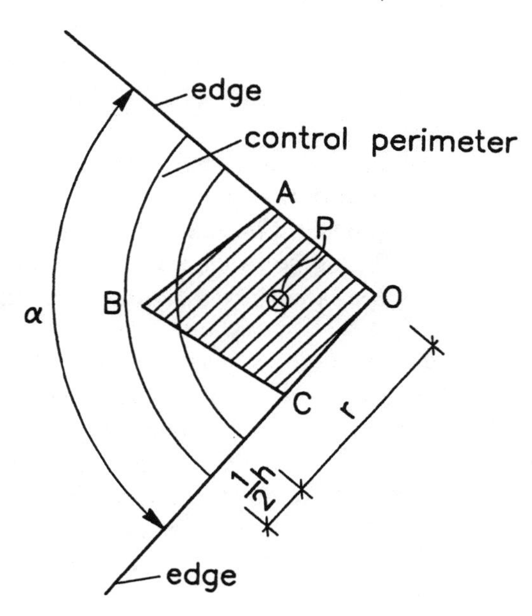

Figure 7.3.8 Concentrated load in a corner with an opening angle α.

Here α is the opening angle of the corner.

We have suggested using an area transformation instead of a perimeter transformation. In the case shown in Fig. 7.3.8 it would be rather obvious how to transform the perimeter ABC to the perimeter of a circle. We would in this case neglect the perimeter along the free edges of the loaded area. However, loaded areas may have many strange forms for which one might encounter difficulties about the parts of the perimeter to include, so an area transformation must be preferred in this case.

The punching shear capacity is

$$P = \alpha\left(r + \frac{1}{2}h\right)h\,\tau_u \tag{7.3.6}$$

In this formula $r + \frac{1}{2}h$ is the radius of the control perimeter.

Since we do not know much about the influence of sliding in initial cracks in corner shear problems, it might be argued that the shear capacity in the control surface τ_u should, on the safe side, be taken as the shear capacity of a nonshear reinforced beam or slab (cf. Section 5.2.6). In fact it is strongly recommended to do so, the very approximate character of the calculation being taken into account. The value of τ_u determined in this way normally will be roughly half the value valid for the punching shear capacity for internal loads (cf. Section 7.2.2).

When using formula (7.3.6) normally only a fraction of the opening angle is inserted because only a part of the load is carried by the usual shear forces (see below).

Let us illustrate the application of this procedure by considering the general solution in Section 6.5.4 for a point load in a corner with an arbitrary opening angle α. The slab is isotropically reinforced corresponding to the yield moments m_p and m_p', respectively.

The plastic solution for failure by yielding is determined by the formulas (6.5.113) and (6.5.114). We will confine ourselves to small reinforcement degrees, so we do not have to consider the limitations imposed by the upper value of the torsional capacity (7.3.2). The introduction of this limitation requires knowledge of the reinforcement directions since the calculation of t_p refers to torsional moments in sections parallel to the reinforcement directions. When the reinforcement directions have been chosen, the yield condition (2.3.65) for higher reinforcement degrees in Section 2.3 makes it possible to investigate if a particular moment field may be resisted and thereby to find out whether or not the maximum value of t_p

imposes restrictions on the moments which can be carried. By assuming small reinforcement degrees we avoid this complication. Fortunately, the greater majority of cases encountered in practice are covered in this way.

Referring to the coordinate systems shown in Fig. 7.3.9, a statically admissible moment field may be taken as follows:

In the angles γ we select

$$m_y = -m_p'$$
$$m_x = m_p' \tan^2 \gamma \tag{7.3.7}$$
$$m_{xy} = 0$$

In the angle β we select

$$m_r = -m_p'$$
$$m_\theta = m_p \tag{7.3.8}$$
$$m_{r\theta} = 0$$

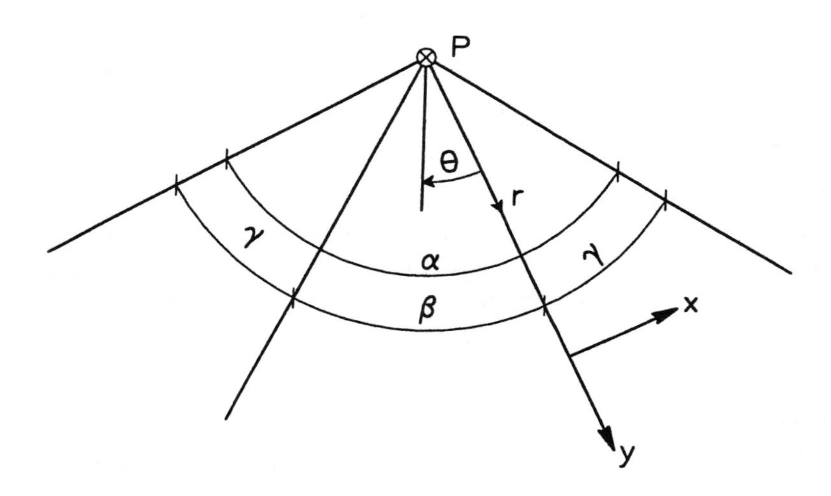

Figure 7.3.9 Concentrated force in a corner with arbitrary opening angle.

It is easily shown that the bending moments along the edges are zero and that the torsional moments along the edges are constant and equal to

$$|m_v| = m_p' \tan\gamma \tag{7.3.9}$$

The two moment fields satisfy the continuity requirements if

$$m_p = m_p' \tan^2\gamma \quad \Rightarrow \quad \tan\gamma = \sqrt{\frac{m_p}{m_p'}} \tag{7.3.10}$$

Thus the load carried in the angles γ is

$$P_\gamma = 2|m_v| = 2m_p' \tan\gamma \tag{7.3.11}$$

The load carried by the moment field (7.3.8) is

$$P_\beta = (m_p + m_p')\beta = (m_p + m_p')(\alpha - 2\gamma) \tag{7.3.12}$$

This part of the load is carried by uniformly distributed shear forces in circumferential sections.

The total load carried is

$$P = P_\gamma + P_\beta \tag{7.3.13}$$

or

$$\begin{aligned} P &= 2m_p' \tan\gamma + (m_p + m_p')(\alpha - 2\gamma) \\ &= 2\sqrt{m_p m_p'} + (m_p + m_p')\left(\alpha - 2\,\mathrm{Arc}\tan\sqrt{\frac{m_p}{m_p'}}\right) \end{aligned} \tag{7.3.14}$$

which is in agreement with (6.5.113).

If $m_p > m_p' \tan^2 \alpha/2$ we will find $\gamma > \alpha/2$ and β will be found negative. Thus in this case we must put $\gamma = \alpha/2$ and the load-carrying capacity is determined by

$$P = 2m_p' \tan\frac{\alpha}{2} \tag{7.3.15}$$

which is in agreement with (6.5.114).

The solution is, in fact, an exact solution when all supports, irrespective of their form, are restrained edges. The type of failure mechanism involved is shown in Fig. 6.5.20 and Fig. 6.5.21, respectively.

To carry the load P_γ (7.3.11) or the load P (7.3.15) edges must be shear reinforced to carry the torsional moment $|m_v| = m_p' \tan\gamma$. When $\beta = 0$ no punching shear analysis has to be carried out. The load is carried exclusively by the Kirchhoff forces.

When β is not zero a punching shear analysis is carried out to control whether the load P_β may be resisted. This may be done by means of formula (7.3.6) replacing α with β. If the load cannot be carried, the dimensions of the loaded area must be increased or shear reinforcement in the angle β must be provided.

The analysis suggested is, of course, conservative. A local punching failure in the angle β is not likely to take place at the same time as the reinforcement is yielding. The analysis may be justified by a kind of lower bound reasoning.

The above statically admissible moment field contains as a special case the limit when the whole force is carried by the usual shear forces. An example has already been considered in Section 7.3.1. When no edge reinforcement is required, nevertheless the edges should be provided with some edge reinforcement. Carrying the reinforcement around the edges also substantially improves the anchorage.

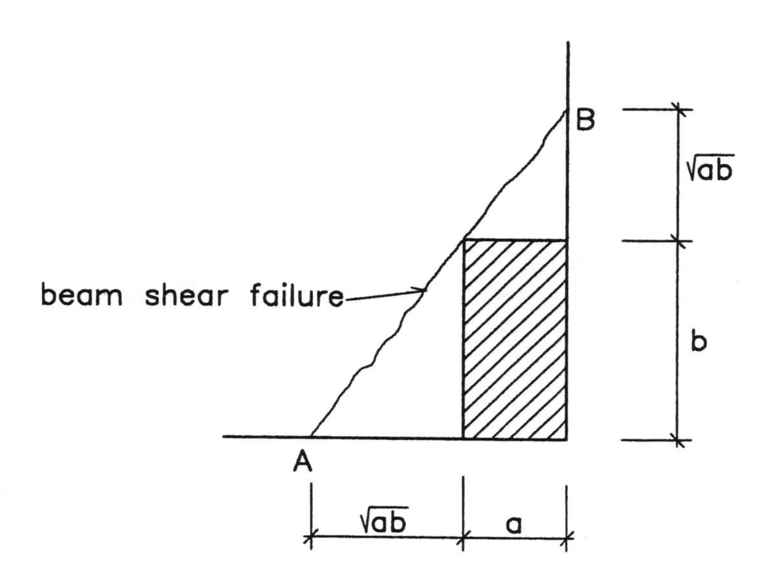

Figure 7.3.10 Beam shear failure in a right angled corner without edge reinforcement.

If no edge reinforcement is provided, the punching shear analysis described should be supplemented by a beam shear analysis if relevant.

A beam shear analysis is illustrated in Fig. 7.3.10 for a rectangular load in a right angled corner. The beam shear failure takes place along a straight line *AB*. The calculation may be carried out approximately by requiring the average shear stress along *AB* to be less than or equal to the shear capacity of a nonshear reinforced beam (cf. Section 5.2.6). *AB* should be chosen as the line with the shortest length touching the loaded area. For a loaded area in the form of a rectangle the line with shortest length is shown in the figure. Normally this analysis does not lead to answers very different from what is found by the method described above, but it is pertinent anyway to have one more check of the load-carrying capacity, the uncertainty of the whole procedure taken into account.

The general edge and corner solution of this section may be used for loads near an edge or near a corner, too. In fact, when the load is near a free edge or a corner, the region between the load and the edge will contribute very little to the load-carrying capacity. The relative enhancement will be of the order of the distance to the edge divided by the distance from the force to the supports. When this ratio is small the analysis is simply performed by imagining the loaded area moved to the edge or to the corner and then perform an analysis as described above. Edge reinforcement is provided as for an edge or corner load.

7.3.5 Eccentric Loading

Edge and corner load-carrying capacities are sensitive to eccentricities. Eccentric loading may be analyzed by the method outlined in Section 7.2.4.

In order to be able to perform the calculation we need to know the load-carrying capacity for concentric loading and for a pure moment loading, respectively. Concentric loading has already been discussed. Pure moment loading will be illuminated by considering some elementary cases.

Fig. 7.3.11 shows a pure moment loading on a rectangular area $b \cdot a$ in a corner of a rectangular slab $\ell_1 \cdot \ell_2$ simply supported along two sides. The slab is isotropically reinforced parallel to the edges in the top and bottom and the corresponding yield moments are equal, i.e., $m_p = m_p'$. The moment vector is parallel to the side ℓ_1 and the sides of the loaded area are assumed parallel to the supported edges. The load may be carried partly by pure torsion in a strip of width b. To include bending action in the strip of width a we consider at first this strip to be supported along part of the edge ℓ_1 and by a uniformly distributed reaction R on the loaded area. The loading on the strip is shown in the figure.

R is found by a moment equation

$$R\left(\ell_2 - \frac{b}{2}\right) = M - 2t_p b \quad \Rightarrow$$

$$R = \frac{M - 2t_p b}{\ell_2 - \dfrac{b}{2}} \cong \frac{M - 2t_p b}{\ell_2} \tag{7.3.16}$$

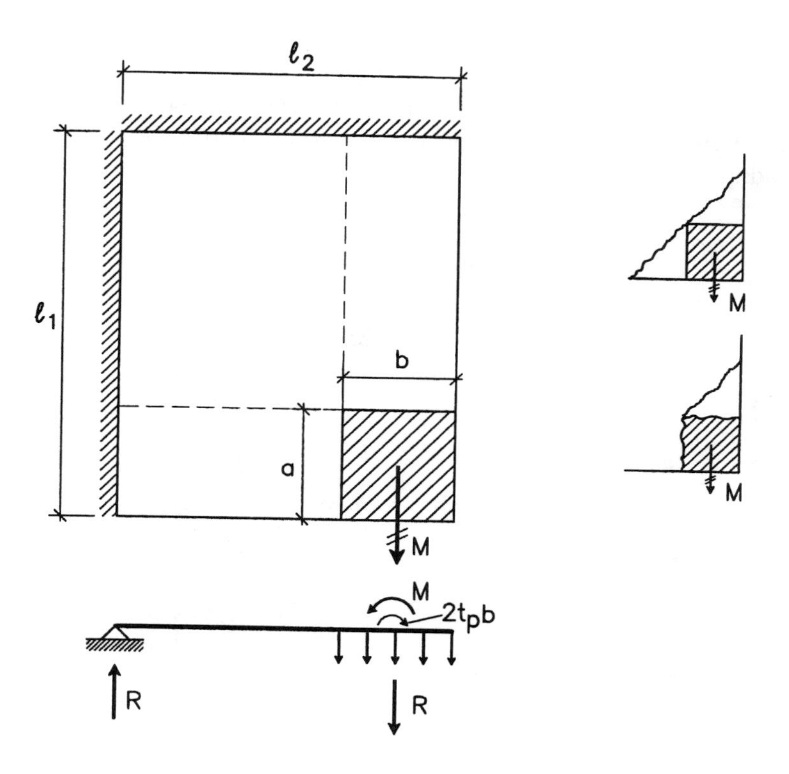

Figure 7.3.11 Pure moment loading in a corner.

having assumed $b << \ell_2$. As usual, t_p is the torsional capacity (see Box 7.3.1). As above, we have included the effect of the vertical shear stresses in the section $b \cdot h$ with pure torsion. The maximum bending moment per unit length in the strip a is

$$m_b = \frac{R(\ell_2 - b)}{a} \cong \frac{R\ell_2}{a} = \frac{M}{a} - 2t_p \frac{b}{a} \tag{7.3.17}$$

Inserting, as in the cases treated in Section 7.2.4, $m_b = t_p$ we find

$$M = M_p = a\left(1 + 2\frac{b}{a}\right)t_p \tag{7.3.18}$$

Strictly speaking this is not the true load-carrying capacity according to the present lower bound analysis because we must add the moment field for a uniform upward-directed load R on the area $a \cdot b$. This load may be carried by a pure torsion field $R/2$. Since

$$\frac{R}{2} = \frac{t_p a}{2\ell_2} \tag{7.3.19}$$

the torsion field might be neglected when $a << \ell_2$. Therefore (7.3.18) approximately furnishes the load-carrying capacity by a pure moment loading in a corner.

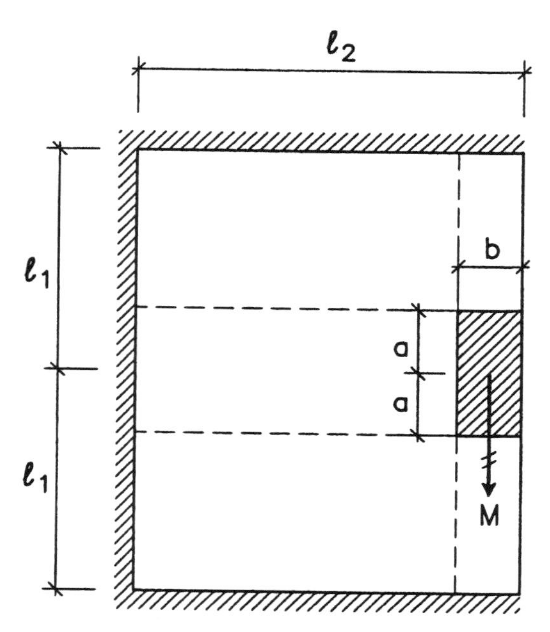

Figure 7.3.12 Pure moment loading at an edge.

It might be instructive to compare the lower bound solution with the results of some simple yield line patterns. Some possibilities are sketched in the figure to the right. The calculations are left for the reader.

Considering next the edge loading case, Fig. 7.3.12 shows a pure moment loading with a moment vector parallel to the side $2\ell_1$. The load-carrying capacity may immediately be found by means of the corner load solution, i.e.,

$$M = M_p = 2a\left(1 + 2\frac{b}{a}\right)t_p \tag{7.3.20}$$

In Fig. 7.3.13 the moment vector is parallel to the sides ℓ_2. The moment may be carried by torsion fields originating from concentrated forces X in the corners of the loaded area. They have, of course, opposite directions above and below the loaded area. In addition, we may have pure torsion in the strip of width $2a$ and bending action in the strip of width b. The loads on the strip b are shown in the figure. A moment equation furnishes

$$R \cdot 2\ell_1 + 2X \cdot 2a = M - 2t_p \cdot 2a \quad \Rightarrow$$

$$R = \frac{M - 2t_p \cdot 2a - 2X \cdot 2a}{2\ell_1} \tag{7.3.21}$$

The maximum bending moment per unit length m_b in the strip of width b will then be

$$m_b = \frac{R(\ell_1 - a)}{b} \cong \frac{R\ell_1}{b} = \frac{M}{2b} - 2t_p\frac{a}{b} - 2X\frac{a}{b} \tag{7.3.22}$$

Since in the strip of width b bending action is superimposed with a torsion field from one concentrated force X we cannot, as before, take m_b equal to t_p. Since the maximum torsional moment outside the strips $2a$ and b is $2X/2 = X = t_p$, the torsion field to be superimposed on the bending action is $t_p/2$. Thus the maximum value of the bending moment m_b may be determined by the yield condition (2.3.62), i.e.,

$$-(t_p - m_b)t_p + \frac{1}{4}t_p^2 = 0 \quad \Rightarrow$$

$$m_b = \frac{3}{4}t_p \tag{7.3.23}$$

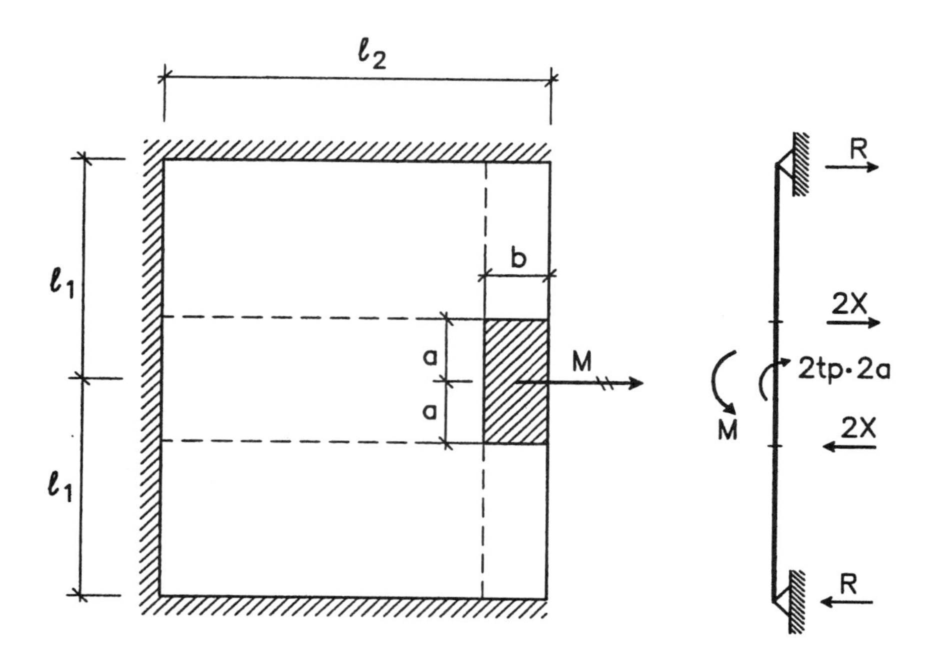

Figure 7.3.13 Pure moment loading at an edge.

We have used the approximation $m_p = m_p' \cong t_p$. Of course, a negative moment $m_b = -3t_p/4$ may also be carried.

Inserting this result together with $X = t_p$ into (7.3.22) we find the load-carrying capacity

$$M = M_p = b\left(\frac{3}{2} + 8\,\frac{a}{b}\right)t_p \qquad (7.3.24)$$

If we had neglected the combined effects of bending and torsion in the strip b we would find the number 2 instead of 3/2 in formula (7.3.24). The effect is thus minor unless the length a is small compared to b.

The results above should not be used unless the reinforcement distances are small compared to the dimensions of the loaded area.

In the cases treated the lengths b should not be less than h because of the vertical shear stresses at the ends of the section $b{\cdot}h$. In the case in Fig. 7.3.13 the lengths a should not be less than h

because we have vertical shear stresses both from the strip with pure torsion and from the torsional fields.

If the lengths a in Fig. 7.3.13 are near the limit the length of the section with pure torsion should be put to $2a - h$ instead of $2a$ because of the vertical shear stresses from the torsional fields. In all cases, if the dimensions are near the limits, the torsional capacity of the strips with pure torsion should be calculated by the method developed for pure torsion of beams. This has already been stated and explained in Section 7.3.5.

It is left for the reader to reestablish the equilibrium solutions under these modified assumptions.

For combined actions the load-carrying capacity is determined by (7.2.44). For eccentricity in only one direction we have formula (7.2.51). Notice that here as well as in the previous cases the eccentricity is measured from the center of gravity of the loaded area and not from the center of gravity of one or another control perimeter. Eccentricities in two directions may be treated by the same method.

In principle it is easy to extend the procedure to cover any edge or corner load on any slab.

The applicability of linear interaction formulas was studied by Larsen [98.1]. It turned out that in the case of edge and corner loads it is important to discern between the cases where the normal force and the moment, so to speak, counteract each other and where they coact. For instance, a downward normal force and the moment vector shown in Fig. 7.3.11 are counteracting each other. Likewise in Fig. 7.3.12, while in Fig. 7.3.13 there will be no dependence on the sign of the moment.

In the case of counteraction the linear interaction formula is far too conservative. Instead, as shown in [98.1], a bilinear interaction diagram may be etablished using the same types of stress fields as above. Then good agreement with tests was obtained.

For the case with coaction no tests exist for corner loads while for the edge load case good agreement was found using linear interaction.

A preliminary conclusion seems to be that for edge and corner loads linear interaction is applicable in the case of coaction and consequently when both signs of the moment may occur for a particular direction of the normal force.

7.4 Concluding Remarks

We have emphasized shear problems related to concentrated forces. But, of course, all shear stresses in a slab should in principle be checked. Generally for slabs without shear reinforcement the limit should be taken as the shear capacity of nonshear reinforced beams and slabs, except, of course, in sections close to concentrated forces (cf. Section 5.2.6). The maximum value of the shear force on a point may be calculated by means of formula (6.1.3) in Chapter 6.

One main result of our analysis is that for concentrated forces at or near a free edge, the edge must be reinforced to carry a shear force equal to the torsional moment. Normally it is not necessary to carry out a shear design of the Kirchhoff edge beam (see below) but if it is done a 45° strut inclination should be used. There is no need for checking the concrete stress. This point is taken care of by the torsion calculation. We repeat in Fig. 7.4.1 two ways of reinforcing the edge. In the figure to the left the edge is reinforced as a beam with closed stirrups. The stirrup width should be of the order of half the slab thickness. However, tests have shown that normally it is enough to reinforce the edge as shown in the figure to the right.

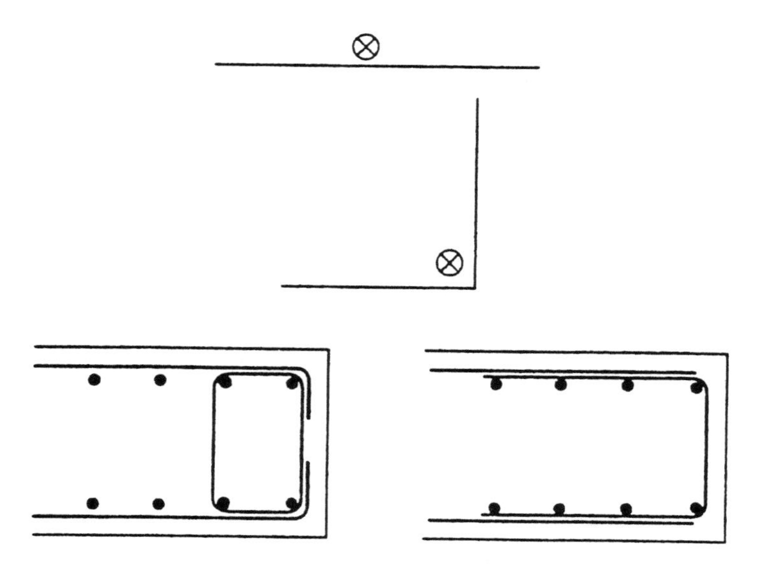

Figure 7.4.1 Edge reinforcement in slabs with concentrated forces at or near a free edge or at or near a corner.

Either U-stirrups are supplied as shown in the figure or the slab reinforcement is carried around the edge. When the reinforcement is parallel and perpendicular to the edge, respectively, no shear calculation is required for the edge beam. The largest reinforcement perpendicular to the edge is carried around the edge or a corresponding number of U-stirrups are supplied. The distance between the bars should not exceed the slab thickness minus two times the maximum cover of the slab reinforcement.

It has been pointed out repeatedly that many problems regarding shear in slabs have not yet been solved. This is demonstrated clearly by the diversity of the code rules available at the moment. We have tried our best to carry the logic of the subject as far as possible in the hope that this will enable the advanced designer to understand the design rules before applying them. Often this is impossible when dealing with code rules alone because they are compromises between many divergent points of view, which is unavoidable when the science of a subject is far from being closed.

It might turn out that shear problems for quasiplastic materials like concrete are so complicated that our culture will never invest enough time and capital to solve them completely. However, we believe that at the present stage of development the subject is sufficiently cleared up to enable the designer to solve most of the problems encountered in practice in a reasonably safe way.

Chapter 8

SHEAR IN JOINTS

8.1 INTRODUCTION

This chapter deals with the shear strength of joints. The first analysis based on the theory of plasticity was performed by B. C. Jensen [75.6; 76.1; 78.12]. A theory of less general character but rendering similar results is the shear-friction theory developed by Mattock and associates [69.5; 72.9; 74.8].

The strength of a disk normally is a plane stress problem (cf. Chapter 4). However, since a joint often introduces a weakness in the structure, failure is inclined to take place in the joint, where lateral (out of plane) strains will be more or less prevented. In most cases, the strength of a joint can, therefore, be treated as a plane strain problem.

The concrete surrounding the joint is identified as a modified Coulomb material. The cohesion is termed c and the angle of friction φ. The compressive strength is f_c and the tensile strength f_t.

For the joint the failure condition is also assumed to be of the modified Coulomb type.

The joint leads to a strength different and normally less than in monolithic concrete. For the cohesion in the joint, the symbol c' is used. The angle of friction is often the same in the joint as in the monolithic concrete. However, in special cases it can be less as well as greater than in monolithic concrete. For the angle of friction in joints the symbol φ' is used. An equivalent compressive strength combining c' and φ' according to (2.1.26) is denoted f_c'. The tensile strength of the joint is denoted f_t'.

8.2 ANALYSIS OF JOINTS BY PLASTIC THEORY

8.2.1 General

The shear problem considered is illustrated in Fig. 8.2.1, where a concrete disk is loaded to shear along line ℓ by two opposite concentrated loads P. Along ℓ one or another kind of joint is supposed to be present, or, alternatively, the disk is supposed to be of monolithic concrete.

Although we know beforehand, according to our assumptions, the shear strength of the joint, it will be instructive to apply the upper bound technique to derive the formulas for the shear strength.

We begin by considering a yield line along ℓ through monolithic concrete, i.e., there is no joint.

8.2.2 Monolithic Concrete

Consider a failure mechanism in the form of a yield line along the line of loading. The relative displacement of the right-hand part to the left-hand part is u, forming the angle α to the yield line (see Fig. 8.2.2). The external work is

$$W_E = Pu\cos\alpha \tag{8.2.1}$$

The dissipation consists of two parts, one from the concrete and one from the reinforcement. The reinforcement bars are perpendicular to the yield line.

As before, the dowel effect of the reinforcement is neglected, which means that the dissipation in the reinforcement is

$$D_R = A_s f_Y u \sin\alpha \tag{8.2.2}$$

where A_s is the reinforcement area and f_Y the yield stress.

From the concrete the contribution is

$$D_C = W_\ell h \tag{8.2.3}$$

where h is the length of the yield line, and W_ℓ is given as a function of α by the formulas of Section 3.4.5, plane stress problems.

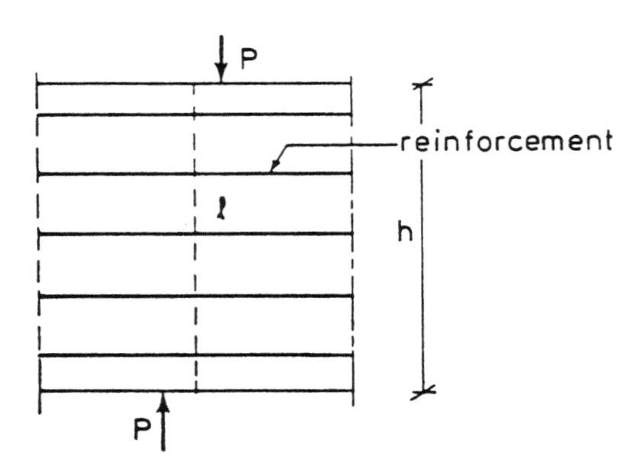

Figure 8.2.1 Specimen subjected to shear.

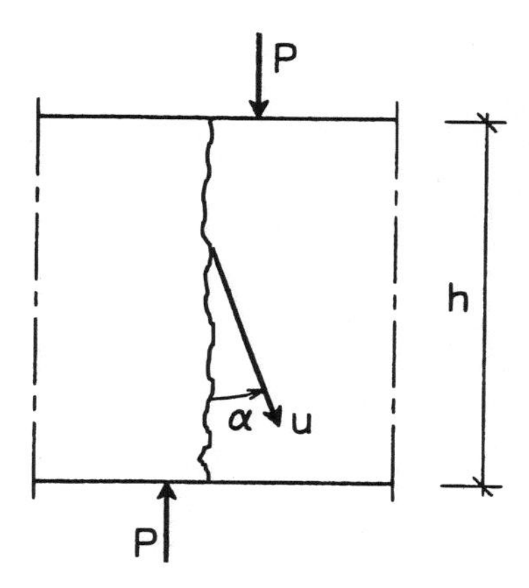

Figure 8.2.2 Failure mechanism in disk subjected to shear.

Consider the following cases.

$\boldsymbol{\alpha} = \mathbf{0}$. The concrete dissipation is determined by (3.4.90). Since the displacement is perpendicular to the reinforcement, $D_R = 0$. The work equation is

$$Pu = \tfrac{1}{2} f_c u h t \tag{8.2.4}$$

t being the thickness of the disk. Consequently

$$\frac{\tau}{f_c} = \frac{1}{2} \tag{8.2.5}$$

where $\tau = P/ht$.

$\mathbf{0} < \boldsymbol{\alpha} < \boldsymbol{\varphi}$. The concrete dissipation is determined by (3.4.90). The work equation is

$$Pu\cos\alpha = \tfrac{1}{2} f_c u(1 - \sin\alpha) ht + A_s f_Y u \sin\alpha \tag{8.2.6}$$

or

$$\frac{\tau}{f_c} = \frac{1 - \sin\alpha}{2\cos\alpha} + \psi \tan\alpha \tag{8.2.7}$$

where the reinforcement degree ψ has been introduced, i.e.,

$$\psi = \frac{A_s f_Y}{htf_c} = r\frac{f_Y}{f_c} \tag{8.2.8}$$

r being the reinforcement ratio. The minimum value of (8.2.7) is

$$\frac{\tau}{f_c} = \sqrt{\psi(1 - \psi)} \tag{8.2.9}$$

which is found to occur for

$$\sin\alpha = 1 - 2\psi \tag{8.2.10}$$

Utilizing that $0 < \alpha < \varphi$ and (8.2.10), we find the following interval, in which (8.2.9) governs the solution:

$$\frac{1 - \sin\varphi}{2} < \psi < \frac{1}{2} \tag{8.2.11}$$

In a $\psi, \tau/f_c$-coordinate system, (8.2.9) represents a circle with radius ½ and the center at (0, ½). Equation (8.2.5) is tangent to (8.2.9) in (½, ½).

$\boldsymbol{\alpha = \varphi}$. The concrete dissipation is given by (3.4.90); that is, the work equation reads

$$Pu\cos\varphi = \tfrac{1}{2} f_c u (1 - \sin\varphi) ht + A_s f_Y u \sin\varphi \tag{8.2.12}$$

or

$$\frac{\tau}{f_c} = \frac{1 - \sin\varphi}{2\cos\varphi} + \psi\tan\varphi = \frac{c}{f_c} + \psi\tan\varphi \tag{8.2.13}$$

Equation (8.2.13) is tangent to (8.2.9) at the point

$$\left(\frac{1 - \sin\varphi}{2}, \frac{\cos\varphi}{2}\right)$$

$\boldsymbol{\alpha > \varphi}$. The concrete dissipation is determined by (3.4.89) or (3.4.83). We get the work equation

$$Pu\cos\alpha = \left[\frac{1}{2} f_c (1 - \sin\alpha) + \frac{\sin\alpha - \sin\varphi}{1 - \sin\varphi} f_t\right] uht \\ + A_s f_Y u \sin\alpha \tag{8.2.14}$$

or

$$\frac{\tau}{f_c} = \frac{1 - \sin\alpha}{2\cos\alpha} + \frac{\sin\alpha - \sin\varphi}{(1 - \sin\varphi)\cos\alpha}\frac{f_t}{f_c} + \psi\tan\alpha \tag{8.2.15}$$

The minimum value is found to be

$$\frac{\tau}{f_c} = \sqrt{\left(\psi + \frac{f_t}{f_c}\right)\left[1 - 2\frac{f_t}{f_c}\frac{\sin\varphi}{1 - \sin\varphi} - \left(\psi + \frac{f_t}{f_c}\right)\right]} \tag{8.2.16}$$

occurring for

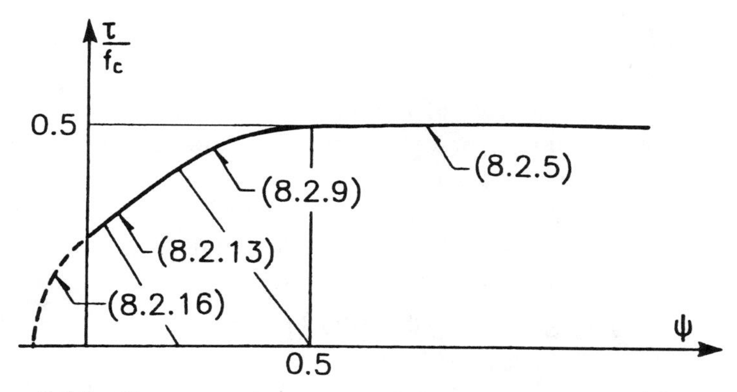

Figure 8.2.3 Shear capacity versus reinforcement degree for monolithic concrete.

$$\sin\alpha = 1 - 2\,\frac{\psi + f_t/f_c}{1 - 2(f_t/f_c)[\sin\varphi/(1-\sin\varphi)]} \tag{8.2.17}$$

(8.2.16) is a circle with radius

$$\frac{1}{2} - \frac{f_t}{f_c}\,\frac{\sin\varphi}{1-\sin\varphi}$$

Equation (8.2.13) is tangent to (8.2.16) at the point

$$\left(\frac{1-\sin\varphi}{2} - (1+\sin\varphi)\,\frac{f_t}{f_c}\,,\ \frac{1+\sin\varphi}{\cos\varphi}\left[\frac{1-\sin\varphi}{2} - \sin\varphi\,\frac{f_t}{f_c}\right]\right)$$

By introducing the parameters ℓ and m (see formulas (3.4.84) and (3.4.85)), (8.2.16) may be written in the more compact form

$$\frac{\tau}{f_c} = \frac{1}{2}\sqrt{\ell^2 - (m - 2\psi)^2} \tag{8.2.18}$$

The load-carrying capacity determined by (8.2.5), (8.2.9), (8.2.13), and (8.2.16) is shown in Fig. 8.2.3. For $\psi < 0$, (8.2.16) is shown by a dotted line, because the solution is without physical meaning in this region.

The small circle (8.2.16) is dependent on the tensile strength f_t of the concrete. For decreasing tensile strengths, the circle approaches the circle (8.2.9), and for $f_t = 0$ it becomes identical to (8.2.9), and (8.2.13) vanishes. In this case the load-carrying capacity is the same as for beams (see Section 5.2.1 on beam shear), where the tensile strength is assumed to be zero.

8.2.3 Joints

Consider now the case of a joint along the line of loading. The joint has the cohesion c' and the angle of friction φ'. From Section 3.4.5 we know that the angle α between the yield line and the relative displacement u is greater than or equal to φ'.

$\boldsymbol{\alpha = \varphi'}$. The concrete dissipation is determined by (3.4.83) valid for plane strain problems. The work equation reads

$$Pu\cos\varphi' = \frac{1-\sin\varphi'}{2} f_c' uht + A_s f_Y u \sin\varphi' \qquad (8.2.19)$$

or

$$\frac{\tau}{f_c'} = \frac{1-\sin\varphi'}{2\cos\varphi'} + \psi'\tan\varphi' = \frac{c'}{f_c'} + \psi'\tan\varphi' \qquad (8.2.20)$$

where we have introduced $\psi' = \psi f_c / f_c'$. Except for the primes, (8.2.20) is identical to (8.2.13).

$\boldsymbol{\alpha > \varphi'}$. Equation (8.2.14) is valid; that is, except for the primes we get the same equation as in the monolithic case. The load-carrying capacity is therefore analogous to (8.2.16); that is,

$$\frac{\tau}{f_c'} = \sqrt{\left(\psi' + \frac{f_t'}{f_c'}\right)\left[1 - 2\frac{f_t'}{f_c'}\frac{\sin\varphi'}{1-\sin\varphi'} - \left(\psi' + \frac{f_t'}{f_c'}\right)\right]} \qquad (8.2.21)$$

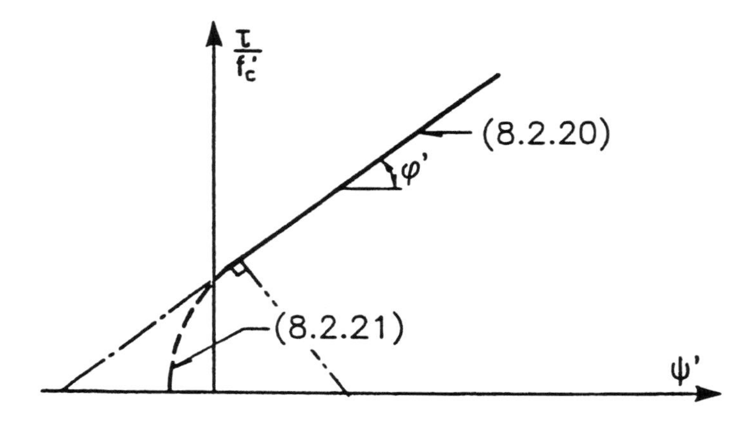

Figure 8.2.4 Shear capacity versus reinforcement degree for joint.

Figure 8.2.4 shows the load-carrying capacity. It is seen that the load-carrying capacity increases rapidly with the degree of reinforcement.

The load-carrying capacity for the monolithic case cannot, of course, be exceeded by a disk with a joint.

8.2.4 Statical Interpretation

The upper bound solutions found above can be given a statical interpretation as shown below. However, it must be emphasized that we are not dealing with true lower bound solutions, since the stress fields considered are defined only along the yield line.

The maximum force in the reinforcement is

$$T = A_s f_Y \tag{8.2.22}$$

If a compression force of the same magnitude is distributed over the whole concrete area between the loads (Fig. 8.2.1), we find a normal stress in the concrete equal to (σ positive for tension)

$$\sigma = -\frac{T}{ht} = -\psi f_c \tag{8.2.23}$$

The average shear stress is given by

$$\tau = \frac{P}{ht} \tag{8.2.24}$$

From geometrical considerations it can be shown that the circular cutoff in the modified Coulomb failure criterion (cf. Fig. 3.4.12) is determined by the equation

$$\left(\sigma + \frac{1}{2}f_c - \frac{f_t}{1 - \sin\varphi}\right)^2 + \tau^2 = \left(\frac{1}{2}f_c - f_t\frac{\sin\varphi}{1 - \sin\varphi}\right)^2 \tag{8.2.25}$$

Inserting (8.2.23) into (8.2.25) we find the shear strength (8.2.16) or (8.2.21) with corresponding primed symbols.

The straight line has the equation

$$\tau = c - \sigma\tan\varphi \tag{8.2.26}$$

Inserting (2.1.27) and (8.2.23), we get (8.2.13) or (8.2.20) with corresponding symbols.

When we assume plane stress, one of the principal stresses is always zero. In the σ, τ-coordinate system the greatest Mohr's circle is therefore the circle through $(\sigma, \tau) = (0,0)$ and $(\sigma, \tau) = (-f_c, 0)$. This circle has the equation

$$\left(\sigma + \frac{1}{2}f_c\right)^2 + \tau^2 = \left(\frac{1}{2}f_c\right)^2 \tag{8.2.27}$$

Inserting (8.2.23), we get (8.2.9).

For degrees of reinforcement greater than $\frac{1}{2}$, the reinforcement does not yield, because the shear strength of the concrete is limited to $\tau/f_c = \frac{1}{2}$, which is expressed in (8.2.5). These statical solutions are thus shown to be identical to the upper bound solutions.

8.2.5 Axial Forces

Compressive forces parallel to the reinforcement have the same effect as adding reinforcement, and tensile forces have the same effect as removing reinforcement. With axial forces, all equations are still valid if we replace ψ by

$$\psi^* = \psi - \frac{N}{htf_c} \tag{8.2..28}$$

N being positive as a tensile force.

For large normal forces the limiting curve for monolithic concrete is governing the load-carrying capacity. The limiting curve is shown in Fig. 8.2.5 ($f_t = 0$). It consists of two Mohr's circles with radii f_c connected by a straight line. We shall not write down the formulas.

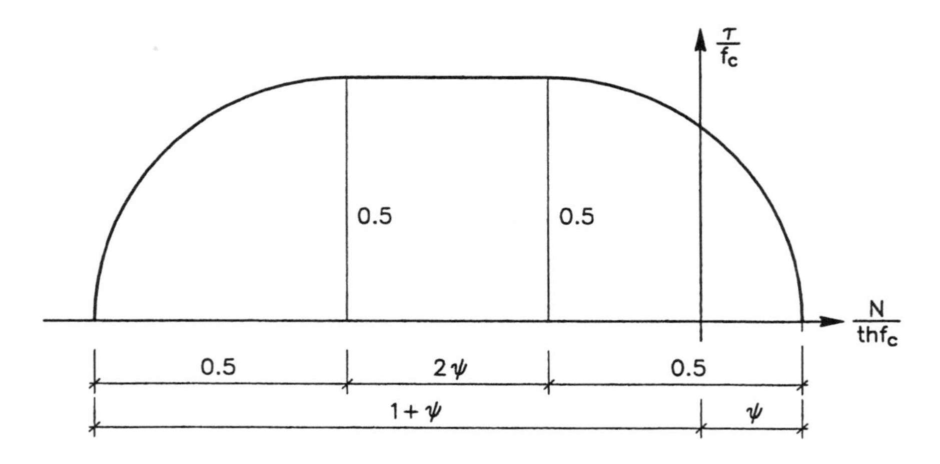

Figure 8.2.5 Limiting curve for monolithic concrete in the case of an external normal force.

8.2.6 Effectiveness Factors

As in many cases where the strength of the concrete is important, the theoretical results are not in agreement with the experimental results unless effective concrete strengths are introduced. The weakness of the joints gives rise to a smaller effectiveness factor than for the monolithic concrete.

Introducing the effectiveness factors in the equations by replacing f_c with νf_c, f_t with ρf_c, f_c' with $\nu' f_c$, and f_t' with $\rho' f_c$, we get, for monolithic concrete,

$$\frac{\tau}{f_c} = \sqrt{(\psi + \rho)\left[\nu - 2\rho\frac{\sin\varphi}{1 - \sin\varphi} - (\psi + \rho)\right]} \tag{8.2.29}$$

valid for

$$\psi \le \nu\frac{1 - \sin\varphi}{2} - \rho(1 + \sin\varphi) \tag{8.2.30}$$

and

$$\frac{\tau}{f_c} = \nu\frac{1 - \sin\varphi}{2\cos\varphi} + \psi\tan\varphi = \frac{\nu c}{f_c} + \psi\tan\varphi \tag{8.2.31}$$

valid for

$$\nu\frac{1 - \sin\varphi}{2} - \rho(1 + \sin\varphi) \le \psi \le \nu\frac{1 - \sin\varphi}{2} \tag{8.2.32}$$

Further,

$$\frac{\tau}{f_c} = \sqrt{\psi(\nu - \psi)} \tag{8.2.33}$$

when

$$\nu\frac{1 - \sin\varphi}{2} \le \psi \le \tfrac{1}{2}\nu \tag{8.2.34}$$

and finally,

$$\frac{\tau}{f_c} = \frac{1}{2}\nu \tag{8.2.35}$$

when

$$\psi \ge \tfrac{1}{2}\nu \tag{8.2.36}$$

For joints,

$$\frac{\tau}{f_c} = \sqrt{(\psi + \rho')\left[\nu' - 2\rho' \frac{\sin\varphi'}{1 - \sin\varphi'} - (\psi + \rho')\right]} \qquad (8.2.37)$$

when

$$\psi \le \nu' \frac{1 - \sin\varphi'}{2} - \rho'(1 + \sin\varphi') \qquad (8.2.38)$$

For greater ψ-values

$$\frac{\tau}{f_c} = \nu' \frac{1 - \sin\varphi'}{2\cos\varphi'} + \psi\tan\varphi' = \frac{c'}{f_c} + \psi\tan\varphi' \qquad (8.2.39)$$

The equations for joints are valid only when the ultimate load is less than the ultimate load for monolithic concrete.

As shown in Section 2.1.3, we may put $\varphi = 37^o$ for monolithic concrete.

The values of ν, ρ, ν', ρ', and φ' can as yet be determined only by means of tests.

In principle, the load-carrying capacity of joints is shown in Fig. 8.2.6 in the case $\varphi = \varphi'$.

It should be noticed that when $\varphi = \varphi'$ then $f_c'/f_c = c'/c = \nu'$ (see Box 8.2.1) where the effectiveness factors for compression are reviewed.

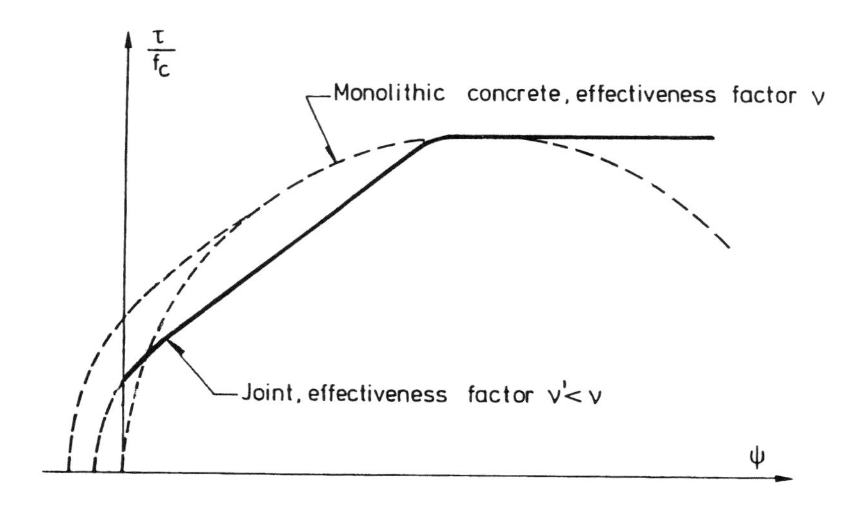

Figure 8.2.6 Shear capacity of joint versus reinforcement degree with due account to the strength of the surrounding concrete.

Virgin material

Cohesion: c

Angle of friction: φ

Compressive strength expressed by c and φ: $f_c = 2c\cos\varphi / 1 - \sin\varphi$

Cohesion expressed by f_c and φ: $c = f_c(1-\sin\varphi)/2\cos\varphi$

For $\varphi = 37^\circ \Rightarrow c = f_c/4$

Monolithic concrete with softening and cracking

Compressive strength: νf_c

Angle of friction: φ (not changed)

Cohesion: $\nu f_c(1-\sin\varphi)/2\cos\varphi = \nu c$

In theoretical formulas f_c is replaced by
νf_c or c by νc keeping φ unchanged

Joints

Cohesion: c'

Angle of friction: φ'

Compressive strength expressed by c' and φ':
$f_c' = 2c'\cos\varphi' / 1 - \sin\varphi'$

Cohesion expressed by f_c' and φ': $c' = f_c'(1-\sin\varphi')/2\cos\varphi'$

$f_c' = \nu' f_c \Rightarrow c' = \nu' f_c(1-\sin\varphi')/2\cos\varphi'$

If $\varphi' = \varphi$: $c' = \nu' c \Rightarrow \nu' = f_c'/f_c = c'/c$

The joint may be characterized either by
c' and φ' or f_c' and φ'. If c' and φ' are known,
for instance from tests

$$\nu' = \frac{c'}{f_c}\frac{2\cos\varphi'}{1-\sin\varphi'} \qquad \left(\varphi' = \varphi \Rightarrow \nu' = \frac{4c'}{f_c}\right)$$

Box 8.2.1 Review of effectiveness factors on compressive strengths for monolithic concrete and joints.

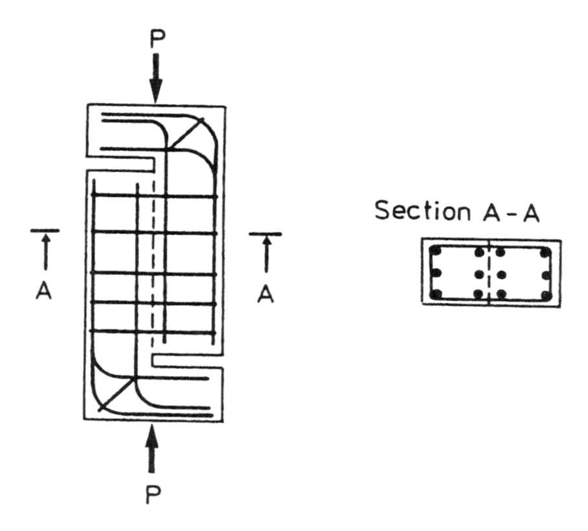

Figure 8.2.7 Test specimen of the push-off type.

We mention a few test results for monolithic concrete.

Hofbeck et al. [69.5] carried out tests on specimens as shown in Fig. 8.2.7. This type of specimen, which has been used in various forms, is called a specimen of the push-off type. One of the series involved monolithic specimens. The results are shown in Fig. 8.2.8 together with (8.2.31), (8.2.33), and (8.2.35).

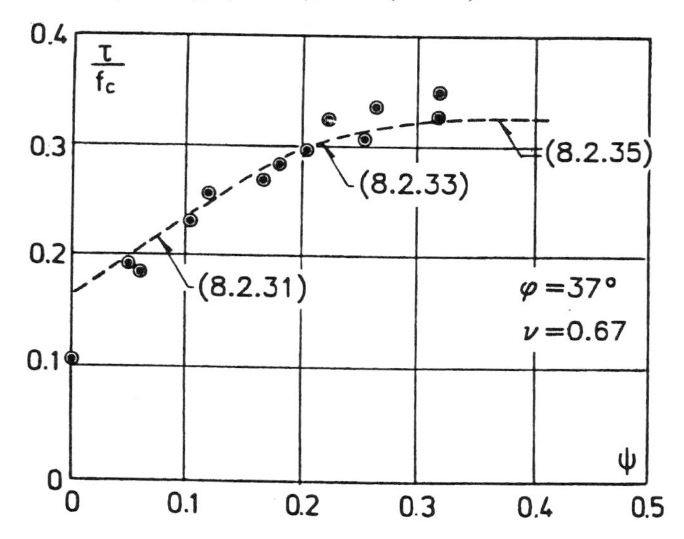

Figure 8.2.8 Test results for monolithic concrete [69.5].

The region where (8.2.29) governs the load-carrying capacity was not covered very well by the tests. Therefore, (8.2.29) is not drawn in the figure. Using $\nu = 0.67$, the equations fit well with the experiments.

The reduced concrete strength must be attributed to softening effects. To obtain failure along the line of force a considerable stress redistribution is likely to take place. This is contrary to specimens with a premade crack, where softening effects seem to be missing (see Section 8.3.2).

The upper limit of the shear stress is thus in these tests $0.5 \cdot 0.67 f_c = 0.33 f_c$. This limit will be used in the graphs in the following, disregarding the fact that it depends on the concrete strength. In actual designs this will be taken into account by using the usual ν-values for monolithic concrete.

8.2.7 Skew Reinforcement

A joint may not always be provided with reinforcement perpendicular to the joint. In Fig. 8.2.9 the bars form the angle β with the joint. The displacement vector u is under the angle α with the joint. Remembering that in a yield line the reinforcement contribution is the displacement component along the bar times the yield force we have for $f_t' = 0$ the work equation $(\varphi' \leq \alpha \leq \pi - \varphi')$

$$Pu \cos\alpha = \frac{1}{2}(1 - \sin\alpha) f_c' u h t + A_s f_Y u \cos(\beta - \alpha) \qquad (8.2.40)$$

Here A_s is the total reinforcement area of the bars crossing the yield line. Introducing the reinforcement ratio and the reinforcement degree, respectively

$$r_{\text{skew}} = \frac{A_s}{ht} \qquad \psi_{\text{skew}} = r_{\text{skew}} \frac{f_Y}{f_c} \qquad (8.2.41)$$

we find

$$\frac{\tau}{f_c} = \frac{\nu'}{2} \frac{1 - \sin\alpha}{\cos\alpha} + \psi_{\text{skew}} \cos\beta + \psi_{\text{skew}} \sin\beta \tan\alpha \qquad (8.2.42)$$

When minimizing with respect to α, the term $\psi_{\text{skew}} \cos\beta$ does not change. Therefore, as long as (8.2.40) is valid (see below) the previous results may be used when ψ is replaced by $\psi_{\text{skew}} \sin\beta$ and when an extra term $\psi_{\text{skew}} \cos\beta$ is added. This result is easily interpreted. $A_s f_Y \cos\beta$ is the yield force component along the joint and $A_s f_Y \sin\beta$ is the yield force component perpendicular to the joint.

When the usual reinforcement ratio r, i.e., the ratio measured in a section perpendicular to the reinforcement bars, is used, we have

$$r = \frac{A_s}{ht\sin\beta} = \frac{r_{\text{skew}}}{\sin\beta} \qquad \psi = \frac{rf_Y}{f_c} = \frac{\psi_{\text{skew}}}{\sin\beta} \tag{8.2.43}$$

and

$$\frac{\tau}{f_c} = \frac{\nu'}{2}\frac{1-\sin\alpha}{\cos\alpha} + \psi\sin\beta\cos\beta + \psi\sin^2\beta\tan\alpha \tag{8.2.44}$$

The term $rf_Y \sin^2\beta$ is the normal stress and $rf_Y \sin\beta \cos\beta$ is the shear stress in the joint section corresponding to a uniaxial stress rf_Y in the reinforcement direction.

When $\beta = \pi/2 + \alpha$ the reinforcement contribution vanishes because then the displacement vector is perpendicular to the bars. For $\beta > \pi/2 + \alpha$ the reinforcement is in compression. Assuming the same yield stress in compression as in tension we find for $\beta > \pi/2 + \alpha$

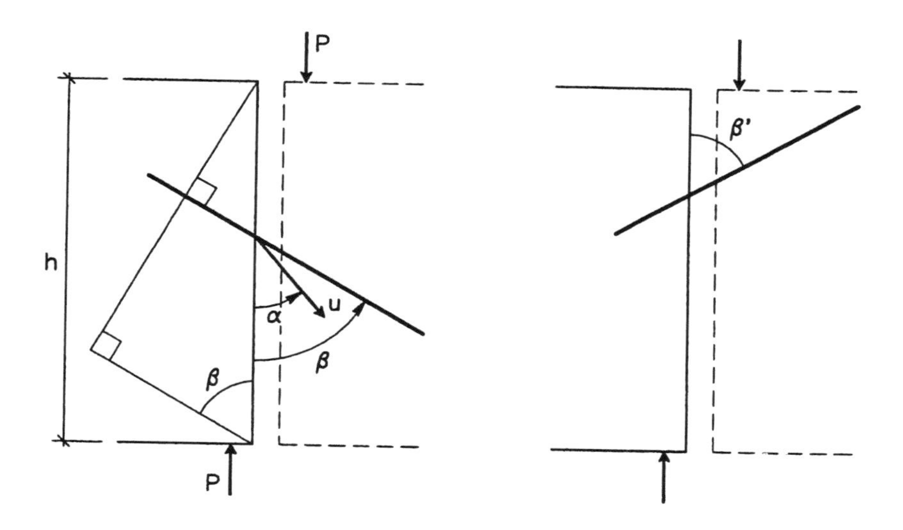

Figure 8.2.9 Joint with skew reinforcement.

$$\frac{\tau}{f_c} = \frac{v'}{2}\,\frac{1-\sin\alpha}{\cos\alpha} + \psi_{skew}\cos\beta' - \psi_{skew}\sin\beta'\tan\alpha \qquad (8.2.45)$$

or

$$\frac{\tau}{f_c} = \frac{v'}{2}\,\frac{1-\sin\alpha}{\cos\alpha} + \psi\sin\beta'\cos\beta' - \psi\sin^2\beta'\tan\alpha \qquad (8.2.46)$$

Here $\beta' = \pi - \beta$ (see Fig. 8.2.9 right).

Let's consider the important case $\alpha = \varphi'$ in more detail. In this case (8.2.44) and (8.2.46) read

$$\frac{\tau}{f_c} = \frac{c'}{f_c} + \psi\sin\beta\cos\beta + \psi\sin^2\beta\tan\varphi' \qquad (8.2.47)$$

$$\frac{\tau}{f_c} = \frac{c'}{f_c} + \psi\sin\beta'\cos\beta' - \psi\sin^2\beta'\tan\varphi' \qquad (8.2.48)$$

$$(\beta > \pi/2 + \varphi')$$

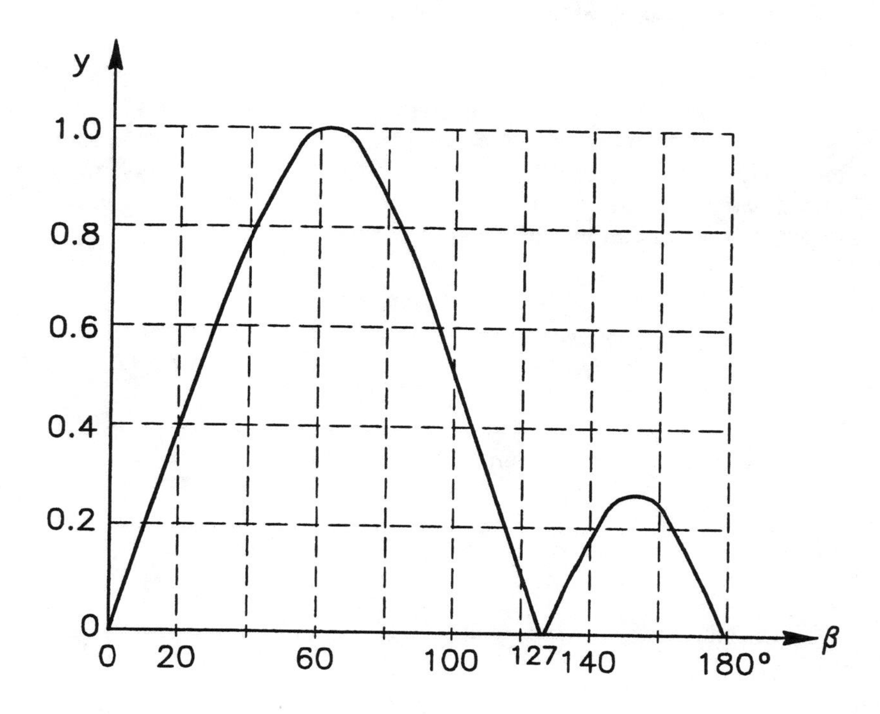

Figure 8.2.10 The function $y = y(\beta)$ versus β, formula (8.2.49).

The function

$$y = y(\beta) = \begin{cases} \sin\beta\cos\beta + \sin^2\beta\tan\varphi' \\ \sin\beta'\cos\beta' - \sin^2\beta'\tan\varphi' \end{cases} \tag{8.2.49}$$

has been drawn in Fig. 8.2.10 for $\varphi = \varphi' = 37^\circ$ $(\tan\varphi' = 0.75)$. It has a maximum, which may be shown to occur for $\beta = \pi/4 + \varphi'/2 = 63^\circ.5$. For $\beta = \pi/2$ we have the usual value $y = \tan\varphi' = 0.75$. It appears that in cases where the reinforcement is in compression the contribution to the load-carrying capacity is small compared with the contribution for more reasonable β-values.

Skew reinforcement in relation to plastic theory was studied by B. C. Jensen [79.21]. He found that when $40^\circ < \beta < 120^\circ$ the plastic theory gave excellent results. Outside this interval the effectiveness factor dropped down. In practical applications of skew reinforcement one should comply with this result.

8.2.8 Compressive Strength of Specimens with Joints

Consider a rectangular disk subjected to pure compression corresponding to the compression stress σ (see Fig. 8.2.11). The disk has a joint forming an angle β with the sections with pure compression. The thickness of the disk is t. If the failure is in the joint the upper part may slide down along the joint. The displacement u of the upper part forms an angle $\alpha \geq \varphi'$ with the joint.

The external work is

$$W_E = \sigma b t u \sin(\beta - \alpha) \tag{8.2.50}$$

The dissipation D_C is determined by (3.4.83). Using (2.1.26) it may be written

$$\begin{aligned} D_C &= \frac{1}{2} f_c' u t \left[1 - \sin\alpha + 2\,\frac{f_t'}{f_c'}\,\frac{\sin\alpha - \sin\varphi'}{1 - \sin\varphi'} \right] \frac{b}{\cos\beta} \\ &= \left[\frac{c'\cos\varphi'(1 - \sin\alpha)}{1 - \sin\varphi'} + f_t'\,\frac{\sin\alpha - \sin\varphi'}{\sin\varphi'} \right] \frac{b}{\cos\beta}\, u t \end{aligned} \tag{8.2.51}$$

From the work equation $W_E = D_C$ we find the load-carrying capacity

$$\sigma = \frac{c'(1 - \sin\alpha)\cos\varphi' + f_t'(\sin\alpha - \sin\varphi')}{\cos\beta(1 - \sin\varphi')\sin(\beta - \alpha)} \tag{8.2.52}$$

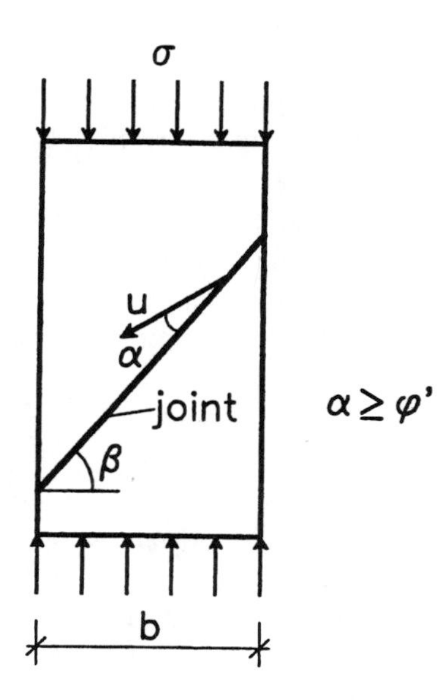

Figure 8.2.11 Specimen with a joint subjected to pure compression.

This expression must now be minimized with respect to α where $\alpha \geq \varphi'$.

If $f_t' = 0$ we find after some calculations that

$$\frac{\pi}{4} + \frac{\varphi'}{2} \leq \beta < \frac{\pi}{2} : \quad \alpha = 2\beta - \frac{\pi}{2} \qquad \sigma = \frac{2c' \cos\varphi'}{1 - \sin\varphi'} = f_c' \quad (8.2.53)$$

$$\beta \leq \frac{\pi}{4} + \frac{\varphi'}{2} : \quad \alpha = \varphi' \qquad \sigma = \frac{c' \cos\varphi'}{\sin(\beta - \varphi') \cos\beta} \quad (8.2.54)$$

The load-carrying capacity can of course not be higher than $\sigma = f_c$, the compressive strength of the concrete outside the joint ($\nu = 1$). Thus $\sigma = f_c$ is the load-carrying capacity for small β-values and for $\beta = \pi/2$, where the joint is in the direction of the compression stress.

The result is illustrated in Fig. 8.2.12. We have used $\varphi = \varphi' = 37^\circ$ and $c' = c/2$, i.e., $\nu' = 1/2$.

The result is rather interesting. For the failure condition used, the slightest inclination of the joint relative to the compression direction leads to a reduced compressive strength. The reason is illustrated in Fig. 8.2.13. Because of the transformations we made in Chapter 2, taking us from the Coulomb sliding failure condition in the form of a straight line to the failure condition in principal stress space, we ended up with a failure condition with a circular transition curve being tangential to the straight line expressing the Coulomb sliding failure criterion and passing through the point on the σ-axis corresponding to the tensile strength. In Fig. 8.2.13 $f_t' = 0$ and σ is positive as compression. It appears that it is impossible to construct a Mohr's circle for uniaxial compression which for large values of the angle β lies outside the Mohr's circle corresponding to the compressive strength f_c'. This strange result was first obtained by B. C. Jensen [76.1]. It was further elaborated in [97.3], from where the results of this section have been taken.

When $f_t' \neq 0$ even very small values of the tensile strength dramatically change the appearance of the curve in Fig. 8.2.12 for large β-values. This is demonstrated in Fig. 8.2.14 where f_t' has been set to $\lambda f_c'$. We have again assumed $\varphi = \varphi' = 37^\circ$ and $\nu' = 1/2$. For $\lambda = 0.05$, i.e., the tensile strength is 5% of the compressive strength f_c' of the joint, we almost get the curve marked $\alpha = \varphi$. This is the compressive strength which is found when the relative displacement is under an angle $\varphi = \varphi'$ with the yield line, formula (8.2.54). The reader may find it instructive to verify that this is also the load-carrying capacity determined by assuming the failure condition in the joint be the Coulomb straight line without a circular transition curve. By expressing the stress in the joint by σ and the angle β and inserting these stresses into the failure condition $\tau = c' + \sigma \tan\varphi'$, formula (8.2.54) is easily obtained.

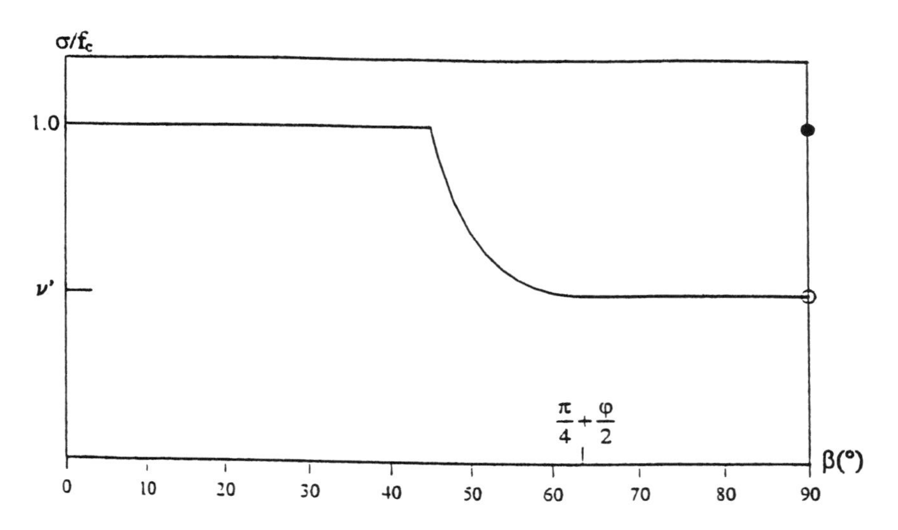

Figure 8.2.12 Compressive strength for a specimen with a joint.

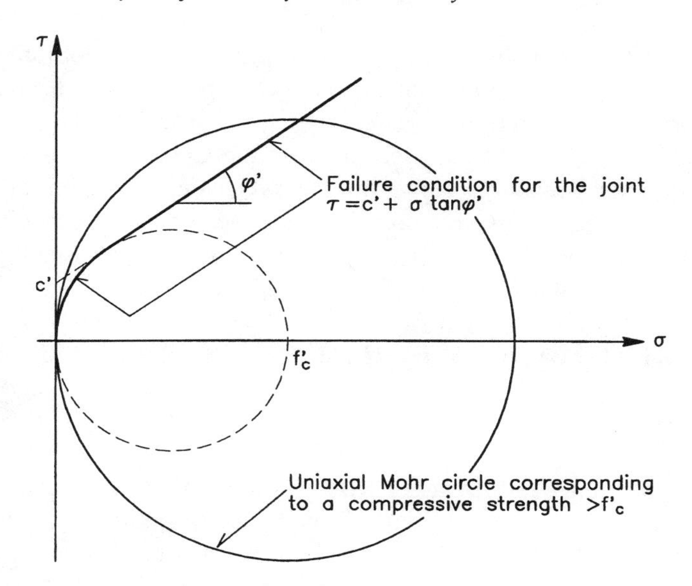

Figure 8.2.13 Failure condition for a joint.

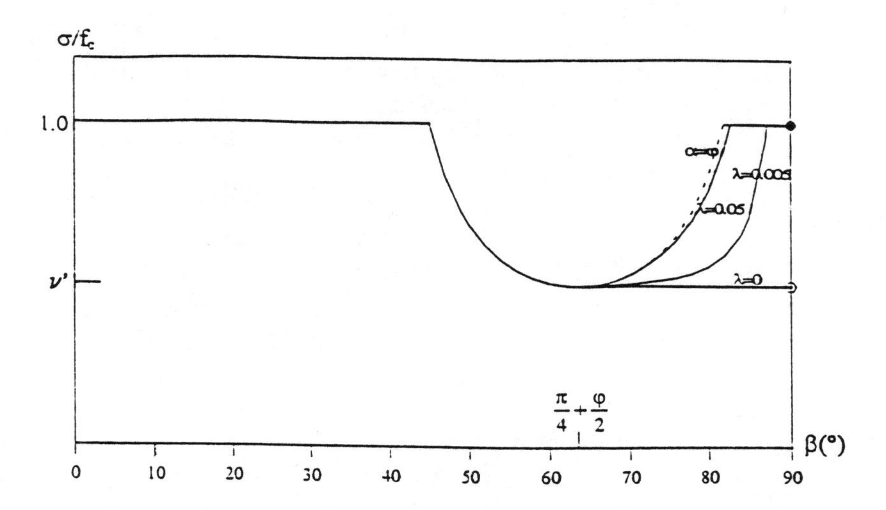

Figure 8.2.14 Compressive strength for a specimen having $f_t' \neq 0$

In this case the minimum value of σ is obtained for $\beta = \pi/4 + \varphi/2$, i.e., when the joint coincides with the yield line in monolithic concrete (cf. Section 2.1.2).

Thus we may conclude that for large values of β the compressive strength reduction due to a joint is strongly dependent on the failure condition for small normal stresses. This part of the yield condition is seldom accurately known. One may use the Coulomb straight line with a circular transition curve on the safe side. The straight line without a circular transition curve is only accurate for sufficiently high compression normal stresses. We return to this important point in Section 8.3.2, where the joint considered is a crack.

8.3 STRENGTH OF DIFFERENT TYPES OF JOINTS

8.3.1 General

In this section we summarize what is known about the strength of the most common types of joints in concrete technology. Probably the most important joint is the crack. Thus we begin with this special joint.

8.3.2 The Crack as a Joint

A crack develops by a separation failure in the cement paste. This failure is very brittle. After cracking in the cement paste the aggregate particles, which are crossed by the crack, are pulled out of the concrete on the weakest side of the crack. The pulling out may be envisaged as a kind of punching failure. This failure is much more ductile than the separation failure through the cement paste and explains the long tail of the stress-strain curve in tension. The force in a crack under tension may vanish only after a crack opening on the surface up to 0.1 to 0.4 mm depending on the aggregate size.

When a crack is subjected to a shear force a substantial shear strength remains, since the aggregate particles prevent sliding through the crack in the cement paste (aggregate interlocking).

The shear strength of a crack is probably one of the most important properties to be studied in concrete mechanics, a property which was surely not envisaged by the inventors of reinforced concrete. The reason why it is so important is that without this property cracked concrete would only be capable of very little redistribution of stresses and plastic theory would be meaningless.

A micromechanical model expressing the above-mentioned features has been developed by Jin-Ping Zhang [97.2]. We shall not enter into the details of the model but confine ourselves to some of the important aspects and results.

When a crack is subjected to shear forces one possibility is that the stresses will be carried by a combination of friction in the above-mentioned punching failure surfaces and, depending on the crack width, shear failure in the cement paste which has not yet suffered a final failure. Another possibility is that new failure surfaces will be created meaning that the failure is partly through uncracked cement paste. The word "partly" is included because, by all means, the crack in the cement paste will be involved in the failure mechanism.

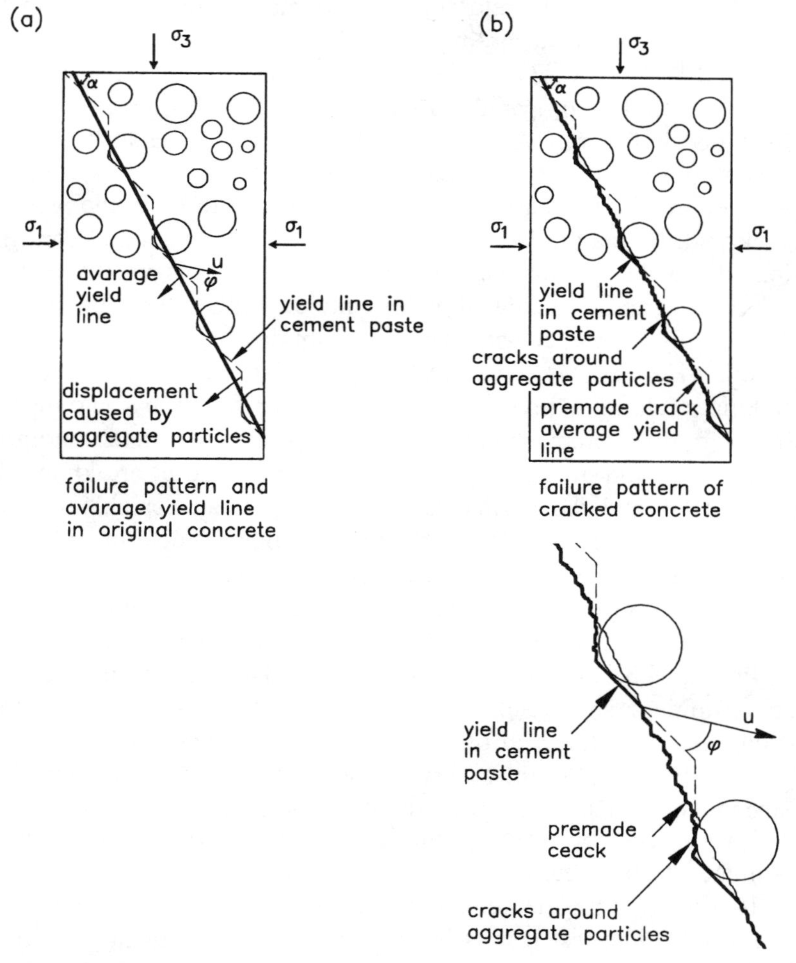

Figure 8.3.1 a) Shear failure pattern and average yield line in original concrete. b) Average yield line along an existing crack.

Fig. 8.3.1a shows a yield line in uncracked concrete. If for simplicity the aggregate particles are considered as having the same size, then the yield line in the cement paste according to the model will zig-zag around the aggregate particles as shown in the figure. Part of the zig-zag line will be true yield lines in the cement paste and part of it will be load induced cracks in the direction of the largest principal compression stress σ_3. Under the assumption of aggregate particles with equal size, the global average yield line will be crossing the local yield lines in the cement paste and the load induced cracks in their middle points.

If now the global yield line is a premade straight crack in the cement paste then, according to this simplified model, it is to be expected that the local yield lines in the cement paste will have only half the length compared to concrete without a crack. This is illustrated in Fig. 8.3.1b. The other half of the local yield lines in the cement paste is now replaced by a local yield line following the premade crack. The part of the yield line going through the premade crack and the load induced cracks will only render a small contribution to the dissipation and may be neglected. This means that the dissipation can be calculated as for concrete without a premade crack by multiplying the cement paste contribution by 0.5. Thus combinations of σ_3 and the transversal confining stress $\sigma_1 < \sigma_3$ which lead to failure along a premade crack may be found by varying the direction of the premade crack and for each direction determining the values of σ_3 and σ_1 leading to failure. When σ_3 and σ_1 are known the shear stress and the normal stress in the premade crack are easily found. In fact the resulting failure condition is determined as an envelope curve because the angle φ (see Fig. 8.3.1) between the local yield line in the cement paste and the displacement u is not a constant for cement paste but varies with the normal stress.

In Fig. 8.3.2 some results are shown. The curves represent the dimensionless shear capacity of a crack τ/f_c for different concrete strengths f_c as a function of the reinforcement degree ψ. It turns out that the lines are slightly curved and that the dimensionless shear capacity is almost independent of the concrete strength.

The curves are only drawn in the region where the model is estimated to be accurate.

In Fig. 8.3.3 the theoretical curves have been compared with some test results. The test results have been obtained by using specimens of the push-off type, one variant of which is sketched in Fig. 8.2.7. If all available test results are included the scatter is very large.

However, if only test results with initial crack widths $w_o < 0.7$ mm are included and if tests where the ultimate strength was not measured are excluded the scatter becomes reasonable. For a complete list of references, the reader is referred to [97.2].

It appears that the agreement between theory and experiment is reasonable.

The dependence of the shear capacity of a crack on the initial crack width has already been illustrated in Fig. 2.1.35. We may assume that when the initial crack width $w_o \leq 0.7$ mm the shear capacity is independent of the crack width.

Finally, in Fig. 8.3.4 the test results have been compared with the modified Coulomb failure condition with circular transition curve. With the notation used in Section 8.2 the Coulomb failure condition has been drawn for $f_t' = 0$, $\varphi = \varphi' = 37°$ and $\nu' = 0.5$, i.e., the angle of friction has the same value as for monolithic concrete and the cohesion c' (and the concrete strength f_c') is half the value for monolithic concrete. This means that $c' = f_c/8$.

It appears that there is a remarkable agreement between the Coulomb failure condition and the test results.

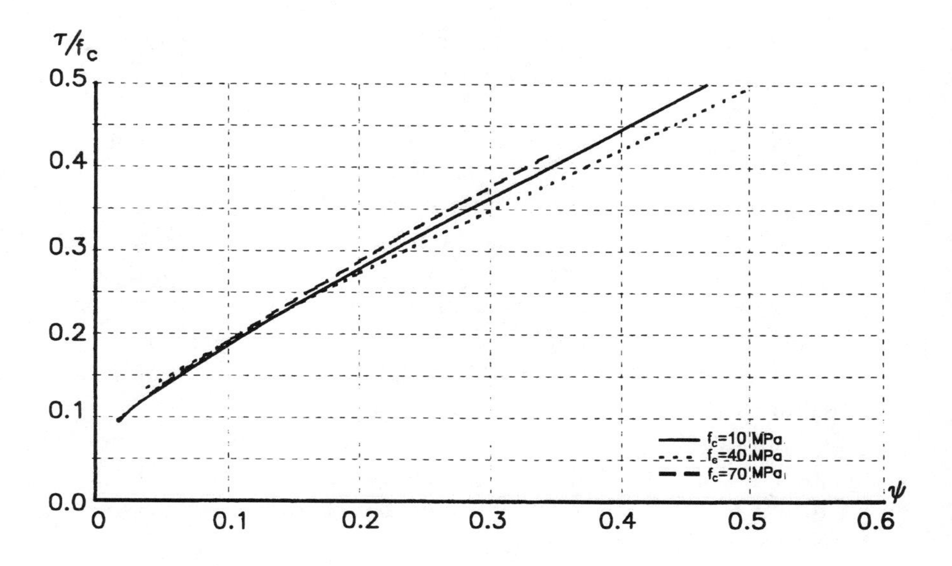

Figure 8.3.2 Shear capacity of a crack.

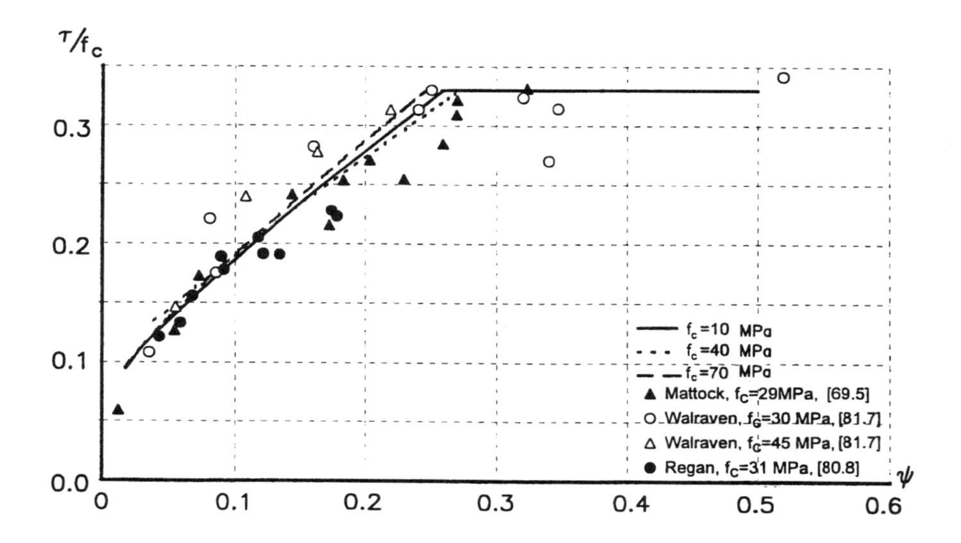

Figure 8.3.3 Shear capacity of cracks, theory compared with experimental results.

Unfortunately there are only a few test results for small values of the reinforcement degree ψ. The important problem about the cohesion in a crack for small normal stresses obviously cannot yet be finally solved. However, there are, on the basis of the model, good reasons to believe that the shear capacity is not zero for a zero normal stress as found from the Coulomb condition with $f_t' = 0$ and a circular transition curve. On the other hand, it clearly seems to be unsafe to assume the straight Coulomb line to be valid for small normal stresses.

In the previous chapters we have used the notation ν_s, the sliding reduction factor, for what we here have denoted ν'. Thus the value $\nu_s = 0.5$ of the sliding reduction factor used before is well founded by both experiment and theory.

However, there is one point further which must be commented upon. In the push-off tests one would expect an influence from the damage caused by the reinforcement bars crossing the crack, i.e., we would expect a resulting effectiveness factor $\nu = \nu_s \nu_o$, ν_o being a result of the local damage due to the stressed, transverse bars. The reason why there is good agreement with test results using $\nu_o = 1$ is probably that most of the tests have been performed with plain bars. This type of bar causes less damage than deformed bars, as mentioned before. A ν_o-value stemming from softening also seems unimportant, contrary to what we found for monolithic concrete.

The model explains very well the effect of the reinforcement crossing a crack. Due to the dilatation in the crack the reinforcement is stressed to tension and not to shear. This means that dowel action is unimportant.

The type of shear stress-displacement curve obtained in push-off tests is illustrated in Fig. 8.3.5, [81.7]. The displacement in the direction of the crack is termed Δ, the crack opening w. It appears that a substantial crack opening takes place during the loading. The ratio between the increments of w and Δ at failure is about 0.6 which is a little lower than expected from the normality condition used in the model.

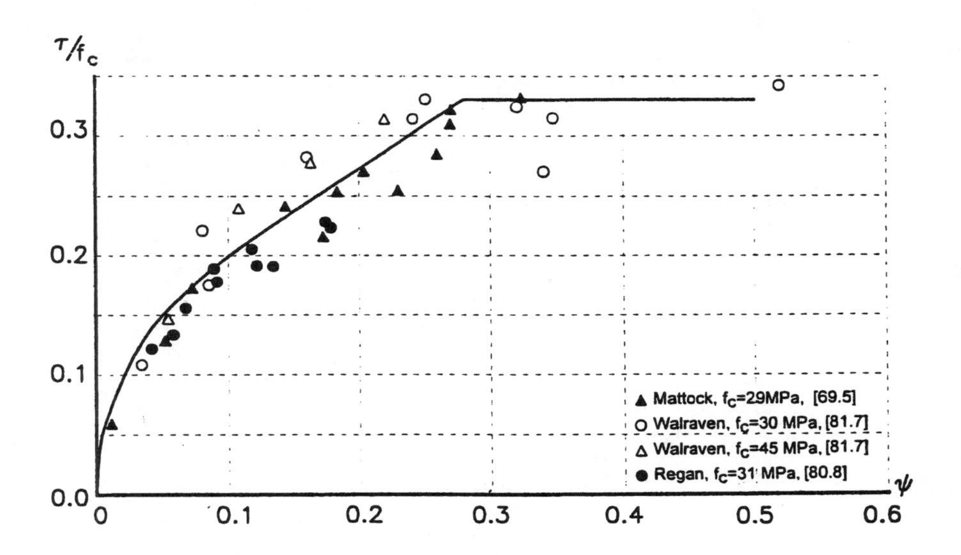

Figure 8.3.4 Shear capacity of cracks. Test results compared with the modified Coulomb failure condition.

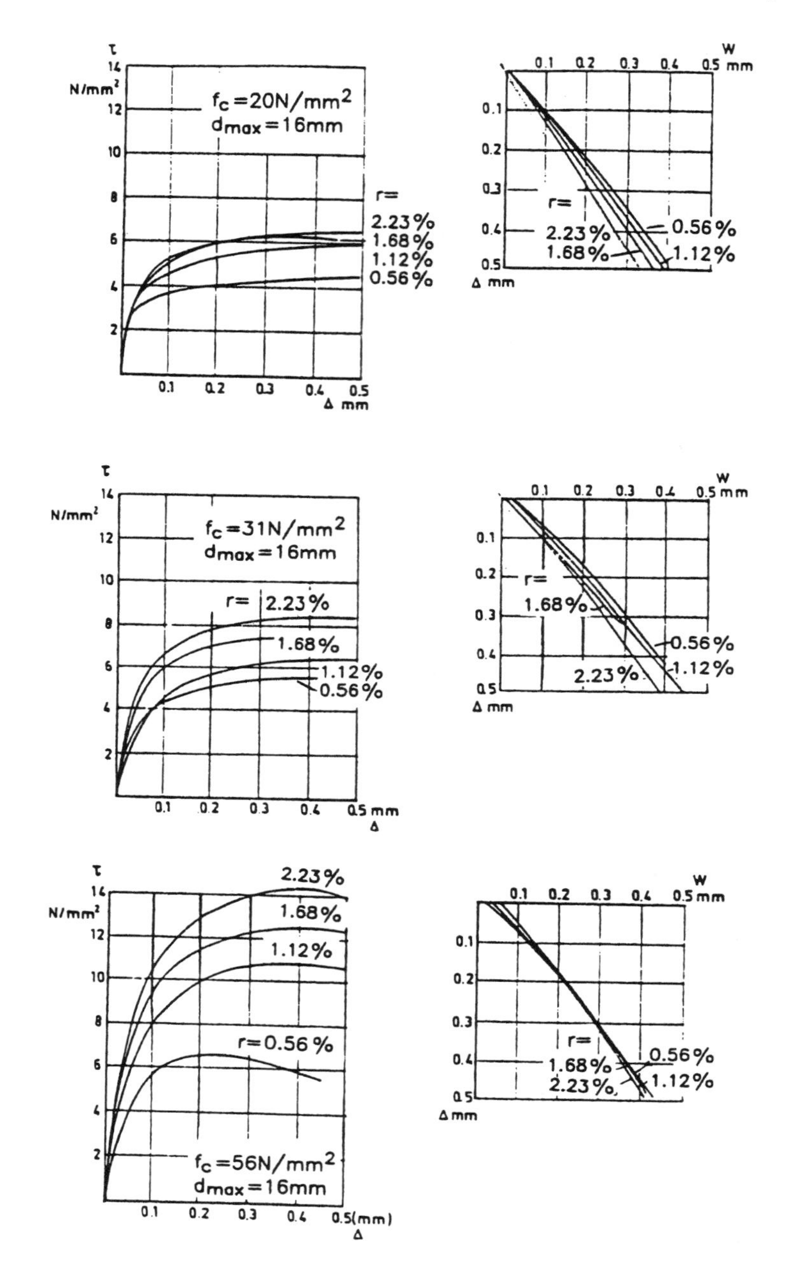

Figure 8.3.5 Shear stress-displacement curves in push-off tests [81.7]. Δ - displacement in the direction of the crack, w - crack opening, f_c - concrete compressive strength, r - reinforcement ratio, d_{max} - maximum aggregate size.

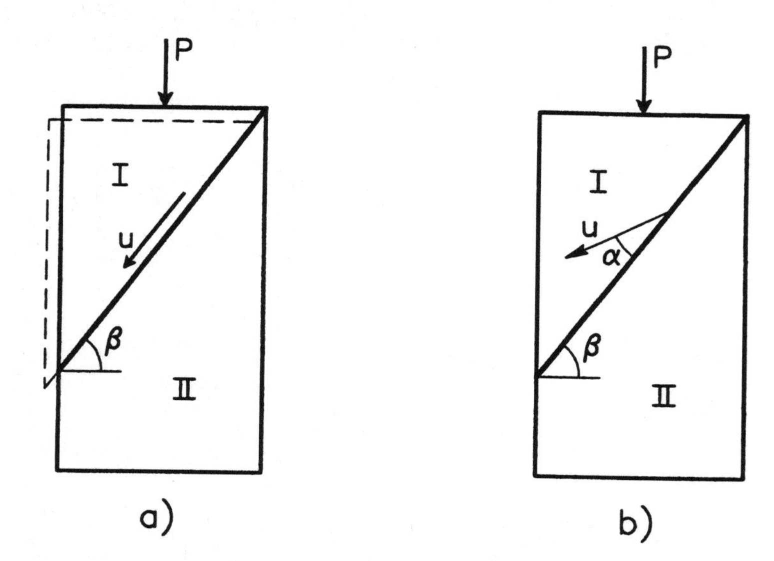

Figure 8.3.6 Sliding of a body I on a body II.

The model cannot differentiate between transverse normal stresses applied by an external normal force and those obtained by a transverse reinforcement. That there is a difference in behavior between these two cases has been demonstrated in [80.8]. However, the difference is small.

Experiments on high strength concrete are reported in [94.9]. For a complete list of references, see [97.2].

We conclude this section by some remarks on friction.

Let's imagine that the width of a crack has been made so wide that no tensile stresses are left in the crack and that the cracked surfaces then are brought as close together as possible. If this joint is subjected to shear forces one might expect that any resistance against shear would stem from pure friction. It seems not to be so.

We recall that when a body I is sliding upon another body II (see Fig. 8.3.6a), the magnitude of the angle β between a line perpendicular to the force P, producing the sliding (our body is weightless), and the surface along which sliding takes place is decisive for whether sliding takes place or not. If $\tan\beta < \mu$, μ being the coefficient of friction, sliding will not occur. If $\tan\beta = \mu$ there is balance between the force component parallel to the sliding surface and the resistance against sliding, which is determined by the force component perpendicular to the sliding surface times the coefficient of friction μ. Thus the body I may slide with uniform velocity down the surface. If $\tan\beta > \mu$ sliding will occur with increasing velocity.

Now a cracked surface is far from being plane. It has a large number of mountains and valleys due to the pulling out of the aggregate particles as mentioned above.

Fig. 8.3.7 shows a number of crack profiles in one and the same specimen which has been cracked by a splitting failure. Profiles opposite to each other have been approximately continuous before failure (it is not entirely correct, the cutting was not particularly accurate in this respect). When studying such a surface it becomes clear that the sliding is not along, say, the average crack plane. Thus Fig. 8.3.6b is more correct in describing the sliding along a cracked surface. In this figure the sliding takes place under an angle α with the average crack plane, the inclination of which is, as before, characterized by the angle β. The condition for no sliding is now changed to $\tan(\beta - \alpha) < \mu$.

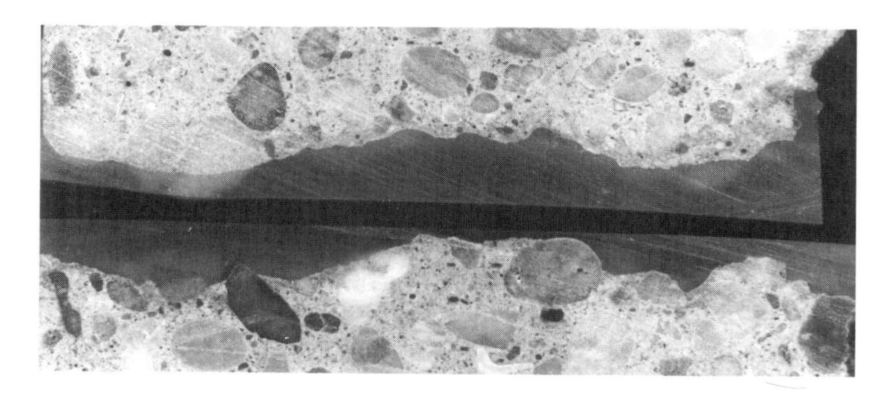

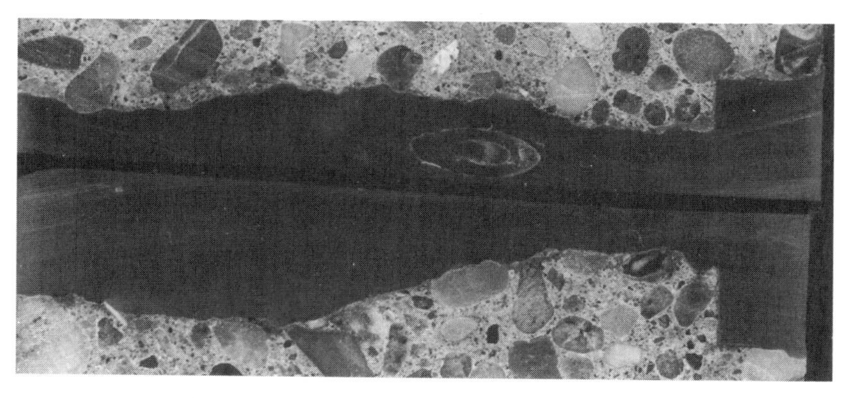

Figure 8.3.7 Crack profiles in concrete from splitting tests.

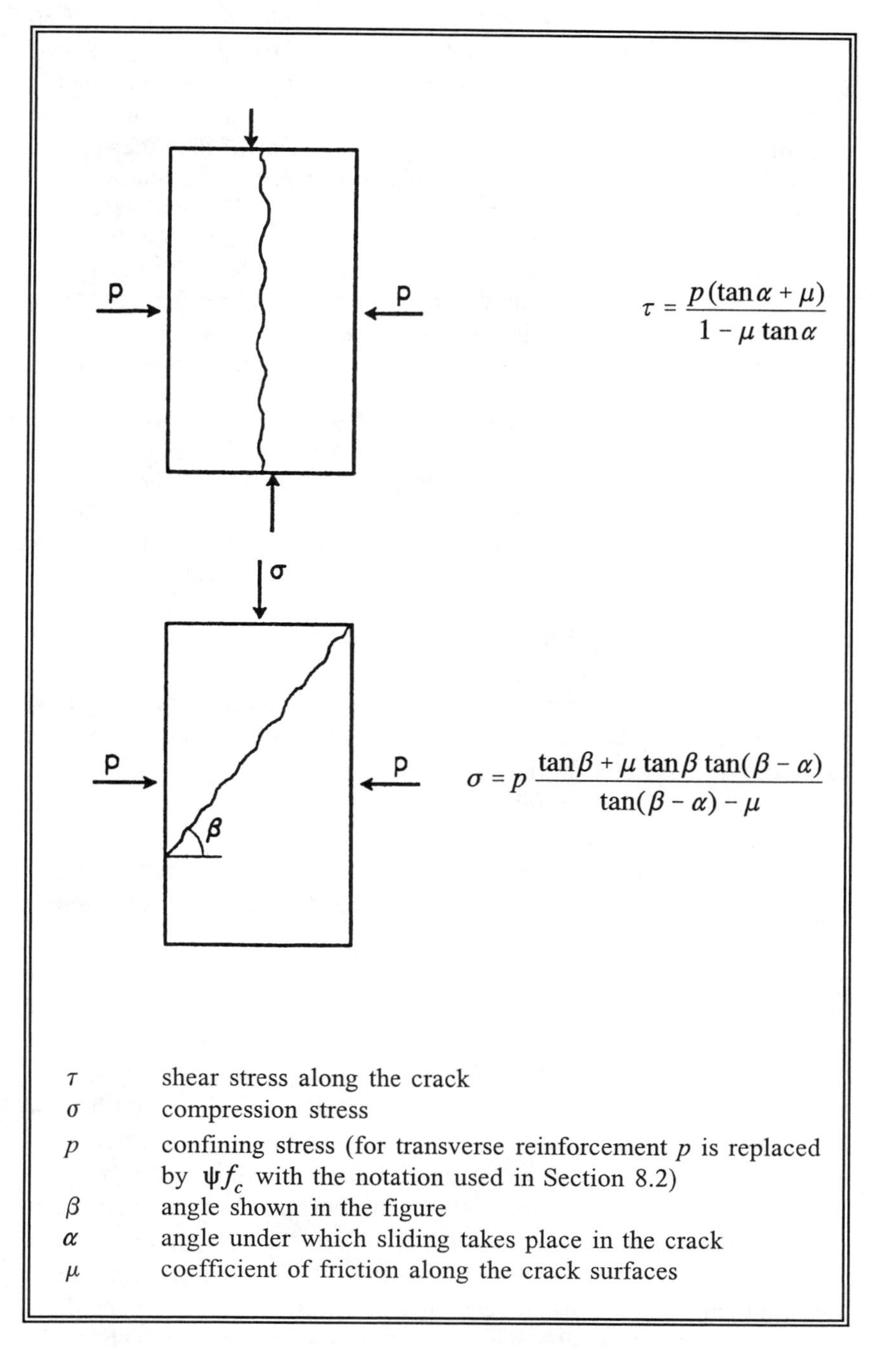

Box 8.3.1 Load-carrying capacity for shear failure along a crack when it is governed by friction in the crack surfaces.

A primitive test carried out with the cracked surface of Fig. 8.3.7 before cutting, showed that the angle α is of order 65°. This value was found at the beginning of the sliding. The angle gradually changed to around 25° but only after a displacement perpendicular to the surface of the order of 1 to 2 mm. This is far more than normally obtained in tests, so in our estimate we may put α to 65°. The order of μ is probably 0.6 (friction research is obviously not prestigious, one is always short of numbers). The condition for no sliding is therefore $\tan(\beta - 65°) < 0.6$, i.e., $\beta < 96°$, which is always fulfilled. Thus it seems that our assumption, that in a shear failure along a crack new failure surfaces must be formed, is justified. Friction along the pull-out surfaces may not have to be considered when the crack widths are sufficiently small.

For larger crack widths friction cannot be excluded at least for small values of confining stress. We shall not enter into this problem here. However, for future use, Box 8.3.1 contains the formulas for the shear capacity in the two elementary cases considered above when friction is governing the load-carrying capacity. They may be useful when the shear capacity for large initial crack widths is going to be studied.

8.3.3 Construction Joints

We have a construction joint when new concrete is cast against a surface of old, already set and more or less hardened concrete. Normally the surface is cleaned to be free from cement slurry and watered to surface-dry condition before casting. The surface may be roughened for instance by water or sand blasting.

A micromechanical model of the fracture in a construction joint is yet to be developed, so we are only able to sketch rather roughly the essential features of such a future model.

Johansen [30.2] was one of the first to design tests with construction joints. He observed that failure took place along the interface between the old and the new concrete, i.e., failure surfaces neither in the old nor in the new concrete were created. The surfaces were completely undamaged. If we take this for granted the strength must be determined solely by the shear bond strength in the interface. This means that the aggregate particles and the hard unhydrated cement grains play no role. Further, since cement paste is a frictionless material under confined conditions, it is seen that the dilatation must be entirely due to the roughness of the joint. Fig. 8.3.8 shows some measurements of the roughness of a concrete surface obtained by casting it against steel, plastic and ply-wood, respectively. The roughness depth defined as twice the amplitude of the "wave" describing the surface profile is of the order 5 to 10 μm. It is smallest in the case of steel and largest in the case of plywood.

It appears from Fig. 8.3.8 that it is possible to define an almost constant sliding direction when a joint with this kind of roughness is subjected to shear. The situation is shown schematically in Fig. 8.3.9. The sliding takes place in the direction forming an angle α with the shear direction. The displacement is along the concrete interfaces to be "climbed" when the concrete part on one side of the joint is moved relative to the other part.

It is likely that only interfaces like AB make a significant contribution to the shear capacity. Along surfaces like BC and CD the displacement vector forms a rather large angle with the surface. This displacement pattern is, of course, not possible if an interface forms an angle larger than α with the shear direction. Then a new failure surface must be created leading to a higher contribution since the failure is through concrete outside the joint. For an interface with a slightly lower inclination than α the contribution will be slightly reduced.

Under the assumption of constant sliding direction, an upper bound calculation immediately furnishes

$$\tau = \frac{c_{\text{bond}}}{\cos^2\alpha} \sum \frac{a_i}{L} + \sigma \tan\alpha \tag{8.3.1}$$

Here τ is the shear capacity of the joint; σ the normal stress, positive as compression; c_{bond} is the cohesion due to the bond between the old and the new concrete; and $\Sigma a_i / L$ is the dimensionless sum of the projections of the interfaces like AB in Fig. 8.3.9 along the length L of the joint. The roughness of the joint thus is sufficiently described by the sum of the projections a_i and the angle α.

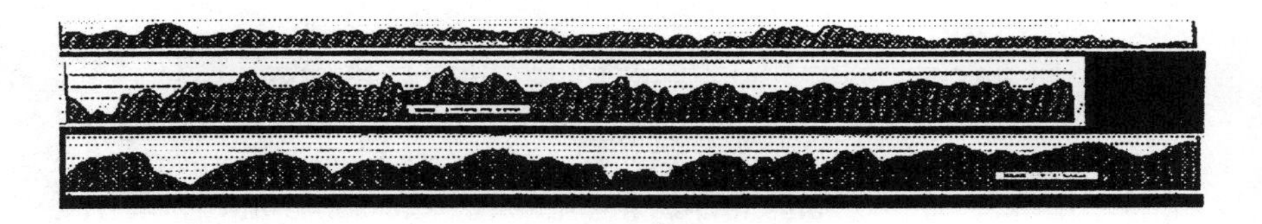

Figure 8.3.8 Roughness of concrete surfaces cast against different materials. The upper one is cast against steel, the one between against plastic and the lower one against plywood. The length measured is 350-400 μm (measurements by Lasse Johansen).

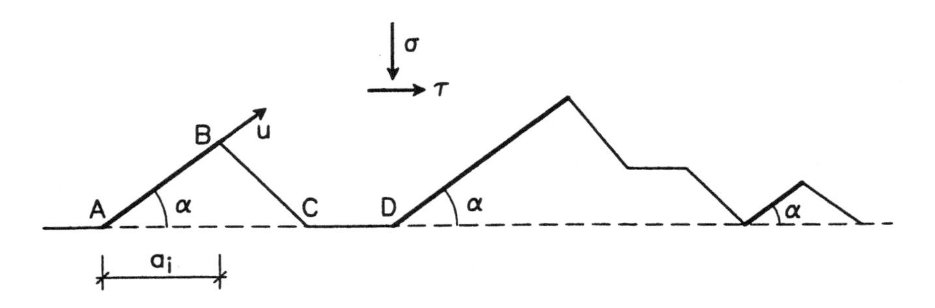

Figure 8.3.9 Shear failure in a construction joint.

The cohesion c' of a construction joint according to this probably highly oversimplified model is therefore

$$c' = \frac{c_{\text{bond}}}{\cos^2\alpha} \sum \frac{a_i}{L} \tag{8.3.2}$$

Experimentally (see below) α is normally around the angle of friction for monolithic concrete 37° and $\Sigma a_i \,/\, L$ is of the order 0.3 to 0.4, the upper limit being 0.5. Since c' is of the order 10% of the concrete strength f_c, c_{bond} is of the order 20% of f_c. Since cement paste is a frictionless material the cohesion of cement paste is half the compressive strength of the cement paste f_{cc}, which is higher than the compressive strength of the concrete. Thus it seems that c_{bond} is only a fraction of the cohesion of cement paste, which is, of course, to be expected.

According to this model the absolute value of the amplitude of the "interface wave" is not an important parameter. The important roughness parameter is the angle α. Notice that increasing α leads to both increasing c' and increasing $\sigma \tan\alpha$, the last term in (8.3.1).

There is no chance that we can rely upon the above model to calculate the strength of a construction joint until more is known about the shear bond strength c_{bond}. So for the time being we must rely on tests. Fig. 8.3.10 shows the test results obtained by Johansen [30.2]. The concrete strength was f_c = 30 MPa. The tests were done by applying a uniaxial stress on prisms in which joints were placed with varying inclination. No special treatment of the joint surface was carried out. The joint surface was cast against a paraffinated steel plate whose surface character is, of course, unknown. The test results lie beautifully on a straight line as they should according to the model. The cohesion $c' = 0.1f_c$ and $\tan\alpha = 0.8$. In the figure Mohr's circle corresponding to a uniaxial stress f_c is shown. In this test method it is, of course, impossible to measure joint strengths lying outside this circle.

Tests of the push-off type were carried out by Houborg and Sørensen [74.12]. Some of their results are shown in Fig. 8.3.11. In the upper curve the joint surface was floated with a steel rail and brushed with a steel brush after about two hours, in the lower one the treatment indicated in the figure was used. In Fig. 8.3.11 the shear strength is plotted as a function of the reinforcement degree ψ.

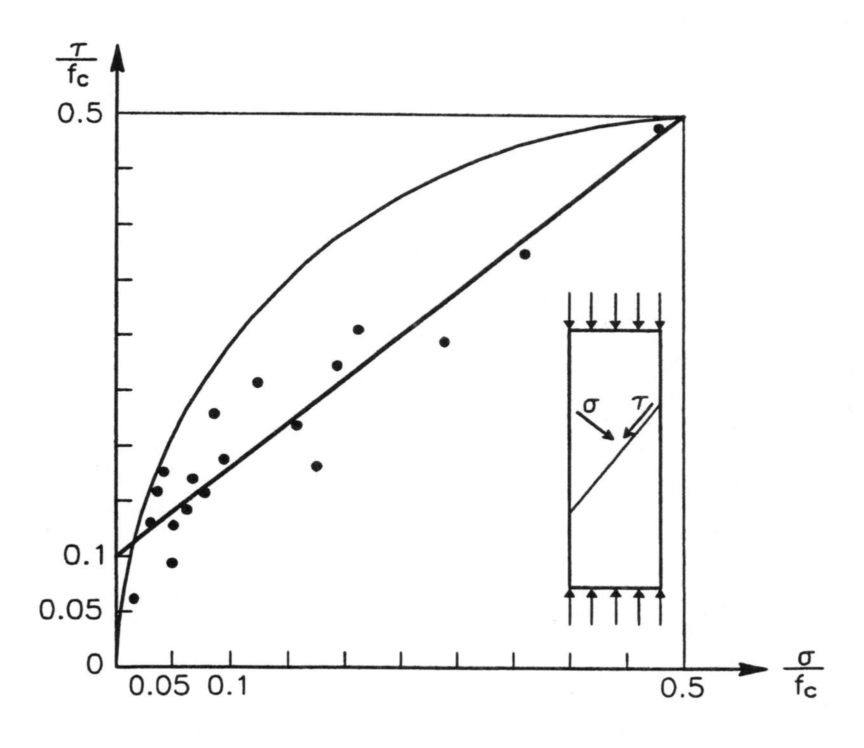

Figure 8.3.10 Construction joint test results by Johansen [30.2].

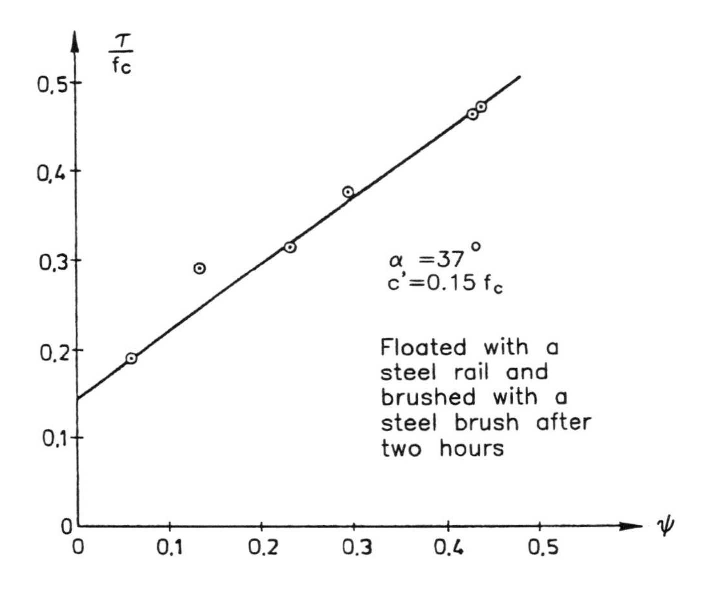

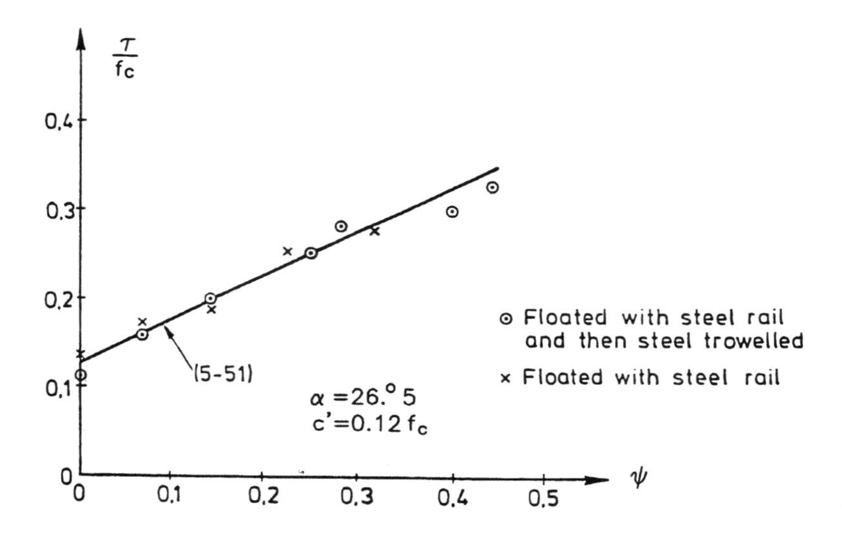

Figure 8.3.11 Construction joint test results by Houborg and Sørensen [74.12].

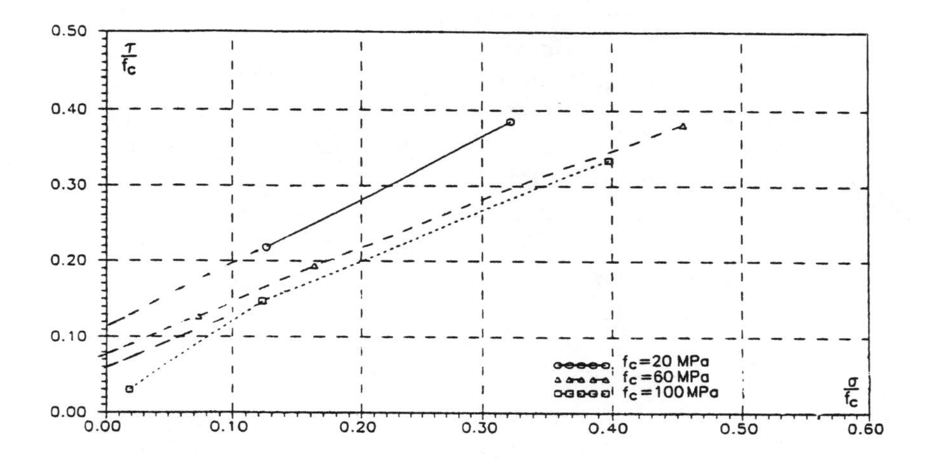

Figure 8.3.12 Construction joint test results by Kaare K.B. Dahl. [94.11].

The upper curve gives $c' = 0.15\,f_c$ and $\tan\alpha = 0.75$ $(\alpha = 37°)$, the lower one gives $c' = 0.12\,f_c$ and $\tan\alpha = 0.5$ $(\alpha = 26.°5)$. It is in agreement with the model that lower α also leads to lower cohesion. It is likely that the treatment leading to the lower curve provides the smoothest surface which will be encountered in practice, i.e., $\tan\alpha = 0.5$ will probably be a lower limit. The test results were converted to an average strength $f_c = 22.5$ MPa.

These tests indicate that the action of a transverse reinforcement in a construction joint is equivalent to an external normal compression stress. The reason why the reinforcement is stressed to tension is not the sliding along the inclined interfaces of the joint, but is due to the formation of cracks in a skew direction to the joint. The sliding in the joint interface takes place at the final failure. Sliding along a surface with a roughness depth of 5 to 10 μm is, of course, not able to stress the reinforcement in tension to any significant level.

Tests including high strength concrete were carried out by Kaare K.B. Dahl [94.11]. Some of his results are shown in Fig. 8.3.12. The three curves drawn are valid for f_c = 20, 60 and 100 MPa, respectively. In this case the concrete in the joint was cast against a plywood board. The tests were, like those of Johansen, carried out on compression specimens.

The results seem to indicate that the cohesion is a decreasing function of the compressive strength. The value of $\tan\alpha$ does not differ much from that obtained in Johansen's tests.

From an engineering point of view most of the above tests will be characterized as tests with a smooth joint. Fortunately for such joints it seems that the cohesion c' may be taken only to be dependent on the compressive strength of the concrete. If the tests of Johansen and Kaare K.B. Dahl are plotted to show c' as a function of f_c, the relationship shown in Fig. 8.3.13 is obtained. The points lie approximately on the curve

$$\frac{c'}{f_c} = \frac{0,55}{\sqrt{f_c}} \quad \text{(smooth joint)} \qquad (f_c \text{ in MPa}) \tag{8.3.3}$$

which is also depicted in the figure. Further the tests show that the value of $\tan\alpha$ may be taken as the usual value for monolithic concrete, i.e., $\tan\alpha = 0.75$ corresponding to $\alpha = 37°$. In terms of plastic theory we thus get, using $c' = \nu' f_c/4$,

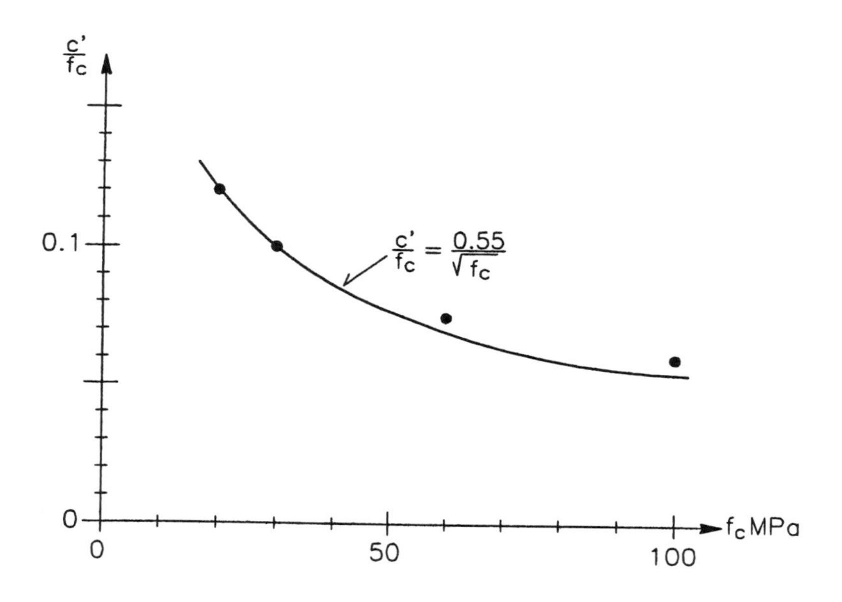

Figure 8.3.13 Cohesion in smooth construction joints as a function of compressive strength.

$$\varphi = \varphi' = 37^\circ \quad \Rightarrow \quad \tan\varphi = \tan\varphi' = 0.75 \tag{8.3.4}$$

$$\nu' = 4 \cdot \frac{0.55}{\sqrt{f_c}} = \frac{2.2}{\sqrt{f_c}} \quad (f_c \text{ in MPa}) \tag{8.3.5}$$

The tests carried out by Houborg and Sørensen indicate that the straight line

$$\tau = c' + 0.75\sigma \tag{8.3.6}$$

or

$$\tau = c' + 0.75 r f_Y \tag{8.3.7}$$

is valid in the limits $\sigma \rightarrow 0$ or $rf_Y \rightarrow 0$, so there is no need for using the circular transition curve in this case. However, it should be noted that this conclusion may not be valid for high strength concrete (cf. Fig. 8.3.12). Whether the decrease in shear strength for low values of σ in this case is due to the tests, which certainly caused some difficulties, or it is real, is not known at the present time.

Tests with smooth joints have been performed by other authors. They do not change the above conclusions. For reference to these tests, see [78.12].

The limitations imposed by the strength of the concrete surrounding the joints are treated as before.

A smooth construction joint is, due to the small roughness depths, extremely sensitive to tensile forces. Even a very small crack width, of the order 5 to 10 μm, will cause the cohesion to vanish. Thus smooth joints should only be considered to have cohesion if tensile stresses in the joint surface are not likely to occur.

If tensile stresses cannot be excluded the joint must be treated as a butt joint (cf. Section 8.3.4).

To remove the sensitivity of the construction joint to cracking, the joint may be given a higher roughness depth. Sufficient depth can normally not be obtained by water or sand blasting but must be obtained using mechanical means, for instance, an airpowered chisel or similar tool [94.11]. Besides improving the roughness depth such a treatment may also improve the length of the projections $\Sigma a_i/L$ in our model.

Kaare K.B. Dahl [94.11] found that the roughness obtained by mechanical treatment by a chisel led to an approximately 50% increase of the cohesion c'. The roughness depth was about 3 mm. His test results are shown in Fig. 8.3.14 for a f_c = 60 and a f_c = 100

MPa concrete. He also found that water blasting even for a f_c = 20 MPa concrete did not improve the cohesion much compared to that of a smooth joint.

Thus for a rough construction joint we have approximately

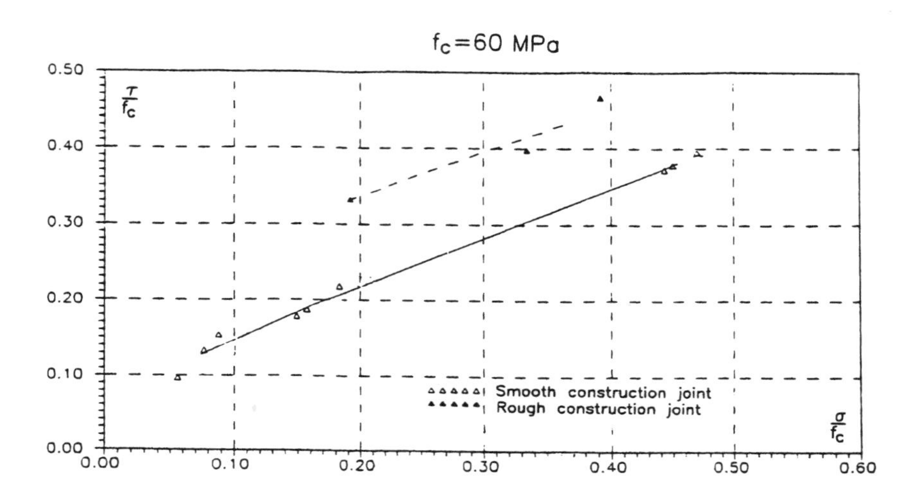

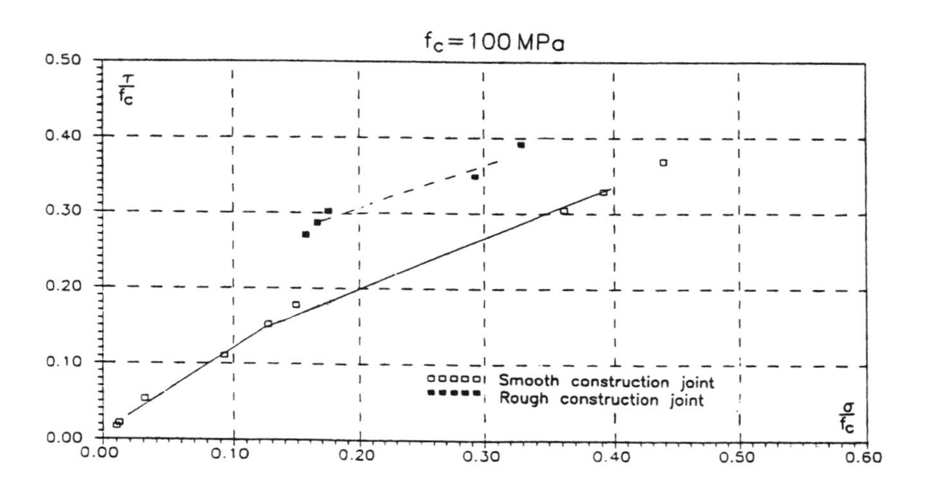

Figure 8.3.14 Test results for rough construction joints compared with those for smooth joints.

$$\frac{c'}{f_c} = \frac{0.83}{\sqrt{f_c}} \quad \text{(rough joint)} \qquad (f_c \text{ in MPa}) \tag{8.3.8}$$

A joint having a roughness depth of some millimeters will also be active in the case of external tensile forces, i.e., the following formula (8.3.9) may in such case be applied to negative values of σ as well.

We may conclude that a smooth construction joint has a shear capacity which may be calculated by the formula

$$\tau = c' + 0.75\left(rf_Y + \sigma\right) \tag{8.3.9}$$

The cohesion c' is determined by the formula (8.3.3). As usual r is the reinforcement ratio of the transverse reinforcement and f_Y is the yield stress. Finally, σ is the normal stress in the joint from external loads, positive as compression. When the external load gives rise to tensile stresses the cohesion should be put to zero and the joint is treated as a butt joint (cf. Section 8.3.4).

Skew reinforcement may be treated as shown in Section 8.2.7.

Regarding the definition of a smooth joint we repeat that it is a joint obtained by casting against normal smooth steel or plywood molds or similar.

For a rough construction joint the cohesion is around 50% higher, cf. formula (8.3.8), than for a smooth joint. A rough construction joint must be obtained by a mechanical treatment. Water or sand blasting is normally not sufficient. The roughness depth must be some millimeters. In a rough construction joint the cohesion may be considered active even for external tensile stresses (σ negative in formula (8.3.9)).

We conclude this section with a remark on the present code rules. The cohesion given in most codes is relatively low compared to the above experimental values. For instance, in Eurocode 2, part 1-3, [93.11] the values given are only about 1/3 of our values for smooth joints. Conservative code rules may be partly justified because the strength of a joint obtained on the site definitely is often lower than that obtained in the laboratory. However, this is true for any strength parameter and there is little basis for claiming that the joint strength is more sensitive to careless site operations than other strength parameters, at least when, for a smooth joint, it is made sure that it will never suffer tensile stresses. Thus the above values are recommended for practical design when the precautions mentioned are taken into account.

8.3.4 Butt Joints

In a butt joint two concrete surfaces are brought into contact without any inter-layer material. The shear strength of a butt joint is solely determined by the coeefficient of friction μ between the surfaces. For a smooth surface obtained by casting against steel or plywood μ is around 0.5 to 0.6. A very small cohesion is also sometimes present, some tenth of 1 MPa. For a rough surface μ may be much larger, up to twice as large depending on the roughness.

Problems involving friction can, strictly speaking, not be treated by plastic theory, since the sliding is practically without dilatation.

Thus the upper bound and lower bound theorems, which are based on the normality condition, cannot rigorously be applied. Nevertheless, engineering judgement combined with plastic theory may often be used to solve problems where friction is involved.

When a butt joint subjected to pure shear is traversed by reinforcement, small loads must be carried by dowel action, since for small loads we have no cracks in a skew direction to the joint and thus there is no elongation in the direction perpendicular to the cracks which can cause elongation in the reinforcement direction. However, as soon as the load is large enough to produce skew cracks, the reinforcement may be stressed to tension, which in turn will cause normal compressive stresses in the concrete in the joint. Failure occurs by sliding without dilatation when the coefficient of friction μ times the normal stress equals the average shear stress in the joint. The load-carrying capacity calculated in this way will normally exceed the load-carrying capacity by dowel action although sometimes only slightly (see below). Thus we will expect the load-carrying capacity to be given by

$$\tau = \mu r f_Y \tag{8.3.10}$$

with notation as before.

In a butt joint there is no cohesion term. If there are normal stresses σ from external loads, positive as compression, we have

$$\tau = \mu\left(r f_Y + \sigma\right) \tag{8.3.11}$$

This equation may also be used for tensile stresses, $\sigma < 0$, as long as they do not, in absolute value, exceed $r f_Y$.

In practical design μ is usually set to 0.5 or 0.6 corresponding to a smooth surface, but it may, as mentioned before, have far larger values for rough surfaces. If important μ must be measured since no reliable method is known of how to relate one or another kind of roughness measure to the coefficient of friction μ.

Tests with cohesionless joints are collected in [78.12], to which the reader is referred.

Let's make a more exact estimate of the load-carrying capacity by dowel action than we did in Section 2.2.1. The purpose is partly to compare it with the above result and partly to supply a useful formula for the designer of precast structures, where dowel action often is utilized.

Consider a butt joint subjected to pure shear (see Fig. 8.3.15). The shear force to be carried is P. The joint is traversed by a reinforcement bar with diameter d and yield stress f_Y. The load is supposed to be carried by dowel action (cf. Section 2.2.1). If, as an approximation, the stresses developed in the concrete are considered uniformly distributed over the diameter d and the lengths ℓ shown in the figure we get a bending moment diagram and a shear force diagram as shown. The lengths ℓ may be calculated by formula (2.2.3) (notice that $\ell = L/2$), i.e.,

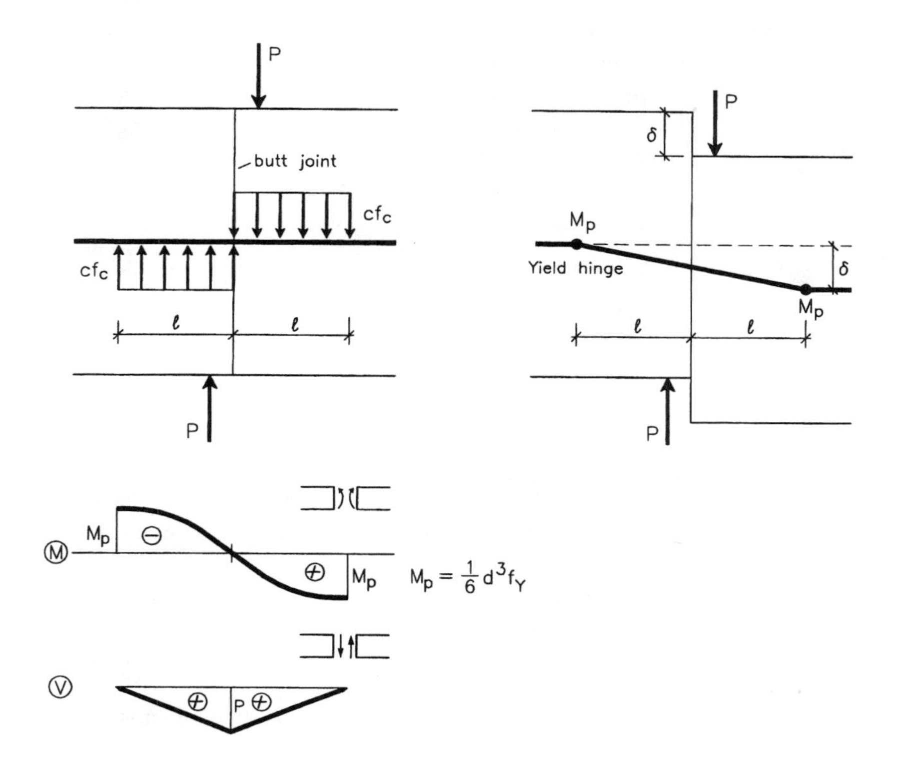

Figure 8.3.15 Dowel action in butt joint.

$$\ell = \frac{d}{\sqrt{3}} \sqrt{\frac{f_Y}{cf_c}} \tag{8.3.12}$$

We have replaced f_c by $c\, f_c$, c being an enhancement factor due to the concentrated loading of the concrete. Thus the load-carrying capacity is

$$P = \ell d c f_c = \frac{d^2}{\sqrt{3}} \sqrt{c f_c f_Y} \tag{8.3.13}$$

The same result would be found by an upper bound calculation for the mechanism shown in the figure to the right.

If a number of bars are crossing the joint we find the shear capacity

$$\frac{\tau}{f_c} = \frac{4}{\pi\sqrt{3}}\, \psi \sqrt{\frac{cf_c}{f_Y}} \tag{8.3.14}$$

As usual τ is the average shear stress over the whole concrete section and ψ is the reinforcement degree.

B. Højlund Rasmussen [63.12] has determined the enhancement factor in some tests. The dowel diameter varied from 15.8 mm to 25.9 mm, f_c from 11 to 44 MPa and f_Y from 225 to 440 MPa. He found that the enhancement factor attained the value $c \cong 5$ when the concrete section was 250 x 300 mm. The load was in the direction of the long side length and the dowel was placed in the center of the section. The specimen was reinforced by closed stirrups along the periphery. If this value of c is adopted, formula (8.3.13) reads

$$P = 1.3\, d^2 \sqrt{f_c f_Y} \tag{8.3.15}$$

and formula (8.3.14) reads

$$\tau = 1.6\, r f_Y \sqrt{\frac{f_c}{f_Y}} \tag{8.3.16}$$

The last formula renders $\tau = 0.5\, rf_Y$ for $f_c = 25$ MPa and $f_Y = 250$ MPa. For $f_c = 25$ MPa and $f_Y = 500$ MPa we get $\tau = 0.36\, rf_Y$. Thus it is seen that dowel action normally gives smaller load-carrying capacity than sliding by friction.

The enhancement factor cannot be calculated by a direct application of formulas for concentrated loads because the reinforcement bars prevent the loaded area from suffering a sliding failure with a displacement component in the reinforcement direction.

Formula (8.3.15) may be used to calculate the load-carrying capacity by dowel action.

Care should be shown when the bar is near a surface or when several bars are used, for instance by fixing the bars by one or another kind of transverse reinforcement.

8.3.5 Keyed Joints

In precast structures the natural way of transferring shear forces between concrete elements is by using a keyed joint. Such a joint is schematically shown in Fig. 8.3.16. A typical shear key *ABCD* has the total depth H and the distance in vertical direction between the keys is Δ. The joint width is a and the length of the joint is L. After the concrete elements have been erected the joint is filled with concrete.

In Fig. 8.3.17 joints, as they may often be found in the precast industry, are shown. The thickness b to be used in the following formulas is indicated in the figure.

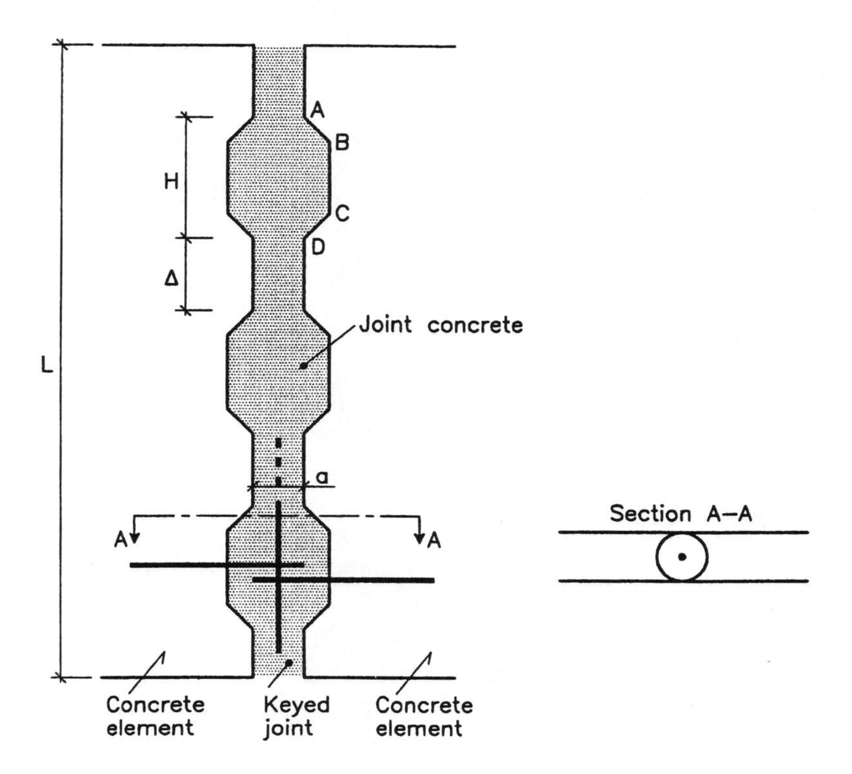

Figure 8.3.16 Keyed joint.

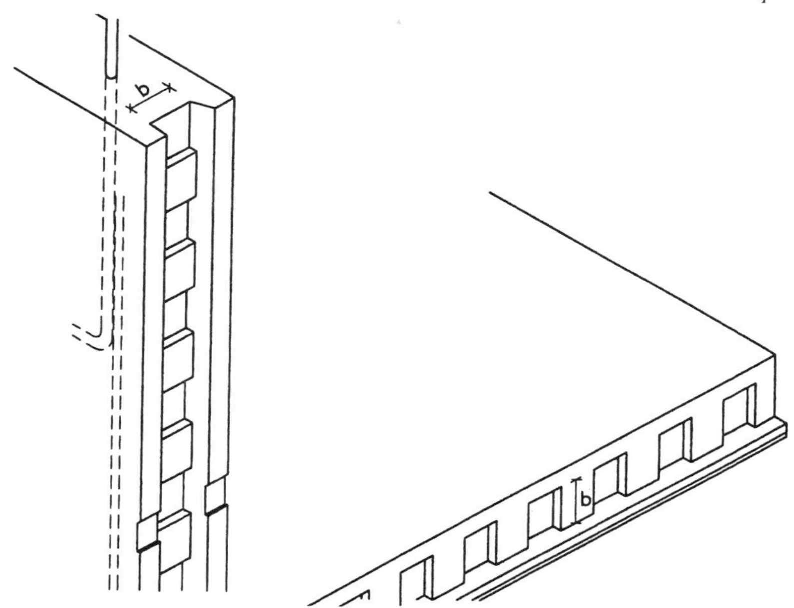

Figure 8.3.17 Panel shear key (left) and slab shear key (right) as they may be found in the precast industry.

Reinforcement may be placed in joints perpendicular to the joint in question. This is the method usually adopted in slabs. In panels the reinforcement may be carried through the joints, for instance, as shown in Fig. 8.3.16. In this case usually a bar through the centers of the circular areas, a so-called locking bar, is provided.

Formally the keyed joint may be treated by adopting suitable values for the cohesion c' and the angle of friction φ' in the plastic solution. Such a treatment may be found in [75.6] and [76.1] and also in the first edition of this book [84.11]. This method is, however, not able to provide the designer with any idea of the mechanical action of a keyed shear joint. In what follows we will try to formulate a more rational theory. The basic work was carried out by Jens Christoffersen [96.1] [97.10] who even pointed out that the empirical formulas may be dangerous to use when the joint parameters are essentially different from those of the tests.

Our departure point will be the assumption that at the final failure most of the interfaces between the precast elements and the joint concrete will be cracked and thus make no contribution to the load-carrying capacity. Tensile stresses due to shrinkage of the joint concrete may have caused cracking in these interfaces already before loading. The only action left for carrying the load will be a strut action from shear key to shear key. The problem is analyzed as a plane stress problem since the joint width may be rather large compared to the thickness.

The strut formulas to be used are summarized in Box 8.3.2. The shaded areas with the horizontal length x_o and the vertical length y_o are under plane hydrostatic pressure f_c . The strut itself is under uniaxial compression f_c. The formula for x_o expressed by y_o may be found by requiring moment equilibrium for the whole strut or by the procedure used in Section 4.8. The only task left is to determine from which shear key to which shear key the struts go.

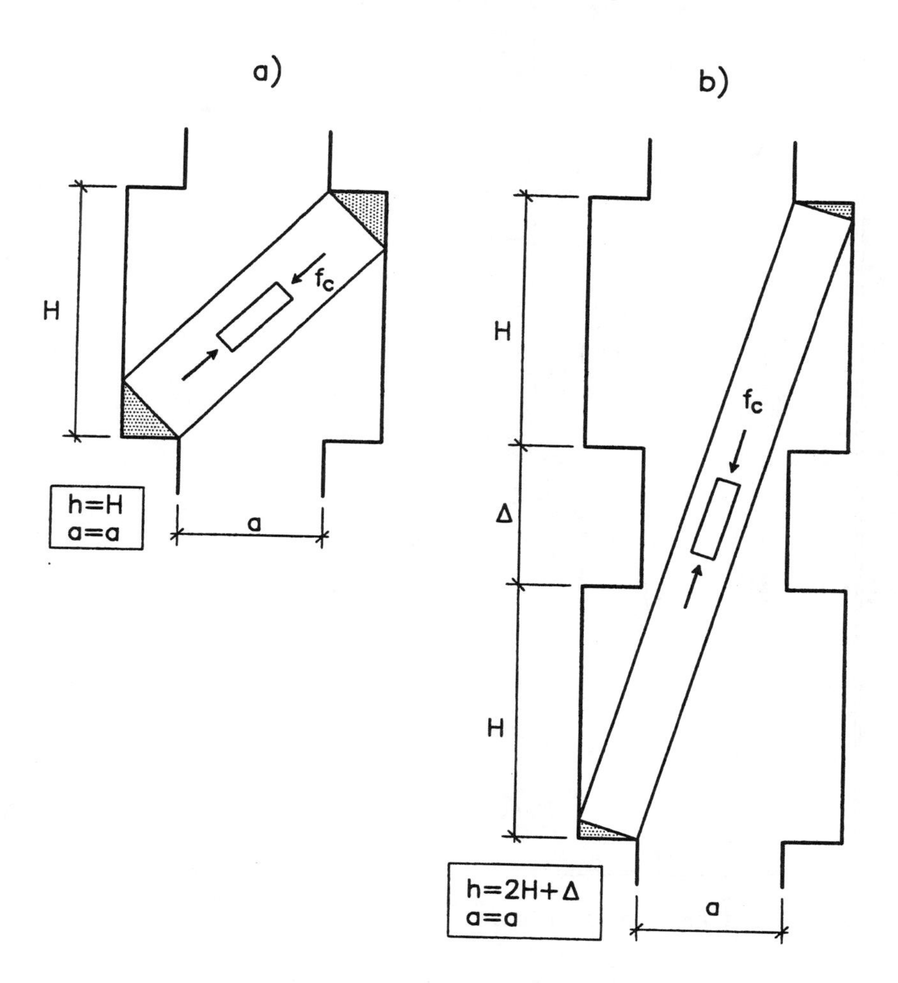

Figure 8.3.18 Strut action between shear keys.

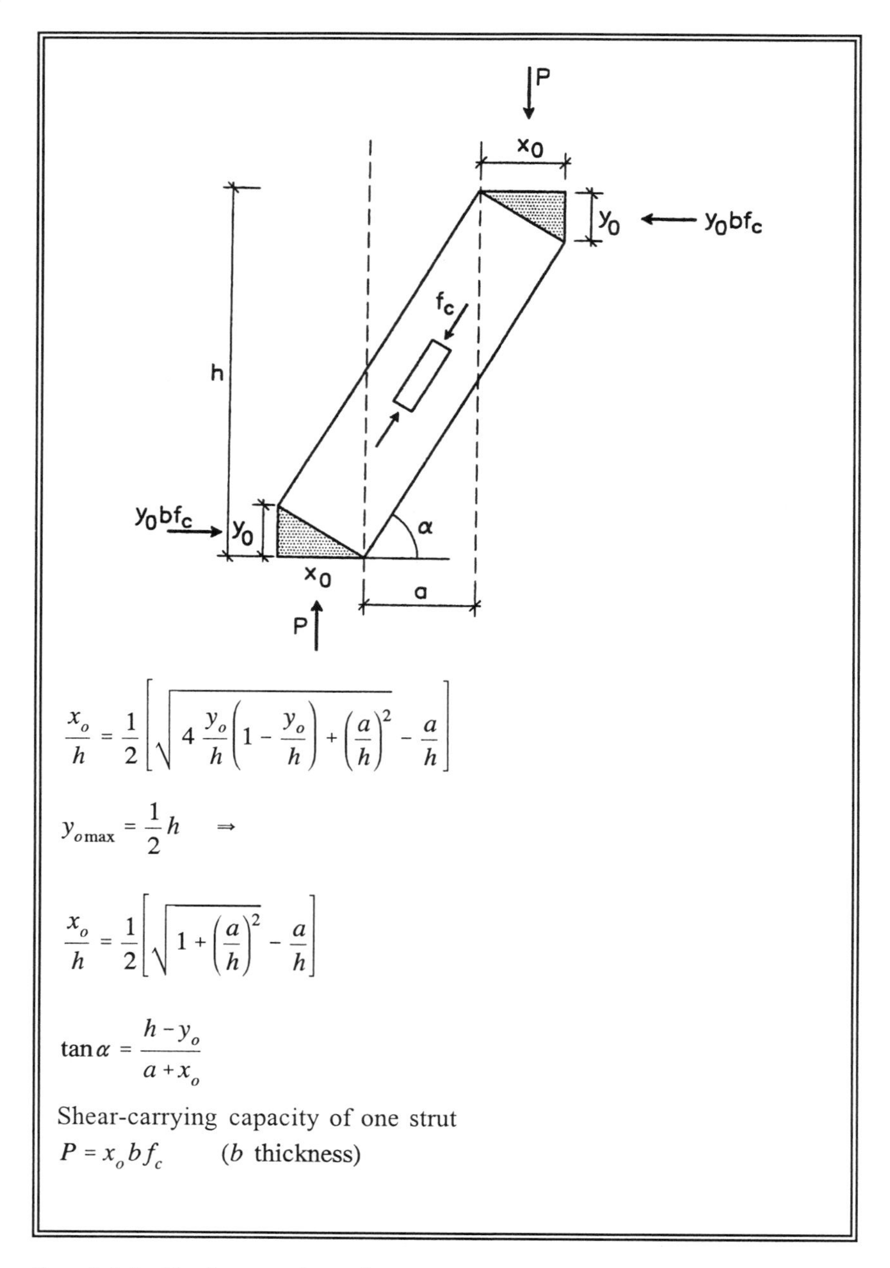

Box 8.3.2 Single strut formulas.

Let us for a moment assume that the geometry of the keys does not impose any restrictions on the strut action. Then it is quite easy to calculate the load-carrying capacity. In Fig. 8.3.18 case a) the struts go between two opposite shear keys and in case b) they go from a shear key in one element to the shear key in the next level in the vertical direction in the adjacent element. When y_o is known the contribution from the strut to the load-carrying capacity may in case a) be found by inserting into the strut formulas $h = H$ and $a = a$. The total load-carrying capacity is the load-carrying capacity for one strut times the number of struts. Similarly in case b) one strut makes a contribution found by inserting into the strut formulas $h = 2H + \Delta$ and $a = a$. Any strut path may be easily treated in the same way.

Formulas may be written for the different cases but they are rather long and we shall avoid them. The highest load-carrying capacity found by calculating all possible strut paths is the best value we can obtain by this method, since essentially we are dealing with a lower bound method. In most cases one of the strut systems shown in Fig. 8.3.18 will govern the load-carrying capacity.

Thus it is a simple procedure to calculate the load-carrying capacity when y_o is known.

If the total reinforcement area available for the joint is A_s and the number of struts is S, then by horizontal projection

$$y_o = \frac{A_s f_Y}{S b f_c} \tag{8.3.17}$$

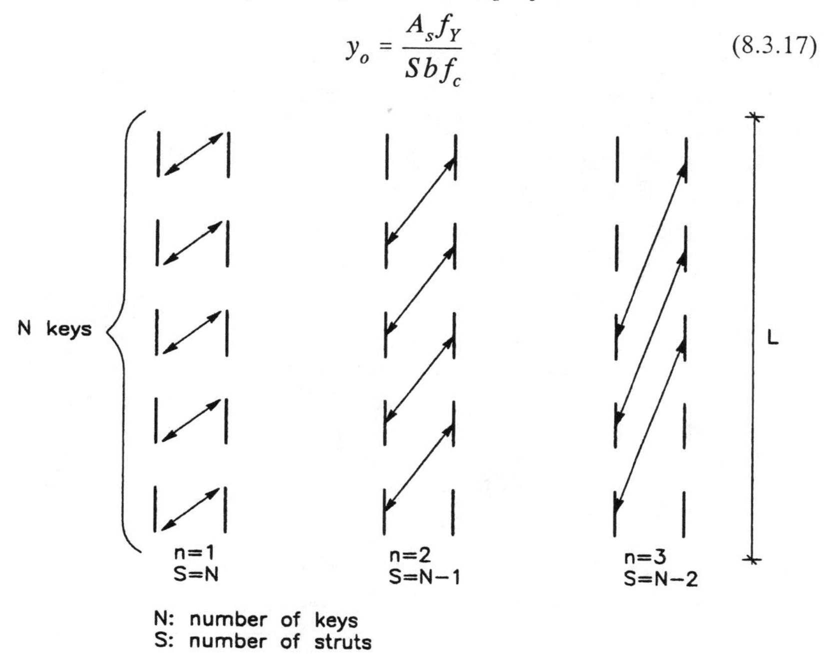

Figure 8.3.19 Notation for different strut paths.

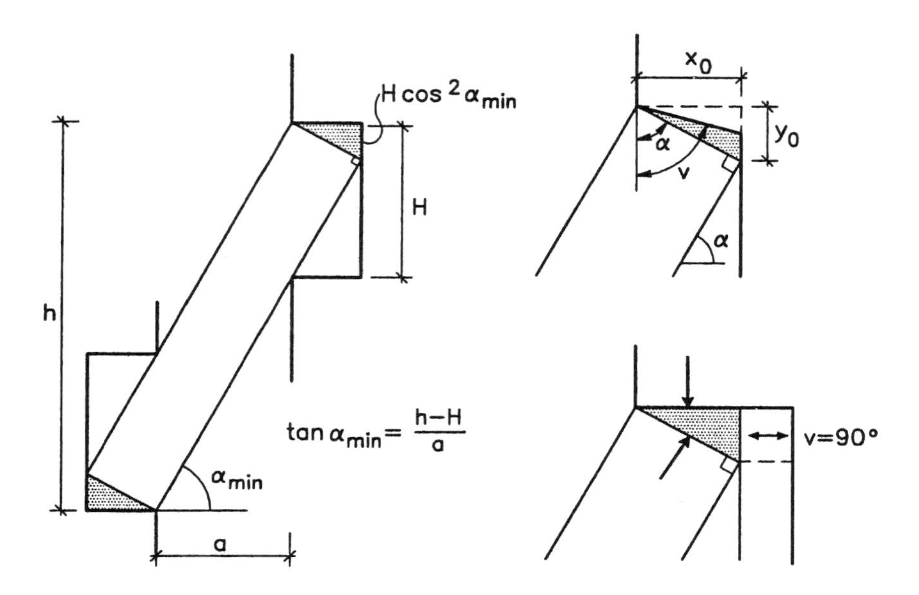

Figure 8.3.20 Key geometry.

If the joint carries a shear force as well as a normal force, the normal force (positive as tension) is subtracted from $A_s f_Y$ in (8.3.17).

The notation used in the following is illuminated in Fig. 8.3.19. The case where the struts go to the opposite key is characterized by the number $n = 1$. When they go to the next key level, $n = 2$, etc. In the figure the number of struts S is also indicated when the total number of keys is N.

Now we treat the limitations imposed by the key geometry. If we imagine the reinforcement to grow from zero, the angle α defined in Box 8.3.2 will decrease. The limiting case is shown in Fig. 8.3.20 (left). The value $\alpha = \alpha_{min}$ is the lowest value which can be accepted. From the figure it is easily seen that

$$\tan\alpha_{min} = \frac{h-H}{a} \tag{8.3.18}$$

where h is the active strut depth for the case considered. The corresponding value of $y_o = y_{omax}$ is found to be

$$y_{omax} = H\cos^2\alpha_{min} \tag{8.3.19}$$

This is the largest value which is allowed in the strut formulas. When the reinforcement is able to supply this value of y_o no further increase of the load-carrying capacity is possible.

In Fig. 8.3.20 (right) the problem of how to design the key is illustrated. It is seen that the angle ν shown in the figure must be larger than the angle α in order that a hydrostatic area can develop. Further the depth of the key should equal x_o to facilitate the formation of the hydrostatic area. In the case $\nu = 90°$ the depth of the key may be larger as illustrated in the figure since a uniaxial stress field may supply the necessary force on the hydrostatic area.

Probably these requirements to the key geometry are not particularly important. They are based on the assumption that only normal stresses can be carried by the key surfaces. This is not so even if these surfaces may be very smooth. Thus the key geometry may be designed with more freedom and the above requirements are only taken as guidelines.

Figure 8.3.21 Initial cracks and strut inclination for low reinforcement degrees.

In Fig. 8.3.22 the theory developed has been compared with a number of test series.[1] Regarding a detailed treatment of the tests and the test data, the reader is referred to [96.1] [97.10].

In the curves τ/f_c has been shown as a function of the reinforcement degree ψ. Here

$$\frac{\tau}{f_c} = \frac{\Sigma P}{bLf_c} \qquad \psi = \frac{A_s f_Y}{bLf_c} \tag{8.3.20}$$

ΣP is the total shear force measured or calculated, bL the total joint area, A_s the total transverse reinforcement available for the joint, f_Y the yield stress of the reinforcement and finally f_c the compressive strength of the joint concrete. In each figure some main data have been shown, e.g., the number N of shear keys and data regarding key geometry.

In some of the test series the scatter was very large, almost identical specimens giving rise to quite different shear capacities. In such cases a mean value for a group with nearly identical reinforcement degrees have been depicted.

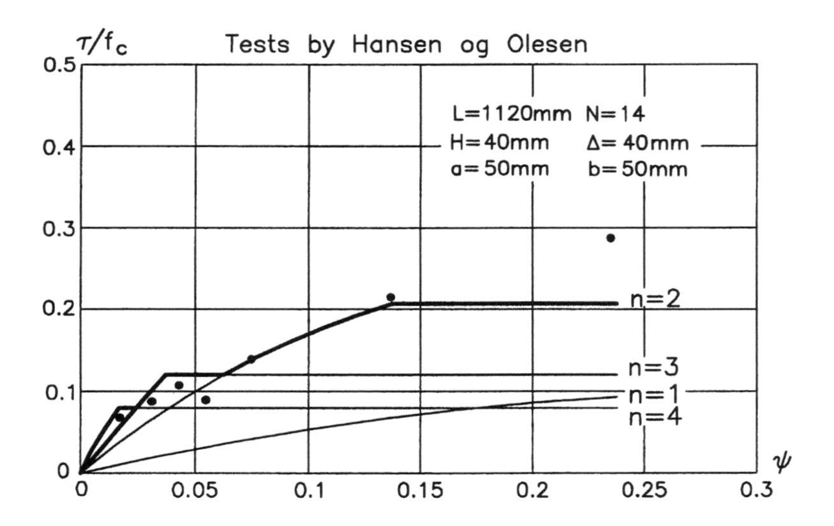

Figure 8.3.22 Test results for keyed joints. The test data may be found in [96.1] [97.10] together with detailed references to the original reports.

[1] The calculations were carried out by Junying Liu.

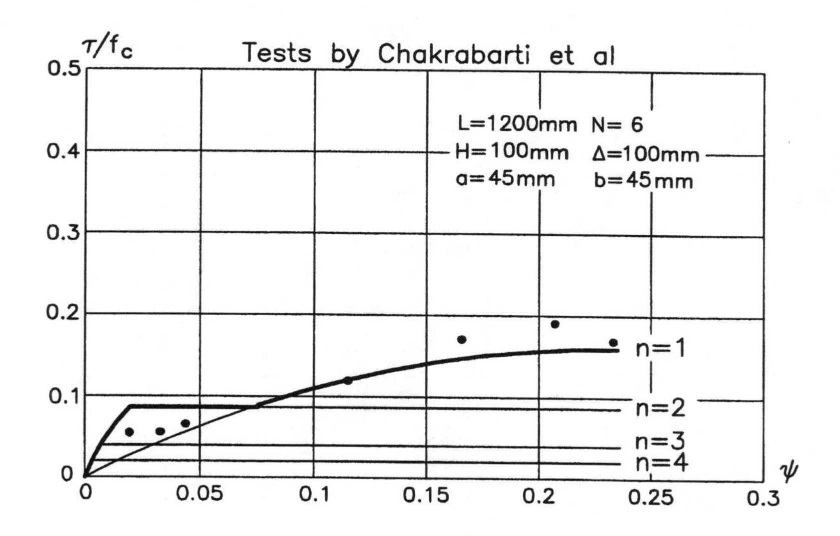

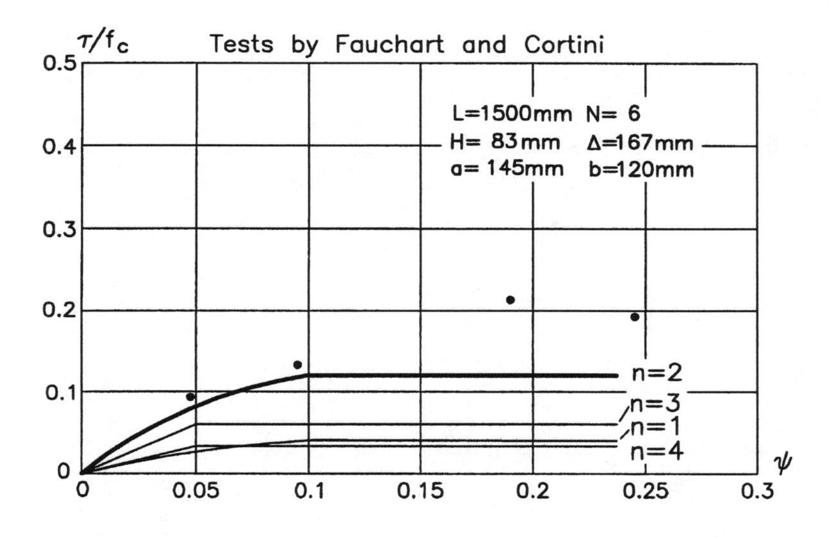

Figure 8.3.22 (continued).

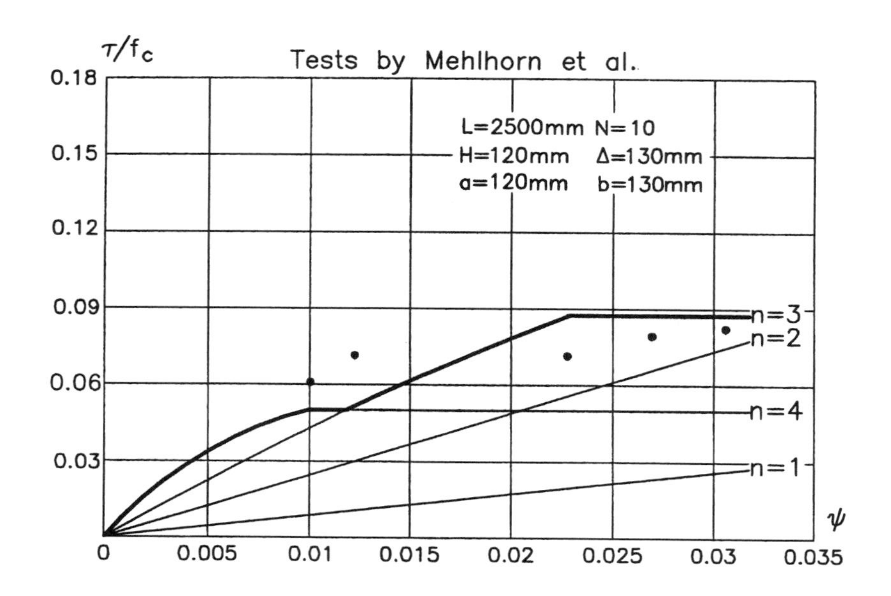

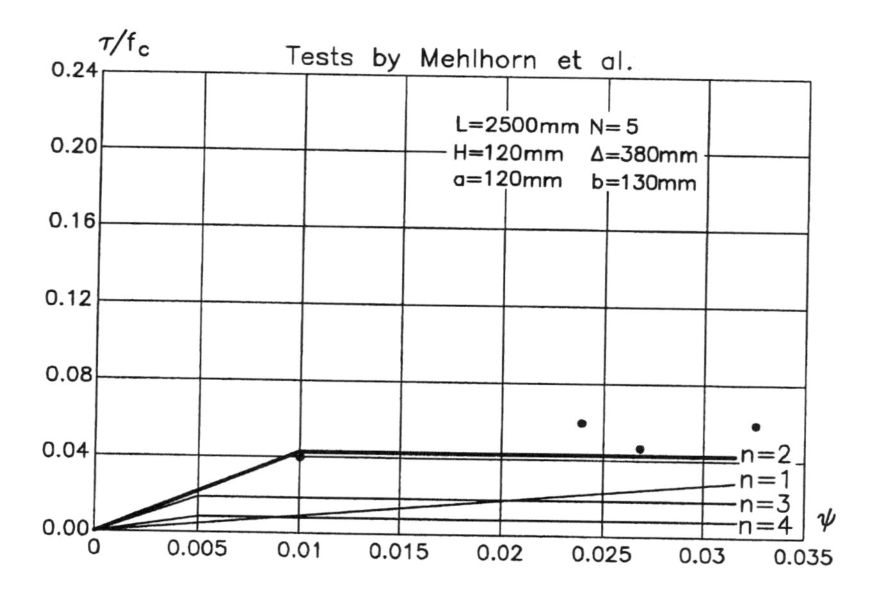

Figure 8.3.22 (continued).

Curves have been drawn for different n-values, the meaning of n being shown in Fig. 8.3.19.

The agreement between theory and tests is not particularly impressive. No reduction factor has been used on the concrete strength f_c, i.e., $\nu = 1$. In some tests there is a tendency that the test points are too low compared with theory for low reinforcement degrees and opposite for higher reinforcement degrees. This may be explained by a reduction of the compressive strength due to cracking (see Fig. 8.3.21). For low reinforcement degrees the strut inclination is rather steep and cracking formed at early loading stages may reduce the compressive strength. If the cracking direction is known, the effective compressive strength for any strut inclination may be calculated by means of formula (4.6.15) with $\nu_o = 1$ and $\nu_s = 0.5$. For pure shear it will be natural to assume an initial crack direction under 45° with the joint. No explanation is offered regarding the high test values for high ψ-values. One reason may be that the limitation on y_o imposed by the condition (8.3.19) may not really be necessary. In some cases, at least, better agreement is obtained if this condition is disregarded.

We conclude that the simple strut and tie model may be used to estimate the load-carrying capacity of keyed shear joints.

Chapter 9

THE BOND STRENGTH OF REINFORCING BARS

9.1 INTRODUCTION

The transfer of stresses between a reinforcing bar and the surrounding concrete is a very complex problem. Only a few attempts to create a rational theory have been made, Ferguson and Briceno [69.7], Tepfers [73.3] and Esfahani and Rangan [96.7]. Hitherto designers have therefore had to rely on empirical rules based on experiments.

A smooth bar may easily be pulled out of a concrete body, the failure being assisted by the elastic contraction of the bar. Hence smooth bars will have to be anchored by mechanical means: hooks or anchor plates.

The pull-out of a rough or deformed bar involves failure in the surrounding concrete, and a mechanism of pure slip is no longer possible. Because of the dilatation of the concrete, the longitudinal displacement of the bar is accompanied by radial deformations of the concrete. The shear stresses along the bar are components of inclined compressive stresses in the concrete leading to circumferential tensile stresses. If the bar is close to the concrete surface, the splitting action will lead to spalling.

The plastic theory of anchorage and splice problems was developed by Hess [78.11] [79.8] [84.7] and Andreasen [89.10]. Some master students have made important contributions, too. They will be referred to later. In Andreasens's work the results of plastic theory were compared with more than 500 test results covering almost all problems of practical interest. Good agreement was found.

In the first edition of this book [84.11], only a brief account of the basic theory could be given. In later work it turned out that applications often led to very complicated equations difficult to use in practical design. Furthermore, the theory ran into difficulties here and there, primarily due to the fact that the tensile strength of concrete is an extremely important parameter in anchorage and splice problems. Some of the difficulties are still with us. Thus we are certainly at the limit of what plastic theory may be supposed to supply.

In the following we put some emphasis on developing simple formulas based on the theory but, as usual, modified by means of the effectiveness factors. It will be shown that plastic theory is an attractive alternative to the fully empirical or semi-empirical approach.

9.2 THE LOCAL FAILURE MECHANISM

Consider a deformed bar with axisymmetrical ribs with depth e and spacing a (see Fig. 9.2.1). A geometrically possible local failure mechanism consists of a sliding failure forming a truncated cone as shown in the figure. The bar is assumed to move a distance u_s in the direction of the bar axis, and the surrounding concrete is assumed to be displaced axisymmetrically a distance u_c in the direction perpendicular to the bar. The failure surface forms the angle γ with the bar axis. The bar is subjected to a force P in the direction of the bar.

The relative displacement u_{cs}, which we formally put to unity, along the failure surface, may be determined as shown in the figure. We shall assume that there is no adhesion between concrete and steel.

If the angle between u_{cs} and u_s is denoted α, we get the external work

$$W_E = P\cos\alpha \tag{9.2.1}$$

The dissipation in the concrete may be determined by (3.4.83). The following expressions are seen to be valid.

For $\gamma \geq \gamma_0$:

$$D = L = \pi(d+e)\frac{e\dfrac{\ell}{a}}{\sin\gamma}\left[\frac{f_c}{2}(1-\sin(\alpha-\gamma)) + f_t\frac{\sin(\alpha-\gamma)-\sin\varphi}{1-\sin\varphi}\right] \tag{9.2.2}$$

where ℓ is the length of the anchorage and d is the diameter of the bar. The significance of the angle γ_0 is shown in Fig. 9.2.1.

For $0 \leq \gamma \leq \gamma_0$:

$$D = L = \pi(d+2e-a\tan\gamma)\,\frac{a}{\cos\gamma}\,\frac{\ell}{a}\left[\frac{f_c}{2}(1-\sin(\alpha-\gamma))\right.$$
$$\left. + f_t\frac{\sin(\alpha-\gamma)-\sin\varphi}{1-\sin\varphi}\right] \tag{9.2.3}$$

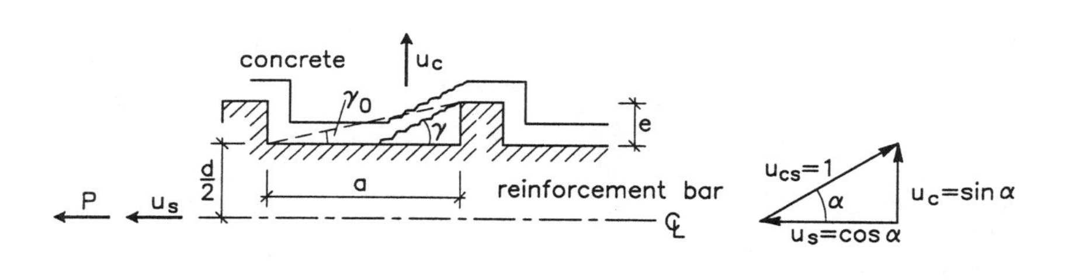

Figure 9.2.1 Local failure mechanism.

It may be shown (see [84.7] [89.10]) that almost always one arrives at the smallest load-carrying capacity when $\alpha = \gamma + \varphi$, i.e., the angle between the displacement vector and the failure surface is the angle of friction φ. Then:

For $\gamma = \alpha - \varphi \geq \gamma_0$:

$$L = \frac{\pi}{2} f_c (d + e) \frac{e\ell}{a} \frac{(1 - \sin\varphi)}{\sin(\alpha - \varphi)} \tag{9.2.4}$$

For $0 \leq \alpha - \varphi \leq \gamma_0$:

$$L = \frac{\pi}{2} f_c [d + 2e - a\tan(\alpha - \varphi)]\,\ell \frac{(1 - \sin\varphi)}{\cos(\alpha - \varphi)} \tag{9.2.5}$$

Notice that the dissipation is now independent of f_t.

To these expressions must be added the dissipation in the surrounding concrete. Further, there may be contributions from transverse reinforcement and the external work may have contributions from other external loads, e.g., a pressure along a support. Designating the total dissipation in the surrounding concrete by S, the contribution from possible transverse reinforcement by B, and assuming that n bars are involved in the failure we get the work equation

$$nP\cos\alpha = nL + S + B \tag{9.2.6}$$

If we divide on both sides with $n\pi d\ell f_c$ we find the average shear stress, the bond strength, per bar τ to be

$$\frac{\tau}{f_c} = \frac{P}{\pi d\ell f_c} = \frac{L}{\pi d\ell f_c \cos\alpha} + \frac{S}{n\pi d\ell f_c \cos\alpha} + \frac{B}{n\pi d\ell f_c \cos\alpha} \tag{9.2.7}$$

The contribution τ_o/f_c from the local failure may be written:
For $\alpha - \varphi \geq \gamma_0$:

$$\frac{\tau_o}{f_c} = \frac{L}{\pi d \ell f_c \cos\alpha} = \frac{1}{2}\frac{d+e}{d}\frac{e}{a}\frac{1-\sin\varphi}{\sin(\alpha-\varphi)\cos\alpha} \tag{9.2.8}$$

For $0 \leq \alpha - \varphi \leq \gamma_0$:

$$\frac{\tau_o}{f_c} = \frac{L}{\pi d \ell f_c \cos\alpha} = \frac{1}{2}\frac{d+2e-a\tan(\alpha-\varphi)}{d}\frac{1-\sin\varphi}{\cos(\alpha-\varphi)\cos\alpha} \tag{9.2.9}$$

With this notation the work equation (9.2.7) becomes:

$$\frac{\tau}{f_c} = \frac{\tau_o}{f_c} + \frac{S}{n\pi d \ell f_c \cos\alpha} + \frac{B}{n\pi d \ell f_c \cos\alpha} \tag{9.2.10}$$

In practice f_c must be replaced by νf_c, ν being the effectiveness factor for compression. Then:
For $\alpha - \varphi \geq \gamma_o$:

$$\frac{\tau_o}{f_c} = \frac{\nu}{2}\frac{d+e}{d}\frac{e}{a}\frac{1-\sin\varphi}{\sin(\alpha-\varphi)\cos\alpha} \tag{9.2.11}$$

For $0 \leq \alpha - \varphi \leq \gamma_o$:

$$\frac{\tau_o}{f_c} = \frac{\nu}{2}\frac{d+2e-a\tan(\alpha-\varphi)}{d}\frac{1-\sin\varphi}{\cos(\alpha-\varphi)\cos\alpha} \tag{9.2.12}$$

f_c now meaning the standard uniaxial compressive strength.

When $\tan\varphi = 0.75$ ($\varphi = 37^\circ$), which is used throughout in the following, we have $\sin\varphi = 3/5$ and $\cos\varphi = 4/5$.

Regarding the dissipation in the local failure mechanism, two cases deserve special attention.

Failure mode 1:

$$\alpha = 45^\circ + \frac{1}{2}\varphi \qquad \gamma = 45^\circ - \frac{1}{2}\varphi \tag{9.2.13}$$

$$\tan\alpha = \tan(45^\circ + \frac{1}{2}\varphi) = \tan 63^\circ.4 = 2$$

$$\frac{\tau_o}{f_c} = \frac{L}{\pi d \ell f_c \cos\alpha} = \frac{d+e}{d}\frac{e}{a} \tag{9.2.14}$$

It may be verified that in this case the compressive strength f_c is reached on the rib area. This is, for practical purposes, a lower limit of the dissipation in the local failure mechanism.

Failure mode 2:

$$\alpha = \varphi \qquad \gamma = 0 \tag{9.2.15}$$

$$\tan\alpha = \tan\varphi = 0.75$$

$$L = \pi(d + 2e)\,\ell\, c\cos\varphi$$

$$= \pi(d + 2e)\,\ell \cdot \frac{f_c}{4} \cdot \cos\alpha \tag{9.2.16}$$

$$\frac{\tau_o}{f_c} = \frac{L}{\pi d\,\ell f_c \cos\alpha} = \frac{1}{4}\,\frac{d + 2e}{d} \tag{9.2.17}$$

In this case the cohesion $c = f_c/4$ is reached along the length a. The failure surface is cylindrical from rib to rib. This is an upper limit of the dissipation in the local failure mechanism.

For practical purposes it may also be useful to know the value of τ_o/f_c when $\tan\alpha$ is about an average value of the limits 2 and 0.75, i.e., $\cong 1.4$.

Failure mode 1a:

$$\alpha = 54^{\circ}.46 \qquad \gamma = 17^{\circ}.6$$

$$\tan\alpha = 1.4 \tag{9.2.18}$$

$$\frac{\tau_o}{f_c} = 1.14\,\frac{d + e}{d}\,\frac{e}{a} \tag{9.2.19}$$

In Table 9.2.1 some typical reinforcement data have been given. There is some variation in the parameter in (9.2.14) while the parameter in (9.2.17) is practically constant for these data.

In the following we will often use the parameters valid for Danish Kam Steel 16 mm. In this case we have for the three failure modes:

Failure mode 1: $$\frac{\tau_o}{f_c} = 0.11 \tag{9.2.20}$$

Failure mode 1a: $$\frac{\tau_o}{f_c} = 0.12 \tag{9.2.21}$$

Failure mode 2: $$\frac{\tau_o}{f_c} = 0.28 \qquad (9.2.22)$$

When f_c is replaced by νf_c we have

Failure mode 1: $$\frac{\tau_o}{f_c} = 0.11\,\nu \qquad (9.2.23)$$

Failure mode 1a: $$\frac{\tau_o}{f_c} = 0.12\,\nu \qquad (9.2.24)$$

Failure mode 2: $$\frac{\tau_o}{f_c} = 0.28\,\nu \qquad (9.2.25)$$

	Danish Kam Steel			Danish Tentor Steel			ASTM A615		
d(mm)	10	16	25	10	16	25	9.5	15.9	25.4
a(mm)	6.5	10.0	15.0	8.5	12.4	18.2	6.7	11.1	17.8
e(mm)	0.6	1.0	1.6	0.6	1.0	1.6	0.4	0.7	1.3
γ_0(°)	5.3	5.7	6.1	4.0	4.6	5.0	3.4	3.6	4.2
$\frac{d+e}{d}\frac{e}{a}$	0.098	0.106	0.113	0.075	0.086	0.094	0.062	0.066	0.077
$\frac{1}{4}\frac{d+2e}{d}$	0.280	0.281	0.282	0.280	0.281	0.282	0.271	0.272	0.276

Table 9.2.1 Reinforcement data.

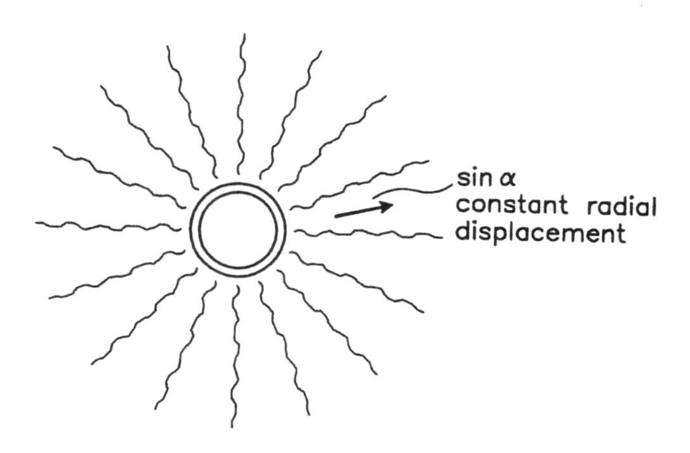

Figure 9.2.2 Axisymmetrical failure of surrounding concrete.

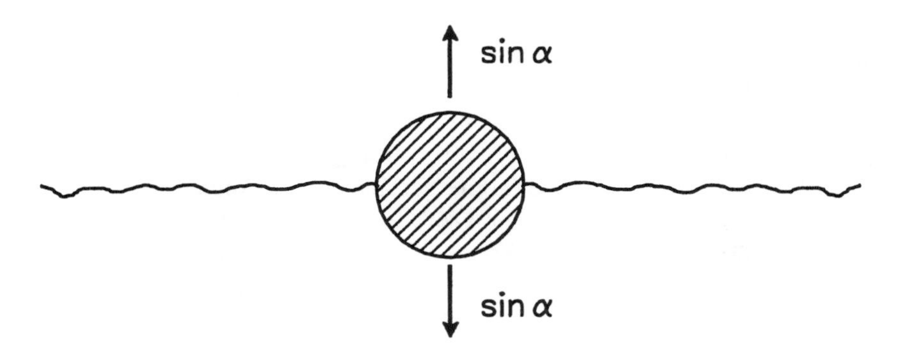

Figure 9.2.3 A single tensile yield line (crack) passing a reinforcement bar.

The ribs are not, in the real world, of the simple form assumed in the above analysis. Normally the rib surface with the depth e is inclined somewhat to the bar axis. Many tests have shown that as long as this surface does not get too flat, i.e., if it is not inclined less than 30-40° with the bar axis, the inclination is of no importance regarding bond strength.

The expressions are strictly correct only if the surrounding concrete is displaced axisymmetrically to the bar axis, which, for instance, is the case if the concrete gets a constant radial displacement corresponding to pure tensile yield lines (cracks) along all radii (see Fig. 9.2.2).

Often the concrete displacement varies around the bar. In the following the surrounding concrete is often subdivided into a number of rigid bodies. In these cases the expressions above have been used in an approximate manner by inserting for each part of the bar surface a value of the displacement complying in the best possible way with the axisymmetrical displacement field for which the expressions are correct.

One would imagine that when we have only one tensile yield line or crack passing a reinforcement bar as shown in Fig. 9.2.3, it would be too inaccurate to use the expressions developed. Certainly a displacement of the surrounding concrete in the direction perpendicular to the yield line violates the normality condition. However, practical experience shows that even in this case we get reasonable results by using the above formulas in a straightforward manner. The explanation may be that, as mentioned, theory has to be modified by the introduction of effectiveness factors. Thus the violation of the normality condition may be compensated for by the use of reduced strengths. Another explanation, probably closer to the truth, may be that since the cement paste is a frictionless material there might exist a complex system of failure surfaces zig-zagging between the aggregate particles and having mainly sliding surfaces in the required displacement direction, these failure surfaces being connected to each other by cracks in the direction of the skew compressive forces emerging from the ribs. The explanation of the true cause must await future careful examinations of the failure surfaces of bars pulled out of concrete.

9.3 FAILURE MECHANISMS

9.3.1 Review of Mechanisms

The only method which can be used to calculate anchor strengths and splice strengths is the upper bound method. Normally a lower bound method will be far too complicated. Since the upper bound solutions in this case cannot be checked by lower bound solutions it is important that a sufficient number of mechanisms are investigated.

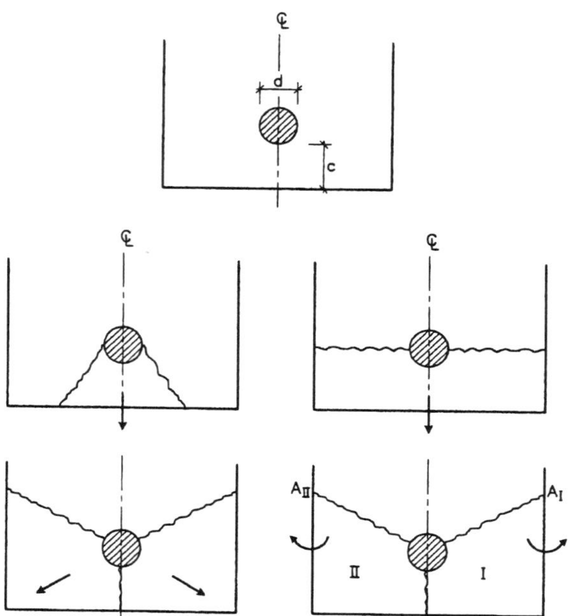

Figure 9.3.1 Failure mechanisms for anchor problems with one bar.

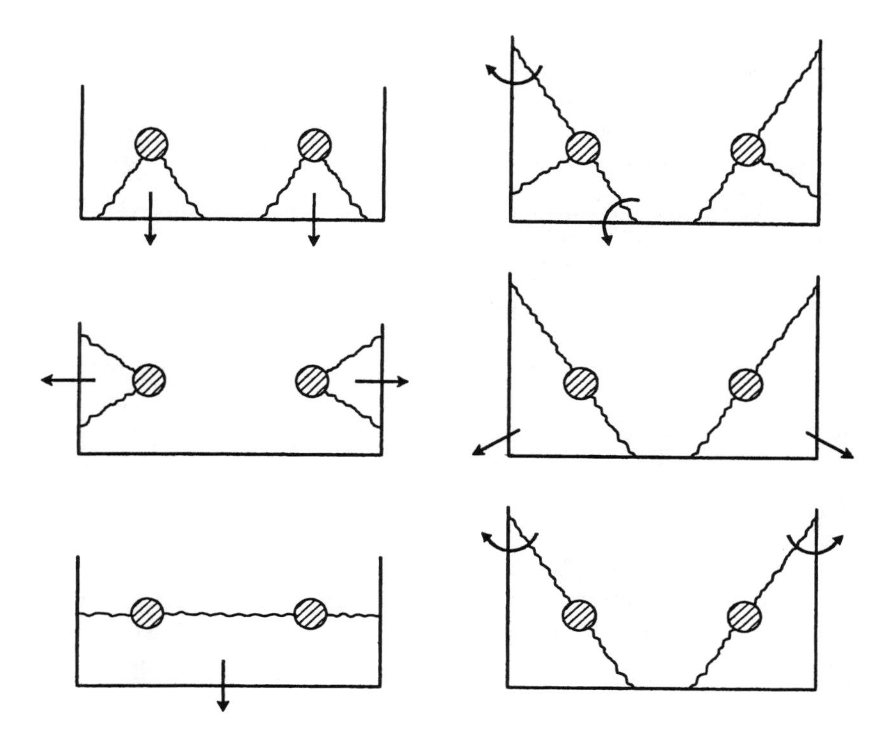

Figure 9.3.2 Failure mechanisms for anchor problems with two bars.

Consider as an example the case where a single bar is anchored. Some geometrically possible failure mechanisms of the surrounding concrete are shown in Fig. 9.3.1. The displacements of the concrete are indicated in the figure. Parts I and II rotate about points A_I and A_{II}, respectively.

The load-carrying capacity τ/f_c becomes a function of the geometrical parameters. As the upper bound theorem is used, τ/f_c has to be minimized with respect to the geometrical parameters.

Fig. 9.3.2 shows some failure mechanisms in the case where two bars have to be anchored.

9.3.2 Splice Strength versus Anchor Strength

When a bar is anchored the bar force is transferred to the surrounding concrete along the anchor length. When a bar force has to be transferred to another bar, the two bars are placed close to each other along a length, the lap length, and the bar force is transferred to the other bar by stresses in the concrete.

The average shear stress in the bar surface along the anchor length is termed the anchor strength. The average shear stress in the bar surface along the lap length is termed the splice strength.

In experiments one usually finds that the anchor strength approximately equals the splice strength under otherwise same conditions. There is not yet any quantification in plastic theory of this fact. We must be content with the following qualitative argument.

As indicated in Example 9.3.1, the splitting tendency, which the reinforcement bars give rise to in the surrounding concrete, may be visualized as shown in Fig. 9.3.3. The shear stresses around the bars are carried by skew compressive stresses emerging from the ribs. These skew forces will make the surrounding concrete feel something like a plane hydrostatic pressure (bursting forces) which again leads to tensile stresses in circumferential direction in the concrete near the bar. In Fig. 9.3.3 (left) a concrete ring around the bar is subjected to such a hydrostatic pressure. This may illustrate the conditions for an anchor bar. Fig. 9.3.3 (right) illustrates in the same qualitative way the conditions in a lap splice. Since the bars are rather close, large shear stresses and therefore large compressive stresses may be transferred directly from bar to bar. These stresses equilibrate each other. Thus a ring-shaped region around the bars does not need to have larger tensile stresses in the splice case than those present in the anchor case, because a large part of the bursting forces are self-equilibrated. Thus at least a primitive argument can be given to explain the equality of the anchor strength and the splice strength. Future research may succeed in quantifying this argument. Here we take it as an experimental fact, which means that in the following usually only anchor bars will be dealt with theoretically.

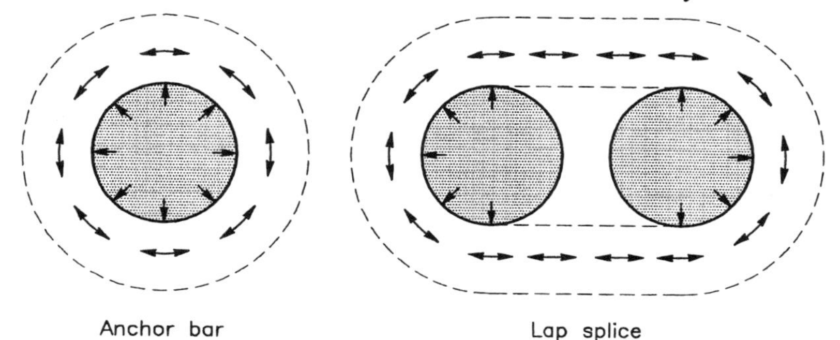

Figure 9.3.3 Ring model for anchor bar and lap splice.

We note in passing that such ring considerations play an important role in the theories of Tepfers [73.3] and of Esfahani and Rangan [96.7].

9.3.3 The Most Important Mechanisms

Fortunately the most important cases encountered in practice may be dealt with by considering only three mechanisms. They are illustrated in Fig. 9.3.4.

In the corner splitting failure mechanism or short, the corner mechanism, an inclined crack intersects the reinforcement bar. The corner is pushed away either by translation or by rotation as shown in the figure. The rotation mechanism has been found to be decisive. Notice the influence of the relative magnitude of the two covers c_1 and c on the rotation point. The bond strength is a symmetrical function of c_1 and c.

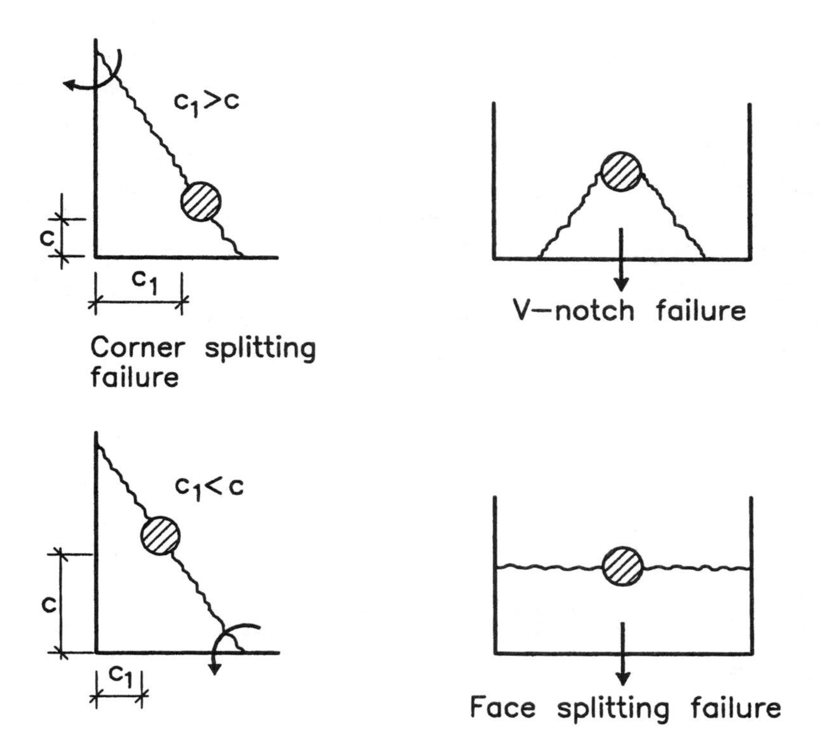

Figure 9.3.4 Important failure mechanisms.

In the *V*-notch failure a triangular area is pushed out in a kind of punching failure. As elaborated further below there are serious doubts whether this mechanism really exists. In fact it will be replaced by a mechanism named the cover bending mechanism.

The third mechanism is the face splitting mechanism. Here the cover as a whole is pushed off.

These mechanisms are analyzed in detail in what follows.

9.3.4 Lap Length Effect

When a plastic analysis is performed along the lines described an extremely important effect is not revealed, that is the strong dependence of the bond strength on the anchor or splice length. The longer the length the lower the bond strength τ/f_c. For some time it was believed (see [89.10]) that this phenomenon could be explained as an end effect as illustrated in Fig. 9.3.5 for a splice. An end failure would lead to a decrease of the bond strength with the length. However, if we imagine the bond strength is determined by a bending test, bending cracks develop at the ends of the splice. Before final failure these cracks are often so wide that no shear transfer can take place, thus excluding the explanation of the length dependence as an end effect. Therefore, we are left with the only possibility: to include the length dependence in the effectiveness factors. The physical explanation must be that the longer the length the more tensile softening takes place in the cracks in the surrounding concrete and the lower the strength. How to include, in the most effective way, the length effect in the effectiveness factors was solved by Christensen [95.12] (cf. Section 9.4).

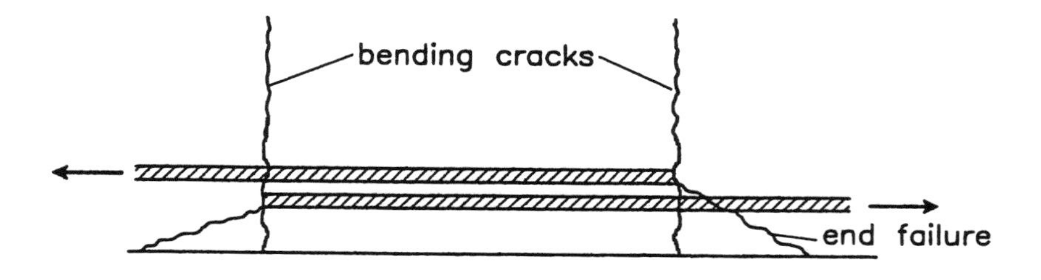

Figure 9.3.5 Hypothetical end effects in a splice.

9.3.5 Development Length

The most important application in practice of the theory is the calculation of the length necessary for an anchor bar or a splice to carry, respectively to transfer, the yield load of the bar. According to our assumptions the two lengths are equal. The length is often termed the development length. If the bond strength of a bar is τ, the development length ℓ, the diameter of the bar d, and the yield strength f_Y, we have

$$\tau \cdot \pi d \ell = \frac{\pi}{4} d^2 f_Y \tag{9.3.1}$$

or

$$\ell = \frac{f_Y d}{4\tau} \tag{9.3.2}$$

Since τ depends on ℓ and the failure mechanism, the equation must be solved by trial and error. The development length is of the order $30d$ to $40d$.

Example 9.3.1

We conclude this section by calculating the bond strength for a bar of length ℓ being pulled out of a cylindrical concrete specimen. The bar axis coincides with the axis of the concrete specimen. The cover is c (Fig. 9.3.6).

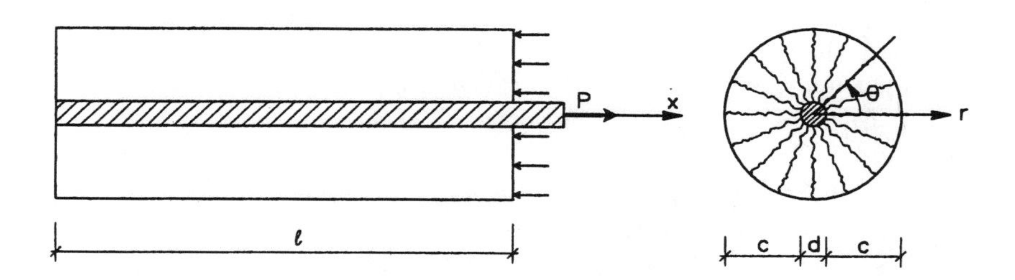

Figure 9.3.6 Pull-out of a bar from a cylindrical specimen.

In one end the bar is acted upon by a tensile force P. The concrete cylinder is supported at the same end by some kind of arrangement able to supply compressive stresses statically equivalent to $-P$. The arrangement must also allow the concrete cylinder to expand in a lateral direction.

Consider a failure mechanism corresponding to a constant radial displacement $u_r = \sin\alpha$. Thus radial cracks will develop.

The strain in the circumferential direction is $\varepsilon_\theta = u_r/r = \sin\alpha/r$. The dissipation in the concrete is then found to be

$$S = \ell f_t \int_0^{2\pi} \int_{d/2}^{c+\frac{d}{2}} \frac{\sin\alpha}{r} r\, d\theta\, dr = \ell f_t \sin\alpha \cdot c \cdot 2\pi$$

The work equation (9.2.7) leads to the bond strength

$$\frac{\tau}{f_c} = \frac{P}{\pi d \ell f_c} = \frac{L}{\pi d \ell f_c \cos\alpha} + 2\frac{c}{d}\frac{f_t}{f_c}\tan\alpha$$

The best upper bound solution is obtained by minimizing τ/f_c with respect to α. Before showing the solution we consider the three failure modes 1, 1a and 2 introduced above, understanding that any choice of α renders an upper bound solution.

Failure mode 1 leads to

$$\frac{\tau}{f_c} = \frac{d+e}{d}\frac{e}{a} + 4\frac{c}{d}\frac{f_t}{f_c}$$

since $\tan\alpha$ in this case is 2.

Failure mode 1a gives

$$\frac{\tau}{f_c} = 1.14\frac{d+e}{d}\frac{e}{a} + 2.8\frac{c}{d}\frac{f_t}{f_c}$$

since $\tan\alpha = 1.4$.

Finally, failure mode 2 provides

$$\frac{\tau}{f_c} = \frac{1}{4}\frac{d+2e}{d} + 1.5\frac{c}{d}\frac{f_t}{f_c}$$

since now $\tan\alpha = 0.75$.

Failure mode 1 governs the load-carrying capacity only for very small values of the cover and may be neglected. Only failure modes 1a and 2 are important.

For small values of the cover the bond strength rapidly increases with the cover. For larger values the increase is much smaller.

Notice that by calculating the average pressure σ on the bar surface according to the free body diagram in Fig. 9.3.7, this pressure multiplied by $\tan\alpha$ immediately gives the second term in the equations for τ.

Now, in practice, effectivenesss factors must be introduced according to the usual scheme:

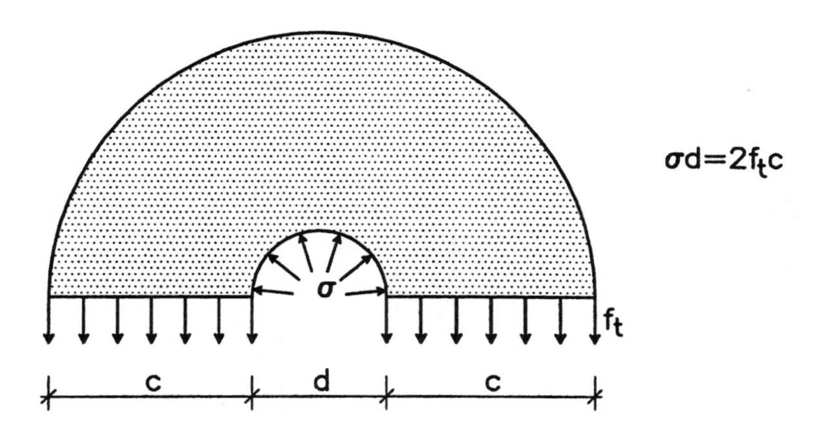

Figure 9.3.7 Free body diagram for calculating the transverse pressure σ.

$$f_c \rightarrow \nu f_c$$

$$f_t \rightarrow \rho f_c \text{ or } \nu_t f_t$$

where f_c and f_t now represent the standard strengths.

Using the data for Danish Kam Steel 16 mm we get for the failure modes 1a and 2, respectively:

$$\frac{\tau}{f_c} = 0.12\,\nu + 2.8\,\rho\,\frac{c}{d}$$

$$\frac{\tau}{f_c} = 0.28\,\nu + 1.5\,\rho\,\frac{c}{d}$$

or

$$\frac{\tau}{f_c} = 0.12\,\nu + 2.8\,\frac{c}{d}\,\frac{\nu_t f_t}{f_c}$$

$$\frac{\tau}{f_c} = 0.28\,\nu + 1.5\,\frac{c}{d}\,\frac{\nu_t f_t}{f_c}$$

The result has been illustrated in Fig. 9.3.8 using some typical values of ν and ν_t, namely $\nu = 0.5$ and $\nu_t = 0.5$. The standard tensile strength f_t is calculated using the usual equation $f_t = \sqrt{0.1 f_c}$ (f_c in MPa). Then $\rho = \nu_t f_t / f_c$ attains the value 0.032 when $f_c = 25$ MPa. This value of ρ has been used in the figure. In practice the effectiveness factors must be calculated taking into account the concrete strength and the ℓ/d ratio (cf. Section 9.4.2).

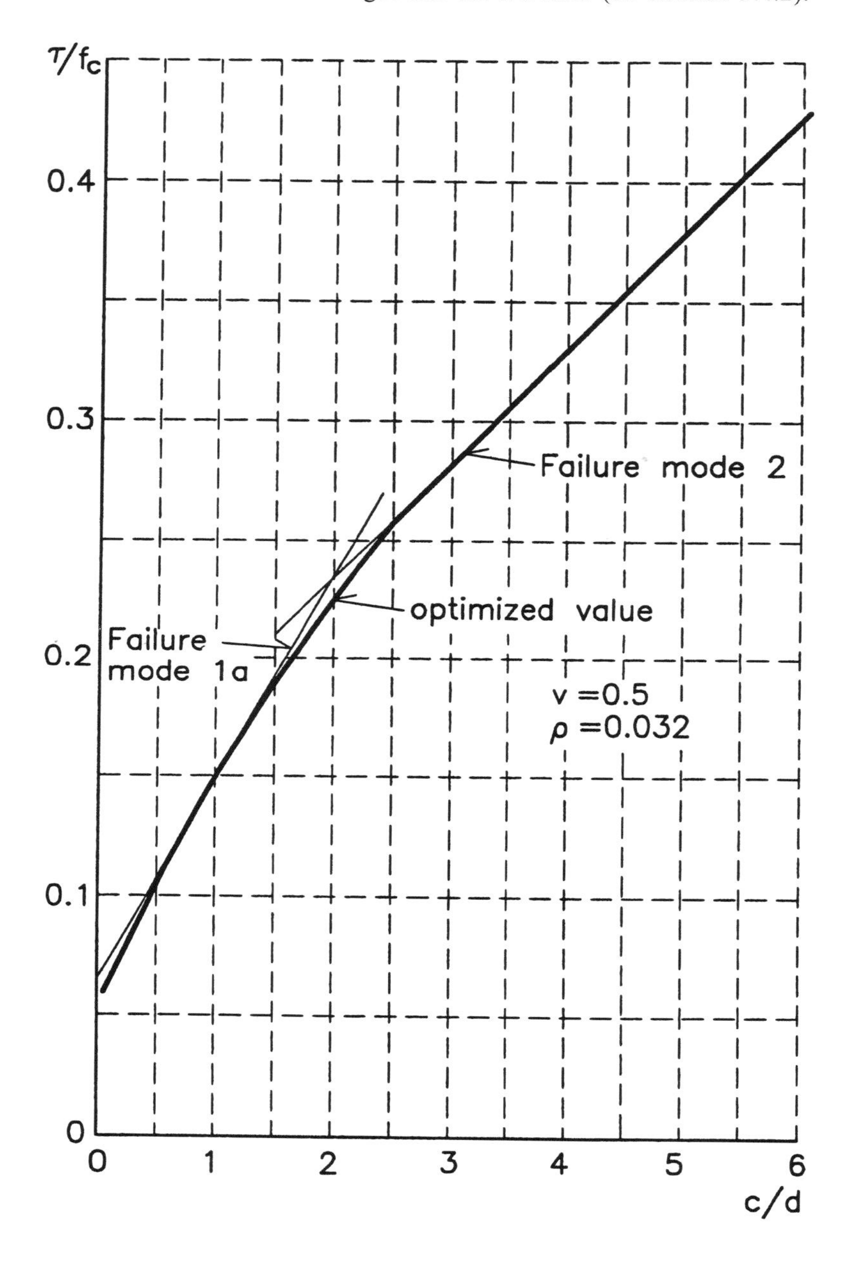

Figure 9.3.8 Pull-out of a bar. Bond strength versus the relative cover.

In the figure we have also shown the result when the bond strength is optimized with respect to α. It appears that failure modes 1a and 2 give an excellent approximation to the correct curve.

The two straight lines approximation using $\tan\alpha = 1.4$ and $\tan\alpha = 0.75$ will often be used below. We refer to this approximation as the procedure of Example 9.3.1.

An approximate statically admissible stress field is easily found. The σ_x-stresses are determined assuming these stresses to be uniformly distributed over the concrete section and to equilibrate the tensile force in the reinforcement. This may be found by assuming uniform shear stress along the bar surface. Then the stresses $\tau_{rx} = \tau_{xr}$ may be determined by considering equilibrium of a body limited by two sections at $x = 0$ and $x = x$ and two cylindrical surfaces at $r = r$ and $r = c + d/2$. Finally, in a thin disk limited by two normal sections at $x = x$ and $x = x + dx$, the stress distribution used in Example 1 in Section 4.5.3 may be used.

The stress distribution arrived at is not fully correct, since there are shear stresses statically equivalent to zero at the ends. These must be removed by introducing some end transition zones.

9.4 ANALYSIS OF FAILURE MECHANISMS

9.4.1 General

In this section the three most important failure mechanisms are analyzed in detail. The effectiveness factors are determined by comparison with a large number of tests.

9.4.2 Corner Failure

The corner failure mechanism is depicted in Fig. 9.4.1. The rotation point is shown on the vertical face which is valid when $c_1 \geq c$, i.e., the side cover is larger than or equal to the bottom cover. If $c_1 < c$, the rotation point lies on the bottom face. The load-carrying capacity is, as mentioned before, a symmetrical function of c_1 and c, so we need only analyze the case $c_1 \geq c$.

There is only one geometrical parameter, the angle β which the yield line makes with the vertical face. If only the triangle limited by the yield line and the vertical and horizontal faces is moving, the displacement of the bar is $\sin\alpha$ perpendicular to the yield line. Therefore the displacement of the triangle will be $2\sin\alpha$ perpendicular to the yield line at the position of the bar (cf. Section 9.2). Thus the angular rotation is

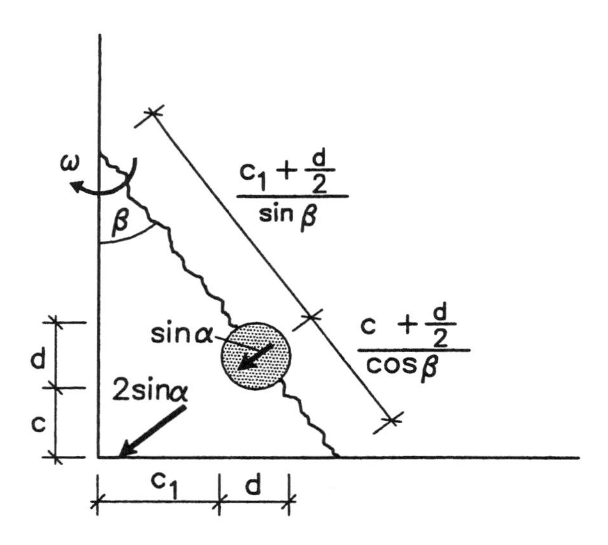

Figure 9.4.1 Corner failure mechanism.

$$\omega = \frac{2\sin\alpha\sin\beta}{c_1 + \frac{d}{2}} \tag{9.4.1}$$

and the dissipation in the yield line is

$$S = \frac{1}{2}\omega\left(\frac{c_1+\frac{d}{2}}{\sin\beta} + \frac{c+\frac{d}{2}}{\cos\beta}\right)^2 \ell f_t - 2\sin\alpha\cdot\ell d f_t$$

$$= \frac{\sin\alpha\sin\beta}{c_1+\frac{d}{2}}\left(\frac{c_1+\frac{d}{2}}{\sin\beta} + \frac{c+\frac{d}{2}}{\cos\beta}\right)^2 \ell f_t - 2\sin\alpha\cdot\ell d f_t \tag{9.4.2}$$

Then

$$\frac{S}{\pi d\,\ell f_c\cos\alpha} = \frac{1}{\pi}\left[\frac{\sin\beta}{\frac{c_1}{d}+\frac{1}{2}}\left(\frac{\frac{c_1}{d}+\frac{1}{2}}{\sin\beta} + \frac{\frac{c}{d}+\frac{1}{2}}{\cos\beta}\right)^2 - 2\right]\frac{f_t}{f_c}\tan\alpha \tag{9.4.3}$$

If $c_1 < c$ the factor $\sin\beta/(c_1/d + 1/2)$ is replaced by $\cos\beta(c/d + 1/2)$.

Now the work equation (9.2.7) or (9.2.10) may be established. It is rather complicated since minimization must be carried out both with respect to α and β. Currently this is easily done numerically by means of standard optimization programs.

Fig. 9.4.2 shows an example of such an optimization. Bar parameters are for Danish Kam Steel d = 16 mm. As before, f_c has been replaced by νf_c and f_t by ρf_c. In the figure ν = 0.5 and ρ = 0.042. The bond strength τ/f_c has been depicted as a function of c_1/d, i.e., as a function of the largest cover. The curve marked $c_1/d = c/d$ shows τ/f_c when the two covers are equal. This curve is very similar to the curve we obtained in Example 9.3.1 (see Fig. 9.3.8). The point determined by the two dotted lines indicates the cover where, in the local failure mechanism, the failure surface becomes a cylindrical surface from rib to rib. The curves marked c/d = 0.5, 1.0, etc. give the bond strength when the two covers are different.

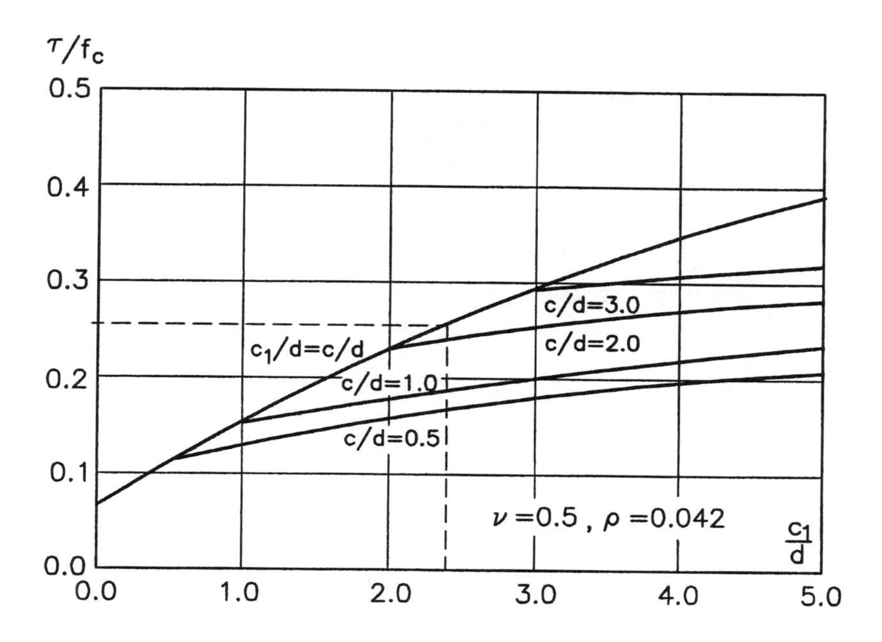

Figure 9.4.2 Bond strength versus cover.

Thus when $c_1 = c$ increases from zero to say $c_1/d = c/d = 1.0$, we follow the curve marked $c_1/d = c/d$ up to the point where $c_1/d = 1.0$. If from that point c is kept constant at $c/d = 1.0$, while c_1 still grows we follow the curve marked $c/d = 1.0$. We notice that the inclination of the curves marked $c/d = 0.5$, 1.0, etc. is much smaller than the inclination of the curve marked $c_1/d = c/d$, i.e., an increase of the smaller cover c has a greater effect on the bond strength than an increase of the larger one.

Fig. 9.4.2 suggests that it may be possible to describe the bond strength at corner failure by an approximate relation of the type

$$\frac{\tau}{f_c} = \alpha_o + \alpha_1 \frac{c}{d} + \alpha_2 \left(\frac{c_1}{d} - \frac{c}{d} \right) \qquad (c_1 \geq c) \tag{9.4.4}$$

The curve in Fig. 9.4.2 marked $c_1/d = c/d$ is consequently approximated by the straight line $\alpha_o + \alpha_1 \, c/d$ and the curves marked $c/d = 0.5$, 1.0, etc. are approximated by straight lines with inclination α_2.

The expression may be written

$$\frac{\tau}{f_c} = \alpha_o + \alpha_2 \left(\left(\frac{\alpha_1}{\alpha_2} - 1 \right) \frac{c}{d} + \frac{c_1}{d} \right) \qquad (c_1 \geq c) \tag{9.4.5}$$

A parameter study carried out by Tolderlund [96.8] showed that with good approximation $\alpha_o \simeq 0.13\,\nu$, $\alpha_2 \cong 0.52\,\rho$ and $\alpha_1/\alpha_2 \cong 4.2$. Thus

$$\frac{\tau}{f_c} \cong 0.13\,\nu + 0.52\,\rho \left(3.2 \frac{c}{d} + \frac{c_1}{d} \right) \qquad (c_1 \geq c) \tag{9.4.6}$$

If we want an expression valid for any combination of c_1 and c, i.e., if we want to relax on the condition $c_1 \geq c$, we must write

$$\frac{\tau}{f_c} \cong 0.13\,\nu + 0.52\,\rho \left(3.2 \frac{c_{\min}}{d} + \frac{c_{\max}}{d} \right) \tag{9.4.7}$$

The formula may also be written

$$\tau = 0.13\,\nu f_c + 0.52\,\nu_t f_t \left(3.2 \frac{c_{\min}}{d} + \frac{c_{\max}}{d} \right) \tag{9.4.8}$$

If we assume that ν may be described by an expression of the usual form

$$\nu = \frac{K_1}{\sqrt{f_c}} \qquad (f_c \text{ in MPa}) \tag{9.4.9}$$

where K_1 is a constant and if we introduce $f_t = \sqrt{0.1 f_c}$ (f_c in MPa) we arrive at

$$\frac{\tau}{\sqrt{f_c}} = 0.13 K_1 + 0.164\, v_t \left(3.2 \frac{c_{\min}}{d} + \frac{c_{\max}}{d} \right) \qquad (9.4.10)$$

The effectiveness factors must be determined by tests.

The length dependence of the shear strength is illustrated in Fig. 9.4.3 by some tests of Tepfers [73.3] on lap splices where corner failure was governing. In this case the bond strength may be excellently described by a power function which is almost a square root dependence. It was suggested by Christensen [95.12] to take the length dependence into account by letting v_t depend on $\sqrt{d/\ell}$, i.e.,

$$v_t = K_2 \sqrt{\frac{d}{\ell}} \qquad (9.4.11)$$

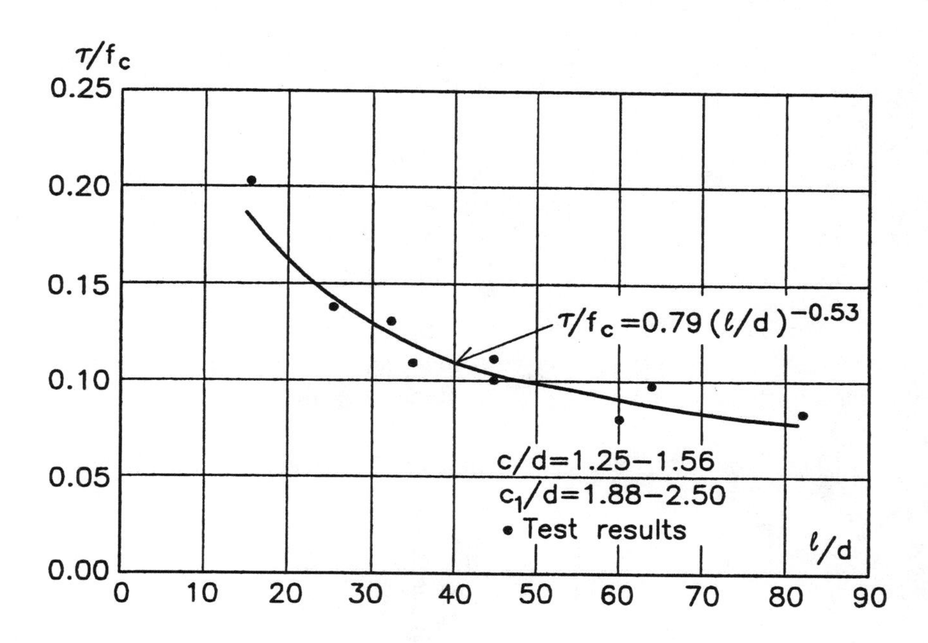

Figure 9.4.3 The length dependence of the bond strength according to tests by Tepfers [73.3] (see also [96.8]).

K_2 being another constant. To include the length dependence in ν_t is quite natural if the physical cause of the length dependence is softening in the tensile yield lines as mentioned in Section 9.3.4.

Our final formula is thus

$$\frac{\tau}{\sqrt{f_c}} = 0.13K_1 + 0.164K_2\sqrt{\frac{d}{\ell}}\left(3.2\frac{c_{\min}}{d} + \frac{c_{\max}}{d}\right) \quad (f_c \text{ in MPa}) \tag{9.4.12}$$

A statistical analysis of 100 tests mainly on splices was carried out by Christensen [95.12] and Tolderlund [96.8]. Tolderlund found the constants

$$K_1 = 1.8 \quad \Rightarrow \quad \nu = \frac{1.8}{\sqrt{f_c}} \leq 1 \quad (f_c \text{ in MPa}) \tag{9.4.13}$$

$$K_2 = 1.9 \quad \Rightarrow \quad \nu_t = 1.9\sqrt{\frac{d}{\ell}} \leq 1 \tag{9.4.14}$$

This ν_t-value leads to

$$\rho = \frac{\nu_t f_t}{f_c} = \frac{\nu_t\sqrt{0.1f_c}}{f_c} = \frac{0.32\,\nu_t}{\sqrt{f_c}} \quad (f_c \text{ in MPa}) \tag{9.4.15}$$

or

$$\rho = \frac{0.61}{\sqrt{f_c}}\sqrt{\frac{d}{\ell}} \quad (f_c \text{ in MPa}) \tag{9.4.16}$$

The mean value was 1.003 and the coefficient of variation 12.7%. Tepfers′ tests (133) were not included because they have, in general, rather large scatter, about 22%. Regarding the test data used, see [96.8].

The parameter intervals covered by the tests are

$$6 \leq \frac{\ell}{d} \leq 80$$

$$0.6 \leq \frac{\text{cover}}{\text{d}} \leq 4.96$$

$$9.5 \leq d \leq 35.8 \text{ mm}$$

$$18 \leq f_c \leq 100 \text{ MPa}$$

The size of the cover probably has a similar effect on ν_t as the length ℓ, but this problem has not yet been studied due to the lack of experiments with very large covers.

For $f_c > 75$ MPa an empirical correction factor mentioned below was applied.

The ν_t-formula becomes meaningless for $\ell \rightarrow 0$. Notice, however, that tests have been carried out down to $\ell = 6d$. The bond strength formulas should not be used for smaller lengths than $5d$ to $6d$.

If K_1 and K_2 found from the tests are inserted into (9.4.12) we find that the bond strength at corner failure is given by the formula

$$\frac{\tau}{\sqrt{f_c}} = 0.23 + 0.31 \sqrt{\frac{d}{\ell} \left(3.2 \frac{c_{\min}}{d} + \frac{c_{\max}}{d} \right)} \quad (f_c \text{ in MPa}) \qquad (9.4.17)$$

In Fig. 9.4.4 theory and tests are compared.

The formula arrived at is quite satisfactory. However, it is not yet perfect. It underestimates bond strengths when the bar diameter is small, about 10-15% when d = 10 mm. It overestimates bond strengths when the diameter is very large, about 10-15% when d = 35 mm. This could be remedied by adding an empirical correction factor which is a function of the diameter only. However, at the present stage of development this does not seem necessary.

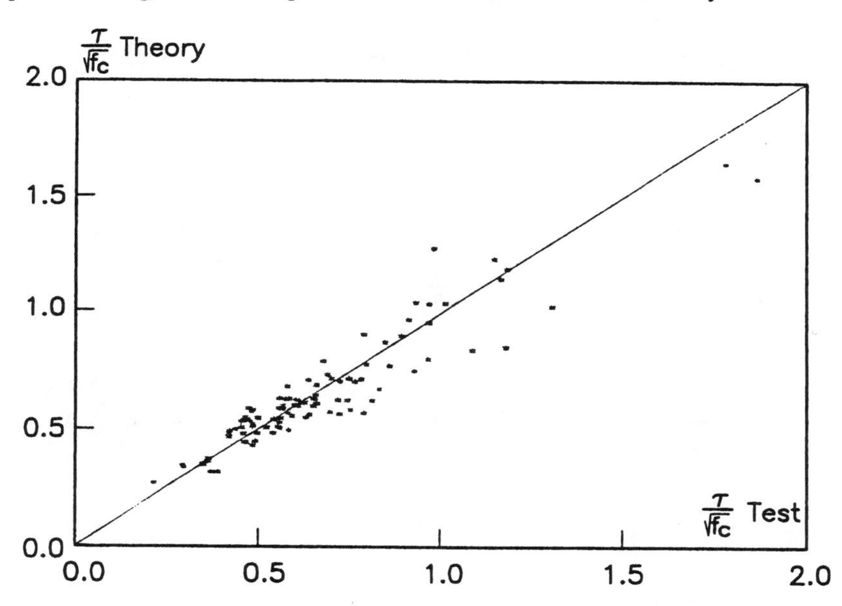

Figure 9.4.4 Formula (9.4.17) compared with 100 test results.

Another defect, and this is more serious, is that the bond strength is overestimated for high strength concrete with microsilica. Why this is so is not yet completely clear. It is well known that concrete with microsilica often suffers from severe microcracking stemming from self-desiccation. The microcracking may be enhanced around reinforcement bars which prevent the shrinkage strains.

A phenomenon like this may explain why microcracking reduces bond strength more than it reduces the compressive strength.

In [95.12] it was suggested to multiply the bond strength by an empirical reduction factor when $f_c > 75$ MPa. This proposal has been included in Box 9.4.1, to which the reader is referred.

There are empirical or semi-empirical formulas with both larger and smaller scatter. Orangun et al. [75.11] suggested a formula which in general gives higher scatter. Darwin et al. [92.23] modified Orangun's formula and the scatter was reduced to about the same level as we have obtained. The most accurate formula seems to be that of Esfahani and Rangan [96.7] with a coefficient of variation as low as 9.3%.

9.4.3 *V*-notch failure

The *V*-notch failure mechanism is shown in Fig. 9.4.5 for anchorage of one bar.

The dissipation in the yield lines is (cf. formula (3.4.83))

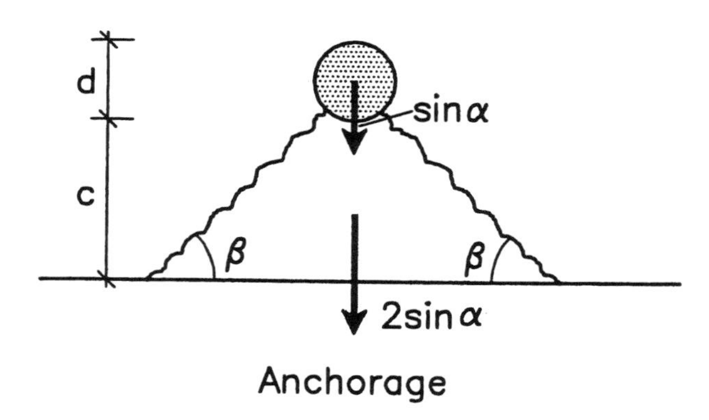

Figure 9.4.5 *V*-notch failure.

$$S = 2\left(\frac{c + \frac{d}{2}}{\sin\beta} - \frac{d}{2}\right)\ell \cdot 2\sin\alpha\left[\frac{f_c(1 - \cos\beta)}{2} + f_t\,\frac{\cos\beta - \sin\varphi}{1 - \sin\varphi}\right] \tag{9.4.18}$$

The yield lines are assumed to intersect at the center of the bar, although it might have been more natural to let them emerge from the outermost point of the bar in the horizontal direction. The mechanism is only possible for $\beta \leq \pi/2 - \varphi$ according to the normality condition.

Now the work equation (9.2.10) may be set up. We get the following bond strength

$$\frac{\tau}{f_c} = \frac{\tau_o}{f_c} + \frac{4}{\pi}\left(\frac{\frac{c}{d} + \frac{1}{2}}{\sin\beta} - \frac{1}{2}\right)\left[\frac{1}{2}(1 - \cos\beta) + \frac{f_t}{f_c}\,\frac{\cos\beta - \sin\varphi}{1 - \sin\varphi}\right]\tan\alpha \tag{9.4.19}$$

The contribution from the local failure τ_o/f_c is given by the formulas (9.2.24) and (9.2.25).

As usual in practice, f_c is replaced by νf_c and f_t by ρf_c or $\nu_t f_t$.

A parameter study shows that ρ has very little influence on the bond strength in this case. Except for very small covers, we have approximately

$$\frac{\tau}{f_c} \cong 0.16\,\nu + 0.32\,\nu\,\frac{c}{d} \tag{9.4.20}$$

According to this equation one would expect the length dependence to be rather small compared to what we found for the corner failure. The reason is that tensile softening cannot play any significant role when ρ does not enter into the formula. Further, if a sliding failure took place in a crack, which is what we would expect, the aggregate interlocking will be almost fully intact along the whole length since the crack widths are always extremely small. This conclusion is at variance with experiments. The experimental bond strength displays the same type of length dependence as we found in the corner failure case.

Consequently, we are led to the conclusion that the V-notch failure does not take place in the manner we have assumed. There must be another mechanism in which tensile softening plays an important role. One such mechanism is shown in Fig. 9.4.6. In this mechanism the cover suffers a kind of bending failure. The displacement of the bar has not been shown in Fig. 9.4.6, a remark valid for the figures below, too. The shaded areas to the left and to the right are under compression, the other shaded areas are under tension. Strictly speaking this kind of compression zone is not allowed in plane strain problems, but this difficulty may be removed by imagining the compression zones to be Rankine fields (cf. Section 3.5). This mechanism will be referred to as the cover bending mechanism.

The dissipation is, when $\delta = 2\sin\alpha$,

$$S = 2\ell f_t\left[\frac{1}{2}\delta x + \frac{1}{2}\frac{\delta}{x}y^2 + \frac{1}{2}\frac{\delta}{x}y^2\frac{f_c}{f_t} - \frac{1}{2}\delta d\right] \tag{9.4.21}$$

Minimizing with respect to the geometrical parameter x we get

$$x = y\sqrt{1 + \frac{f_c}{f_t}} \tag{9.4.22}$$

Thus

$$S = 2\ell f_t\left[y\sqrt{1 + \frac{f_c}{f_t}} - \frac{1}{2}d\right]\cdot 2\sin\alpha \tag{9.4.23}$$

The work equation (9.2.10) leads to the following bond strength τ / f_c

$$\frac{\tau}{f_c} = \frac{\tau_o}{f_c} + \frac{2\ell f_t\left[y\sqrt{1 + \frac{f_c}{f_t}} - \frac{1}{2}d\right]\cdot 2\sin\alpha}{\pi d\ell f_c\cos\alpha} \tag{9.4.24}$$

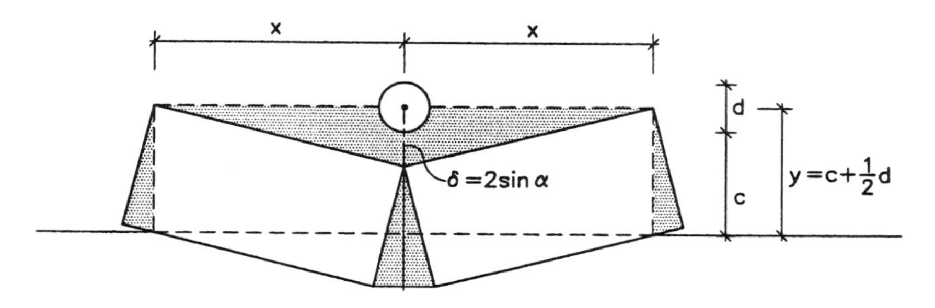

Figure 9.4.6 Cover bending mechanism.

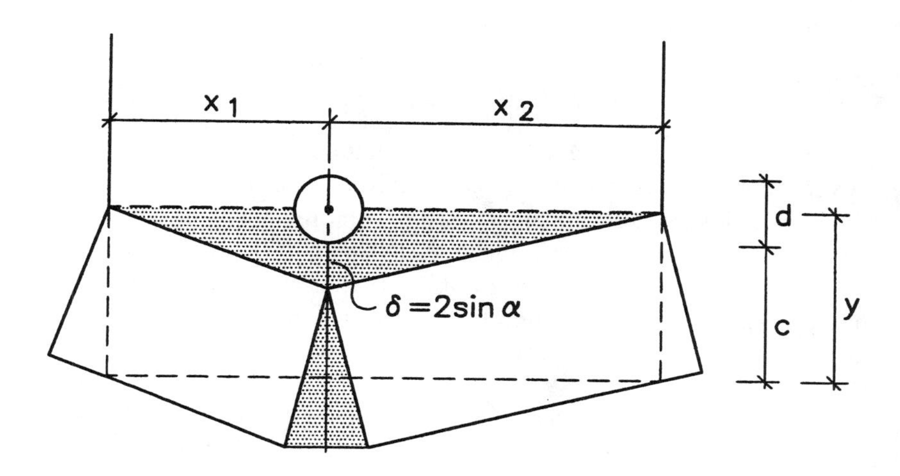

Figure 9.4.7 Cover bending mechanism, bars near two free surfaces.

Replacing f_c by νf_c and f_t by ρf_c we get

$$\frac{\tau}{f_c} = \frac{\tau_o}{f_c} + \frac{2}{\pi}\left[2\left(\frac{c}{d} + \frac{1}{2}\right)\sqrt{1 + \frac{\nu}{\rho}} - 1\right]\rho \tan\alpha \tag{9.4.25}$$

having inserted $y = c + \frac{1}{2}d$. The square root $\sqrt{1 + \nu/\rho}$ is of the order 3 to 5. A conservative estimate is 3. A more accurate value might be found by using the ν-value (9.4.13) and the ρ-value (9.4.16), but such an apparent accuracy is probably not in place.

We thus have

$$x \cong 3y \tag{9.4.26}$$

By using the procedure of Example 9.3.1 we find the approximate formulas

$$\frac{\tau}{f_c} = \min \begin{cases} 0.12\,\nu + 0.89\left(6\dfrac{c}{d} + 2\right)\rho \\ 0.28\,\nu + 0.48\left(6\dfrac{c}{d} + 2\right)\rho \end{cases} \tag{9.4.27}$$

Introducing the ν- and the ρ-values referred to above we find

$$\frac{\tau}{\sqrt{f_c}} = \min \begin{cases} 0.22 + 0.54\left(6\frac{c}{d} + 2\right)\sqrt{\frac{d}{\ell}} & (f_c \text{ in MPa}) \\ 0.50 + 0.29\left(6\frac{c}{d} + 2\right)\sqrt{\frac{d}{\ell}} & (f_c \text{ in MPa}) \end{cases} \tag{9.4.28}$$

When the bars to be anchored or spliced are near free surfaces the cover bending mechanism, of course, has to be modified. In Fig. 9.4.7 the relevant mechanism is shown when the bar is near two free surfaces. Then the compressive zones cannot develop, so only the tensile zones make their contribution to the dissipation. With the notation explained in the figure we find by a procedure completely analogous to the above the following bond strength

$$\frac{\tau}{\sqrt{f_c}} = \min \begin{cases} 0.22 + 0.54\left[\frac{1}{2}\frac{x_1}{d} + \frac{1}{2}\frac{y^2}{x_1 d} + \frac{1}{2}\frac{x_2}{d} + \frac{1}{2}\frac{y^2}{x_2 d} - 1\right]\sqrt{\frac{d}{\ell}} \\ 0.50 + 0.29\left[\frac{1}{2}\frac{x_1}{d} + \frac{1}{2}\frac{y^2}{x_1 d} + \frac{1}{2}\frac{x_2}{d} + \frac{1}{2}\frac{y^2}{x_2 d} - 1\right]\sqrt{\frac{d}{\ell}} \end{cases}$$

(Fig. 9.4.7) (f_c in MPa) (9.4.29)

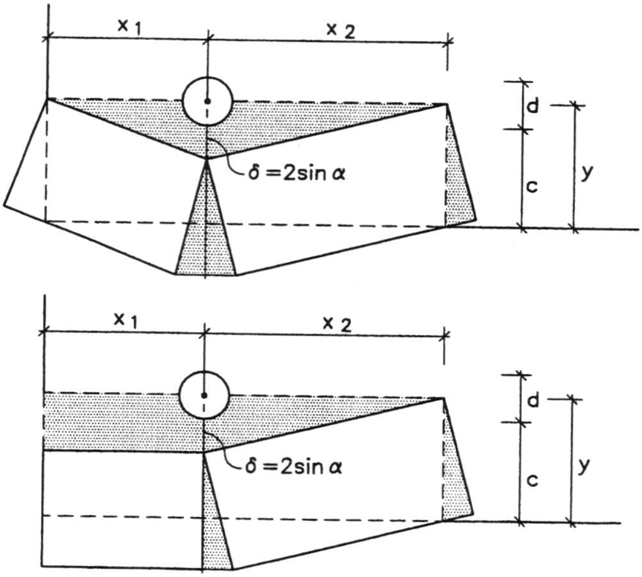

Figure 9.4.8 Cover bending mechanism, bar near a free surface.

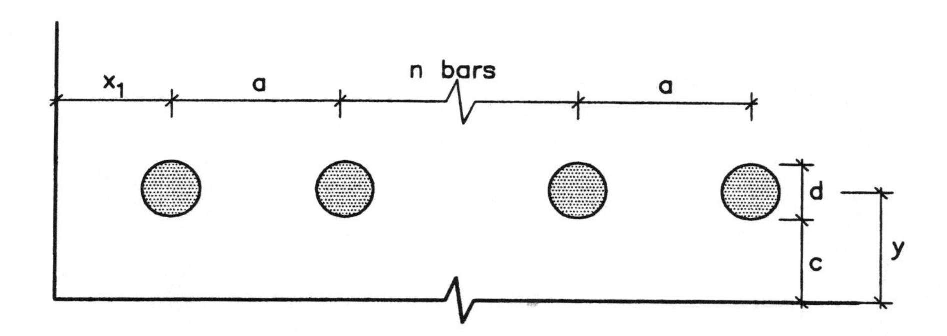

Figure 9.4.9 Cover bending mechanism, a number of bars near one free surface (mechanism not shown).

The formula is only valid if $x_1 \geq y$ and $x_2 \geq y$. If $x_1 < y$ and $x_2 > y$ the first two terms in the square bracket are replaced by x_1/d. In this case the cover part to the left in Fig. 9.4.7 is translating in the vertical direction instead of rotating. Similarly, if $x_1 > y$ and $x_2 < y$, the last two terms are replaced by x_2/d. If $x_1 < y$, and $x_2 < y$ the terms in the square bracket read $x_1/d + x_2/d$. In the last mentioned case, in fact, the mechanism is the face splitting mechanism (cf. Section. 9.4.4).

In Fig. 9.4.8 the bar is near only one surface so one compression zone may develop. In this case we find for the mechanism with rotation of the cover near the free surface

$$\frac{\tau}{\sqrt{f_c}} = \min \begin{cases} 0.22 + 0.54\left[\dfrac{1}{2}\dfrac{x_1}{d} + \dfrac{1}{2}\dfrac{y^2}{x_1 d} + \dfrac{3y}{d} - 1\right]\sqrt{\dfrac{d}{\ell}} \\ 0.50 + 0.29\left[\dfrac{1}{2}\dfrac{x_1}{d} + \dfrac{1}{2}\dfrac{y^2}{x_1 d} + \dfrac{3y}{d} - 1\right]\sqrt{\dfrac{d}{\ell}} \end{cases} \qquad (9.4.30)$$

(Fig. 9.4.8) $\quad$ (f_c in MPa)

This formula is valid when the distance x_1 exceeds the distance y.

If $x_1 < y$, the second mechanism shown in Fig. 9.4.8 is more dangerous. In this mechanism the cover near the free surface suffers no rotation but a translation parallel to the free surface.

We get

$$\frac{\tau}{\sqrt{f_c}} = \min\begin{cases} 0.22 + 0.54\left[\dfrac{x_1}{d} + 3\dfrac{y}{d} - 1\right]\sqrt{\dfrac{d}{\ell}} \\ 0.50 + 0.29\left[\dfrac{x_1}{d} + 3\dfrac{y}{d} - 1\right]\sqrt{\dfrac{d}{\ell}} \end{cases} \tag{9.4.31}$$

(Fig. 9.4.8) (f_c in MPa)

It is interesting to notice the similarity between the formulas in (9.4.31) and the formula (9.4.17) valid for the corner failure case.

There are, of course, numerous variants of the cover bending mechanism. They may be investigated along the previous lines in the individual cases. We deal with only one more case, the one shown in Fig. 9.4.9. Here n bars are near one free surface. The distance between the bars is a (center to center).

We find

$$\frac{\tau}{\sqrt{f_c}} = \min\begin{cases} 0.22 + \dfrac{0.54}{n}\left[\dfrac{1}{2}\dfrac{x_1}{d} + \dfrac{1}{2}\dfrac{y^2}{x_1 d} + \dfrac{(n-1)a}{d} + \dfrac{3y}{d} - n\right]\sqrt{\dfrac{d}{\ell}} \\ 0.50 + \dfrac{0.29}{n}\left[\dfrac{1}{d}\dfrac{x_1}{d} + \dfrac{1}{2}\dfrac{y^2}{x_1 d} + \dfrac{(n-1)a}{d} + \dfrac{3y}{d} - n\right]\sqrt{\dfrac{d}{\ell}} \end{cases}$$

(Fig. 9.4.9) (f_c in MPa) (9.4.32)

As in the previous case, when x_1 is less than y, the cover near the free surface will translate instead of rotate, and we get

$$\frac{\tau}{\sqrt{f_c}} = \min\begin{cases} 0.22 + \dfrac{0.54}{n}\left[\dfrac{x_1}{d} + \dfrac{(n-1)a}{d} + \dfrac{3y}{d} - n\right]\sqrt{\dfrac{d}{\ell}} \\ 0.50 + \dfrac{0.29}{n}\left[\dfrac{x_1}{d} + \dfrac{(n-1)a}{d} + \dfrac{3y}{d} - n\right]\sqrt{\dfrac{d}{\ell}} \end{cases} \tag{9.4.33}$$

(Fig. 9.4.9) (f_c in MPa)

The structure of the formulas for the cover bending mechanisms is often evident, so they may be written down without any calculation. If the cover near a free surface translates, there is an x_1/d-term and/or an x_2/d-term. If it rotates ($x_1 > y$ or $x_2 > y$), there is a term $\frac{1}{2}x_1/d + \frac{1}{2}y^2/x_1 d$ and/or a term $\frac{1}{2}x_2/d + \frac{1}{2}y^2/x_2 d$, the factor $\frac{1}{2}$ being due to the rotation.

When there are a number of translating cover parts, like in the case of Fig. 9.4.9, we have a term $(n - 1)a/d$, i.e., the number of translating parts times their length over d. Thus a formula valid when the distances a between the bars or the splices are varying may immediately be written down. When there is rotation with a compression zone each zone contributes $3y/d$. Finally the number of bars n involved is subtracted from the terms in the square bracket of the formulas.

In Box 9.4.1 one more common case has been dealt with (see formula (6)).

When the formulas are applied to splices, the geometrical values to be used may be estimated in each individual case. The y-distances should, of course, always be the smallest cover plus $d/2$. The x_1- and x_2-distances may be taken as the cover plus $d/2$ or d.

There are not very many tests reported in the literature which can be used to verify the cover bending formulas. In [96.8] 19 tests which presumably exhibited this kind of failure were picked up. Furthermore, 4 new splice tests of the type shown in Fig. 9.4.10 were completed.

All tests were splice tests, 5 with one splice and 14 with two splices in the bottom of the test beams.

The tests show good agreement with theory. The 14 tests with two splices gave a mean value theory/test of 0.9 and a coefficient of variation 14%. The 5 tests with one splice gave a mean value 1.02. The 4 tests of the type shown in Fig. 9.4.10 gave a mean value 1.36 because two tests showed a very low and unrealistic bond strength.

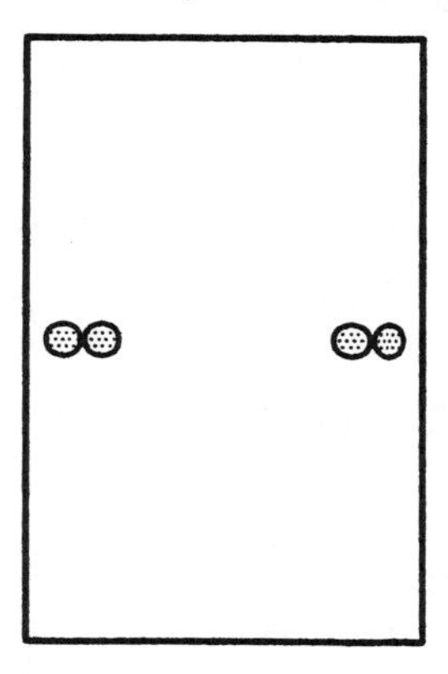

Figure 9.4.10 Splice tests reported in [96.8].

The 14 tests which were done by Tepfers [73.3] with two splices in the bottom of the test beams are remarkable in one respect, since some of the tests were done with zero cover. The cover bending formulas reproduced even these tests rather well. The test bond strength was as high as $\sim 0.5\sqrt{f_c}$ (f_c in MPa). The designer should, of course, not be encouraged to use such values of the cover.

The parameter intervals covered by the tests are

$$6.3 \le \frac{\ell}{d} \le 39$$

$$0 \le \frac{c}{d} \le 1.56$$

$$10 \le d \le 35.8 \text{ mm}$$

$$16.9 \le f_c \le 42.5 \text{ MPa}$$

Although the number of tests is small, we may conclude that the cover bending formulas seem to represent reality reasonably well.

9.4.4 Face Splitting Failure

Now we turn to the third of the three important failure mechanisms. As shown in Fig. 9.4.11, in this case the whole cover spalls off. The size of the cover is unimportant, only the width b enters into the dissipation formula. If there are n bars or splices along the yield line we get the dissipation

$$S = (b - nd)\ell f_t \cdot 2\sin\alpha \tag{9.4.34}$$

The work equation (9.2.10) thus renders the bond strength per bar

$$\frac{\tau}{f_c} = \frac{\tau_o}{f_c} + \frac{2}{\pi}\left(\frac{b}{nd} - 1\right)\frac{f_t}{f_c}\tan\alpha \tag{9.4.35}$$

Using the procedure of Example 9.3.1, the ν-value (9.4.13) and the ρ-value (9.4.16), we arrive at the formula

$$\frac{\tau}{\sqrt{f_c}} = \min\begin{cases} 0.22 + 0.54\sqrt{\dfrac{d}{\ell}\left(\dfrac{b}{nd} - 1\right)} \\ 0.50 + 0.29\sqrt{\dfrac{d}{\ell}\left(\dfrac{b}{nd} - 1\right)} \end{cases} \quad (f_c \text{ in MPa}) \tag{9.4.36}$$

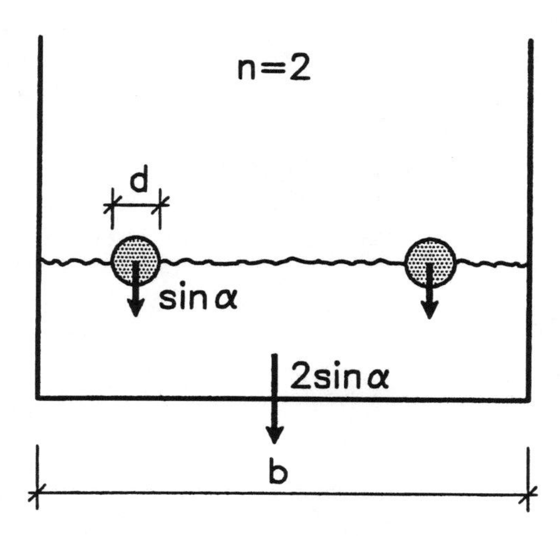

Figure 9.4.11 Face splitting failure.

In [96.8] 41 tests are collected which presumably exhibited a face splitting failure. They were all splice tests. Since we do not distinguish between anchor and splice strength, n in formula (9.4.36) is the number of splices, not the number of bars.

The tests show a remarkably good agreement with theory. The mean value theory/test is 0.98 and the coefficient of variation 13.7%.

The parameter intervals covered by the tests are

$$5 \leq \frac{\ell}{d} \leq 6.0$$

$$92 \leq b \leq 914 \text{ mm}$$

$$1 \leq n \leq 6$$

$$8 \leq d \leq 35.8 \text{ mm}$$

$$19 \leq f_c \leq 52 \text{ MPa}$$

We may conclude that the face splitting formula (9.4.36) seems to be reliable.

9.4.5 Concluding Remarks

The reader may have noticed that for the cover bending failure and the face splitting failure the formulas given are purely theoretical. We have determined the ν-value and the ρ- or ν_t-value for the corner failure and have applied them without modification to the other two failure modes. The good results are quite remarkable since we are dealing with plastic theory applied to problems where the tensile strength of concrete is a governing factor. The reason for such an unexpected success is probably that the crack widths in an anchor failure are very small, which means that there is a good possibility of retaining an essential part of the tensile strength while the failure mechanism is developing.

The most important formulas are reviewed in Box 9.4.1.

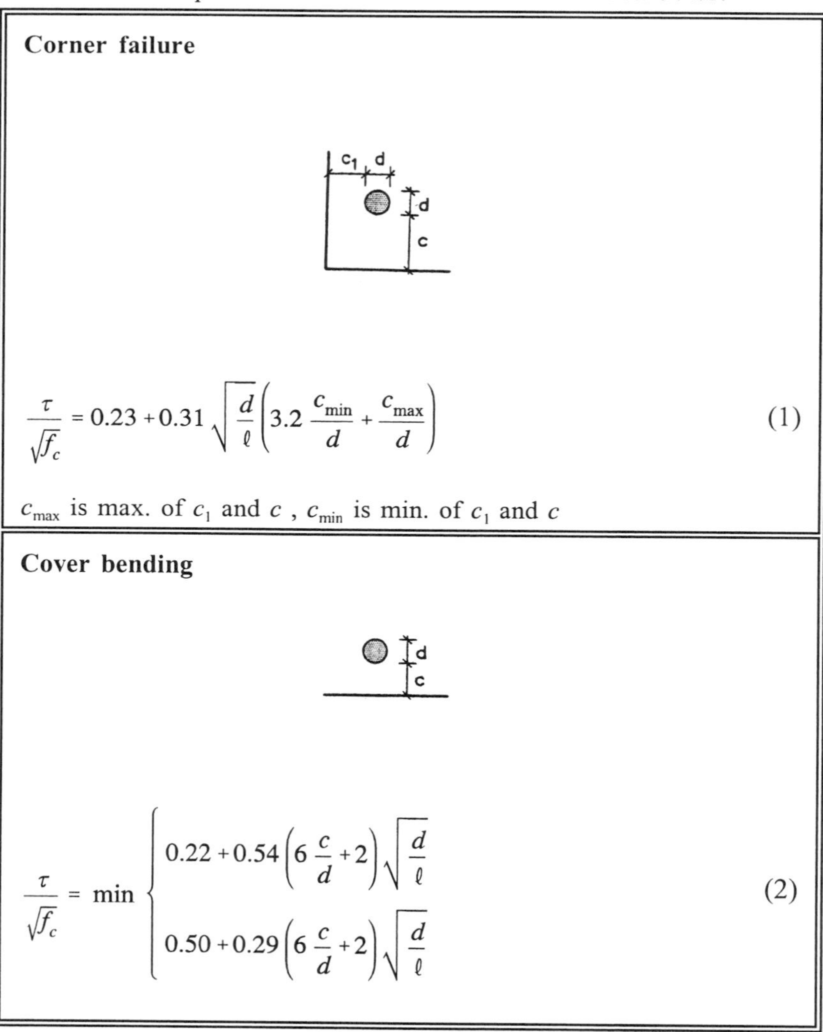

Box 9.4.1 Review of bond strength formulas.

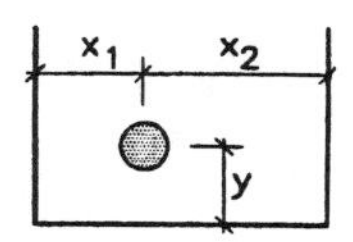

$x_1 \geq y \quad x_2 \geq y$

$$\frac{\tau}{\sqrt{f_c}} = \min \begin{cases} 0.22 + 0.54\left[\frac{1}{2}\frac{x_1}{d} + \frac{1}{2}\frac{y^2}{x_1 d} + \frac{1}{2}\frac{x_2}{d} + \frac{1}{2}\frac{y^2}{x_2 d} - 1\right]\sqrt{\frac{d}{\ell}} \\ 0.50 + 0.29\left[\frac{1}{2}\frac{x_1}{d} + \frac{1}{2}\frac{y^2}{x_1 d} + \frac{1}{2}\frac{x_2}{d} + \frac{1}{2}\frac{y^2}{x_2 d} - 1\right]\sqrt{\frac{d}{\ell}} \end{cases} \tag{3}$$

If $x_1 < y$, $x_2 > y$, the first two terms in "[ ]" are replaced by x_1/d.
If $x_1 > y$, $x_2 < y$, the two terms no. 3 and 4 in "[ ]" are replaced by x_2/d.
If $x_1 < y$, $x_2 < y$, face splitting governs.

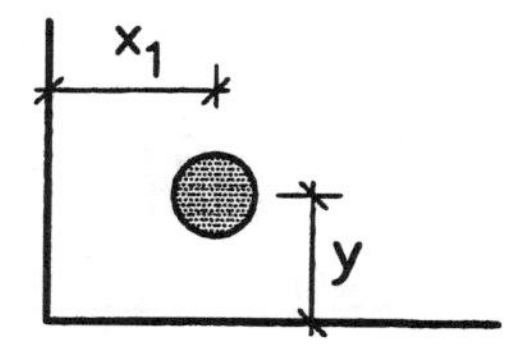

$x_1 \geq y$

$$\frac{\tau}{\sqrt{f_c}} = \begin{cases} 0.22 + 0.54\left[\frac{1}{2}\frac{x_1}{d} + \frac{1}{2}\frac{y^2}{x_1 d} + \frac{3y}{d} - 1\right]\sqrt{\frac{d}{\ell}} \\ 0.50 + 0.29\left[\frac{1}{2}\frac{x_1}{d} + \frac{1}{2}\frac{y^2}{x_1 d} + \frac{3y}{d} - 1\right]\sqrt{\frac{d}{\ell}} \end{cases} \tag{4}$$

$x_1 < y$ The first two terms in "[ ]" are replaced by x_1/d

Box 9.4.1 (continued).

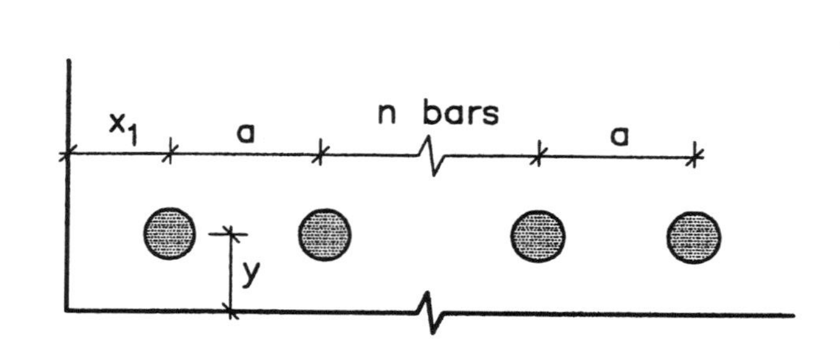

$x_1 \geq y$

$$\frac{\tau}{\sqrt{f_c}} = \min \left\{ \begin{array}{l} 0.22 + \dfrac{0.54}{n}\left[\dfrac{1}{2}\dfrac{x_1}{d} + \dfrac{1}{2}\dfrac{y^2}{x_1 d} + \dfrac{(n-1)a}{d} + \dfrac{3y}{d} - n\right]\sqrt{\dfrac{d}{\ell}} \\ 0.50 + \dfrac{0.29}{n}\left[\dfrac{1}{2}\dfrac{x_1}{d} + \dfrac{1}{2}\dfrac{y^2}{x_1 d} + \dfrac{(n-1)a}{d} + \dfrac{3y}{d} - n\right]\sqrt{\dfrac{d}{\ell}} \end{array} \right. \qquad (5)$$

$x_1 < y$ The first two terms in "[ ]" are replaced by x_1/d.

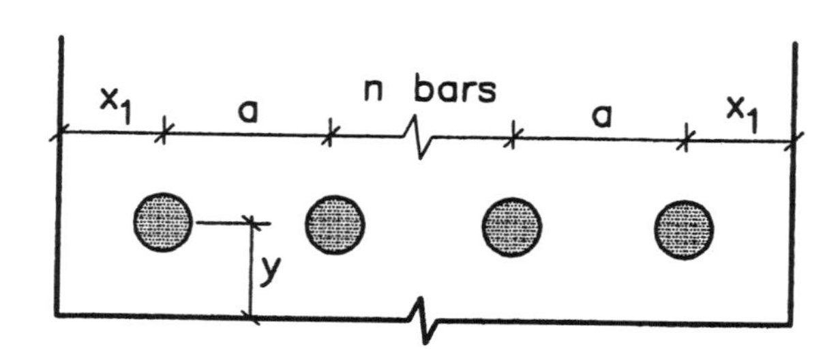

$x_1 \geq y$

$$\frac{\tau}{\sqrt{f_c}} = \left\{ \begin{array}{l} 0.22 + \dfrac{0.54}{n}\left[2\left(\dfrac{1}{2}\dfrac{x_1}{d} + \dfrac{1}{2}\dfrac{y^2}{x_1 d}\right) + \dfrac{(n-1)a}{d} - n\right]\sqrt{\dfrac{d}{\ell}} \\ 0.50 + \dfrac{0.29}{n}\left[2\left(\dfrac{1}{2}\dfrac{x_1}{d} + \dfrac{1}{2}\dfrac{y^2}{x_1 d}\right) + \dfrac{(n-1)a}{d} - n\right]\sqrt{\dfrac{d}{\ell}} \end{array} \right. \qquad (6)$$

$x_1 < y$ The first term in "[ ]" is replaced by $2\,x_1/d$.

Box 9.4.1 (continued).

Face splitting

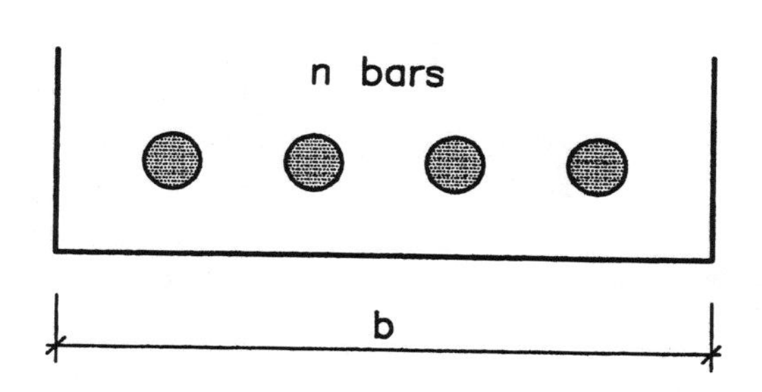

$$\frac{\tau}{\sqrt{f_c}} = \min\begin{cases} 0.22 + 0.54\sqrt{\dfrac{d}{\ell}\left(\dfrac{b}{nd} - 1\right)} \\ 0.50 + 0.29\sqrt{\dfrac{d}{\ell}\left(\dfrac{b}{nd} - 1\right)} \end{cases} \quad (7)$$

Notation

τ	bond strength in MPa per bar
f_c	concrete compressive strength in MPa
d	bar diameter
ℓ	anchor or splice length
n	number of bars or splices

For splices the distances x_1, x_2, a, etc. may be measured to and from the centers of the spliced bars.

For high strength concrete, $f_c > 75$ MPa, $\tau/\sqrt{f_c}$ is multiplied by $2 - f_c/75$.

Box 9.4.1 (continued).

9.5 ASSESSMENT OF ANCHOR AND SPLICE STRENGTH

We now have in our hands some important formulas for the assessment of the anchor or splice strength of an assembly of bars.

The force distribution in a number of anchor bars or lap splices is, in a real structure, determined by the whole set of statical, geometrical and constitutive equations. In a test one might apply a prescribed force to each individual bar and measure the force as failure progresses in each bar and the force carried by the whole assembly for different force distributions. In a test one might also prescribe the bond slip distribution, for instance, by fixing each bar to a rigid plate and moving the plate in different ways to give rise to different bond slip distributions. Probably the last way of performing a test would imitating what is going on in most real structures in the best way.

Normally one bar or a group of bars in an assembly is weaker than the remaining bars. For instance, bars near the faces are normally weaker than bars farther from the faces. The weak bars will reach their failure load first and the load carried by these bars will start to fall down due to the tensile softening in the surrounding concrete. The assembly may still carry a higher total force if the stronger bars are able to compensate for the softening of the weaker bars. The failure will normally be a kind of progressive failure. In such a failure the highly stressed elements start to fail, other elements take over the forces and begin to fail and so on.

Thus, we realize that the load-carrying capacity of an assembly of bars cannot be calculated by assuming a completely plastic behavior. We must use a conservative approach, which may involve the following steps.

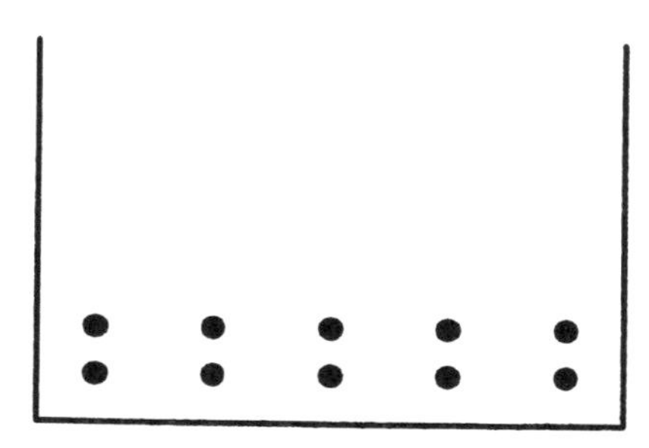

Figure 9.5.1 Assembly of bars in two layers.

First a number of failure mechanisms are selected and the most dangerous one is identified. If the force distribution is approximately uniform the load-carrying capacity of the assembly may be taken as the lowest load-carrying capacity per bar times the total number of bars.

If the load distribution over the bars is not uniform, we shall assume that the load distribution is determined by a scaling factor μ. Again a number of failure mechanisms are selected. If a failure mechanism involves n bars the failure load per bar for this mechanism times n is set equal to the sum of the forces, expressed by μ, in these n bars. In this way different values of the scaling factor μ are determined and the lowest one is assumed to govern the load-carrying capacity of the whole assembly.

In most cases it is only necessary to consider mechanisms involving bars near the faces.

For instance, in the case shown in Fig. 9.5.1, one has only to deal with corner failure, cover bending along the horizontal face, cover bending along the vertical faces and face splitting through the bottom layer. If all bars are equal in size and have the same anchor length, face splitting through the second layer will give the same load-carrying capacity per bar as face splitting through the bottom layer. Of course, if there are more bars in the second layer face splitting through these bars must be dealt with.

One or another kind of cover bending mechanism involving bars from both layers is normally not dangerous if the bar distances, required by practical experience, are used. However, more research is needed to deal with bars in a second or third layer in a fully satisfactory manner.

The procedure suggested is illustrated below in some simple cases.

Example 9.5.1

a)

We begin on a small scale. Fig. 9.5.2 shows a case with only one bar, $d = 14$ mm. The covers and other relevant distances are given in the figure. Three mechanisms are selected as indicated in the figure. The cover bending mechanism is without compression zones since these would be found at a distance $3 \cdot 27 = 81$ mm $> x_1 = x_2$ from the bar. The compressive strength of the concrete equals $f_c = 24$ MPa. The anchor length is $\ell = 120$ mm.

This anchor length is much smaller than the development length normally required (cf. Section 9.3.5). This short anchor length and the particular section have been chosen because they were used by Rathkjen [72.7] in a test series which we are going to refer to extensively. It will be instructive for the reader to go through the calculations using an anchor length of, for example, $30d$ = 420 mm.

Corner failure

We refer to the formulas in Box 9.4.1. Formula (1) applies.

$$\frac{\tau}{\sqrt{f_c}} = 0.23 + 0.31\sqrt{\frac{14}{120}}\left(3.2 \cdot \frac{20}{14} + \frac{63}{14}\right) = 1.19$$

$$\tau = 1.19\sqrt{24} = 5.83 \text{ MPa}$$

$$\frac{\tau}{f_c} = \frac{5.83}{24} = 0.24$$

Cover bending

Since $x_1 = x_2 > y$, formula (3) applies.

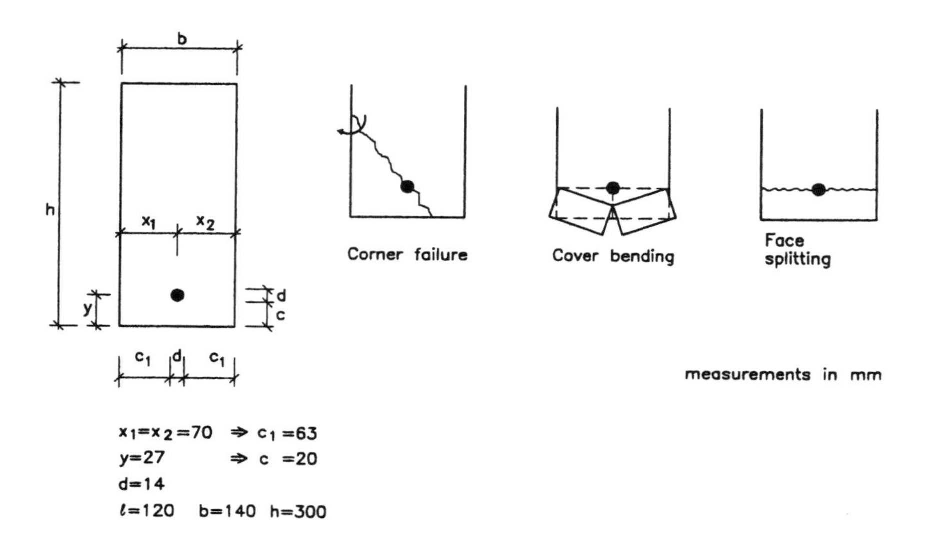

Figure 9.5.2 Section with one bar.

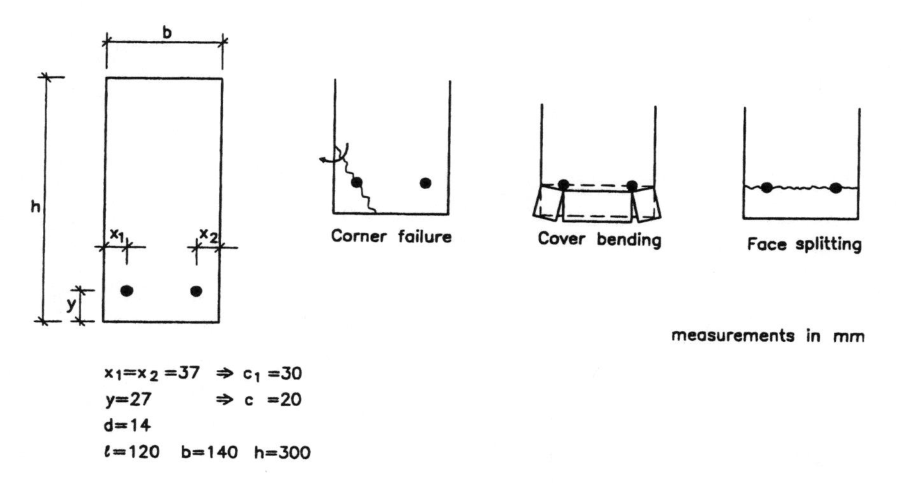

Figure 9.5.3 Section with two bars.

$$\frac{\tau}{\sqrt{f_c}} = \min \begin{cases} 0.22 + 0.54\left[\frac{1}{2}\frac{70}{14} + \frac{1}{2}\frac{27^2}{70 \cdot 14} + \frac{1}{2}\frac{70}{14} + \frac{1}{2}\frac{27^2}{70 \cdot 14} - 1\right]\sqrt{\frac{14}{120}} = 1.09 \\ 0.50 + 0.29\left[\frac{1}{2}\frac{70}{14} + \frac{1}{2}\frac{27^2}{70 \cdot 14} + \frac{1}{2}\frac{70}{14} + \frac{1}{2}\frac{27^2}{70 \cdot 14} - 1\right]\sqrt{\frac{14}{120}} = 0.97 \end{cases}$$

$\Rightarrow$

$$\tau = 0.97\sqrt{24} = 4.75 \text{ MPa}$$

$$\frac{\tau}{f_c} = \frac{4.75}{24} = 0.20$$

Face splitting

Formula (7) applies with $n = 1$

$$\frac{\tau}{\sqrt{f_c}} = \min \begin{cases} 0.22 + 0.54\sqrt{\frac{14}{120}}\left(\frac{140}{14} - 1\right) = 1.88 \\ 0.50 + 0.29\sqrt{\frac{14}{120}}\left(\frac{140}{14} - 1\right) = 1.39 \end{cases}$$

$\Rightarrow$

$$\tau = 1.39\sqrt{24} = 6.81 \text{ MPa}$$

$$\frac{\tau}{f_c} = \frac{6.81}{24} = 0.28$$

Thus cover bending is slightly more dangerous than corner failure. The bond strength is

$$\tau = 4.75 \text{ MPa}$$

and the total force, which may be carried, is

$$T = \pi \cdot 14 \cdot 120 \cdot 4.75 = 25.1 \cdot 10^3 \text{ N} = 25.1 \text{ kN}$$

b)

Same problem as in a) except now we have two bars at the corners (Fig. 9.5.3). The mechanisms considered are shown in the figure.

Corner failure

Formula (1) applies.

$$\frac{\tau}{\sqrt{f_c}} = 0.23 + 0.31 \sqrt{\frac{14}{120}} \left(3.2 \cdot \frac{20}{14} + \frac{30}{14} \right) = 0.94$$

$$\tau = 0.94\sqrt{24} = 4.61 \text{ MPa}$$

$$\frac{\tau}{f_c} = \frac{4.61}{24} = 0.19$$

Cover bending

The mechanism is shown in Fig. 9.5.3. As mentioned previously we encourage the reader to practice writing down the formulas for cover bending directly. Formula (6) applies. It is written for half the beam width.

$$\frac{\tau}{\sqrt{f_c}} = \min \begin{cases} 0.22 + 0.54 \left[\frac{1}{2} \cdot \frac{37}{14} + \frac{1}{2} \frac{27^2}{37 \cdot 14} + \frac{70-37}{14} - 1 \right] \sqrt{\frac{14}{120}} = 0.84 \\ 0.50 + 0.29 \left[\frac{1}{2} \cdot \frac{37}{14} + \frac{1}{2} \frac{27^2}{37 \cdot 14} + \frac{70-37}{14} - 1 \right] \sqrt{\frac{14}{120}} = 0.84 \end{cases}$$

$\Rightarrow$

$$\tau = 0.84\sqrt{24} = 4.12 \text{ MPa}$$

$$\frac{\tau}{f_c} = \frac{4.12}{24} = 0.17$$

Face splitting

Formula (7) applies with $n = 2$.

$$\frac{\tau}{\sqrt{f_c}} = \min \begin{cases} 0.22 + 0.54\sqrt{\dfrac{14}{120}\left(\dfrac{140}{2 \cdot 14} - 1\right)} = 0.96 \\ 0.50 + 0.29\sqrt{\dfrac{14}{120}\left(\dfrac{140}{2 \cdot 14} - 1\right)} = 0.90 \end{cases}$$

$\Rightarrow$

$$\tau = 0.90\sqrt{24} = 4.41 \text{ MPa}$$

$$\frac{\tau}{f_c} = \frac{4.41}{24} = 0.18$$

Thus cover bending is slightly more dangerous than corner failure and face splitting.

The bond strength per bar is

$$\tau = 4.12 \text{ MPa}$$

and the total force, which may be carried, is

$$T = 2 \cdot \pi \cdot 14 \cdot 120 \cdot 4.12 = 43.5 \cdot 10^3 \text{ N} = 43.5 \text{ kN}$$

Notice that the bond strength per bar is lower than in a).

c)

Same problem as in a), except now we have three bars along the bottom face (Fig. 9.5.4). The mechanisms considered are shown in the figure.

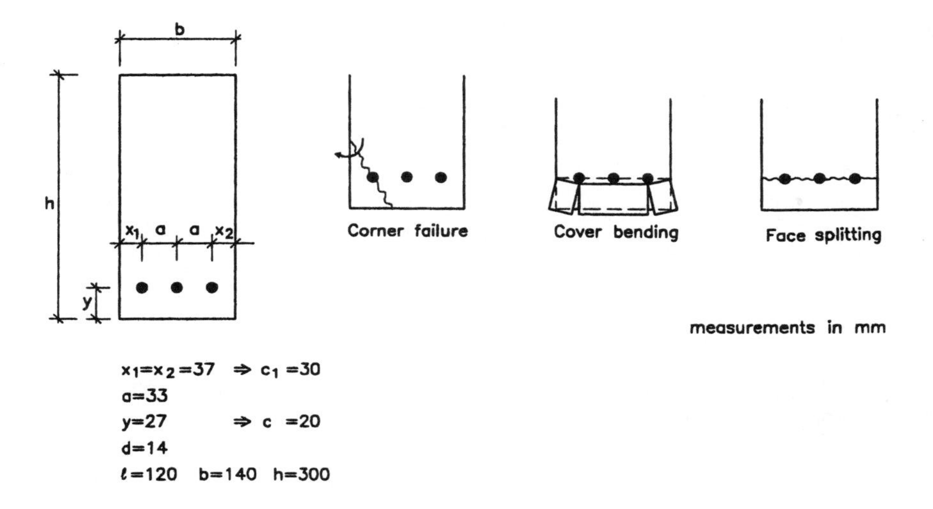

Figure 9.5.4 Section with three bars.

Corner failure

The result is the same as in b), i.e.,

$$\tau = 4.61 \text{ MPa}$$

$$\frac{\tau}{f_c} = \frac{4.61}{24} = 0.19$$

Cover bending

Formula (6) applies with $n = 3$.

$$\tau = \min\begin{cases} 0.22 + \dfrac{0.54}{3}\left[2\left(\dfrac{1}{2}\dfrac{37}{14} + \dfrac{1}{2}\dfrac{27^2}{37\cdot 14}\right) + \dfrac{2\cdot 33}{14} - 3\right]\sqrt{\dfrac{14}{120}} = 0.57 \\ 0.50 + \dfrac{0.29}{3}\left[2\left(\dfrac{1}{2}\dfrac{37}{14} + \dfrac{1}{2}\dfrac{27^2}{37\cdot 14}\right) + \dfrac{2\cdot 33}{14} - 3\right]\sqrt{\dfrac{14}{120}} = 0.69 \end{cases}$$

$\Rightarrow$

$$\tau = 0.57\sqrt{24} = 2.79 \text{ MPa}$$

$$\frac{\tau}{f_c} = \frac{2.79}{24} = 0.12$$

Face splitting

Formula (7) applies with $n = 3$.

$$\frac{\tau}{\sqrt{f_c}} = \min\begin{cases} 0.22 + 0.54\sqrt{\dfrac{14}{120}}\left(\dfrac{140}{3\cdot 14} - 1\right) = 0.65 \\ 0.50 + 0.29\sqrt{\dfrac{14}{120}}\left(\dfrac{140}{3\cdot 14} - 1\right) = 0.73 \end{cases}$$

$\Rightarrow$

$$\tau = 0.65\sqrt{24} = 3.18 \text{ MPa}$$

$$\frac{\tau}{f_c} = \frac{3.18}{24} = 0.13$$

Thus cover bending is again the most dangerous mechanism. The bond strength per bar is

$$\tau = 2.79 \text{ MPa}$$

and the total force which may be carried is

$$T = 3 \cdot \pi \cdot 14 \cdot 120 \cdot 2.79 = 44.2 \cdot 10^3 \text{ N} = 44.2 \text{ kN}$$

Notice the substantial decrease in bond strength per bar compared to the case with two bars in b).

d)

Same problem as in c), except now we have added two bars in a second layer (Fig. 9.5.5). The mechanisms are the ones considered in c) plus the cover bending mechanism shown in Fig. 9.5.5. Notice that in this case the lowest part of the cover will translate instead of rotate as in the previous cases. In this cover bending mechanism a compression zone might possibly develop.

Formula (5) applies with $n = 2$.

$$\frac{\tau}{\sqrt{f_c}} = \min \begin{cases} 0.22 + \dfrac{0.54}{2}\left[\dfrac{27}{14} + \dfrac{28}{14} + 3 \cdot \dfrac{37}{14} - 2\right]\sqrt{\dfrac{14}{120}} = 1.13 \\ 0.50 + \dfrac{0.29}{2}\left[\dfrac{27}{14} + \dfrac{28}{14} + 3 \cdot \dfrac{37}{14} - 2\right]\sqrt{\dfrac{14}{120}} = 0.99 \end{cases}$$

$\Rightarrow$

$$\tau = 0.99\sqrt{24} = 4.85 \text{ MPa}$$

$$\frac{\tau}{f_c} = \frac{4.85}{24} = 0.20$$

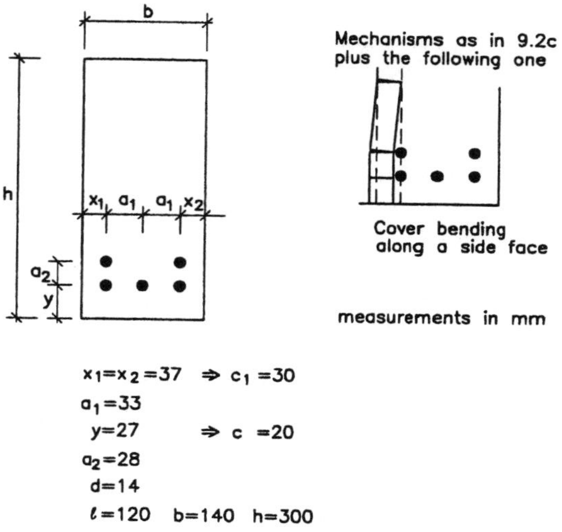

Figure 9.5.5 Section with five bars.

We must also consider a cover bending mechanism where no compression zone develops but where the upper rotating part of length 300-27-28 = 245 mm extends to the top of the beam. Formula of type (6) with $n = 2$ applies.

$$\frac{\tau}{\sqrt{f_c}} = \min \begin{cases} 0.22 + \dfrac{0.54}{2}\left[\dfrac{27}{14} + \dfrac{28}{14} + \dfrac{1}{2}\dfrac{245}{14} + \dfrac{1}{2}\dfrac{37^2}{245 \cdot 14} - 2\right]\sqrt{\dfrac{14}{120}} = 1.22 \\ 0.50 + \dfrac{0.29}{2}\left[\dfrac{27}{14} + \dfrac{28}{14} + \dfrac{1}{2}\dfrac{245}{14} + \dfrac{1}{2}\dfrac{37^2}{245 \cdot 14} - 2\right]\sqrt{\dfrac{14}{120}} = 1.04 \end{cases}$$

$\Rightarrow$

$$\tau = 1.04\sqrt{24} = 5.09 \text{ MPa}$$

$$\frac{\tau}{f_c} = \frac{5.09}{24} = 0.21$$

Combining with the mechanisms from c) we see that the most dangerous one is the cover bending mechanism in c).

Thus the bond strength per bar is

$$\tau = 2.79 \text{ MPa}$$

and the total force, which may be carried, is

$$T = 5 \cdot \pi \cdot 14 \cdot 120 \cdot 2.79 = 73.6 \cdot 10^3 \text{ N} = 73.6 \text{ kN}$$

Example 9.5.2

Consider a slab reinforced by $d = 12$ mm bars at a distance 125 mm. The cover is $c = 20$ mm and the anchor length is $\ell = 432$ mm. The concrete strength is $f_c = 35$ MPa.

Only two mechanisms are relevant. The cover bending mechanism involving only one bar and the face splitting mechanism involving all bars (see Fig. 9.5.6).

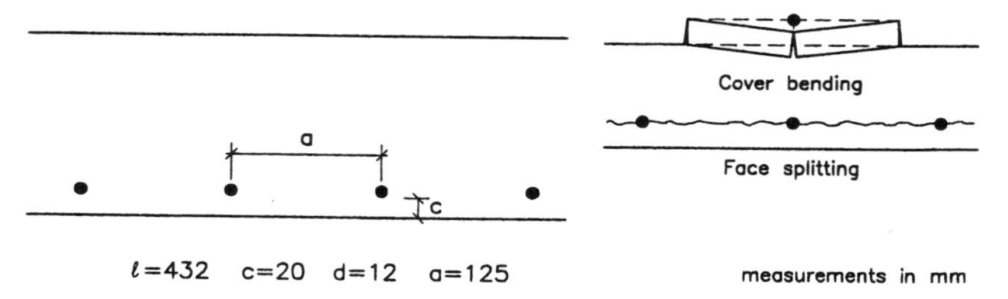

Figure 9.5.6 Slab reinforced along one face.

<u>Cover bending</u>

Formula (2) applies.

$$\frac{\tau}{\sqrt{f_c}} = \min \begin{cases} 0.22 + 0.54\left(6 \cdot \dfrac{20}{12} + 2\right)\sqrt{\dfrac{12}{432}} = 1.30 \\ 0.50 + 0.29\left(6 \cdot \dfrac{20}{12} + 2\right)\sqrt{\dfrac{12}{432}} = 1.08 \end{cases}$$

$\Rightarrow$

$$\tau = 1.08\sqrt{35} = 6.39 \text{ MPa}$$

$$\frac{\tau}{f_c} = \frac{6.39}{35} = 0.18$$

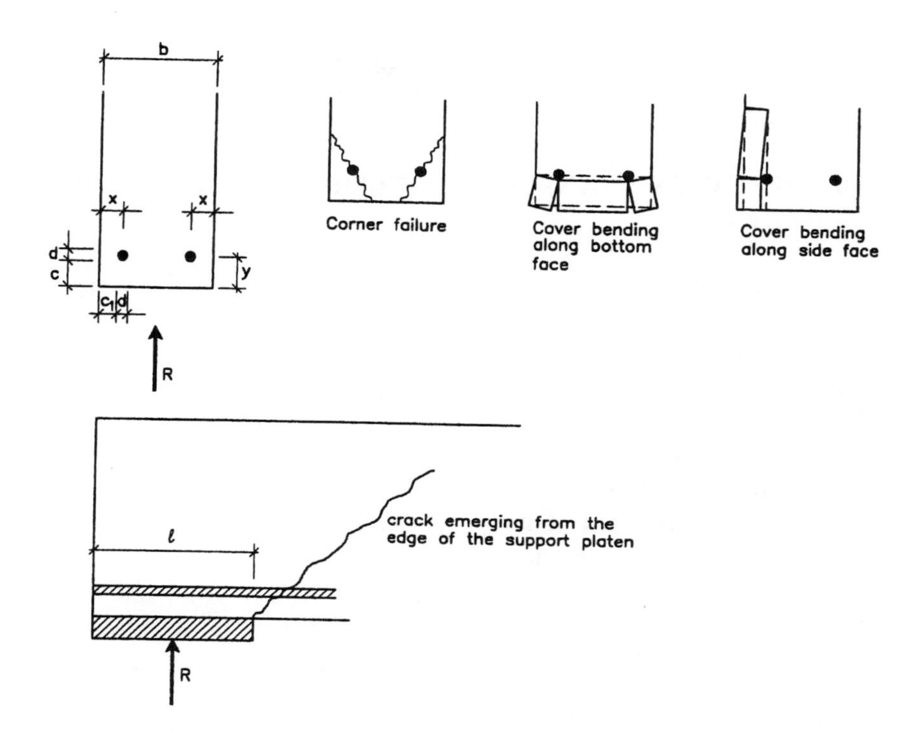

Figure 9.6.1 Anchorage failure at a support.

Face splitting

A length 125 mm is available for each bar. Formula (7) renders:

$$\frac{\tau}{\sqrt{f_c}} = \min \begin{cases} 0.22 + 0.54\sqrt{\dfrac{12}{432}\left(\dfrac{125}{12} - 1\right)} = 1.07 \\ 0.50 + 0.29\sqrt{\dfrac{12}{432}\left(\dfrac{125}{12} - 1\right)} = 0.96 \end{cases}$$

$\Rightarrow$

$$\tau = 0.96\sqrt{35} = 5.68 \text{ MPa}$$

$$\frac{\tau}{f_c} = \frac{5.68}{35} = 0.16$$

Face splitting is governing the load-carrying capacity. The force, which may be carried by one bar, is

$$T = \pi \cdot 12 \cdot 432 \cdot 5.68 = 92.5 \cdot 10^3 \text{ N} = 92.5 \text{ kN}$$

Since the sectional area of the bars is 113 mm^2 this load corresponds to a stress σ_s = 819 MPa, i.e., for most yield strengths used in practice the anchor length 432 mm is far too large.

9.6 EFFECT OF TRANSVERSE PRESSURE AND SUPPORT REACTION.

If concrete mechanics in some distant future is reduced to a branch of mathematics, a section like this will be very short. Only one statement would be needed, i.e., if a transverse pressure or other external loads is present along the anchor or splice length, the work done by these forces must be included in the work equation.

Things are not yet that easy.

The most important case encountered in practice is the pressure supplied by the reaction at a support along which the reinforcement is anchored. We will confine ourselves to this case.

In the first edition of this book [84.11] a number of solutions were presented. In these solutions the reaction was supposed to deliver a uniform pressure along the anchorage and the work done by this uniform pressure was included in the work equation. However, as pointed out in [84.7], such a procedure is only correct if the pressure, i.e., both its distribution and size, does not change appreciably when the anchor failure takes place. In other words, the deformations at final failure must be so small that they do not substantially change the support pressure. This assumption is not very likely to be fulfilled. Furthermore, the solutions obtained in this way often show too small an enhancement of the bond strength when the support pressure is increased, at least for relatively high values of the support pressure. This assumption has therefore been discarded in recent calculations, where the alternative, namely to assume the support platen to be rigid, has been applied. In what follows we only deal with this case. As there is no general theory yet, the problem is dealt with by calculating some important examples seen from the designers′ point of view. Once again, therefore, a rectangular section with two corner bars is treated (see Fig. 9.6.1).

A cover bending mechanism along the bottom face is only likely to occur at very small support pressures since the work done by the reaction in this mechanism is relatively large (see below).

In the cover bending mechanism along the side face, shown in the figure, the reaction does no work at all, so one would imagine this to be particularly relevant for large values of the support pressure. However, such pressures are seldom reached before another limit, treated below, governs. One more reason is that the area, which has to be cracked in this mechanism, is greatly enhanced since this area is limited, not by a vertical section through the edge of the support platen, but by a skew crack surface as the one shown in the figure.

Thus in most cases we are left with one important mechanism, the corner failure mechanism (cf. Section 9.4.2).

In the corner failure mechanism the rotation point may lie either at a vertical face or at the bottom face depending on the relative magnitudes of the covers c_1 and c. If $c_1 > c$, the rotation point is on a vertical face; if $c_1 < c$, the rotation point is on the bottom face in the case with no support pressure. However, when a support pressure is applied the rotation point at a rather small pressure jumps from the bottom face to a vertical face so in most cases we may assume the rotation point to be on a vertical face.

Instead of establishing the work equation for a whole beam with anchorage failure at the support we consider the tensile force in the reinforcement which has to be anchored and the reaction R to be the active forces. The result of the two procedures will be the same if a mechanism for the whole beam, which only gives rise to dissipation in the anchorage zone, is geometrically possible. We shall assume this condition to be fulfilled, which is a conservative assumption. We have been tacitly assuming the same thing in all the previous cases.

In Fig. 9.6.2 the corner failure mechanism is once again depicted. The dissipation in the yield line making the angle β with the vertical face has been given in Section 9.4.2. We assume that corner failure takes place in both corners at the same time. The dissipation in one yield line is given by the formulas:

Rotation point on vertical face:

$$S = \ell f_t \left[\frac{\sin\beta}{x} \left(\frac{x}{\sin\beta} + \frac{y}{\cos\beta} \right)^2 - 2d \right] \sin\alpha \tag{9.6.1}$$

Rotation point on bottom face:

$$S = \ell f_t \left[\frac{\cos\beta}{y} \left(\frac{x}{\sin\beta} + \frac{y}{\cos\beta} \right)^2 - 2d \right] \sin\alpha \tag{9.6.2}$$

Here $x = c_1 + d/2$ and $y = c + d/2$.

The work done by the reaction R is negative. When it is transferred to the dissipation side of the work equation we get the following positive contribution U:

Rotation point at vertical face:

$$U = \omega\,(x + y\tan\beta)\,R = \frac{2\sin\alpha\sin\beta}{x}\,(x + y\tan\beta)\,R \tag{9.6.3}$$

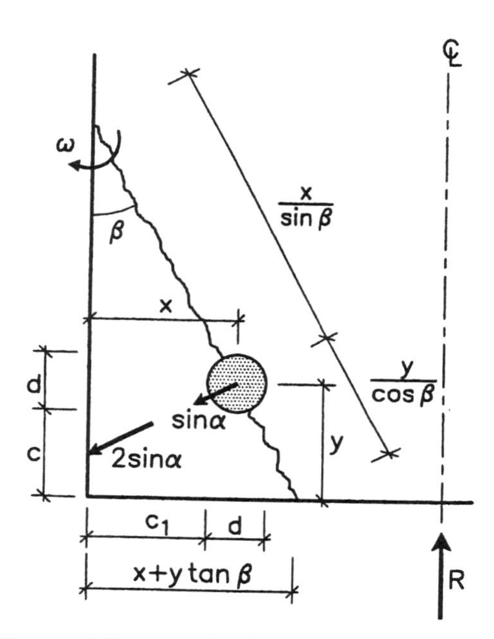

Figure 9.6.2 Corner failure mechanism.

Rotation point at bottom face:

$$U = \frac{2\sin\alpha\cos\beta}{y}(x + y\tan\beta)R \qquad (9.6.4)$$

The reader may notice that we have disregarded any contribution from friction between the support platen and the concrete bottom face.

The work equation (9.2.10) now renders the bond strength of one bar (notice that the work done by the reaction must be shared by two bars):

$$\frac{\tau}{f_c} = \frac{\tau_o}{f_c} + \frac{1}{\pi}\left[B\frac{f_t}{f_c} + \frac{b}{d}C\frac{r}{f_c}\right]\tan\alpha \qquad (9.6.5)$$

To shorten the formulas we have introduced the notation:

Rotation point on vertical face:

$$B = \frac{\sin\beta}{\dfrac{x}{d}}\left(\frac{\dfrac{x}{d}}{\sin\beta} + \frac{\dfrac{y}{d}}{\cos\beta}\right)^2 - 2 \qquad (9.6.6)$$

$$C = \sin\beta\left(1 + \frac{\dfrac{y}{d}}{\dfrac{x}{d}}\tan\beta\right) \qquad (9.6.7)$$

Rotation point on bottom face:

$$B = \frac{\cos\beta}{\dfrac{y}{d}}\left(\frac{\dfrac{x}{d}}{\sin\beta} + \frac{\dfrac{y}{d}}{\cos\beta}\right)^2 - 2 \qquad (9.6.8)$$

$$C = \cos\beta\left(\frac{\dfrac{x}{d}}{\dfrac{y}{d}} + \tan\beta\right) \qquad (9.6.9)$$

$$r = \frac{R}{b\ell} \qquad (9.6.10)$$

The parameter r is the average pressure over the support area.

By replacing f_c with νf_c and f_t with ρf_c we get

$$\frac{\tau}{f_c} = \frac{\tau_o}{f_c} + \frac{1}{\pi}\left[B\rho + \frac{b}{d}C\frac{r}{f_c}\right]\tan\alpha \tag{9.6.11}$$

where τ_o/f_c is determined by the formulas (9.2.24) and (9.2.25).

The procedure of Example 9.3.1 leads to the final formula

$$\frac{\tau}{f_c} = \min\begin{cases} 0.12\,\nu + 0.45\left[B\rho + \dfrac{b}{d}C\dfrac{r}{f_c}\right] \\ 0.28\,\nu + 0.24\left[B\rho + \dfrac{b}{d}C\dfrac{r}{f_c}\right] \end{cases} \tag{9.6.12}$$

The ν-value is determined by (9.4.13) and the ρ-value by (9.4.16).

In fact (9.6.12) provides an alternative formula to (9.4.17) by setting $r/f_c = 0$.

This alternative formula may be written

$$\frac{\tau}{\sqrt{f_c}} = \min\begin{cases} 0.22 + 0.27\,B\sqrt{\dfrac{d}{\ell}} \\ 0.50 + 0.15\,B\sqrt{\dfrac{d}{\ell}} \end{cases} \tag{9.6.13}$$

$$x \ge y \quad : \quad B = \frac{\sin\beta}{\frac{x}{d}}\left(\frac{\frac{x}{d}}{\sin\beta} + \frac{\frac{y}{d}}{\cos\beta}\right)^2 - 2 \tag{9.6.14}$$

$$x < y \quad : \quad B = \frac{\cos\beta}{\frac{y}{d}}\left(\frac{\frac{x}{d}}{\sin\beta} + \frac{\frac{y}{d}}{\cos\beta}\right)^2 - 2 \tag{9.6.15}$$

An approximate value of β may be determined by the following formulas (9.6.21) and (9.6.22). The first one is valid when $x \ge y$, the second one when $x < y$.

When comparing these formulas with the tests available we find almost the same coefficient of variation as we found using the approximate formula (9.4.17). However, when the value of β is important the alternative formulas may be useful.

When $r/f_c \neq 0$, τ/f_c determined by (9.6.12) must be minimized with respect to β. Account must be taken of the fact that the position of the rotation point is not known beforehand. For $x < y$ the rotation point will lie on the bottom face for $r = 0$ and for small values of r but, as mentioned before, when r is increased it jumps rather fast to the vertical face. So in the main range of the r-values the formulas valid for the rotation point on the vertical face govern the bond strength.

If corner failure does not govern the bond strength for $r = 0$, there is a short interval of small r-values where another mechanism governs the bond strength. Normally this interval is of little practical interest but for the sake of completeness we give the formulas for the bond strength increment when cover bending along the bottom face or face splitting is the most dangerous mechanism for $r = 0$. In both cases we have

$$U = 2\sin\alpha \cdot R \tag{9.6.16}$$

Thus the bond strength increment is

$$\frac{\Delta\tau}{f_c} = \frac{U}{2\pi d \ell f_c \cos\alpha} = \frac{1}{\pi}\frac{b}{d}\tan\alpha\,\frac{r}{f_c} \tag{9.6.17}$$

The factor $\tan\alpha$ is either 1.4 or 0.75 depending on which bond strength formula is valid in the actual case. If the first formula written in the bond strength formulas is valid, $\tan\alpha = 1.4$; if the second one is valid, $\tan\alpha = 0.75$.

For later use we mention that if the work done by the reaction must be shared by n bars, (9.6.17) reads

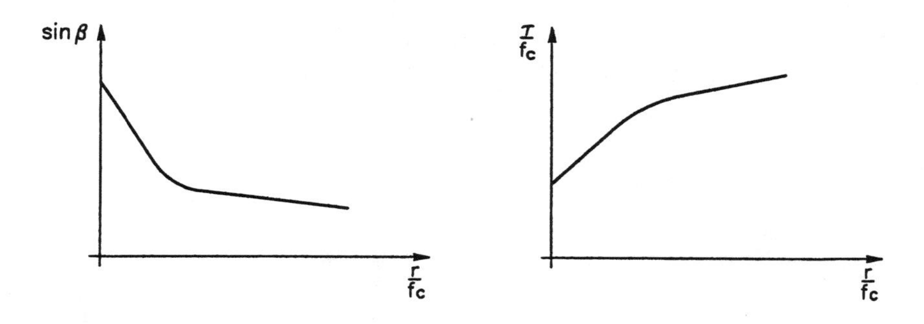

Figure 9.6.3 Schematical performance of $\sin\beta$ and τ/f_c versus r/f_c.

$$\frac{\Delta\tau}{f_c} = \frac{2}{\pi n}\frac{b}{d}\tan\alpha\,\frac{r}{f_c} \tag{9.6.18}$$

Introduction of the effectiveness factors does not change (9.6.17) and (9.6.18).

Instead of minimizing (9.6.12) with respect to β the true solution may be approximated by a number of straight lines found by choosing a suitable number of β-values. Each choice of β gives rise to a straight line τ/f_c versus r/f_c. For small values of r, rotation points on the vertical as well as on the bottom face must be dealt with if $x < y$. If corner failure does not govern the bond strength for $r = 0$, a straight line (9.6.17) leaving the τ/f_c-axis at the point corresponding to the bond strength for $r = 0$ is added. The envelope curve corresponding to the set of straight lines is normally a good approximation of the true curve.

A particularly simple case is obtained when the following three conditions are fulfilled, i.e., $x \geq y$ (rotation point always on vertical face), the corner failure governs the bond strength for $r = 0$ and the second formula in (9.6.12) is valid for all r-values. Then only two straight lines may be used with good approximation.

This is due to the fact that the appearance of the optimized value of, say $\sin\beta$, as a function of r/f_c and the appearance of τ/f_c as a function of r/f_c are as shown schematically in Fig. 9.6.3.

The value of $\sin\beta$ drops down rather fast to an almost constant value and it is seen that the τ/f_c-curve may be very closely approximated by two straight lines. The initial value of β, i.e., the value for $r/f_c = 0$, may be found by minimizing the parameter B. When the derivative with respect to β, using formula (9.6.6) for B, is set equal to zero we find the equation

$$-\frac{x^*\cos\beta}{\sin^2\beta} + \frac{2y^*\sin\beta}{\cos^2\beta} + \frac{y^{*2}}{x^*}\,\frac{\cos^3\beta + 2\sin^2\beta\cos\beta}{\cos^4\beta} = 0 \tag{9.6.19}$$

Here $x^* = x/d$ and $y^* = y/d$. This equation may be written

$$-\frac{x^*}{\tan^2\beta} + 2y^*\tan\beta + \frac{y^{*2}}{x^*}\left(1 + 2\tan^2\beta\right) = 0 \tag{9.6.20}$$

Parameter studies show that the initial value is often around $\sin\beta \sim 0.55$ or $\tan\beta \sim 0.66$. The last term in (9.6.20), which is small, may then be approximated by putting $\tan\beta \sim 0.66$, i.e., $1 + 2\tan^2\beta \sim 1.87 \sim 2.8\tan\beta$. Then (9.6.20) is easily solved with respect to $\tan\beta$ to render

$$\tan\beta \cong \sqrt[3]{\frac{\dfrac{x^*}{y^*}}{2+2.8\dfrac{y^*}{x^*}}} = \sqrt[3]{\frac{\dfrac{x}{y}}{2+2.8\dfrac{y}{x}}} \quad (x \geq y) \tag{9.6.21}$$

This formula which is valid for $x \geq y$ determines the initial value of β, i.e., the value of β for $r = 0$.

When we use formula (9.6.13) we might also be interested in the β-value when the rotation point is on the bottom face. We find

$$\tan\beta = \sqrt[3]{\frac{2+2.8\dfrac{x}{y}}{\dfrac{y}{x}}} \quad (x < y) \tag{9.6.22}$$

which is valid for $x < y$. Notice that the two formulas for β do not give the same β-value for $x = y$. However, the bond strengths will be the same.

For large values of r the rotation point is always, as mentioned, on the vertical face and the angle β is rather small, $\sin\beta$ being around 0.2. In this case the terms in the square bracket of formula (9.6.5) may be approximated in the following way.

$$\frac{\tau}{f_c} \cong \frac{\tau_o}{f_c} + \frac{1}{\pi}\left[\left(\frac{\dfrac{x}{d}}{\sin\beta} + 2\frac{y}{d} - 2\right)\frac{f_t}{f_c} + \frac{b}{d}\sin\beta\,\frac{r}{f_c}\right]\tan\alpha \tag{9.6.23}$$

If this expression is minimized with respect to $\sin\beta$ we find

$$\sin\beta = \sqrt{\frac{xf_t}{br}} \tag{9.6.24}$$

The formula may be used to find an approximate value of β in the region where it may be considered to be almost constant. It is done by choosing an appropriate value of r, say in the middle of the actual range of r.

By replacing f_c with νf_c and f_t with ρf_c the formulas (9.6.21) and (9.6.22) are not changed but (9.6.24) is changed to

$$\sin\beta = \sqrt{\frac{x\,\rho}{b\dfrac{r}{f_c}}} \tag{9.6.25}$$

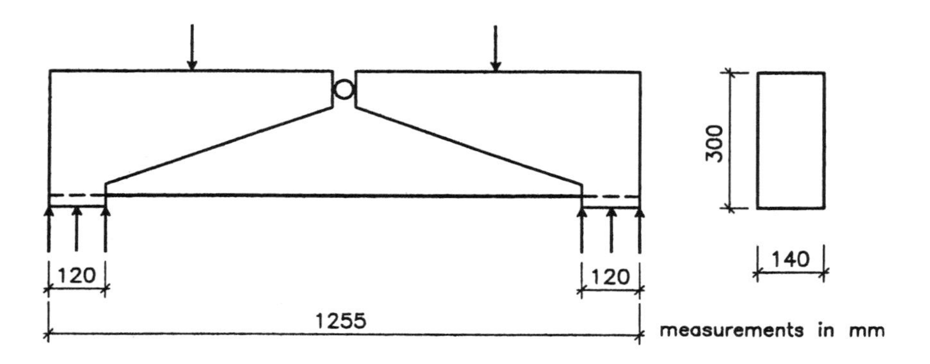

Figure 9.6.4 Test specimen used by Rathkjen [72.7].

When two β-values have been determined by (9.6.21) and (9.6.25), B is calculated for each β-value by (9.6.6) and C is calculated for each β-value by (9.6.7). Then the second formula in (9.6.12), which has been assumed to be valid, provides two straight lines τ/f_c versus r/f_c.

In Fig. 9.6.5 the theory has been compared with some test results by Rathkjen [72.7]. Rathkjen's test beams are shown schematically in Fig. 9.6.4. By varying the positions of the two symmetrically applied loads different support reactions were obtained. In the figure both the optimized value of the bond strength and the two straight lines approximation have been shown. It may be verified that the conditions for using only two straight lines are fulfilled. In formula (9.6.25) r/f_c has been set equal to 0.2.

It appears that there is an excellent agreement between theory and experiment. Other test results have been collected in [89.10], but they have not yet been compared with the present variant of the plastic theory.

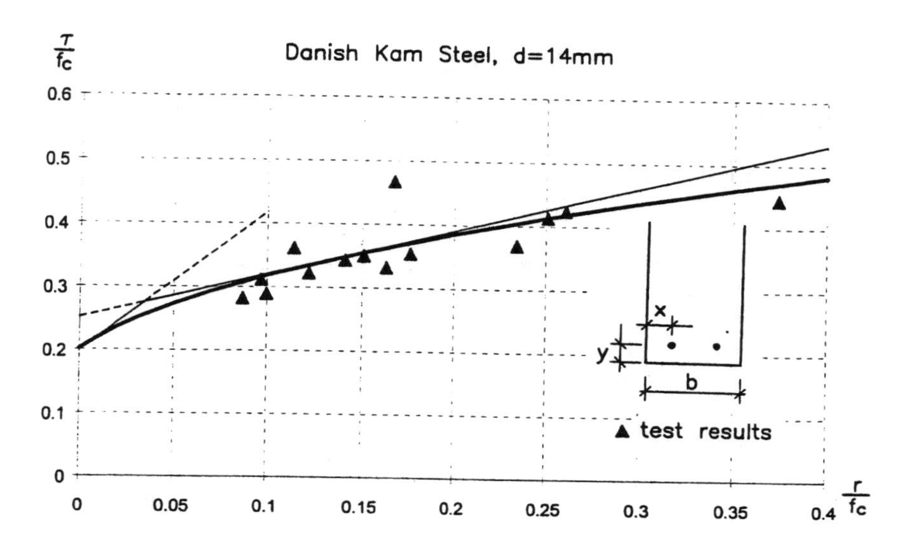

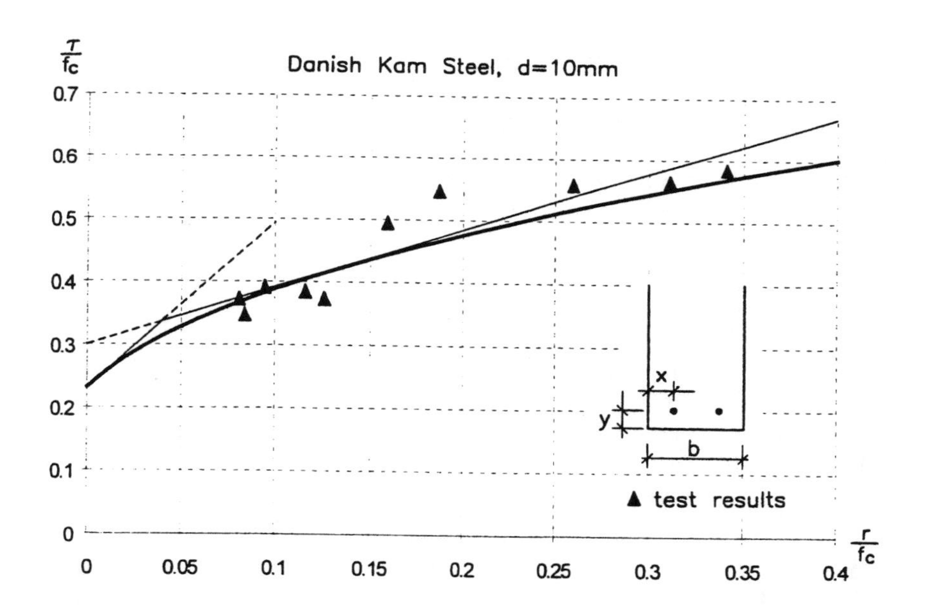

Figure 9.6.5 Bond strength versus transverse pressure.
Data, upper figure: Danish Kam Steel d = 14 mm, 2 corner bars, x = 37 mm, y = 27 mm, b = 140 mm, ℓ = 120 mm, f_c = 21.2 MPa.
Data, lower figure: Danish Kam Steel, d = 10 mm, 2 corner bars, x = 37 mm, y = 27 mm, b = 140 mm, ℓ = 120 mm, f_c = 20.1 MPa.
In both figures test results from [72.7] are compared with optimized values and the two straight lines approximation.

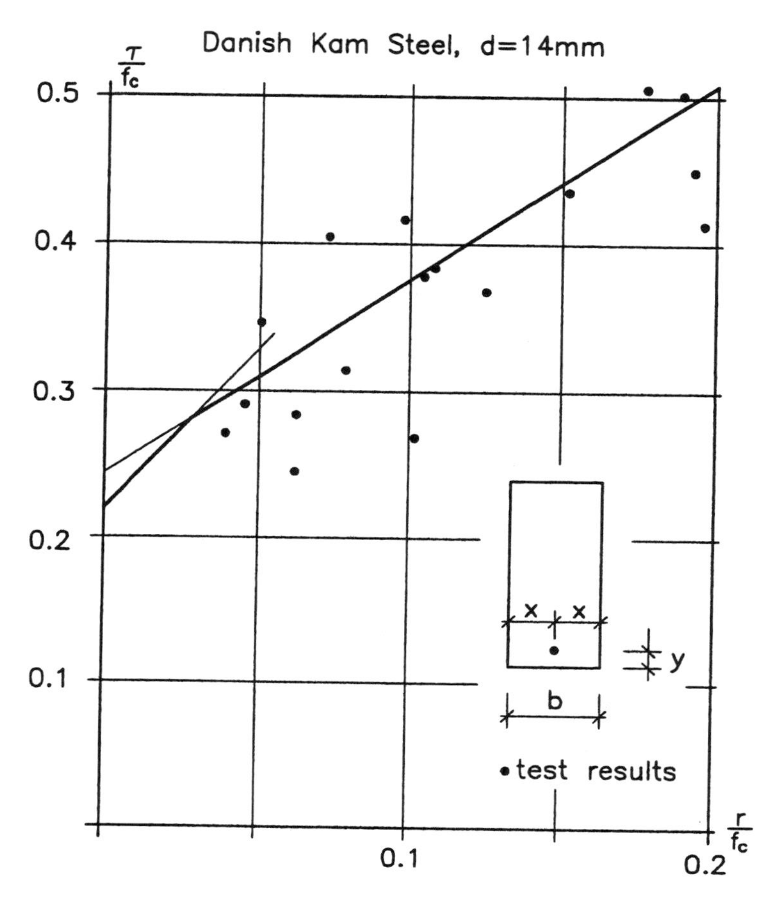

Figure 9.6.6 Bond strength versus transverse pressure. Test results from [72.7]. Only the straight lines approximation is shown.
Data: Danish Kam Steel, d = 14 mm, x = 70 mm, y = 37 mm, b = 140 mm, ℓ = 120 mm, f_c = 23.0 MPa.

We would not expect the formulas developed for corner bars to be valid for one single bar in the middle of the section. If a corner failure of the type considered takes place, the symmetry assumed is broken and when symmetry is preserved, a more complicated failure mechanism would be required.

Nevertheless, in Fig. 9.6.6 we have compared some test results with theory for one bar. These tests are also from [72.7]. The formulas are applied in a straightforward manner assuming a one-sided corner failure to take place. It appears that there is good agreement between theory and tests.[1] In the figure only the straight lines approximation has been plotted. In formula (9.6.25) $r/f_c = 0.1$ has been inserted.

Much remains to be learned before the effect of a support pressure on the bond strength of an assembly of bars can be predicted in a reliable way. Let's use a few lines to discuss in which direction improved knowledge may take us.

Consider as a simple example the anchorage of three bars in one layer at a support (see Fig. 9.6.7). Except possibly for very small support pressures it is likely that the corner failure will be governing as in the case of two corner bars. To evaluate the contribution to the load-carrying capacity from the third bar the following primitive arguments may be thought about. On the way to corner failure the beneficial pressure supplied by the reaction is likely to be more and more concentrated around the corner bars. Thus the third bar, regarding bond strength, is likely to benefit relatively little from the support pressure. In fact, if the width of the beam is large enough, one might even imagine that the anchor failure around the third bar will be almost independent of the support reaction. A complex failure mechanism may be envisaged that consists of a modified corner failure mechanism mixed with a failure around the third bar. Another possibility is a progressive failure that begins with a corner failure and, as concrete material around the corner bars suffers tensile softening, a secondary failure involving the third bar takes place.

It appears that even in such a simple example the conditions at failure may be rather complex.

To make possible a simple estimate we must convert to extremely simplified assumptions.

Let's assume that the load-carrying capacity of the assembly has been determined by the method suggested in Section 9.5, which is, of course, also strongly simplified. Let's further assume that the most dangerous mechanism is not a cover bending failure along the vertical faces (see the mechanism to the right in Fig. 9.6.1), in which the reaction does no work at all. If such a mechanism is the most dangerous one the effect of the support pressure should a priori be set at zero.

Our departure point will be the formula (9.6.12) valid for two corner bars. According to this formula the total force in the two bars at failure will be

[1] In the first edition of this book [84.11] results from the same tests were plotted. Some of the original results were excluded because in some tests the force in the bar was not measured, only calculated, which turned out not to be accurate enough. However, here all test results are included.

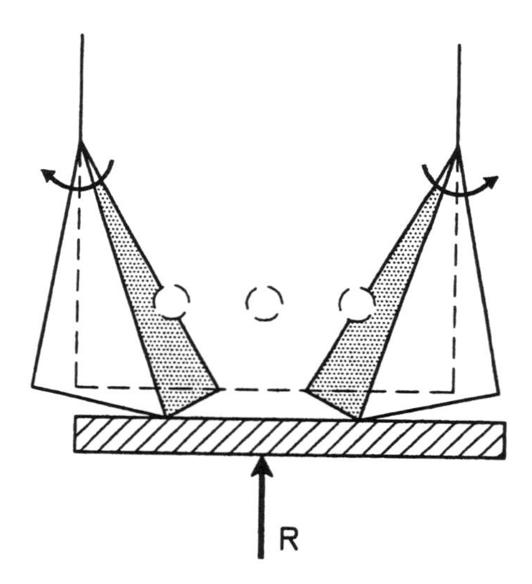

Figure 9.6.7 Bond strength of three bars anchored at a support (displacements of bars not shown).

$$T = 2\pi d \ell f_c \left[K_1 \nu + K_2 B \rho + K_2 \frac{b}{d} C \frac{r}{f_c} \right] \tag{9.6.26}$$

where K_1 and K_2 are constants.

Now we revert to the straight lines approximation. The first two terms in the square bracket represent the bond strength for $r = 0$. Therefore we may write

$$T = T(r = 0) + 2\pi d \ell f_c \cdot K_2 \frac{b}{d} C \frac{r}{f_c} \tag{9.6.27}$$

A conservative estimate of the total force T which may be carried by an assembly of n bars almost suggests itself. It is obtained by assuming that the only contribution to the total load-carrying capacity from the support pressure stems from the two corner bars. Thus formula (9.6.27) is assumed to be valid also for n bars, $T(r = 0)$ now meaning the total force in the n bars when $r = 0$.

Formula (9.6.27) may be rewritten to give the dimensionless bond strength per bar by dividing on both sides with $n\pi d\ell f_c$. We then get

$$\frac{\tau}{f_c} = \frac{\tau}{f_c}(r = 0) + \frac{2b}{nd} K_2 C \frac{r}{f_c} \tag{9.6.28}$$

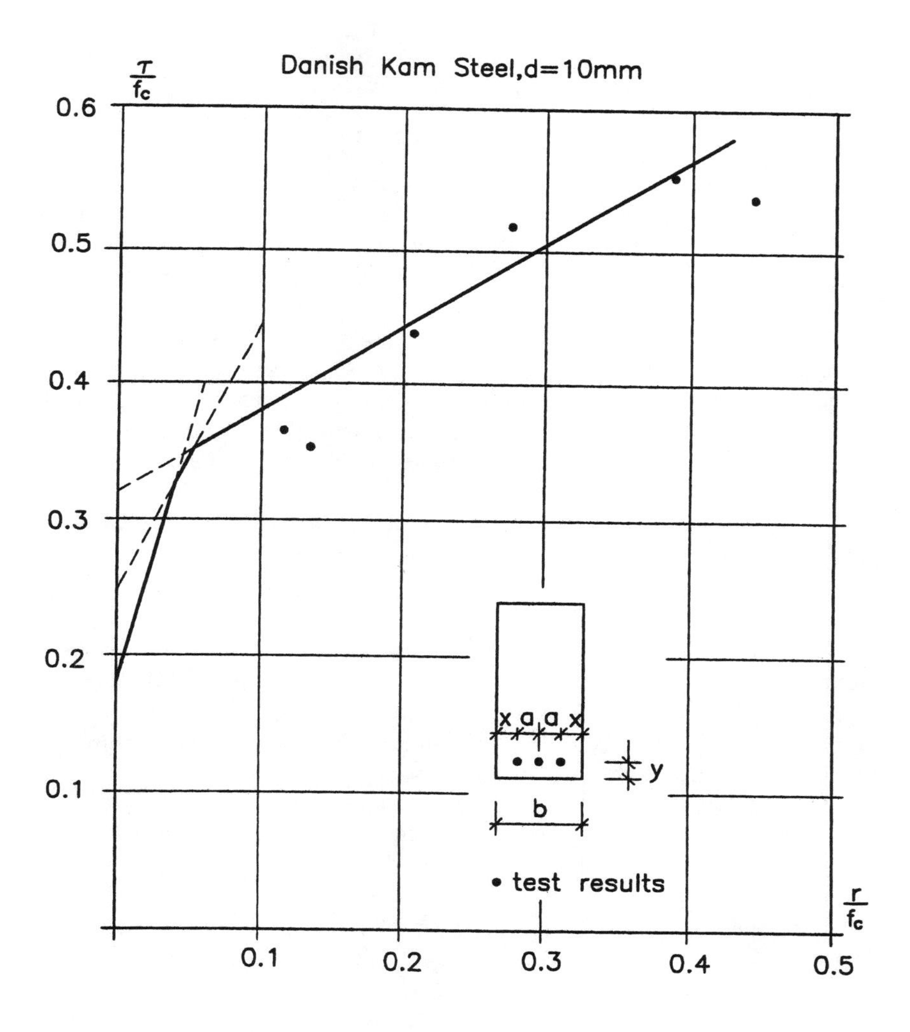

Figure 9.6.8 Bond strength versus transverse pressure. Theory compared with tests from [72.7]. Data: Danish Kam Steel, 3 bars, $d = 10$ mm, $x = 37$ mm, $a = 33$ mm, $y = 27$ mm, $b = 140$ mm, $\ell = 120$ mm, $f_c = 17.7$ MPa.

This formula expresses in words that the contribution to bond strength from the reaction is shared evenly by the total number of bars.

Thus the calculation may proceed according to the previous lines with only one change, namely that b/d is replaced by $2b/nd$. The values of B and C should be calculated for the two corner bars case.

Since the whole procedure is endowed with a number of approximations it will be natural to simplify further. This might be achieved by using the two straight lines approximation and the second formula in (9.6.12), which has the smallest K_2-value. Then

$$\frac{\tau}{f_c} = 0.28\,\nu + 0.24 B \rho + 0.24 \cdot \frac{2b}{nd} C \frac{r}{f_c} \qquad (9.6.29)$$

One straight line is found by determining β by formula (9.6.21) and B and C by (9.6.6) and (9.6.7), respectively. The other straight line is found by determining β by formula (9.6.25). B and C are determined by the same formulas.

It may, of course, happen that the bond strength of the assembly of bars in question is not governed by corner failure for $r = 0$, but by cover bending along the bottom face or by face splitting. Then a straight line with an inclination determined by (9.6.18) and leaving the τ/f_c-axis at the point corresponding to the correct bond strength for $r = 0$ is added to the set of two straight lines already found.

To see whether such a primitive approach bears any resemblance to reality, the bond strength per bar has been calculated in a case with three bars, $d = 10$ mm (see Fig. 9.6.8). The calculation is again compared with test results from Rathkjen [72.7].

When the bond strength is calculated for $r = 0$ as in Example 9.5.1 it is found to be $\tau/f_c = 0.17$ and it is governed by cover bending. The value of $\tan\alpha$ is 1.4. With these numbers the straight line in Fig. 9.6.8 valid for very small values of r has been determined. When calculating β by means of formula (9.6.25), $r/f_c = 0.2$ has been used.

The agreement between our approximate formula and the tests is remarkable. However, much more work needs to be done before a straightforward use of the procedure in practical design can be recommended.

A quick, rough estimate of the contribution from the reaction R to the anchor strength is found by putting $K_2 = 0.24$ and $C \sim \sin\beta$, where $\sin\beta$ is determined by formula (9.6.25). Then the total force increment T_R due to the reaction will be (cf. formula (9.6.27))

$$T_R = 2\pi d\,\ell f_c \cdot 0.24 \cdot \frac{b}{d} \sqrt{\frac{x\rho}{b\dfrac{r}{f_c}}}\,\frac{r}{f_c} \qquad (9.6.30)$$

or

$$T_R = 1.5 \sqrt{\frac{x\rho}{b\dfrac{r}{f_c}}}\,R \qquad (9.6.31)$$

The effectiveness factor ρ is determined by formula (9.4.16). T_R is normally around 25-40% of R.

An estimate of the total force T which may be carried by bars anchored along a support is then

$$T = T_o + T_R \tag{9.6.32}$$

Here T_o is the total force which may be carried by all the bars anchored at the support at $r = 0$. T_o is determined as described in Section 9.5. This estimate first of all does not include the fast enhancement of the force near $r = 0$ for small increments of r.

We conclude this section by reminding ourselves about the importance of limiting the support pressure. This problem was treated in Section 4.8.

In Fig. 9.6.9 we have repeated the situation at a simple support. When the total force in the bars to be anchored is T, the reaction is R, then the average support pressure must be limited to (cf. formula (4.8.14)),

$$r \le \frac{f_c}{1 + \left(\frac{T}{R}\right)^2} \tag{9.6.33}$$

When the number of bars at the support is large this limit very often determines the failure load rather than the bond strength.

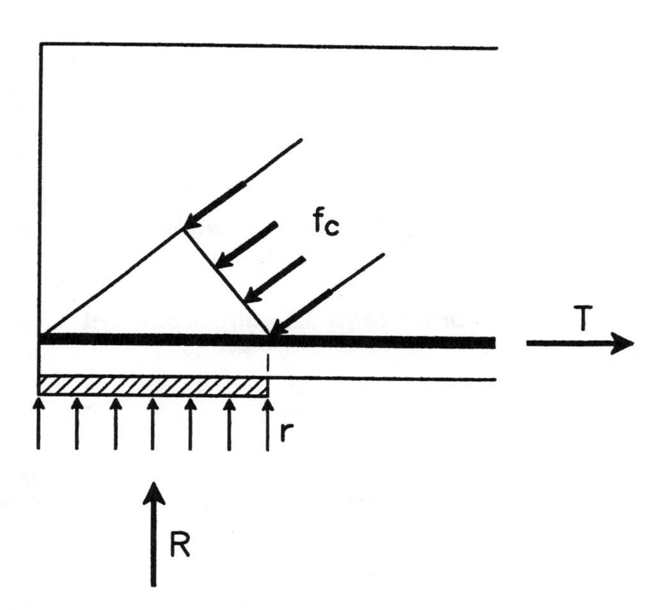

Figure 9.6.9 Conditions at a simple support.

9.7 EFFECT OF TRANSVERSE REINFORCEMENT

9.7.1 General

Reinforcement of one or another kind should always, if practically feasible, be present around anchor bars or splices. Reinforcement is a means to enhance the bond strength as well as the ductility. Adding transverse reinforcement may change the bond failure from being without warning and very violent to be warned by large crack widths and be very ductile.

Some of the difficulties in the bond strength theory that is still with us are related to the effect of transverse reinforcement.

The first question to be answered is whether transverse reinforcement may be supposed to yield before the maximum load is reached. The answer is that the reinforcement does not yield before the peak load is reached for small amounts of transverse reinforcement. If more reinforcement is added the reinforcement may yield. The range where there is no yielding would seem outside the realm of plastic theory. However, it turns out that a simple approach exists which preserves all we have developed up to now but works with a stress in the reinforcement far below the yield stress.

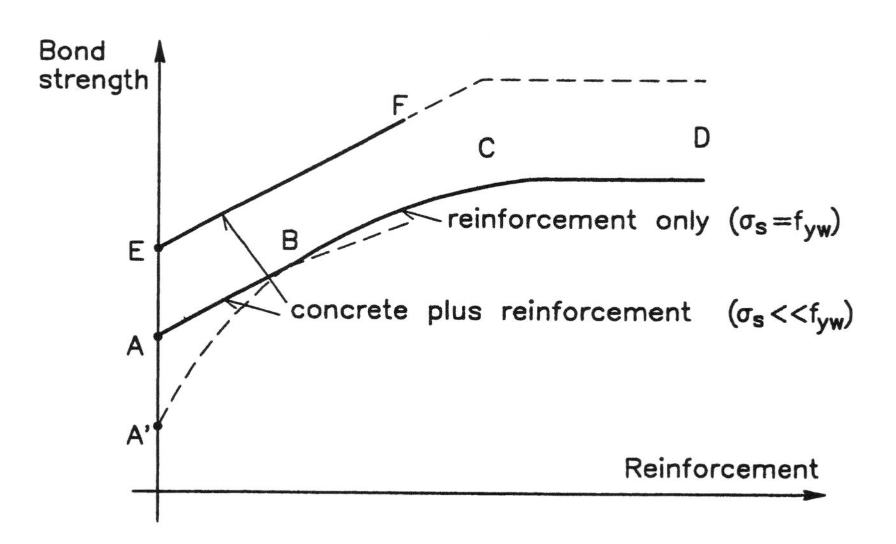

Figure 9.7.1 Bond strength versus the amount of transverse reinforcement (schematical).

Yielding in transverse reinforcement normally only takes place when the tensile stresses in the cracks in the surrounding concrete have disappeared almost completely. For low reinforcement degrees the bond strength is reached before the stage of yielding. In this region the concrete tensile stress contribution and a reinforcement contribution may be added. For larger reinforcement degrees the bond strength may be reached after yielding of the transverse reinforcement. Consequently, in this region the concrete tensile stress contribution cannot be added to the reinforcement contribution. Thus the bond strength as a function of some measure of the amount of transverse reinforcement may be as shown schematically in Fig. 9.7.1.

For normal anchor lengths and no transverse pressure we may follow a curve *ABCD* when the transverse reinforcement is increased. Along *AB*, the bond strength, when calculated by adding the concrete tensile strength contribution and a reinforcement contribution, is higher than the bond strength calculated by taking into account only the reinforcement contribution and assuming the yield strength f_{Yw} in the transverse reinforcement to be reached. We have a brittle failure but after the brittle failure the load falls down to a point on the curve *A'B* which is part of the bond strength curve *A'BCD* valid when the transverse reinforcement is yielding and when the concrete tensile strength contribution is disregarded. If we increase the reinforcement beyond *B* we have yielding in the transverse reinforcement and no concrete tensile strength contribution at the peak load. The failure is ductile. The bond strength calculated by adding the two contributions assuming a small reinforcement stress is lower than the bond strength found by taking into account only the reinforcement contribution. When further increasing the reinforcement it seems that an upper limit *CD* exists (cf. Section 9.7.3).

When the anchor length is small or when a transverse pressure is present we may follow a curve *EF*. All the way through, this curve lies above the curve *ABCD*, i.e., the reinforcement does not yield at the peak load. The failure is brittle. In this case probably one or another kind of upper limit always exists, too (cf. the end of Section 9.7.2).

When the bond strength is calculated by disregarding the tensile strength of the surrounding concrete we still have a contribution from the local failure. Therefore, the curve named "reinforcement only" does not emerge from the origin but from a point corresponding to the local failure contribution.

9.7.2 Transverse Reinforcement does not Yield

To get some understanding of the behavior of reinforcement around anchor bars and splices let us start on a small scale and look at the simple case of a tensile bar. In what follows an important situation is when the bar is underreinforced, i.e., the yield force of the bar is smaller than the load-carrying capacity of the concrete section in tension. In this case only one or a few cracks will be formed because the yield force in the bar is not able to crack adjacent sections.

The load as a function of the total elongation δ of the bar is schematically shown in Fig. 9.7.2. Along *AB* we have classical, almost linear-elastic, behavior. The stress in the bar is $n = E_s/E_c$ times the stress in the concrete, E_s and E_c being the Young's modulus of steel and concrete, respectively. At *B*, cracking occurs. By further elongation the stress in the concrete in the cracked section gradually decreases while the stress in the bar increases. Normally the increase of the bar force cannot compensate for the decrease of the concrete stress so we get a softening branch *BC*. Along *CD* the bar force is still growing in the linear-elastic range while the concrete tensile stress has almost vanished. Along *DC* the steel stress at yielding is the only contribution to the total force of the bar.

Thus it is rather evident that when the concrete sections around the bars are underreinforced the only effect of adding reinforcement is to raise point *B* by a certain amount.

If anchor or splice failures are required to be completely ductile, the branch *DE* in Fig. 9.7.2 must lie above point *B* and then the contribution from the tensile strength of the surrounding concrete cannot be added to the reinforcement contribution.

The stress in the reinforcement at point *B* in Fig. 9.7.2 may be estimated to be

$$\sigma_s = \frac{E_s}{E_c} f_t = n f_t \tag{9.7.1}$$

The tensile strength of the concrete f_t in this situation is evaluated by using a realistic value without any effectiveness factor. Such a value may be found by using formula (2.1.44).

Since n is of the order 10 and f_t is of the order 3 MPa we would expect a tensile stress in the steel of the order σ_s = 30 MPa.

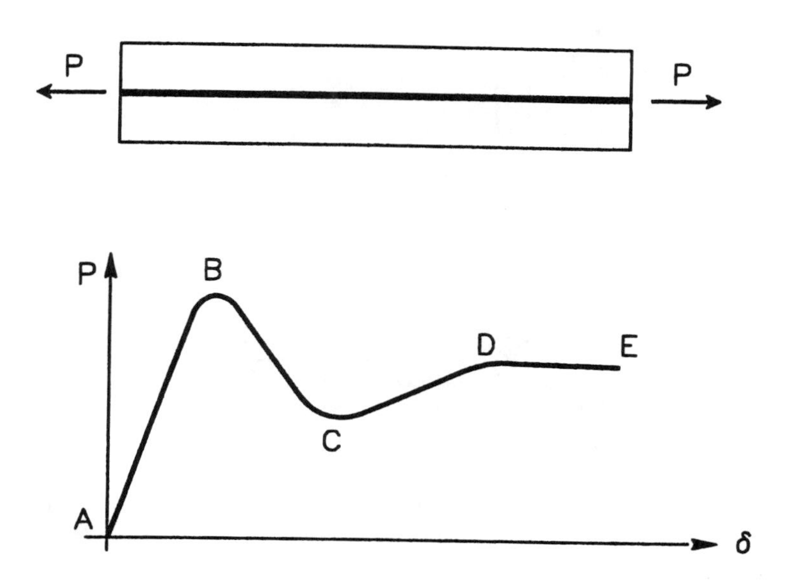

Figure 9.7.2 Force-elongation relationship for an underreinforced tensile bar.

A primitive way of including a reinforcement contribution in the work equations for bond strength mechanisms would then be to add a term $\sigma_s A_s \delta$ for each bar crossing a yield line. Here $\sigma_s \approx 30$ MPa according to our estimate, A_s is the sectional area of the bar and δ is the relative displacement in the bar direction. Thus we consider σ_s to be a pseudo yield stress, being approximately constant during tensile softening.

This procedure may underestimate the reinforcement contribution somewhat because we have used the conditions prevailing at the very onset of cracking. As soon as cracking occurs the steel stress will increase to some extent. We consider the steel stress at cracking as an empirical parameter, which will be denoted $\sigma_s = f_{scr}$, whereby the terms to be added in the work equations will be of the form $f_{scr}\ A_s\ \delta$. This is certainly a very strange form of plastic theory, but it seems to be the only way at present.

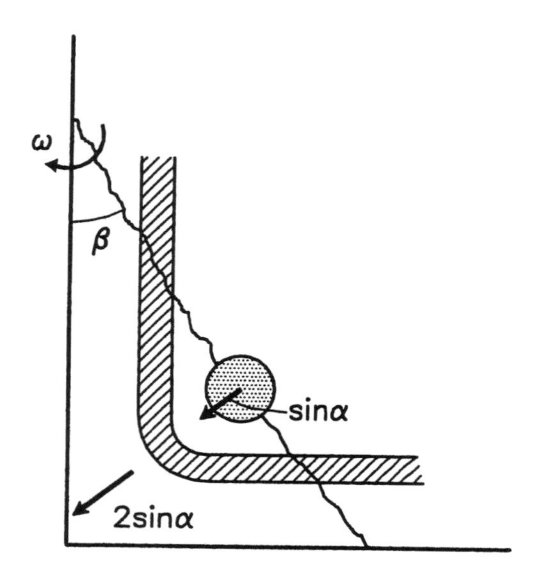

Figure 9.7.3 Corner bar surrounded by normal stirrups.

Since in most tests with surrounding reinforcement the corner failure has been decisive we may solve two problems by considering that. First, we may illustrate the procedure suggested above and second, we may find a formula which may be calibrated according to experiments.

Fig. 9.7.3 shows a very common case where the transverse reinforcement is a normal stirrup. The contribution to the dissipation is

$$B = f_{scr} \sum A_s \cdot 2 \sin\alpha (\cos\beta + \sin\beta) \tag{9.7.2}$$

Here ΣA_s is the total sectional area of the stirrups along the anchor length or the splice. Therefore, ΣA_s may be calculated as the number of stirrups along the anchor or splice length times the sectional area of one stirrup. The angle β has been used before to characterize the corner failure. We have assumed the stirrups to be close to the bar, so that the relative displacement components along the stirrup legs may be evaluated at the bar center.

We find the same result for rotation around a point at the bottom face.

Now the work equation (9.2.10) furnishes the following bond strength increment due to the stirrups

$$\Delta\tau = \frac{B}{\pi d \ell \cos\alpha} = \frac{2}{\pi} f_{scr} (\cos\beta + \sin\beta) \tan\alpha \frac{\sum A_s}{d\ell} \tag{9.7.3}$$

Strictly speaking the bond strength derived from the work equation with all terms present should be minimized. However, such an apparent accuracy is not in place here due to the uncertainties in f_{scr}. Thus it will be sufficient to use the optimized values for the case without transverse reinforcement. Then β is determined by the formulas (9.6.21) or (9.6.22) depending on the relative size of the geometrical parameters x and y. To find the value of $\tan\alpha$ we must determine which formula in (9.6.13) is valid. If the upper one is valid, $\tan\alpha = 1.4$; if the lower one is valid, $\tan\alpha = 0.75$. In many practical cases the first formula is governing, i.e., $\tan\alpha = 1.4$. To get an estimate of $\Delta\tau$ we may further realize that the sum $\cos\beta + \sin\beta$ does not vary much in cases without transverse pressure. It may be put to 1.35 in the β-region of practical interest. Then

$$\Delta\tau \cong 1.20 f_{scr} \frac{\sum A_s}{d\ell} \tag{9.7.4}$$

There are about 70 tests in the literature dealing with splices with transverse reinforcement. They are collected in [95.12]. However, many of these tests are with transverse reinforcement in the form of spirals, one spiral normally surrounding each splice. Since this form of transverse reinforcement has not become popular we will not treat it here.

The evaluation of f_{scr} by means of the tests gives rise to certain difficulties because the reinforcement degree of most of the test specimens is too large to be in the region of interest.

Thus we must adhere to the above estimate and to the few tests on anchor bars at a support mentioned below. They show that the round value

$$f_{scr} \approx 40 \text{ MPa} \tag{9.7.5}$$

may be used provisionally.

Then the estimate (9.7.4) with a rounded-off value reads

$$\Delta\tau \approx 50 \frac{\sum A_s}{d\ell} \text{ MPa} \tag{9.7.6}$$

In fact a formula of this type was suggested many years ago by Tepfers [73.3].

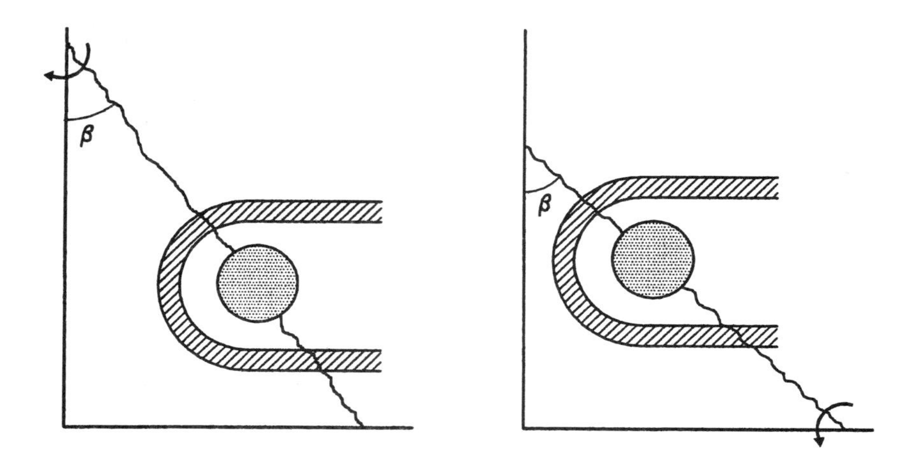

Figure 9.7.4 Corner bar surrounded by *U*-stirrups.

In Fig. 9.7.4 the transverse reinforcement is formed by *U*-stirrups. The rotation point is shown on the vertical face as well as on the bottom face. For simplicity we assume the yield lines intersect the stirrups on the horizontal part of the legs. Then in both cases the dissipation is

$$B = 4 f_{scr} \sum A_s \sin\alpha \cos\beta \tag{9.7.7}$$

which means that

$$\Delta\tau = \frac{B}{\pi d \ell \cos\alpha} = \frac{4}{\pi} f_{scr} \cos\beta \tan\alpha \frac{\sum A_s}{d\ell} \tag{9.7.8}$$

The angle β depends on the distances x and y shown in Fig. 9.6.2. If $x \geq y$, formula (9.6.21) applies; if $x < y$, formula (9.6.22) applies. The value of $\tan\alpha$ is either 1.4 or 0.75 depending on whether the first formula in (9.6.13) or the second one is valid. Since $\cos\beta$ will normally be around 0.8 we get for $\tan\alpha = 1.4$ the estimate

$$\Delta\tau = 57 \frac{\sum A_s}{d\ell} \text{ MPa} \quad (\tan\alpha = 1.4) \tag{9.7.9}$$

and for $\tan\alpha = 0.75$

$$\Delta\tau = 31 \frac{\sum A_s}{d\ell} \text{ MPa} \quad (\tan\alpha = 0.75) \tag{9.7.10}$$

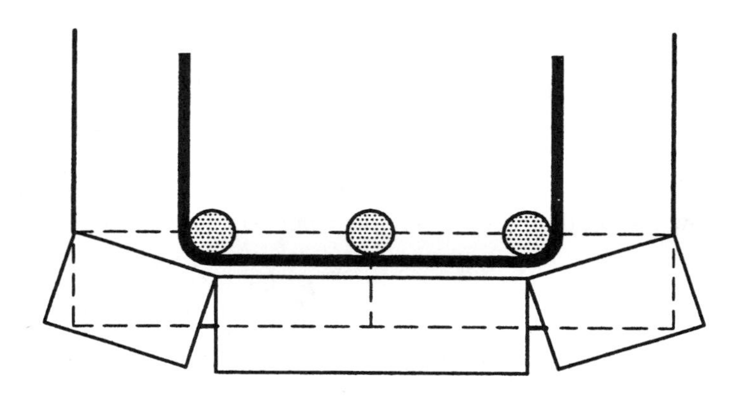

Figure 9.7.5 Cover bending mechanism involving transverse reinforcement in the form of normal closed stirrups.

Since in most cases without transverse pressure $\tan\alpha$ = 1.4, the *U*-stirrups are normally slightly more effective than the standard stirrups. With the accuracy we are working with here the difference is, however, so small that it will be hard to detect it experimentally. Since the normal stirrups are effective in carrying vertical shear in the section as well, they are practically always preferred for the *U*-stirrups.

When cover bending is decisive the dissipation in the surrounding reinforcement is calculated by including, for each intersection between a failure surface and a bar leg, a term of the form $f_{scr}\ A_s\ \delta$. When the surrounding reinforcement is close to the bar and in case of cover bending along the bottom face the contribution from the vertical yield lines may be neglected. With cover bending along a vertical face the contribution from the horizontal yield lines may be neglected. Thus with normal stirrups having vertical and horizontal legs we only have to count the number of intersections between the stirrup legs and the horizontal failure surfaces when we have cover bending along the bottom and the number of intersections between the stirrup legs and the vertical failure surfaces when we have cover bending along a vertical face. Thus the dissipation in the reinforcement is

$$B = 2f_{scr}\sum A_s \sin\alpha \qquad (9.7.11)$$

In this formula ΣA_s is the number of intersections between the stirrup legs and the horizontal or the vertical failure surfaces along the anchor or splice length times the sectional area of one stirrup leg. In the case shown in Fig. 9.7.5 with one closed stirrup ΣA_s equals twice the number of stirrups along the anchor or splice length times the sectional area of the stirrup.

If n bars or splices are involved in the failure mechanism we find the following bond strength increment per bar

$$\Delta\tau = \frac{B}{n\pi d\ell\cos\alpha} = \frac{2}{\pi n} f_{scr} \tan\alpha \frac{\sum A_s}{d\ell} \qquad (9.7.12)$$

Here we may insert the value f_{scr} = 40 MPa. The value of $\tan\alpha$ depends on whether the first or the second formula, respectively, in the expressions for the cover bending bond stress is valid. If the first formula is valid, $\tan\alpha = 1.4$; if the second one is valid, $\tan\alpha = 0.75$.

Since face splitting may be considered a special case of cover bending, formula (9.7.12) is valid for this case as well.

The above formulas should, of course, only be used in the transition region (the region AB in Fig. 9.7.1), i.e., the region of transverse reinforcement ratios leading to higher load-carrying capacity than that found by assuming yielding in the transverse reinforcement and neglecting the concrete tensile strength contribution. The transition point must be determined by calculating the bond strength in two ways, one found by adding the concrete tensile strength contribution and the reinforcement contribution with $f_{scr} \cong 40$ MPa and one found by neglecting the concrete tensile strength contribution and assuming yielding in the transverse reinforcement. How to perform the last mentioned calculation is discussed in the following section.

When the anchor length is small or when a transverse pressure is present the transition point may not exist for the transverse reinforcement ratios which are practically feasible (the curve EF in Fig. 9.7.1).

The formulas in this section should not at present be used for higher reinforcement ratios than those approximately satisfying the condition

$$\frac{\sum A_s}{d\ell} < 0.04 \qquad (9.7.13)$$

The Rathkjen tests [72.7] mentioned previously also included some specimens with transverse reinforcement. Two corner bars were anchored at the support. In one case normal stirrups were used as transverse reinforcement and in another case small closed stirrups surrounding both bars were used. The last mentioned case is treated above as the *U*-stirrup case.

The angle β may in this case be determined by formula (9.6.25). Since $f_c \cong 20$ MPa, $d = 14$ mm and $\ell = 120$ mm we have according to (9.4.16) $\rho = 0.047$. The average pressure r/f_c was $\cong 0.3$ and the geometrical parameters b and x were 140 mm and 37 mm, respectively. Then

$$\sin\beta = \sqrt{\frac{37 \cdot 0.047}{140 \cdot 0.3}} = 0.203 \tag{9.7.14}$$

$$\Rightarrow \quad \cos\beta = 0.979 \tag{9.7.15}$$

According to formula (9.7.3) we have for the normal stirrup case

$$\Delta\tau = \frac{2}{\pi} \cdot 40(0.979 + 0.203) \cdot 0.75 \frac{\sum A_s}{d\ell} = 22.6 \frac{\sum A_s}{d\ell} \text{ MPa} \tag{9.7.16}$$

The calculations performed in relation to Fig. 9.6.5 show that we are dealing with failure mode 2, i.e., $\tan\alpha = 0.75$.

Thus

$$\frac{\Delta\tau}{f_c} = \frac{22.6}{20} \frac{\sum A_s}{d\ell} = 1.13 \frac{\sum A_s}{d\ell} \tag{9.7.17}$$

In the *U*-stirrups case formula (9.7.8) renders

$$\Delta\tau = \frac{4}{\pi} \cdot 40 \cdot 0.979 \cdot 0.75 \frac{\sum A_s}{d\ell} = 37.4 \frac{\sum A_s}{d\ell} \tag{9.7.18}$$

or

$$\frac{\Delta\tau}{f_c} = \frac{37.4}{20} \frac{\sum A_s}{d\ell} = 1.87 \frac{\sum A_s}{d\ell} \tag{9.7.19}$$

The two straight lines representing $\Delta\tau/f_c$ versus $\Sigma A_s/d\ell$ have been drawn in Fig. 9.7.6. The intersection point with the τ/f_c axis is at $\tau/f_c = 0.45$, since this is the bond strength we would expect according to Fig. 9.6.5 when $\Sigma A_s = 0$.

However, an average line through the test results intersects the τ/f_c axis at $\tau/f_c \cong 0.52$. The difference may probably be explained as an enhancement of the effectiveness factors due to the presence of the transverse reinforcement. More tests are needed to quantify this enhancement before it can be used for design purposes.

Notice that in Fig. 9.7.6 the limiting value of the support pressure (9.6.33) has also been shown.

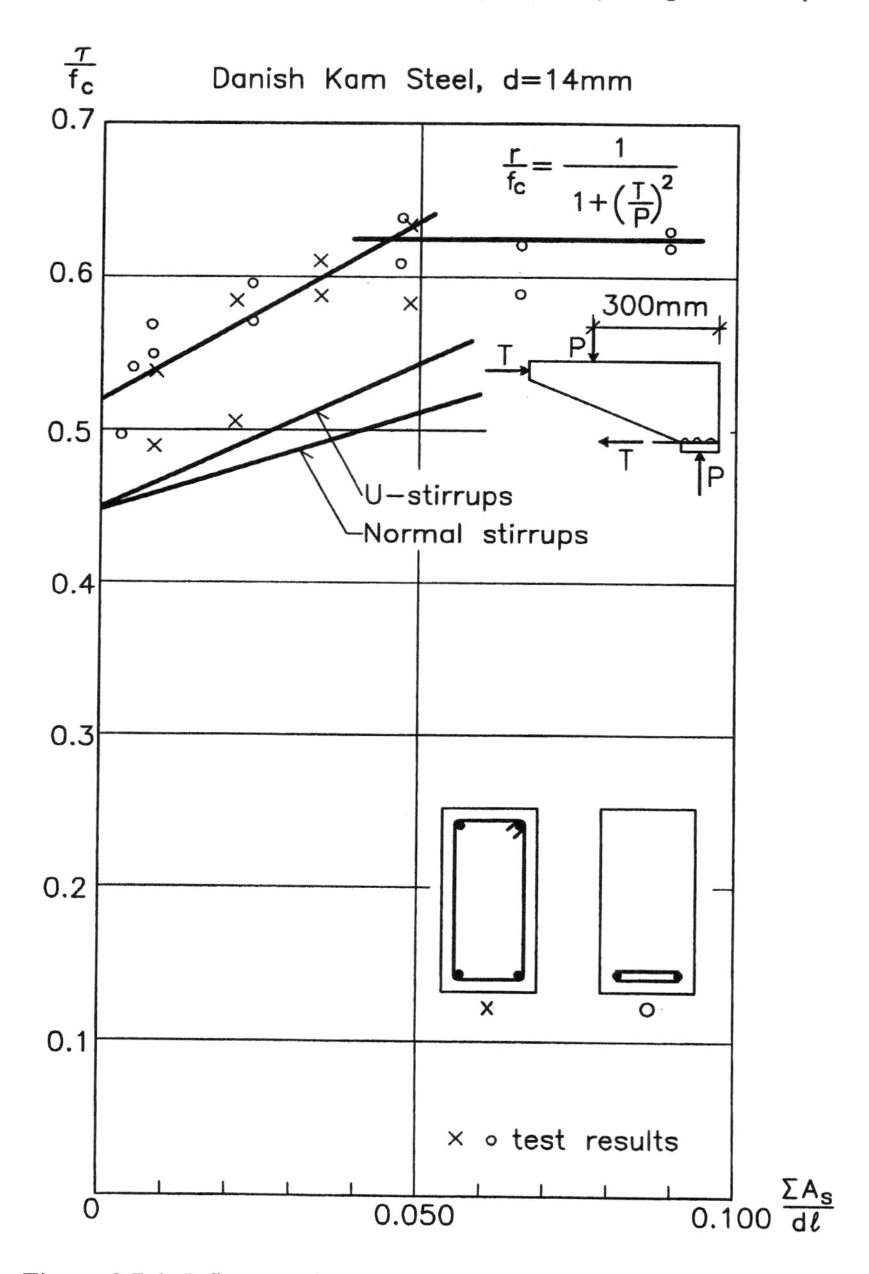

Figure 9.7.6 Influence of transverse reinforcement and pressure on bond strength. Test results from [72.7]. Data: Danish Kam Steel, d = 14 mm, 2 corner bars, geometry as in Fig. 9.6.5. Stirrups: plain wires or bars of mild steel, $f_c \cong 20$ MPa.

Although the scatter in the tests is considerable the trend is predicted quite correctly.

9.7.3 Transverse Reinforcement Yields

The prototype for discussing bond strength has become the corner bar, so let´s begin with that again. As mentioned previously, when the transverse reinforcement yields no contribution from the concrete, tensile strength can be present at the same time.

Fig. 9.7.7 shows a situation where the concrete in a corner has spalled off. The vanishing tensile strength is illustrated by simply removing from the figure the concrete spalled off. The stirrups have been shown in contact with the longitudinal bar, but the following reasoning is also valid when there is a small concrete volume between the stirrup and the bar.

In such a situation it is rather evident that there is no meaning to calculating the dissipation in the transverse reinforcement in the way done before, where the relative displacements in the yield lines were governing the stirrup dissipation. Rather it will be natural to assume that the stirrup part in contact with the longitudinal bar follows the displacements of the bar. This means, in fact, that the dissipation in the stirrups is normally half the value used before.

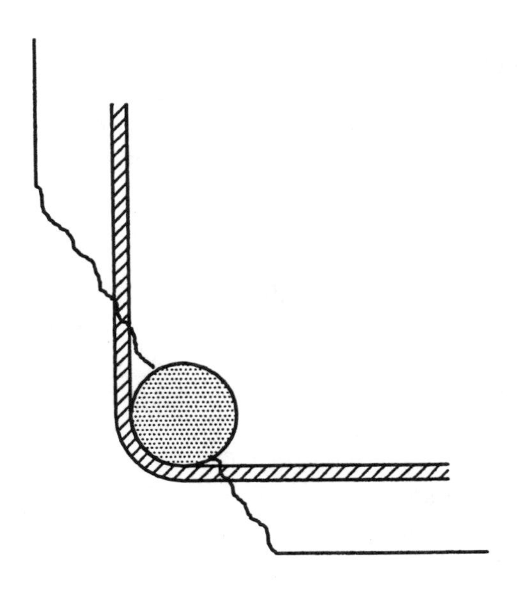

Figure 9.7.7 Concrete spalling at a corner.

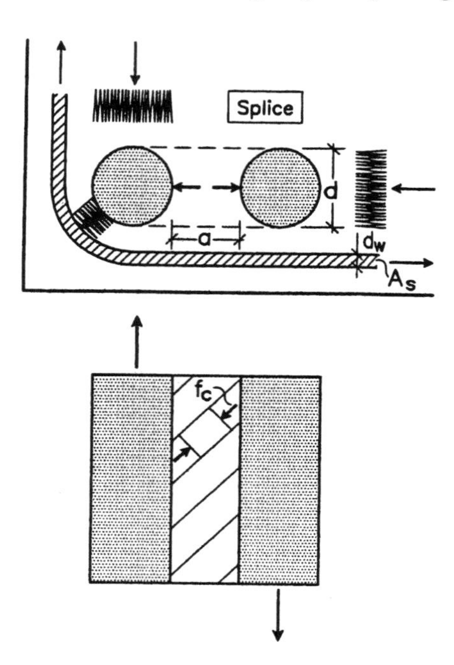

Figure 9.7.8 Beam model for a splice.

Now we might proceed as before with no change in the local failure dissipation. If this is done and the theory is compared with tests the results are rather dubious compared with what we have been used to. The most essential points where such a theory fails are that the variation of bond strength with concrete strength is poorly predicted and, even more important, there seems to be an upper limit of bond strength which is not predicted at all. Thus we are, with some reluctance, inclined to leave the procedure which has been guiding us so well up to now.

It turns out that an alternative model is possible. This has the further advantage of possessing a simplicity which one would not even dream about.

The alternative model is a simple beam model. It is most transparent in the case of a splice.

In this model we assume that the concrete volume $d\ell a$ between the bars acts as a small beam (cf. Fig. 9.7.8). The notation is almost as before: d meaning the diameter of the longitudinal bars, ℓ the splice length, and a the clear distance between the bars. In this concrete volume we have skew, uniaxial compressive stresses f_c. The statics of the problem is schematically illustrated in the figure.

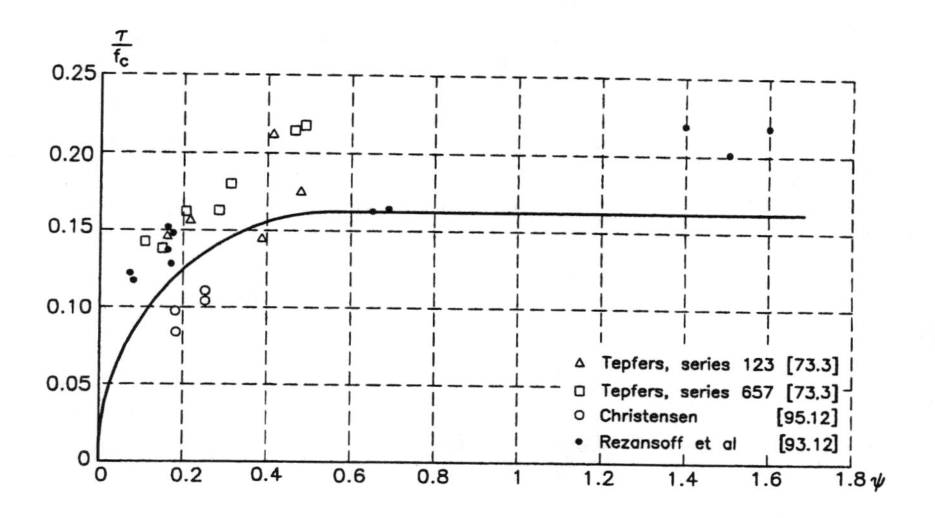

Figure 9.7.9 Bond strength versus transverse reinforcement degree. No effectiveness factor introduced.

The surrounding stirrups act as a transverse shear reinforcement of the small beam. The shear reinforcement degree of the beam is

$$\psi = \frac{\sum A_s}{d\ell} \frac{f_{Yw}}{f_c} \tag{9.7.20}$$

The term ΣA_s means, as previously in this chapter, the number of stirrups along the splice length ℓ times the sectional area of one stirrup. The yield strength of the transverse reinforcement has been denoted f_{Yw}.

The shear strength of the beam may be determined by formula (5.2.18). Transformed to bond strength we get

$$\frac{\tau}{f_c} = \begin{cases} \dfrac{1}{\pi}\sqrt{\psi(1-\psi)} & \psi \le 0.5 \\ \dfrac{1}{\pi}\cdot 0.5 = 0.16 & \psi > 0.5 \end{cases} \tag{9.7.21}$$

This simple model automatically gives an upper limit of the bond strength, namely $0.16\, f_c$.

If the test results available are compared with (9.7.21) the picture shown in Fig. 9.7.9 appears. This does not seem very promising.

There is a large group of tests lying far above (9.7.21) and a small group lying far below. The first group comprised normal strength concrete, f_c being around 25 MPa; and the second group was done with high strength concrete, f_c being about 70 MPa. But we quickly realize that it would be rather peculiar if the effective concrete compressive strength in the small beam equals the uniaxial standard compressive strength. Thus we modify (9.7.21) by replacing f_c with νf_c whereby we find

$$\frac{\tau}{f_c} = \begin{cases} \frac{1}{\pi}\sqrt{\psi(\nu - \psi)} & \psi \le 0.5\,\nu \\ 0.16\,\nu & \psi > 0.5\,\nu \end{cases} \tag{9.7.22}$$

In this formula f_c now means the standard compressive strength and ψ is determined by (9.7.20), inserting f_c as the standard compressive strength.

If we set

$$\nu = \frac{6.5}{\sqrt{f_c}} \quad (f_c \text{ in MPa}) \tag{9.7.23}$$

and plot the test results, the picture shown in Fig. 9.7.10 appears. This is much more satisfactory, in fact the agreement is remarkably good considering the simplicity of the model.

Notice that when using formula (9.7.23), ν-values > 1 are accepted. That we may have $\nu > 1$ is obviously due to the confinement left even when the corner has spalled off. Naturally, the size of the local beam may also play a role.

Thus a simple beam model quite accurately describes the bond strength when the transverse reinforcement yields.

The parameter intervals covered by the tests are

$$6.3 \le \frac{\ell}{d} \le 63.8$$

$$16 \le d \le 29.9 \text{ mm}$$

$$4 \le d_w \le 16 \text{ mm}$$

$$1.0 \le \frac{\text{cover}}{d} \le 2.5$$

$$21.2 \le f_c \le 77 \text{ MPa}$$

$$330 \le f_{Yw} \le 580 \text{ MPa}$$

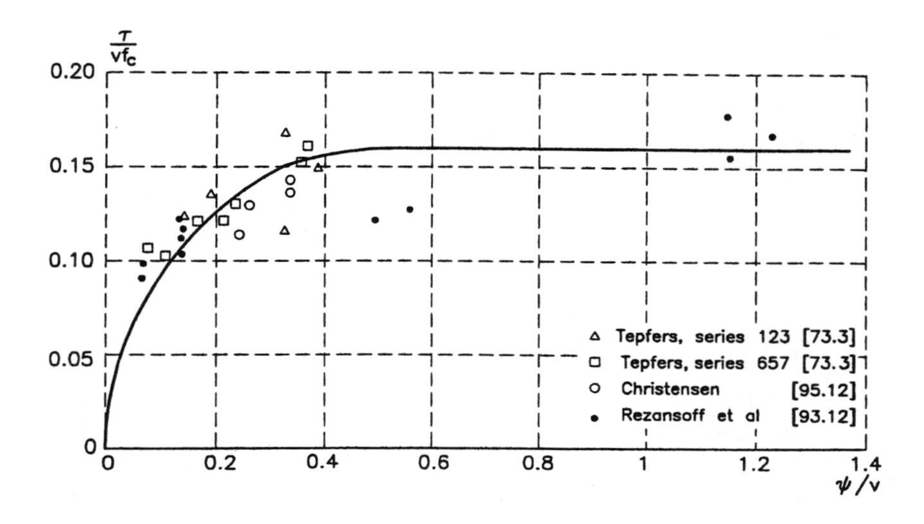

Figure 9.7.10 Bond strength versus reinforcement degree. Effectiveness factor introduced.

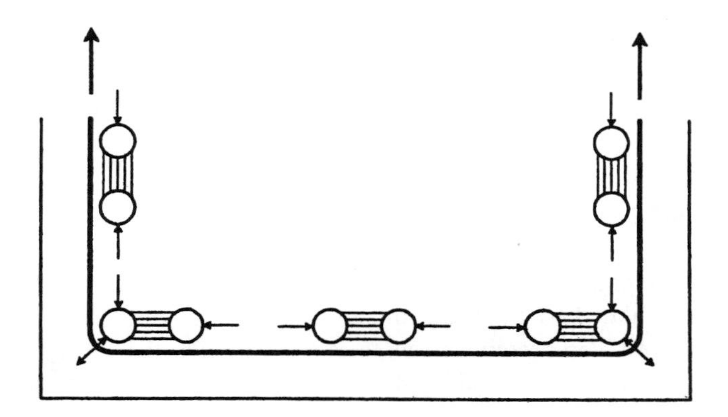

Figure 9.7.11 Beam model applied to a number of bar splices.

Here d_{sw} means the diameter of the transverse reinforcement.

The beam model is easily extended to the case of several splices. In Fig. 9.7.11 the local beams are shown as shaded regions. One stirrup suffices to activate all the local beam strengths according to formula (9.7.22), because the compressive forces outside the local beams equilibrate each other. This means that the local beam reinforcement degree (9.7.20) may be calculated as before. If the stirrup diameter and the stirrup distance have been calculated, for instance by a global beam shear analysis, the local beam shear reinforcement degree can be calculated as well, and then the bond strength may be found by using formula (9.7.22) (cf. Example 9.7.1). Finally the necessary splice length may be determined by means of formula (9.3.2).

The procedure suggested leads to a very ductile structural element and is preferred for a calculation utilizing the concrete tensile strength.

Notice that according to the beam model the splice strength is independent of the cover thickness and bar distances. Further, the length only enters into the reinforcement degree calculation. No special length correction, for instance in the formula for the effectiveness factor, is necessary.

The beam model is only relevant for bars near the faces because it rests on the assumption that the confinement is largely reduced due to the spalling off of the cover. Bars farther away from the faces probably do not need extra stirrups to equilibrate the forces from the local beam action, but more tests are needed to quantify this.

It also remains for future research to deal with anchor bars at the stage where the transverse reinforcement is yielding. Anchor bars may turn out to be stronger than splices in this case, but until more is learned, anchor bars should be treated as splices when using the beam model.

Example 9.7.1

The development length for a splice in the beam section shown in Fig. 9.7.12 is calculated by the beam model.

The concrete compressive strength is $f_c = 25$ MPa, the yield strength of the longitudinal bars is $f_Y = 550$ MPa and the yield strength of the stirrups is $f_{Yw} = 230$ MPa. The beam width is $b = 300$ mm, the diameter of the longitudinal bars is $d = 16$ mm and the diameter of the stirrups is $d_w = 7$ mm. The stirrup center distance is $s = 90$ mm.

The reinforcement ratio r_w in a horizontal section is

$$r_w = \frac{2 \cdot \frac{\pi}{4} d_w^2}{bs} = \frac{2 \cdot \frac{\pi}{4} \cdot 7^2}{300 \cdot 90} = 0.285\%$$

Thus $\Sigma A_s/d\ell$ to be used for calculating the reinforcement degree (9.7.20) in the beam model is

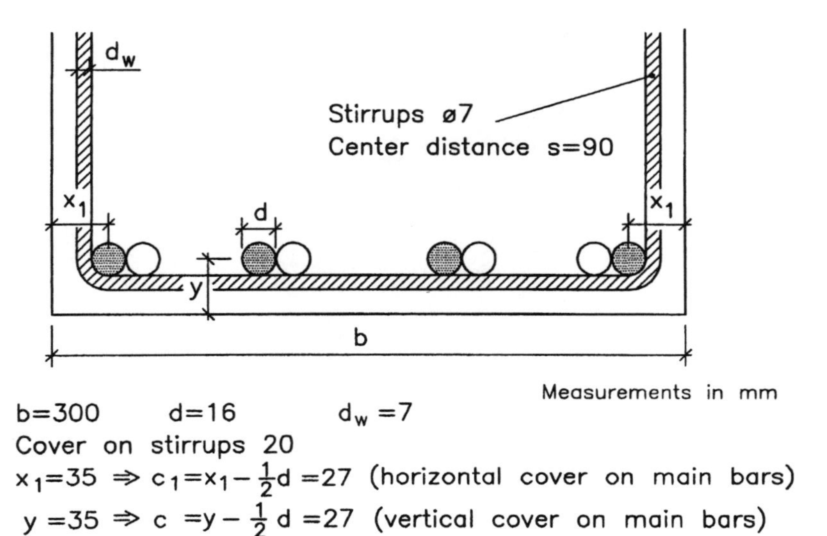

Figure 9.7.12 Beam section with four splices.

$$\frac{\sum A_s}{d\ell} = \frac{1}{2} r_w \frac{b}{d} = \frac{1}{2} \cdot 0.285 \cdot 10^{-2} \frac{300}{16} = 0.0267$$

and

$$\psi = \frac{\sum A_s}{d\ell} \frac{f_{Yw}}{f_c} = 0.0267 \frac{230}{25} = 0.246$$

According to (9.7.23)

$$\nu = \frac{6.5}{\sqrt{25}} = 1.3$$

Since $\psi < 0.5\ \nu = 0.65$ the bond strength is, formula (9.7.22),

$$\frac{\tau}{f_c} = \frac{1}{\pi} \sqrt{0.246(1.3 - 0.246)} = 0.162$$

$$\Rightarrow$$

$$\tau = 0.162 \cdot 25 = 4.05 \text{ MPa}$$

Then the development length is (see formula (9.3.2)),

$$\frac{\ell}{d} = \frac{550}{4 \cdot 4.05} = 34 \quad \Rightarrow \quad \ell = 34 \cdot 16 = 544 \text{ mm}$$

The bond strength of the bars, without stirrups, but with the same position of the bars in the concrete section is governed by face splitting. Formula (7) in Box 9.4.1 renders

$$\frac{\tau}{\sqrt{f_c}} = \min \begin{cases} 0.22 + 0.54\sqrt{\dfrac{1}{34}\left(\dfrac{300}{4 \cdot 16} - 1\right)} = 0.56 \\ 0.50 + 0.29\sqrt{\dfrac{1}{34}\left(\dfrac{300}{4 \cdot 16} - 1\right)} = 0.68 \end{cases}$$

$$\Rightarrow$$

$$\tau = 0.56\sqrt{25} = 2.80 \text{ MPa}$$

The reinforcement contribution, if the reinforcement does not yield, is determined by formula (9.7.12). The $\Sigma A_s/d\ell$-value to be inserted into this formula is seen to be $2 \cdot 0.0267 = 0.0534$. Thus

$$\Delta\tau = \frac{2}{\pi \cdot 4} \cdot 40 \cdot 1.4 \cdot 0.0534 = 0.48 \text{ MPa}$$

We have used $\tan\alpha = 1.4$ since the first formula in formula (7) in Box 9.4.1 governs the bond strength.

The bond strength found by adding the concrete tensile strength contribution and a reinforcement contribution is then

$$\tau = 2.80 + 0.48 = 3.28 \text{ MPa}$$

This is lower than the value $\tau = 4.05$ MPa found by assuming yielding of the stirrups and adding no concrete tensile strength contribution.

Thus the bond strength is $\tau = 4.05$ MPa, and the above calculation of the development length is correct.

9.8 Concluding Remarks

The reader who has followed us up to this point may agree that the plastic theory applied to bond is now in a shape where practical applications may begin. Thus more accurate development lengths may be provided than those furnished by the extremely crude empirical rules used in practical design.

However, when this has been said, it must be added that there is much more to be learned. To learn more requires more tests, first of all regarding bars which are not near a concrete face, but also regarding the effects of transverse pressure and transverse reinforcement.

These words are not only the last ones in the present chapter but the last ones in the book. Once again the hope is expressed that all the words written may be useful, not only for the designer for whom the book is written, but also that they may serve as an inspiration for future engineering research.

REFERENCES

Besides works directly referred to in the text, this section includes a number of important, related books and papers listed in order of publication year.

[1638.1] **Galileo, G.**, Discorsi e dimostrazioni matematiche interno à due nuove scienze attenenti alla mecanica & i movimenti locali, Leyden, Elsevirs, 1638 (English translation: Dialogue concerning two new sciences, New York, Dover, 1950, 294 pp.).

[1776.1] **Coulomb, C. A.**, Essai sur une application des règles de maximis & minimis à quelgues problèmes de statigue, relatifs a l'architecture, Mémoires de Mathèmatique & de Physique présentés à l'Academie Royale des Sciences, Vol. 7, 1776, pp. 343-382, (English translation: Note on an application of the rules of maximum and minimum to some statical problems, relevant to architecture, Heyman [72.1], pp. 41-74).

[1899.1] **Ritter, W.**, Die Bauweise Hennebique, Schweiz. Bauztg., Vol. 33, Nos. 5, 6, and 7, Feb. 1899, pp. 41-43, 49-52, 59-61.

[03.1] **Kötter, F.**, Die Bestimmung des Druckes an gekrümmten Gleitflächen, Sitzungsber., Kgl. Preuss. Akad. Wiss, Berlin, 1903.

[06.1] **Mörsch, E.**, Die Scher- und Schubfestigkeit des Eisenbetons, Beton Eisen, Vol. 5, No. XI, 1906, pp. 289-290.

[12.1] **Mörsch, E.**, Der Eisenbetonbau, seine Theorie und Anwendung, 4th ed., Stuttgart, 1912, 710 pp.

[19.1] **Suenson, E.**, Jernbetonarmering under en vinkel med normalkraftens retning (Reinforcement at an angle to the normal force), Ingeniøren, 1919, pp. 422.

[20.1] **Nielsen, N. J.**, Beregning af spændinger i plader (Calculation of stresses in slabs), Copenhagen, 1920.

[20.2] **Prandtl, L.**, Über die Härte plastischer Körper, Nachr. Ges. Wiss., Göttingen, 1920, pp. 74-85.

[21.1] **Ingerslev, Å.**, Om en elementær beregningsmetode af krydsarmerede plader (On an elementary method of calculation for two-way slabs), Ingeniøren, Vol. 30, No. 69, 1921, pp. 507-515.

[22.1] **Suenson, E.**, Eisenbetonbewehrung unter einem Winkel mit der Richtung der Normalkraft, Beton Eisen, Vol. 21, 1922, pp. 145-149.

[22.2] **Mörsch, E.**, Der Eisenbetonbau, seine Theorie und Anwendung, 5th ed., Stuttgart, 1922 (see also 6th ed., Stuttgart, 1929).

[23.1] **Ingerslev, Å.**, The strength of rectangular slabs, J. Inst. Struct. Eng., Vol. 1, No. 1, 1923, pp. 3-14.

[23.2] **Leitz, H.**, Eisenbewehrte Platten bei allgemeinem Biegungszustande, Bautechnik, Vol. 1, 1923, pp. 155-157, 163-167.

[23.3] **Carathéodory, C. and Schmidt, E.**, Über die Hencky-Prandtlschen Kurven, Z. Angew. Math. Mech., Vol. 3, 1923, pp. 468-475.

[23.4] **Hencky, H.**, Über einige statisch bestimmten Fälle des Gleichgewichts in plastischen Körpern, Z. Angew. Math Mech., Vol. 3, 1923, pp. 241-251.

[25.1] **Leitz, H.**, Die Drillungsmomente bei kreuzweise bewehrten Platten, Bautechnik, Vol. 3, 1925, pp. 717-719.

[25.2] **Leitz, H.**, Zum stand der Berechnung kreuzweise bewehrter Platten, Bauingenieur, Vol. 6, 1925, pp. 920-924.

[26.1] **Leitz, H.**, Über die Anwendung der Elastizitätstheorie auf kreuzweise bewehrten Beton, Beton Eisen, Vol. 25, 1926, pp. 240-245.

[26.2] **Marcus, H.**, Die Grundlagen der Querschnittbemessung kreuzweise bewehrter Platten, Bauingenieur, Vol. 7, 1926, pp. 577-582.

[28.1] **von Mises, R.**, Mechanik der plastischen Formänderung von Kristallen, Z. Angew. Math. Mech., Vol. 8, 1928, pp. 161-185.

[28.2] **Johansen, K. W.**, Om virkningen af bøjler og skråjern i jernbetonbjælker (On the action of stirrups and bent-up bars in reinforced concrete beams), Dan. Naturvidensk. Samf., A, No. 17, Copenhagen, 1928.

[28.3] **Richart, F. E., Brandtzæg, A. and Brown, R. L.**, A study of the failure of concrete under combined compressive stresses, Univ. Ill., Eng. Exp. Stat., Bull. 185, 1928.

[28.4] **Nadai, A.**, Plastizität und Erddruck, Chap. 6 in Handbuch der Physik, Vol. VI, Berlin, Springer, 1928.

[29.1] **Wagner, H.**, Ebene Blechwandträger mit sehr dünnen Stegblech, Z. Flugtech. Motorluftschiffahr, Vol. 20, Nos. 8-12, Berlin, 1929.

[30.1] **Leitz, H.**, Bewehrung von Scheiben und Platten, Int. Congr. Bet. Eisenbet., 1930, Vol. 1, pp. 115-123.

[30.2] **Johansen, K. W.**, Styrkeforholdene i støbeskel i beton (Strength of concrete construction joints), Bygningsstat. Medd., Vol. 2, 1930, pp. 67-68.

[31.1] **Johansen, K. W.**, Beregning af krydsarmerede jernbetonpladers brudmoment (Calculation of the rupture moment of two-way slabs), Bygningsstat. Medd., Vol. 3, No. 1, 1931, pp. 1-18.

[31.2] **Bay, H.**, Über den Spannungszustand in hohen Trägern und die Bewehrung von Eisenbetontragwändern, Stuttgart, Konrad Wittwer, 1931.

[31.3] **Nadai, A.**, Plasticity, New York, McGraw-Hill, 1931.

[32.1] **Johansen, K. W.**, Bruchmomente der kreuzweise bewehrten Platten, Mem. Ass. Int. Ponts Charp., Vol. 1, 1932, pp. 277-295.

[32.2] **Johansen, K. W.**, Nogle pladeformler (A number of slab formulas), Bygningsstat. Medd. Vol. 4, No. 4, 1932, pp. 77-84.

[33.1] **Dischinger, F.**, Die Ermittlung der Eisenanlagen in wandartigen Trägern, Beton Eisen, Vol. 32, 1933, pp. 237-239.

[34.1] **Brice, L. P.**, Détermination expérimentale et tracée de la courbe de résistance intrinsèque du béton, Sci. Ind. (Constr. Trav. Pub.), Jan. 1934, pp. 5-13.

[35.1] **Leon, A.**, Über die Scherfestigkeit des Betons, Beton Eisen, Vol. 34, No. 8, 1935, pp. 130-135.

[37.1] **Johansen, K. W.**, Bøjningsfri spændingstilstande i skaller (Stress fields in shells without bending), Bygningsstat. Medd., Vol. 9, 1937, 1938-39, pp. 61-84.

[38.1] **Gvozdev, A. A.**, Opredelenie velichiny razrushayushchei nagruzki dlya staticheski neopredelimykh sistem, preterpevayushchikh plasticheskie deformatsii, Moscow/Leningrad, Akademia Nauk SSSR, 1938, pp. 19-38 (English translation: The determination of the value of the collapse load for statically indeterminate systems undergoing plastic deformation, Int. J. Mech. Sci., Vol. 1, 1960, pp. 322-333.

[42.1] **Suenson, E.**, Byggematerialer, 3. bind, Natursten (Building materials, Vol. 3, Rock), Copenhagen 1942

[43.1] **Johansen, K. W.**, Brudlinieteorier (Yield line theories), Copenhagen, Gjellerup, 1943, 189 pp.

[48.1] **Dorn, J. E.**, The effect of stress state on the fracture strength of metals, in Fracturing of Metals, Cleveland, Ohio, American Society of Metals, 1948, pp. 32-50.

[49.1] **Gvozdev, A. A.**, Raschet nesushchei spasobnosti konstruktsii po metodu predelnogo ravnoveiya (Analysis of the load-carrying capacity of structures by the method of limiting equilibrium), Moscow, Stroiizdat, 1949, 280 pp.

[49.2] **Lundgren, H.**, Cylindrical shells, Copenhagen, The Danish Technical Press, The Institution of Danish Civil Engineers, 1949.

[49.3] **Johansen, K. W.**, Pladeformler (Slab formulas), Copenhagen, Polyteknisk Forening, 1949.

[49.4] Structural Research Laboratory, Denver, Colo., Report SP-23, 1949.

[50.1] **Drucker, D. C.**, Some implications of work hardening and ideal plasticity, Q. Appl. Math. Vol. 7, 1950, pp. 411-418.

[50.2] **Hill, R.**, The mathematical theory of plasticity, Oxford, Clarendon Press, 1950, 356 pp.; 2nd ed., 1956.

[50.3] **Nadai, A.**, Theory of flow and fracture of solids, Vol. I, New York, McGraw-Hill, 1950.

[50.4] **Lee, E. H.**, On stress discontinuities in plane plastic flow, Proc. Symp. Appl. Math., Vol 3, 1950, pp. 213-228.

[51.1] **Hill, R.**, On the state of stress in a plastic-rigid body at the yield point, Philos. Mag. (7), Vol. 42, 1951, pp. 868-875.

[51.2] **Prager, W. and Hodge, P. G., Jr.**, Theory of perfectly plastic solids, New York, Wiley & Sons, London, Chapman & Hall, 1951 (see also the Dover publication, New York, 1968).

[52.1] **Drucker, D. C., Greenberg, H. J. and Prager, W.**, Extended limit design theorems for continuous media, Q. Appl. Math., Vol. 9, 1952, pp. 381-389.

[52.2] **Prager, W.**, The general theory of limit design, Proc. 8th Int. Congr. Theor. Appl. Mech., Istanbul, 1952, Vol. II, pp. 65-72.

[52.3] **Hill, R.**, A note on estimating the yield-point loads in a plastic-rigid body, Philos. Mag. (7), Vol. 43, 1952, pp. 353-355.

[52.4] **Drucker, D. C. and Prager, W.**, Soil Mechanics and plastic analysis or limit design, Q. Appl. Math., Vol. 10, 1952, pp. 157-165.

[52.5] **Balmer, G.**, A general analytical solution for Mohr's envelope, ASTM Proceedings, 1952, pp. 1260-1271.

[53.1] **Hillerborg, A.**, Armering av elasticitetsteoretiskt beräknade plattor, skivor och skal (Reinforcement in slabs, disks and shells calculated according to the theory of elasticity), Betong, 1953, No. 2, pp. 101-109.

[53.2] **Cowan, H. J.**, The strength of plain, reinforced and prestressed concrete under the action of combined stresses, with particular reference to the combined bending and torsion of rectangular sections, Mag. Concr. Res., Vol. 5, 1953, pp. 75-86.

[53.3] **Rausch, E.**, Drillung (Torsion), Schub and Scheren im Stahlbetonbau, 3rd ed., Düsseldorf, Deutscher Ingenieur Verlag GmbH, 1953.

[53.4] **Bishop, P. W.**, On the complete solution to problems of deformation of a plastic-rigid material, J. Mech. Phys. Solids, Vol. 2, 1953, pp. 43-53.

[53.5] **Brinch Hansen, J.**, Earth pressure calculation, Copenhagen, Danish Technical Press, 1953.

[54.1] **Shield, R. T.**, Stress and velocity fields in soil mechanics, J. Math. Phys., Vol. 33, No. 2, 1954, pp. 144-156.

[54.2] **Brice, L.-P.**, Etude des conditions de formation des fissures de glissement et de décohésion dans les solides, Travaux, June 1954, pp. 475-506.

[55.1] **Ockleston, A. J.**, Load tests on a three-story reinforced concrete building in Johannesburg, Struct. Eng., Vol. 33, No. 10, 1955, pp. 304-322.

[55.2] **Rosenbleuth, E.**, Shell reinforcement not parallel to principal stresses, J. Am. Concr. Inst., Sept. 1955, pp. 61-71.

[55.3] **Bresler, B. and Pister, K. S.**, Failure of plain concrete under combined stresses, Proc. ASCE, Vol. 81, Reprint No. 674, Apr. 1955, 17 pp.

[55.4] **Wood, R. H.**, Studies in composite construction, Part II, National Building Studies, Research Paper No. 22, London, 1955.

[56.1] **Baker, A. L. L.**, The ultimate-load theory applied to the design of reinforced and prestressed concrete frames, London, Concrete Publications, 1956, 91 pp.

[56.2] **Hillerborg, A.**, Jämnviktsteori för armerade betongplattor (Theory of equilibrium for reinforced concrete slabs), Betong, Vol. 41, No. 4, 1956, pp. 171-181.

[56.3] **Falconer, B. H.**, Theory of stresses induced in reinforced concrete by applied two-dimensional stress, J. Am. Concr. Inst., Sept. 1956, pp. 277-293.

[56.4] **Girkmann, K.**, Flächentragwerke, Wien, Springer, 1956.

[56.5] **Drucker, D. C. and Shield, R. T.**, Design for minimum weight, Proc. 9th Int. Congr. Appl. Mech., Brussels, Vol. V, 1956, pp. 212-222.

[56.6] **Elstner, R. C. and Hognestad, E.**, Shearing strength of reinforced concrete slabs, J. Am. Concr. Inst., Proc., Vol. 53, July 1956, pp. 29-58.

[56.7] **Powell, D. S.**, The ultimate strength of concrete panels subjected to uniformly distributed loads, Thesis presented to Cambridge University, England, in fulfillment of the requirements for the degree of Doctor of Philosophy, 1956.

[57.1] Proc. 2nd Symp. Concr. Shell Roof. Constr., Oslo, 1957, (Discussion by Johansen, K. W., Vol. 2, pp. 270, 336).

[57.2] **Mansfield, E. H.**, Studies in collapse analysis of rigid-plastic plates with a square yield diagram, Proc. R. Soc. A, Vol. 241, 1957, pp. 311-338.

[58.1] **Johansen, K. W.**, Brudbetingelser for sten og beton (Failure criteria for rock and concrete), Bygningsstat. Medd., Vol. 29, No. 2, 1958, pp. 25-44.

[58.2] **McHenry, D. and Karni, J.**, Strength of concrete under combined tensile and compressive stress, J. Am. Concr. Inst., Apr. 1958, pp. 829-839.

[58.3] **Theimer, O. F.**, Hilfstafeln zur Berechnung wandartiger Stahlbetonträger, Berlin, Wilhelm Ernst & Sohn, 1958.

[58.4] **Haythornthwaite, R. M. and Shield, R. T.**, A note on the deformable region in a rigid-plastic structure, J. Mech. Phys. Solids, Vol. 6, 1958, pp. 127-131.

[58.5] **Bresler, B. and Pister, K. S.**, Strength of concrete under combined stresses, J. Am. Concr. Inst., Sept. 1958, pp. 321-345.

[58.6] **Ockleston, A. J.**, Loading test on reinforced concrete slabs spanning in two directions, Portland Cement Institute, Paper No. 6, Oct. 1958.

[59.1] **Hillerborg, A.**, Strimlemetoden (The strip method), Stockholm, Svenska Riks-byggen, 1959.

[59.2] **Base, G. D.**, Some tests on the punching shear strength of reinforced concrete slabs, London, Cement and Concrete Association, Technical Report No. TRA 321, July 1959.

[59.3] **Prager, W.**, An introduction to plasticity, Reading, Mass., Addison-Wesley, 1959, 148 pp.

[59.4] **Hodge, P. G., Jr.**, Plastic analysis of structures, New York, McGraw-Hill, 1959.

[60.1] **Sawczuk, A. and Rychlewski, J.**, On yield surfaces for plastic shells, Arch. Mech., Stosow., Vol. 12, 1960, pp. 29-52.

[60.2] **Nylander, H.**, Knutkrafter vid brottlinjeteorin (Nodal forces in the yield line theory), Swedish Cement and Concrete Research Institute at the Royal Institute of Technology, Bull. No. 42, 1960.

[60.3] **Kinnunen, S. and Nylander, H.**, Punching of concrete slabs without shear reinforcement, Stockholm, Royal Institute of Technology, Transactions, No. 158, 1960, 112 pp.

[60.4] **Mansfield, E. H.**, An analysis of slabs supported along all edges, Concr. Constr. Eng., Vol. 55, No. 9, 1960, pp. 333-340.

[61.1] **Drucker, D. C.**, On structural concrete and the theorems of limit analysis, Publ. Int. Assoc. Bridge Struct. Eng., Vol. 21, 1961, pp. 49-60.

[61.2] **Ziegler, H.**, On the theory of the plastic potential, Q. Appl. Math., Vol. 19, 1961, pp. 39-44.

[61.3] **Save, M.**, On yield conditions in generalized stresses, Q. Appl. Math., Vol. 19, 1961, pp. 259-267.

[61.4] **Paul, B.**, A modification of the Coulomb-Mohr theory of fracture, J. Appl. Mech., Vol. 28, June 1961, pp. 259-268.

[61.5] **Wood, R. H.**, Plastic and elastic design of slabs and plates, London, Thames & Hudson, London, 1961.

[61.6] **Moe, J.**, Shear strength of reinforced concrete slabs and footings under concentrated loads, Portland Cement Association, Development Department, Bull. P 47, April 1961.

[62.1] **Johansen, K. W.**, Yield line theory, London, Cement and Concrete Association, 1962.

[62.2] **Nielsen, M. P.**, Plasticitetsteorien for jernbetonplader (Theory of plasticity for reinforced concrete slabs), Licentiatafhandling, Danmarks Tekniske Højskole, Copenhagen, 1962.

[62.3] **Nielsen, M. P.**, On the calculation of yield line patterns with curved yield lines, Proc. Symp. Use Computers Civil Eng., Lisbon, 1962, Vol. I, Paper No. 22 (see also RILEM Bull., Vol. 19, 1963, pp. 67-74).

[63.1] **Massonnet, C. E. and Save, M. A.**, Calcul plastique des constructions, Vol. II: Structure spatiales, Brussels, Centre Belgo-Luxemburgeois d'Information de l'Acier, 1963 (English edition: Plastic analysis and design of plates, shells and disks, Amsterdam, North-Holland, 1972).

[63.2] **Nielsen, M. P.**, Flydebetingelser for jernbetonplader (Yield conditions for reinforced concrete slabs), Nord. Betong, Vol. 7, No. 1, 1963, pp. 61-82.

[63.3] **Nielsen, M. P.**, Yield conditions for reinforced concrete shells in the membrane state, IASS Symp. Non-classical Shell Problems, Warsaw, 1963, Proc., Amsterdam, North-Holland Publishing Company, 1964.

[63.4] **Nylander, H.**, Knutkrafter vid brottlinjeteorin (Nodal forces in the yield line theory), Nord. Betong, Vol. 7, No. 1, 1963, pp. 45-60.

[63.5] **Sawczuk, A. and Jaeger, T.**, Grenztragfähigkeitstheorie der Platten, Berlin, Springer, 1963.

[63.6] **Kuyt, B.**, Overhet wapenen van platen en schijven en twee richtingen (Reinforcement in slabs and disks in two directions), Ingenieur, Vol. 75, No. 24, 1963, pp. Bt43-46.

[63.7] **Ziegler, H.**, Some extremum principles in irreversible thermodynamics with application to continuum mechanics, in Progress in Solid Mechanics, Vol. IV (eds. Sneddon, I. N. and Hill, R.), Amsterdam, North-Holland, 1963, pp. 93-193.

[63.8] **Baus, R. and Tollaccia, S.**, Calcul à la rupture des dalles en béton armé et étude expérimentale du critère de rupture en flexion pure, Ann. Inst. Tech. Bat. Trav. Pub., Sept. 1963, 16th year, No. 189, Sér. Béton Armé (71), pp. 870-894.

[63.9] **Janas, M.**, Limit analysis of non-symmetric plastic shells by a generalized yield line method, IASS Symp. Non-classical Shell Problems, Warsaw, 1963; Proc., Amsterdam, North-Holland Publishing Company, 1964.

[63.10] **Nielsen, M. P.**, Exact solutions in the plastic plate theory, Bygningsstat. Medd., Vol. 34, No. 1, 1963, pp. 1-28.

[63.11] **Christiansen, K. P.**, The effect of membrane stresses on the ultimate strength of the interior panel in a reinforced concrete slab, Struct. Eng., Vol. 41, 1963, pp. 261-265.

[63.12] **Rasmussen, B. Højlund**, Betonindstøbte tværbelastede boltes og dornes bæreevne (Load carrying capacity of transversely loaded bolts and dowels imbedded in concrete), Bygningsstat. Medd., Vol. 34, No. 2, 1963, pp. 39-55.

[64.1] **Wolfensberger, R.**, Traglast und optimale Bemessung von Platten, Zurich, Ph.D. dissertation, 1964.

[64.2] **Kuyt, B.**, Zur Frage der Netzbewehrung von Flächentragwerken, Beton- Stahlbetonbau, Vol. 59, No. 7, 1964, pp. 158-163.

[64.3] **Nielsen, M. P.**, Limit analysis of reinforced concrete slabs, Copenhagen, Acta Polytech. Scand., Civil Eng. Build. Constr. Ser. No. 26, 1964, 167 pp.

[64.4] **Nielsen, M. P.**, On the formulation of linear theories of thin shells, Bygningsstat. Medd., Vol. 35, No. 2, 1964, pp. 37-77.

[64.5] **Nielsen, J.**, Kræfter i gitterflader (Forces in trusses), Nord. Betong, Vol. 8, No. 4, 1964, pp. 465-484.

[64.6] **Park, R.**, Ultimate strength of rectangular concrete slabs under short-term uniform loading with edges restrained against lateral movement, Proc. Inst. Civil Eng., Vol. 28, 1964, pp. 125-150.

[64.7] **Park, R.**, The ultimate strength and long-term behavior of uniformly loaded, two-way concrete slabs with partial lateral restraints at all edges, Mag. Concr. Res., Vol. 16, No. 48, 1964, pp. 139-152.

[64.8] **Sawczuk, A.**, On initiation of the membrane action in R-C plates, J. Mec., Vol. 3, No. 1, 1964, pp. 15-23.

[64.9] **Kupfer, H.**, Erweiterung der Mörsch'schen Fachwerkanalogie mit Hilfe des Princips vom Minimum der Formänderungsarbeit, CEB, Bull. Inf., No. 40, June 1964, pp. 44-57.

[64.10] **Nielsen, M. P.**, En tilnærmelsesformel for den rektangulære plade (A design formula for rectangular slabs), Copenhagen, 1964, (unpublished).

[64.11] **Szmodits, K.**, Scheibenbemessung auf Grund des Traglastverfahrens, Acta Tech., Acad. Sci. Hung., Vol. 46, 1964, pp. 371-389.

[64.12] **Drucker, D. C.**, On the postulate of stability of material in the mechanics of continua, J. Mec., Vol. 3, 1964, pp. 235-249.

[64.13] **Sawczuk, A.**, Large deflections of rigid-plastic plates, Proc., 11th Int. Congress on Applied Mechanics, 1964, pp. 224-228.

[65.1] **Green, A. E. and Naghdi, P. M.**, A comment on Drucker's postulate in the theory of plasticity, Berkeley, Calif., University of California, Institute of Engineering Research, Report No. AM-65-4, Apr. 1965, 10 pp.

[65.2] **Kemp, K. O.**, The yield criterion for orthotropically reinforced concrete slabs, Int. J. Mech. Sci., Vol. 7, 1965, pp. 737-746.

[65.3] **Nielsen, M. P.**, A new nodal force theory: recent developments in yield-line theory, Mag. Concr. Res., Spec. Publ., May 1965, pp. 25-30.

[65.4] **Kemp, K. O.**, The evaluation of nodal and edge forces in yield-line theory: recent developments in yield-line theory, Mag. Concr. Res., Spec. Publ., May 1965, pp. 3-12.

[65.5] **Wood, R. H.**, New techniques in nodal force theory for slabs: recent developments in yield-line theory, Mag. Concr. Res., Spec. Publ., May 1965, pp. 31-62.

[65.6] **Jones, L. L.**, The use of nodal forces in yield-line analysis: recent developments in yield-line theory, Mag. Concr. Res., Spec. Publ., May 1965, pp. 63-74.

[65.7] **Morley, C. T.**, Equilibrium methods for least upper bounds of rigid-plastic plates: recent developments in yield-line theory, Mag. Concr. Res., Spec. Publ., May 1965, pp. 13-24.

[65.8] **Møllmann, H.**, On the nodal forces of the yield line theory, Bygningsstat. Medd., Vol. 36, 1965, pp. 1-24.

[65.9] **Ceradini, G. and Gavarini, C.**, Calcolo a rottura e programmazione lineare, G. Genio Civile, Jan.-Feb. 1965.

[65.10] **Sawczuk, A.**, Membrane action in flexure of rectangular plates with restrained edges, Am. Concr. Inst. Spec. Publ., SP-12, 1965, pp. 347-358.

[65.11] **Taylor, R. and Hayes, B.**, Some tests on the effect of edge restraint on punching shear in reinforced concrete slabs, Mag. Concr. Res., Vol. 17, No. 50, Mar. 1965.

[65.12] **Granholm, H.**, A general flexural theory of reinforced concrete with particular emphasis on the inelastic behaviour of concrete and reinforcement, Stockholm, Almqvist & Wiksell, 1965, 209 pp.

[66.1] **Morley, C. T.**, The minimum reinforcement of concrete slabs, Int. J. Mech. Sci., Vol. 8, No. 4, 1966, pp. 305.

[66.2] **Morley, C. T.**, On the yield criterion of an orthogonally reinforced concrete slab element, J. Mech. Phys. Solids, Vol. 14, 1966, pp. 33-47.

[66.3] **Gavarini, C.**, I theoremi fondamentali del calcolo a rottura e la dualita in programmazione lineare, Ing. Civile, Vol. 18, 1966.

[66.4] **Sacchi, G.**, Contribution à l'analyse de limite des plaques minces en béton armé, Dissertation, Faculté Polytechnique de Mons, 1966.

[66.5] **Janas, M. and Sawczuk, A.**, Influence of positions of lateral restraints on carrying capacity of plates, CEB Bull. No. 58, 1966, pp. 164-189.

[66.6] **Courant, R. and Hilbert, D.**, Methods of mathematical physics, Vol. I, New York, Interscience Publishers, 1966, 561 pp.

[66.7] **Leonhardt, F. and Walther, R.**, Wandartige Träger, Deutscher Ausschuss für Stahlbeton, Heft 78, Berlin 1966.

[67.1] **Reckling, K. A.**, Plastizitätstheorie und ihre Anwendung auf Festigkeitsprobleme, Berlin, Springer-Verlag, 1967, 361 pp.

[67.2] **Nielsen, M. P.**, Om forskydningsarmering i jernbetonbjælker (Shear reinforcement in beams), Bygningsstat. Medd., Vol. 38, No. 2, 1967, pp. 33-58 (see also discussion with Sørensen, H. C., Vol. 40, No. 1, 1969, pp. 55-63).

[67.3] **Lenschow, R. J. and Sozen, M. A.**, A yield criterion for reinforced concrete slabs, J. Am. Concr. Inst., May 1967.

[67.4] **Lenkei, P.**, On the yield criterion for reinforced concrete slabs, Arch. Inz. Ladowej, Vol. 13, No. 1, 1967, pp. 5-11.

[67.5] **Massonnet, C. E.**, Complete solutions describing the limit state of reinforced concrete slabs, Mag. Concr. Res., Vol. 19, No. 58, 1967, pp. 13-32

[67.6] **Morley, C. T.**, Yield line theory for reinforced concrete slabs at moderately large deflections, Mag. Concr. Res., Vol. 19, No. 61, Dec. 1967, pp. 212-222.

[68.1] **Armer, G. S. T.**, The strip method: A new approach to the design of slabs, Concrete, Vol. 2, No. 9, 1968, pp. 358-363.

[68.2] **Wood, R. H. and Armer, G. S. T.**, The theory of the strip method for design of slabs, Proc. Int. Civil Eng., Vol. 41, 1968, pp. 285-311.

[68.3] **Lampert, P. and Thürlimann, B.**, Torsionsversuche an Stahlbetonbalken, Zurich, Eidgenössische Technische Hochschule, Institut für Baustatik, Bericht No. 6506-2, June, 1968.

[68.4] **Wood, R. H.**, Some controversial and curious developments in the plastic theory of structures, in Engineering Plasticity (eds. Heyman, J. and Leckie, F. A.), Cambridge, Cambridge University Press, 1968.

[68.5] **Calladine, C. R.**, Simple ideas in the large-deflection plastic theory of plates and slabs, Engineering Plasticity (eds. Heyman, J. and Leckie, F. A.), Cambridge, Cambridge University Press, 1968, pp. 93-127.

[68.6] **Janas, M.**, Large plastic deflections of reinforced concrete slabs, Int. J. Solids Struct., Vol. 4, No. 1, 1968, pp. 61-74.

[69.1] **Kachanov, L. M.**, Osnovy teorii plastichnosti, Moscow, Izdatelstvo Nauka, 1969 (English translation: Fundamentals of the theory of plasticity, Moscow, Mir Publishers, 1974, 445 pp.).

[69.2] **Nielsen, M. P.**, Om jernbetonskivers styrke, Copenhagen, Polyteknisk Forlag, 1969, 254 pp. (English edition: On the strength of reinforced concrete discs, Acta Polytech. Scand., Ci-70, Copenhagen, 1971, 261 pp.).

[69.3] **Chen, W. F. and Drucker, D. C.**, Bearing capacity of concrete blocks or rock, J. Eng. Mech. Div., Proc. ASCE, Vol. 95, No. EM4, Aug. 1969, pp. 955-978.

[69.4] **Lampert, P. and Thürlimann, B.**, Torsions-Biegeversuche an Stahlbetonbalken, Zurich, Eidgenössische Technische Hochschule, Institut für Baustatik, Bericht Nr. 6506-3, June, 1969.

[69.5] **Hofbeck, J. A., Ibrahim, I. O. and Mattock, A. H.**, Shear transfer in reinforced concrete, J. Am. Concr. Inst., Proc., Vol. 66, Feb. 1969, pp. 119-128.

[69.6] **Kupfer, H., Hilsdorf, H. K. and Rüsch, R.**, Behaviour of concrete under biaxial stresses, J. Am. Concr. Inst., Proc., Vol. 66, No. 8, Aug. 1969, pp. 656-666.

[69.7] **Ferguson, P. M. and Briceno, E. A.**, Tensile lap splices - Part 1: Retaining wall type, varying moment zone, University of Texas at Austin, Center for Highway Research, Research Report 113-2, July 1969 (see also Part 2, Research Report 113-3, Apr. 1971).

[69.8] **Demorieux, J. M.**, Essai de traction-compression sur modèles d'ame de poutre en beton armé, Annales de l'institut technique du bâtiment et des traveaux publics, No. 258, Juin 1969, pp. 980-982.

[70.1] **Bræstrup, M. W.**, Yield-line theory and limit analysis of plates and slabs, Mag. Concr. Res., Vol. 22, No. 71, 1970, pp. 99-106.

[70.2] **Lampert, P.**, Bruchwiderstand von Stahlbetonbalken unter Torsion und Biegung, Zurich, Eidgenössische Technische Hochschule, Institut für Baustatik, Bericht Nr. 26, Jan. 1970.

[70.3] **Bræstrup, M. W.**, Yield lines in discs, plates, and shells, Copenhagen, Technical University of Denmark, Structural Research Laboratory, Report No. R-14, 1970, 54 pp.

[70.4] **Nielsen, M. P.**, Brudstadieberegninger for jernbetonkonstruktioner (Ultimate limit state calculations for reinforced concrete structures), DBF-Publikation No. 4, Copenhagen, Dansk Betonforening, 1970.

[70.5] **Mast, P. E.**: Stresses in flat plates near columns, J. Am. Concr. Inst., Oct. 1970.

[70.6] **Hobbs, D. W.**, Strength and deformation properties of plain concrete subject to combined stress. Part 1: Strength results obtained on one concrete, Cement and Concrete Association, Report 42.451, Nov. 1970.

[71.1] **Nielsen, M. P.**, Kombineret bøjning og vridning af jernbetonbjælker (Combined bending and torsion of reinforced concrete beams), Aalborg, Danmarks Ingeniørakademi, Bygningsafdelingen, Ren & Anvendt Mekanik, Report No. R-7103, 1971.

[71.2] **Nielsen, M. P.**, Bestemmelse af armering i jernbetonskaller (Design of reinforcement in reinforced concrete shells), Aalborg, Danmarks Ingeniørakademi, Bygningsafdelingen, Ren & Anvendt Mekanik, Report No. R-7104, 1971.

[71.3] **Lampert, P. and Thürlimann, B.**, Ultimate strength and design of reinforced concrete beams in torsion and bending, IABSE, Publ. 31-I, 1971, pp. 107-131.

[71.4] **Hopkins, D. C. and Park, R.**, Test on a reinforced concrete slab and beam floor with allowance for membrane action, ACI, SP-30-10, 1971.

[71.5] **Hung, T. Y. and Navy, E. G.**, Limit strength and serviceability factors in uniformly loaded, isotropically reinforced two-way slabs, ACI, SP-30-14, 1971.

[71.6] **Brotchie, J. F. and Holley, M. J.**, Membrane action in slabs, ACI, SP-30-16, 1971.

[72.1] **Heyman, J.**, Coulomb's Memoir on Statics: An essay on the history of civil engineering, Cambridge, Cambridge University Press, 1972, 212 pp.

[72.2] **Fox, E. N.**, Limit analysis for plates: A simple loading problem involving a complex exact solution, Philos. Trans. R. Soc., Vol. 272A, 1972, pp. 463-492.

[72.3] **Johansen, K. W.**, Yield-line formulae for slabs, London, Cement and Concrete Association, 1972.

[72.4] **Anderheggen, E. and Knöpfell, H.**, Finite element limit analysis using linear programming, Int. J. Solids Struct., Vol. 8, 1972, pp. 1413-1431.

[72.5] **Chan, H. S. Y.**, The collapse load of reinforced concrete plates, Int. J. Numm. Meth. Eng., Vol. 5, 1972, pp. 57-64.

[72.6] **Bäcklund, J.**, Membraneffect i armerade betongplattor - en litteraturöversikt (Membrane effect in reinforced concrete slabs - a survey), Chalmers Tekniska Högskola, Inst. f. Konstruktionsteknik, Betongbyggnad, Report 72:1, 1972, 14 pp.

[72.7] **Rathkjen, A.**, Forankringsstyrke af armeringsjern ved bjælkeunderstøtninger (Bond strength of reinforcement bars at beam supports), Aalborg, Danmarks Ingeniørakademi, Bygningsafdelingen, Ren & Anvendt Mekanik, Report No. R-7203, 1972.

[72.8] **Izbicki, R. J.**, General yield condition: I. plane deformation, Bull. Acad. Pol. Sci., Ser. Sci. Tech., Vol. 20, No. 7-8, 1972.

[72.9] **Mattock, A. H. and Hawkins, N. M.**, Shear transfer in reinforced concrete - recent research, PCI Journal, Mar.-Apr. 1972, pp. 55-75.

[73.1] **Bäcklund, J.**, Finite element analysis of nonlinear structures, Göteborg, Chalmers Tekniska Högskola, Dissertation, 1973.

[73.2] Dansk Ingeniørforening: Norm for betonkonstruktioner. Dansk Standard DS 411. Vejledning til norm for betonkonstruktioner. Bilag til Dansk Standard, DS 411, København, Teknisk Forlag, Normstyrelsens Publikationer, 2. udgave, Dec. 1973 (English edition: Structural use of concrete. Danish standard, DS 411. Supplementary guide to code of practice for the structural use of concrete. Supplement to Danish standard, DS 411, Copenhagen, Teknisk Forlag, Normstyrelsens Publikationer, 1976).

[73.3] **Tepfers, R.**, A theory of bond applied to overlapped tensile reinforcement splices for deformed bars, Göteborg, Chalmers Tekniska Högskola, Division of Concrete Structures, Publ. 73:12, 1973.

[73.4] **Kupfer, H.**, Das Verhalten des Betons unter mehrachsigen Kurzzeitbelastung unter besonderer Berücksichtigung der zweiachsigen Beanspruchung, Deutscher Ausschuss für Stahlbeton, No. 229, 1973, 105 pp.

[73.5] **Jensen, B. C.**, Koncentrerede belastninger på uarmerede betonprismer (Concentrated loads on unreinforced concrete prisms), Bygningsstat. Medd., Vol. 44, No. 4, Dec. 1973, pp. 89-111.

[74.1] **Hillerborg, A.**, Strimlemetoden (The strip method), Stockholm, Almqvist & Wiksell, 1974.

[74.2] **Mitchell, D. and Collins, M. P.**, Diagonal compression field theory - a rational model for structural concrete in pure torsion, J. Am. Concr. Inst., Vol. 71, Aug. 1974, pp. 396-408.

[74.3] **Fox, E. N.**, Limit analysis for plates: The exact solution for a clamped square plate of isotropic homogeneous material obeying the square yield criterion and loaded by uniform pressure, Philos. Trans. R. Soc., Vol. 277A, 1974, pp. 121-155.

[74.4] **Pedersen, H.**, Optimum design of thin concrete plates, Proc. Int. Symp. Discr. Meth. Eng., CISE-SEGRATE, Milan, 1974, pp. 374-389.

[74.5] **Dragosavic, M. and van den Beukel, A.**, Punching shear, Heron, Vol. 20, No. 2, 1974, 48 pp.

[74.6] **Hansen, K., Kavyrchine, M., Melhorn, G., Olesen, S. Ø., Pume, D. and Schwing, H.**, Design of vertical keyed shear joints in large panel buildings, Build. Res. Pract., July-Aug. 1974.

[74.7] **Brøndum-Nielsen, T.**, Optimum design of reinforced concrete shells and slabs, Copenhagen, Technical University of Denmark, Structural Research Laboratory, Report No. R-44, 1974.

[74.8] **Mattock, A. H.**, Shear transfer in concrete having reinforcement at an angle to the shear plane, Detroit, Am. Concr. Inst., Spec. Publ. SP-42, Vol. I, 1974, pp. 17-42.

[74.9] **Cederwall, K., Hedman, O. and Losberg, A.**, Shear strength of partially prestressed beams with pretensioned reinforcement of high grade deformed bars, Detroit, Am. Concr. Inst., Spec. Publ. SP-42, Vol. I, 1974, pp. 215-230.

[74.10] **Rajendran, S. and Morley, C. T.**, A general yield criterion for reinforced slab elements, Mag. Concr. Res., Vol. 26, No. 89, 1974, pp. 212-270.

[74.11] **Hobbs, D. W.**, Strength and deformation properties of plain concrete subject to combined stress. Part 2: Results obtained on a range of flint gravel aggregate concrete, Cement and Concrete Association, Report 42.497, July 1974.

[74.12] **Houborg, J. and Sørensen, A. B.**, Støbeskel (Casting joints), M.Sc. thesis, Technical University of Denmark, 1974.

[75.1] **Martin, J. B.**, Plasticity: Fundamentals and general results, Cambridge, Mass., MIT Press, 1975, 931 pp.

[75.2] **Chen, W. F.**, Limit analysis and soil plasticity, Developments in Geotechnical Engineering, Vol. 7, New York, Elsevier, 1975, 638 pp.

[75.3] **Nielsen, M. P. and Bræstrup, M. W.**, Plastic shear strength of reinforced concrete beams, Bygningsstat. Medd., Vol. 46, No. 3, 1975, pp. 61-99.

[75.4] **Nielsen, M. P.**, Beton 1, Del 1, 2 og 3 (Concrete structures 1, Part 1, 2 and 3), Aalborg/Copenhagen, 1975, 678 pp.

[75.5] **Hess, U.**, Udtrækning af indstøbte inserts (Pull-out strength of anchors), Copenhagen, Danmarks Ingeniørakademi, Bygningsafdelingen, Ren & Anvendt Mekanik, Report No. 75:54, Jan. 1975, 25 pp.

[75.6] **Jensen, B. C.**, On the ultimate load of vertical, keyed shear joints in large panel buildings, 2nd Int. Symp. on Bearing Walls, Warsaw, 1975, Individual Reports, pp. 187-195 (Also: Copenhagen, Technical University of Denmark, Institute of Building Design, Report No. 108, 8 pp.).

[75.7] **Fernando, J. S. and Kemp, K. O.**, The strip method of slab design: Unique or lower-bound solutions?, Mag. Concr. Res., Vol. 27, No. 90, 1975, pp. 23-29.

[75.8] **Jensen, B. C.**, Lines of discontinuity for displacements in the theory of plasticity of concrete, Mag. Concr. Res., Sept. 1975, pp. 143-150.

[75.9] **Jensen, B. C. and Nielsen, M. P.**, Om spalteforsøget og om koncentrerede kræfter på uarmerede betonprismer, (On the splitting test and concentrated loads on non-reinforced concrete blocks), Nord. Betong, No. 3, 1975, pp. 9-13.

[75.10] **Birke, H.**, Kupoleffekt vid betongplattor (Dome effect in concrete slabs), Inst. for Byggnadsstatik, Royal Technical University, Stockholm, 1975.

[75.11] **Orangun, C. O., Jirsa, J. O. and Breen, J. E.**, The strength of anchor bars: A reevaluation of test data on development length and splices, Center for Highway Research, The Univ. of Texas at Austin, Research Report 154-3F, 1975.

[76.1] **Jensen, B. C.**, Nogle plasticitetsteoretiske beregninger af beton og jernbeton, Ph.D. thesis, Copenhagen, Technical University of Denmark, Institute of Building Design, Report 111, 1976, 115 pp. (English edition: Some applications of plastic analysis to plain and reinforced concrete, Report No. 123, 1977, 119 pp.).

[76.2] **Grob, J. and Thürlimann, B.**, Ultimate strength and design of reinforced concrete beams under bending and shear, Int. Assoc. Bridge Struct. Eng., Mem., Vol. 36, No. II, 1976, pp. 105-120.

[76.3] **Bræstrup, M. W., Nielsen, M. P., Bach, F. and Jensen, B. C.**, Shear tests on reinforced concrete T-beams, Ser. T, Copenhagen, Technical University of Denmark, Structural Research Laboratory, Report No. R-75, 1976, 114 pp.

[76.4] **Müller, P.**, Failure mechanisms for reinforced concrete beams in torsion and bending, Int. Assoc. Bridge Struct. Eng., Mem., Vol. 36, No. II, 1976, pp. 147-163.

[76.5] **Rozvany, I. N. and Hill, R. D.**, The theory of optimal load transmission by flexure, Adv. Appl. Mech., Vol. 16, 1976, pp. 184-308.

[76.6] **Bræstrup, M. W., Nielsen, M. P., Jensen, B. C. and Bach, F.**, Axisymmetric punching of plain and reinforced concrete, Copenhagen, Technical University of Denmark, Structural Research Laboratory, Report No. R-75, 1976, 33 pp.

[76.7] **Jensen, B. C. and Bræstrup, M. W.**, Lok-tests determine the compressive strength of concrete, Nord. Betong, Vol. 20, No. 2, Mar. 1976, pp. 9-11.

[76.8] **Kærn, J. and Jensen, L. F.**, Gennemlokning af beton (Punching shear in concrete), Copenhagen, Danmarks Ingeniørakademi, Bygningsafdelingen, Ren & Anvendt Mekanik, Report No. 76:78, May 1976, 34 pp.

[76.9] **Clark, L. A.**, The provision of tension and compression reinforcement to resist in-plane forces, Mag. Concr. Res., Vol. 28, No. 94, Mar. 1976, pp. 3-12.

[76.10] **Lyngberg, B. S.**, Ultimate shear resistance of partially prestressed reinforced concrete I-beams, J. Am. Concr. Inst., Proc., Vol. 73, Apr. 1976, pp. 214-222.

[76.11] **Nielsen, M. P.**, Punching shear resistance according to the CEB-Model Code, ACI/CEB/FIP/PCI Symp., Philadelphia, Mar.-Apr. 1976.

[76.12] **Thürlimann, B.**, Shear strength of reinforced and prestressed concrete beams - CEB approach, ACI/CEB/FIP/PCI Symp., Philadelphia, Mar.-Apr. 1976.

[76.13] **Thürlimann, B.**, Torsional strength of reinforced and prestressed concrete beams - CEB approach, ACI/CEB/FIP/PCI Symp., Philadelphia, Mar.-Apr. 1976.

[77.1] **Bræstrup, M. W., Nielsen, M. P. and Bach, F.**, Plastic analysis of shear in concrete, Z. Angew. Math. Phys., Vol. 58, 1978, pp. 3-14 (also: Copenhagen, Technical University of Denmark, DCAMM Report No. 120, May, 1977, 24 pp.).

[77.2] **Ottosen, N. S.**, A failure criterion for concrete, J. Eng. Mech. Div., No. EM4, Aug. 1977.

[77.3] **Bach, F., Nielsen, M. P. and Bræstrup, M. W.**, Forskydningsforsøg med jernbetonbjælker (Shear tests on reinforced concrete beams), Copenhagen, Technical University of Denmark, Structural Research Laboratory, Internal Report No. I-49, 1977, 19 pp.

[77.4] **Roikjær, M., Nielsen, M. P., Bræstrup, M. W. and Bach, F.**, Forskydningsforsøg med spændbetonbjælker uden forskydningsarmering (Shear tests on prestressed concrete beams without shear reinforcement), Copenhagen, Technical University of Denmark,

Structural Research Laboratory, Internal Report No. I-57, 1977, 15 pp.

[77.5] **Morley, C. T. and Gulvanessian, H.**, Optimum reinforcement of concrete slab elements, Proc. Inst. Civil Eng., Part 2, Vol. 63, 1977, pp. 441-454.

[77.6] **Jensen, J. F., Bræstrup, M. W., Bach, F. and Nielsen, M. P.**, Præfabrikerede sandwichelementer af letbeton (Prefabricated lightweight concrete sandwich elements), Copenhagen, Technical University of Denmark, Structural Research Laboratory, Internal Report No. I-54, 1977, 28 pp.

[77.7] **Marti, P. and Thürlimann, B.**, Fliessbedingung für Stahlbeton unter Berüchsichtigung der Betonzugfestigkeit (Yield criteria for reinforced concrete including the tensile strength of concrete), Beton-Stahlbetonbau, Vol. 72, No. 1, 1977, pp. 7-12.

[77.8] **Robinson, J. R. and Demorieux, J.-M.**, Essai de modeles d'ame de poutre en double Té. Annales de l'Institut Technique du Bâtiment et des Traveaux Publics, No. 354, Oct. 1977, Serie: Beton No. 172, pp 77-95.

[78.1] **Nielsen, M. P., Bræstrup, M. W., Jensen, B. C. and Bach, F.**, Concrete plasticity, beam shear - shear in joints - punching shear, Danish Society for Structural Science and Engineering, Special Publication, Lyngby 1978, 129 pp.

[78.2] **IABSE**, Plasticity in reinforced concrete, Introductory report, Zurich, Int. Assoc. for Bridge and Struct. Eng., Reports of the Working Commissions, Vol. 28, Oct. 1978, 172 pp.

[78.3] **Jensen, J. F., Nielsen, M. P., Bræstrup, M. W. and Bach, F.**, Nogle plasticitetsteoretiske bjælkeløsninger (Some plastic solutions concerning the load-carrying capacity of reinforced concrete beams), Copenhagen, Technical University of Denmark, Structural Research Laboratory, Report No. R-101, 1978, 50 pp.

[78.4] **Collins, M. P.**, Reinforced concrete in flexure and shear, Proc. Mark W. Huggins Symp. Struct. Eng., University of Toronto, Sept. 1978.

[78.5] **Collins, M. P.**, Towards a rational theory for RC members in shear, J. Struct. Div., ASCE, Vol. 104, No. ST4, Apr. 1978, pp. 649-666.

[78.6] **Jensen, J. F., Jensen, V., Christensen, H. H., Nielsen, M. P., Bræstrup, M. W. and Bach, F.**, On the behaviour of cracked reinforced concrete beams in the elastic range, Copenhagen, Technical University of Denmark, Structural Research Laboratory, Report No. R-103, 1978.

[78.7] **Pedersen, C., Jensen, J. F., Nielsen, M. P. and Bach, F.**, Opbøjet længdearmering som forskydningsarmering (Bent-up bars as shear reinforcement), Copenhagen, Technical University of

Denmark, Structural Research Laboratory, Report No. R-100, 1978.

[78.8] **Thürlimann, B.**, Plastic analysis of reinforced concrete beams, IABSE Colloquium, Copenhagen, 1979, Plasticity in Reinforced Concrete, Introductory Report, Zurich, Int. Assoc. for Bridge and Struct. Eng., Reports of the Working Commissions, Vol. 28, Oct. 1978, pp. 71-90.

[78.9] **Müller, P.**, Plastische Berechnung von Stahlbetonscheiben und -Balken (Plastic analysis of walls and beams of reinforced concrete), Institut für Baustatik und Konstruktion, ETH Zurich, Bericht Nr. 83, Stuttgart, Birkhäuser Verlag, 1978, 160 pp.

[78.10] **CEB-FIP**, Code Modèle CEB-FIP pour les structures en béton, Paris, Comité Euro-International du Béton, 1978, 336 pp.

[78.11] **Hess, U., Jensen, B. C., Bræstrup, M. W., Nielsen, M. P. and Bach, F.**, Gennemlokning af jernbetonplader (Punching shear of reinforced concrete slabs), Copenhagen, Technical University of Denmark, Structural Research Laboratory, Report No. R-90, 1978, 63 pp.

[78.12] **Jensen, B. C.**, Ultimate strength of joints, RILEM/CEB/CIB Symp. Mech. & Insulation Properties of Joints of Precast Reinforced Concrete Elements, Athens, Sept. 1978, Vol. I, pp. 223-240, Vol. III, pp. 279-280.

[78.13] **Hansen, K. E., Bryder, K. L. and Nielsen, M. P.**, Armeringsbestemmelse i jernbetonskaller (Reinforcement design in shells), Copenhagen, Technical University of Denmark, Structural Research Laboratory, Report No. R-91, 1978.

[78.14] **Nielsen, M. P., Hansen, L. P. and Rathkjen, A. R.**, Mekanik 2.2, del 1 og del 2, Rumlige deformations- og spændingstilstande (Structural Mechanics 2.2, Parts 1 and 2, 3-dimensional strain and stress fields), 2nd edition, 1978.

[78.15] **Marti, P.**, Plastic analysis of reinforced concrete shear walls, IABSE Colloquium, Plasticity in Reinforced Concrete, Copenhagen 1979, Introductory Report, Vol. 28, Oct. 1978, pp. 51-69.

[78.16] **Jensen, J. F., Pedersen, C., Bræstrup, M. W., Bach, F. and Nielsen, M. P.**, Rapport over forskydningsforsøg med 6 spændbetonbjælker (Shear tests with 6 prestressed concrete beams), Copenhagen, Technical University of Denmark, Structural Research Laboratory, Report No. R-102, 1978, 28 pp.

[78.17] **Bræstrup, M. W.**, Punching shear in concrete slabs, IABSE Colloquium, Copenhagen, 1979, Session III, Plasticity in Reinforced Concrete, Introductory Report, Vol. 28, Oct. 1978, pp. 115-136.

[78.18] **Chen, W. F.**, Constitutive equations for concrete, IABSE Colloquium, Copenhagen, 1979, Session I, Plasticity in Reinforced Concrete, Introductory Report, Vol. 28, Oct. 1978, pp. 11-34.

[78.19] **Nielsen, M. P. and Bræstrup, M. W.**, Shear strength of prestressed concrete beams without web reinforcement, Mag. Concr. Res., Vol. 30, No. 104, Sept. 1978.

[78.20] **Nielsen, M. P., Bræstrup, M. W. and Bach, F.**, Rational analysis of shear in reinforced concrete beams, Int. Assoc. Bridge Struct. Eng., Proc. P-15/78, May 1978, 16 pp.

[78.21] **Wang, P. T., Shah, S. P. and Naaman, A. E.**, Stress-strain curves of normal and lightweight concrete in compression, J. Am. Concr. Inst., Proc., Vol. 75, No. 11, Nov. 1978, pp. 603-611.

[78.22] **Fernando, J. S. and Kemp, K. O.**, A generalized strip deflection method of reinforced concrete slab design, Proc. Inst. Civil Eng., Part 2, Vol. 65, 1978, pp. 163-174.

[78.23] **Marti, P.**, Anwendung der Theorie starr-ideal plastischer Körper auf Stahlbetonplatten, Neuntes Forschungskolloquium des Deutschen Ausschusses für Stahlbeton (DAfStb), Institut für Baustatik und Konstruktion, ETH Zurich, Bericht No. 85, Nov. 1978, pp. 1-4.

[78.24] **Marti, P.**, Plastische Berechnungsmethoden, Vorlesungsautographie, Abteilung für Bauingenieurwesen, ETH Zurich, 1978.

[79.1] **IABSE**, Plasticity in reinforced concrete, Final report, Zurich, Int. Assoc. for Bridge and Struct. Eng., Reports of the Working Commissions, Vol. 29, Aug. 1979, 360 pp.

[79.2] **Exner, H.**, On the effectiveness factor in plastic analysis of concrete, IABSE Colloquium, Copenhagen, 1979, Session I, Plasticity in Reinforced Concrete, Final Report, Vol. 29, Aug. 1979, pp. 35-42.

[79.3] **Pedersen, H.**, Optimization of reinforcement in slabs by means of linear programming, IABSE Colloquium, Copenhagen, 1979, Session IV, Plasticity in Reinforced Concrete, Final Report, Vol. 29, Aug. 1979, pp. 263-272.

[79.4] **Roikjær, M., Pedersen, C., Bræstrup, M. W., Nielsen, M. P. and Bach, F.**, Bestemmelse af ikke-forskydningsarmerede bjælkers forskydningsbæreevne (Load-carrying capacity of beams without shear reinforcement), Danmarks tekniske Højskole, Afdelingen for Bærende Konstruktioner, Rapport Nr. I-62, 1979, 44 pp.

[79.5] **Pedersen, C.**, Shear in beams with bent-up bars, IABSE Colloquium, Copenhagen, 1979, Session II, Plasticity in Reinforced Concrete, Final Report, Vol. 29, Aug. 1979, pp. 79-86.

[79.6] **Gurley, C. R.**, The bimoment method for Hillerborg slabs, IABSE Colloquium, Copenhagen, 1979, Session III, Plasticity in Reinforced Concrete, Final Report, Vol. 29, Aug. 1979, pp. 153-157.

[79.7] **Kærn, J.**, The stringer method applied to discs with holes, IABSE Colloquium, Copenhagen, 1979, Session II, Plasticity in Reinforced Concrete, Final Report, Vol. 29, Aug. 1979, pp. 87-93.

[79.8] **Hess, U.**, The anchorage strength of reinforcement bars at supports, IABSE Colloquium, Copenhagen, 1979, Session V, Plasticity in Reinforced Concrete, Final Report, Vol. 29, Aug. 1979, pp. 309-316.

[79.9] **Morley, C. T.**, Punching shear failure of hollow concrete spheres, IABSE Colloquium, Copenhagen, 1979, Session III, Plasticity in Reinforced Concrete, Final Report, Vol. 29, Aug. 1979, pp. 167-174.

[79.10] **Nielsen, M. P.**, Some examples of lower-bound design of reinforcement in plane stress problems, IABSE Colloquium, Copenhagen,, 1979, Session V, Plasticity in Reinforced Concrete, Final Report, Vol. 29, Aug. 1979, pp. 317-324.

[79.11] **Clyde, D. H.**, Direct design by concrete flow, IABSE Colloquium, Copenhagen, 1979, Session V, Plasticity in Reinforced Concrete, Final Report, Vol. 29, Aug. 1979, pp. 325-332.

[79.12] **Collins, M. P.**, Investigating the stress-strain characteristics of diagonally cracked concrete, IABSE Colloquium, Copenhagen, 1979, Session I, Plasticity in Reinforced Concrete, Final Report, Vol. 29, Aug. 1979, pp. 27-34.

[79.13] **Collins, M. P.**, Reinforced concrete members in torsion and shear, IABSE Colloquium, Copenhagen, 1979, Session II, Plasticity in Reinforced Concrete, Final Report, Vol. 29, Aug. 1979, pp. 119-130.

[79.14] **Bræstrup, M. W.**, Effect of main steel strength on shear capacity of reinforced concrete beams with stirrups, Copenhagen, Technical University of Denmark, Structural Research Laboratory, Report No. R-110, 1979.

[79.15] **Clyde, D. H.**, Nodal forces as real forces, IABSE Colloquium, Copenhagen, 1979, Session III, Plasticity in Reinforced Concrete, Final Report, Vol. 29, Aug. 1979, pp. 159-166.

[79.16] **Jensen, B. C.**, Reinforced concrete corbels - Some exact solutions, IABSE Colloquium, Copenhagen, 1979, Session V, Plasticity in Reinforced Concrete, Final Report, Vol. 29, Aug. 1979, pp. 293-300.

[79.17] **Jensen, J. F.**, Plastic solutions for reinforced concrete beams in shear, IABSE Colloquium, Copenhagen, 1979, Session II, Plasticity in Reinforced Concrete, Final Report, Vol. 29, Aug. 1979, pp. 71-78.

[79.18] **Bryder, K. L.**, Optimization of reinforcement in discs by means of linear programming, Bygningsstat. Medd., Vol. 50, No. 4, 1979, pp. 63-77.

[79.19] **Fredsgård, S. and Kirk, J.**, Brudstadieberegninger med edb (Collapse analysis by numerical methods), Lyngby, ADB-Udvalget, Rap. Nr. R7911, Oct. 1979.

[79.20] **Nielsen, M. P. and Bach, F.**, A class of lower bound solutions for rectangular slabs, Bygningsstat. Medd., Vol. 50, No. 3, 1979, pp. 43-58.

[79.21] **Jensen, B. C.**, Om forskydning i støbeskel (Shear in casting joints), Bygningsstat. Medd., Vol. 50, No. 4, 1979, pp. 63-77.

[79.22] **Cookson, P. J.**, A general yield criterion for orthogonally reinforced concrete slab elements, IABSE Colloquium, Copenhagen, 1979, Session I, Plasticity in Reinforced Concrete, Final Report, Vol. 29, Aug. 1979, pp. 43-50.

[79.23] **Cookson, P. J.**, Generalized yield lines in reinforced concrete slabs, J. Struct. Mech., Vol. 7, No. 1, 1979, pp. 65-82.

[80.1] **Bach, F., Nielsen, M. P. and Bræstrup, M. W.**, Shear tests on reinforced concrete T-beams, Series V, U, X, B, and S, Copenhagen, Technical University of Denmark, Structural Research Laboratory, Report No. R-120, 1980, 86 pp.

[80.2] **Campbell, T. I., Chitnuyanondh, L. and Batchelor, B. D.**, Rigid-plastic theory v. truss analogy method for calculating the shear strength of reinforced concrete beams, Mag. Concr. Res., Vol. 32, No. 110, Mar. 1980, pp. 39-44.

[80.3] **Bjørnbak-Hansen, J. and Krenchel, H.**, Undersøgelse af let konstruktionsbeton til bærende konstruktioner (Investigation of structural lightweight aggregate concrete), Copenhagen, Technical University of Denmark, Structural Research Laboratory, Internal Report No. I-65, 1980, 40 pp.

[80.4] **Bræstrup, M. W.**, Dome effect in RC slabs: Rigid-plastic analysis, J. Struct. Div., Proc. ASCE, Vol. 106, No. ST6, Paper No. 15501, June 1980, pp. 1237-1253.

[80.5] **Bræstrup, M. W. and Morley, C. T.**, Dome effect in RC slabs: Elastic-plastic analysis, J. Struct. Div., Proc. ASCE, Vol. 106, No. ST6, 1980, pp. 1255-1262.

[80.6] **Marti, P.**, Zur plastischen Berechnung von Stahlbeton, Institut für Baustatik und Konstruktion, ETH Zurich, Bericht Nr. 104, Oct. 1980, 176 pp.

[80.7] **Nielsen, M. P. and Bach, F.**, Beregning af forskydningsarmering efter diagonaltrykmetoden (Calculation of shear reinforcement according to the diagonal compression field theory), Bygningsstat. Medd., Vol. 51, No. 3-4, 1980, pp. 75-139.

[80.8] **Hamadi, Y. D. and Regan, P. E.**, Behaviour of normal and light weight aggregate beams with shear cracks, The Struct. Eng., Vol. 58B, No. 4, Dec. 1980, pp. 71-79.

[81.1] **Bach, F. and Nielsen, M. P.**, Nedreværdiløsninger for plader (Lower bound solutions for slabs), Copenhagen, Technical

University of Denmark, Structural Research Laboratory, Report No. R-136, 1981, 76 pp.

[81.2] **Jensen, J. F.**, Plasticitetsteoretiske løsninger for skiver og bjælker af jernbeton (Plastic solutions for disks and beams of reinforced concrete), Copenhagen, Technical University of Denmark, Structural Research Laboratory, Ph.D. thesis, Report No. R-141, 1981, 153 pp.

[81.3] **Marti, P.**, Discussion of Proc. Paper 15705, "Concrete reinforcing net: Safe design" by Bazant, Z. P., Tsubaki, T. and Belytschko, T. B. (Sept. 1980), J. Struct. Div., ASCE, Vol. 107, No. ST7, July 1981, pp. 1391-1393.

[81.4] **Thürlimann, B. and Marti, P.**, Plastizität im Stahlbeton, Vorlesungsunterlagen, Abteilung für Bauingenieurwesen, ETH Zurich, 1981.

[81.5] **Marti, P.**, Gleichgewichtslösungen für Flachdecken, Schweizer Ingenieur und Architekt, Vol. 99, 1981, pp. 799-809.

[81.6] **Jensen, J. F., Bach, F., Rasmussen, J. and Nielsen, M. P.**, Jernbetonplader med hjørnelast (Reinforced concrete slabs with corner loads), Department of Structural Engineering, Technical University of Denmark, Report I, No. 68, 1981.

[81.7] **Walraven, J. C. and Reinhardt, H. W.**, Theory and experiments on the mechanical behaviour of cracks in plain and reinforced concrete subjected to shear loading, Heron, Vol. 26, No. 1A, 1981, 68 pp.

[81.8] **Damkilde, L. and Kirk, J.**, RUPTUS, et program til brudstadieberegninger (RUPTUS, a program for failure analysis), ADB-udvalget, Rapport 8115, 1981.

[82.1] **Jensen, B. C.**, On the ultimate load of reinforced concrete corbels, DIALOG 1:82, Miscellaneous Papers in Civil Engineering (25th anniversary of Denmark's Engineering Academy), Danish Engineering Academy, Department of Civil Engineering, 1982, pp. 119-137.

[82.2] **Marti, P. and Clyde, D. H.**, Discussion of "An upper-bound rigid-plastic solution for the shear failure of concrete beams without shear reinforcement" by Kemp, K. O. and Al Safi, M. T. (Vol. 33, No. 115), Mag. Concr. Res., Vol. 34, No. 119, June 1982, p. 96.

[82.3] **Marti, P.**, Strength and deformations of reinforced concrete members under torsion and combined actions, Comité Euro-International du Béton, Bulletin d'Information No. 146, Jan. 1982, pp. 97-138.

[82.4] **Vecchio, F. and Collins, M. P.**, The reponse of reinforced concrete to in-plane shear and normal stresses, The Department of Civil Engineering, University of Toronto, Canada, Mar. 1982, 332 pp.

[82.5] **Nishimura, A., Fujii, M., Miyamoto, A. and Yamada, S.**, Shear transfer at cracked section in reinforced concrete, Trans. of the Japan Concrete Institute, Vol. 4, 1982, pp. 257-268.

[82.6] **Christiansen, K. P.**, Membrane effect in concrete slabs, DIALOG, Civil Engineering Academy of Denmark, Lyngby, Aug. 1982.

[82.7] **Nielsen, M. P. and Christensen, S. B.**, Post-buckling strength of steel plate girders subjected to shear, Bygningsstat. Medd., Vol. 53, No. 3, 1982, pp. 97-122.

[82.8] **Chen, W. F.**, Plasticity in reinforced concrete, McGraw-Hill Book Company, 1982.

[83.1] **Feddersen, B. and Nielsen, M. P.**, Opbøjet spændarmering som forskydningsarmering (Bent-up prestressed reinforcement as shear reinforcement), Department of Structural Engineering, Technical University of Denmark, Report No. R-160, 1983, 20 pp.

[83.2] **Exner, H.**, Plasticitetsteori for Coulomb materialer (Theory of plasticity for Coulomb materials), Ph.D. thesis, Department of Structural Engineering, Technical University of Denmark, Report No. R-175, 1983.

[83.3] **Feddersen, B. and Nielsen, M. P.**, Revneteorier for enaksede spændingstilstande (Crack width theories for uniaxial stress fields), Department of Structural Engineering, Technical University of Denmark, Report No. R-162, 1983, 37 pp.

[83.4] **Feddersen, B. and Nielsen, M. P.**, Revneteori for biaksiale spændingstilstande (Crack width theory for biaxial stress fields), Department of Structural Engineering, Technical University of Denmark, Report No. R-163, 1983, 24 pp.

[83.5] **Nielsen, M. P. and Feddersen, B.**, Effektivitetsfaktoren ved bøjning af jernbetonbjælker (The effectiveness factor of reinforced concrete beams in bending), Department of Structural Engineering, Technical University of Denmark, Report No. R-173, 1983, 123 pp.

[83.6] **Feddersen, B. and Nielsen, M. P.**, Effektivitetsfaktoren ved vridning af jernbetonbjælker (The effectiveness factor of reinforced concrete beams in pure torsion), Department of Structural Engineering, Technical University of Denmark, Report No. R-174, 1983, 84 pp.

[83.7] **Marti, P.**, Discussion of "Is the 'Staggering concept' of shear design safe?", by Hsu, T. T. C. (Nov.-Dec. 1982), J. Am. Concr. Inst., Vol. 80, Sept.-Oct. 1983, pp. 445-446.

[83.8] **Marti, P.**, Discussion of Proc. Paper 17457, "Concrete plate reinforcement: Frictional limit design" by Bazant, Z.P. and Lin, C. (Nov. 1982), J. Struct. Div., ASCE, Vol. 109, No. ST9, Sept. 1983, pp. 2231-2233.

[83.9] **Marti, P.**, Plasticity in reinforced concrete - potential and limitations, W. Prager Symp. on Mechanics of Geomaterials: Rocks, Concretes, Soils, Northwestern University, Evanston, Illinois, 11-15 Sept. 1983, Preprints, pp. 657-660.

[83.10] **Thürlimann, B., Marti, P., Pralong, J., Ritz, P. and Zimmerli, B.**, Anwendung der Plastizitätstheorie auf Stahlbeton, Institut für Baustatik und Konstruktion, ETH Zurich, 1983, 252 pp.

[83.11] **Marti, P.**, Über die Bedeutung von Gleichgewichtsbetrachtungen im Massivbau, Schweizer Ingenieur und Architekt, Vol. 101, No. 7, 1983, pp. 184-185.

[83.12] **Exner, H.**, Betonbjælkers bøjningsbæreevne (Bending capacity of concrete beams), Department of Structural Engineering, Technical University of Denmark, Report No. R-176, 1983.

[83.13] **Bræstrup, M. W. and Nielsen, M. P.**, Plastic design methods of analysis and design, Section 20, Kong, F. K., Evans, R. H., Cohen, E. and Roll F. (eds.), Handbook of Structural Concrete, Pitman Advanced Publishing Program, 1983.

[83.14] **Christiansen, K. P.**, Ultimate strength of concrete slabs with horizontal restraints, Bygningsstat. Medd., Vol. LIV, No. 3, 1983.

[83.15] **Eyre, J. R. and Kemp, K. D.** A graphical solution for predicting the increase in strength of concrete slabs due to membrane action, Mag. Concr. Res., Vol. 35, No. 124, Sept. 1983.

[83.16] **Christiansen, K. P. and Frederiksen, V. T.**, Experimental investigation of rectangular concrete slabs with horizontal restraints, Mat. and Struct., Vol. 16, No. 93, May-June 1983, pp. 179-192.

[83.17] **Christiansen, K. P. and Frederiksen, V. T.**, Tests on rectangular concrete slabs with horizontal restraints on three sides only, Institute of Building Design, Nordic Concr. Res., 1983.

[84.1] **Feddersen, B., Hess, U., Exner, H. and Nielsen, M. P.**, Gennemlokning af huldækelementer (Punching shear strength of prestressed hollow core elements), Department of Structural Engineering, Technical University of Denmark, Report No. R-182, 1984, 37 pp.

[84.2] **Feddersen, B. and Nielsen, M. P.**, Plastic analysis of reinforced concrete beams in pure bending and pure torsion, The Danish Society of Structural Science and Engineering, Vol. 55, No. 2, 1984, 24 pp.

[84.3] **Feddersen, B. and Nielsen, M. P.**, Cracks in reinforced concrete subjected to uniaxial and biaxial stress fields, The Danish Society of Structural Science and Engineering, Vol. 55, No. 4, 1984, 20 pp.

[84.4] **Jensen, B. C.**, Armerede betonkonsollers bæreevne (Ultimate load of reinforced concrete corbels), Bygningsstat. Medd., No. 1, 1984, pp. 1-36.

[84.5] **Marti, P.**, The use of truss models in detailing, Developments in Design for Shear and Torsion, ACI-ASCE Committee 445, Shear and Torsion, Mar. 1984, pp. 49-55.

[84.6] **Kærn, J. C.**, Numerisk brudstadieberegning af stift, plastiske materialer (Numerical failure analysis of rigid plastic materials), Department of Structural Engineering, Technical University of Denmark, Report No. R-181, 1984.

[84.7] **Hess, U.**, Plasticitetsteoretisk analyse af forankring og stød af forkammet armering i beton (Plastic analysis of anchorage and splicing of ribbed bars in concrete), Department of Structural Engineering, Technical University of Denmark, Report No. R-184, 1984.

[84.8] **Bræstrup, M. W.**, Discussion of "Deformation and failure in large-scale pullout tests" by Stone, W. C. and Carino, N. J., J. Am. Concr. Inst., Vol. 81, No. 5, Sept.-Oct. 1984, pp. 525-526.

[84.9] **Rasmussen, J., Christensen, S. B., Exner H. and Nielsen, M. P.**, Plasticitetsteoretisk analyse af beregningsmetoder for jernbetonbjælkers forskydningsbæreevne (Analysis of plastic solutions for the shear design of reinforced concrete beams), Department of Structural Engineering, Technical University of Denmark, Report No. R-137, 1984.

[84.10] **Bellotti, R. and Ronzoni, E.**, Results of tests carried out on cylindrical concrete specimens subjected to complex stress state: A critical analysis, Int. Conf. on Concrete under Multiax. Cond., RILEM, Press de l'Universite Paul Sabatier, Toulouse, May 1984, pp. 53-74.

[84.11] **Nielsen, M. P.**, Limit Analysis and Concrete Plasticity, 1st ed., Prentice-Hall, Inc., Englewood Cliffs, New Jersey, 1984.

[84.12] **FIP Recommendations**, Practical design of reinforced and prestressed concrete structures, Thomas Telford Ltd., London 1984.

[85.1] **Feddersen, B. and Nielsen, M. P.**, Opgaver i styrkeberegning af beton og jernbeton (Problems in Limit Analysis and Concrete Plasticity), Department of Structural Engineering, Technical University of Denmark, Report No. F-102, 1985, 25 pp.

[85.2] **Feddersen, B. and Nielsen, M. P.**, Besvarelser til opgaver i styrkeberegning af beton og jernbeton (Solutions to problems in Limit Analysis and Concrete Plasticity), Department of Structural Engineering, Technical University of Denmark, Report No. I-81, 1985, 125 pp.

[85.3] **Feddersen, B. and Nielsen, M. P.**, Opbøjet spændarmering som forskydningsarmering (Bent-up prestressed reinforcement as shear reinforcement), The Danish Society of Structural Science and Engineering, Vol. 56, No. 1, 1985, 24 pp.

[85.4] **Andreasen, B. S. and Nielsen, M. P.**, Armering af beton i det tredimensionale tilfælde (Reinforcement of concrete in the 3-dimensional case), Bygningsstat. Medd., Dansk Selskab for Bygningsstatik, Vol. 56, No. 2-3, 1985, pp. 25-79.

[85.5] **Marti, P.**, Basic tools of reinforced concrete beam design, J. Am. Concr. Inst., Vol. 82, No. 1, Jan.-Feb. 1985, pp. 46-56. Discussion and Closure in Vol. 82, No. 6, Nov.-Dec. 1985, pp. 933-935.

[85.6] **Brynjarsson, A. Ø.**, Forskydningsarmerede jernbetonplader (Shear reinforced concrete slabs), M.Sc. thesis, Department of Structural Engineering, Technical University of Denmark, 1985.

[85.7] **Maier, J. and Thürlimann, B.**, Bruchversuche an Stahlbetonscheiben, Institut für Baustatik und Konstruktion, ETH Zürich, Bericht Nr. 8003-1, January 1985.

[85.8] **Andreasen, B. S.**, Trykmembranvirkning i betonplader (Compression membrane action in concrete slabs), M.Sc. thesis, Department of Structural Engineering, Technical University of Denmark, July 1985.

[85.9] **Regan, P. E. and Bræstrup, M. W.**, Punching shear in reinforced concrete - A state of art report, CEB Bull. d'Information, No. 168, Jan. 1985.

[86.1] **Andreasen, B. S.**, The bond strength of reinforcing bars at supports, Int. Symp. on Fundamental Theory of Reinforced and Prestressed Concrete, Nanjing, China, Sept. 1986, pp. 387-397.

[86.2] **Andreasen, B. S.**, The bond strength of deformed reinforcing bars, Chalmers University of Technology, Division of Concrete Structures, Publikation 86:1, pp. 44-58.

[86.3] **Andreasen, B. S. and Nielsen, M. P.**, Dome effect in reinforced concrete slabs, Department of Structural Engineering, Technical University of Denmark, Report No. R-212, 1986.

[86.4] **Jensen, B. C.**, Ultimate load of reinforced concrete corbels, Int. Symp. on Fundamental Theory of Reinforced and Prestressed Concrete, Nanjing, China, Sept. 1986, Vol. II, pp. 554-561.

[86.5] **Marti, P.**, Staggered shear design of concrete bridge girders, Int. Conf. on Short and Medium Span Bridges, Ottawa, Ontario, Canada, Aug. 17-21, 1986, Proc., Vol. 1, pp. 139-149.

[86.6] **Marti, P.**, Staggered shear design of simply supported concrete beams, J. Am. Concr. Inst., Vol. 83, No. 1, Jan.-Feb. 1986, pp. 36-42.

[86.7] **Marti, P.**, Truss models in detailing, Concrete International: Design and Construction, Vol. 7, No. 12, Dec. 1985, pp. 66-73. Discussion and Closure in Vol. 8, No. 10, Oct. 1986, pp. 66-68.

[86.8] **Bræstrup, M. W.**, Shear strength of RC beams with distributed loading - a coinciding lower bound plastic solution, Int. Symp. on

Fundamental Theory of Reinforced and Prestressed Concrete, Nanjing, China, Sept. 1986, Vol. 1, pp. 195-202.

[86.9] **Hauge, L.**, Koncentreret last på beton (Concentrated loads on concrete), M.Sc. thesis, Department of Structural Engineering, Technical University of Denmark, 1986.

[86.10] **Madsen, K.**, Lokalt brud i beton (Local failure in concrete), Sektionen for Konstruktionsmekanik, Danmarks Ingeniørakademi, Bygningsafdelingen, 1986.

[86.11] **van Mier, J. G. M.**, Fracture of concrete under complex stress, Heron, Vol. 31, No. 3, 1986.

[87.1] **Feddersen, B.**, Brudberegning af beton og jernbeton ved hjælp af plasticitetsteorien (Calculation of the load-bearing capacity of concrete and reinforced concrete by use of the theory of plasticity), Dansk Beton (Journal of Danish Concrete), Vol. 4, No. 2, 3 and 4, 1987.

[87.2] **Ichinose, T. and Takiguchi, K.**, Shear deformation mode of reinforced concrete beam, J. Struct. Eng., ASCE, Vol. 113, No. 4, 1987, pp. 689-703.

[87.3] **Cerruti, L. M. and Marti, P.**, Staggered shear design of concrete beams: Large scale tests, Canadian Journal of Civil Engineering, Vol. 14, No. 2, Apr. 1987, pp. 257-268.

[87.4] **Nielsen, M. P.**, Minimumsarmering (Minimum reinforcement), Dansk Beton (Journal of Danish Concrete), Vol. 4, No. 2, 1987.

[87.5] **Schlaich, J., Schäfer, K. and Jennewein, M.**, Toward a consistent design of structural concrete, PCI Journal, May/June 1987, pp. 75-150.

[88.1] **Andreasen, B. S.**, Anchorage of ribbed reinforcing bars, Ph.D. thesis, Department of Structural Engineering, Technical University of Denmark, Report No. R-238, 1988.

[88.2] **Andreasen, B. S.**, Anchorage tests with ribbed reinforcing bars in more than one layer at a beam support, Ph.D. thesis, Department of Structural Engineering, Technical University of Denmark, Report No. R-239, 1988.

[88.3] **Nielsen, M. P., Andreasen, B. S. and Chen, G.W.**, Dome effect in reinforced concrete slabs, 11th ACMSM, University of Auckland, 1988, 9 pp.

[88.4] **Marti, P.**, New concepts of shear design, VSL Symposium 1988, Lugano, 8 pp.

[88.5] **Bræstrup, M. W.**, Gennemlokning af betonplader. Normregler, beregninger og forsøgsresultater (Punching of concrete slabs. Code rules, analyses and test results). Bygningsstat. Medd., Vol. 59, No. 3-4, 1988, pp. 113-156.

[88.6] **Bræstrup, M. W.**, Discussion of "Concrete-to-concrete friction" by Tassios, T. P. and Vintzeleou, E. N., J. Struct. Eng., Vol. 114, No. 12, Dec. 1988, pp. 2824-2827.

[88.7] **Bræstrup, M. W.**, Non-linear design - A historical review, FIP Sym. on Practical Application of Non-linear Design of Prestressed Concrete Structures, Jerusalem, Israel, Sept. 1988, Proc. pp. 227-236.

[88.8] **Bræstrup, M. W.**, Discussion of "Ultimate strength of slab-column connections" by Alexander, S. D. B. and Simmonds, S. H., ACI Struct. Jour., Vol. 85, No. 2, Mar.-Apr. 1988, pp. 1606-1626.

[88.9] **Levi, F. and Marro, P.**, Shear tests up to failure of beams made with normal and high strength concrete, Special Discussion Session, Design Aspects of High Strength Concrete, Sept. 1988, Dubrovnik.

[88.10] **Madsen, K.**, Koncentreret last på beton (Concentrated loads on concrete), Dansk Beton (Journal of Danish Concrete), No. 1, 1988.

[88.11] **Harder, N. A.**, Brudlinieteoriens knudekræfter (The nodal forces of the yield line theory), Bygningsstat. Medd., Vol. 59, No. 2, 1988, pp. 81-111.

[88.12] **Chen, G.W.**, Plastic analysis of shear in beams, deep beams and corbels, Department of Structural Engineering, Technical University of Denmark, Report No. R-237, 1988.

[89.1] **Feddersen, B., Nielsen, M.P. and Olsen, D. H.**, Opgaver i styrkeberegning af beton og jernbeton (Problems in Limit Analysis and Concrete Plasticity), Department of Structural Engineering, Technical University of Denmark, Report No. F-117, 1989, 44 pp.

[89.2] **Feddersen, B.**, Betonbjælkers forskydningsbæreevne under hensyntagen til buevirkning (Shear capacity of concrete beams taking account of arch effects), The Danish Society of Structural Science and Engineering, Vol. 60, No. 1, 1989, 48 pp.

[89.3] **Andreasen, B. S. and Nielsen, M. P.**, Dome effect in reinforced concrete two-way slabs, Bygningsstat. Medd., Dansk Selskab for Bygningsstatik, Vol. 60, No. 3-4, 1989, pp. 79-120.

[89.4] **Marti, P., Leesti, P. and Khalifa, W. U.**, Torsion tests on reinforced concrete slab elements, J. Struct. Eng., ASCE, Vol. 113, No. ST5, May 1987, pp. 994-1010. Discussion and Closure in Vol. 115, No. ST2, Feb. 1989, pp. 495-497.

[89.5] **Marti, P. and Kong, K.**, Response of reinforced concrete slab elements to torsion, J. Struct. Eng., ASCE, Vol. 113, No. ST5, May 1987, pp. 976-993. Discussion and Closure in Vol. 115, No. ST2, Feb. 1989, pp. 493-495.

[89.6] **Marti, P.**, Size effect in double-punch tests on concrete cylinders, ACI Materials Journal, Vol. 86, No. 6, Nov.-Dec. 1989, pp. 597-601.

[89.7] **Bræstrup, M. W.**, Punching of reinforced concrete slabs: Code rules, plastic analysis, test results, Nordic Concrete Research, Publication No. 8, 1989, pp. 24-48.

[89.8] **Yin, X.Q., Ottosen, N. S., Thelandersson, S. and Nielsen, M. P.**, Review of constitutive models for concrete, Commission of the European Communities, Nuclear Science and Technology, Shared Cost Action, Reactor Safety Programme 1985-87, Final Report ISPRA, Nov. 1989.

[89.9] **Bigom, L.**, Forskydningsbæreevne af højstyrkebeton (Shear capacity of high strength concrete), M.Sc. thesis, Department of Structural Engineering, Technical University of Denmark, 1989.

[89.10] **Andreasen, B. S.**, Anchorage of ribbed reinforcing bars, Department of Structural Engineering, Technical University of Denmark, Report R, No. 238, 1989.

[89.11] **Andreasen, B. S.**, Anchorage tests with ribbed reinforcing bars in more than one layer at a beam support, Department of Structural Engineering, Technical University of Denmark, Report R, No. 239, 1989.

[90.1] **Feddersen, B.**, Jernbetonbjælkers bæreevne -rene og kombinerede påvirkninger (Load-bearing capacity of reinforced concrete beams subjected to pure and combined actions), Ph.D. thesis, Department of Structural Engineering, Technical University of Denmark, Report No. R-251 and R-252, 1990, 450 pp.

[90.2] **Feddersen, B. and Nielsen, M. P.**, Dimensionering af betonbjælkers vederlag (Design of the supports in reinforced concrete beams), Dansk Beton (Journal of Danish Concrete), Vol. 7, No. 3 and 4, 1990.

[90.3] **Andreasen, B. S.**, Forankring i beton (Anchorage in concrete), Part 1, Dansk Beton (Journal of Danish Concrete), No. 4, Nov. 1990.

[90.4] **Naganuma, K. and Yamaguchi, T.**, Tension stiffening model under in-plane shear stress, Summaries of Technical Papers of Annual Meeting, Architectural Institute of Japan, 1990, pp. 649-650 (in Japanese).

[90.5] **Olsen, D. H. and Nielsen, M. P.**, Ny teori til bestemmelse af revneafstande og revnevidder i betonkonstruktioner (New theory for crack width design in concrete structures), Department of Structural Engineering, Technical University of Denmark, Report No. R-254, 1990.

[90.6] **Olsen, D. H., Chen, G.W. and Nielsen, M. P.**, Plastic shear solutions of prestressed hollow core concrete slabs, Department of Structural Engineering, Technical University of Denmark, Report No. R-257, 1990.

[90.7] **Chen, G.W. and Nielsen, M. P**, Shear strength of beams of high strength concrete, Department of Structural Engineering, Technical University of Denmark, Report No. R-258, 1990.

[90.8] **Chen, G.W., Nielsen, M. P. and Janos, K.**, Ultimate load carrying capacity of unbonded prestressed reinforced concrete beams, Department of Structural Engineering, Technical University of Denmark, Report No. R-259, 1990.

[90.9] **Chen, G.W. and Nielsen, M. P.**, A short note on plastic shear solutions of reinforced concrete columns, Department of Structural Engineering, Technical University of Denmark, Report No. R-260, 1990.

[90.10] **Dahl, K. K. B.**, Preliminary state-of-the-art report on multiaxial strength of concrete, Department of Structural Engineering, Technical University of Denmark, Report No. R-262, 1990.

[90.11] **Bræstrup, M. W.**, Discussion of "Relationship of the punching shear capacity of reinforced concrete slabs with concrete strength" by Gardner, N. J., ACI Struct. Jour., Vol. 87, No. 6, Nov.-Dec. 1990, pp. 743-744.

[90.12] **Bræstrup, M. W.**, Plastic analysis as a design tool: Shear strength of deep beams and corbels, Int. Conf. on Bridge, Amirkabir, University of Technology, Tehran, 1-3 May, 1990.

[90.13] **Nielsen, M. P.**, Shear and torsion, Eurocode 2, Editorial Group, 1st Draft, Oct. 1990 (unpublished).

[90.14] **Nielsen, M. P.**, Plastic analysis of slabs, Eurocode 2, Editorial Group, 1st Draft, Mar. 1990 (unpublished).

[90.15] **Nielsen, M. P.**, Beton 1, Del 2 - Tværsnitsundersøgelse i brugs- og brudstadiet (Concrete Structures 1, Part 2, Analysis of sections in the serviceability and the ultimate limit state), 2nd preliminary edition, Department of Structural Engineering, Technical University of Denmark, Lyngby 1990.

[90.16] **Nielsen, M. P.**, Beton 1, Del 3 - Bøjning med forskydning, To-aksede spændingstilstande, søjler og bjælkesøjler, konstruktive regler, resumésamling (Concrete Structures 1, Part 3, Bending and shear, plane stress fields, columns and beam columns, design guidelines, summaries), 2nd preliminary edition, Department of Structural Engineering, Technical University of Denmark, Lyngby 1990.

[90.17] **Olsen, N. H.**, The strength of overlapped deformed tensile reinforcement splices in high strength concrete, Department of Structural Engineering, Technical University of Denmark, Report No. R-234, 1990.

[90.18] **Olsen, N. H.**, Design proposal for high strength concrete sections subjected to flexural and axial loads, Department of Structural Engineering, Technical University of Denmark, Report No. R-233, 1990.

[90.19] **Madsen, K.**, Ny beregningsregel for koncentreret last på uarmeret beton (New design rule for concentrated loads on unreinforced concrete), Dansk Beton (Journal of Danish Concrete), No. 3, 1990.

[90.20] **Muttoni, A.**, Die Anwendbarkeit der Plasticitätstheorie in der Bemessung von Stahlbeton, Institut für Baustatik und Konstruktion, ETH, Zurich, Bericht Nr. 176, Juni 1990.

[91.1] **Feddersen, B., Olsen, D. H. and Nielsen, M. P.**, Besvarelser til opgaver i styrkeberegning af beton og jernbeton, 2. udgave (Answers to Problems in Limit Analysis and Concrete Plasticity), 2nd edition, Department of Structural Engineering, Technical University of Denmark, Report No. I-101, 1991, 189 pp.

[91.2] **Andreasen, B. S.**, Forankring i beton, (Anchorage in concrete), Part 2, Dansk Beton (Journal of Danish Concrete), No. 1, Feb. 1991.

[91.3] **Chen, G.W., Andreasen, B. S. and Nielsen, M. P.**, Membrane action tests of reinforced concrete square slabs, Department of Structural Engineering, Technical University of Denmark, Report No. R-273, 1991.

[91.4] **Andreasen, B. S.**, Cast-in-place thread bars, Int. Symp. on Concrete Engineering, Southeast University, Nanjing, China, Proc. Vol. 3, 18-20 Sept. 1991, pp. 1487-1492.

[91.5] **Andreasen, B. S.**, Anchorage of multilayered reinforcement at supports, Nordic Concrete Research, The Nordic Concrete Federation, 1991.

[91.6] **Ichinose, T.**, Interactions between bond and shear in R/C interior joints, Am. Concr. Inst., Spec. Publ. SP 123-14, 1991, pp. 379-399.

[91.7] **Ichinose, T. and Yokoo, Y.**, A shear design procedure of reinforced concrete beam with web opening, Proc. of the Japan Concrete Institute, Vol. 13, No. 2, 1991, pp. 303-308.

[91.8] **Takeda, T., Yamaguchi, K. and Naganuma, K.**, Reports on tests of nuclear prestressed concrete containment vessels, Concrete shear in earthquake, Edited by Hsu, T. C. C. and Mau, S. T., Elsevier Applied Science (Transactions of the International Workshop on Concrete Shear in Earthquake), 1991, pp. 163-172.

[91.9] **Yamaguchi, T. and Naganuma, K.**, Experimental study on mechanical charateristics of reinforced concrete panels subjected to in-plane shear force, J. Struct. and Constr. Eng., Architectural Institute of Japan, No. 419, Jan. 1991, pp. 77-86 (in Japanese).

[91.10] **Naganuma, K. et al.**, Proposal and verification of nonlinear analysis model for reinforced concrete panels, Report of Obayashi Corporation Technical Research Institute, No. 42, Feb. 1991, pp. 9-14 (in Japanese).

[91.11] **Naganuma, K. et al.**, Nonlinear analytical model for reinforced concrete panels under in-plane stresses, J. Struct. and Constr. Eng., Architectural Institute of Japan, No. 421, Mar. 1991, pp. 39-48 (in Japanese).

[91.12] **Naganuma, K., Yoshioka, K. and Omote, Y.**, Three dimensional nonlinear finite element analysis of reinforced concrete beam-column joint, Computer Applications in Civil and Building Engineering (Proc. of the 4th Int. Conf. on Computing in Civil and Building Engineering), July 1991, pp. 219-226.

[91.13] **Takeda, T., Yamaguchi, T. and Naganuma, K.**, An analytical model of reinforced concrete panel under in-plane shear stress, Transactions of the 11th International Conference on Structural Mechanics in Reactor Technology (SMiRT 11), H14/3, Aug. 1991, pp. 413-418.

[91.14] **Naganuma, K.**, Effects of compressive deterioration of cracked concrete on analytical results of shear walls with high strength concrete, Summaries of Technical Papers of Annual Meeting, Architectural Institute of Japan, 1991, pp. 435-436 (in Japanese).

[91.15] **Marti, P.**, Dimensioning and detailing, IABSE Colloquium 'Structural Concrete', Stuttgart 1991, IABSE Reports, Vol. 62, Zurich, 1991, pp. 411-443.

[91.16] **Marti, P.**, Discussion of "Stirrup stresses in reinforced concrete beams" by Hsu, T. T. C. (Sept.-Oct. 1990), ACI Struct. Jour., Vol. 88, No. 4, July-Aug. 1991, pp. 517-519.

[91.17] **Marti, P.**, Discussion of "Behavior of reinforced concrete membranes with compatible stress and cracking" by Fialkow, M. N. (Sept.-Oct. 1990), ACI Struct. Jour., Vol. 88, No. 4, July-Aug. 1991, pp. 519-522.

[91.18] **Marti, P.**, Design of concrete slabs for transverse shear, ACI Struct. Jour., Vol. 87, No. 2, Mar.-Apr. 1990. Discussion and Closure in Vol. 88, No. 1, Jan.-Feb. 1991, pp. 117-118.

[91.19] **Yin, X.Q.**, Constitutive equations and their application in finite element analysis, Department of Structural Engineering, Technical University of Denmark, Report No. R-269, 1991.

[91.20] **Andreasen, B. S. and Nielsen, M. P.**, Arch effect in reinforced concrete one-way slabs, Department of Structural Engineering, Technical University of Denmark, Report No. R-275, 1991.

[91.21] **Bræstrup, M. W.**, Discussion of "Punching shear behavior of slabs with varying span-depth ratios" by Lovrovich, J. S. and McLean, D. I., ACI Struct. Jour., Vol. 88, No. 4, July-Aug. 1991, pp. 515-516.

[91.22] **Nielsen, M. P.**, Concrete beam shear design according to Eurocode 2, Int. Symp. on Concrete Engineering, Southeast University, Nanjing, China, Additional Vol., 18-20 Sept. 1991, pp. 1949-1954.

[91.23] **EC 2, Eurocode 2**, Design of concrete structures - Part 1, General rules for buildings, European Prestandard ENV 1992-1-1:1991, European Committee for Standardisation TC 250, Brussels, 1991.

[91.24] **Rønfeldt, B.**, Bøjning af højstyrkebetonbjælker (Bending of high strength concrete beams), M.Sc. thesis, Department of Structural Engineering, Technical University of Denmark, 1991.

[91.25] **Rangan, B. V.**, Web crushing strength of reinforced and prestressed concrete beams, ACI Struct. Jour., Vol. 88, No. 1, Jan.-Feb. 1991, pp. 12-16.

[91.26] **Sigurdsson, T. G.**, Gennemlokning i beton ved excentrisk belastning (Punching shear by eccentric loading), M.Sc. thesis, Department of Structural Engineering, Technical University of Denmark, July 1991.

[91.27] **Collins, M. P. and Mitchell, D.**, Prestressed concrete structures, Prentice-Hall, Englewood Cliffs, New Jersey, 1991.

[91.28] **Schäfer, K., Schlaich, J. and Jennewin, M.**, Strut-and-tie modelling of structural concrete, 'Structural Concrete', IABSE Colloquium, Stuttgart 1991, IABSE Reports, Vol. 62, Zurich, 1991.

[91.29] **Jirsa, J. O., Breen, J. E., Bergmeister, K., Barton, D., Anderson, R. and Bouadi, H.**, Experimental studies of nodes in strut-and-tie models, 'Structural Concrete', IABSE Colloquium, Stuttgart 1991, IABSE Reports, Vol. 62, Zurich 1991, pp. 525-532.

[91.30] **Jensen, H. E.**, State-of-the-art report for revnet betons styrke (State-of-the-art report on the strength of cracked concrete), High Performance Concretes in the 90s, Report 7.1, Aug. 1991.

[92.1] **Andreasen, B. S.**, Cast-in-place bonded anchors, Nordic Concrete Research, The Nordic Concrete Federation, No. 12, Feb. 1992.

[92.2] **Andreasen, B. S.**, Anchorage of ribbed reinforcing bars, Int. Conf. on Bond in Concrete, Riga, Latvia, Oct. 1992, Proc., Vol. 1, pp. 18-27.

[92.3] **Madsen, K., Møller, N. and Rosetzsky, T.**, The shear strength of rough construction joints in normal and high-strength concrete, Miscellaneous Papers in Civil Engineering, 35th Anniversary of the Danish Engineering Academy, Copenhagen, 1992, pp. 69-80.

[92.4] **Jensen, B. C.**, Carrying capacity of reinforced concrete corbels, Int. Conf. on Concrete, Tehran, Iran, Nov. 1992, Proc., pp. 274-287.

[92.5] **Ichinose, T.**, A shear design equation for ductile R/C members, Earthquake Engineering & Structural Dynamics, Vol. 21, No. 3, 1992, pp. 197-214.

[92.6] **Ichinose, T. and Yokoo, Y.**, Effects of sparseness of shear reinforcement on shear strength of R/C beam, J. Struct. and

Constr. Eng., Architectural Institute of Japan, No. 437, 1992, pp. 97-103 (in Japanese).

[92.7] **Ichinose, T. and Yokoo, Y.**, Shear strength of R/C beams with spiral reinforcement, J. Struct. and Constr. Eng., Architectural Institute of Japan, No. 441, 1992, pp. 85-91 (in Japanese).

[92.8] **Watanabe, F. and Ichinose, T.**, Strength and ductility design of RC members subjected to combined bending and shear, Concrete shear in earthquake, Elsevier Applied Science, Publishers Ltd., 1992, pp. 429-438.

[92.9] **Naganuma, K.**, Method and applicability of nonlinear analysis for reinforced concrete shear walls, J. Struct. and Constr. Eng., Architectural Institute of Japan, No. 431, Jan. 1992, pp. 7-16 (in Japanese).

[92.10] **Naganuma, K. et al.**, Three dimensional nonlinear finite element analysis of a reinforced concrete column subjected to horizontal load under high axial load, Part 1 - Comparison of plane stress analysis with three dimensional analysis, Part 2 - Quantification of confinement effect of lateral reinforcement, Summaries of Papers of Annual Meeting, Architectural Institute of Japan, 1992, pp. 1065-1068 (in Japanese).

[92.11] **Marti, P. and Meyboom, J.**, Response of prestressed concrete elements to in-plane shear forces, ACI Struct. Jour., Vol. 89, No. 5, Sept.-Oct. 1992, pp. 503-514.

[92.12] **Marti, P.**, Truss models, Topical Volume: Cast-in-place-concrete in tall building design and construction, Council on Tall Buildings and Urban Habitat, 1992, Chap. 5, Sect. 5.1, pp. 67-81.

[92.13] **Marti, P.**, State-of-the-art of membrane shear behavior - European work, Concrete Shear in Earthquake, Edited by Hsu, T. C. C. and Mau, S. T., Elsevier Applied Science, London, 1992, pp. 187-195.

[92.14] **Dahl, K. K. B.**, Uniaxial stress-strain curves for normal and high strength concrete, Department of Structural Engineering, Technical University of Denmark, Report No. R-282, 1992.

[92.15] **Dahl, K. K. B.**, The calibration and use of a triaxial cell, Department of Structural Engineering, Technical University of Denmark, Report No. R-285, 1992.

[92.16] **Dahl, K. K. B.**, A failure criterion for normal and high strength concrete, Department of Structural Engineering, Technical University of Denmark, Report No. R-286, 1992.

[92.17] **Dahl, K. K. B.**, A constitutive model for normal and high strength concrete, Department of Structural Engineering, Technical University of Denmark, Report No. R-287, 1992.

[92.18] **Jensen, H. E.**, State-of-the-art rapport for revnet betons styrke (State-of-the-art report on the strength of cracked concrete),

Department of Structural Engineering, Technical University of Denmark, Report No. R-295, 1992.

[92.19] **Bræstrup, M. W.**, Discussion of "Web crushing strength of reinforced and prestressed concrete beams" by Rangan, B. V., ACI Struct. Jour., Vol. 89, No. 1, 1992, p. 115.

[92.20] **Bræstrup, M. W.**, Discussion of "Ultimate shear force of structural concrete members without transverse reinforcement derived from a mechanical model" by Reineck, H.-H., ACI Struct. Jour., Vol. 89, No. 4, 1992, pp. 476-477.

[92.21] **Bach, F., Thorsen, T. S. and Nielsen, M.P.**, Load carrying capacity of structural members subjected to alkali-silica reaction, The 9th Int. Conf. on Alkali - Aggregate Reaction in Concrete 1992, Chameleon Press Limited Conference Proceedings, London 1992.

[92.22] **Rasmussen, L. J. and Ibsø, J. B.**, Vridning af armerede normal- og højstyrkebetonbjælker (Torsion of normal and high strength reinforced concrete beams), High Performance Concretes in the 90s, Department of Structural Engineering, Technical University of Denmark, Report 8.4, June 1992.

[92.23] **Darwin, D. McCabe, S. L., Idun, E. K. and Schoenekase, S. P.**, Development length criteria: Bars not confined by transverse reinforcement, ACI Struct. J., Vol. 89, No. 6, Nov.-Dec. 1992, pp. 709-720.

[93.1] **Feddersen, B.**, Armerede betonbjælker påvirket til kombineret forskydning, normalkraft og bøjning (Load-bearing capacity of reinforced concrete beams subjected to combined shear, normal force and bending), Dansk Beton (Journal of Danish Concrete), Vol. 10, No. 2, 1993.

[93.2] **Sugiura, M. and Ichinose, T.**, Shear strength analysis of RC beams with web openings using truss model, Proc. of the Japan Concrete Institute, Vol. 15, No. 2, 1993, pp. 305-310 (in Japanese).

[93.3] **Watanabe, F., Nishiyama, M. and Muguruma, H.**, Strength and ductility of high strength concrete beams subjected to combined bending and shear, Utilization of High Strength Concrete, Symp. in Lillehammer, Norway, June 20-23, Proc. Vol. 1, 1993, 8 pp.

[93.4] **Naganuma, K.**, Analytical study on shear strength of reinforced concrete shear walls, Jour. of Struct. and Constr. Eng., Architectural Institute of Japan, May 1993, pp. 107-117 (in Japanese).

[93.5] **Naganuma, K.** Applicability of finite element method for strength estimation of reinforced concrete shear walls, Summaries of Technical Papers of Annual Meeting, Architectural Institute of Japan, 1993, pp. 311-312 (in Japanese).

[93.6] **Sigrist, V. and Marti, P.**, Discussion of "Design model for structural concrete based on the concept of the compressive force path" by Kotsovos, M. D. and Bobrowski, J. (Jan.-Feb. 1993), ACI Struct. Jour., Vol. 90, No. 6, Nov.-Dec. 1993, pp. 699-700.

[93.7] **Sigrist, V. and Marti, P.**, Versuche zum Verformungsvermögen von Stahlbetonträgern, Institut für Baustatik und Konstruktion, ETH Zurich, Bericht IBK No. 202, Nov. 1993, 90 pp.

[93.8] **Marti, P.**, Dimensioning and detailing of structural concrete, Civil Engineering Committee of the Polish Academy of Sciences Section of Concrete Structures, Conf. on Analytical Models and New Concepts in Mechanics of Structural Concrete, Bialystok, Poland, May 6-8, 1993.

[93.9] **Hsu, T. T. C.**, Unified theory of reinforced concrete, CRC Press, 1993.

[93.10] **CEB-FIP** Model Code 1990, Bulletin d'Information No. 213/214, Comité Euro-International du Beton, Lausanne, May 1993.

[93.11] **Eurocode 2,** Part 1-3, Precast concrete elements and structures, ENV 1992 1-3, Draft 1993.

[93.12] **Rezansoff, T. Akanni, A. and Sparling, B.**, Tensile lap splices under static loading: A review of the proposed ACI 318 code provisions, ACI Struct. J., Vol. 90, No. 4, July-Aug. 1993, pp. 374-384.

[93.13] **Ibell, T. J. and Burgoyne, C. J.**, An experimental investigation of the behaviour of anchorage zones, Mag. Concr. Res., Vol. 45, Dec. 1993, pp. 281-292.

[93.14] **Gupta, A. and Rangan, B. V.**, High-strength concrete structural walls under inplane vertical and horizontal loads, Proc. 3rd Int. Conf. on utilization of high-strength concrete, Lillehammer, Norway, June 1993, pp. 177-183.

[94.1] **Feddersen, B.**, The theory of plasticity for concrete and reinforced concrete - a consistent computation tool for practical use, The Danish Society of Structural Science and Engineering, Vol. LXV, No. 2-3-4, Dec. 1994, 19 pp.

[94.2] **Ichinose, T. and Ogura, N.**, Shear strength analysis of RC beam with two web openings, Proc. of the Japan Concrete Institute, Vol. 16, No. 2, 1994, pp. 551-556 (in Japanese).

[94.3] **Naganuma, K. et al.**, Proposal and verification of nonlinear analysis method for reinforced concrete shear walls, Report of Obayashi Corporation Technical Research Institute, No. 48, Feb. 1994, pp. 9-16 (in Japanese).

[94.4] **Marti, P.**, Schubbemessung von Voutenträgern mit geneigten Spannliedern (Shear design of variable-depth girders with draped prestressing tendons), XII FIP Congress & Exhibition, Washington D.C., May 29 - June 2, 1994, Proc., Prestressed Concrete in Switzerland.

[94.5] **Bræstrup, M. W.**, Discussion of "Shear strength of haunched beams without shear reinforcement" by MacLeaod, I. A. and Houmsi, A., ACI Struct. Jour., Vol. 91, No. 6, 1994, pp. 729-730.

[94.6] **Bræstrup, M. W.**, Concrete plasticity - The Copenhagen shear group 1973-79, Bygningsstat. Medd., Vol. 65, No. 2-3-4, 1994, pp. 33-87.

[94.7] **Bræstrup M. W.**, Test validation of structural concrete punching shear formulae, Workshop on Development of EN 1992 in Relation to New Research Results and to the CEB-FIP Model Code 1990, Prague, Czech Republic, 1994, Proc., pp. 25-36.

[94.8] **Liu, W. Q., Jensen, H. E., Ding, D. J. and Nielsen, M. P.**, Experimental report on the mechanical behavior of cracks in reinforced concrete subjected to shear loading, High Performance Concretes in the 90s, Report 7.7, 1994.

[94.9] **Liu W. Q., Jensen, H. E., Ding, D. J. and Nielsen, M. P.**, Experimental research on the shear transfer of cracks in reinforced concrete, High Performance Concretes in the 90s, Report 7.8, Mar. 1994.

[94.10] **Zhang, J.-P.**, Strength of cracked concrete, Part 1 - Shear strength of conventional reinforced concrete beams, deep beams, corbels, and prestressed reinforced concrete beams without shear reinforcement, Department of Structural Engineering, Technical University of Denmark, Report No. R-311, 1994.

[94.11] **Dahl, K. K. B.**, Construction joints in normal and high strength concrete, Department of Structural Engineering, Technical University of Denmark, Report No. R-314, 1994.

[94.12] **Hansen, T. C.**, Triaxial tests with concrete and cement paste, Department of Structural Engineering, Technical University of Denmark, Report No. R-319, 1994.

[94.13] **Ashour, A. F. and Morley, C. T.**, The numerical determination of shear failure mechanisms in reinforced concrete beams, The Structural Engineer, Vol. 72, No. 23 & 24/6, Dec. 1994, pp. 395-400.

[94.14] **Gupta, A. and Rangan, B. V.**, High strength concrete shear walls, Australian Structural Engineering Conference, Sydney, Sept. 1994.

[94.15] **Damkilde, L., Olsen, J. F. and Poulsen, P. N.**, A program for limit state analysis of plane, reinforced concrete plates by the stringer method, Bygningsstat. Medd., Vol. 65, No. 1, 1994, pp. 1-26.

[94.16] **Anderheggen, E., Despot, Z., Steffen, P. N. and Tabatabai, S. M. R.**, Computer-aided dimensioning of reinforced concrete wall and flat slab structures, Struct. Eng. Int., Vol. 4, No. 1, Feb. 1994, pp. 17-22.

[94.17] **Anderheggen, E., Despot, Z., Steffen, P. N. and Tabatabai, S. M. R.**, Reinforced concrete dimensioning based on element nodal forces, J. Struct. Eng. (ASCE), Vol. 120, No. 6, June 1994, pp. 1718-1731.

[94.18] **Dahl, K. K. B.**, Construction Joints in Normal and High Strength Concrete, High Performance Concretes in the 90s, Department of Structural Engineering, Technical University of Denmark, Report 7.9, March 1994.

[94.19] **Ibell, T. J. and Burgoyne, C. J.**, A plasticity analysis of anchorage zones, Mag. Concr. Res., Vol. 46, March 1994, pp. 39-48.

[94.20] **Ibell, T. J. and Burgoyne, C. J.**, A generalised lower-bound approach to the analysis of anchorage zones for prestressed concrete, Mag. Concr. Res., Vol. 46, June 1994, pp. 133-143.

[95.1] **Sakata, H., Ichinose, T. and Kamiya, N.**, Shear strength analysis of RC beams with web opening and diagonal reinforcement, Proc. of the Japan Concrete Institute, Vol. 17, No. 2, 1995, pp. 601-606 (in Japanese).

[95.2] **Iida, M. and Ichinose, T.**, Biaxial shear strength of RC column and three dimensional shear failure, Proc. of the Japan Concrete Institute, Vol. 17, No. 2, 1995, pp. 863-868 (in Japanese).

[95.3] **Ogura, N. and Ichinose, T.**, Analysis of concrete confinement by circular hoops, Concrete under Severe Condition, E & FN Spon, Vol. 2, 1995, pp. 1501-1510.

[95.4] **Ichinose, T. and Hanya, K.**, Three dimensional shear failure of R/C beams, Concrete under Severe Condition, E & FN Spon, Vol. 2, 1995, pp. 1737-1747.

[95.5] **Ichinose, T.**, Truss mechanism in critical region of RC member, J. Struct. and Constr. Eng., Architectural Institute of Japan, No. 475, 1995 (in Japanese).

[95.6] **Naganuma, K. et al.**, Analysis of reinforced concrete panels subjected to reversed cyclic loads, Part 1 - Modelling of hysteretic characteristics of concrete, Part 2 - Simulation of pure shear tests of panels and torsion tests of cylinders, Summaries of Technical Papers of Annual Meeting, Architectural Institute of Japan, 1995, pp. 823-826 (in Japanese).

[95.7] **Naganuma, K.**, Stress-strain relationship for concrete under triaxial compression, J. Struct. and Constr. Eng., Architectural Institute of Japan, No. 474, Aug. 1995, pp. 163-170 (in Japanese).

[95.8] **Bræstrup, M.W.**, Truss model for shear design of structural concrete members with unbonded tendons, Bygningsstat. Medd., Vol. 66, No. 1, 1995, pp. 17-22.

[95.9] **Lorenzen, K. G.**, Koncentreret last på beton (Concentrated loads on concrete), M.Sc. thesis, Department of Structural Engineering, Technical University of Denmark, 1995.

[95.10] **Anderheggen, E., Despot, Z., Steffen, P. N. and Tabatabai, S. M. R.**, Linearisierte Fliessbedingungen für ein finites Element aus Stahlbeton: Theoretische grundlagen, IBK, ETH Zürich, 1995.

[95.11] **Anderheggen, E., Despot, Z., Steffen, P. N. and Tabatabai, S. M. R.**, Finite elements and plasticity theory: Integration in optimum reinforcement design, Proc. of the 6th Int. Conference on Computing in Civil and Building Engineering, Berlin, July 1995, pp. 653-660.

[95.12] **Christensen, L. K.**, Forankring af armeringsjern (Anchorage of reinforcing bars), M.Sc. thesis, Department of Structural Engineering, Technical University of Denmark, 1995.

[95.13] **Despot, Z.**, Methode der finiten Elemente und Plasticitätstheorie zur Bemussung von Stahlbetonscheiben, Institut für Baustatik und Konstruktion, ETH Zürich, Bericht Nr. 215, December 1995.

[96.1] **Christoffersen, J.**, Ultimate capacity of joints in precast large panel concrete buildings, Ph.D. thesis, Department of Structural Engineering, Technical University of Denmark, 1996.

[96.2] **Jagd, L.**, Non-linear FEM analysis of 2D concrete structures, Ph.D. thesis, Department of Structural Engineering, Technical University of Denmark, 1996.

[96.3] **Gupta, A.**, Behaviour of high strength concrete structural walls, Thesis presented for the award of the Degree of Doctor of Philosophy, Curtin University of Technology, April 1996.

[96.4] **Anderheggen, E., Steffen, P. N. and Tabatabai, S. M. R.**, Finite elements and plasticity theory: Application in practice-oriented dimensioning of reinforced concrete structures, Proc. of the 15th Congress of IABSE, Copenhagen, Denmark, June 1996, pp. 1219-1220.

[96.5] **Anderheggen, E.**, On plasticity theory and reinforced concrete computer-aided design, Proc. of the 15th Congress of IABSE, Copenhagen, Denmark, June 1996, pp. 889-894.

[96.6] **Hoang, L. C. and Nielsen, M. P.**, Continuous reinforced concrete beams - stress and stiffness estimates in the serviceability limit state, Department of Structural Engineering and Materials, Technical University of Denmark, Report R, No. 12, 1996.

[96.7] **Esfahani, M. R. and Rangan, B. V.**, Studies on bond between concrete and reinforcing bars, Curtin Univ. of Technology, Research Report No. 1/96, 1996.

[96.8] **Tolderlund, C.**, Forankring af armeringsjern (Anchorage of reinforcing bars), M.Sc. thesis, Department of Structural Engineering, Technical University of Denmark, 1996.

[96.9] **Jagd, L. K.**, Non-linear seismic analysis of RC shear wall, Technical University of Denmark, Department of Structural Engineering and Materials, Report R, No. 5, 1996.

[96.10] **Steffen, P.**, Elastoplastische Dimensionierung von Stahlbetonplatten mittels finiter Bemessungselemente und linearer Optimierung, Institut für Baustatik und Konstruktion, ETH Zürich, Bericht Nr. 220, Juni 1996.

[97.1] **Muttoni, A., Schwartz, J. and Thürlimann, B.**, Design of concrete structures with stress fields, Birkhäuser Verlag, Basel, Switzerland, 1997.

[97.2] **Zhang, J.-P.**, Strength of cracked concrete, Part 2 - Micromechanical modelling of shear failure in cement paste and in concrete, Department of Structural Engineering and Materials, Technical University of Denmark, Report R, No. 17, 1997.

[97.3] **Zhang, J.-P.**, Strength of cracked concrete, Part 3 - Load carrying capacity of panels subjected to in-plane stresses, Department of Structural Engineering and Materials, Technical University of Denmark, Report R, No. 18, 1997.

[97.4] **Hoang, L. C.**, Shear strength of non-shear reinforced concrete elements, Part 1 - Statically indeterminate beams, Department of Structural Engineering and Materials, Technical University of Denmark, Report R, No. 16, 1997.

[97.5] **Liu, J. Y.**, Plastic theory applied to shear walls, Ph.D. thesis, Department of Structural Engineering and Materials, Technical University of Denmark, 1997.

[97.6] **Ibell, T. J., Morley, C. T. and Middleton, C. R.**, The assessment in shear of existing concrete bridges, Proc. FIP Symp. '97, The concrete way to development, Johannesburg, March 1997, pp. 529-536.

[97.7] **Hoang, L. C.**, Shear strength of non-shear reinforced concrete elements, Part 2 - T-beams, Department of Structural Engineering and Materials, Technical University of Denmark, Report R, No. 29, 1997.

[97.8] **Hoang, L. C.**, Shear strength of non-shear reinforced concrete elements, Part 3 - Prestressed hollow-core slabs, Department of Structural Engineering and Materials, Technical University of Denmark, Report R, No. 30, 1997.

[97.9] **Hoang, L. C.** Shear strength of lightly reinforced concrete beams, Department of Structural Engineering and Materials, Technical University of Denmark, Report R (to be published).

[97.10] **Christoffersen, J.**, Ultimate capacity of joints in precast large panel concrete buildings, Department of Structural Engineering and Materials, Technical University of Denmark, Report R, No. 25, 1997.

[97.11] **Nielsen, M. P.**, Beton 2, del 1, 2 og 3 (Concrete Structures 2, part 1, 2, and 3), Department of Structural Engineering and Materials, Technical University of Denmark, 1997 (to be published).

[97.12] **Nielsen, M. P.**, Beton 1, del 1, Materialer (Concrete Structures, part 1, Materials), Deptartment of Structural Engineering, Technical University of Denmark, 1997 (to be published).

[97.13] **Jagd, L. K.**, Non-linear FEM analysis of 2D concrete structures, Department of Structural Engineering and Materials, Technical University of Denmark, Report R, No. 31, 1997.

[97.14] **Zhang, J.-P.**, Diagonal cracking and shear strength of reinforced concrete beams, Mag. Concr. Res., Vol. 49, No. 178, 1997, pp. 55-65.

[97.15] **Lorenzen, K. G. and Nielsen, M. P.**, Koncentreret last på beton (Concentrated loads on concrete), Department of Structural Engineering and Materials, Technical University of Denmark, Report R, No. 34, 1997.

[97.16] **Liu, J. Y.**, Plastic theory applied to shear walls, Department of Structural Engineering and Materials, Technical University of Denmark, Report R, No. 36, 1997.

[97.17] **Christiansen, M. B. and Nielsen, M. P.**, Modelling tension stiffening in reinforced concrete structures - rods, beams and disks, Department of Structural Engineering and Materials, Technical University of Denmark, Report R, No. 22, 1997.

[97.18] **Mitrofanov, V. P. and Artsev, S. I.**, Reinforced concrete bar systems with optimal steel expenditure: Design by limit equilibrium method with taking into account of shear forces, The Industrialization of Building: Planning, design and construction, Proc. Int. Scientific Conf., INDIS'97 and CIB W-63, Novi Sad, Yogoslavia, Nov. 1997, Vol. 1, pp. 257-265.
[The paper contains further references to the work by V. P. Mitrofanov and associates at Poltava State Technical University, Ukraine].

[98.1] **Larsen, M. S.**, Eccentrical load on slab-column connections - determination of the load-carrying capacity, M.Sc. thesis, Department of Structural Engineering and Materials, Technical University of Denmark, 1998.

Added in proof:

[98.2] **Ibell, T. J., Morley, C. T. and Middleton, C. R.**, An upper-bound plastic analysis for shear, Mag. Concr. Res., Vol. 50, No. 1, 1998, pp. 67-73.

[98.3] **Gupta, A. and Rangan, B. V.**, High-strength concrete (HSC) structural walls, ACI Struct. Journal, Vol. 95, No. 2, March-April 1998.

[98.4] **Esfahani, M. R. and Rangan, B. V.**, Local bond strength of reinforcing bars in normal and high-strength concrete (HSC), ACI Struct. Journal, Vol. 95, No. 2, March-April 1998.

[98.5] **Esfahani, M. R. and Rangan, B. V.**, Bond between normal strength and high strength concrete (HSC) and reinforcing bars in splices in beams, ACI Struct. Journal, Vol. 95, No. 3, May-June 1998.

[98.6] **Hoang, L. C. and Nielsen, M. P.**, Plasticity approach to shear design, Cement and Concrete Composites 20, Elsevier Science Ltd., 1998 (to be published).

INDEX

A

Affinity theorems
  concrete stress field 87
  disks 289
  slabs 608
α- and β-lines 173
Anchor length; see development length
Anchor strength 774
Angle of friction 24
Arbitrary cross-section, design of shear reinforcement 450
Arrows to indicate load bearing direction 590
Axial and transversal strains in uniaxial compression 155
Axial force, effect of; see normal force, effect of
Axisymmetrical mechanism at bond failure, discussion 771

B

Bauschinger's area ratio 217
Beams in bending 365
  yield (failure) moment 367
  bending with normal force and compression reinforcement 371
Beams in shear
  maximum shear capacity 376, 381, 389, 391
  general upper bound method 385
  influence of longitudinal reinforcement 393
  beams without shear reinforcement 401
  beams with normal force 428
  beam columns 434
  variable depth 435
  bent-up bars 435
  inclined prestressing reinforcement 435
  lightly reinforced beams 441
  strong flanges 450
  arbitrary cross-section 450
Beams in torsion 452
  corner problems 457
  rectangular sections 459
  combined bending, shear, and torsion 466

Beam model for bond strength 839
Bending and torsional moments in a yield line 491
Bond strength formulas 798
Bond strength in joints 741
Bond strength per bar, definition 767
Bond strength, reinforcing bars 765
  local failure mechanism 766
  bond strength per bar, definition 767
  reinforcement data 770
  failure mechanisms 772, 775
  splice strength versus anchor length 773
  lap length effect 776
  development length 777
  pull-out of a bar 777
  analysis of failure mechanisms 781
  corner failure 781
  V-notch failure 788
  cover bending failure 790
  face splitting failure 796
  bond strength formulas 798
  assessment of anchor and splice strength 802
  effect of transverse pressure and support reaction 812
  effect of transverse reinforcement 828
  beam model, bond strength 839
Boundary conditions
  three dimensional 143
  disks 240
  slabs 482, 486, 499
Boundary (edge) reinforcement, slabs 607, 708
Bredt's formula 452
Butt joint 750

C

Catenary curve 653
Circular fan 325, 498
Coefficient of friction (in failure condition) 24
Cohesion, definition 23
Cohesion in joint 741, 746, 749
Collapse by yielding 9
Collapse load 9

Combined bending, shear, and torsion 466
Combined loading, slabs 566
Compatibility conditions
  disks, rectangular coordinates 241
  disks, polar coordinates 241
  slabs 479
Compressive membrane effect 626
Compressive strength
  definition 26
  cement paste 34, 43
  concrete 26, 52
  disks 76
  with regards to initial cracks 276, 726
  of specimens with joints 726
Concentrated loading, punching 650
Concentrated loads (unreinforced concrete)
  strip load 198
  point load 207
  design formulas 230
  effect of reinforcement 231
  edge and corner loads 232
  group action 235
  size effects 237
  effect of friction 226
  upper limit of load-carrying capacity 230
Concentrated loads on beams near supports 418
Concentrated loads on slabs 514, 519, 553, 561, 584, 586, 588, 598, 602, 603, 618, 646
Concentric loading, punching 650
Conical failure 27
Constitutive equations
  Coulomb materials 145
  modified Coulomb materials 153
  disks 242
  slabs 487
Construction joints 740
Control perimeter 666
Control surface 650
Corner failure 775, 781
Corner lever 558
Coulomb's failure criterion, various formulas 31

Coulomb material 24
Coulomb material, modified 24
Counterpressure, effect of 676
Cover bending failure 790
Crack as a joint 730
Cracking moment, beam 403
Curved yield lines 203
Curved strut solution 436
Cylinder compressive strength of concrete 52

D

Degree of reinforcement
 disks 76
 bending 366, 391
 shear 376
Design examples, bond strength 803, 844
Design examples, slabs 569, 591, 593, 595, 603, 618
Design formulas, concentrated loads 230
Development length 777
Deviation from elastic solution
 beams 19
 reinforcement design 139
 disks 359
Diagonal compression field
 disks 336
 beams 373
Diagonal tension field (Wagner) 378
Difference expressions 287
Discontinuity lines
 definition 14
 plain concrete 160
 slabs 475, 479
Disk loaded through a corbel (half joint) 295
Disks 239
 definition 239
 statical conditions 239
 geometrical conditions 241
 constitutive equations 242
 yield zones 246
 exact solutions 248

effective compressive strength 262
lower bound solutions 285
stringer method 290
shear zone solution 300
strut and tie models 306
strut and tie systems 311
more refined models 325
shear walls 328
homogeneous reinforcement 351
combination of homogeneous and concentrated reinforcement 355
deep disks 357
relation to elastic theory 359
Dissipation formulas
plain concrete 166
Rankine zone 181
Prandtl zone 183
disks 244
slabs 490
Dowel action 73, 751
Drucker's stability postulate 13

E

Eccentric loading, punching 668, 702
Edge and corner loads 232, 686, 695
Effect of prestressing
beams with shear reinforcement 400
beams without shear reinforcement 412
inclined prestressing reinforcement 435
Effect of reinforcement, concentrated loads 231
Effective depth 98, 365
Effective strengths; see effectiveness factors
Effectiveness factors
definitions 52
transformation of theoretical formulas 70
disks 89, 127, 262, 306, 320
slabs 117, 643
struts 320
shear walls 342
beams in bending 368, 403

beams in shear 397
beams without shear reinforcement 405, 411, 412, 465
membrane action in slabs 643
punching shear 662, 663
torsion, edge and corner loads 692
joints 719, 721, 733, 746, 749
bond strength 786, 842
Elastic theory
beams 16
reinforcement design 138
disks 359
Empirical tensile strength formulas, concrete 51
Equilibrium conditions
three dimensional, Cartesian coordinates 143
three dimensional, cylinder coordinates 143
disks, rectangular coordinates 239
disks, polar coordinates 240
slabs, rectangular coordinates 475
slabs, polar coordinates 475
Equilibrium method, slabs 525
Equivalent stresses 76
Euler equation 652
Exact solutions
definition 16
plain concrete 187, 190, 191, 195, 201
disks 248
beams 384, 385, 389, 392
slabs 503
Extent of top reinforcement, slabs 560
External yield line 528
Extremum principles 9

F

Face splitting failure 775, 796
Failure criteria
concrete 21
Coulomb materials 23
modified Coulomb materials 23
simplified failure condition for concrete 49
cracks 275, 282

Failure load 10
Failure mechanism, geometrically possible, definition 11
Failure patterns in uniaxial compression 188
Failure surface, punching 657, 659, 661
Flanges, influence of 450
Flat slab, design example 603
Flow rule
  general 4
  at an apex 7
  at an edge 7
Formulas, Coulomb failure condition 31
Friction angle 24
Friction, sliding failure involving 737
Frictional hypothesis 21

G

Generalized yield lines 167
Generalized strains
  general 1
  slabs 477
Generalized stresses
  general 1
  slabs 475
Geometrical conditions
  three dimensional, Cartesian coordinates 144
  three dimensional, cylinder coordinates 144
  disks, rectangular coordinates 241
  disks, polar coordinates 241
Geometrically possible displacement field, definition 9, 11
Grashof stress distribution 466
Group action, concentrated loads 235

H

Half joint; see disk loaded through a corbel
Hankinson's formula 94
Hole, reinforcement around 298, 590
Homogeneous stress field solutions 259, 351, 355

I

Increments of plastic strain 9
Influence of stirrup spacing 446
Ingerslev's solution 547
Internal cracking 53, 263
Internal forces in slabs 473
Internal yield lines 528
Isotropic disk 77
Isotropic slab 105, 111

J

Jin-Ping Zhang models
  cement paste 33
  concrete 37
  beams without shear reinforcement 401
  crack 730
Joints 711
  plastic theory 711
  monolithic concrete 712
  joints 716
  axial forces 718
  skew reinforcement 723
  the crack as a joint 730
  construction joints 740
  butt joints 750
  dowel action 751
  keyed joints 753

K

Key geometry 758
Keyed joint 753
Kirchoff assumptions 477
Kirchoff boundary condition 483
Kirchoff edge beam 484, 686
Kötter's equations 175

L

Lap length 774
Lap length effect 776
Limit analysis, definition 10
Limits on cotθ (κ) in beams 425
Line load on a free edge of a slab 552, 595
Linearized yield conditions 624
Load induced microcracking 53
Load-carrying capacity 9
Local failure mechanism, bond strength 766
Longitudinal reinforcement degree 391
Lower bound solution, definition 16
Lower bound solutions, plane stress problems
  general theory 285
  stringer method 290
  shear zone solution 300
  strut and tie models 306
  refined models 325
  shear walls 328
  homogeneous reinforcement solutions 351
  combination of homogeneous reinforcement and concentrated reinforcement solutions 355
Lower bound solutions, slabs
  rectangular slabs with various support conditions 572
  strip method 587
  concentric internal load 672
  eccentric internal load 673
  corner load 683
  edge load 695
  eccentric edge and corner load 702
Lower bound theorem 10

M

Macrocracking 53, 58
Maximum shear capacity, beams 376, 381, 389, 391
Maximum work hypothesis 2
Membrane action, rough estimate of load-carrying capacity 647
Membrane action, slabs 625
Microcracking; see internal cracking

Microcracking parameter χ 56, 268
Minimum reinforcement rules
  concentrated loads 232
  disks 280
  beams 400
Minimum reinforcement solutions
  beams 17
  slabs 622
Mises', von flow rule 4
Modified Coulomb material 24
Mohr's circle, applied to homogeneous stress field solutions 259
Mohr's failure envelope 22

N

Navier stress distribution 466
Negative yield line 539, 614
Nodal forces
  type 1 531
  type 2 533
  orthotropic slabs 616
Normal force, effect of
  shear walls 328
  beams in bending 371
  beams in shear 413, 428
Normal horizontal restraint, slabs 643
Normality condition 4
Normally reinforced beams
  bending 366
  shear 393, 394
Notation for boundary conditions in slabs 487
Notation for loading on slabs 476
Numerical methods, slabs 623

O

One-way slab with a hole 590
Orthogonal reinforcement 75
Overreinforced beam
  bending 366
  shear 391, 401

Orthotropic disk 86
Orthotropic slab 112

P

Panel shear key 754
Plain concrete 143
Plastic deformation 1
Plastic strains, flow rule 4
Plastic strains at collapse 14
Point load 207
Position of dangerous crack in beams 407
Positive yield line 501
Prandtl solution, semi-infinite body 193
Prandtl zone 178
Prestressing; see effect of
Principal axes 77
Principal curvatures 478
Principal moments 474
Principal sections (directions)
  general 29
  plane stress 29
  slabs 474, 478
Principal stresses 25
Proportional loading 11
Pull-out of a bar 777
Punching shear 649
  control surface 650
  concentric loading 650
  upper bound solution 658
  failure surface 657, 659, 661
  ultimate load, concentric loading 663, 666
  locally reduced thickness 665
  variable thickness 665
  control perimeters 666
  eccentric loading, internal loads 668
  ultimate load, eccentric internal loads 676
  counterpressure 676
  shear reinforcement, internal loads 676, 680, 681
  failure mechanisms involving shear reinforcement, internal loads 679

corner load 686
torsional yield moment 688
torsion tests 689, 693
edge load 695
general case of edge and corner loads 696
eccentric edge and corner loads 702
Push-off specimen 722

R

Rankine zone 177
Reaction, effect on bond strength 812
Reactions in slabs, upper bound method 565
Regions with different crack sliding behavior 407
Reinforcement data 770
Reinforcement degree
disks 76
bending 366, 391
shear 376
Reinforcement design, combined bending, torsion, and shear in beams 466
Reinforcement design, disks
exact solutions 248
homogeneous reinforcement, bending 257
minimum reinforcement 280
supplementary reinforcement to avoid sliding in initial cracks 284
stringer method 290
disk or beam loaded through a corbel (half joint) 295
frame corner 296
reinforcement around a hole 298
shear zone solution 300
shear zone solution, bending 301
the single strut 306
strut and tie systems 311
circular fan 325
shear walls 328
homogeneous reinforcement solutions 351
combination of homogeneous and concentrated reinforcement 355
disks with small reinforcement degrees 356

deep disks 357
design according to elastic theory 359
Reinforcement design, general 120
shear 131
skew reinforcement 132
slabs 135, 607
shells 137
three dimensional case 137
using elastic theory 138
against sliding in initial cracks 283
around a hole 298, 590
Reinforcement design, shear in beams
shear reinforced beams 423
normal forces 428
variable depth 435
bent-up bars 435
inclined prestressing reinforcement 435
lightly reinforced beams 441
strong flanges 450
arbitrary cross-section 450
Reinforcement design, slabs
exact solutions 504, 505
corners in simply supported slabs 525
upper bound solutions 525, 614, 632
effect of corner levers 558
extent of top reinforcement 558, 560
lower bound solutions 571
strip method 587
around a hole 590
including membrane effect 634, 642
shear reinforcement, punching, concentric loading 676
edge and corner reinforcement, punching 683, 708
Reinforcement design, torsion in beams 452
Reinforcement ratio 138, 366
Rigid horizontal restraint, slabs 643
Rigid part at collapse 13
Rigid plastic first-order theory 625
Rigid plastic material 1
Rotation of reinforcement bar in shear strain field 74
Rough joint 749

S

Safe stress distribution 10
Separation failure 24
Separation resistance 24
Shear capacity, beams
  transverse reinforcement 376, 381
  general upper bound method 385
  inclined reinforcement 389
  nonshear reinforced beams 391, 401
  influence of longitudinal reinforcement 393
  lightly shear reinforced beams 443
  finite stirrup spacing 447
Shear capacity, disks 77, 265
Shear capacity, joints
  plastic theory 716
  shear capacity of a crack 730
  construction joints 740
  butt joints 750
  keyed joints 753
Shear capacity, slabs; see punching shear
Shear failure mechanism observed in tests 384
Shear-friction theory 711
Shear key 753
Shear span 373
Shear strength of concrete, definition 28
Shear stresses in reinforcement bars 72, 751
Shear stresses, limits in slabs 708
Shear walls 328
  strut solution with web reinforcement 329
  diagonal compression field solution 336
  modification of diagonal compression field 340
Simple affine case 490, 608, 612
Size effects
  concentrated loads 237
  cracking moment 404
Skew reinforcement 132, 723
Slab shear key 754
Slabs 473
  definition 473
  internal forces 473

transformation of moments 474
transformation of shear forces 474
yield zones 493
statically admissible moment fields 506
exact solutions 503
upper bound solutions 525
work equation method and equilibrium method 525
nodal forces 530
combined actions 566
design examples 569, 591, 593, 595, 603, 618
lower bound solutions 571
strip method 587
column support, concentrated loads 598
orthotropic slabs 608
optimum reinforcement solutions 622
numerical methods 623
membrane action 625
Sliding failure 24
Sliding failure involving friction 737
Sliding in initial cracks
general 59
influence of crack width 63
strength reduction in disks 274
implications on design 278
plastic solutions 281
nonshear reinforced beams 401
Sliding reduction factor 59, 734
Sliding resistance 23
Slipline theory 168
stress field 168
Rankine zone 177
Prandtl zone 178
strain field 178
displacement conditions 186
Smooth joint 746
Snap-through action in slabs 626
Softening in concrete
compression 63, 262
tension 69
Splice strength 774
Splitting tensile strength 51, 202

Square yield condition 105
Square yield locus 52
Statically admissible failure zones 175
Statically admissible stress field 10
Stationary moment field, slabs 528
Stiffness in cracked state
  plane stress 140
  torsion 457
Stirrup spacing; see influence of stirrup spacing
Strain increments 9
Strain tensor
  general 144
  slabs 476
Strains below the yield point 14
Stress-strain (deformation) relation
  concrete, compression 63
  concrete, tension 68
  reinforcement 71
Stress tensor
  general 143
  slabs 477
Stringer method 290
Strip load 198
Strip method 587
Structural concrete strength 52
Strut and tie models 306, 753
Strut formulas 306, 756
Support
  limiting value of support stress 312
  concentrated loads on beams near supports 418

T

Tensile membrane effect 626
Tensile strength formulas, concrete 51
Tensile strength in design 421
Tensile strength of concrete, definition 27
Tensile strength of disks 76
Tension cut-off 22
Tension yield load 5
Torsion capacity of rectangular sections 459

Torsion, yield moment
slabs 102, 688
beams 456
Torsion tests, slabs 689, 693
Total stresses 76
Transformation formulas for moments in slabs 474
Transformation formulas for shear forces in slabs 474
Transformation of a crack into a yield line 403
Transformation of theoretical formulas by introduction of effectiveness factors 70
Transversal and axial strains in uniaxial compression 155
Transverse pressure, effect on bond strength 812
Transverse reinforcement, effect on bond strength 828

U

Underreinforced concrete bar 830
Uniaxial compression, axial and transversal strains 155
Uniaxial stress and strain 93
Uniqueness theorem 12
Upper bound solutions
definition 16
slope 195
strip load 198
point load 207
beams in bending 367
beams in shear 381
isotropic slabs 525
orthotropic slabs 614
punching 658
bond strength of reinforcing bars 765

V

V-notch failure 775, 788
Variable θ solutions 436
Virtual work
three dimensional 145
disks 241
slabs 480
von Mises; see Mises, von

W

Web crushing criterion 375
Work equation 12
Work equation, isotropic slabs 540
Work equation method, isotropic slabs 525
Work equation, orthotropic slabs 617

Y

Yield conditions
 definition 1
 disks 70, 265
 slabs 98
 pure torsion in slabs 100, 688
 plain concrete 145
 beams in torsion 455
 beams, combined bending and torsion 470
Yield lines
 plain concrete 160
 dissipation formulas, plain concrete 166
 generalized 167
 curved 203
 dissipation formulas, disks 245
 slabs 479, 497
Yield load 10
Yield moment
 beams in bending 5, 366
 beams in torsion 456
 slabs in torsion 102, 688
Yield stress 1
Yield surface 3
Yield zones 246, 493

Z

Zhang, Jin-Ping; see Jin-Ping Zhang